Natural Hazards and Disasters

Natural Hazards and Disasters

From Avalanches and Climate Change to Water Spouts and Wildfires

VOLUME 1: Natural Hazards and Aid Organizations
Bimal Kanti Paul, Editor

An Imprint of ABC-CLIO, LLC
Santa Barbara, California • Denver, Colorado

Copyright © 2021 by ABC-CLIO, LLC

All rights reserved. No part of this publication may be reproduced, stored in a retrieval system, or transmitted, in any form or by any means, electronic, mechanical, photocopying, recording, or otherwise, except for the inclusion of brief quotations in a review, without prior permission in writing from the publisher.

Library of Congress Cataloging-in-Publication Data

Names: Paul, Bimal Kanti, editor.
Title: Natural hazards and disasters : from avalanches and climate change to water spouts and wildfires / Bimal Kanti Paul, editor.
Description: Santa Barbara : ABC-CLIO, 2021. | Includes bibliographical references and index. | Contents: v. 1. Natural hazards and aid organizations — v. 2. Natural disasters.
Identifiers: LCCN 2020019064 (print) | LCCN 2020019065 (ebook) | ISBN 9781440862151 (v. 1 ; hardcover) | ISBN 9781440862168 (v. 2 ; hardcover) | ISBN 9781440862137 (set : hardcover) | ISBN 9781440862144 (ebook)
Subjects: LCSH: Natural disasters—Encyclopedias.
Classification: LCC GB5014 .N389 2021 (print) | LCC GB5014 (ebook) | DDC 363.3403—dc23
LC record available at https://lccn.loc.gov/2020019064
LC ebook record available at https://lccn.loc.gov/2020019065

ISBN: 978-1-4408-6213-7 (set)
        978-1-4408-6215-1 (vol. 1)
        978-1-4408-6216-8 (vol. 2)
        978-1-4408-6214-4 (ebook)

25 24 23 22 21   1 2 3 4 5

This book is also available as an eBook.

ABC-CLIO
An Imprint of ABC-CLIO, LLC

ABC-CLIO, LLC
147 Castilian Drive
Santa Barbara, California 93117
www.abc-clio.com

This book is printed on acid-free paper ∞

Manufactured in the United States of America

Contents

VOLUME 1

Guide to Related Topics   xi

Preface   xv

Introduction   xvii

Natural Hazards   1

Avalanches   1

Blizzards   11

Climate Change   22

Coastal Erosion   33

Desertification   43

Droughts   53

Earthquakes   64

Erosion   75

Expansive Soils   87

Extinction   97

Floods   107

Hail   119

Hurricanes   129

Ice Storms   139

Landslides   151

Lightning   162

Salinization   172

Storm Surges   181

Subsidence   189

Temperature Extremes   199

Tornadoes   210

Tsunamis   220

Volcanic Activity   230

Waterspouts   240

Wildfires   250

Aid Organizations   263

ActionAid International   263

American Red Cross   267

Catholic Relief Services (CRS)   272

Concern Worldwide   276

Cooperative for Assistance and Relief Everywhere   281

Doctors Without Borders   286

International Federation of Red Cross and Red Crescent Societies (IFRC)   290

International Organization for Migration (IOM)   295

Islamic Relief Worldwide   299

Lutheran World Federation   304

Mennonite Central Committee (MCC)   309

Oxfam International   314

Pan American Health Organization (PAHO)   319

Refugees International   323

Save the Children   328

United Nations Children's Fund (UNICEF)   332

World Food Program (WFP)   337

World Health Organization (WHO)   341

World Vision International (WVI)   346

VOLUME 2

Guide to Related Topics   xi

Natural Disasters   1

Bam Earthquake, Iran, 2003   1

Bangladesh Flood, 1998   6

Bengal Famine, 1943–1944   10

Bhola Cyclone, Bangladesh, 1970   14

Big Thompson Canyon Flash Flood, Colorado, 1976   18

Black Saturday Bushfires, Australia, 2009   23

BP Deepwater Horizon Oil Spill, United States, 2010   27

Brisbane and Queensland Flood, Australia, 2011   32

Buffalo Blizzard, New York, 1977   36

California Drought, 2012–2016   40

Chicago Heat Wave, Illinois, 1995   45

Chi-Chi Earthquake, Taiwan, 1999   50

Christchurch Earthquake, New Zealand, 2010–2011   54

Colombia Floods, 2010–2011   59

Colorado Flood, United States, 2013   64

Cyclone Gorky, Bangladesh, 1991   69

Cyclone Nargis, Myanmar, 2008   73

Cyclone Pam, Vanuatu, 2015   77

Cyclone Phailin, India, 2013   82

Cyclone Sidr, Bangladesh, 2007   87

The Dust Bowl, 1930s   91

East African Drought, 2011–2012   96

Edmonton Tornado, Canada, 1987   101

European Heat Wave, 2003   105

Eyjafjallajökull Eruption, Iceland, 2010   109

Grand Forks Flood, North Dakota, 1997   114

Great Ice Storm of 1998, Canada 118

Great Kanto Earthquake, Japan, 1923 122

Great Mississippi River Flood, United States, 1993 127

Gujarat Earthquake, India, 2001 132

Haiti Earthquake, 2010 137

Heat Wave and Wildfires, Russia and Eastern Europe, 2010 141

Hurricane Andrew, United States and the Bahamas, 1992 146

Hurricane Charley, United States, 2004 150

Hurricane Galveston, United States, 1900 155

Hurricane Harvey, Texas and Louisiana, 2017 159

Hurricane Ike, United States, 2008 164

Hurricane Irma, Florida, 2017 168

Hurricane Katrina, United States, 2005 173

Hurricane Maria, Puerto Rico, 2017 179

Hurricane Matthew, United States, 2016 183

Hurricane Mitch, Central America, 1998 188

Hurricane Stan, Guatemala, 2005 193

Indian Ocean Tsunami, 2004 197

Iowa Flood, United States, 2008 202

Izmit/Marmara Earthquake, Turkey, 1999 207

Johnstown Flood, Pennsylvania, 1889 211

Joplin Tornado, Missouri, 2011 216

Kashmir Earthquake, Pakistan, 2005 221

Kerala Floods, India, 2018 225

Kobe Earthquake, Japan, 1995 230

Loma Prieta Earthquake, California, 1989 235

Mexico City Earthquakes, Mexico, 1985 239

Millennium Drought, Australia, 2001–2012 244

Mozambique Flood, 2000 248

Nepal Earthquakes, 2015 253

Pakistan Flood, 2010 258

Contents

Sichuan Earthquake, China, 2008   262

Sulawesi Earthquake and Tsunami, Indonesia, 2018   267

Summer Floods, United Kingdom, 2007   272

Superstorm Sandy, United States, 2012   276

Tangshan Earthquake, China, 1976   281

Thomas Fire, California, 2017–2018   286

Tohoku Earthquake and Fukushima Tsunami, Japan, 2011   290

Tri-State Tornado, United States, 1925   295

Tropical Storm and Floods, Yemen, 2008   299

Typhoon Haiyan (Yolanda), Philippines, 2013   304

Valdivia Earthquake, Chile, 1960   309

Vietnam Flood, 1999   313

Yangtze River Flood, China, 1931   317

Bibliography   323

About the Editor and Contributors   331

Index   335

Guide to Related Topics

VOLUME 1

CLIMATOLOGICAL
Climate Change
Desertification
Droughts
Ice Storm
Temperature Extremes
Wildfires

GEOPHYSICAL
Avalanches
Coastal Erosion
Earthquakes
Erosion
Expansive Soils
Extinction
Landslides
Subsidence
Tsunamis
Volcanic Activity

HUMANITARIAN AGENCIES
ActionAid International
American Red Cross
Catholic Relief Services (CRS)
Concern Worldwide
Cooperative for Assistance and Relief Everywhere
Doctors Without Borders
International Federation of Red Cross and Red Crescent Societies (IFRC)
International Organization for Migration (IOM)
Islamic Relief Worldwide
Lutheran World Federation
Mennonite Central Committee (MCC)
Oxfam International
Pan American Health Organization (PAHO)
Refugees International
Save the Children
United Nations Children's Fund (UNICEF)
World Food Program (WFP)
World Health Organization (WHO)
World Vision International (WVI)

HYDROLOGICAL
Floods
Salinization
Storm Surges
Waterspouts

METEOROLOGICAL
Blizzards
Hail
Hurricanes
Lightning
Tornadoes

VOLUME 2

CLIMATOLOGICAL
Bengal Famine, 1943–1944
Black Saturday Bushfires, Australia, 2009
California Drought, 2012–2016
Chicago Heat Wave, Illinois, 1995
The Dust Bowl, 1930s
East African Drought, 2011–2012
European Heat Wave, 2003
Great Ice Storm of 1998, Canada
Heat Wave and Wildfires, Russia and Eastern Europe, 2010
Millennium Drought, Australia, 2001–2012
Thomas Fire, California, 2017–2018

GEOPHYSICAL
Bam Earthquake, Iran, 2003
Chi-Chi Earthquake, Taiwan, 1999
Christchurch Earthquake, New Zealand, 2010–2011
Eyjafjallajökull Eruption, Iceland, 2010
Great Kanto Earthquake, Japan, 1923
Gujarat Earthquake, India, 2001
Haiti Earthquake, 2010
Indian Ocean Tsunami, 2004
Izmit/Marmara Earthquake, Turkey, 1999
Kashmir Earthquake, Pakistan, 2005
Kobe Earthquake, Japan, 1995
Loma Prieta Earthquake, California, 1989
Mexico City Earthquakes, Mexico, 1985
Nepal Earthquakes, 2015
Sichuan Earthquake, China, 2008
Sulawesi Earthquake and Tsunami, Indonesia, 2018
Tangshan Earthquake, China, 1976
Tohoku Earthquake and Fukushima Tsunami, Japan, 2011
Valdivia Earthquake, Chile, 1960

HYDROLOGICAL
Bangladesh Flood, 1998
Big Thompson Canyon Flash Flood, Colorado, 1976
BP Deepwater Horizon Oil Spill, United States, 2010
Brisbane and Queensland Flood, Australia, 2011
Colombia Floods, 2010–2011
Colorado Flood, United States, 2013
Grand Forks Flood, North Dakota, 1997
Great Mississippi River Flood, United States, 1993
Iowa Flood, United States, 2008
Johnstown Flood, Pennsylvania, 1889
Kerala Floods, India, 2018
Mozambique Flood, 2000
Pakistan Flood, 2010

Summer Floods, United Kingdom, 2007

Vietnam Flood, 1999

Yangtze River Flood, China, 1931

METEOROLOGICAL

Bhola Cyclone, Bangladesh, 1970

Buffalo Blizzard, New York, 1977

Cyclone Gorky, Bangladesh, 1991

Cyclone Nargis, Myanmar, 2008

Cyclone Pam, Vanuatu, 2015

Cyclone Phailin, India, 2013

Cyclone Sidr, Bangladesh, 2007

Edmonton Tornado, Canada, 1987

Hurricane Andrew, United States and the Bahamas, 1992

Hurricane Charley, United States, 2004

Hurricane Galveston, United States, 1900

Hurricane Harvey, Texas and Louisiana, 2017

Hurricane Ike, United States, 2008

Hurricane Irma, Florida, 2017

Hurricane Katrina, United States, 2005

Hurricane Maria, Puerto Rico, 2017

Hurricane Mathew, United States, 2016

Hurricane Mitch, Central America, 1998

Hurricane Stan, Guatemala, 2005

Joplin Tornado, Missouri, 2011

Superstorm Sandy, United States, 2012

Tri-State Tornado, United States, 1925

Tropical Storm and Floods, Yemen, 2008

Typhoon Haiyan (Yolanda), Philippines, 2013

Preface

When I set out to compile *Natural Hazards and Disasters: From Avalanches and Climate Change to Water Spouts and Wildfires* in 2017, I intended to complete the project within two years. As most projects go, however, it took longer than I expected. Volume 1 includes 25 individual types of hazards, ranging from landslides to climate change, with each entry averaging 5,000 words, including the establishment of each hazard in a global context. With few exceptions, most of the entries in volume 1 are arranged as follows: introduction, physical dimensions (e.g., magnitude, duration, frequency, seasonality, areal extent, and dispersion), geographic distribution, impacts (i.e., deaths, injuries, and property and infrastructure damage), preparedness and response, prevention and mitigation strategies, and conclusion (see the section "Guide to Related Topics"). Volume 1 also highlights 19 of the world's major international aid agencies, including Doctors without Borders, Catholic Relief Services, the Lutheran World Federation, the Mennonite Central Committee (MCC), ActionAid International, the American Red Cross, Concern Worldwide, and the International Organization for Migration. Each one of these entries is 2,000 words long.

Volume 2 describes 70 natural disasters that have occurred long ago, such as the 1889 Johnstown Flood, Pennsylvania, or recently, such as the 2018 Kerala (India) flood. Each entry is 2,000 words long and contains a description of the event (including physical dimensions), causes (if appropriate), preparedness, warning and evacuation, impact, response and relief efforts, recovery and reconstruction, and conclusion (see the section "Guide to Related Topics").

I contacted over 100 young, emerging, and established scholars in the multidisciplinary field of natural hazards and disasters, and 55 of those scholars agreed to write one or more entries for this work. I was appreciative of the responses of my fellow hazard and disaster researchers. As a book review editor of *The Professional Geographer* (2011–2012), I knew that academicians are reluctant to contribute to nonrefereed writings, but with minimal exceptions, contributors to this work are experts in the hazard field in general or specific hazards and disasters in particular. Most of them contributed entries for both volumes based on their firsthand knowledge about the specific disasters, including personal field surveys after selected natural disasters and published papers in refereed journals on these specific topics. In addition to primary sources, the contributors also included materials from secondary sources. Several university students are also contributors in the volumes.

The extreme events selected for volume 2 include a range of disasters that have occurred since 1889 because relevant and reliable information for many disasters that occurred before that time is not readily available. Several internet sources list the 25, 10, or 5 deadliest natural disasters in history, including nonnatural disasters, such as the Black Death in 1348, Great White Plague of the 1600s, the North American Smallpox Epidemic in 1775, and Spanish Influenza in 1918. However, this work focuses on natural hazards and disasters only.

One or a combination of physical characteristics, as well as the number of fatalities and injuries, the geographic area or people affected, and monetary damage (in millions of dollars), were considered when selecting the natural disasters for volume 2. A distinction was made between developed and developing countries in applying these criteria, particularly for the number of fatalities and injuries and monetary damage. For example, although the death of 10 people is a relatively significant figure for developed countries, disasters cause significantly more fatalities in developing countries due to overpopulation, widespread poverty, and a lack of resources to adequately prepare citizens for extreme natural events.

Similarly, in monetary terms, disaster damage of $1 million has different effects in developed and developing countries. In addition, economic thresholds are less meaningful over time. Therefore, the inflation factor must be seriously considered when selecting natural disasters that occurred over a relatively long period of time. This work includes both high- and low-magnitude events in order to understand their damaging power. Overall, although an attempt was made to maintain geographic balance when selecting extreme events for these volumes, a majority of extreme events were deliberately selected from the United States.

Although both volumes are written for high school students, they are ideal reference texts for undergraduate and graduate courses on natural hazards and disasters. My hope is that they will provide an indispensable resource for all levels of students, as well as lay readers and hazard and disaster experts, disaster managers and planners, and international humanitarian organizations involved in responding and recovering during postdisaster periods. Comprehensive coverage of each topic will introduce readers to major worldwide natural hazards and disasters that have occurred recently and in distant past.

I have received unconditional support, cooperation, and help from several colleagues, friends, and students from my own and other institutions. I am particularly thankful to Dr. Charles W. Martin (professor and head, Department of Geography and Geospatial Sciences, Kansas State University), Dr. Jeffrey Smith (Kansas State University), Avantika Ramekar (Kansas State University), Dr. Luke Juran (Virginia Tech), Dr. Mitchel Stimers (Compita Consulting, Junction City, Kansas), and Dr. Jennifer Trivedi (University of Delaware). I would also like to thank Patrick Hall, development editor, Barbara Patterson, project coordinator, ABC-CLIO, and Jaishree Thiyagarajan, editorial project manager, Amnet, for their guidance and patience during the successful completion of this project. Finally, my deepest gratitude goes to my wife, Anjali Paul; daughters, Anjana and Archana; and son, Rahul, for their love, encouragement, and support over the years and particularly during the time I was busy in editing these two volumes.

Introduction

Though the terms *natural hazard* and *natural disaster* are often used interchangeably, there is a recognized distinction between the two. Natural hazards are threats to people and have the potential to kill and injure them as well as cause considerable damage to their property and environment. In contrast, natural disasters occur when the potential turns into reality (Alexander 2000). But not all hazards necessarily become disasters. If the actual event is not large enough and/or does not affect (harm) people, it is not considered a disaster, but rather remains a hazard. For example, an earthquake of considerable strength that occurred on an uninhabited island cannot be classified as a disaster because it does not affect people.

Traditionally, a natural hazard or disaster has been considered a freak event of nature or an "Act of God" that is punishment for a breach in religious laws. For example, after Hurricane Katrina damaged the Gulf Coast on August 28, 2005, many religious groups within the United States claimed that the hurricane was sent to punish the residents of New Orleans, a city known for disorderly festivals. The view of the natural hazards and disasters as punishment from God disregards any role humans may play in causing them or in intensifying or reducing their impact. In truth, however, humans do play a role in the occurrence of natural hazards and disasters. For example, floods can occur from excessive rainfall, whose effects are exacerbated by human actions (e.g., intensive use of land or construction of dykes along river banks, or from a combination of both). For example, Houston, Texas, experienced a devastating flood in 2017, which was caused by record-breaking quantities of rain dropped by Hurricane Harvey and intensified by the city's sprawl and poorly regulated residential growth. All this took place after paving of the sawgrass prairie reduced the ground's capacity to absorb rainfall. To further compound the issue, sometimes it is not easy to distinguish between "natural variables" and "human actions:" for instance, what if a variability in rainfall, which seems "natural" enough, itself is caused by human actions?

Natural hazards and disasters may not happen just to us but often may happen because of us. Recent definitions of natural hazards and disasters directly acknowledge the role of humans in causing and exacerbating such events (Montz et al. 2017). Many contemporary hazard researchers widely believe that natural hazards and disasters are less acts of nature and more the result of human actions. Others maintain that these events are the outcome of human failures to adequately prepare for them. Furthermore, contemporary researchers argue that "natural" hazards or disasters are misleading terms, as such events tend to have a social impact in terms of who is affected most. These researchers assert instead that natural hazards or disasters are "Acts of Humans." This perspective is useful in mitigating disaster impact and reducing frequency or intensity.

Moreover, natural hazards and disasters are generally defined in a negative way. However, such extreme events can have some positive consequences. For example, forest fires have beneficial ecological consequences. Similarly, flood water recharges groundwater and deposits fertile sediment in agricultural areas, and ash produced by volcanic eruptions increases soil fertility. Nevertheless, these benefits tend to become apparent months or years after an extreme event. For this reason, some researchers maintain that extreme events should be understood as neutral (Puleo and Sivak 2013). These events cause damage to places, people, and environment, but over time they establish a new order that is often better than that predisaster. The researchers claim that: "London would not be the same city it is today had it not been for the fire of 1666, nor would San Francisco have undergone the same pattern of development and expansion it did had it not suffered an earthquake in 1906" (Puleo and Sivak 2013, 458).

Next, natural hazards and disasters arise from interrelated causes and their classification is often difficult. Despite their technical difference, as discussed earlier, "hazard" will be used hereafter for the sake of simplicity to include both hazards and disasters. On the basis of the nature of physical processes involved, some researchers categorized natural hazards into four classes: (1) meteorological (e.g., tornadoes, windstorms, ice storms, and tropical cyclones/hurricanes/typhoons), (2) geological (e.g., earthquakes and tsunamis, volcanic eruptions, landslides, and subsidence), (3) hydrological (e.g., floods, droughts, and wildfires), and (4) extraterrestrial (e.g., meteorites) (Montz et al. 2017).

It is important to mention that an extreme event may fit within more than one of the preceding groups. For example, large-scale volcano eruptions (a geological event) can block incoming sunlight, potentially enough to trigger cold waves (a meteorological event). An earthquake (a geological event) can cause tsunamis (a hydrological event) that occur under sea water, as evidenced by the 2004 Indian Ocean Tsunami (IOT) or the 2011 Japan Earthquake and Tsunami. Further complicating the issue is the fact that one disaster often leads to other disasters. For example, earthquakes produce landslides, liquefaction, tsunami, coastal flooding, and fire.

Also, the disaster that initiated other disasters is called the primary disaster, and the disasters initiated by the primary disaster are often referred to as secondary or collateral disasters. Both primary and secondary disasters are called cascading disasters. Secondary disasters often cause far more damage and problems than a primary disaster (Montz et al., 2017). For example, the 9.0-magnitude Japan earthquake of March 11, 2011, killed more than 18,000 people. But only about 100 people died as a direct result of the earthquake, which originated off the Pacific coast of Tohoku. However, the massive underwater earthquake triggered a tsunami with about 30-foot (10 m) high waves that damaged the Fukushima Dai'ichi nuclear reactors. Subsequent melt down of the reactors led to the evacuation of

200,000 people from the surrounding area. Thus, the term *secondary* does not refer to scale but rather sequencing. However, for many cascading disasters, it may be difficult to allocate all fatalities or losses to a single event. When more than one hazard event impacts the same area, a multiple hazard situation arises. These different hazard events may occur at the same time or may be spaced in time.

Each natural hazard has unique characteristics and shares characteristics with other hazards. Knowing these characteristics is essential to understand their ability to cause damage and destruction, determine how great an impact these extreme events will have on people or affected-areas, or assess what measures and response are appropriate for reducing the impact of each hazard. An understanding of these characteristics is also necessary to detect similarities among natural hazards and to undertake appropriate disaster management practices.

Among all physical characteristics or dimensions of natural hazards, the most important one is magnitude. Although it varies over time, magnitude indicates the strength of an extreme event at a specific location or area. Generally, the greater the magnitude of the hazard, the greater the destructive power of the event. Magnitude is expressed in different ways for different hazards. For example, tornado magnitude is expressed in terms of the Enhanced Fujita (EF) Scale, which was introduced in 2007. It replaced the Fujita Scale which was used between 1971 and 2006. Based on extent of damage, the EF Scale ranges from 0 (light damage) to 5 (incredible damage). Earthquake magnitude, on the other hand, is expressed by the Richter Scale and it ranges from 1 through 10; the higher the number, the more severe the earthquake impact.

Duration is another important physical dimension of natural hazards; it describes how long an event persists in terms of seconds (e.g., earthquakes and landslides), minutes (e.g., tornadoes and tsunamis), hours and days (e.g., flash floods and hurricanes), weeks and months (river floods), and years (e.g., drought and climate changed-induced sea level rise). It is relatively easy to measure, making possible a degree of generalizability across hazards.

Frequency describes how often a natural disaster of a given magnitude occurs over a given period of time. This can be expressed in different ways for different disasters. For floods, frequency is generally expressed in terms of a recurrence interval or return period. For this event, a return period of 20 years suggests that in any given year, a flood of that magnitude has a 1 in 5 (20 percent) chance of occurring. It is therefore important to understand the relationship between frequency of an event and the magnitude of the event. Generally, larger events occur less frequently than smaller events. Another physical property of hazards is their seasonality. Thus, certain hazards occur more frequently during a given season. For example, blizzards and snowstorms are clearly winter phenomena in areas with temperate climate, while heat waves and tropical cyclones are typically summer events.

Areal extent and spatial distribution/dispersion are also two important dimensions of hazards. The former refers to the area over which a hazard event occurs. For example, a tornado generally affects a small area, while a drought affects large geographic areas. Meanwhile, dispersion refers to the distribution of a hazard across a geographic area in which the hazard occurs. A particular hazard does not occur in all geographic locations. For example, hurricanes occur in the Gulf of Mexico and on the Atlantic coasts of the United States, but not in the Pacific. Similarly, snowstorms occur only in temperate latitudes, not in tropical areas.

The remaining two important physical dimensions of hazards are rate of onset and diurnal factor. The former is the speed at which an extreme event converts from its first appearance to peak strength, or the length of time between the first appearance of the event and its peak. There are several types of rate of onset: very rapid (e.g., earthquakes, landslides, and tornadoes), or fairly slow (e.g., hurricanes and river floodings), to extremely slow (e.g., droughts, climate change, desertification, deforestation, and subsidence). Rapid-onset disasters are also called sudden-onset disasters and slow-onset disasters are called "creeping" disasters. Finally, the diurnal factor refers to the occurrence of an event at a specific time of day or night. For example, thunderstorms usually occur during the late afternoon in response to heat buildup during the day. In contrast, several disasters such as floods, droughts, or earthquakes can occur anytime. Duration, frequency, seasonality, and diurnal factors represent temporal characteristics of natural hazards. The impact of a particular disaster depends on the combination of several of its physical characteristics.

Bimal Kanti Paul

Further Reading

Alexander, D. 2000. *Confronting Catastrophe: New Perspective on Natural Disasters.* New York: Oxford University Press.

Montz, B. E., G. A. Tobin, and R. R. Hagelman, III. 2017. *Natural Hazards: Explanation and Integration.* New York: The Guilford Press.

Paul, B. K. 2011. *Environmental Hazards and Disasters: Contexts, Perspective and Management.* Hoboken, NJ: Wiley Blackwell.

Puleo, T. J., and H. Sivak. 2013. Introduction: The Ambivalence of Catastrophe. *Geographical Review* 103: 458–468.

Natural Hazards

Avalanches

Mass movement events constitute a class of disasters in which large amounts of material overcome the cohesive forces holding the mass in place and move downslope. Avalanches are one type of mass movement, with two major subtypes: snow avalanches and rock avalanches. Snow avalanches are also called snowslides and are composed of snow, although it is common for a snow avalanche to quickly pick up other forms of debris once the slide is initiated. Rock avalanches, as the name suggests, are composed of rocks of varying sizes and may also be referred to as debris slides or rockslides, but they should not be confused with landslides (a large mass of cohesive soil flow) or rockfalls (airborne rocks, falling in such a way as to not be in contact with the ground), which are two different types of mass movement event. Snow and rock avalanches exhibit several common characteristics in their formation and motion, such as the reliance on gravity, friction, moisture, and relief, and even in the way in which they flow. Rock avalanches can be affected by a process called fluidization, whereby air cannot escape from spaces between particles, which can cause the entire flow to act more like a fluid than a mass of solid rocks. Climatic factors, including most notably temperature and precipitation, play key roles in the stability of slopes and in the potential for slope failure; because of this, the study of global climate change is paramount in understanding the changing dynamics and frequencies of this natural disaster type.

PHYSICAL CHARACTERISTICS OF AVALANCHES

Both snow and rock avalanches share several common characteristics, with the most basic being that sliding can be broadly defined as the occurrence when gravity and pressure (specifically water pressure in the case of rock avalanches) exceed the shear strength of a block. First, material on a slope is necessary for a slide to be initiated; as such, avalanches are seen only in areas with significant relief. As material gathers on the slope over time, a constant struggle between cohesion and gravity takes place over the basal plane, which can also be thought of as the "base plane" of the material on the slope. Cohesion is the ability of one object to successfully remain in place in relation to another through static friction, which is the force of friction present between two bodies at rest. Gravity works to pull material downslope, but if the force of static friction between the two objects is greater than the force of gravity attempting to pull the object situated on top of the underlying surface, then the object on top will move downward. Regardless of whether the slope is covered in snow or soil (and rock), this relationship is at the center of the mechanics controlling downslope movement.

The speed at which material moves is also directly related to gravity and slope. With steeper slopes, more potential energy is stored in the material at rest, and once a slide is initiated, the material in motion will naturally contain more kinetic energy. Thus, with steeper slopes come quicker acceleration and overall higher speeds as the material makes its way down the grade. Moisture also plays a role in initiating a slide, as does the presence (or absence) or vegetation and its root systems. The addition of moisture may work to decrease the force of static friction, thus allowing a portion of the mass to break away and slide downslope. Weather conditions, specifically precipitation, are important common factors. For rock avalanches, heavy rains can often serve as the catalyst to a slide event, while for snow avalanches, a heavy snowfall event may add enough weight to an existing slab to initiate movement. Jarring or some other sudden application of force may be enough to allow the material to overcome the cohesive forces and begin moving. Such initiators could include geologic events such as earthquakes and volcanic eruptions or human-caused triggers such as recreational activities (e.g., snowmobiling and skiing, although these typically are associated with snow avalanche initiation). Military exercises have also been known to initiate snow avalanches, more specifically artillery explosions.

THE TWO PRIMARY FACTORS IN ROCKSLIDES

Land masses at rest on a slope pose little danger; it is when two primary actors are factored in, and become drivers of the slope's changing conditions, that mass movement events become a threat. For the dry mass movement event considered first, rockslides, there are two primary factors: slope overloading (also referred to as oversteepening) and moisture content. As with all other mass movement types, the relationship between the slope's cohesive ability and gravity is another vital component to consider.

Overloading

The balance of a slope may be altered in one of two ways: the addition of more material (load) to the slope and/or the increase of the steepness of the slope. Adding more load to a slope means that the weight from the overburden material has increased, and with that comes increased downward pressure on the slope, which is at some angle usually greater than 45 degrees but less than 60 degrees. With this increased pressure, the potential for failure is increased. There are two ways in which load can be increased; one is the more obvious, adding more material to the middle or upper portions of the slope, and the other is to remove material from the bottom, termed the toe, of the slope. Each of these occurrences not only will increase the force of gravity on the slope but may also increase the coefficient of static friction; thus, adding more load does not translate directly into slope failure. Overloading can result from human intervention, most typically building in high-relief areas, which can create the demand for material removal or addition, or by natural means such as erosion. Many problems tend to arise in areas where people desire to build in high areas for scenically driven reasons, as areas along waterbodies that are perched along high vistas exist in that condition due to natural forces that once altered can have unintended consequences. Problems of slope failure due to overburdening are common in coastal areas, where slope toes are placed under more stress due to the increased load (structures) on or at the top of the slope.

Moisture Content

Water content in a slope helps provide cohesion for that slope, but the addition of too much water will eventually trigger a slide. Dry sand has a gentler angle of repose (the angle at which the structure will fail) than does wet soil. For this reason, some water content is necessary in order to maintain stability on the slope. However, precipitation will add water content to a slope, and if added in large enough quantities, it can significantly weaken the slope by invading pores in the soil, which increases pore pressure while simultaneously adding more weight (load) to the slope, which will ultimately cause the slope to fail. Humans can also add water to slopes through direct infiltration or raising of the local water table via leaky municipal pipes, swimming pools, lawn care habits, and agricultural irrigation. The removal of vegetation, whether natural or human caused, can further cause more water to be taken up by slopes, since plant life both consumes water and evaporates it back into the atmosphere, and both functions serve to divert water away from the soil. Further, thaw-induced rockslides are commonly initiated when shear strength of a slope is compromised by the creation of fractures from thawing ice.

COMPONENTS OF SNOW AVALANCHES

Avalanches have three major components: snowpack, terrain, and weather. Each of these elements works together to create the specific conditions that allow for the formation of avalanche-prone sections of snow-covered hills and mountains. For all types of mass movement, gravity is an important consideration, as it is the interplay between the slope's ability to resist gravity and the pull of gravity that initiates a slide of any type. Other mechanical factors are also important, but any amount of outside influence will not result in a slide unless the ability of the material to stay in place (cohesion) is overcome by gravity. Thus, the three elements of avalanches all relate to formation and the potential to slide.

Snowpack

Concerning snow avalanches, the condition and characteristics of the snow are obviously important; the overall collection of snow on a slope is referred to as snowpack, and standing snowpack is a body of snow that has persisted from one season to the next. These are layers that can build up over several winter seasons, with layers from subsequent seasons having different characteristics than those from previous accumulations. These characteristics include tensile strength (the ability to withstand stretching), ductile strength (the ability to bend or change shape without breaking), shear strength (material breaking away at an angle), a coefficient of friction, grain size and type, and temperature. Typically, a weak layer situated below a cohesive layer is a key factor in determining whether an avalanche event will occur.

Terrain

As with rockslides, terrain plays a critical role; however, for snow avalanches, the concept of collection of snow is slightly different. While rock avalanches are occurrences in which the actual material is part of the slope slides, in a snow avalanche, the slope acts as the collecting surface only and does not itself slide during the avalanche event. During snowfall, surfaces that are too gentle (typically 25 degrees or less) may collect snow, but they are not steep enough to allow for gravity to move mass amounts of the material downslope. Conversely, slopes that are too steep (typically 60 degrees or more) simply will not collect adequate volumes of snow to pose a danger. The steepness of the slope must be such that it will allow for the collection of enough snow to eventually slide while providing a steep enough angle to allow for gravity to greatly accelerate the slide once it is initiated. The starting point of the slide, usually on slopes of 30–45 degrees, the pathway the slide follows (the track), and the runout (where the snow travels over flat ground and eventually comes to rest) are all directly influenced and controlled by the element of terrain.

Weather

Meteorological conditions can be highly variable over short distances on the earth's surface, owing to changes in elevation, proximity to large water bodies, and latitudinal position. Snowpack is the primary physical component of a slope during a pre-avalanche and is heavily dependent on weather conditions to develop its characteristics. Areas of high altitude (colder) and latitude (more precipitation in the form of snow) are typically those with the proper weather conditions for standing snowpack to develop. Due to the moderating effect of large water bodies, areas inland are typically those that see less stability in avalanche-prone slopes, while coastal locations tend to be more stable. Cycles of freezing and thawing as days progress also play a critical role, resulting in surfaces that are relatively unstable during the day as thaw occurs compared to nighttime conditions when freezing (and thus stability) is more common, producing loose crystals called hoarfrost that can result in additional weakness on the slope.

TYPES OF SNOW AVALANCHE

There are several different ways to classify snow avalanches. The slab avalanche occurs when a large block, or slab, of snow breaks away from the larger deposit of snow and moves downslope. It is this particular type of slide that is responsible for the majority of recreational skiing deaths. The slab avalanche will be characterized by a crown fracture at the top and flank (lateral) fractures along the sides, which demarcate the area of the slide that was entrained or captured in the slide. A fracture will also appear at the bottom and is termed the stauchwall. Wet snow avalanches are typically low-speed events, due to the fact that the mass is heavy with moisture. However, the relatively low speed (up to 25 mph, or 40 kph) cannot be underestimated, as the mass of the flow supplants the lack of speed in terms of its ability to deliver a great deal of force to objects in its path, such as structures. Powder snow avalanches, by contrast, move as fast as about 186 mph (300 kph), building great force due to those high speeds, and can run out over long flat distances. A unique type of snow avalanche is the ice avalanche, which as suggested by the name is composed of pieces of ice of varying size usually resulting from breaking away from a glacier. As is often repeated through urban myth, but is not true, avalanches are not triggered by sound.

CLASSIFICATION OF SNOW AVALANCHES

Rockslides have no specific classification scheme, but attempts have been made to categorize its counterpart—the snow avalanche. While there is no universally agreed-upon classification scheme for snow avalanches, there are several ways in which the events are categorized. In Europe, the European Avalanche Risk Table adopted in 1993 is employed to assign a risk level of 1–5 (with 1 being the lowest risk and 5 being the highest) for slopes and describes the snow stability as well as the avalanche risk in general. The European Avalanche Size Table serves a similar function, but it does not assess risk level; rather, it categorizes avalanches into five types—small, medium, large, very large, and extremely large—based on increasing size, runout, and potential damage. In North America, Canadians and Americans use the North American Public Avalanche Danger Scale, which, like the European version, ranks risk potential ascending from 1 to 5 as well as provides a classification of avalanche size via tables describing the destructive potential (CAIC n.d.).

MAJOR AVALANCHE EVENTS

Deaths from mass movement events are difficult to classify by exact type, but most are caused by landslides, which, as indicated, is not an interchangeable term with *rockslide*, *debris avalanche*, or *rock avalanche*. Comparatively, few people die each year as a result of landslides or rockslides in the United States (up to 50 annually), but hundreds and possibly thousands more die worldwide (many deaths go unreported in developing countries) due largely to slides triggered by volcanic activity (USGS n.d.). It is much more common for larger groups of less-affluent people to live near volcanic hazards in developing countries than in developed countries; thus, marginalized groups are disproportionally affected by avalanches.

Although landslides, as an event, can boast of many events in which more than 1,000 people have been killed, avalanches cannot lay claim to a similar record. There have been just three snow avalanches where the number of people killed went into the thousands, several dozen with death tolls in the hundreds, and many events with 100 fatality counts. Most notably on this list are two events resulting on the same mountain range, Huascarán, in Peru. In 1960, a snow avalanche occurred that claimed the lives of up to 4,000 people and destroyed nine towns at the base of Nevados Huascarán. This event, however, would pale in comparison to an earthquake-induced slide in the same area a few years later.

On May 31, 1970, the Ancash earthquake triggered a debris avalanche made of ice and rock that tumbled down from the highest peak in Peru, an event that was predicted by two American scientists eight years earlier, at which time they were told to retract the statement or face arrest; they left the country. Moving at over 100 mph (160 kph) and launching one-ton boulders as much as about 2.4 miles (4 km) into the runout area, the mass ultimately affected over 25,00 square miles (64,000 square km) and killed an estimated 18,000–22,000 people (Anonymous 1970), but estimates range as high as 25,000 killed. The massive rush of air in front of the flow was said to be strong enough to knock people over and, as it contained small rock particles, strip bare skin. A total of approximately 80,000 people died from the event as a whole, earthquake deaths included. The town on Yungay was almost completely destroyed, with only 350 of the town's 25,000 surviving. The government of Peru dedicated the site as a national cemetery, which also means study in the form of an excavation is forbidden, and the modern-day city of Yungay is located within 1 mile north of the disaster site.

Human-induced slides are also responsible for some of the largest events in history. On what is now known as White Friday, December 13, 1916, saw of these major slides. The Dolomite Mountains in northeastern Italy hosted several encampments for soldiers during World War I (1914–1918), and it is thought that each side was purposefully firing artillery shells above enemy locations in order to destabilize the slopes and cause slides, easily destroying the wooden barracks that made up the camps. Aiding in these efforts was the heavy snowfall in the Alps that winter, which made ripe the conditions for avalanches due to increased snowpack. Both Austrian and Italian positions were caught in such slides, killing an estimated 2,000 soldiers, although some estimates place the figure as high as

10,000 total for the month of December 1916, making it the second worst avalanche disaster in history.

The only other known event to kill more than 1,000 people was the 1618 Plurs, Switzerland, avalanche. Called the Rodi avalanche, it struck and completely covered the community of Plurs, killing 2,427 people. Occurring as a series of approximately 650 avalanches, the Alps along the Switzerland/Austria border were home to what is known as the Winter of Terror, which saw 265 people killed during the winter season. Also near the middle of the twentieth century, Blons, Austria, was destroyed by two avalanches that occurred within nine hours of one another. The first killed 85 people, and the second killed an additional 115, including many people who were attempting to rescue those buried by the first event. The Blons/Montclav avalanche event is the worst in Austria's history. In 1979, unusually heavy snow in the Lahaul Valley in northern India caused several avalanches in the state of Himachal Pradesh, killing 200. Other notable events include 150 killed by events on Mount Kazvek, North Ossetia, Russia; 310, 201, and 172 people killed in avalanches in Afghanistan in the years 2015, 2012, and 2010, respectively; and 138 people killed in avalanches in the Japanese village of Mitsumata. In 2012, 138 lives were lost to events in Pakistan, and an additional 102 died in 2010; 125 people died in 2002 because of slides in Russia.

The deadliest avalanche in the United States occurred in 1910, in Wellington, Washington, which killed 96 people. Eleven feet (3.4 m) of snow fell over the course of nine days in February of that year, trapping a train at a depot in Spokane. On February 28, a massive slab broke loose, having been initially loosened by rain and warmer weather that followed the storms but ultimately severed from the slope by a lightning strike. The resulting avalanche threw the train off the tracks and into Tye River Valley below. Amazingly, 23 people survived. The only other avalanche in U.S. history to come near the Wellington event in terms of lives lost was the Palm Sunday disaster, which occurred on the Chilkoot Trail, Yukon, Alaska, on April 3, 1898, killing 65. Canada's worst avalanche disaster took place on March 4, 1910. The Rogers Pass Avalanche happened in close proximity, both in time and space, to the Wellington, Washington, event and resulted from the same system of winter storms that brought so much snow to that area. A work crew of 63 men were attempting to clear the tracks from a previous slide when a slab from the aptly named and nearby Avalanche Mountain broke loose, killing 62 of the crewmen.

AVALANCHES AND HUMANS

Avalanches can, and often do, occur without warning, and the conditions that can lead to a slide will vary by geographic location and can even vary considerably within small areas of just a few square miles. While attempts are being made to rely on sensors to detect motion precursors to such events, they remain notoriously rapid in their onset when compared to many other hazard types such as heatwaves, droughts, and hurricanes. This disaster type is not the largest contributor to deaths annually as a result of natural forces, but there have been a few events in history that have claimed thousands of lives. Preparing and planning for avalanches can pose different problems depending on what type is being considered. People do not live in high-altitude, snow-covered areas in the same numbers as those who live in more temperate areas. The simplest solution to preventing death and destruction due to slope failure is to simply choose not to build in the most slide-prone areas, although this is a multidimensional concept, consisting of more than just idea of choice. Prevention of snow avalanches can include such measures as using explosives to trigger small slides before large buildups can occur and installing snow nets or other barriers.

Surviving an avalanche is a grim prospect. Deaths result from avalanches in two major ways: blunt force trauma from collisions with objects, the person either colliding with the object or being hit by falling debris, or suffocation. In the case of snow avalanches, hypothermia may be the primary cause of death if the victim is located in an air pocket, reducing the possibility of dying by suffocation. The time in which a person can expect to survive a snow avalanche while waiting for rescue is very short, 20–30 minutes depending on the conditions of the snow, and only slightly longer if a person is trapped under a soil- or rock-based slide. The best advice concerning avalanche safety is to simply avoid areas that pose the highest risk. If traveling into areas known to harbor a higher threat of avalanches, people should either travel with guides who know the area or at least become as familiar with the local conditions as possible before venturing into backcountry, where rescues are more difficult.

AVALANCHES AND CLIMATE CHANGE

The far-reaching effects of climate change are connected to virtually every physical system on the planet. From the rainforests to the deserts, no ecosystem or biome will be able to escape from some change related to an altered climate system. While it may seem counterintuitive to connect a mass movement event such as landslides, rockfalls, rock/debris slides, and snow avalanches to climate, as discussed earlier, weather (and thus its longer temporal counterpart, climate) is closely linked to such disaster event types. Rockslides and snow avalanches are different in several key ways, and as such, climate change dynamics are affecting the actualization of those events differently. Still, there are some similarities, mostly related to load, either snowpack for snow avalanches or oversteepening for rockslides. Examined in its most basic form, the addition of more weight in the form of more moisture will increase the load on a slope, causing a slide to be more likely.

With a warming world, the warmer atmosphere will hold more moisture, which means more precipitation throughout the year, which in turn results in higher load weight. Increased temperatures will also be common contributors to the frequency of these event types (Sidle and Ochiai 2006). The similarities essentially end there, however, as soil-based disasters incorporate the element of moisture differently than snow avalanches. Further, the diurnal cycle of temperature changes, and the seasonal freeze-thaw cycle will be present in locations that house the potential for both rockslides and snow avalanches, but the mechanical processes that those cycles enact on the two different types of slide are markedly different. It should be noted, though, that records of slides are relatively incomplete, with few events recorded previous to the period (roughly) 1850–1900, and with those sparse records comes the reality that prediction of future frequency of slide events may prove a formidable task, with advanced modeling, including coupled ocean-atmosphere models, becoming an important tool in the study of these phenomena. A further challenge presented here is the high degree of variability exhibited on slopes even within the same region, and by extension, globally; modeling efforts will need to be sensitive to local conditions if they are to have any hope of providing realistic and useable predictions for slide occurrence.

Rockslides and Climate Change

The most prominent structural difference between a slope prone to rockslides and one prone to snow avalanches is obviously the material that the slope is composed of: rockslides require rock and snow avalanches require snow, and those two natural occurrences are not the same. Rocks, when subjected to the freeze-thaw cycle, experience a process termed *frost wedging* (which is sometimes interchanged with the term *freeze-thaw cycle* in a general sense). When water freezes, it can expand up to 9 percent of its original volume, and when this occurs in small spaces and cracks in rocks, the rock will slowly be broken apart into smaller units. However, if a large block of rock breaks away from a vertical surface, the free-falling rocks can eventually crash into the slope, initiating a rockslide. Additional moisture during winter months may cause accelerated frost wedging, which could result in more rockslides. Additionally, during the summer months, a similar process occurs, although without the help of expanding water. At very small scales, rock expands as it warms and can cause large sections to break away and tumble down slopes (Collins and Stock 2016).

One of the most apparent aspects of global climate change is global warming, and it is this warming in high-altitude areas that may warm rocks enough to cause an increase in frequency of slides, as heat-driven expansion will act as a catalyst for such events. Further, the availability of moisture is of concern, as more moisture on a slope could prove to be enough extra load to initiate a slide; with a warmer world, more precipitation will come on the ground (Seneviratne et al. 2012). In addition, increased spring runoff from increased winter precipitation could impact load on a slope, as rockslides are more likely to happen during periods of greater runoff.

Snow Avalanches and Climate Change

Relating climate change to the incidence of snow avalanches is similar to the relationship between climate change and rockslides in that the addition of moisture to the atmosphere will result in more snow during the winter months and, consequently, more weight on slopes, which can lead to more rapid and frequent slope destabilization. Additionally, with more heat in the atmosphere during winter months, more and earlier snow melting, including glacial melting, will occur, which may work to initiate more slide events through faster melting. During the summer months, more rain may be present in areas where persistent snowpack could become destabilized, resulting in slides. This is of special concern in glacial regions, where the addition of more meltwater has the capacity to function as a lubricant to the bottom of the glacier, which could lead to collapses like the one seen in Tibet in 2016, which killed nine people. Thus, the people can think of glaciers in high-altitude regions to be basically stuck to high-relief rock walls or rock walls with weak structures like various shale types; when the underside of a glacier of this type (or an area at the toe) becomes unstable, enormous amounts of snow and ice will slide into the valley below.

A lesser-known hazard resulting from avalanches in mountainous areas are glacial outburst floods. This occurs when a large slide moves into a valley and temporarily dams the flow of a river. With the newly formed ice dam in place, the water behind the impoundment will increase in the same way lakes form behind artificially created (by humans) dams. But the temporary nature of an ice dam makes it very unstable and dangerous, even if human settlements are far removed from the location. Once the buildup behind the dam is too great for the ice wall to hold in place, the structure fails, and an outburst flood is the result. (Failure can also result from erosion, seismic events, additional avalanches, or volcanic eruptions.) From the study of climate change, it seems that higher elevations are warming at a faster rate than other areas of the world. Given this point, understanding the potential magnitude and frequency of snow avalanches is an important area of study to mitigate against high-loss events of this type.

CONCLUSION

Rockslides and avalanches are caused by the failure of the slope to hold a mass of material in place. In general, the two types are similar in that they are both engaged in a fight against gravity, whereby the shear strength of the slope must hold in order to remain stable. When the forces of gravity overcome those of static friction, cohesion, and pressure, a downslope movement in the form of a slide will occur. While the actual slide may appear to be the result of the interplay of those forces alone, the dynamics of slope stability also heavily depends on factors related to meteorological conditions. Precipitation and temperature are key elements affecting whether or not the slope will remain stable. The addition of moisture can act to destabilize soil- and rock-based slopes; freeze-thaw cycles can split rocks, causing falls leading to slides, and the removal of ice during spring thaw months can reduce the shear strength of slopes enough to cause sliding. For mountainous areas prone to snow avalanches, the addition of more snow will add additional weight to the slope, leading to destabilization, and changing temperature and precipitation patterns can cause nonuniform development of alternating layers of snow, resulting in weaker cohesion between two planes.

In the context of global natural disaster threats, rockslides and snow avalanches have historically not been major contributors to annual death and damage tolls. However, with global climate change emerging as the most prescient problem in recent times, more attention will undoubtedly be paid to these disaster types. Few events in history have killed more than 1,000 people, and none of those were directly related to climate (the three deadliest events in history were induced either seismically or by humans). Understanding of these phenomena must continue to advance to predict them better in light of the changing conditions that affect environments.

Mitchel Stimers

Further Reading
Anonymous. 1970. Environmental Disaster of Nature and Man: The Peru Earthquake: A Special Study. *Bulletin of the Atomic Scientists* 26(8): 17–19.
CAIC (Colorado Avalanche Information Center). n.d. Avalanche Danger. https://avalanche.state.co.us/forecasts/help/avalanche-danger/, accessed August 12, 2019.
Collins, B. D., and G. M. Stock. 2016. Rockfall Triggering by Cyclic Thermal Stressing of Exfoliation Fractures. *Nature Geoscience* 9: 395.
Seneviratne, S. I., N. Nicholls, D. Easterling, C. M. Goodess, S. Kanae, J. Kossin, Y. Luo, J. Marengo, K. McInnes, M. Rahimi, M. Reichstein, A. Sorteberg, C. Vera, and X. Zhang. 2012. Changes in Climate Extremes and Their Impacts on the Natural Physical Environment. In *Managing the Risks of Extreme Events and Disasters to Advance Climate Change Adaptation*, edited by C. B. Field, V. Barros, T. F. Stocker, D. Qin, D. J. Dokken, K. L. Ebi, M. D. Mastrandrea, K. J. Mach, G.-K. Plattner, S. K. Allen, M. Tignor, and P. M. Midgley. A Special Report of Working Groups I and II of the Intergovernmental Panel on Climate Change (IPCC), 109–230. Cambridge, UK: Cambridge University Press.
Sidle, C. R., and H. Ochiai. 2006. *Landslides: Processes, Prediction, and Land Use*. Water Resources Monograph. Series 18. Washington, DC: American Geophysical Union, Washington, DC.
USGS (United States Geologic Survey). n.d. How Many Deaths Result from Landslides Each Year? https://www.usgs.gov/faqs/how-many-deaths-result-landslides-each-year?qt-news_science_products=0#qt-news_science_products, accessed August 16, 2019.

Blizzards

Mid-latitude cyclones and winter storms are significant atmospheric hazards with large variations in the intensity, timing, and location. Blizzards are the most dangerous of winter storms, as they combine very cold temperatures with strong winds and blowing snow and can spread hazardous conditions across wide geographic areas. Elements of a blizzard for a definitional purpose include a minimal wind speed and falling or blowing snow over an extended period as defined by a governmental agency, such as the U.S. National Weather Service.

Winter storm events that include heavy snow, ice storms, frigid temperatures, and blizzards have received widespread media attention in recent years. Since 2012, the Weather Channel (U.S.) has named significant winter storms each season analogous to the naming conventions of tropical cyclone basins; the meteorological offices of the United Kingdom and Ireland followed a similar format starting in 2013. Some North American winter weather episodes, such as the Arctic air mass outbreaks in early 2014 (as well as 2019) and the mid-Atlantic snowstorms of 2010, earned dubious monikers such as the "Polar Vortex" and "Snowmageddon." The heightened media interest coupled with above-average winter storm events has spurred investigation into whether or not this apparent increase in severe winter weather is part of an overall upward trend and/or a changing climate (Coleman and Schwartz 2017).

Atmospheric hazards cause significant disruption and damage to various socioeconomic sectors with annual losses in the billions. Average winter storm losses in the United States are around $1.2 billion per year. Winter storms may adversely impact agriculture (including livestock), transportation and utilities infrastructure, building integrity (e.g., roof collapse), commerce, school and work schedules, and human health. Winter storm impacts are most likely in high mid-latitudes with snow and subfreezing weather, conditions that may last for several months in the winter, and in lower mid-latitude locations (e.g., the southeastern United States), where snow and extreme cold are rarer and locals are less prepared. The "Superstorm" of 1993 and the New England Blizzard of 1978 were two of the top weather events in the United States in the twentieth century. In the twenty-first century, the Groundhog Day Blizzard of 2011 and the Eastern winter storm of January 2014 are two examples of winter events that resulted in multi-billion dollar losses and collectively resulted in over 50 fatalities (Coleman and Schwartz 2017).

DEFINITION AND HISTORICAL USAGE

Before entering the English lexicon as a winter weather occurrence in the late nineteenth century, a blizzard referred to a violent or forceful strike. American frontiersman Davy Crockett used the term *blizzard* in the 1830s to mean a blast, such as from a musket, or barrage of words in a tremendous argument (OED 2019). The first documented description of a winter storm matching blizzard characteristics was by mariner Henry Ellis (1721–1806) along the Hudson Bay in 1746. Ellis described a storm with strong northeast winds and intense cold that was filled with fine, hard particles of snow, although his meaning at the time was ambiguous (Wild 1997; Fitzhugh 1928). David Ludlum reported that the word was first used in the United States to describe an 1870 winter storm in Iowa (Ludlum 1968); however, other sources suggest that blizzards were synonymous with blinding snow or snow squalls in South Dakota and Kansas a decade earlier (Wild 1997; OED 2019). American meteorologist William Ferrel (1817–1891) speculated that blizzard derived from the German *blitzartig*, or "lightning like," to describe powerful snowstorms common to the German-settled Dakotas (Fitzhugh 1928). By the 1880s, blizzard was in general used to describe intense winter storms with poor visibility from strong wind and blowing snow (Schwartz and Schmidlin 2002).

Winter storms may range from a moderate snowfall that lasts a few hours to a blizzard that may last several days with impacts from the community to international level. Generally, snowfall and low temperatures define winter storms. Regional and national differences exist on definitions of a winter storm, as much depends on the local terrain (e.g., large water body proximity, elevation, mountains), storm frequency, and societal adaptation (e.g., snow removal methods). For example, the U.S. National Weather Service (Schwartz 2001) has established some key winter weather terms:

- Snow Flurries: Light snow falling for short durations with only light dusting or no accumulation
- Snow Showers: Snowfalls at varying intensities for brief temporal periods with some accumulation possible
- Snow Squalls: Brief, intense snow showers accompanied by strong, gusty winds with possibly significant accumulation
- Blowing Snow: Wind-driven snow
- Sleet: Rain that freezes into ice pellets before reaching the ground
- Freezing Rain: Rain that falls on a surface with the temperature below freezing point and forms an ice coating
- Winter Storm Outlook: Winter storm conditions possibly 48–60 hours in advance of winter storm
- Winter Storm Watch: Possibility of blizzard, heavy snow, freezing rain, or heavy sleet within 12–36 hours
- Winter Storm Warning: Combination of heavy snow, heavy freezing rain, or heavy sleet expected within 6–24 hours
- Blizzard Warning: Sustained or gusty winds of 35 mph (56.3 kmph) or more, falling or blowing snow creating visibilities at or below 1/4 mile (0.4 km) for at least three hours
- Lake Effect Snow Warning: Lake effect snow expected
- Wind Chill Warning: Wind chill temperatures below -34 degrees Fahrenheit (-36.67 degrees Celsius)
- Wind Chill Advisory: Wind chill temperatures between -20 degrees Fahrenheit and -34 degrees Fahrenheit (-28.87 and -36.67 degrees Celsius)
- Winter Weather Advisories: Issued for accumulations of snow, freezing rain, freezing drizzle, and sleet

Definitions for heavy snow are often subjective and/or vary with location. In the United States, for example, northern Ohio terms *heavy snow* as six inches (15.24 cm) or more snow in a 24-hour period, while in southern and central Ohio, five inches (12.7 cm) or more in the same period is considered "heavy snow" (Schwartz 2001). Snow intensity (and blizzard classification) is also based on visibility. The U.S. National Weather Service (NWS) for instance describes snow intensity as "light" with visibility greater than 0.5 miles (800 m), "moderate" with visibility ranging between 0.25 and 0.5 miles (400–800 m) and "heavy" with visibility less than 0.25 miles (400 m) (Schwartz 2001). Heavy snow intensity periods and diminished visibility are defining characteristics of blizzards.

An extreme form of winter storms is the blizzard that combines strong winds with falling or blowing snow to cause the potential of low visibility, deep snowdrifts, and extreme wind chill. No universally accepted definition of a blizzard exists, but measures often include a threshold for sustained winds and duration of reduced visibility from blowing snow. As defined by the U.S. National Weather Service (NWS 2018), blizzards have winds over 35 miles per hour (16 km per second) and falling or blowing snow causing visibility less than 0.25 miles (400 m) for at least three hours. An additional blizzard criterion of temperatures below 20 degrees Fahrenheit (-7 degrees Celsius) and 10 degrees Fahrenheit (-12 degrees Celsius) or less for severe blizzards was used for many years but was abandoned in the 1970s, and presently, many locations use the definition established by the U.S. National Weather Service with some regional variations (Schwartz and Schmidlin 2002). Environment Canada issues blizzard warnings when winds 25 miles per hour (11 m per second) or greater are anticipated to reduce visibility to 0.25 miles (400 m) or less from blowing or falling snow for at least four hours; however, regional definitions exist for locations north of the tree line (e.g., Northwest and Nunavut Territories). In the United Kingdom, the Met Office defines blizzards based on moderate to heavy snowfall, diminished visibility (0.12 miles, or 200 m), and sustained winds of 25 miles per hour (13 per second) or greater yet without a time period stipulation.

PHYSICAL FORMATION

Components of a winter storm are cold air, moisture, and a lifting mechanism such as a warm air mass ascending over a cold air mass (i.e., a frontal boundary). Winter storms are associated with mid-latitude cyclones, areas of relatively low pressure surrounded by counterclockwise (in the Northern Hemisphere) spinning surface winds that rotate inward toward the center of the system. The counterclockwise motion brings contrasting air masses (temperature, moisture, or both) together where a front forms. Winter cyclones begin as a disturbance along a stationary front separating cold and warm air. Eventually, pressure falls due to upper-level circulation characteristics (e.g., divergence aloft, positive vorticity advection), and counterclockwise circulation develops around the area of low pressure.

Development of a blizzard usually requires a strong, deep low-pressure system to form during the cold season. Intense low-pressure systems and blizzards commonly occur in regions with a strong winter thermal gradient whereby temperatures change rapidly over a short distance, such as along the polar front representing the boundary between colder, drier air masses poleward and warmer, humid air masses equatorward. Rapid changes in temperature, and thereby pressure, produce strong winds and blizzard conditions, and when coupled with falling snow or loose, previously fallen snow, they reduce surface visibility. Strong pressure gradients during winter are frequent along the backside (poleward side) of mature mid-latitude cyclones, edge of polar anticyclones, and boundary of major terrain shifts (e.g., continental coasts). Mountains can also play a role in enhancing uplift and increasing wind speeds.

Although blizzards are typically associated with powerful storm systems and significant snowfall accumulation, ground blizzards occur when high winds blow previously fallen snow. Ground blizzards are most common in large areas of flat, open terrain without significant vegetation to impede snowdrifts and diminish wind speeds. Depending on the wind orientation (horizontal or vertical), the blowing snow can be concentrated near the ground with relatively clear skies above or can reach hundreds of feet height. Blizzards in Canadian High Arctic, for instance, are often the ground-blizzard variety, a location with a strong winter pressure gradient, no trees or topographic obstructions, and perpetual light or powdery snow cover (Stewart et al. 1995).

GEOGRAPHIC DISTRIBUTION

Blizzards take place on all continents, from the poles to the high-altitude locations of the tropics, but they are most frequent in North America, central Asia, and Antarctica. Flatter terrain areas of continental centers during the Northern Hemisphere boreal winter with persistent snow cover are high blizzard activity zones. The northern interiors of Eurasia and North America are dominated by polar or Arctic air masses, and topographic obstructions for reducing wind speeds are minimized. Known locally as *purgas* or *poorgas*, central Russian blizzards have frigid temperatures and strong northeasterly winds around the Siberian High, which blow surface snow along the flat tundra into ground blizzards. Mongolia (and northern China) even have a term for winters with frequent blizzards and severe cold coupled with extreme livestock loss—the *dzud*, or "the white death"—such as the one that occurred during the 2008–2010 winters. The deadliest recorded blizzard in the world occurred in Iran when intermittent blizzard conditions over the course of several days in early February 1972 buried the country in 10–26 feet (3–8 m) of snow and resulted in over 4,000 deaths (Associated Press 1972).

In the contiguous United States and central Canada, the highest blizzard frequency occurs in the northern Great Plains and surrounding region (e.g., North and South Dakota, Minnesota, and Iowa) positioned on the northwest or backside of Colorado Lows and Alberta Clippers, two major winter storm tracks. As this type of low-pressure system tracks eastward, the strong northerly winds and frigid temperatures move in behind the low center, thus lifting the fresh, loose snowfall that has not had sufficient time to become compacted. The southern Canadian Prairie provinces (Alberta, Saskatchewan, and Manitoba) produce approximately five blizzards annually each winter of the Alberta clipper variety (Stewart et al. 1995). Blizzards are also common along the northeastern coasts of the United States and Canada with northeasters that often produce heavy snowfall from the added moisture influx from the North Atlantic. The strong thermal contrast that develops from the movement of continental polar air masses across the warm Gulf Stream current makes the North American Eastern seaboard a prime area for rapid intensification of mid-latitude cyclones (explosive cyclogenesis or "bombogenesis") with strong winds and likely blizzard development. Yet the most repeated winter storm and blizzard activity in North America arises along the Alaskan and British Columbia coasts that are frequented by strong low-pressure systems around the Bering Sea and Gulf of Alaska.

Localized blizzard conditions also occur in regions with large topography changes that induce fast gravity-induced air currents known as katabatic winds. The ice-capped regions of Greenland and Antarctica are notorious for high-velocity coastal winds because of air descending from the high-elevation, snow-packed interior. The interaction of subpolar low-pressure systems, especially during the transition seasons, produces a strong pressure gradient around these katabatic winds and lifts surface snow along with any light falling snow. Around Antarctica and other complex terrain regions, storm systems can also channel strong winds through mountain valleys and ice sheet crevasses. Based on early twentieth-century expeditions (e.g. Douglas Mawson), the shorelines of Cape Denison, Antarctica, earned the moniker of "home of the blizzard" for recurrent whiteout conditions and sustained winds in excess of 80 miles per hour (35 m per second). The European Alps and Pyrenees also produce blizzards and avalanches from strong northerly winds flowing within the mountain valleys with regional names such as the boulbie (southwestern France).

U.S. Climatology

Based on *Storm Data* reports, the conterminous United States recorded 714 blizzards of various sizes, intensities, and durations during 1959–1960 and 2013–2014. The average number of blizzards per season (September through August) is 13.0. Seasonal blizzard frequency ranged from 1 blizzard in the 1980–1981 season to 32 blizzards in the 2007–2008 season. Based on three-month intervals, blizzard frequencies as expected are highest during the boreal winter peak (December through February) at 8.3. Late season blizzards (March through May) occur twice as often as early season blizzards (September through November), averaging 3.1 per season compared with 1.5. Summer blizzards were only reported twice, occurring in June and July 2002 (Coleman and Schwartz 2017).

Blizzard activity in the United States is strongly concentrated in the northern Great Plains, particularly in the Dakotas and western Minnesota (or the "blizzard zone"). Nearly all 119 counties in North Dakota and South Dakota and 39 counties in western Minnesota average at least one blizzard or more per year. Blizzard totals remain relatively high on the blizzard zone margins. Eastern Minnesota, central Iowa, Nebraska, northwestern Kansas, eastern Colorado, southeastern Wyoming, and portions of Idaho and Montana have 27–48 blizzards reported. Frequencies decrease to 12–16 blizzards for southern Wisconsin, upper Michigan, and the remainder of Idaho, Montana, Iowa, and Colorado (Coleman and Schwartz 2017).

Outside the blizzard zone and its periphery, blizzard totals are generally between 1 and 11 blizzards per county with higher frequencies generally in the northern and high-altitude locations with a few exceptions. A secondary blizzard (geographic) maximum occurs in eastern Maine (Hancock and Washington counties) with 27–48 blizzards. Higher regional blizzard totals (n = 12–26) also occur for counties in coastal New England, interior Maryland (Garrett County), and lakeside New York (Erie County). With the exception of northwestern Georgia (Atlanta), blizzards are overall absent in the southeast and low-elevation areas in western states (Arizona, California, Nevada, and Washington) (Coleman and Schwartz, 2017).

United States' Blizzard Variability by Decade

The geographic distribution of blizzards by decade in the United States displays distinct periods of concentrated blizzard events contrasted with phases of more widespread activity. Spatially, blizzards in the 1960s and 1980s are more focused in the northern Great Plains blizzard zone with secondary activity around the extreme northeast coast. Although the blizzard activity is also more concentrated in the 2000s, blizzard reports shift westward from the Dakota's centered blizzard zone to the mountainous regions of the West (Rockies, Cascades, and Sierra Nevada) with no secondary geographic blizzard peak.

In comparison, blizzard reports from the 1970s are prevalent throughout the North and are shifted south of their mean position to include states such as New Mexico, Oklahoma, and North Carolina where severe winter weather events are rare. However, the 1970s had some severe winters (e.g., the great Midwest blizzard in January 1978), especially around the Great Lakes region; a similar severe winter cycle in this region did not occur again until the 2010s. In the 1990s, blizzards were more active in the west and the entire northeast corridor. Blizzard activity trends for the 2010s demonstrate a trend toward more geographically extensive blizzards reports with current activity highly spread throughout the central United States and Rockies (Coleman and Schwartz 2017).

IMPACTS

Fatalities, Injuries, Economic, Infrastructure

The intense blend of wind and snow from blizzards causes a general cessation of routine societal activities and severe damage to socioeconomic infrastructures. Transportation issues include closures of highways, railroads, and airports, often stranding thousands of travelers and commuters, such as during the February 2015 northeast U.S. blizzard. Retail sales and employee absenteeism can be impacted with reduced business and lost wages, which potentially causes financial losses to business and industry. Extensive drifting snow on roofs can lead to structural collapse due to extreme snow loads, as occurred during the collapse of the Metrodome, the Minnesota Vikings' stadium in Minneapolis, during a December 2010 blizzard. Ranchers in South Dakota experienced devastating impacts from an October 2013 blizzard when an estimated 70,000 livestock perished in the storm and area damages totaled $1.7 billion (Coleman and Schwartz 2017). The "Storm of the Century" for Canada was the March 1971 Montreal storm that combined a record 17 inches (43 cm) of snow with 68 miles per hour (109 km per hour) winds to create two-story snowdrifts, a 10-day power outage, and, notably, cancelation of a Montreal Canadiens hockey game for only the second time in franchise history; yet, the Montreal snowfall record was broken (18 inches, or 45 cm) with a similar severe winter storm in December 2012 (Perreaux and d'Aliesio 2012). Although these examples highlight the adverse impacts associated with blizzards, literature on winter storm preparedness often fails to adequately discuss the dangers with these most severe winter storms and areas prone to blizzard occurrence.

In postindustrial nations where blizzards are more common, financial and indirect losses (e.g., unemployment) from natural disasters are often much larger than direct human impacts. In the United States, for example, the National Centers for Environmental Information (NCEI) of the National Oceanic and Atmospheric Administration (NOAA) (2019) summarizes deaths, injuries, and damage costs attributed to various billion-dollar weather events. Based on NCEI statistics from 1980 through 2018, winter storms overall have less societal impact than severe storms (e.g., tornadoes), tropical cyclones, floods, droughts and, more recently, wildfires. Winter storms, which include blizzards, account for only 2.8 percent or approximately $47.3 billion of the $1,670 billion in losses (2018 adjusted) from all natural disasters during this period; however, total monetary losses are heavily inflated by a few tropical cyclone events. Hurricanes Katrina (2005), Harvey (2017), Maria (2017), Sandy (2012), and Irma (2017) accounted collectively for damages over $500 billion or about one-third of all losses from 1980 to 2018. In comparison, the costliest (and deadliest) winter storm was the March 1993 East Coast blizzard, dubbed the "Storm of the Century," resulting in nearly $10 billion in losses (and 270 deaths) from hurricane force winds coupled with heavy snowfall that created extensive wind and water damage. Other top damaging winter storms that involved blizzard conditions impacting the Northeast and/or Midwest include billion-dollar events in January 1996 ($5.2 billion), December 1992 ($4.5 billion), February 2015 ($3.2 billion), and March 2018 ($2.2 billion).

Disaster Declarations

For locations to receive a federal disaster declaration in the United States, a three-phase process through a bureaucratic hierarchy is followed. First, the local jurisdiction determines the financial and humanitarian needs that cannot be met and then requests emergency assistance from the state. The state governor may declare a state of emergency and invoke state disaster plan protocols and utilize state resources. Second, the governor may request federal aid if local and state government resources combined are deemed insufficient and meet the guidelines of the Stafford Act for disaster relief and emergency assistance. Third, the Federal Emergency Management Agency (FEMA) conducts a preliminary damage assessment and gives a recommendation to the president. Between 1953 and 2014, the number of major disaster declarations totaled 2,202, nearly all atmospheric-related hazards (Coleman and Schwartz 2017).

In comparison with mesoscale events (i.e., tornado outbreaks) and tropical cyclones, blizzards comprised a small proportion of federal disaster declarations and are usually a subset of the more generalized "winter storm" category. Between the 1959–1960 and 2013–2014 seasons, federal disaster declarations due to blizzards totaled 57, displaying a visible increase from the mid-1990s that coincided with the increase in reported blizzard activity (Coleman and Schwartz 2017). Over half of the blizzard declarations (n=33) occurred in the twenty-first century alone with the active 2009/2010 blizzard season over the mid-Atlantic and Great Plains receiving the highest individual season frequency (n=9). In comparison, the disaster declarations in earlier decades (n=24) were highly concentrated in the late 1970s and late 1990s and largely confined to the Dakotas and northeastern Michigan (Iosco County). The first federally declared blizzard declaration occurred in the 1974–1975 season with the January 10–12, 1975, blizzard that impacted the Dakotas, Minnesota, Iowa, Nebraska, and Missouri.

Storm Data (SD), a U.S. weather events database, attribute 711 fatalities from blizzards between the 1959–1960 and 2013–2014 seasons with an average life loss of approximately one individual per event (Coleman and Schwartz 2017). Fatalities ranged from a minimum (and most common value) of no fatalities per storm to a maximum of 73 deaths associated with the late January Midwest blizzard in 1978, although these totals do not include indirect fatalities that put the 1993 Superstorm event as the deadliest winter storm event at 270 deaths. Although blizzard frequencies increased substantially in the twenty-first century, blizzard fatalities actually decreased compared with earlier decades. From the 1959–1960 season through the 1999–2000 season, blizzards accounted for 679 fatalities with an average of 1.55 deaths per event compared with only 32 fatalities with an average of 0.12 for the 2000–2001 through 2013–2014 seasons. Reported injuries yielded similar results. All blizzard events for the study period totaled 2,044 injuries with a mean of 2.87 per blizzard. As with fatalities, most events had zero reported injuries. The maximum number of injuries reported (n=426) occurred with the March 1993 Superstorm over the densely populated East Coast. The period from 1959–1960 through 1999–2000 seasons yielded 2,011 injuries with a mean of 4.59 injuries per event while the 2000–2001 through 2013–2014 seasons had only 33 injuries averaging 0.12. Although more recent (since the late 1990s) SD reports make a distinction between direct and indirect causation, indirect fatalities and injuries from winter events (e.g., snow-related vehicles crashes) are significantly underestimated in the overall SD database (Coleman and Schwartz 2017).

EMERGENCY MANAGEMENT

Preparedness, Response, Mitigation

Whether the event is a hurricane, tornado, wildfire, or blizzard, a natural disaster requires adequate preparation. Events such as hurricanes usually have warning times greater than a day, while tornadoes and wildfires can have a range of outlooks from over a day to only a few minutes. Winter storms systems usually have ample warnings (days or more) so individuals can be prepared for the event. Even though blizzards are the most extreme of winter storms, similar preparedness actions can help assure safety of citizens. Some of the phenomena associated with a blizzard include cold temperatures, high winds, and heavy snow. Blizzard impacts are hours (reduced visibility and drifting snow) to days or even weeks (societal disruptions). The probability of power and communication loss is high, which could mean no heat from sources requiring electricity. Other issues include vehicle accidents and health issues such as heart attacks (from overexertion), hypothermia, and carbon monoxide poisoning. Those at a greater risk include those with illnesses, young children, and the elderly.

Individuals should also monitor weather watches and warnings from official sources from their national and local government, news, and radio agencies. In the United States, a NOAA weather radio gives alerts (texts or reverse 911) from local officials when blizzard conditions and other weather hazards are imminent. Another information option is signing up for alert and news through the Emergency Alert System (EAS) on television or radio. Due to the snow and high winds, visibility is very limited, and driving and other travel means should be avoided. Other preparedness actions include making sure structures have proper insulation, wrapping water pipes to avoid freezing, caulking window and door opening, weather stripping, working smoke and carbon monoxide detectors with battery backups, and having a properly maintained heating system. If there is a generator, it should be in working order. Generators need to be kept away from living areas to avoid carbon monoxide poisoning.

Other preparedness actions include having proper supplies such as food, water, medicine, and other items necessary if unable to leave the house for several days. Nonperishable foods (e.g., canned items, bread, crackers, dried goods) are recommended in case of power outages. In addition to bottled water, fill the bathtubs with water for an extra supply. Avoid melting snow for drinking water unless absolutely necessary as snow often has chemicals that are not easily removed even by the boiling process. If grills or camp stoves are used, be sure to have proper ventilation or avoid altogether as the fumes are lethal. An emergency kit for the house is highly recommended. Some of the items in the kit are extra batteries for flashlights and radios, alternative methods for charging devices such as phones, food, water, blankets, and jackets. Battery-powered lanterns and lamps are preferable over candles, a leading cause of home fires. Additionally, pets will need to be placed indoors and proper food and water supplies ensured.

Besides having an emergency kit for the house, another kit is recommended for the vehicle. Items included are blankets, water, food, flashlights, charging device, jumper cables, shovel, warm clothes, rope, compass, and a candle. The fuel tank should also be full to prevent the fuel lines from freezing and to have energy for the heating system. If stranded in a vehicle, do not go out in blizzard conditions as visibility can be near zero. Only run the heater and engine around 10–15 minutes per hour and make sure the snow does not cover the exhaust pipe. The candle will help heat up the interior and melt snow for drinking water. When trapped in a vehicle for an extended period of time, make continuous movements to increase circulation and body warmth. If you must leave the vehicle for a short distance, then attach a rope to yourself and the vehicle and utilize road flares and a compass so you can find your way back to the vehicle.

Since blizzards are not preventable and are an endemic natural hazard to some regions, mitigation measures may minimize blizzard impacts. Response during a blizzard is common sense. Stay indoors and keep out of the severe winter weather. If one does find it necessary to go outside, be aware of the symptoms of frostbite and hypothermia. Treat immediately if necessary. Check on neighbors, especially if they are elderly or have other special needs.

CONCLUSION

Blizzards do not always cause the same magnitude of impact of financial and humanitarian loss like large tropical cyclones or tornado outbreaks do, but they do cause significant societal disruptions. The transportation and economic sectors are heavily impacted. As with other hazards, the number of federally declared disasters in the United States and elsewhere associated with blizzards has increased. Locations with better preparedness and warning systems have also seen precipitous decline in fatalities and injuries.

Robert M. Schwartz and Jill S. M. Coleman

NOTE

Portions of this entry have been adapted from Coleman, J. S., and R. M. Schwartz. 2017. An Updated Blizzard Climatology of the Contiguous United States (1959–2014): An Examination of Spatiotemporal Trends. *Journal of Applied Meteorology and Climatology* 56: 173–187.

Further Reading

Associated Press. 1972. Missing Put at 6,000 in Iranian Blizzard. *New York Times*. February 11, 1972. https://www.nytimes.com/1972/02/11/archives/missing-put-at-6000-in-iranian-blizzard.html, accessed January 20, 2019.

Coleman, J. S., and R. M. Schwartz. 2017. An Updated Blizzard Climatology of the Contiguous United States (1959–2014): An Examination of Spatiotemporal Trends. *Journal of Applied Meteorology and Climatology* 56: 173–187.

Department of Homeland Security. 2017. Snowstorms and Extreme Cold. https://www.ready.gov/winter-weather, accessed August 3, 2018.

Fitzhugh, T. C. 1928. What Is a Blizzard? *Bulletin of the American Meteorological Society* 9: 55.

Ludlum, D. M. 1968. *Early American Winters II 1821–1870 Historical Monograph.* Boston: American Meteorological Society.

NOAA (National Oceanic and Atmospheric Administration). 2019. U.S. Billion-Dollar Weather and Climate Disasters. https://www.ncdc.noaa.gov/billions/, accessed February 9, 2019.

NWS (National Weather Service). 2018. National Weather Service Expanded Winter Weather Terms. https://www.weather.gov/bgm/WinterTerms, accessed August 8, 2018.

OED (Oxford English Dictionary). 2019. Blizzard. Oxford University Press. https://public.oed.com/updates/February, accessed February 8, 2019.

Perreaux, L., and R. D'Aliesio. 2012. Montrealers Take Snow in Stride as Storm Smashes 1971 Record. *The Globe and Mail*. https://www.theglobeandmail.com/news/national/montrealers-take-snow-in-stride-as-storm-smashes-1971-record/article6758046/, accessed December 27, 2012 (and updated May 9, 2018).

Schwartz, R. M. 2001. Geography of Blizzards in the Conterminous United States, 1959–2000. PhD Dissertation, Kent State University, Kent, OH.

Schwartz, R. M., and T. W. Schmidlin, 2002. Climatology of Blizzards in the Conterminous United States, 1959–2000. *Journal of Climate* 15: 1765–1772.

Stewart, R. E., D. Bachand, R. R. Dunkley, A. C. Giles, B. Lawson, L. Legal, S. T. Miller, B. P. Murphy, M. N. Parker, B. J. Paruk, and M. K. Yau. 1995. Winter Storms over Canada. *Atmosphere-Ocean* 33: 223–247.

Wild, R. 1997. Historical Review on the Origin and Definition of the Word Blizzard. *Journal of Meteorology* 22: 331–340.

Climate Change

There are numerous hazards that become worse with a change in climate. Climate, according to the Intergovernmental Panel on Climate Change (IPCC), "is usually defined as the average weather." The formal IPCC definition goes on to discuss statistical variability (the entire suite of weather that impacts a location) and a time window to differentiate weather from climate that extends from one month to millions of years. Climate normals have been established by the World Meteorological Organization (WMO) as the statistical data for 30-year intervals that are updated every decade. It is common to characterize the climate of a given location with a presentation of the normal values.

According to the IPCC, climate change "refers to a change in the state of the climate that can be identified (e.g., using statistical tests) by a change in the average and/or the variability that persists for an extended period, typically several years or longer" (Stocker 2014, 126). Decades of research addressing the earth's climate system confirm that emissions of carbon dioxide from fossil fuel burning accompanied by burning associated with land-use changes have produced a climate system imbalance. The amount of carbon dioxide in the atmosphere is increasing over time at a rate that continues to grow by as much as 2.0 percent each year. Vegetation and the oceans remove (sequester) carbon dioxide from the atmosphere, but not at the same rate that human actions are adding carbon dioxide and other greenhouse gases (GHGs) to the atmosphere. The growth in the amount of GHGs in the atmosphere by more than 40 percent since the start of the industrial revolution has triggered a change in the behavior of the climate system. An increase in GHGs is similar to adding a blanket (or two) to keep the surface of the planet warmer.

Emissions of GHGs at a pace that exceeds natural abilities to remove the GHGs have broken the balance between incoming energy from the sun and outgoing thermal infrared radiation. These anthropogenic (resulting from human influence) changes have caused a shift in climate. One measure of the state of the earth's climate is the average planetary temperature. The IPCC Special Report indicates an increase of 1.8 degrees Fahrenheit (1.0 degrees Celsius) in the mean global temperature compared with that in 1850–1900 due to changes in the planetary energy budget (Allen et al. 2018). The increase in temperature has contributed to other changes in the earth's physical systems, such as more extreme precipitation events, ice sheet melting, sea-level rise, and ocean acidification. Since the 1980s, the area of sea ice cover and thickness of the sea ice in the Arctic Ocean have shrunk dramatically, and the world's land-based glaciers have experienced considerable losses in volume. The shrinking thickness of mountain glaciers and the calving of ice sheets have provided rich visual evidence of the impact of global warming. Melting of glaciers and ice sheets has contributed to sea-level rise. Although the changes in sea-level rise are not uniform all over the globe, long-term tide gauges and short-term satellite data showed an increase of about 7 inches (18 cm) at an overall rate of 0.07 inch per year (1.8 mm per year) for the global average sea level in the twentieth century, with a more recent twenty-first century value of 0.2 inch per year (5.0 mm per year). For more information, see Figure 1.1 in the IPCC report at https://www.ipcc.ch/report/ar5/syr/.

Ocean acidification is another indicator of the changes in the earth system. Phytoplankton in the oceans act as a vegetative pool to uptake and store large amounts of carbon dioxide, but that sequestration results in an increase in carbonic acid. Ocean acidification impacts marine life and the jobs and economies associated with those who rely on food from ocean ecosystems. Since the preindustrial era (1765), the average pH of the ocean's surface water has dropped from 8.25 to 8.14 (a 30 percent increase in hydrogen ion concentration).

CAUSES OF CLIMATE CHANGE

A variety of natural and human-made factors influence the climate system. In the past, the earth has been through both warm and cold phases because of a combination of natural internal and external forces. Recent changes in climate (since the mid-twentieth century), however, cannot be explained with an understanding of the natural variability of the climate system alone. Research studies summarized by the IPCC have concluded that humans are progressively influencing the climate system by burning fossil fuels and land-use changes such as deforestation and agricultural activities (Stocker 2014). These human-induced changes to the climate system have increased the amount of GHGs in the atmosphere and created an energy imbalance for the climate system, with more heat being stored in the oceans and the near-surface atmosphere. Anthropogenic global warming (AGW) is a phrase used to describe the result of the human-induced change to the earth's climate system.

Drivers of Climate Change

Understanding climate change involves identifying the external and internal drivers that can change the energy flows within the atmosphere. Internal drivers include the slow movement of lithospheric plates with associated mountain building, volcanic eruptions, and energy and GHG exchanges between the oceans and the atmosphere (such as with an El Niño event in the tropical Pacific Ocean). External drivers include changes in the receipt of solar energy and meteorite impacts. During the last 2.5 million years, adjustments in the tilt of the earth's axis, the shape of the earth's orbit around the Sun, and the orientation of the earth's North Pole axis have changed the geographic distribution of solar energy received, and these shifts have produced a number of cold glacial periods and shorter warm interglacial intervals.

The interconnectedness of the many climate system components (including the atmosphere, oceans, ice sheets, and vegetation cover) is very complex with numerous feedbacks. Overall, the amount of incoming short-wave or solar radiation and outgoing long-wave or thermal radiation controls the temperature of the earth's atmosphere and oceans. Slightly over two-thirds (69 percent) of the incoming radiation from the Sun is absorbed by the earth's oceans and land surfaces and by gases and particulates in the atmosphere. The remaining solar radiation (31 percent) is reflected back into space. The reflectivity (also known as albedo) of the earth's surface, oceans, clouds, and atmospheric particles determines the amount of solar energy that is unused in the earth system and returned to space. Ice sheets, snow cover, and clouds play a critical role in increasing reflectivity and cooling temperatures during an ice age, as they have a greater albedo compared to the darker oceans and land surface.

Earth loses some energy through reflected sunlight, but a much greater amount of energy is lost, associated with emitted long-wave or thermal infrared radiation moving out to space. Energy emitted by the earth's surface and the atmosphere helps to cool temperatures and keep the system in balance. However, most of the infrared energy flowing away from the earth's surface is absorbed by clouds and GHGs leading to a warming of the lower atmosphere. Naturally, the most important GHG has been water vapor. If there were no water vapor and clouds in the earth's lower atmosphere, the average global temperature would be well below freezing rather than a value of approximately 59 degrees Fahrenheit (15 degrees Celsius) today.

The effect of long-wave or thermal energy loss can be observed at night when there are clear skies and the air is dry; the local near-surface air temperature drops quickly with the energy lost from emitted thermal radiation. At other times or places, when it is humid and cloudy, evening temperatures do not drop as fast. For the planet as a whole, clouds and GHGs (including water vapor, carbon dioxide, methane, and nitrous oxide) absorb most of the emitted thermal energy moving away from the surface. This capture of outgoing thermal radiation warms the gases and clouds in the atmosphere. The molecules in the atmosphere then radiate energy in all directions, some out to space helping to cool the planet and balance the solar input, with other emissions directed downward where this recycled thermal energy can be absorbed and warm the earth's oceans and land surface.

The recent increase in GHGs in the atmosphere has shifted the earth's temperature away from an equilibrium, therefore altering the climate system. The impact of each component of the earth's energy budget can be quantified by its radiative forcing (RF). Satellite measurements of the total solar flow since 1973 shows an approximately 11-year cycle of slight changes in solar output that varies with the number of sunspots. However, the changes in RF from the sun cannot be used to explain the increase in global temperature and the climate changes that have occurred in the last two centuries. Major volcanic eruptions can impact RF, but the impact lasts only a few years. The June 1991 eruption of Mount Pinatubo resulted in a massive injection of gases and ash into the stratosphere, which influenced climate and resulted in more beautiful sunsets. Stratospheric injection of sulfur dioxide produces sulfate aerosols that can linger for several years in the stratosphere, impacting RF by increasing the albedo and cooling the planet for a few years. Planetary cooling from the Mount Pinatubo eruption reached about 0.9 degrees Fahrenheit (0.5 degrees Celsius), and the effect lasted three years. Some climate engineering proposals, called solar radiation management (SRM), suggest that one strategy to offset the impact of warming related to increased GHGs is to add reflective aerosols to the stratosphere. For a number of scientific, political, and moral reasons, many scholars think that SRM is unwise.

To date, the shifts in temperature associated with anthropogenic climate change have followed a generally linear upward trend since about 1975. There is a concern among climate scientists who model the earth system that forcing may reach a level, called a tipping point, where the system response jumps to a new level. Scientific work is currently addressing the idea of an anticipated surprise and a major shift in the earth's climate.

Emissions of GHGs

GHGs are the gases that absorb thermal radiation and "trap heat" in the lower atmosphere. Carbon dioxide, methane, and nitrous oxide, along with water vapor, are the principal GHGs. Prior to the industrial era (1750), the atmospheric carbon dioxide concentration of approximately 280 parts per million (ppm) had changed only slightly during the Holocene (the last 11,000 years). However, carbon dioxide emissions have increased exponentially in the industrial era. Burning fossil fuels including gas, oil, and coal; deforestation; and cement manufacturing are primary sources of the carbon dioxide growth in the atmosphere since the industrial era began. Methane, which contributes about 17 percent to the impact of GHGs, is currently over 1850 parts per billion (ppb) and has a growing atmospheric concentration. A variety of natural and anthropogenic sources affect the atmospheric concentration of methane, including natural wetlands, animal husbandry, rice agricultural, mining, and natural gas extraction.

Methane is more effective than carbon dioxide in absorbing thermal radiation, but it has a shorter lifetime in the atmosphere. Nitrous oxide is also an important anthropogenic GHG with a smaller concentration (just over 330 ppb) and a 6 percent contribution to the impact of GHGs. Agricultural fertilizers, biomass burning, industrial activities, and vehicle exhaust are anthropogenic sources of nitrous oxide. Other GHGs include chlorofluorocarbons (CFCs), halons, hydrochlorofluorocarbons (HCFCs), hydrofluorocarbons (HFCs), perfluorocarbons (PFCs), and sulfur hexafluoride. Together, these mostly human-made gases are called halocarbons. Water vapor is also a very strong GHG, accounting for about half of the total greenhouse warming effect, but scientists do not have a long-term record of changes in atmospheric water vapor concentrations. Since temperature controls the amount of water in the lower atmosphere, global warming provides more energy to add more water vapor into the atmosphere, and the warming effect becomes intensified.

CLIMATE SENSITIVITY AND FEEDBACKS

Scientists identify the amount of temperature change with a doubling of the atmospheric carbon dioxide concentration as climate sensitivity. Depending on the climate model used in the analysis, sensitivity values range from 2.7 to 8.1 degrees Fahrenheit (1.5 to 4.5 degrees Celsius). Recent research suggests that as the amount of carbon dioxide in the atmosphere continues to increase, the sensitivity also increases. Climate feedbacks can be either reinforcing/amplifying (positive) or regulating (negative). Any substantial change in the climate system may initiate a feedback with a warming or cooling effect. For example, higher temperatures facilitate more evaporation of water from oceans, and that adds more water vapor into the atmosphere. Water vapor is a GHG, but more water vapor leads to more clouds that also help absorb outgoing thermal radiation. While more clouds do increase the planetary albedo, the net effect is an increase in lower atmospheric and ocean temperatures. The complex interactions among temperature changes, evaporation, and clouds identify an amplifying feedback. An increase in water vapor in the lower atmosphere will provide some storms with more precipitable water, and this has led to an increase in the flood hazard.

An additional compound effect of warmer temperatures is greater evaporation from soils and increased transpiration by plants. These adjustments on the land surface can trigger a greater risk for agricultural and hydrologic drought, especially in semiarid and desert regions. Hotter and drier conditions can lead to increases in both the heat wave and wildfire hazards.

Areas with considerable snow and/or ice cover are subject to a different amplifying feedback. In this case, warmer conditions help melt away the snow or ice, and this produces a darker surface that will absorb more of the incoming solar and reflected thermal energy. With more energy absorbed to warm the surface and the air above, more snow and ice will melt. This reinforcing impact is currently at work with the loss of Arctic sea ice. During the onset of glacial periods, this reinforcing feedback works in the other direction. As the temperature cools, more snow and/or ice stays on the surface to reflect the incoming energy, and the increase in albedo helps lower the temperature. With cooler temperatures, snow accumulates, allowing glaciers and ice sheets to grow.

CLIMATE HAZARDS

Climate is characterized by not only the changes in average temperature and precipitation but also the shifts in the severity, frequency, and duration of extreme events such as droughts, heat waves, cold waves, storms, and flooding events. Changes in extreme climate events are expected as a result of global warming with economic and security impacts that include human health, food production, water quality and quantity, ecosystems, and infrastructure (Mora et al. 2018).

Warming

Increased air and ocean temperatures have occurred for much of the world. Since the late nineteenth century, the global mean temperature has shown a warming of 1.8 degrees Fahrenheit (1.0 degrees Celsius) with warming over land areas approximately twice as high as warming over the oceans. The United States (U.S.) National Climatic Assessment reported an increase in the U.S. average temperature by a rate of 1.23 degrees Fahrenheit (0.68 degrees Celsius) since 1895 for the 48 contiguous states with the greatest increase since 1970 (Vose et al. 2017). National and regional studies have documented temperature increases for India, China, Australia, and Africa (Stocker 2014). *Arctic amplification* is a term used to describe the faster rate of temperature increase for land areas in polar environments. Temperatures in Alaska warmed at a faster rate than the rest of the United States, with an increase of 1.7 degrees Fahrenheit (0.93 degrees Celsius) in a comparison of recent values with those from the mid-twentieth century (Vose et al. 2017). Sea surface temperature and marine air temperature have also changed significantly since the late nineteenth century (Stocker 2014). Patterns of temperature change have not been similar over all oceanic regions because of the variability patterns of sea surface temperature in the Pacific Ocean (for example, the El Niño Southern Oscillation (ENSO) cycle) and due to hemispheric asymmetry between the North Atlantic and South Atlantic Oceans (Stocker 2014). Warmer temperatures have increased the likelihood of heat waves (both land based and marine) and the societal impacts associated with hot temperatures.

Precipitation

The increase in atmospheric humidity as a result of increased evaporation has intensified precipitation in humid regions and caused more flooding. Conversely, in some regions warmer temperatures can lead to intensified drying and droughts. The statement "Wet gets wetter, dry gets drier" is a summary of the impact of climate change on water resources. Generally speaking, precipitation is likely to decrease in drier subtropical latitudes while the wetter latitudes expect an increase in precipitation (Held and Soden 2006; Stocker 2014).

Heavy precipitation (large precipitation amounts) or precipitation extremes are rare events when precipitation is higher than a fixed threshold like the 95th or 99th percentile. A 95th percentile threshold means that in 95 percent of the time at that location, the precipitation total was lower than the threshold amount. Across the world, there is an overall increasing trend in the frequency and intensity of extreme precipitation events since 1951. The immediate effect of extreme precipitation is an increase in the prospect of flooding especially in urban areas where less vegetation and more impervious surfaces force rainwaters to run off rapidly.

Flooding

A flood is a high flow or water inundation that may cause serious damage. Floods are amplified by either weather-related or human-caused features. Heavy or prolonged precipitation events can be followed by floods, especially when soils are saturated. Thunderstorms, storm surges from hurricanes, and even rapid snowmelt can trigger floods. Humans impact flooding by activities such as land-use change that help move water more rapidly into rivers and streams. Replacing the natural vegetative systems with impermeable surfaces decreases infiltration of water into the soil and consequently accelerates surface runoff or overland flow. In addition, structural failures of dams and altered drainage behavior, such as the creation of concrete channels, are human activities that can produce a flood. The IPCC reports that the factors that contribute to floods such as heavier rains and higher sea levels have been influenced by climate change, and the U.S. National Climate Assessment reports that flood frequency has increased in the Mississippi River Valley, Midwest, and Northeast and that coastal flooding has doubled in a matter of decades.

Drought

Drought depends on different natural factors for different geographic regions; however, most scientists link changes in droughts to higher temperatures and higher rates of evaporation and plant transpiration. It is hard to analyze drought trends on a global scale; however, there is evidence of longer and more intense droughts in some regions around the world since the 1950s, especially in the Mediterranean region (Stocker 2014).

Storms

In addition to heavy rainfall, storm damage can occur related to strong winds, blowing dust, hailstones, lightning, blowing and drifting snow, and the related impacts on human activities and structures. Currently, there is no robust evidence of an increase in the frequency of tropical storms across the world. However, the intensity of tropical cyclones seems to be increasing, with evidence of stronger storms since the 1970s in the North Atlantic Ocean (Stocker 2014). For extratropical storms that are produced by moving disturbances in the jet stream, the numbers of events seem to be decreasing, and there is documentation of a poleward shift in storm tracks for the Northern Hemisphere since the 1950s.

Extreme Heat and Heat Waves

Since the middle of the twentieth century, warm temperature extremes and related hazards have become more common while cold temperature extremes have decreased across the globe. With global warming, heat waves are becoming longer, more frequent, and more intense in North America, Europe, and Australia (Meehl and Tebaldi 2004; Tavakol et al. 2020b). Scientific studies indicate that preexisting dry soils tend to amplify heat wave characteristics (Alexander 2011; Stocker 2014). A heat wave can amplify the hazardous impacts of a drought when the phenomena occur simultaneously (Tavakol et al. 2020a). Oceanic or marine heat waves have increased by a factor of three since 1980.

Wildfire

Wildfires are dramatic hazards driven by dynamic factors including weather, fuels, and people. Wildfires are more problematic during a drought and heat wave because the atmospheric conditions dry the fuel. An extended drought can increase vegetation mortality and raise the risk of an out-of-control wildfire. Hot and dry conditions can also favor insect pests that injure or kill the trees. As summer temperatures increase significantly, climate change may have a nonuniform impact on the fire regimes in many regions of the globe.

FUTURE CLIMATE CHANGE

Scientists who address what future climate conditions are likely use scenarios and climate models to help address questions related to possible futures. These scenarios differ regarding future GHGs emissions and what strategies are used to scrub or remove the GHGs from the atmosphere. Given the importance of GHG concentrations in climate models, the different scenarios are called Representative Concentration Pathways (RCPs). The IPCC has used four RCPs that address differences in population growth, energy use and fuel sources, economic activity, and other social factors. At current emission rates, the IPCC estimates that carbon dioxide may double by midcentury or may even triple further into the future. Considering the influence of GHGs on the climate system, some aspects of the climate will respond rapidly (near-term changes) and other characteristics will emerge at slower rates (a time lagged response or long-term changes). In future climate change research, the ideas from Shared socioeconomic pathways (SSPs) scenarios are available for incorporation into model runs.

Near Term

Near-term climate change is very important to decision-makers who develop strategies to adapt to the shifting conditions. Near-term predictions for the next couple of decades (until 2035) show an additional increase of 0.5 degrees Fahrenheit to 1.3 degrees Fahrenheit (0.3 degrees Celsius to 0.7 degrees Celsius) in global mean surface temperature, if no major volcanic eruptions occur (Stocker 2014). Results vary due to differences across the four RCP scenarios and which climate models are used. The temperature increase will impact the water cycle and the atmospheric circulation of the earth system. Warmer temperatures will shrink the Arctic sea ice cover, reduce the area of snow-covered ground, and melt permafrost over northern high latitudes. Changes to the earth's climate system will change the salinity of the oceans with increases projected for the tropical and subtropical Atlantic and a decrease over the western tropical Pacific. Near-term climate change will impact extreme events as well. Expected shifts include more hot days and warm nights, fewer cold days and nights, and a greater probability of heavy precipitation over most land areas.

Long Term

Unlike near-term projections, long-term climate change studies address the RCPs and also rely more on an understanding of external climate forcing. These studies have greater uncertainty; nevertheless, if the concentration of GHGs continues to rise, greater climate changes than already observed are expected in the future, and tipping points become more of a concern. Warming is not expected to be regionally uniform. A greater increase in temperature is expected over land areas than the oceans, especially for polar areas. Projected temperature changes will influence the frequency and magnitude of extreme temperature events with warmer hot extremes and fewer cold extremes. A temperature increase will intensify evaporation, and as a result, precipitation is expected to increase globally.

IMPACTS OF CLIMATE CHANGE

The effects of changing climate are already having an economic impact, and the cost of not addressing the problem will be greater the longer humanity waits to address the problem. Impacts are occurring across most economic sectors, such as agriculture, health, and transportation, and supporting resources, including biodiversity, freshwater, and marine resources.

Food Safety and Natural Ecosystems

Food security and agricultural activities, including growing livestock and harvesting fisheries, are extremely dependent on environmental conditions. Yield reductions, including quantity and nutritional quality, and increased death rates for important species are increasingly likely with the hotter temperature extremes associated with climate change. Challenges for farmers and ranchers have increased with the changing frequency and severity of extreme weather, including conditions that exceed the physiological tolerance levels of the plants or animals (Mora et al. 2018). For example, livestock are sensitive to extreme heat, which can increase the cost of production or result in mortality.

Stresses in natural ecosystems include more frequent and damaging impacts from insect pests. Heat and drought hazards increase the likelihood that beetles or other forest pests will kill large areas of trees. Dry fuels enable fires to grow more rapidly and affect larger areas. In the oceans, warmer waters are causing geographic shifts in the distribution, quantity, and quality of important commercial species and impacting coral reefs with more frequent stressful bleaching events (Mora et al. 2018).

Freshwater Resources

Quality and quantity of water resources have been negatively impacted by climate change. As melting glaciers and extreme precipitation events disrupt the water cycle, the year-round availability of clean freshwater is changing. In addition to changing flood hazard frequency (especially in urban areas), increased runoff from agricultural lands is transporting more sediment and fertilizers to coastal zones, increasing the frequency of the hypoxia hazard or low oxygen dead zones. With more glacier meltwaters and the expansion from warmer ocean temperatures, coastal fresh groundwater resources are being impacted by saltwater intrusion. And as sea levels rise, hazardous "clear-sky" (high tides) flooding events with high tides are becoming more frequent.

Human Health and the Economy

Young children and the elderly are at greatest risk during temperature extremes, especially heat waves. As temperatures warm, the areas impacted by tropical diseases, such as malaria, are expanding. In addition to death and disease, mental health issues are likely to be influenced by climate change (Mora et al. 2018). Changing environments will have profound impacts on jobs, labor productivity, and the local and global economy. For developing countries, climate change impacts include a reduction in gross domestic product (GDP). Major events, such as a heat wave, drought, or wildfire, have impacts on the local, national, and perhaps global economy. Examples include the 2010 Pakistan flood, where about 88 percent of householders reported up to 50 percent loss of income, and the 2010 heat wave and drought in Russia that damaged grain production and caused an increase in the global price of wheat (Field 2014). Wildfires, which are more frequent and worse due to climate change, are increasingly becoming costlier with the 2018 wildfire season in California producing losses exceeding $24.5 billion (NOAA 2019).

MITIGATION AND ADAPTATION OPTIONS

Mitigation and adaptation are two different responses to climate change. Mitigation involves reducing the sources of emission (like getting a car that gets more miles per gallon), preventing emissions (replacing electricity from fossil fuels with solar or wind power), or increasing the sinks that store GHGs (such as planting trees). An objective of mitigation activities is to first stabilize and then reduce the level of GHGs in the atmosphere. An energy-intensive lifestyle generally increases the cost of mitigation.

Adaptation is an adjustment that attempts to decrease the risk of and vulnerability to the harmful impacts of climate change. Adaptation efforts build capacity for nations to cope with climate change impacts. Adaptation strategies include physical and structural, social, and institutional or political options. Success of adaptation strategies depends on available knowledge along with access to a source of funding and technology (Field 2014). Developed countries tend to have greater access to the needed resources whereas developing countries are more likely to suffer the impacts of climate change. Categories of adaptation include spontaneous versus planned, with planned adaptation more likely to help build resilient communities.

International programs that address climate change can include either mitigation or adaptation or both. Programs tend to think about the global problem but act to do what will work at the local scale. For over a decade, the UN-REDD Programme has enabled efforts to reduce emissions from deforestation and forest degradation and increase the amount of carbon in forests while promoting local sustainability. In order to be successful, this program needs to address many factors including lifestyles, livelihood and poverty, culture, and behavior.

CONCLUSION

Accumulation of GHGs in the earth's atmosphere from human activities has changed the earth's climate system. Surface air temperature and ocean temperature have gone up and are expected to rise during the twenty-first century. Extreme weather events have become an increasing concern. As the frequencies of floods, droughts, heat waves, and other weather-related hazards increase, other aspects of the earth system are responding with a loss in polar sea ice, shrinking thickness of glaciers and ice sheets, sea level rise, and ocean acidification. Climate change is affecting human health, agriculture, water management, natural ecosystems, and infrastructure. The future magnitude of climate change and related hazards depends on the trajectory of global GHG levels. With no major reduction in GHGs emission, the earth system could be pushed toward tipping points we cannot predict.

Ameneh Tavakol, Vahid Rahmani, and John A. Harrington Jr.

Further Reading

Alexander, L. 2011. Climate Science: Extreme Heat Rooted in Dry Soils. *Nature Geoscience* 4: 12–13.

Allen, M. R., O. P. Dube, W. Solecki, F. Aragón-Durand, W. Cramer, S. Humphreys, M. Kainuma, J. Kala, N. Mahowald, Y. Mulugetta, R. Perez, M. Wairiu, and K. Zickfeld. 2018. Framing and Context. In *Global Warming of 1.5°C. An IPCC Special Report on the Impacts of Global Warming of 1.5°C above Pre-industrial Levels and Related Global Greenhouse Gas Emission Pathways, in the Context of Strengthening the Global Response to the Threat of Climate Change, Sustainable*

Development, and Efforts to Eradicate Poverty, edited by V. Masson-Delmotte, P. Zhai, H.-O. Pörtner, D. Roberts, J. Skea, P. R. Shukla, A. Pirani, W. Moufouma-Okia, C. Péan, R. Pidcock, S. Connors, J.B.R. Matthews, Y. Chen, X. Zhou, M. I. Gomis, E. Lonnoy, T. Maycock, M. Tignor, and T. Waterfield. Cambridge, UK: IPCC.

Field, C. B. 2014. *Climate Change 2014—Impacts, Adaptation and Vulnerability: Regional Aspects*. Cambridge, UK: Cambridge University Press.

Held, I. M., and B. J. Soden. 2006. Robust Responses of the Hydrological Cycle to Global Warming. *Journal of Climate* 19: 5686–5699.

Makowski, K., M. Wild, and A. Ohmura. 2008. Diurnal Temperature Range over Europe between 1950 and 2005. *Atmospheric Chemistry and Physics* 8: 6483–6498.

Meehl, G. A., and C. Tebaldi. 2004. More Intense, More Frequent, and Longer Lasting Heat Waves in the 21st Century. *Science* 305: 994–997.

Melillo, J. M. 2014. *Climate Change Impacts in the United States, Highlights: US National Climate Assessment*. Washington, DC: U.S. Global Change Research Program.

Mora, C., D. Spirandelli, E. C. Franklin, J. Lynham, J., M. B. Kantar, W. Miles, C. Z. Smith, K. Freel, J. Moy, L. V. Louis, E. W. Barba, K. Bettinger, A. G. Frazier, J. E. Colburn IX, N. Hanasaki, E. Hirabayashi, W. Knorr, C. M. Little, K. Emanuel, J. Sheffield, J. A. Patz, and C. L. Hunter. 2018. Broad Threat to Humanity from Cumulative Climate Hazards Intensified by Greenhouse Gas Emissions. *Nature Climate Change* 8: 1062–1071.

Mouillot, F., and C. B. Field. 2005. Fire History and the Global Carbon Budget: a 1× 1 Fire History Reconstruction for the 20th Century. *Global Change Biology* 11: 398–420. https://doi.org/10.1111/j.1365-2486.2005.00920.x

NOAA (National Oceanic and Atmospheric Administration). 2019. Billion-dollar Weather and Climate Disasters. https://www.ncdc.noaa.gov/billions, accessed September 3, 2019.

Stocker, T. F. 2014. *Climate Change 2013: The Physical Science Basis: Working Group I Contribution to the Fifth Assessment Report of the Intergovernmental Panel on Climate Change*. Cambridge, UK: Cambridge University Press.

Tavakol, A., V. Rahmani, and J. Harrington Jr. 2020a. Changes in the Frequency of Hot, Humid Days and Nights in the Mississippi River Basin. *International Journal of Climatology*. 1–16. https://doi.org/10.1002/joc.6484.

Tavakol, A., V. Rahmani, and J. Harrington Jr. 2020b. Evaluation of Hot Temperature Extremes and Heat Waves in the Mississippi River Basin. *Atmospheric Research*. 239: 104907. https://doi.org/10.1016/j.atmosres.2020.104907

Vose, R. S., D. R. Easterling, E. E. Kunkel, A. N. LeGrande, and M. F. Wehner. 2017. Temperature Changes in the United States. In *Climate Science Special Report: Fourth National Climate Assessment, Volume I*, edited by D. J. Wuebbles, D. W. Fahey, K. A. Hibbard, D. J. Dokken, B. C. Stewart, and T. K. Maycock, 185–206. Washington, DC: U.S. Global Change Research Program.

Coastal Erosion

Coastal erosion is the removal of coastal land, including sand, rock, soil, or other sediments by a range of processes including waves, tides, storm surge, tsunamis, rain, wind, subsidence (i.e., land sinking), mass wasting processes, and others, some of which can themselves also cause other sediment removal or hazards such as landslides. Mass wasting processes are related to the movement of large amounts of sediment, sometimes due to gravity in general and at other times due to other related processes like landslides or rockfalls, and the sediment that is dropped is then itself often moved by waves or currents over time. Such erosion processes are part of the natural environment, part of periodic cycles that occur over time periods ranging from days to decades, particularly in areas dominated by sandy sediment, like beaches and barrier islands in some areas. Seventy percent of sandy beaches in the world experience some level of regular coastal erosion, and other coastlines, like Arctic bluffs, also experience it to varying degrees (Frederick et al. 2016; Leatherman et al. 2000).

Humans also contribute to coastal erosion through steps like construction, dredging, and other manipulations of existing lands. Each of these processes may impact the land and sediments involved by removing them, adding to them, or just moving them from one specific area to another, or they may impact the larger systems involved through steps like rerouting existing waterways, thereby disrupting the previously described natural cycles of erosion and sediment movement. Further complications arise as a result of varying systems that can cause coastal erosion to be either a slow or rapid onset hazard. Both forms are worldwide problems, potentially affecting communities in a range of areas and countries. Twenty percent of the world's population lives within 15.5 miles (25 km) of the coastline, and 40 percent lives within 62 miles (100 km), an area that includes just 20 percent of land in the world (Williams et al. 2018). Given the large number of people who live and work along coastlines, knowing how coastal erosion works, how humans improve or worsen coastal erosion, and what risks are associated with coastal erosion is imperative to the safety of the natural environment and human populations, something that is especially true in an era of climate change and rising sea levels.

Despite these concerns, scientific understanding of coastal erosion remains a work in progress. Researchers are still working to develop consistently successful mathematical and conceptual models of coastal erosion and on gathering data on where coastlines are and about their erosion processes over time, both in the short and long terms. Further, experts understand some forms of erosion better than others. For example, there is a relative lack of understanding around problems related to Arctic coastal erosion, despite it being roughly one-third of the global coastline with ongoing concerns over the rapidity of coastal erosion increasing along those shores (Frederick et al. 2016).

Coastal erosion can be problematic for several reasons, particularly, (1) the loss of land, including potential land used by local populations, and (2) the loss of buffering beaches or landforms that help reduce the impact of flooding and storms, especially in communities along these coasts. The former potentially impacts local homes and businesses in these areas, causing additional economic problems for these communities. The second may, in time, result in increased costs as well, by increase in not only the impact of such hazards on coastal cities but also associated costs like insurance rates and transforming building codes.

Ongoing issues with coastal erosion are also tied to concerns regarding disasters, hazards, and climate change, further complicating the larger situation and potential approaches to minimizing coastal erosion. Understanding these ongoing transformations of and potential future increase of coastal erosion requires not only an understanding of ongoing and future environmental, weather, and climate patterns but also research into their history. Coastal erosion in some areas has been rapidly increasing in recent years, a process that is at least in part seemingly driven by climate change. Such increased amount and speed of coastal erosion raises additional concerns about the process; the current impacts it is having on land, people, and communities; and potential future increases in the amount of erosion and the areas affected. In addition, problems with coastal erosion also can result in increasing associated financial costs, as well as the "social and cultural integrity" of areas such as the Alaskan coast (Frederick et al. 2016).

CAUSES AND TYPES

Coastal erosion comes at two main speeds: slow onset and rapid onset. Fully understanding the extent and effects of coastal erosion on an area, if studied via mathematical and conceptual modeling or by comparing and contextualizing historical data, requires an understanding of the rate of change and the extent of it. This also requires an attempt to better appreciate the potential ongoing nature of the erosion process and the future extent of coastal erosion in ongoing and changing conditions.

Slow-onset coastal erosion takes anywhere from years to centuries to progress, caused by the gradual erosion of the coastline by regular waves, water currents, and other processes. Rapid-onset coastal erosion occurs over days or weeks and may be caused by hazards such as storms, storm surges, coastal flooding, or, even in some cases, tsunamis. With the scale of hazards causing rapid-onset coastal erosion, like tsunamis, the location of eroding lands can transform in addition to the scale and speed of erosion, resulting in normally nonaffected lands being eroded away or even the undercutting of cliffs. In addition, rapid-onset coastal erosion triggered by hazards like tsunamis or cyclones may cause more damage in days or less than slow-onset coastal erosion can cause in decades. And despite the rapidity with which the initial erosion happens, the recovery process still generally takes much longer.

Either slow onset or rapid onset may also be the by-product of human activity, caused either by the disruption of land areas or systems that contribute to ongoing slow-onset erosion or by the rapid onset of land movement, such as when entire areas of land are removed or moved to reroute water ways. Such human activity may, in turn, open up more areas to (or speed up) the process of naturally occurring erosion. This may also cause other problems such as increased cyclone or flooding impacts by disrupting natural impediments to hazards such as barrier islands or beaches.

Different forms of land, including beaches, bluffs, etc., are subject to different rates of erosion, depending on variables such as the materials they are made of, the coarseness of the sediment in that area, their height, their slope, their orientation relative to the ocean, how sheltered they are, temperatures of materials involved, and/or the age of the surface being affected. In addition, these issues also shape whether or not coastal erosion can be reversed once it has happened.

Unconsolidated sediments—like beaches—are looser accumulations of materials. They are among the most vulnerable to coastal erosion and some of the areas that have the most rapid processes of erosion. This can result in not only direct coastal erosion but also things like shifting of dunes. However, in coastal erosion of noncohesive and unconsolidated sediments like sand, at least some of the damage can be repaired, although sometimes only temporarily, by restoring the sediment. With cohesive and consolidated sediments that are more solid and bound together, like Arctic bluffs, the process of coastal erosion often takes more time generally, although not always, occurring more gradually. Despite this, the damage of coastal erosion once it occurs is largely irreversible.

Naturally occurring landforms may minimize coastal erosion on the main coast of an area, such as barrier islands. This minimizing of coastal erosion occurs both with slow-onset and rapid-onset forms, sheltering more mainland areas from both the slow and consistent wearing down of sediment by wind and water, as well as from the more rapid and severe impacts of natural hazards such as storms, cyclones, and tsunamis. Research has shown that in comparable areas, land sheltered by barrier islands may exhibit less coastal erosion than other nearby areas (Frederick et al. 2016). However, it is also worth noting that in some cases naturally occurring varied sediment that forms in contact with one another can either act as a buffer against further erosion or cause more erosion. As Hurst et al. note, where beaches and cliffs meet, the effects may vary. If the beaches are stable, they can provide a barrier to dissipate the impact of waves and minimize cliff erosion. But if the beach sediment is more mobile, the abrasion caused by moving sands may actually contribute to cliff erosion (Hurst et al. 2016).

SEA LEVEL RISE AND CLIMATE CHANGE

Rising sea levels are tied to rising global temperatures and climate change more broadly. The rise of global mean temperature by a half degree Celsius throughout the twentieth century is likely responsible for a nearly 8-inch (203 mm) global sea level rise in that same period (Leatherman et al. 2000). By some predictions, it could be as much as an increase of over 38 inches (98 cm) by the end of the twenty-first century (Williams et al. 2018). This remains an ongoing problem. Greenhouse gases are predicted to further increase global temperatures in the twenty-first century, causing accelerated rises in sea levels, which will, in turn, result in even more severe increases in coastal erosion. And these issues are likely to have extremely serious impacts not only on the world in general and specific coastal regions but also especially on areas like low-lying barrier islands. Damage to these low-lying barrier islands is likely to have a cascading effect, opening up new areas that they had protected to increasing levels of coastal erosion. Understanding these issues also requires knowledge of long-term changes to both the shoreline and sea levels, including rates of sea level rise.

Sea level rise associated with climate change and increased impacts of varying hazards like flooding, storm surge, and cyclones may be contributing to increasing coastal erosion in key locations. What makes this relationship even more problematic is a relatively minimal amount of sea level rise can produce significantly larger effects on coastal erosion. One study found that it took just a sustained sea level rise of 3.9 inches (10 cm) to produce 49.2 feet (15 m) of erosion, noting that "such an amount is more than an order of magnitude greater than would be expected from a simple response to sea level risk through inundation of the shoreline" (Leatherman et al. 2000, 55). Making this even more complicated is the fact that such changes may be masked by other causes of erosion, such as coastal engineering projects (Leatherman et al. 2000).

Rising sea levels tied to climate change produce increased coastal erosion in different ways. On one level, they simply replicate ongoing coastal erosion issues, leading to erosion tied to wave action and other naturally occurring processes. On another level, they allow for more powerful waves to both reach further inland and redistribute that land further offshore, transforming the otherwise naturally occurring movement of noncohesive sediments in particular. In fact, by the year 2050, global sea level is estimated to increase by nearly 8 inches (20 cm), and in some specific areas, this increase may be as high as 16 inches (40 cm), which will then result in up to 197 feet (60 m) of coastal erosion in some areas (Leatherman et al. 2000).

MEASUREMENTS AND TRACKING

Measuring current rates of coastal erosion may be complicated by factors such as what datasets are available, and in some cases, data on the materials that make up a coastline or their exact location, slope, etc. may not exist or may not be entirely accurate (Udo and Takeda 2017). Researchers have attempted to calculate previous and ongoing coastal erosion in order to determine future projections of potential coastal erosion, using mathematical and conceptual models (Frederick et al. 2016; Udo and Takeda 2017). When trying to predict future erosion, such efforts are complicated by issues such as changing hazards, increasing wave height, or sea level rise, each of which may have unforeseen impacts on coastal erosion or whose future numbers may be difficult to quantify, leaving researchers working with a range of potential options or needing to choose to exclude such factors.

Further complicating such modeling is the consideration for the types of materials involved. Most coastal erosion modeling has focused on noncohesive sediments like sand and mostly in temperate climates. Nevertheless, coastal erosion cannot generally be modeled the same way when, for example, we consider Arctic coastal bluffs made of "ice-bonded silts or clays" (Frederick et al. 2016). Gaining knowledge of how rocky coastlines or cliffs erode is also complicated by the fact that it is difficult to tell what their pre-erosion states were. However, understanding the process of coastal erosion and the type, amount, and location of change in coastlines in general and on specific forms like cliffs is also crucial to awareness of the risks involved and the potential impact on natural systems and human development. And some models can be problematic in what they do or do not consider.

The Bruun model or rule, for example, predicts that over the long-term coastal erosion can be 50–200 times the magnitude of sea level rise in the area. It is based on a two-dimensional model of coastal responses to rising sea levels. In the model, rising sea levels result in sediments being moved further inland, which in turn cause erosion on the coastline as those sediments are moved, leading to the argument that increasing sea levels enable waves and water during storms to be able to access land further up the beach and remove it entirely (Leatherman et al. 2000).

The Bruun model has been both widely used and rejected because it does not factor in issues like sediment transport in all contexts or wave climates. While some tests seem to verify the model, others seem to fully disprove it, making the ongoing use of the model complicated. The model fails in more dynamic areas like the regions around barrier islands, a particularly problematic failure given that the failure of barrier islands can allow additional storm damage in the coastal erosion cycle, leading to rapid onset and increasing levels of coastal erosion.

In the case of situations like Arctic coastal erosion, specific forms of conceptual modeling may also be of use in understanding the impacts of various problems and long-term processes. For example, the use of conceptual models of block failure may help researchers better understand the role of falling blocks of ice in erosion of coastal bluffs in the Arctic (Frederick et al. 2016). Work on this has revealed that rather than adding additional materials at the base of a bluff that might buffer them against erosion, the way the fallen blocks themselves erode and move leaves the bluffs as susceptible to ongoing issues with coastal erosion (Frederick et al. 2016).

Beyond mathematical and conceptual modeling, recent history has seen efforts toward geodetic surveys, increasing since the mid-nineteenth century. Such efforts may lead to clearer data on what land exists at various points in history, allowing for more direct tracking of loss to coastal erosion and other factors. However, this process is also somewhat difficult, as it may take years or even decades to gather enough data to clarify if slow-onset erosion is occurring or to determine the "normal" state of the coastline, a process that is made even more complicated by human intrusions into the area and attempts to minimize erosion, which may obscure long-term issues. Moreover, in many areas historical knowledge of coastal sediment, mapping, and erosion may be lacking, with few accurate historical records being available from more than 150 years ago. This lack of data also makes it even more important that we continue to gather as much data as possible now and move forward to better understand how coastal erosion changes coastlines.

IMPACTS

Coastal erosion may have a wide array of impacts, depending on where it is happening. As previously discussed, coastal erosion may lead to the reduction of sediments along beaches, barrier islands, cliffs, and a range of other land types. These impacts may also have a cascading effect, opening up new areas to increasing coastal erosion or flooding, the shifting of sediments into new areas that may transform the way existing sediments and waters move throughout environments, or even have potential effects on different species that live in affected areas.

Coastal erosion may also affect people and communities along the coastline. Coastal areas in general are hugely important to humans and their settlements and are spaces of extensive cultural and economic exchanges. In addition, large numbers of people live in coastal areas, along with the associated homes, business, and infrastructure built, developed, and maintained by those human populations, many of whom are critically tied to their continued living in the area. Coastal erosion in general or increasing amounts of it in certain areas can also be tied in to increase in flooding, causing additional destruction or problems for people in many areas.

Already issues seem to be appearing in various coastlines, as illustrated in examples like research into Arctic coastlines, which point to increasing coastal erosion and larger scale issues with a transforming coastal environment tied to issues like sea level rise and problems associated with climate change. Research-based estimates suggest that as much as 86 percent of native villages on Alaskan river or sea coastlines are more affected by flooding and coastal erosion (Frederick et al. 2016). Translated from percentages to lived experiences, this means that (1) archaeological, landfill, and other sites have been undermined or damaged; (2) infrastructure tied to people's daily lives and economic networks like those related to oil and gas extraction are threatened; and (3) homes are falling into the water, undermined by erosion, and when combined with floodwaters, calls have increased for and work has begun on the potential relocation of entire communities (Frederick et al. 2016).

In areas with sandy beaches, coastal erosion may lead to eroded beaches and, subsequently, may increase the exposure of human-built structures like lighthouses or other buildings to direct wave damage or flooding. Coastal erosion on cliffs can also directly affect infrastructure, homes, businesses, and other construction at the top of the cliffs even when erosion is primarily at their base, a process that can undercut both the cliffs themselves and the development on top of them. And even processes further away can play a role, as human transformation of environments further inland and on the coasts can impact which sediments end up where and how sediments move, including potentially reshaping preexisting natural processes. These issues happen at both a more local level and a more global level, raising problems related to human activity like dam construction, which has resulted in a worldwide reduction in sediments moving to the sea.

The impacts of coastal erosion may affect not only the present existence of coastal communities but also their future while simultaneously threatening to potentially uncover elements of their past that may cause problems for inhabitants of coastal areas. While ongoing problems with coastal erosion may affect existing and future infrastructure, a lack of more complete knowledge about potential future coastal erosion may result in new construction, development, and relocation being placed in harm's way and simply pushing problems related to construction and infrastructure further down the road. Beyond this, if current and/or future erosion interferes with buried chemicals, radiological materials, or other potential hazards, such as those existing around the Arctic coastline, new threads may emerge and impact people's health and well-being (Frederick et al. 2016).

RESPONSES

There are a wide range of responses to coastal erosion, some of which focus on the land itself and others that focus on human uses of the land. Williams et al. (2018) argue that historically much of the decision making in response to coastal erosion was influenced by economic issues and/or a cost-benefit analysis including both potential economic losses and gains with different approaches to coastal erosion management. However, they also note that this is shifting with more work being done on so-called managed retreat, efforts to manage a retreat from spaces being impacted by coastal erosion rather than ongoing efforts to stay on that land (Williams et al. 2018). Ultimately, planning for management of coastal areas requires an understanding of coastal erosion, including natural processes and human impacts on the region.

In an effort to maintain ongoing use of land susceptible to coastal erosion, people turn to a range of different responses that allow for ongoing use. Many such efforts can be classified into three larger categories: (1) defense, (2) adaptation, or (3) managed retreat/realignment that is often combined with other approaches like the creation of setback zones, buy-back programs, or environmental assessments and regulations (Williams et al. 2018). Each of these approaches includes both benefits and potential risks, both to humans and their built environment and to the local and larger natural environment.

Defense includes practices such as building things like levees or seawalls to keep water off areas of land being used by humans, although these can be both costly and can result in temporary solutions. The results of these efforts vary widely, based both on the approach used and its context. Some may provide short-term protection, but that protection can be reduced over time as issues like erosion of the barriers themselves, problems with barrier maintenance (see, for example, the damage to levees in New Orleans, Louisiana, both predating and following Hurricane Katrina in 2005), or shifting of the impact of coastal erosion to other nearby areas as water and sediment are simply rerouted.

Adaptation involves using tools such as elevated homes and buildings and related changes to building codes to avoid direct impacts of rising water and erosion, as well as transforming or attempting to repair eroded land as with wetland restoration or growing salt-tolerant crops in areas of saltwater erosion. This may be problematic in some areas and for some populations, in part due to the associated costs of such efforts. For example, in coastal areas where residents own or live on small plots of land, building up can be costly, particularly for disabled or elderly residents, as it can require either the additional costs of structures like elevators to make buildings accessible or the additional costs, if even possible, of additional land to provide the space for accessible ramps instead of stairs.

Managed retreat or managed realignment, including potentially the development of sacrificial zones, done by processes like moving structures inland or rebuilding further inland while demolishing, letting existing buildings in the affected areas degrade, or wholesale abandonment of certain areas. This method has been used historically, but with increasing problems with sea level rise, we are seeing it used more frequently and more broadly. The potential costs and complexities of managed retreat/realignment can vary widely across different areas, due to a range of issues such as (1) the cost of land in the specific area in question; (2) legal or other requirements to compensate local landowners or others who may be affected, including issues like zoning or land use restrictions or regulations; (3) the costs of dismantling or destroying human infrastructure and buildings in a way that minimizes or eliminates potential pollution; (4) what defenses may be needed to minimize coastal erosion or inland hazard or weather problems; (5) the coast and availability of experts versed in such efforts; (6) the amount of monitoring needed for the long-term project; (7) long-term relocation plans and the speed with which they need to happen; (8) stabilization of land and sediment being retreated from; (9) modification of existing developments and infrastructure, including both areas further inland that may be affected and those that will remain in areas of coastal erosion; and (10) the specifics of local areas, including incomes, costs, environment, available land, etc. (Williams et al. 2018; Frederick et al. 2016).

If done correctly, managed retreat/realignment can reduce the impact of coastal erosion on both natural landscapes and human infrastructure and development, and in many cases, these types of efforts are very specifically managed to allow ongoing natural processes that may protect the environment and minimize coastal erosion themselves to continue. However, this approach can be problematic in some areas where coastal access is prioritized for a variety of reasons. For example, in the Arctic, interior areas range from less accessible than coastal areas to entirely inaccessible by the majority of people looking to live in or access the general area. In such areas, managed retreat may not be a workable or affordable option.

Such examples of potential responses to coastal erosion are also a question of scale. Primary and initial interventions to coastal erosion often happen or at least begin at the local level, responding to issues like coastal erosion on local beaches. However, previous discussions of issues like the worldwide reduction in sediment movement due to structures like dams also open up the potential space for a larger scale discussion and response, including conversations about and research on coastal erosion and sediment movement at a regional, national, or even global level. Beyond this, while local-level response is critical to understanding and reacting to coastal erosion processes, it is also vital to keep in mind how coastal erosion is tied into larger systems, with sediments coming from or moving to other areas. Responses to coastal erosion in one location can impact coastal erosion or build up in other areas, although often unintentionally. In addition, managing and responding to coastal erosion must be done by looking not only to the sea but also to inland areas that may also have an impact via development and sedimentary movement.

CONCLUSION

Coastal erosion is an ongoing process, triggered not only by natural processes and hazards such as water and waves, storms, cyclones, or tsunamis but also by human action, including sometimes actions taken to mitigate the impact of coastal erosion. Moreover, in the face of ongoing rising sea levels and climate change, it is not just a process that is not going away, but it's potentially a growing problem for a range of peoples, communities, natural habitats, and ecosystems. Tracking and modeling coastal erosion is complicated, regardless of whether it's done with mathematical or conceptual modeling or data collection and comparisons, and results that reflect current and future problems require a long-term perspective. Nevertheless, it is crucial that researchers understand coastal erosion and the ways in which it is increasing and transforming landscapes and that policy makers and communities be engaged in such understandings to enable researchers both to reflect on human activity in coastal erosion and to help them minimize future effects and problems. In short, coastal erosion occurs naturally, but is affected by human activity and decision making on both a global and local scale. Knowing about it and its potential effects on people, their homes and communities, and the environment is crucial to addressing problems related to it and taking the best steps to reduce its impacts on people and people's impacts on it.

Jennifer Trivedi

Further Reading

Frederick, J. M., M. A. Thomas, D. L. Bull, C. A. Jones, and J. D. Roberts. 2016. *The Arctic Coastal Erosion Problem*. Sandia Report SAND 2016-9762. Albuquerque, NM: Sandia National Laboratories.

Geoscience Australia. 2019. Coastal Erosion. http://www.ga.gov.au/scientific-topics/hazards/coastalerosion, accessed April 1, 2019.

Hurst, M. D., D. H. Rood, M. A. Ellis, R. S. Anderson, and U. Dornbusch. 2016. Recent Acceleration in Coastal Cliff Retreat Rates on the South Coast of Great Britain. *PNAS* 113(47): 13336–13341.

Leatherman, S. P., K. Zhang, and B. C. Douglas. 2000. Sea Level Rise Shown to Drive Coastal Erosion. *Eos* 81(6): 55–57.

Senevirathna, E.M.T.K., K.V.D. Edirisooriya, S. P. Uluwaduge, and K.B.C.A. Wijerathn. 2018. Analysis of Causes and Effects of Coastal Erosion and Environmental Degradation in Southern Coastal Belt of Sri Lanka Special Reference to Unawatuna Coastal Area. 7th International Conference on Building Resilience; Using scientific knowledge to inform policy and practice in disaster risk reduction, ICBR2017, November 27–29, 2017, Bangkok, Thailand. *Procedia Engineering* 212: 1010–1017.

Udo, K., and Y. Takeda. 2017. Projections of Future Beach Loss in Japan Due to Sea-Level Rise and Uncertainties in Projected Beach Loss. *Coastal Engineering Journal* 59(2): 1740006-1–1740006-16.

Williams, A. T., N. Rangel-Buitrago, E. Pranzini, and G. Anfuso. 2018. The Management of Coastal Erosion. *Ocean & Coastal Management* 156: 4–20.

Desertification

Experts have found the concept of desertification difficult to define, but it can be broadly understood as a disturbance in the functioning of an ecosystem in terms of its ability to manage water, energy, and nutrients (Veron et al. 2006). While the moniker of this disaster may lend itself to be interpreted as an expansion of desert, desertification is more accurately defined as the conversion of certain ecosystems into desert-like terrain, which may include considering whether or not the area can be brought back into productivity should less-arid conditions return at some point. Consequently, desertification may be seen as overall land degradation, as opposed to the conversion of land into desert. Due to their closely related conditions, semiarid and arid lands (drylands, which cover approximately 40 percent of the earth's land area) are most at risk of shifting to desert conditions, and the process of desertification has been called by an United Nations (UN) official as "the greatest environmental challenge of our time" (Carrington 2010). Desertification needs to be understood under the lens of several overlapping concepts, but it can be categorized under two canopies: (1) natural systems, and the natural variability of climate systems, including how continentality and topography play a role in defining climate regions, and (2) human influence on and use of the natural environment. While the natural operations of the earth's climate system create the conditions at some initial state, humans have done a great deal of work to alter those natural systems, which in turn alter the world in which humans must live.

As human settlements expand due to increasing population, more food is required, which demands more, and often more intense, agricultural and rangeland practices. Mismanagement and overuse of land, including the collection of biomass to be used as fuel, leads to degradation of soil, which in turn can radically alter the ability of the land to support plant and animal life. Taken in the context of global climate change, semiarid and arid regions that see significant pressure placed on already sparse natural resources coupled with increasing temperatures may set the stage for rapidly advancing desertification. Increasing temperatures will lead to prolonged periods of heat and drought in semiarid and arid areas, and less frequent but more intense period of rainfall will greatly contribute to erosion.

Desertification will most likely be accompanied by political and civil conflicts over dwindling resources, including water, which is also placed at risk due to desertification. The UN recognized this decades ago and in 1977 at the UN Conference on Desertification, held in Nairobi, Kenya, convened the first international meeting dedicated to examining desertification. Other major gatherings include the 1993 Conference to Combat Desertification, the 2006 International Year of the Desert and Desertification, and the 2013 UN Convention to Combat Desertification. Many regional, national, and international management plans have been implemented to combat this threat. The two key drivers of desertification are climate and land management practices, which lead to soil degradation or the shift of soil from productive to unproductive in terms of its ability to support plant life, which forms the basis for the global food chain. Understanding the role of climate with respect to desertification can be broadly accomplished by gaining a basic knowledge of how climates in these regions are classified and the ways in which humans use the environmental services available to them in these regions.

PHYSICAL CHARACTERISTICS OF DESERTIFICATION

As classified by the Koeppen-Geiger system, semiarid, or steppe (*BS*), climates are situated between humid (*C and some subgroups*) and arid/desert (*B; BWh, hot desert; BSh, hot semiarid*) climate types. The primary factor is lack of precipitation, with the second letter following *B* assigned as either *S* (steppe) or W (desert) to further indicate the level of precipitation typically found in the defined or delineated region. A third letter may be added to dictate hotter (*h*) or colder (*k*) regions. Within these groups and subgroups, different ecological communities can also exist; called biomes, the vegetation found within these divisions can play a crucial role in fending off desertification, due to the highly variable nature of plant types in dealing with drought conditions. There are several classification systems for biomes, but most identify, in some fashion, shrubland, scrubland, desert grasslands, and/or savanna as vegetation types typically found in arid or semiarid climate regions. More specifically defined regions termed *ecoregions* were developed by Robert Bailey (1989), and they can further assist policy makers in identifying appropriate management principles to apply in a given area.

MAJOR PROCESSES IN DESERTIFICATION

Soil Degradation

The importance of land-use issues cannot be overstated in terms of the ability of humans to effectively manage and efficiently use available arable land in light of increasing population and the need for productive soil for agriculture as well as the production of biomass for energy needs, largely in developing countries. Not only does the production of biomass increase pressure on land use, but also when used as a primary fuel source, biomass releases greenhouse gases (GHGs) into the atmosphere, further compounding the climate change problem, thus further advancing desert areas (scientists call this a "positive feedback loop," which is a steady move away from the original condition or state of a given system). While simple to describe in general terms, it should be noted that this issue is scale dependent, in that large variations in extent and intensity of land use for the growth of biomass to be used as fuel exist over relatively small areas of inhabited land; however, according to the Intergovernmental Panel on Climate Change (IPCC 2019), approximately one-quarter of the earth's ice-free land suffers from soil degradation due to human action, most notably, erosion, discussed in the following paragraph.

Erosion is a key component of soil degradation. The forces of water and wind, as well as human activities, work to detach soil from the column and remove it, sometimes to great distances. The eroded soil is deposited in a new location. Erosion by water is referred to as fluvial erosion and can occur in either established river channels or overland, as is typical during flood events, but can also be caused by poorly managed irrigation practices, while wind-based erosion is called Aeolian (or eolian) erosion and is wholly a natural occurrence. However, the ability of wind to remove soil is greatly increased in areas under tillage. The IPCC reports that erosion rates are anywhere from 10 times higher (in areas of no tillage) to 100 times higher (conventional tillage) than soil formation rates in those areas. So while the agent Aeolin erosion may be a natural process in and of itself, it is one greatly exacerbated by human action and mismanagement of natural resources (IPCC 2019).

Removal of soil by water can be classified into five major types: (1) rill erosion, (2) gully erosion, (3) sheet erosion, (4) stream channel erosion, and (5) splash erosion. Rill erosion involves the formation of a small channel in a slope, which in turn can lead to the second type listed, gully erosion. The process and action of these two types is similar, but can be separated by thinking of rill erosion as the precursor to gully erosion. Sheet erosion is a slow process, occurring over many years when small and generally uniform layers of soil are removed. Stream channel erosion occurs in channels with continuous flow and on relatively flat gradients. Erosion can take place through the natural flow of the stream as the banks are undercut, by the action of scouring the bed of the stream or by flood events. Splash erosion can occur with either nature or humans as the agent: when nature driven, large and heavy raindrops can land with enough force to dislodge soil particles, and when human driven, the agent is irrigation, but the physical process is the same. In considering one of the predicted effects of global climate change, more intense rainfall, splash erosion may pose more of a threat as the climate continues to warm; the same can be stated for stream channel erosion, in that stream systems are likely to see an increase in flood events due to climate change.

Vegetation Degradation (Deflation)

Soil degradation, namely, erosion, is caused mostly by vegetative degradation, or "deflation," which is the loss of vegetation that normally would work to stabilize soil in place, thus fending off the forces of fluvial or Aeolian erosive agents. This is another example of a positive feedback loop when topsoil erosion is also considered, as the loss of vegetation can lead to deflation, which in turn leads to more erosion, and in a final turn, more vegetation loss can occur. The relationship between vegetation and soil is a key factor in the process of desertification, as the degradation or loss of one leads to the other and begins the positive feedback loop. When deflation occurs in arid areas, it can be exceedingly difficult to reverse the process, as soils in arid locations contain very few nutrients, very little topsoil, and very little parent material below the surface, which could lead to rapid redevelopment of the soil. Illustrating the interconnectedness of these processes, the lack of vegetation at the surface, which is one of the major contributors to productive soil formation, further stalls the process of soil regeneration in deflated areas. Deflation may also lead to the increase in invasive species in an area, which tend to have extremely destructive effects on local plant and animal life, which in turn could remove additional vegetation, making the problem of desertification in the area worse.

Salinization

Salinization is the third major process in soil degradation and occurs when saline water used in irrigation slowly adds water-soluble salts to the soil column, greatly inhibiting the ability of organisms within the soil to respire and decompose normally and the ability of plants to uptake nutrients and flourish in that column. Irrigation-driven salinization occurs more readily in areas with high evaporation rates (arid regions) and, thus, poses more of a threat to advancing desertification in these areas, as the salt content cannot be washed out by rainfall and, due to lessened evaporation, has a tendency to remain situated in the soil column. The application of excessive water can lead to what is called waterlogging, which in turn can lead to salinization. Salts can also accumulate naturally in soil as a result of several natural forces. Chemical or mechanical weathering of parent material will add salts to the soil column as that parent material breaks down, especially in areas containing parent material high in carbonate minerals. If the area was once submerged under a saline water body, the remaining salts will affect the salinity of the soil in the present. Coastal areas are also naturally affected by saltwater, as sea level rise can saturate groundwater, causing seepage of saline water into local soil columns.

Soil Compaction

Removing pore space in a soil column is the result of soil compaction. This occurs when pressure is exerted at the surface, thus pushing the soil into a more densely packed unit. Rain can work to compact soil, but this fourth major process in desertification is largely driven by human action. Farm machinery moving over soil works to add pressure to the surface, which compacts soil. While machinery is more often associated with mechanical agriculture in the developed countries, and lower-impact subsistence farming (in terms of machinery) is associated with the developing world, the presence of machine-based agriculture in arid areas cannot be ignored. Any presence of mechanized agriculture in areas already under threat from desertification carry more weight, as arid soils are more easily compacted. The use of land as grazing pastures for livestock also presents an opportunity for soil to compact under the weight of the animal. Additionally, connecting back to deflation, the presence of livestock affects two of the four major processes of desertification, as livestock removes vegetation for food, resulting in less vegetative cover.

GLOBAL CLIMATE CHANGE AND DESERTIFICATION

Since desertification features climate as one of its two primary drivers, it follows that large-scale changes in the earth's climate systems will invariably affect the overall process of arid and semiarid lands shifting to drier conditions. Two of the major areas of interest within climate change studies are temperature and precipitation regimes. The latter regimes is largely defined the biomes under threat of desertification. However, increasing temperatures, on average, will continue to put pressure on already-arid regions. The human driver of increasing land use, leading to more intense uses of available land, will work in concert with climate change to compound the problem in already at-risk areas. According to the IPCC, as of 2019, humans were using anywhere between one-quarter and one-third of the land's ability to produce food, feed for animals, fiber used in the production of goods like clothes, wood used in construction, and wood (biomass fuel) for energy (IPCC 2019).

The productive capacity of a given parcel of land is called net primary production (NPP), and it is the sum of the ability of all vegetation in that area to photosynthesize the sun's energy into plant life. With climate change continuing to advance desertification, more land that would have been available for NPP is taken out of production, which adds to pressure in other areas. Further, vegetation and land act as a sink for GHGs, absorbing some of the excess added to the atmosphere; thus, it follows that with more land being converted to desert, which is largely devoid of vegetation, less land is available to absorb GHGs, meaning they will remain in the atmosphere, further fueling a positive feedback loop. Heat and heat-related events connected to climate change will also play an important role in desertification. Longer periods of drought in areas already prone to extended dry periods will greatly add to the problems of vegetation loss, topsoil erosion, and subsequent Aeolian erosion. Further, dust storms may become more frequent and widespread, as more desert surface area will mean more exposed material that can be transported by Aeolian processes.

MAJOR DESERTIFICATION PROBLEM AREAS

Globally, not all regions, biomes, or other classifications for location are under threat of desertification. Primarily, the semiarid and arid regions will be most affected by desertification, for reasons discussed earlier, including the already-dry nature of these biomes and the looming threat of climate changes predicted to occur in those areas, such as increases in temperature and increased evaporation of water coupled with decreased precipitation. Populations in Africa and Asia are projected by the IPCC (2019) to be the most vulnerable to desertification, but the recent report also identifies several other regions of less aridity as under threat from increased wildfires and crop yield decline due to advancing desertification. Identified here are two major regions that are currently experiencing the most intense desertification and are likely to continue to face the threat for the foreseeable future.

The Sahel Region, Africa

The region known as the Sahel, situated between the Sahara Desert to the north and equatorial Africa to the south, is a transition zone of several hundred miles in width (north to south) and spans the east-west length of the African Continent. Also referred to as a biogeographic zone or an ecoclimatic zone, the Sahel encompasses parts of 14 African nations and covers approximately 1.8 million square miles (3.1 million square km). The region sees very little rainfall, from four to eight inches (100–200 mm) annually, depending on location, and is primarily flat in relief, with the exception of some isolated mountain ranges and plateaus. During the year, the Sahel sees a prolonged season accompanied by a short rainy season. With the prediction of shorter but more intense rainfall, one of the primary threats to the region is erosion due to fluvial processes, which may work to exacerbate already-occurring desertification across the region. Further issues related to soil degradation include the prevalence of livestock throughout the region. Bands of nomadic peoples typically move from north to south during the rainy season and into the dry season, respectively. This puts pressure on the surface and results in soil compaction, lessening the ability of the soil column to support plant life. Although tempting to place the blame on a changing climate as the culprit driving human-environment interaction in the region, in 2006, the United Nations Environmental Programme (UNEP) stated in their report titled *Climate Change and Variability in the Sahel Region: Impacts and Adaptation Strategies in the Agricultural Sector* that climate cannot be the only consideration (Kandji et al. 2006, 8): "Rainfall variability is a major driver of vulnerability in the Sahel. However, blaming the 'environmental crisis' on low and irregular annual rainfall alone would amount to a sheer oversimplification and misunderstanding of the Sahelian dynamics. Climate is nothing but one element in a complex combination of processes that has made agriculture and livestock farming highly unproductive. Over the last half century, the combined effects of population growth, land degradation (deforestation, continuous cropping and overgrazing), reduced and erratic rainfall, lack of coherent environmental policies and misplaced development priorities, have contributed to transform a large proportion of the Sahel into barren land, resulting in the deterioration of the soil and water resources." Desertification is a major threat to this region, with the onset of large dust storms, inhabitable land, lack of wood for building and burning, and general overpopulation that are all major problems faced by the countries that make up this area.

The Gobi Desert, China and Mongolia

Much like problems facing inhabitants of the Sahel region, the increasing threats of heat, lack of rainfall, and soil-related issues are becoming important. Overpopulation leading to increased pressure on the drylands is causing ecosystem services to fail in their support of that population. Since the late 1970s, the Chinese government has recognized the problem of desertification and has been working to combat it. Much as populations in countries throughout Europe as well as the United States have experienced, rapid growth of the population and subsequent building and expansion took its toll on the environment. The removal of productive land to make way for encroaching urbanization left huge swaths of land exposed to the forces of erosion. In 1978, the Three-North Shelterbelt Project was launched, with the aim of planting millions of trees along the roughly 3,000 miles (4,800 km) of the desert in northern China. The project has planted 66 billion trees, and planting trees in this belt is set to terminate in 2050 (Petri 2017).

Other Problem Areas

In South America, desertification is a growing problem, most notably in Argentina and Brazil. In Argentina, approximately 75 percent of the land area is affected by desertification, touching 30 percent of the population. The major cause of desertification is the overreliance on the land by masses of impoverished citizens. Intense pressure on farm and rangeland and the conversion of forestlands to those uses have been widespread. The poor across several regions of the country utilize the land for the harvesting of firewood and for practice of unsustainable sheep grazing and short-term and intense agriculture. Current projects, managed by the Argentinian government, are attempting to address the problem by educating populations on best practices concerning land use and by providing economic incentives to practice those methods.

Another South American nation facing the problem of desertification is Brazil. In 2007, the Brazilian National Conference on Fighting Desertification was held to address the issue. The key pressure on forestlands in Brazil, as related to desertification, is the practice of deforestation, which can occur on both small and large scales. Typically, small-scale deforestation takes place at the hands of subsistence farmers or small ranch operations. Overly intense use of tropical lands is sustainable at any given location for usually less than five years, as tropical soils (oxisols or ultisols) cannot regenerate fast enough; the result is soils exposed to erosion and subsequently, desertification. Large-scale operations cause the same type of damage but complete the task over much larger areas, and in less time, especially when logging is the extractive industry operating in the area. Forest fires also play a role in this system and as such require careful management. Australia as well as northern India, particularly the Himalayan Region, are also global problem areas for desertification, for mostly the same reasons as discussed concerning Africa, China, and South America.

DESERTIFICATION AND HUMAN SYSTEMS

Agriculture

Improper application of agricultural practices, in all areas affected by desertification, is a major cause of the problem. Wherever human settlements are found, the need to set aside arable land for agricultural purposes is a must. Humans, however, have typically not examined their agricultural practices in regards to the long-term sustainability of the land. In the United States, the Soil Conservation Service (SCS), which was the forerunner to the modern National Resource Conservation Service (NRCS, renamed in 1994), was formed in 1933 to begin addressing the problems associated with the misuse of soil. That agency marks the first attempt by a large nation to systematically work to protect and preserve soil, which is the basis for productive agricultural land, as a resource. Since 1933, many nations have followed the lead of the United States, forming similar agencies.

The main threat to soil, in relation to agricultural use, is that soil takes decades or centuries to regenerate from parent bedrock material into useable soil. As such, the mismanagement of soil at the hands of irresponsible agriculturalists creates a scenario where once soil has lost its productive capabilities, that land parcel cannot return to productivity for several generations. In light of this, the mismanagement of agricultural lands left barren is in turn driving desertification in countries where government agency-lead intervention is not a priority or possibility. By examining processes related to the application of fertilizer, saline water used for irrigation, rangeland management, and proper tillage techniques will aid in better preserving arable lands for continued productive use.

Food Security

Food security is another major issue facing humanity in light of desertification and its relationship to agriculture. The problem is two-fold, in that while population across the world is increasing, food production will need to expand rapidly to meet the demand, but it is under threat of reducing crop yields, in part due to desertification of land (especially in more vulnerable, less-developed countries). According to the UNCCD (2019), half of the world's occupants of arid lands live in poverty and represent approximately three-quarters of a billion people, all of whom are at substantially elevated risk of malnutrition and undernutrition due to projected food insecurities. This will likely lead to migration out of areas under risk and into areas of lesser risk, which in turn will put more pressure on the agricultural production systems of those areas, thus increasing the level of food insecurity. Climate change will also play a role in reducing food security, with crop yields reducing and lower growth rates in livestock. The prevalence and increase of agricultural pests and disease are also likely to increase in areas experiencing desertification. The trend of food insecurity is expected to continue as more frequent and higher-magnitude weather events, including those connected to desertification, unfold.

Worsening weather in dryland areas will likely result in damage or destruction to precious crops, furthering the problem. Biodiversity or the range of crop types will also be affected. Globally, humans get their food from very few types of plants and animals, an effect of monoculture (the practice of farming or ranch operations to concentrate on one type of product). With an overreliance on a small number of species, reduction in biodiversity poses a major threat to food security, as any one disease could greatly damage or eliminate that species. With continued desertification, the potential for such an event is increased. Urban expansion will also affect desertification in the context of food security; as more land is demanded for urban areas, cropland is converted for that use, which can exert further pressure on food production in that area.

SECONDARY NATURAL DISASTERS

Desertification is an example of natural disaster, but with the increase of this event, secondary disasters, or disasters that result from the initial disaster event, may become more common in areas experiencing desertification. Drought and heat waves are commonly associated with arid and semiarid regions and will likely become more common and intense as climate change causes the global temperature average to increase. Exemplifying the concept of a positive feedback loop, these secondary disaster types will certainly work to further desertify regions affected by them. Drought is a major contributor to famine; there have been dozens of major famines, killing hundreds of thousands of people, a majority of which were in African nations. With increasing population in the Sahel and sub-Saharan Africa, the prospect of further desertification leading to famine is a grim one, with potentially millions of people at risk of death by starvation. Similarly, heat waves are a major disaster type, one that will create larger vulnerable populations in light of the creation of more desert-like conditions.

Dust storms are also likely to increase in frequency and magnitude. As more surface area of the earth is exposed, it stands to reason that more available material is presented to be swept up in windborne dust storms. Forest regions in humid biomes that experience increased wildfire activity may work to increase or begin the process of desertification. In this example, desertification is the secondary disaster. In the western United States, desertification processes are being initiated by such increased wildfire incidence (Neary 2018), termed *fire-induced desertification*. This process is connected to climate change, in that the conditions that are thought to produce increased wildfire activity (increased fuel/dry fuel, mainly) are driven largely by increased temperatures and less precipitation. Evidence that fire seasons are starting earlier and lasting longer will only work to increase this risk. Fire-induced desertification can also be initiated by "wildfires" resulting from human negligence during an interaction with the forest environment.

CONFLICT

The Index of Human Insecurity (IHI) was designed to estimate the likelihood of conflict among humans based on socioeconomic considerations as well as four resource-based considerations: (1) safe water, (2) energy imports, (3) arable land, and (4) soil degradation (Lonergan et al. 2000). Considerations one and two are outside the scope of this topic, but three and four are directly related to desertification, as discussed in detail earlier. The increase in degraded soil in a given area means simply that food production will be maximized or be able to support the present population. Additionally, arable land (clearly directly related to soil degradation) further drives a population toward instability. The considerations of lack of arable land and degrading soil can directly threaten survival; it then follows that desertification as a disaster type is a major threat to human survival in areas that experience this slow-onset disaster type.

CONCLUSION

Desertification is the process by which land moves toward a desert-like state. Arid and semiarid lands have naturally occurring climate types that support biomes containing both plant and animal life. When suited to such a region, one largely devoid of precipitation and is host to temperatures higher than all other climate types and regions across the planet, its inhabitants can flourish. The application, however, of additional pressures, both anthropogenic and natural, can work together or separately to alter the functioning of an ecosystem in such a way that the climate type of semiarid may be converted to arid over time, or an already-arid region may simply become more arid. Global climate change is playing a major role in advancing desertification in many areas on the planet, including the Sahel Region in North Africa, parts of sub-Saharan Africa, North and South America, South Asia and East Asia. With increasingly warm temperatures, semiarid areas are likely to see further stressors on vegetation, which in turn may result in deflation of vegetative cover, exposing more surface soil to erosion. Precipitation regimes, forecasted to be lessened in volume and frequency but heightened in intensity, will work to further erode that newly exposed surface material, which in turn will create a difficult avenue of recovery for the affected vegetation types.

Humans are also playing a major role in the advancement of desertification across the planet. While the stress of climate change is partially driving the need to migrate to more productive areas, what people do with those arable lands is also contributing to the problem. Overuse of resources through intensive agriculture and rangeland uses is causing soil quality to degrade in many areas; once the soil has been depleted, the land is no longer arable, and subsistence-based lifestyles must be followed elsewhere. Education concerning these detrimental practices is one of the major solutions to solving the problems presented by desertification, but it remains to be seen as to whether or not education and economic incentives alone will be enough to address the issues. Humans will, when placed at risk, do what is needed to survive; in the case of desertification, though, the very actions taken to survive in the short term may ultimately create (or continue to exacerbate) the conditions in the long term that will not be in the best interest of the planet. Desertification is a major problem facing many countries, and with the considerations of both natural and human systems as the drivers for this disaster type, contriving solutions to combat conversion of land to desert may not prove simple.

Mitchel Stimers

Further Reading
Bailey, R. G. 1989. Explanatory Supplement to Ecoregions Map of the Continents. *Environmental Conservation* 16(4): 307–309.
Carrington, D. 2010. Desertification Is Greatest Threat to Planet, Expert Warns. *The Guardian.* December 16, 2010. https://www.theguardian.com/environment/2010/dec/16/desertification-climate-change, accessed September 5, 2019.
IPCC (Intergovernmental Panel on Climate Change). 2019. *Climate Change and Land.* New York: UNEP.
Kandji, S. T., L. Verchor, J. Mackeneu. 2006. *Climate Change and Variability in the Sahel Region: Impacts and Adaptation Strategies in the Agricultural Sector.* Nairobi, Kenya: UNEP.
Lonergan, S., K. Gustavson, and B. Carter. 2000. *The Index of Human Insecurity.* Victoria, BC, Canada: Department of Geography, University of Victoria.
Neary, D. G. 2018. Wildfire Contribution to Desertification at Local, Regional, and Global Scales. In *Desertification,* edited by V. R. Squires and A. Ariapour, 199–222. Hauppauge, NY: Nova Science Publishers.
Petri, A. E. 2017. China's "Great Green Wall" Fights Expanding Desert. *National Geographic.* April 21, 2017. https://www.nationalgeographic.com/news/2017/04/china-great-green-wall-gobi-tengger-desertification/, accessed September 5, 2019.
UNCCD (United Nations Convention to Combat Desertification). 2019. The Global Land Outlook: East Africa Thematic Report: Responsible Land Governance to Achieve Land Degradation Neutrality. https://www.unccd.int/publications/global-land-outlook-east-africa-thematic-report-responsible-land-governance-achieve, accessed September 7, 2019.
Veron, S. R., J. M. Paruelo, and M. Oesterheld. 2006. Assessing Desertification. *Journal of Arid Environments* 66(4): 751–763.

Droughts

Droughts are slow-onset disasters and less dramatic than other extreme natural events; however, they can last for considerable periods, even for several years or decades as in, for example, the Great Plains region of North America and southern margin of the Sahara Desert (Sahel) of Africa. Because of its long duration, climatologists call drought a "creeping disaster." Effects of this natural event are not felt at once, but they slowly take hold in an area and tighten their grip over time. Others compare drought to a python, which slowly and inexorably squeezes its prey to death. Such a drought is generally defined in terms of its impact rather than its genesis. Droughts occur in nearly every part of the world as well as almost all climatic regimes, but with varying frequency. They also occur in both dry and wet seasons and thus are a much more complex phenomenon than a routine dry season. However, droughts should not be confused with aridity. In desert or arid regions, rain is rare, and temperatures are high. Therefore, the lack of rain is the characteristic feature of the climate of such a region. Drought is thus an occasional phenomenon in the region.

In terms of areal extent, droughts are widespread compared to other natural disasters such as earthquakes and tornadoes; therefore, they affect more people across a wider area. With growing population, people are increasingly forced to settle in marginal land, and thus, the areas subject to drought are expanding over time. Since the 1970s, areas affected by drought have doubled. Although droughts may begin any time of the year, they are seasonal in certain places. Their impacts may range from mere local inconvenience to the economic and political breakdown of a nation (WMO 1975). Often droughts are broken by heavy rainfall and floods, particularly in semiarid and arid regions. This association between droughts and floods is linked to the El Niño-Southern Oscillation (ENSO).

DROUGHT DEFINITIONS AND TYPES

The definitions of drought vary considerably, and providing a universally acceptable definition is a difficult task for two reasons. Scholars from different disciplines such as agronomy, climatology, geology, geography, hydrology, and meteorology define drought from their own disciplinary perspectives. For example, agronomists may define drought in terms of plant growth, while meteorologists would focus on precipitation decline (Jedd et al. 2018). Also, based on climate history, each place needs its own operational definition. Thus, drought is interpreted differently depending on the concerned area and its climate. For example, in southern Canada, a drought is any period when no rain occurs in 30 consecutive days. In Australia, such a definition is meaningless because most of the country receives no rainfall for several months. However, drought effects do differ spatially, which therefore calls for many definitions. In fact, research in the early 1980s uncovered more than 150 published definitions of drought.

The National Drought Mitigation Center (NDMC) of the University of Nebraska–Lincoln defines drought in terms of deficient precipitation in a given area or region: "Drought is when a shortfall in precipitation creates a shortage of water, whether it's for crops, utilities, municipal water supplies, recreation, wildlife or other purposes" (NDMC 1996, i). In Australia, drought is usually defined as annual rainfall amounts being in the lower 10 percent of those recorded. Thus, a drought is not absolute in the sense of there being a total lack of rainfall during a period when it is most needed. In general, rainfall does occur, but it is either too late or too little for satisfactory crop growth or other relevant human activities.

In essence, drought is defined as an extended period of deficient rainfall relative to the average annual rainfall for a region. However, drought definitions abound, which are characterized in terms of either lack of rain over an extended time period or measuring impacts such as agricultural losses. Apart from region-specific definitions, some define droughts in the context of impacts of such events. However, droughts are widely defined according to meteorological, agricultural, hydrological, and socioeconomic criteria with each sector focusing on a particular type of drought (Wilhite and Glantz 1985).

Meteorological Drought

The most common type of drought is a meteorological drought, which is usually based on long-term precipitation departures from normal, but there is no consensus regarding the threshold of the deficit or the minimum duration of the lack of precipitation that makes a dry spell an official drought. This category of drought is noticed when a region receives less than adequate or no rainfall in a season when precipitation is more common. Thus, the definition of meteorological drought is usually region-specific and based on a clear understanding of regional climatology. This is so because the atmospheric conditions that result in deficiencies of precipitation are highly variable from region to region. For example, in the United Kingdom a (meteorological) drought occurs if daily precipitation totals of less than 0.01 inch (0.25 mm) are accumulated for 15 consecutive days. Although annual rainfall in the United Kingdom is not comparable to that of tropical countries, the country has close to 170 average number of annual rainy days. In the United States, if an area receives only 30 percent or less precipitation than normal for at least 21 days, it is called a (meteorological) drought.

In other examples, for meteorological drought, annual rainfall should be less than 7 inches (180 mm) in Libya, and in India, this type of drought is defined as actual rainfall being deficient by more than twice the mean deviation or annual rainfall being less than 75 percent of normal. The definition used in Bali, Indonesia, is a period of six days without rain (NDMC 1996). Notably, the entire country of Indonesia experiences a tropical rainforest type of climate, where rainfall occurs almost every afternoon. Thus, in a region characterized by year-round precipitation, meteorological drought is defined based on the number of days with precipitation less than some specified threshold. On the other hand, in regions where rainfall is characterized by a seasonal pattern as in India and Pakistan, this type of drought is generally defined in the context of actual precipitation departures from average amounts on monthly, seasonal, or annual time scales. However, normally, meteorological measurements are the first indicators of drought.

Agricultural Drought

An agricultural drought is essentially tied to other types of drought, but it emphasizes soil moisture deficits. This type of drought occurs when there is not adequate soil moisture to meet the needs of a particular crop at a specific time. A deficit of rainfall in areas during critical periods of growth of crops can result in serious crop damage and hence shortfall of total production. Thus, stages of crop growth are closely correlated to prevailing weather patterns and conditions. However, the onset of agricultural drought is often illusive or almost unnoticeable because subsoils, especially when fine textured, have considerable moisture-storage capacity.

Most of the remaining available definitions of drought are, in fact, definitions of agricultural droughts. In the context of Bangladesh, a predominantly agrarian country, Hugh Brammer defines drought as "a period when soil moisture supply is less than what is required for satisfactory crop growth during a season when crops normally are grown," namely, drought that affects normal agricultural operations and is caused by lack of rainfall (Brammer 1987, 21). Brammer's definition is similar to one provided by Australian geographer R. L. Heathcote, who defined agricultural drought "as a shortage of water harmful to man's agricultural activities. It occurs as an interaction between agricultural activity (i.e., the demand) and natural events (i.e., the supply), which results in a water volume or quality inadequate for plant and/or animal needs" (Heathcote 1974, 128–129). Heathcote's definition is broader than Brammer's definition because he considered agriculture as a whole, not just crops.

Agricultural droughts are generally defined as a temporary reduction in soil moisture availability at a given area below its annual or seasonal average by a certain percentage. Different methods are used to estimate moisture availability, but it is usually dependent on the amount of total precipitation minus the combined amounts of evaporation and runoff. Agricultural drought is typically evident after meteorological drought has been confirmed but before a hydrological drought. Moreover, agriculture is usually the first economic sector to be affected the most by drought, and it can result in greater economic, social, and political repercussions than any other types of droughts.

Hydrological Drought

Hydrological drought results in a marked reduction of natural surface water flow or ground water levels. When only reduction in ground water levels occurs, it is often called "aquifer drought." Distinctively, hydrological drought reflects significant reduction of water in reservoirs, lakes, or streams. This is considered the drinking water type of drought. Conditions for hydrologic drought are built over extended periods of time. As it takes a longer time for reservoirs or streams to become depleted, this corresponds to longer replenishing periods. Hydrological drought is measured by a lack of any type of precipitation (including snowfall) that leads to shortfalls in surface or subsurface water supply. The frequency and severity of this type of drought are often defined on a watershed or river basin scale. Although all droughts originate with a deficiency of precipitation, hydrologists are more concerned with how this deficiency plays out through the hydrologic system.

The impacts of the hydrological drought are usually noticed well after the signs of meteorological or agricultural drought because it takes so long for these conditions to affect the streamflow, groundwater, and reservoir levels. For this reason, the reduction of water in reservoirs may not affect hydroelectric power production or recreational uses at the beginning of drought onset. Also, water in reservoirs and rivers is often used for multiple and competing purposes (e.g., flood control, irrigation, recreation, navigation, hydropower, and wildlife habitat). Accordingly, competition for use of water in these storage systems escalates during hydrological drought, resulting in increased conflicts between water users as well as neighboring states/countries if they share a particular river course or aquifer.

Socioeconomic Drought

Finally, a socioeconomic drought is a product of the previous three, but it usually occurs when water shortage starts to affect people as a result of shortage of food and water supplies. This type of drought happens when the supply of water does not meet the demand and thus affects people's ability to eat, drink, or wash. Of course, they are also not able to irrigate field crops, and their hydroelectric power supply is significantly reduced because power plants are dependent on stream flow. Along with people, livestock also suffer from lack of water and forage. Thus, the effects of socioeconomic droughts are widespread.

A new type of drought, ecological drought, has emerged, which captures how this event impacts an ecosystem. Environmentalists now widely recognize that drought can lead to increases in wildfire and insect outbreaks; local species extinctions; forest diebacks; and altered rates of carbon, nutrient, and water cycling, all of which can have profound consequence for ecosystems. Ecological drought therefore is defined as "a prolonged and widespread deficit in naturally available water supplies—including changes in natural and managed hydrology—that create multiple stresses across ecosystems" (Earth Observatory 2000).

Several countries (e.g., France, Ireland, the United Kingdom, and the United States) also recognize different categories of drought (e.g., environmental, water supply, and physiological). For example, the U.S. National Weather Service (NWS) uses the term *absolute drought* when no measurable rain is recorded for at least 15 consecutive days. NWS defines a "partial drought" after at least 29 consecutive days with a rainfall total averaging less than 0.01 inch (0.2 mm) per day. Similarly, a "dry spell" is defined as a period of 15 or more consecutive days with less than 0.04 inch (0.8 mm) of rainfall (Ebert 2000).

DROUGHT INDICES

Several drought indices are used to make forecast about droughts. Although forecasters use more than one index to make their decisions concerning occurrence of droughts, the most widely used is the Palmer Drought Severity Index (PDSI). The PDSI is a measure of both the intensity and magnitude of any drought, and the index is prolific in the United States. It was developed by Wayne Palmer in 1965, and it focuses on the supply and demand concept of water balance for an area with demand referring to the needs of plants to maintain equilibrium. The index uses precipitation, temperature, and soil moisture data to find variances in temporary climatic conditions. It also considers the rates of evapotranspiration and recharge rates in the flora of an area or a region.

The PDSI uses a value of "0" to show normal conditions, while negative values are representative of drought conditions in the soil. This standardized index ranges from -10 (dry) to +10 (wet). While the Palmer Index is a fairly accurate measure of aridity in soil moisture, it has several limitations. For instance, it generalizes soil types, and certain types of surface runoff (e.g., snow or ice melt) are not taken into consideration. Moreover, it needs historical data for a long period of time to establish trends, besides which the index is mainly suited for agriculture and does not consider specifically the impacts of droughts.

CAUSES OF DROUGHT

Droughts are caused by many factors that are broadly divided into two types: physical and anthropogenic (caused by human activities). The underlying physical cause of most droughts reflects changes in large-scale atmospheric circulation patterns and the locations of warm high-pressure systems. For example, in North America, rotating air masses often prevent rain-bearing westerlies from penetrating the normal west-to-east progression of weather systems. When these "blocking systems" persist for a long period of time, they exacerbate drought conditions. In the United Kingdom, however, drought occurs when the displacement of mid-latitude depressions blocks movement of high-pressure systems. Then again, droughts in Africa result from the failure of the inter-tropical convergence zone to move sufficiently from the equator.

In most tropical regions, droughts are frequently associated with global patterns of sea surface temperature anomalies such as El Niño. During an El Niño, the warm equatorial Pacific Ocean warms the overlying atmosphere, which leads to changes in large-scale atmospheric circulation patterns and increased probabilities of drought in many parts of the world, including Australia and tropical countries of Asia (Earth Observatory 2000). When an El Niño event in the Pacific Ocean coincides with an Indian Ocean Dipole (IOD), a counterpart of the Pacific Ocean El Niño, monsoon circulation becomes weak over South and Southeast Asia, leading to severe drought conditions in these regions.

Research suggests a strong association between the shifting ocean currents and droughts for two reasons: the vast heat-storage capacity of the oceans and the continuing exchange of energy between the world's oceans and atmosphere (World

Meteorological Organization 1975). Research further suggests that droughts are linked with sunspots and the 18.6-year lunar cycles. Thus, some scientists consider droughts to be cyclic in semiarid regions as well as in the temperate latitudes such as in the Great Plains of the United States. This is helpful for predicting a drought event in a timely way. The cycle and ESNO together account for 45 percent of variance in rainfall amounts. This means that the remaining percentage accounts for other anthropogenic and atmospheric causes.

Human activities are also responsible for causing droughts. Removing vegetation cover, unwise farming practices, overgrazing marginal pasture lands, and cutting down trees—all these activities increase trans-evaporation and lessen the ability of the soil to hold water, leading to drought. Deforestation can also influence the occurrence of dry conditions since it reduces a forest's watershed potential. Additionally, overuse of sensitive soils and lowering of the water table by excessive pumping cause dry spells. Human activities have also emitted greenhouse gases into the atmosphere, resulting in global warming; in fact, some of the worst droughts in recent decades in the Sahel of Africa have been associated with global warming and climate change. Overall, the continued rise of the world's average temperatures has resulted in dry spells and exacerbated drought conditions, which in turn have led to forest fires and wildfires.

IMPACTS

The effects of drought are long lasting and widespread and have devastating effects on the environment, economy, and society as a whole. Droughts can affect these sectors in various ways: They can reduce surface water levels, which destroy natural and wildlife habitats; they can destroy the entire food chain and change the ecosystem as many aquatic and other wildlife dependent on water bodies die or become endangered; they can create unsuitable conditions for plants and vegetation cover to survive, often killing fruit trees; they can lead to increased fire danger and fire events that burn not only plants and animals but also residential structures; they can also stunt the growth of trees and kill seedlings. The devastating Colorado wildfire of 2012, largely caused by drought, was responsible for five fatalities and an estimated $450 million in losses. This does not include the costs of fighting the fires (Ryan and Doesken 2012). The lumber industry also suffers major losses due to drought as evidenced by the 1988 drought that damaged U.S. forests at a cost of nearly $5.2 billion (Changnon et al. 2007).

Clearly, drought affects climate, rocks, and fertile soils, leading to desertification by reducing soil quality and causing soil erosion and moisture depletion, which compromises soil microbial activities. As a result, soil quality is lowered because of minimized organic activity and the continued dry spell, which kills soil organisms. The end result is dry and cracked earth, resulting in devalued soil, rendering it therefore unsuitable for agricultural practices.

The economic impacts of drought are varied, so, understandably, the agricultural sector is the most affected by this event. Because prices of agricultural inputs and products during drought periods soar, farmers have to spend more money on irrigating crops and buying fodder for their farm animals. Furthermore, rangeland and extensive irrigated pasturelands are affected by droughts as often people need to drill new wells to provide adequate water to crops and livestock. Unable to spend additional money on irrigation and forage, farmers often have to sell their livestock at distress prices. Because of a lack of sufficient water, yields of certain crops decrease. For example, corn production was 45 percent below average during the 1988 drought in the United States (Changnon et al. 2007).

Agriculture-related industries and businesses incur heavy financial loss because famers are unable to buy farm equipment and modern agricultural inputs. Owners of grain elevators also experience decreased revenues because of reduced production. Other businesses such as marinas and landscapers also incur losses. For example, unprecedented low river levels create major problems with navigation. During the 1988 drought, low flows on the lower Mississippi River initially restricted and then stopped barge traffic. This along with higher shipping costs resulted in loss to the navigational transport sector of nearly $1 billion (Changnon et al. 2007). Similarly, ethanol production is also impacted greatly because of reduced corn production. During the 2012 drought, a considerable number of ethanol plants in several states in the United States reduced production or even closed during the drought.

However, decline in crop yields not only negatively affects farm income but also leads to crop failure and famine in developing countries and reduces agricultural employment due to the delay in sowing and transplanting crops. Employment opportunities are further reduced because of the reduced need for weeding and subsequent reduced crop harvest. Moreover, drought conditions have negative impacts on both human and animal health. Because of scarcity of water, people in developing countries are not able to bathe regularly or are forced to bathe in dirty water, causing skin diseases. Most importantly, people die from malnutrition and undernutrition.

Economic loss in developed countries occurs because droughts affect wildlife, fishing, and hunting from which these countries earn a significant proportion of public income, both in license fees and sales tax from equipment. Thus, tourism and outdoor sports are impacted negatively by drought conditions. For example, there were about 1,000 fewer pheasant hunters in South Dakota during the 2012 drought than in 2011; pheasant hunting is a strong anchor of the economy of this state (Edwards and Todey 2012). During the 1988 drought in several states in the United States, the tourism industry lost an estimated $400 million (Changnon et al. 2007). Another negative consequence is that droughts provide favorable conditions for the spread of epizootic hemorrhagic disease (EHD) in deer in the country.

During drought, people and businesses who are dependent on hydroelectric power pay higher energy costs due to the inability to sustain hydropower generation, which therefore often leads to more coal-fired generation that increases not only utility costs but also air pollution. High temperature during droughts also leads to major increases in use of air-conditioning, which in turn leads to increased cost of grid energy, which leads to economic losses both for energy industries and businesses (Changnon et al. 2007).

Social implications are possibly the most often felt effects of drought. Since water scarcity is high, water quality significantly depreciates. This means the availability of clean water for drinking purposes, sanitation, and cleaning may not be sufficient. Droughts also increase the concentration of chemicals and solid particles or impurities in surface waters, causing rapid increase in incidence of waterborne diseases. People also suffer more from illnesses like asthma, mental and heat stress, and anxiety; also, excessive heat causes deaths, particularly among the elderly. Furthermore, during droughts, costs for water and sewage treatments are not only increased, but also such water becomes difficult to treat. Drought also creates interstate, inter-province, or inter-country conflicts over water supplies.

Hunger, anemia, malnutrition, and deaths of people and livestock are caused by drought in stricken areas, which see abnormal increases in food grain prices, and lack of employment affects the food entitlement of rural people in particular, especially in developing countries. Also, wild animals are often forced to migrate to new locations for water and food, perhaps making those locations vulnerable and endangered because of new threats. This leads to loss of biodiversity and disruption of natural ecosystems. Apart from animals and wildlife, people are also forced to migrate to areas not affected by drought. In fact, this type of disaster causes more displacement of populations than floods, hurricanes, and earthquakes put together.

RESPONSE

In agrarian countries, crop adjustments usually constitute the principal focus of risk-aversion strategies adopted by drought victims since crop losses mean acute shortage in their food stock. Therefore, growers make several agricultural adjustments during the period to reduce their losses. In such a period, farmers usually depart from normal cropping patterns in order to minimize their losses. In the case of early drought (March-April), farmers in Bangladesh and other South Asian countries usually devote relatively more land to cultivate dry crops such as millets. These crops are grown either in pure stands or in combinations of two or more crops in the same field, which is called intercropping. Gap filling, a method of replanting a crop in a patch of land where seedlings have died, is also practiced by farmers to mitigate crop losses from drought. This practice is very effective if drought occurs in the early stage of crop growth. In the case where the drought persists, farmers generally cultivate water-efficient and fast-maturing crops and dispose of a variety of their personal assets such as livestock, land, and utensils. They also rely on loans and donations from different sources.

In developed countries, particularly Australia, farmers rely heavily upon technology to mitigate drought effects. Land management practices there are directed toward conservation of soil moisture and crop protection. Practices that promote rainfall penetration into the subsoil include mulching, tillage, and construction of contour banks and furrows that minimize runoff. When drought lasts for years, ranchers are forced to cull, sell, or transport herds to areas not affected by droughts. Drought-affected farmers in the United States often buy and transport hay from neighboring or distant states. In some areas, special provisions are required to exempt hay-hauling truckers from highway load size limits to permit some oversized loads to be delivered, making hay slightly more affordable (Ryan and Doesken 2012). Crops, meanwhile, are also protected from drought by aerial spraying of herbicides to control weed growth and aerial sowing of seed to prevent soil compaction by heavy farm machinery. Computer management models are also used for maximizing crop yields through climatic forecast and monitoring of moisture and crop conditions. To guard against potential loss of crops from drought, farmers in developed countries usually buy crop insurance (Kellner 2012).

Also in developed countries, water restrictions, which take several forms, are enforced both in rural and urban areas. For example, lawn-watering restrictions are imposed along with requests to conserve water and limit the number of showers per week and the number of times individuals wash their cars per week. Burn bans of grass and pasture lands are also implemented along with campfires or fireworks bans during drought periods. Furthermore, boating, canoeing, and kayaking activities are greatly reduced in rivers, lakes, and reservoirs. This is because blue-green algae often develop in lakes and reservoirs, which restricts access to these resources for safety reasons. Compounding the problems, combination of high winds and ongoing dry conditions causes dust storms during droughts, reducing visibility, altering air quality, and making driving hazardous (Jedd et al. 2018). Therefore, authorities in drought-affected areas often issue air quality alerts and even force closure of local roads since smoke- and dust-induced air creates breathing problems.

Many communities in the United States and other developed countries open public cooling facilities for the elderly to escape the heat during a drought. Community officials and faith-based organizations distribute fans to lower-income residents as well as to public schools. Often, school hours and days are adjusted to avoid drought impacts or even closed during peak drought periods. If school buildings are not equipped with cooling systems, community or city authorities install such systems. Authorities also help low-income households cover utility bills, and community-based organizations help individuals both psychologically and financially. State and federal governments provide funds for agricultural relief such as the federal government providing $4.3 billion in drought assistance to 26 states in 1988 (Changnon et al. 2007).

In developing countries, friends, neighbors, and relatives of nondrought affected areas may aid victims by providing food, seeds, cash, loans, and clothing. Similarly, governments and nongovernmental organizations (NGOs) also provide support to those attempting to cope with losses while the national government often provides formal loans to drought survivors. Bilateral and foreign donor agencies and international communities often provide emergency aid to alleviate the devastation caused by droughts. Working against quick recovery, as a creeping disaster, drought effects develop slowly and therefore do not receive much national or international media attention.

DROUGHT'S WINNERS

Notably, drought impacts are not always necessarily negative. Construction companies and airlines operate with fewer delays when there is no rain, and railroads carry shipments diverted from barges. Because of increased demand, utility companies are able to sell more power during droughts. Major beneficiaries are some agricultural producers of drought-affected areas and most farmers in nondrought areas in developed countries. Some farmers take advantage of a large increase in crop prices and reductions in supply of irrigation water to gain financial benefit from increased prices that offset irrigation costs.

Although parks and recreational facilities are often closed for severe droughts, the dry condition at the early stage lures many more campers than usual to public parks. Public swimming pools and water parks also enjoy a brisk business. Another positive effect of low water is the great fishing opportunity that anglers experience due to the fish being concentrated in areas of deeper water. In some instances, dry conditions can produce favorable foraging, such as for shorebirds, by exposing additional mudflats.

Above all, like other natural disasters, droughts provide a "window of opportunity" to repair and improve infrastructure associated with agriculture practices and water and electricity supply. Other benefits include the following: marginal lands being converted to Conservation Reserve Program (CRP) lands, increased drought awareness, renewed interests in conservation, and adaptation of modern irrigation systems to use less water.

CONCLUSION

In developed countries, many climate-related products are now available to predict and monitor drought conditions. Almost all states in the United States use the multi-sensor precipitation estimator, for instance. The availability of high-resolution temporal precipitation data makes monitoring the departures of precipitation rates from normal relatively easy. In addition, soil moisture models and remote sensing images are useful. However, drought-prone developing countries typically lack this technology, which international agencies and donor countries can sometimes supply it.

Proactive action plans and proper planning are required to deal with substantial reduction of the impacts of drought and timely recovery from the event. Although most drought-prone communities in developed countries have these plans, they need to be revised frequently as lessons are learned from each drought. Given that climate change is expected to trigger droughts, plans for such events are essential for drought-prone countries. Besides, there is a need for a comprehensive drought-monitoring system to track each event and provide early warning of emerging droughts and for a network of people who can assess the evolving effects of drought. Available studies in both developed and developed countries reveal that public awareness and education campaigns can further be improved in order to generate appropriate response to future drought events. In addition to current ways of disseminating drought information, social media could be used to target a much wider audience for future drought events.

Bimal Kanti Paul

Further Reading

Brammer, H. 1987. Drought in Bangladesh: Lessons for Planners and Administrators. *Disasters* 11(1): 21–29.

Changnon, S. A., K. E. Kunkel, and D. Changnon. 2007. *Impacts of Recent Climate Anomalies: Losses and Winners.* Champaign: Illinois Department of Natural Resources and University of Illinois at Urbana-Champaign.

Earth Observatory. 2000. Drought: The Creeping Disasters. August 28, 2000. https://earthobservatory.nasa.gov/features/DroughtFacts/drought_facts.php, accessed December 26, 2018.

Ebert, C. H. V. 2000. *Disasters: An Analysis of Natural and Human-Induced Hazards.* Dubuque, IA: Kendall Hunt.

Edwards, L. M., and D. Todley. 2012. South Dakota. In *From Too Much to Too Little: How the Central U.S. Drought of 2012 Evolved Out of One of the Most Devastating Floods on Record in 2011*, edited by Brian Fuchs, Deborah Wood, and Dee Ebbeka, 84–99. Lincoln: Drought Mitigation Center Faculty Publications, University of Nebraska.

Heathcote, R. L. 1974. Drought in South Australia. In *Natural Hazards: Local, National, Global*, edited by Gilbert F. White, 128–136. New York: Oxford University Press.

Jedd, T. D. Bathke, D. Gill, B. K. Paul, N. Wall, T. Bernadi, J. Petr, A. Mucia, and M. Wall. 2018. Tracking Drought Perspectives: A Rural Case Study of Transformation Following an Invisible Hazard. *Weather, Climate and Society* 10(4): 653–672.

Kellner, O. 2012. Indiana. In *From Too Much to Too Little: How the Central U.S. Drought of 2012 Evolved Out of One of the Most Devastating Floods on Record in 2011*, edited by Brian Fuchs, Deborah Wood, and Dee Ebbeka, 29–364. Lincoln: Drought Mitigation Center Faculty Publications, University of Nebraska.

NDMC (National Drought Mitigation Center). 1996. *What Is Drought*. Lincoln, NE: NDMC.

Ryan, W., and N. Doesken. 2012. Colorado. In *From Too Much to Too Little: How the Central U.S. Drought of 2012 Evolved Out of One of the Most Devastating Floods on Record in 2011*, edited by Brian Fuchs, Deborah Wood, and Dee Ebbeka, 20–24. Lincoln: Drought Mitigation Center Faculty Publications, University of Nebraska.

Wilhite, D. A., and M. H. Glantz. 1985. Understanding the Drought Phenomenon: The Role of Definitions. *Water International* 10(3): 111–120.

WMO (World Meteorological Organization). 1975. *Drought and Agriculture*. Technical Note No. (WMO-No. 392). Geneva, Switzerland.

Earthquakes

An earthquake is the sudden shaking of the ground, resulting most commonly from an abrupt release of energy stored in the earth's crust. Among all natural disasters, it is the deadliest. The Emergency Events Database (EM-DAT) of the Center for Research on the Epidemiology of Disasters (CRED) reported that

Table 1 Selected parameters of earthquakes, 1996–2015

Parameter	Entire Period	Yearly Average
Frequency	525 (3)[3]	26
Number of people killed	748,621 (1)	37,422
Number of people affected[1]	360,644 (4)	18,032
Total damage[2]	590,307 (2)	29,515

[1] In thousand.
[2] In millions of dollars.
[3] Within parentheses reflect rank among all natural disasters.
Sources: IFRC (2007 and 2016).

worldwide 1.35 million people died from natural disasters from 1996 to 2015. Of these deaths, earthquakes claimed 748,621 lives, that is, on average 37,422 deaths per year (see Table 1). This represents 55.6 percent of disaster deaths in the preceding period, which means more deaths than for all other natural disasters combined. EM-DAT further claimed that 7,056 natural disasters occurred between 1996 and 2015. With a frequency of 26 per year, earthquakes, including tsunamis, rank third after floods and windstorms (see Table 1). Despite yearly fluctuations, the frequency of earthquakes remained broadly constant throughout this period (CRED, USAID, and UNISDR 2016).

EARTHQUAKE MYTHS

After the theory of plate tectonics was widely accepted in the middle of the twentieth century, people started to believe that earthquakes occur along the plate boundaries. Before that, inhabitants of earthquake-prone areas came up with many metaphors to explain causes of earth-shaking events. For example, the Japanese in the seventeenth century thought that the country lies on top of an underground giant catfish, called Namazu, and he was guarded by the god Kashima. The catfish liked to play pranks and could only be restrained by the god. When Kashima became tired, Namazu would use this moment to wiggle his tail, causing an earthquake. Meanwhile, Mongolians once believed that the world sat on the back of a gigantic frog and that earthquakes occurred whenever the frog moved. Similarly, the Chinese maintained that the earth rode on the back of a giant ox; when the ox moved, the earth shook (Brumbaugh 1999).

According to an Indian legend, the earth was held up by four elephants that stood on the back of a giant turtle, which was balanced on top of a cobra. Earthquakes occurred when any of these animals moved. Siberian folklore stated that a god named Tuli carried the earth on a dog sled, and when the dogs had fleas and often scratched, it caused earthquakes. As in Asia, much folklore existed in Africa that explained the cause of earthquakes. People in West Africa, for example, claimed that a giant, who used to sit facing east, carried the earth on his head. Thus, earthquakes originated when he turned toward other directions. Some people in Mozambique still believe that the earth is a living creature and that it sometimes gets sick with fever and chills and shakes, which causes earthquakes (Kott 2017).

In Greek mythology, Poseidon, the god of the sea, always carried a trident, and whenever he struck the ground with it, he triggered earthquakes. In contrast, the early Greek philosophers maintained that earthquakes were caused by movements of gases trying to escape from underground. Elsewhere, the Scandinavians used to believe the god Loki was responsible for earthquakes. As a punishment for killing his brother, Loki was tied to a rock in an underground cave. "A serpent would drip poison down on him, which was caught in a bowl by Loki's sister. When she emptied the bowl, Loki had to twist and turn to avoid the poison, thus causing earthquakes" (Brumbaugh 1999, 3). Finally, on the South and North American continents, the Aztec and the Native Americans also blamed the gods for the occurrence of earthquakes.

SEISMIC WAVES AND DEPTH

In contrast with the myths, scientists have discovered that sudden shaking of the ground is caused by seismic waves, which in turn are caused by stresses that build up in the earth's crust over time. The waves produce vibrations in the crust when the stress is great and travel in all directions from the point of disturbance through the earth's layers, including its elastic layers. In essence, a seismic wave transfers energy from one place to another without transferring solid, liquid, or gas matter. However, when the location coincides with a fault, a fracture, or a crack where rupture occurs, it is called the focus or hypocenter. The point vertically above the focus on the earth's surface is the epicenter. The depth of focus from the epicenter is known as focal depth, which is an important parameter in determining the amount of potential damage caused by an earthquake. Distance from epicenter to any point of interest is called epicentral distance.

Two types of seismic waves exist and are differentiated by speed and way of travel. During an earthquake, the wave released may be a "P" (primary) or an "S" (secondary) wave. P-waves are longitudinal in nature, and the vibrations they create are along the same direction as the direction of travel. They travel through solid, liquid, and gas, and they force the materials in their paths to compress and expand in the direction of wave travel. Because P-waves travel faster than S-waves, the former waves reach a seismometer first. P-waves are also known as compressional waves. S-waves, on the other hand, travel from side to side, perpendicular to the wave path, and they cannot propagate in a liquid. S-waves are also called transverse waves. Both waves can be destructive (Abbot 2008).

Earthquakes can occur anywhere between the earth's surface and about 466 miles (750 km) below the surface. Thus earthquake depth range is divided into three zones: shallow (0-44 miles, or 0-70 km), intermediate (44-186 miles, or 70-300 km), and deep (186-466 miles, or 300-750 km). In general, the term *deep-focus earthquakes* applies to earthquakes deeper than 44 miles (70 km), all of which are localized within great slabs of lithosphere that are sinking into the earth's mantle. In contrast, the most damaging earthquakes originate from shallow depths. The strength of shaking from an earthquake decreases with increasing distance from the earthquake's source, so the strength of shaking at the surface from an earthquake that occurs at 311 miles (500 km) deep is considerably less than if the same earthquake had occurred at 12 miles (20 km) deep. This is because seismic waves from deep earthquakes have to travel farther to the surface, losing energy along the way, while the waves from shallow earthquakes travel shorter distances, losing less energy and hence become more damaging than those associated with deeper earthquakes.

The following examples show how deadly a shallow earthquake can be in relation to one with an intermediate depth. An earthquake in Italy on August 24, 2016, measuring 6.2 on the Richter scale, originated at a depth of 2.5 miles (4 km) from the earth's surface and caused 299 deaths and more than 388 injuries. On the same day and in the same year, an earthquake of magnitude 6.8 struck Myanmar. The estimated depth was 52.3 miles (84.1 km). Tremors from the earthquake were felt in neighboring Bangladesh, India, and Thailand. According to reports, several temples in the nearby ancient city of Bagan were damaged, but only four people died (AFP 2016).

A number of smaller sized earthquakes take place before and after a large earthquake or main shock and are related to it in both time and space. Those occurring before the large one are called foreshocks, and the ones after are called aftershocks. Foreshock activity has been detected for about 40 percent of all moderate to large earthquakes and for about 70 percent of events of magnitude larger than 7.0. They occur any time before the main shock, ranging from minutes to days or even years. For example, the 2002 Sumatra earthquake is considered a foreshock of the 2004 Indian Ocean earthquake.

An example of a much shorter gap between the main shock and foreshocks was the 1960 Chilean earthquake that occurred on May 22. Within twenty-four hours of the main shock, Chile experienced three foreshocks. The first foreshock measured 8.3 magnitude, and the second and third foreshocks measured magnitudes of 7.1 and 7.5, respectively. With a magnitude of 9.5, the main shock was the largest earthquake recorded in the twentieth century (BSL 2015). But the main shock did not cause the greatest death toll. This is because the third foreshock, which struck only 30 minutes before the main shock, sent people from their homes to the street. It saved them effectively, as their abandoned houses ultimately were destroyed during the main shock (Brumbaugh 1999). Thus, foreshocks provide early warnings of a main earthquake, but in areas where foreshocks are fairly common, there is no way of distinguishing a foreshock from an independent earthquake. Some large earthquakes show no foreshock activity at all. In fact, there is no evidence of foreshock activity for most of the historic earthquakes in the Pacific Northwest.

Aftershocks are earthquakes that are triggered by a main shock and occur because the earth has to readjust to the new stress condition produced by the main shock. The United States Geological Survey (USGS) claims that aftershocks can continue over weeks, months, or years. In general, the larger the main shock, the larger and more numerous the aftershocks, and the longer they will continue. For example, 421 aftershocks of a magnitude larger than 3.9 occurred up to December 31, 2015, in Nepal following the country being struck by a 7.8 magnitude earthquake on April 25, 2015. Four aftershocks had a magnitude greater than 6.0, including one of magnitude 7.3 on May 12 (Shrestha et al. 2016). Worldwide, the probability that an aftershock will follow within three days of a main shock is somewhere just over 6 percent, which implies that about 94 percent of earthquakes will not have a foreshock. The USGS further claims that in California, about half of the main shocks were preceded by foreshocks; the other half were not.

MEASURES OF MAGNITUDE

Magnitude is a quantitative measure of the relative size of the earthquake or the amount of energy released at the source of the earthquake. An instrument called a seismograph measures the magnitude of earthquakes. This measure is based on the amplitude of a seismic wave determined at a certain wave period (in seconds) or frequency (in hertz). The amplitude is the vertical distance between the highest and lowest points produced by the waves. The greater the amplitude of a wave on a seismograph, the greater the energy released by the earthquake (Coch 1995). However, the size of an earthquake can range from so weak that it cannot be felt to one violent enough to toss people around and destroy large areas. Generally, the smaller the earthquake magnitude, the greater their numbers and vice versa (Abbott 2008). The National Earthquake Information Center (NEIC) in Golden, Colorado, records an average of 20,000 earthquakes every year around the world, which translates to about 50 a day. In addition, millions of earthquakes that are too weak to be recorded by seismographs are estimated to occur every year.

Several measures of earthquake magnitude are available, but the best measure is the Richter scale, developed by Charles Richter in 1935. The scale is based on two principles: (1) at the same distance from the epicenter, larger earthquakes exert greater shaking of the earth and thus have a greater wave amplitude than do smaller earthquakes, and (2) for a given earthquake, seismographs at farther distances have smaller wave amplitude than those at close distances. Accordingly, the Richter scale estimates the amount of energy released by an earthquake, ranging from 1 (micro) to 9 (great). Since the Richter scale uses a base-10 logarithmic scale, each whole number increase in magnitude implies 10 times higher waveform amplitude and about 31 times higher energy released. Earthquakes are often classified into different groups based on their size (see Table 2). For earthquakes larger than 7.5 magnitude, seismologists do not use the Richter scale because the scale measures the amplitude for a one-second period. For larger earthquakes, the amplitude of a seismic wave needs to be measured for longer periods (anywhere from 20 to 100 seconds). As an alternative, scientists use the surface area of rupture along the fault and the amount of slip (displacement) on the fault, the latter being the seismic moment.

Most of the energy released by a given earthquake goes into generating of heat and fracturing rocks, and only a small amount of it goes into the seismic waves

Table 2 Earthquake magnitude and intensity comparison

Group	Magnitude	MMI Scale	Description of MMI Scale
Micro	1–2	I	Not felt
Minor	3	II–III	II—Felt only by a few persons at rest
			III—Felt quite noticeably by persons indoors
Light	4	IV–V	IV—Felt indoors by many, outdoors by few
			V—Felt by nearly every one
Moderate	5	VI–VII	VI—Felt by all
			VII—Damage negligible to poorly constructed buildings
Strong	6	VIII	Slight damage
Major	7	IX–X	IX—Considerable damage
			X—Some well-built wooden structures destroyed
Great	8 and higher	XI–XII	XI—Serious damage in areas several hundred miles across
			XII—Total damage

Source: Compiled from various sources.

that shake the ground and travel large distances. There is also a direct relationship between magnitude of earthquakes and their impacted areas. Generally, a larger earthquake affects more people in a larger area than does a smaller earthquake. A similar relationship holds true for duration of earthquakes or of the shaking of the earth's surface. Thus, a longer duration earthquake can more greatly increase the amount of damage to buildings than can a smaller duration earthquake (Abbott 2008).

The strength of shaking produced by earthquakes at a certain location is called intensity. It is a qualitative measure of the actual shaking at a location during an earthquake and is assigned a Roman Capital Numeral, ranging from I (least perceptible) to XII (most severe). Several intensity scales exist, but the one most commonly used in the United States is the Modified Mercalli Intensity (MMI) scale. MMI scales are based on three features of shaking: human experience of the event, extent of damage to structures, and changes to natural surroundings.

GEOGRAPHIC DISTRIBUTION OF EARTHQUAKES

Earthquakes can strike at any time and in any part of the world. But in the areas that correspond to active tectonic zones, the frequency of earthquakes is greater than anywhere else. Thus, earthquakes are not evenly distributed; in certain areas, they are more frequent and intense, whereas in other places, like Australia, they are extremely rare. They are concentrated in three main belts: the circum-Pacific earthquake belt, the Mediterranean-Asia earthquake belt, and the submerged Mid-Atlantic Ridge. The former belt corresponds with the "Ring of Fire" and includes a region that encircles the vast Pacific Ocean. This belt extends as an isostatically sensitive zone along the rim of the Pacific Ocean from the west coasts of the North and South Americas, including the coasts of Alaska, the Aleutian Islands, Japan, the Philippines, Indonesia, and New Zealand.

This belt hosts over 80 percent of the world's shallow and medium-depth earthquakes and 100 percent of its deep earthquakes. The most earthquake-prone countries of this belt include Chile, Mexico, the United States, Japan, and Indonesia. The circum-Pacific earthquake belt is home to 452 volcanoes, which is over 75 percent of the world's active and dormant volcanoes. The belt exists along boundaries of tectonic plates, where plates of mostly oceanic crust are sinking (or subducting) beneath another plate. Earthquakes in these subduction zones are caused by slip between plates and rupture within plates.

The second belt extends from Gibraltar to the east via the Atlas Mountains, the Alps, the Pyrenees, the Apennines, the Balkan mountains, the mountain chains of Asia Minor, the Caucasus, the Hindu Kush range, and the Himalayan mountains. One of its branches passes through Mongolia and Lake Baikal, and another branch extends to Myanmar and the islands of Indonesia, where it meets the Circum-Pacific belt north of Australia. About 17 percent of the world's largest earthquakes are located in this region, including the 9.1 magnitude Indonesian earthquake in 2004, which generated deadly tsunami waves and killed over 230,000 people. The next year, another destructive earthquake of 7.6 magnitude occurred in Pakistan that killed over 80,000 people. Finally, the third earthquake belt follows the submerged mid-Atlantic Ridge, which is the meeting place of two tectonic plates that are spreading apart (a divergent plate boundary). Most of the mid-Atlantic Ridge is deep underwater and far from human development, except for Iceland, which sits directly over the mid-Atlantic Ridge. Other seismic areas include Northern Africa and the Rift Valley areas of the Red Sea and the Dead Sea.

TYPES OF EARTHQUAKES

According to their mode generation, there are four different types of earthquakes: tectonic, volcanic, collapse, and explosion. The type of earthquake depends on the region where it occurs and the geologic makeup of that region. Tectonic earthquakes are the most common and of tremendous social significance because they pose the greatest hazard. This type of earthquake occurs when rocks in the earth's crust break suddenly due to various geologic forces created by movement of tectonic plates. A second well-known type is called a volcanic earthquake, which is any earthquake associated with volcanic eruptions. The magnitude of this type of earthquake is usually weak; the highest magnitude of a volcanic earthquake ever recorded, 5.5 on the Richter scale, was during the eruptions of Mount St. Helens in the United States in 1981.

Earthquakes and volcanic eruptions do not necessarily accompany each other in all tectonic regions. For example, they need not occur together in the San Andreas fault or the Himalayas, where two plates are moving together along a transform fault and a collision zone, respectively. In contrast, they are associated together in mid-ocean ridges and near deep-ocean trenches. At the subduction zone, that is, along the trenches, as the surface rocks bend and thrust downward, they become strained and finally fracture, thus producing earthquakes (Bolt 2004). Specifically, there are two types of volcanic earthquakes: volcano-tectonic earthquakes and long period earthquakes. The earthquakes that occur due to injection or withdrawal of magma between the stressed rocks are called volcano-tectonic earthquakes. In these types, magma exerts pressure on the tectonic plates until this magma breaks the rocks, causing tremors. After the breakdown of rocks, magma starts flowing toward the surface of the earth. A long period earthquake, on the other hand, occurs with the injection of magma into the surrounding rocks, which happens due to pressure changes among the layers of the earth. These types of volcanic activity provide early warning of imminent volcanic eruption.

Collapse earthquakes are small and weak and occur in caverns and mines. Sometimes, underground blasts (rock breaking) in the mines cause the collapse of mines, which produces seismic waves. Collapse earthquakes also are sometimes produced by massive landslides (Bolt 2004).

Finally, an explosion earthquake is the result of the detonation of a nuclear and/or chemical device. For instance, fracking and underground nuclear explosions are known to produce minor- to moderate-magnitude earthquakes. Some past underground nuclear explosions have been so large that they sent seismic waves throughout the earth's interior and even have produced moderate-sized earthquakes. Another example of an explosive earthquake was the result of the terrorist jet planes' attacks on the World Trade Center in New York on September 11, 2001. The collapse of the north tower produced an earthquake of magnitude 2.4 on the Richter scale in Manhattan (Bolt 2004). Moreover, meteor impacts can also cause earthquakes. Both collapse and explosive earthquakes are human-made earthquakes, which can be stopped.

Depending on plate movements, earthquakes take one of three main forms: convergent boundary, divergent boundary, or transform fault. Earthquakes originate at convergent plate boundaries where two plates meet together, and one plate is forced over another plate whereupon the densest plate will subduct beneath the other plate. For example, the 1960 Chile earthquake originated at the ocean-continental convergence zone, where the Nazca Plate off the coast of Chile subducted below the South American Plate. Convergent plate boundaries produce earthquakes all around the Pacific Ocean basin. Thus, earthquakes in Japan are caused by ocean-ocean convergence. Meanwhile, the Philippine Plate and the Pacific Plate subduct beneath the oceanic crust on the North American or Eurasian plates. This complex plate tectonics situation creates a chain of volcanoes, some of which are Japanese islands, and as many as 1,500 earthquakes annually. Subduction zone earthquakes are among the most powerful, primarily because so much surface area is affected, resulting in a much greater seismic moment, as well as a greater-than-average amount of slip.

Next, divergent plate boundaries are locations where plates are forced away from each other. This kind is common in oceans where new floors are created.

The Mid-Atlantic Ridge is the classic example of this type of plate boundary. Another classic example is the East Africa Rift Valley, which is subject to frequent earthquakes. It has not been completely rifted and is still above sea level, but it is occupied by several large lakes. The earthquakes that occur along these zones are relatively small.

Lastly, when two tectonic plates slide past each other, the place where they meet is a transform or lateral fault. The San Andreas Fault in California is one of the best examples of lateral plate motion. The faults along the San Andreas Fault zone produce around 10,000 earthquakes a year. New Zealand also has a transform plate boundary, where about 20,000 earthquakes occur a year. Often, deadly earthquakes occur at transform plate boundaries because transform faults have shallow focus earthquakes as the plates (or fault) meet near the surface.

EARTHQUAKE PREDICTION

Given their devastating potential, there is great interest in predicting the location, the magnitude, and the date and time when earthquake might occur. However, currently no reliable method exists to predict individual earthquakes. Fortunately, the world's largest earthquakes do have a clear spatial pattern, and prediction is possible for this type. Most large earthquakes occur on long fault zones around the margin of the Pacific Ocean. If a fault segment is known to have broken in a past large earthquake, recurrence time and probable magnitude can also be estimated based on fault segment size, rupture history, and strain accumulation. In the absence of scientific methods to predict earthquakes, a number of precursors that might precede a large earthquake are used for warning of an impending earthquake.

For instance, for centuries there have been anecdotal accounts of anomalous animal behavior preceding earthquakes. The Chinese in ancient times relied upon unusual animal behavior as an indication of a coming tremor. The behavior included snakes crawling out from their holes, dogs barking ceaselessly, birds circling and not roosting, animals not entering their barns, horses running wildly, and yaks rolling over repeatedly (Brumbaugh 1999). It is believed that animals' unusual behavior is a response to P-waves, which travel through the ground about twice as fast as the S-waves that cause the most severe shaking. Animals are more sensitive to small ground motions produced by a P-wave that is too small for a human to feel, but people feel the ground shaking when S-waves later arrive.

One of the most notable successful earthquake predictions based on anomalous animal behavior was for the Haicheng, China, earthquake, which occurred on February 4, 1975. Such behavior was reported just a few days before the tremor, and Chinese authorities issued an evacuation warning 5-10 hours before the 7.3 magnitude earthquake. In addition, in the preceding months, changes in land elevation and in ground water levels and many foreshocks were also indicators for issuing the warning. This warning saved many lives as one million people evacuated the buildings in Haicheng and the surrounding areas even though they stayed outside in the winter cold. The earthquake destroyed or severely damaged 90 percent of the buildings in Haicheng. Most earthquakes do not have such obvious precursors. For example, there was a lack of anomalous animal activity before the 7.6 magnitude 1976 Tangshan, China, earthquake, which killed about 250,000 people (Brumbaugh 1999).

Changes in the P velocities can be used as an earthquake precursor in a seismic area as seismographs and chronometers can detect change in rock properties before an earthquake. In such a case, the speed of seismic waves might also vary, and hence, travel time of a P wave from one side to the other would change. Seismographic stations are designed to measure the travel time very precisely and easily. In the 1970s, measuring P-wave speed was considered a likely breakthrough when Russian seismologists reported observing such changes in the region of a subsequent earthquake. However, now geologists doubt the accuracy of this method.

The release of radioactive gas radon from underground rock into the atmosphere along active fault zones, particularly from deep wells, can be a useful predictor of earthquakes. Radon can be easily detected, and it is sensitive to short-term fluctuations. The release of this gas is due to preseismic stress or fracturing of rock. Reports of spikes in the concentrations of radon prior to a major earthquake seem to support this theory. For example, scientists found a tenfold increase in radon concentration 20 miles (30 km) away from the epicenter, nine days before the 1995 Kobe earthquake in Japan. However, there is at least one problem with this predictor. Radon release may occur for nonseismic reasons. Therefore, a strong correlation is not yet established (Bolt 2004).

Variations in the electrical conductivity of the rocks in an earthquake zone and magnetic phenomena also have been attributed to precursory stress and strain changes that precede earthquakes. Therefore, on first inspection, this anomaly seems to be a reliable earthquake precursor. A few field experiments to check this property in fault zones have been made in China, Japan, and the former Soviet Union, but the results were not conclusive. Even the study of the closely monitored 2004 Parkfield earthquake, California, found no evidence of precursory electromagnetic signals of any type; further study showed that earthquakes with magnitudes less than 5 do not produce significant transient signals. Ultimately, the International Commission on Earthquake Forecasting for Civil Protection (ICEF) has failed to provide evidence in support of these precursors. The variations in electrical resistance before earthquakes quite often interfere with human-made electrical systems.

As noted, foreshocks do indicate an imminent large earthquake; thus, an increased rate of small earthquakes in an area precedes a large earthquake. This proved to be true in several cases, such as the Oroville earthquake in California, which occurred on August 1, 1975. In June 1975, a few small earthquakes occurred to the southwest of the Oroville reservoir, which were followed by an earthquake of magnitude 5.9 near Oroville on August 1. There are other examples of such foreshocks before the main shock in China and Italy (Bolt 2004). However, there are two problems with this predictor of earthquakes: not all large earthquakes are preceded by foreshocks, and the time frame for foreshocks or indeed for shock sequence often extends over a year. There are other predictors of earthquakes, but none of them are definitive.

EARTHQUAKE PREPAREDNESS

Earthquakes strike without warning, and they cause buildings to collapse and heavy items to fall, resulting in deaths, injuries, and property damage. Household preparedness is key to reduce deaths and injuries from earthquakes. One needs to prepare in advance to protect family, pets and livestock, and home. Before the earthquake, each household should develop an emergency plan and identify and discuss evacuation routes with family members. Households should store food and water, assemble first aid kit along with a working flashlight, buy earthquake insurance, and consider a retrofit of any building with structural issues that would make it vulnerable to collapse during an earthquake.

The following listed directions need to be followed by a person who is inside a building during an earthquake:

- Move as little as possible—most injuries during earthquakes occur because of people moving around.
- Try to protect head and torso.
- If in bed, stay there, turn face down, and cover head with a pillow.
- Stay indoors until the shaking stops.
- If leaving a building after the shaking stops, take stairs rather than an elevator in case of aftershocks, power outages, or other damage.
- Be aware that smoke alarms and sprinkler systems frequently go off in buildings during an earthquake, even if there is no fire.
- If gas is noticeable, get out of the house and move as far away as possible.
- Before leaving any building, check to make sure that no falling could threaten safety.

While outdoors, a person should follow the instructions listed here:

- Find a clear spot and drop to the ground. Stay there until the shaking stops.
- Try to get as far away from buildings, power lines, trees, and streetlights as possible.
- If in a vehicle, pull over to a clear location and stop. Avoid bridges, overpasses, and power lines if possible.
- Stay inside with seatbelt fastened until the shaking stops.
- After the shaking has stopped, drive on carefully, avoiding bridges and ramps that may have been damaged.
- If a power line falls on the vehicle, do not get out. Wait for assistance.
- If in a mountainous area or near unstable slopes or cliffs, be alert for falling rocks and other debris as well as landslides.

CONCLUSION

Earthquakes are not only the deadliest natural disasters, but they also produce numerous secondary disasters such as tsunamis, fires, avalanches, and land and mud slides. Some of these secondary disasters cause human deaths. Often, it is difficult to separate deaths contributed by earthquakes from those caused by their secondary disasters. In many instances, deaths associated with earthquakes do not result from the ground shaking, but are linked to the destruction of human-made structures such as buildings and roads. Thus, governments of earthquake-prone countries should enforce building standards to ensure that each building is structurally sound in the event of significant tremors.

Bimal Kanti Paul and Avantika Ramekar

Further Reading

Abbott, P. L. 2008. *Natural Disasters*. Boston: McGraw Hill Higher Education.

AFP (American Free Press). 2016. Myanmar Struck by 6.8-Magnitude Earthquake. August 24, 2016. https://www.theguardian.com/world/2016/aug/24/myanmar-struck-by-6-8-magnitude-Earthquake, accessed August 23, 2019.

Bolt, B. A. 2004. *Earthquakes*. New York: W.H. Freeman and Company.

Brumbaugh, D. S. 1999. *Earthquakes: Science and Society*. Upper Saddle River, NJ: Prentice Hall.

BSL (Berkeley Seismology Lab). 2015. Today in Earthquake History: Chile 1960. May 22, 2015. https://seismo.berkeley.edu/blog/2015/05/22/today-in-earthquake-history-chile-1960.html, accessed August 12, 2019.

Coch, N. K. 1995. *Geohazards: Natural and Human*. Englewood Cliffs, NJ: Prentice Hall.

CRED (Center for Research on the Epidemiology of Disasters), USAID (United States Assistance for International Development), and UNISDR (United Nations Office for Disaster Risk Reduction). 2016. *The Human Cost of Natural Disasters 2015: A Global Perspective*. Brussels, Belgium: CRED.

IFRC (International Federation of Red Cross and Red Crescent Societies). 2007. *World Disaster Report—Focus on Discrimination*. Geneva, Switzerland: IFRC.

IFRC (International Federation of Red Cross and Red Crescent Societies). 2016. *World Disaster Report—Resilience, Saving Lives Today, Investing for Tomorrow*. Geneva, Switzerland: IFRC.

Kott, A. 2017. Earthquake Myths from around the World. http://uraha.de/de/?p=390&lang=en, accessed August 21, 2019.

Shrestha, A. B., S. R. Bajracharya, J. S. Kargel, and N. R. Khanal. 2016. *The Impact of Nepal's 2015 Gorkha Earthquake-Induced Geohazards*. Kathmandu, Nepal: ICIMOD.

Erosion

Erosion is the entrainment (pick up and carry along) of sediment and soil particles by the geomorphic agents of wind, water, ice, and gravity. Erosion is distinguished from weathering, which is the decomposition (by chemical processes) and disintegration (by physical processes) of bedrock. Erosional processes pick up the products of weathering and transport them away. When the rate of erosion outpaces rates of weathering, soil can become thin or disappear altogether, leaving bedrock exposed and/or leading to land degradation (Schaetzl and Thompson 2015). Soil contains an enormous reservoir of carbon and other essential nutrients for life, so erosion limits the growth of natural vegetation and crops and diminishes soil biodiversity (Beach et al. 2019).

It is possible to rank the importance of erosional agents worldwide by comparing the amount of sediment delivered to the oceans by each agent: 90 percent by running water, 7 percent by glaciers, 1 percent by groundwater solution, 1 percent by ocean waves along coastlines, and 1 percent by wind and volcanoes combined. The relative magnitude of erosion varies with ecoregions, but it's always a battle between the magnitude and frequency of erosional agents on the one hand versus the resistance of sediments and soil on the other hand, largely controlled by sediment size, bedrock hardness, steepness of slopes, and vegetation. The maximum erosion rates of natural areas exist in semiarid mountains, where vegetation is relatively sparse, but the action of water and wind proceeds without strong control by vegetation.

Mountains are particularly dynamic regions because they experience ferocious rates of both uplift and erosion. In the Himalayas, for example, rates of uplift have been reported as ranging from 1.6 to 65.6 feet (0.5–20 m) per 1,000 years. Erosion rates have been estimated at rates up to 65.6 feet (20 m) per 1,000 years. Erosion rates correlate directly with slope steepness. Rivers are downcutting at rates up to 72.2 feet (22 m) per 1,000 years. In some areas of the Himalayas, erosion has been so high that it has induced uplift (Marston 2008). The Southern Alps of New Zealand are another example of a mountain range with extreme rates of uplift and erosion. Furthermore, the rates of erosion will be higher on the wetter side of a mountain range, which induces more uplift on that side. Therefore, the windward south-facing flank of the Himalayas erodes and uplifts faster than the drier north-facing rain shadow side. The elevation of Mt. Everest has been increasing over time, but it is not clear how much of this is due to increased precision in measurement as opposed to the erosional-induced uplift effect. As a mountain increases elevation due to uplift, the rate of erosion also increases to the point where some geomorphologists believe a steady state is achieved where uplift and erosion are in balance.

EROSION: NATURAL RATES AND RATES ACCELERATED BY HUMAN ACTIVITIES

Erosion occurs naturally, but it's widely regarded as a global environmental problem and can be altered, usually accelerated by cropland agriculture, irrigation and agricultural drains, dams and levees, grazing, forest removal, conversion of shrubland to grassland, road building, wildfires, mining, urbanization, off-road vehicle use, and other activities (Goudie and Viles 2016).

Cropland agriculture disturbs the soil when it is plowed and tilled before planting. If crops are aggressively pulled from the soil, as with sugarcane, the soil is disturbed and exposed to erosional processes until land cover is reestablished. The erosion from sugarcane fields has been so high on the Hawaiian Islands that coral reefs have been smothered by the sediment. Agricultural soils can be protected over time by conservation tillage techniques, including contour plowing and use of grassed waterways that are left untilled and used to conduct water off fields. Some societies have reduced vulnerability to erosion by incorporating agricultural terraces into farming on steep land slopes (Marston 2008; Beach et al. 2019). Nevertheless, most of the world's farmland has experienced moderate to severe erosion from running water and wind (Chappell et al. 2015).

Grazing can accelerate soil erosion where grasses are eaten low, and grass cover is reduced over time by livestock numbers that exceed the carrying capacity of the rangeland. Grassland areas have been largely converted to croplands or rangeland for grazing. Livestock and animals such as elk are notorious for trampling stream channel banks and grazing on bank vegetation, which can lead to bank erosion. In an interesting turn of events, the introduction of wolves into the northern ranges of Yellowstone National Park has reduced elk populations that had previously overgrazed the willows on streambanks, causing erosion. In recent decades, those banks are recovering.

Forest removal can lead to increased slope failures as the soil-binding effect of tree roots is lost and trees cease to break the energy of falling raindrops. Land surface erosion can also occur on forest roads or where logs are skidded upslope to landing points. When forest or shrubland is converted for rangeland or cropland, through clearing and/or fire, the new vegetation cover provides less interception of rain, and runoff is usually increased. This is a particular problem in Brazil where the savanna-type vegetation, known locally as the cerrado, has been cleared for soy bean production (e.g., in the state of Mato Grosso) and for ranching grazing (e.g., in the state of Goias).

Land surface erosion can be accelerated by wildfires. Fires are sometimes set deliberately to reduce logging residue after timber harvest and, sometimes illegally, to clear the forest for farming or ranching. Moreover, extensive and severe wildfires pose a significant hazard in forested areas worldwide. Intense heat creates water-repellant soils as soil carbon is vaporized and recondenses, forming an impermeable seal at or near the soil's surface. Subsequent rain or snowmelt cannot infiltrate the soils, so it will end up as runoff instead, leading to more erosion. If the wildfire does not burn needles on coniferous tress, and those needles eventually fall to the forest floor, the soils will be protected from rainsplash erosion. On especially steep slopes, the runoff on water repellant soils can lead to dangerous mudflows.

Off-road vehicles (ORVs) such as dune buggies cause little effect in open sand areas and can provide an enjoyable form of recreation for many. However, in areas where ORVs reduce vegetation cover, sediment can be eroded and erosional scars left behind. One form of ORVs would be military maneuvers with tanks and other armed vehicles. Scars are still visible in the Saharan Desert from World War II tank battles. Tank tracks exert little pressure on the soil's surface, but when tanks pivot and turn, the soil is churned and made ready for running water and wind to remove it (Marston 1986).

Urbanization can lead to high rates of erosion during construction processes when vegetation has been removed and soil is being regraded for structures. Concrete, asphalt, and watertight rooftops combine to increase running water. Special efforts (e.g., silt fences and straw bales) are needed to retain sediment on-site. Use of quick-growing vegetation or artificial turf will provide temporary cover to afford some erosion control.

Climate change can affect erosion by shifting the balance between resisting factors, such as vegetation, and the magnitude and frequency of the driving forces of water, wind, waves, and gravity. Erosion is a hazard in the sense that the lost soil cannot be replaced in a time period relevant to human occupation. Rates of soil erosion have been reported to have reached their highest levels in the last 500 million years, leading to the loss worldwide of perhaps half of all topsoil in the last 150 years (Beach et al. 2019). Erosion causes a direct hazard for human infrastructure (e.g., roads, bridges, and homes) and can restrict or eliminate land-use activities. One of the greatest challenges facing geomorphologists is separating erosion caused by human activities from erosion that would have occurred without human interference.

Geomorphologists estimate that natural processes move 111 billion tons per year of sediment, whereas human activities move 129–134 billion tons per year (Rozsa 2007). Beach et al. (2019, 149) write, "We are left with a tension between what we know has been a history of extreme soil erosion in many places around the world and the history of soil conservation and evolving sustainability as cultures developed myriad soil conserving and enhancing features and sustainable agroecosystems." Examples of very degraded soil can be observed around the world, but especially in the Great Plains and Southwest portions of the United States, Central America, the Andes, Iceland, Sahel region of Africa, eastern Europe, Madagascar, much of India, eastern China, and southeast Asia.

EROSION BY RUNNING WATER: SHEETWASH, RILLS, GULLIES, PIPES

Running water causes erosion on hillslopes and in rivers. On hillslopes, running water causes erosion through the processes of rainsplash, sheetwash, rills, gullies, and pipes. Raindrops cause erosion by detaching soil particles and transporting them downslope, a process that grows in importance as vegetation and ground cover (vegetation litter) decreases. Raindrops may also compact and seal the soil surface, which then leads to lower infiltration, greater runoff, and accelerated erosion by other processes. When the rate of rain supplied to the soil's surface exceeds the infiltration rate, water begins to accumulate on the surface, filling small depressions. In the presence of a sloping land surface, runoff begins in an unconcentrated form termed *sheetflow*. Surface runoff can also be created when subsurface flow is conducted to the surface. Regardless of the source of the sheetflow, erosion is caused when the sheetflow detaches soil particles and carries them downslope. Sheetflow erosion is mitigated in part or in whole by the presence of litter on the soil surface, which can absorb some of the water, as well as by microtopography, which creates roughness that slows the flow. Sheetflow is easily concentrated into small rills, with width and depth measured in centimeters, which can lead to further erosion. Rills are small enough that they can be plowed under by modern farm soil tilling equipment, but in the absence of any control, they can reach a length of many meters as they grow in the downslope direction. If rills grow in width and depth to a meter or more, a gully is created. Gullies cannot be destroyed by farm equipment. The head of the gully commonly migrates upslope over time.

Subsurface tunnels, termed *pipes*, can form in the soil and cause significant amounts of erosion. The conditions favorable to pipe formation are as follows: highly variable precipitation (which leads to wet-and-dry cycles), expansive soils (which are especially susceptible to shrink and swell), reduction in vegetation cover (which leads to the loss of the soil-binding effect of plant roots), presence of an impermeable subsoil horizon (which helps concentrate water flow above it), and the presence of a dispersible soil layer (in which clays are easily removed by water). Gophers can create tunnels independent of all of the preceding factors, providing the pipes for concentrated subsurface running water. Piping is very pervasive in badland areas and can, in fact, be the most important process of erosion.

Rainsplash, sheetwash, rills, gullies, and pipes are common processes in badlands areas; steep, angular topography with low vegetation cover; silty soils; and in semiarid or arid climates, where rainfall is relatively rare but intense when it does occur. Badlands are perhaps the most erodible hillslopes by running water. Because the steep slopes of badlands preclude grazing by animals, badlands

Badlands viewed from Zabriskie Point, Death Valley. (Courtesy Richard Marston)

provide an example of a landscape where extremely high rates of erosion can occur even without direct human interference (Marston and Dolan 1999). Hoodoos provide another example of landscapes made skeletal by erosion.

EROSION BY RUNNING WATER: RIVERS

Rivers are perhaps the most widespread and dynamic erosional agents on earth. Rivers adjust their dimensions as a result of upstream controls (e.g., area and shape of the drainage basin, hillslope gradient, vegetation cover, and land use) and downstream controls (e.g., rise or drop in base level of erosion). Chorley et al. (1984, 278) point out, "As rivers adjust, they affect not only their tributary streams, but also the adjacent hillslopes by undercutting or by aggradation." This process is especially significant in the Nepal Himalayas, where rivers undercut the toe of hillslopes during the monsoon season and during glacial lake outburst floods. The large volumes of sediment from streamside slope failures cannot entirely be transported by the floods, so channel aggradation occurs. In subsequent floods, the water in the elevated streambed attacks the stream adjacent hillslopes to an even greater extent, further accelerating slope failures. This is an example of a positive feedback in science, when an environmental change (streamside slope failures) causes other changes (channel aggradation) that reinforce the direction of initial change (streamside slope failures). No engineering solution exists for these aggrading streams of the Nepal Himalayas as they move further away from equilibrium. Rivers in flood cause a great deal of erosion and modify landforms, but over decades and longer time frames, the frequent flow events of low- to moderate-magnitude cause greater amounts of erosion.

Rivers erode by three processes: hydraulic action, abrasion, and solution (also termed *corrosion*). The strength of hydraulic action depends on the shear stress, which is a function of the depth of the flow and the channel gradient; the latter affects velocity. Thus, deep, fast-moving rivers have the greatest erosion potential. Shallow rivers exert little erosional effect, even if the velocity is high. Rivers that are deep, but slow, also cause little erosion. The effective channel gradient will be lowered by the presence of features that cause a vertical drop in stream elevation, such as waterfalls and log steps. Water creates a local plunge pool below these fall obstructions, but the greater effect is to decrease the average gradient of the stream and thereby decrease erodibility. Water that is deep and moving downstream in a steep gradient can move clay, silt, sand, gravel, cobbles, and boulders as shear stress progressively increases. Abrasion is caused by the particles that are already in motion striking bedrock and eroding it by chipping or scraping. Running water in rivers also dissolves bedrock, especially where bedrock is highly soluble or with other weak and young rock types. When comparing the solid (suspended load and bedload) and dissolved load being carried by rivers, the dissolved load can be a minor component in arid regions but can be dominated over the solid load in tropical rivers.

Rivers cause significant amounts of erosion by horizontal and vertical shifting of their channels. When viewed from above (plan view), rivers rarely follow a straight course, unless they are guided by a geologic fault or in a narrow canyon. Most rivers take a plan shape that is either meandering or braided (see Table 1). Sinuosity refers to the deviation from a straight course, so it's used to measure the amount of meandering. Meanders migrate downvalley over time unless controlled by human-made levees or rip-rap placed on channel banks. This migration of meandering streams is inconvenient for many types of land use in the floodplain. Many examples exist around the world where meandering rivers have been straightened ("channelized") for the purpose of increasing human use of floodplains. However, this usually leads to increased flooding and erosion downstream. The percentage of silt and clay in unprotected banks affects the erodibility of those banks. Clay and silt strengthen the banks because of the cohesion between particles of that size, whereas gravel and cobbles are larger but lack cohesiveness. When a river erodes banks of gravel and cobbles, the banks collapse, but the gravel and cobbles are not transported far. Floodplains comprising clay and silt commonly host meandering rivers, whereas floodplains composed of gravel and cobbles commonly host braided rivers.

Uplift or subsidence over geologic time can change the valley gradient; steeper valleys provide more erosional energy to a river. Shear stress, sediment sizes, and sinuosity can change not only during individual storms but also over longer time

Table 1 Characteristics of river patterns and their relative stability

Characteristic	Meandering	Braided
Plan view shape		
Valley gradient	Moderate	High
Shear stress	Low to moderate	High
Sediment sizes	Low	High
Sinuosity at bankfull stage	High	Low
Width-to-depth ratio	Moderate	High
Percent of silt and clay in banks	Moderate	Low
Horizontal stability	Low	High

Sources: Table modified from Schumm (1981, 24) and Knighton (1998, Table 5.10). Meandering river image (U.S. National Archives, 412-DA-8086). Braided river image (Dimitry B./Flickr).

intervals in response to shifts in land cover and human activities (e.g., construction of upstream dams). In some cases, streams metamorphose from a meandering to braided shape or vice versa. Thus, the relative stability of the channel can change, as expressed by lateral migration, vertical downcutting (channel incision), or vertical accretion (channel aggradation). The erosional consequences for any streamside activity can be hazardous due to bridges undercutting, loss of farmland, altered flood regime, and more. The change in erosional regime will also affect floodplain vegetation and the wildlife that depend on it (Marston et al. 2005).

EROSION BY GLACIERS

Glaciers and ice sheets store about 84 percent of the world's freshwater and have modified a significant portion of earth's surface in the past. At present, glaciers cover about 10 percent of earth's surface (most of this is in Antarctica and Greenland), but erosion by glaciers is still important in areas where glaciers are active. Glaciers create hazards by surging in velocity over short time periods and through glacial outburst floods. Outburst floods can create massive amounts of erosion, pose hazards to infrastructure, and alter landforms in the valley bottom. Even though most glaciers have been retreating in recent decades, the ice still moves downhill, so it causes erosion by two processes: abrasion and plucking. In addition, the meltwater from glaciers causes erosion. Ice by itself is too soft to erode bedrock, except perhaps for soapstone. Rocks that protrude out of the base of the ice abrades the rock, creating striations and grooves. Smaller particles of sand, silt, and clay are carried along between the ice and bedrock, and these finer sized sediments can actually polish the bedrock. The magnitude of glacial abrasion depends on the concentration of rock debris in the basal ice, the hardness of rock fragments compared to the hardness of bedrock, plus the velocity and thickness of the ice. Abrasion rates have been measured from only 0.04 inch (1 mm) per year in basalt (a resistant rock) with slow ice to 1.3 inches (34 mm) per year in marble (a relatively nonresistant rock) with thick, fast-moving ice. Glaciers also erode by lifting large blocks of bedrock that have been previously loosened by weathering. The moving ice "plucks" the blocks from the bedrock beneath the glacier and moves the blocks down the valley. Glaciers in temperate regions on earth can erode bedrock at rates between 0.33 and 32.8 feet (0.1 to 10 m) per 1,000 years, much higher than cold-ice polar glaciers.

The glacial erosional processes of abrasion and plucking combine to create an impressive array of landforms: cirques, horns, cols, aretes, glacial troughs, fjords. The sediment that was eroded and transported by glaciers may be deposited in an equally impressive variety of moraines, drumlins, and other features. The meltwater beneath and beyond the glaciers moves sediment downslope, where it can be left as depositional landforms or transported to the oceans. This is especially important with the terminal edge of the Greenland ice sheet and with tidewater glaciers elsewhere in the world. Glacial meltwater deposits sometimes comprise useful well-sorted sand and gravel resources for mining, whereas deposits left

Glacial striations and glacial polish on partly weathered granite, Juneau Icefield, Alaska. (Courtesy Richard Marston)

directly by the ice in moraines are poorly sorted (i.e., contain sizes from clay to boulders), so they are not economical to mine.

EROSION BY GROUNDWATER

As discussed earlier, rivers erode by dissolving rocks and shallow subsurface water erodes by piping. Groundwater also erodes and causes significant erosion by these same two processes, occasionally leading to hazardous sinkholes (or dolines). Groundwater is found at some depth below earth's surface, where water fills all available pore spaces, termed *aquifers*. This zone of saturation is quite unlike the shallow water moving through pipes near the ground surface. In locations where water can move vertically from the earth's surface directly to the water-bearing strata, the groundwater is unconfined, and the top of the zone of saturation is the water table. If the groundwater moves laterally to a point where impervious strata separate the saturated strata from the surface, the aquifer is termed a *confined aquifer*. In a confined aquifer, the groundwater is under pressure, and water in wells will rise to an elevation above the top of the confined aquifer. This elevation is known as the piezometric surface when mapped in three dimensions. The ease with which groundwater can move is measured by the permeability of the rock or sediments and the gradient of the water table or piezometric surface.

Fast-moving, relatively fresh groundwater (low in dissolved solids) can erode soluble types of bedrock, notably limestone and gypsum. A surprisingly large

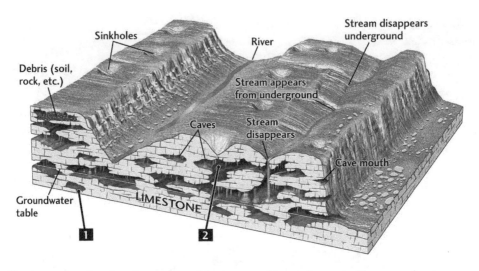

Karst topography, showing erosional features and hazards created by groundwater solution. (U.S. Geological Survey)

percentage of earth's surface, 15 percent, is covered by soluble rocks. When groundwater dissolves bedrock, openings are created, which attract more and more water and thereby grow through time. In this manner, horizontal channels can enlarge into caverns. If the water table drops, the caverns can become filled with air. It is then possible for the overlying layers to collapse into the cavern, creating a sinkhole. Another process by which a sinkhole can be formed is by steady enlargement of a vertical solution cavity. Solution dolines occur where a dense network of joints lie at shallow depths below earth's surface. Because fractures (joints) in the bedrock constitute a likely passageway for groundwater, sinkholes often form at the intersection of different joint sets.

Where groundwater solution (and surface water solution) dominates a landscape, karst topography is created, with unaffected, resistant towers of rock separated

Sinkhole formed in Winter Park, Florida, in 1981. (U.S. Geological Survey)

by low-lying topographic areas, often marking geologic joint patterns. Many caves and passageways are present below the surface. Karst topography is best formed where one finds thick, extensively fractured and soluble rocks, coupled with high rainfall. The process is the solution by groundwater is accelerated by the presence of vegetation, which forms carbon dioxide when decomposing. The carbon dioxide combines with water to form a weak carbonic acid, which is more corrosive than water. Limestone denudation is highest in cold-wet climates and lowest in cold-dry and warm-dry climates. Erosion by running water on earth's surface is greatly reduced in karst landscapes.

EROSION BY WIND

By comparing the density of sediment acted on by moving water and moving air, one can appreciate that wind erosion is comparatively less important than erosion by running water on a global basis. Rocks immersed in water have a density of only about 1.6 times the density of water, but rocks' density is approximately 2,000 times the density of air. Thus, moving air exerts little pressure compared to running water. Nevertheless, wind can be an important erosional agent where one finds sparse vegetation, high winds, a high supply of fine-sized, unconsolidated sediment, and a lack of moisture. Where vegetation is sparse, roots are not effective in binding soil and sediment in place. Without sufficient vegetation, decaying plant matter (humus) is lacking in the soil, which would otherwise help to bind particles together. Finally, vegetation creates aerodynamic roughness for the wind, slowing it down near the surface. Wind erosion reaches maximum effectiveness in arid and semiarid regions, coastal areas, zones where glacial meltwater deposits are widespread and unvegetated, and on damaged or abandoned agricultural lands. It is perhaps surprising that open sand covers less than 1 percent of the desert surface in the southwest United States, but 22–38 percent of the Sahara and deserts of Western Australia.

The two processes of wind erosion are deflation and abrasion. Deflation is the direct removal of loose particles by the wind. Clay is very cohesive, so it's difficult to erode by deflation. Silt is easily eroded, and much of it can be suspended by turbulent, persistent winds and become dust storms. Dust storms have been responsible for major soil degradation in regions with a history of poor agricultural practices, and dust poses major hazards for drivers and aircraft. The second type of wind erosion is termed *abrasion*, and as with running water and glaciers, abrasion is the erosion caused by particles already in transport. Sand in transport is capable of eroded bedrock, creating pits, grooves, or polish as evidence. Automobile drivers report sandblast erosion of paint after encountering windblown sand. When sand is eventually deposited, an amazing array of sand dunes can be formed, depending on the strength of the wind, supply of sand, and effects of vegetation. The deposition of sand can pose a hazard to farmers or transportation.

Wind erosion is accelerated during hot-dry periods (droughts) and where overgrazing and poor farming techniques combine to reduce vegetation cover. Furthermore, as lake beds shrink, dust is easily generated, a process that has been observed in the Aral Sea, Owens Lake of California, and playas around the world.

Grooves in exposed limestone bedrock caused by wind abrasion, Samalayuca Sand Dunes, Chihuahua, Mexico. (Courtesy Richard Marston)

Goudie and Viles (2016) have outlined the human causes of increased dust generation: construction sites, dirt roads, feedlots, fires, military activity, mine tailings, disturbance of soil crusts, and overgrazing. In the 1930s, the southern Great Plains of the United States experienced a period known as the Dust Bowl, when severe wind erosion because of extended drought and farming in areas of inadequate moisture led to the abandonment of many farms and induced out-migration of those farmers to other regions of the country. Similar examples have been observed in other semiarid regions in all continents.

It is worth noting that prolonged dust storms can leave silt-sized deposits known as loess. As mentioned earlier, silt is the particle size most easily eroded by running water. Therefore, loess-covered areas are especially subject to erosion by running water and wind if vegetation cover is reduced. In winter, cold winds from Tibet generate dust that has mantled portions of the Middle Mountains of Nepal, and these soils have been heavily eroded where disturbed by humans.

CONCLUSION

The loss of sediment and soil by erosional processes is a hazard at the local scale, regional scale, continental scale, and global scale. Running water, glaciers, groundwater, and wind entrain sediment and soils, leading to loss of land and posing danger to human activities and infrastructure. Geomorphologists study the magnitude, frequency, causes, and mitigation measures for erosional hazards. Clearly, natural processes will continue to cause erosion and move significant volumes of sediment and soil, but the human impact is growing. The decline of civilizations through history has been linked, in part, to soil erosion. The dilemma is that soil is eroded much faster than it is formed, so human activities must adopt strategies that can mitigate soil erosion in whole or in part. Erosion by glaciers and groundwater cannot be mitigated by humans, so it is a matter of adjusting to the hazards created by these erosional agents. Civilization can become sustainable over the long term if soil conservation measures are enacted that maintain and improve vegetation cover against the expanded effects of human alteration of the landscape and climate change.

Richard Marston

Further Reading

Beach, T., S. Luzzadder-Beach, and N. P. Dunning. 2019. Out of the Soil: Soil (Dark Matter Biodiversity) and Societal "Collapses" from Mesoamerica to Mesopotamia and Beyond. In *Biological Extinction: New Perspective*, edited by P. Dasgupta, P. Raven, and A. McIvor, 138–174. Cambridge, UK: Cambridge University Press.

Chappell, A., J. Baldock, and J. Sanderman. 2015. The Global Significance of Omitting Soil Erosion from Soil Organic Carbon Cycling Schemes. *Nature Climate Change* 6: 187–191.

Chorley, R. J., Schumm, S. A., and D. E. Sugden. 1984. *Geomorphology*. London, UK: Methuen.

Goudie, A. S. and H. A. Viles. 2016. *Geomorphology in the Anthropocene*. Cambridge, UK: Cambridge University Press.

Knighton, D. A. 1998. *Fluvial Forms and Processes: A New Perspective*. 2nd ed. New York: Routledge.

Marston, R. A. 1986. Maneuver-caused Wind Erosion Impacts, South-central New Mexico. In *Aeolian Geomorphology*, edited by W. G. Nickling, 273–290. London, UK: Allen and Unwin.

Marston, R. A. 2008. Land, Life, and Environmental Change in Mountains. *Annals of the Association of American Geographers* 98(3): 507–520.

Marston, R. A. and L. S. Dolan. 1999. Effectiveness of Sediment Control Structures Relative to Spatial Patterns of Upland Soils Loss in an Arid Watershed, Wyoming. *Geomorphology* 31: 313–323.

Marston, R. A., J. D. Mills, D. R. Wrazien, R. Bassett, and D. K. Splinter. 2005. Effects of Jackson Lake Dam on the Snake River and its Floodplain, Grand Teton National Park, Wyoming, USA. *Geomorphology* 71(1–2): 79–98

Rozsa, P. 2007. Attempts at Qualitative and Quantitative Assessment of Human Impact on the Landscape. *Geografia Fisica e Dinamica Quaternaria* 30: 233–238.

Schaetzl, R. J. and M. L. Thompson. 2015. *Soils*. 2nd ed. New York: Cambridge University Press.

Schumm, S. A. 1981. Evolution and Response of the Fluvial System, Sedimentological Implications. *Society of Economic Paleontologists and Mineralogists Special Publication* 31: 19–29.

Expansive Soils

Expansive soils are a class of soils that have a tendency to expand when wet. Alternatively, they typically contract when moisture is removed, such as through evaporation. The degree of expansion depends primarily on the amount and type of mineral present that absorb water. Certain clay minerals include a crystalline structure that has the capacity for the inclusion of water molecules. Some clay minerals more readily absorb water molecules than others, so a spectrum of absorptive and, therefore, expansive minerals exists.

The geographic distribution of expansive soils relates to the distribution of certain soil orders, which generally relate to climate factors along with a degree of weathering and more. Certain soil orders, which are the broadest classification of soil types, are more likely to contain clays that cause soils to shrink and swell, but not all clay-rich soils have this property. In general, soil orders containing less weathered soils are more likely to contain minerals that shrink and swell. For instance, the soil order vertisols are characterized by their capacity to shrink and swell. The prefix *vert-* refers to the fact that these soils appear inverted due to the presence of topsoil at depth. Topsoil washes into large cracks that are formed during a dry spell followed by a rain event so that, over time, the topsoil layer is more dominant at depth.

In the United States, 20 percent of land is underlain by soils with high shrink-swell capacity (Brady and Weil 2002), including large swaths of the states of North and South Dakota, deposited as glaciers receded thousands of years ago. From the upper Midwest to Florida, expansive soils with high moderate shrink-swell capacity underlie much of the landscape but usually make up less than half of all soils. In coastal regions, deposits from marine sediments, lake sediments, or marine-continental sources may have significant expansive properties. Areas of expansive soils with climates characterized by wet and dry seasons, such as those found in parts of Texas and California in the United States, experience the greatest challenges with infrastructure, in general.

Expansive soils are also sometimes called high shrink-swell soils. Classified as a natural hazard, on an annual basis and in a typical year, expansive soils cause significantly more damage on an annual basis in the United States than floods, tornadoes, hurricanes, and earthquakes combined (King 2016). If at all possible, it is best to avoid building structures on these soils as they do not provide a firm, regular surface on which to build. In areas where it is difficult to avoid building on expansive soils, certain mitigating enhancements can be employed to reduce the likelihood that the soil will expand or contract. Engineers continue to develop techniques to mitigate or even correct issues associated with shrink-swell soils.

Measurement of the capacity of a soil to expand when moisture is added is known as the expansive index. Other tests may be combined, such as the plastic limit and liquid limit to determine a site's suitability for building. Further details on soil composition and expandability involve using x-ray diffraction to determine which minerals are present in a soil. Testing of soil typically occurs before a construction project commences to determine if remediation or mitigating procedures need to be included in the engineer's design.

Expansive soils, because they can reduce or even eliminate flow of liquid through soils (i.e., low permeability), prove to be useful for this very property. In certain cases, such as at the bottom and sides of a dump site for garbage or other unwanted items, expansive soils provide a useful barrier. It protects surrounding soil and groundwater from contamination as long as it maintains a certain level of moisture. Besides landfills, common uses of soils containing high levels of swelling clays include sewage lagoons, artificial ponds, and industrial waste lagoons. Expansive soils may be quite useful or destructive, depending on the situation.

CLAY MINERALS THAT EXPAND

Soils comprise sand, silt, and clay particle sizes of varying proportions. For a soil to be expansive, it must contain smaller clay-sized particles with specific properties. In addition, the medium silt-sized soil particles can weather, or break down, into clay-sized particles, to also have expansive properties. These properties relate to soil components even smaller than clay called clay colloids, which have electrically charged surfaces. Different types of clay colloids have different physical and chemical properties, explaining the ability of certain mineral structures to attract and hold water.

Because of the high potential for chemical activity in the colloid fraction of soil, it often determines the level of a soil's fertility in addition to other properties such as expansiveness. The soil colloid's small size (less than 1 µ, or 0.0000001 m) accounts for large surface areas where chemical exchange can occur. Clay colloids also have internal surface area, where substitution of anions and cations occurs. Anions are negatively charged, such the hydroxide ion, OH^-, and nitrate, NO_3^-, while cations are positively charged, such as magnesium, Mg^+, and calcium, Ca^{2+}. For the area of a hectare (about the size of a football field) that is about 5 feet (1.5 m) deep, the surface area could be as large as the entire surface of the United States (Brady and Weil 2002).

For most colloidal surfaces, negative charges predominate, and so cations in the soil solution are loosely attracted or adsorbed due to electrostatic attraction (attraction of opposite charges). The water molecule, H_2O, does not have an overall charge, so it is considered to be neutral. However, it is slightly positive near the two hydrogens and slightly negative near the oxygen, making it a polar molecule (have a separation of charge). Because of this property, water can be adsorbed by soil colloids at either positively or negatively charged sites. However, the expansible nature of certain soils is determined by the presence or absence of particular types of soil colloids.

Three of the four main classes of soil colloids are clays, with the fourth soil colloid being organic colloids, also known as soil humus. Of the three soil colloids, the primary category related to expansive soils are the crystalline silicate clays. These silicate clays contain two types of building blocks (crystals), the first of which is the silica tetrahedra, SiO_4^{+4}. One silicon atom sits at the center of a pyramid-like structure while one oxygen sits at the apex and the remaining three oxygens form the other points of the tetrahedron. These building blocks fit together in many different ways including sheets, chains, or lattices. A tetrahedral sheet forms when these silicate tetrahedra interlock in sheets. The tetrahedral sheet makes up one of the primary mineral structures for expansive soils.

The second crystal building block important in expansive soils is the octahedral sheet, which has a building block of an eight-sided crystalline solid, or octahedron. In this case, an aluminum or magnesium atom sits at the center of six oxygen atoms. These sheets, which are horizontally interlocked, form layers of clay minerals.

Some silicate clays known for having expansive properties comprise one octahedral sheet sandwiched between two tetrahedral sheets, which is how they get the name 2:1-Type Silicate Clays. The ratio refers to the types and relative amounts of sheets present. The two 2:1-Type Silicate Clays groups with expansive properties include smectite and vermiculite. The smectite group expands because of interlayer expansion, where water molecules enter between layers in relatively dry clay, pushing the layers apart. Before water enters, layers are loosely bonded by weak oxygen-oxygen bonds and oxygen to cation linkages. Within the soil solution, dissolved cations are strongly attracted to the negative sea of oxygen atoms between layers. Because of this attraction, the cation along with the water enters the space between layers.

The internal surface area greatly exceeds the external surface area when considering the space between layers of the mineral crystal sheets. However, smectite clays fall within the "very fine" textural category, which means that the clay particles are very small. Water is also attracted to the outside surfaces of clay particles, so the small size of smectites means that significant water is also adsorbed on the exterior surfaces of the clay minerals (Schaetzl and Anderson 2005).

Most common among the smectite group is montmorillonite, but other clays in this group include beidellite, nontronite, and saponite (Brady and Weil 2002). Bentonite also consists of mainly smectite clays. The significant presence of these clays in a soil makes the soil an undesirable surface for building upon. On the other hand, because the clays can reduce soil permeability (the capability of fluids to pass through), these clays can be highly sought after for certain environmental applications, such as containment of contaminated fluids.

More common than the smectite clays is the vermiculite clay group. Vermiculites differ from smectites due to the predominance of aluminum inside the octahedral sheet rather than magnesium. Since aluminum's charge of +1 replaces magnesium's charge of +2, the clays have large quantities of negative charge. Again, cations within the soil solution adsorb into the interlayer spaces along with water. However, water molecules among vermiculite clays found between layers act more as bridges holding layers together rather than expanding wedges. Because of this, vermiculites have limited expansive properties compared to smectites (Brady and Weil 2002).

SOIL ORDERS INCLUDING EXPANSIVE SOILS

More weathered soils contain higher proportions of clays made up of nonexpanding iron and aluminum oxides. Therefore, younger or less developed soils will more likely contain smectite or vermiculite clays. Soil orders more likely to contain smectite, in order of increasing weathering intensity, include aridisols, vertisols, mollisols, and alfisols.

Aridisols occur in arid environments, so weathering processes associated with rare precipitation events, wind, and extreme temperature changes dominate. Aridisols cover more than 12 percent of global land surface and 9 percent of the United States, but challenges associated with the soil order generally focus around the lack of precipitation and presence of hardened layers rather than issues with its tendency to shrink or swell.

By contrast, vertisols are defined by the presence of expansive clays, which causes the inverted appearance of soil layers, or horizons. Vertisols, by definition, contain more than 30 percent of expansive clays to a depth of 3.3 feet (1 m) or more. They typically occur in subhumid to semiarid climates where grasses make up the dominant natural vegetation type.

Often, vertisols occur where the climate pattern includes a dry season, causing the shrinking of clays in the soil. Dramatic shrinking leads to wide cracks in the ground surface, which is a diagnostic characteristic for this soil order. At the surface of the cracked landscape, granules of soil form, which can slough off into the cracks so that top soil is not relocated deeper into the soil profile. The name reflects that the soil is inverted through the prefix "vert" (Brady and Weil 2002).

Vertisols typically form on flat to gentling sloping areas. Since they have low permeability, low slopes assist the soils in accessing water that falls as precipitation (Schaetzl and Anderson 2005). Some of the deepest vertisols occur in shallow depressions, where clay sediments that have eroded from nearby slopes settle. The temporary ponds keep the smectite clays wet until the pond dries up and cracks form.

Like vertisols, mollisols also form under grasslands, but they are characterized by a deep humus-rich top layer. Expansive clays are found in this top layer, which gives a mollisol its characteristic granular structure that is excellent for agriculture. Clays generally dominate deeper in the soil horizon in mollisols and are found in a smaller proportion than in vertisols, so they are less prone to drying and cracking. Mollisols cover about 22 percent of U.S. lands and only about 7 percent of the world (Brady and Weil 2002).

The final soil order where expansive soils are regularly found includes alfisols, an order farther along on the soil weathering continuum. Because of a wetter climate that supports deciduous forests, these soils contain high levels of silicate clays below the surface. The clays have migrated down through the soil profile, mainly due to higher levels of precipitation compared to the other soil orders highlighted here. These silicate clays accumulate at depth, forming a thick silicate clay layer. Alfisols make up about 14 percent of land area in the United States and 10 percent globally.

GEOGRAPHIC DISTRIBUTION

The geographic distribution of expansive soils relates to the spatial distribution of soil orders highlighted earlier. According to the Natural Resources

Conservation Service (NRCS), in the United States, Aridisols dominate in some western states such as Nevada and in southern Arizona and New Mexico. This soil order can also be found in west and south Texas, parts of California, Wyoming, and Colorado and in south central Washington state due to the rain shadow effect (NRCS n.d.). Other than the rain shadow effect, many of these areas experience extensive periods of dry weather associated with the dominance of subtropical high pressure. Vertisols, although less common, significantly increase risk to infrastructure due to their high shrink-swell capacity. These soils, found in pockets of the Dakotas, Minnesota, California, Texas, and along the Mississippi River, require monitoring and maintenance to mitigate associated hazards. Vertisols, since they have colloidal clays that contribute to soil fertility, are sometimes utilized for agriculture as long as enough rainfall or irrigation water is available to keep them from drying and shrinking.

Mollisols, a highly valuable soil order for agriculture, cover a significant portion of the United States (about 22 percent) but are less common globally. They are most common in the Great Plains, but extend from parts of Indiana in the east to the coastal states in the west. They also occur in central Florida over limestone-rich areas. These areas in particular can suffer from both land subsidence and expansive soils.

Alfisols cover less of the United States than mollisols, but have a larger land area globally (7 percent for mollisols compared to 10 percent of land area for alfisols). In the United States, these soils occur in much of the Mississippi River basin including the Ohio River and its tributaries. The more weathered soils may shrink and swell, but when used for rangeland, as many are in Oklahoma and Texas, the adapted grassland vegetation suffers little as a result of expansive soils. However, in other regions where more structures are built with associated infrastructure, such as in the central valley of California, expansive soils can cause many challenges and expensive damage. Globally, expansive soils follow similar spatial patterns associated with climate, degree of weathering, and other local factors, such as parent material, the material from which soil forms. For example, in parts of India's Deccan Plateau, vertisols dominate, having formed on gently sloping plains. Vertisols go by the local term *Tirs* in Algeria and Morocco. In parts of Africa dominated by grasslands, especially in semiarid regions, similar soils are widespread. The nation with the greatest spatial area of vertisols is Australia, home to more than 197 million acres (80 million ha) of the soil order, an area larger than the state of Texas (Schaetzl and Anderson 2005).

NATURAL HAZARD

Expansive soils can swell up to 10 percent or more by volume as they become wet. According to the Colorado Geological Survey (CGS), a sample of pure montmorillonite can swell up to 15 times its dry volume (CGS n.d.). The force exerted by swelling clay minerals can exert pressures of greater than 20,000 lbs./sq. feet (960 kPa). This pressure, when applied to a building's foundation or other structure, can cause enormous damage. Some subsurface soils, when left undisturbed, retain a fairly constant moisture level, so the volume also remains steady. However, if the soil is exposed to drying or more moisture during or after construction, the soil will of course shrink or swell in response.

The key to successfully managing to build a stable structure on expansive soils involves first knowing that the soil type exists at the building site. In cases where expansive soils are not detected early on in the planning phase before onset of construction, improper design accommodations, faulty construction, and even inappropriate landscaping may lead to expensive repairs and ongoing maintenance issues (CGS n.d.).

In 1973, Jones and Holtz (1973) estimated that $2.3 billion in damages occur each year due to damage to houses, other buildings, pipelines, and roads, more than twice the damage from floods, hurricanes, tornadoes and earthquakes. In 2019, the equivalent value is over $13.6 billion annually. At the time of their study, they reported that over 250,000 homes were built on expansive soils each year, and 60 percent of these would experience only minor damage. However, 10 percent of the homes would experience significant damage, in some cases damage that is beyond repair.

Jones and Holtz (1973) go on to make the point that one in 10 people is affected by floods whereas one in five is affected by expansive soils. More recently, the American Society of Civil Engineers (ASCE) estimates that one-quarter of all U.S. homes experience some damage from expansive soils. Many people, never having heard of expansive soils, wrongly attribute damage caused by soil volume changes to poor construction or normal degradation as infrastructure ages.

The cycle of expansion and contraction of soils may not be noticeable year to year. However, when the stress occurs repeatedly to a structure or roadway, the repetitive stress causes damage to worsen over time. Even though expansive soils are classified as a geologic hazard, most home insurance policies do not cover damage caused by expansive soils (King 2016). Costs of mitigation to reduce impacts or cost of repairs can even exceed the value of a home, especially if not addressed at an early stage of impact.

While it is possible to safely build on expansive soils, the building plan must address methods of maintaining a stable moisture content. Maintaining the moisture levels of the soil means that the volume of the soil will be stable. The procedure for ensuring a successful build on expansive soils involves these steps: (1) test the soil to identify presence and type of expansive soil, (2) design a strategy for maintaining a constant soil moisture level, (3) use strategies during construction to minimize soil moisture changes, and (4) employ the designed strategy for maintaining the moisture level after construction ceases (King 2016).

MEASUREMENT

To identify the presence of swelling clay minerals and associated volume changes, several approaches can be used. Named after a German engineer who developed the system, Atterberg limits measure the shrinkage limit, plastic limit, and liquid limit of a soil. Starting with a dry, clayey soil, as it becomes wetter, its behavior and consistency change dramatically. The shrinkage limit occurs when a soil has adsorbed enough water to the point where it becomes crumbly. As more water is added, it sticks together better and begins behaving like a plastic mass. The point at which this occurs is the plastic limit (PL). Finally, on increasing water to the sample, the soil maintains its plastic consistency until it turns into a thick liquid, which is its liquid limit (LL) (Brady and Weil 2002).

Soil engineers use these limits of a soil to determine the suitability of a site for different uses, such as its potential as a building site. Using the liquid limit (LL) and plastic limit (PL) values, the Plasticity Index (PI) of a soil can be calculated: PI = LL – PL. Soils with a plasticity index greater than about 25 are usually expansive clays, so they are not suitable for building sites or roadways (Brady and Weil 2002).

Another measure of the expansiveness of soils is the coefficient of linear extensibility (COLE). The process involves taking a sample of soil moistened to its plastic limit and molded into the shape of a bar with length LM. Next, the bar of soil is allowed to air dry, shrinking to a length of LD (Brady and Weil 2002). The COLE represents the percentage reduction in length of the bar of soil:

$$COLE = \frac{LM - LD}{LM} \times 100.$$

The COLE value of an expansive soil provides a clear idea of the suitability of a site for buildings, roads, etc. Other measures exist for determining site suitability for different uses, but they take into account additional soil factors, such as organic matter content. For example, the U.S. Corps of Engineers and the U.S. Bureau of Reclamation established such a system known as the Unified System of Classification that combines soil texture, Atterberg limits, and organic matter content (Brady and Weil 2002).

ENGINEERING SOLUTIONS

Once a land manager identifies the presence of an expansive soil at a particular site, several engineering solutions can be employed depending on cost, time, and site constraints. The first and most obvious solution is to control moisture changes (Marcus 2017). While expensive systems exist that monitor moisture and apply water as needed, some people simply monitor the ground around their homes and apply water near the foundation before too much drying occurs. However, the effectiveness of this approach is called into question since significant drying has likely already occurred before water is applied.

A second approach involves building more structurally sound (i.e., stronger and heavier) foundations, which reduce the soil's capacity to swell since it cannot overcome the weight of the overlying structure. Third, some engineers remove the expansive soil from a site, replacing it with a nonexpansive substrate. However, this solution can be costly. Related to these approaches is the process of burying the expansive material or capping it with a material to help maintain its moisture levels (Marcus 2017).

In some cases, the shrinking and swelling only provides a nuisance, so walls and doors can be installed or adjusted in a way that adapts to slight shifts in the ground. This approach plans for shifts and helps to avoid or at least reduce regular damage repairs. For example, drywall can be installed with gaps between the roof and the floor so that shifts do not cause cracks in the wall (Marcus 2017).

Finally, a newer less expensive option is to mix expansive soils with nonexpansive materials at the site. This newer approach reduces expansiveness and, when swelling occurs, creates a more even change rather than an unpredictable pattern of shifting (Srinivas et al. 2016).

ENVIRONMENTAL USES

Since expansive soils swell to the point of becoming virtually impermeable, this property can be useful for certain environmental uses. These soils, particularly those high in smectites, provide effective barriers against the movement of water and any fluid contaminants (Brady and Weil 2002). For instance, in a region with soils that drain easily, smectite can be placed at the bottom and sides of a pond so that it will hold water. Expansive clays are also used to seal sewage lagoons and industrial waste lagoons and even as the bottom barrier for landfills. While using expansive clays to prevent downward flow is common, the material also helps prevent upward movement of contaminants in water quality monitoring wells. Wetted smectite (bentonite) provides an impenetrable plug at the top of a bore hole. The plug keeps contaminants from flowing upward, forcing them into the sampling tube. Furthermore, it keeps contaminants from entering from above and contaminating the sampling tube (Brady and Weil 2002).

CONCLUSION

Expansive soils contain clay minerals that change volume, expanding when wet and shrinking as they dry. Certain types of clay mineral structures are more likely to adsorb water molecules than others. Clays that contain an octahedral sheet of crystals sandwiched between two tetrahedral sheets, or 2:1-type silicate clays, can be expansive or nonexpansive.

Expansive 2:1-type clays fall into the smectite or vermiculite group, the second of which is more common. Smectite clays adsorb water both externally and internally, called interlayer adsorption. Water between layers wedges the clay sheets apart, causing expansion. Vermiculite clays also adsorb water between layers, but it serves as a connecting bridge between layers, causing much less expansion.

Certain soil orders contain more expansive clays than others. In general, younger, less-weathered soils are more likely to contain expansive clays. In addition, these types of clays typically develop in flat to gently rolling regions in climates where grasses dominate. The soil orders commonly associated with expansive soils include aridisols, vertisols, mollisols, and alfisols. However, vertisols are characterized by their high shrink-swell capacity, so they are associated with geologic hazards related to expansive soils.

The geographic distribution of expansive soils relates to the distribution of the earlier-mentioned soil order in the United States and globally. While each U.S. state experiences varying levels of issues associated with expansive soils, certain states, such as those with more vertisols, experience greater impacts. Texas, for instance, has wide swaths of vertisols passing through major metropolitan areas, which wreaks havoc on buildings and infrastructure. Australia is home to the greatest spatial area of vertisols, covering a region larger than the state of Texas. Expansive soils can shrink substantially when they are dry and expand dramatically when wetted. During this process, the soils can exert enormous pressures on building foundations and roadways, for instance, causing extensive damage. Damage from expansive soils in the United States annually exceeds damage caused by flooding, tornadoes, hurricanes, and earthquakes combined, in a typical year. Unfortunately, while the cost to state and local governments, businesses, and homeowners is great, few people anticipate the cost since it's a slow process that can cause damage over time. Most homeowners' insurance policies do not cover damage caused by expansive soils.

Before beginning a building project, it's advisable to test the Atterberg Limits, Plasticity Index, or COLE, which are all measures of the capacity of clay to change volume due to moisture changes. With this information in hand, proper building and maintenance practices can be employed to mitigate or avoid damages caused by soil expansion or contraction. Many possible engineering approaches exist, both to address issues of existing buildings and to prepare a site before a construction project commences.

While expansive soils are estimated to cause billions of dollars in damage on an annual basis, there are other applications for the material. Since smectite clays become quite impermeable to fluids, they can be used to arrest the flow of water or contaminants. Smectite (bentonite) clay is often used to seal the bottom of ponds, as boundaries of landfills, in sewage lagoons, and more. Even though they are generally a nuisance in most cases, expansive soils do play a role in protecting environmental quality.

Trisha Jackson

Further Reading

Brady, N. C., and R. R. Weil. 2002. *The Nature and Properties of Soils*. Upper Saddle River, NJ: Prentice Hall.

CGS (Colorado Geological Survey). n.d. Swelling Soils. https://web.archive.org/web/20140801014146/coloradogeologicalsurvey.org/geologic-hazards/swelling-soils/, accessed May 26, 2019.

Jones, J. D., and W. G. Holtz. 1973. Expansive Soils—The Hidden Disaster. *Civil Engineering* 43(8): 49–51.

King, H. M. 2016. Expansive Soil and Expansive Clay. Geology.com. https://geology.com/articles/expansive-soil.shtml, accessed May 26, 2019.

Marcus, M. 2017. Expansive Soil Geologic Hazards—7 Things You Need to Know to Protect Your Investment. June 28, 2017. https://www.globest.com/sites/partnerESI/2017/06/28/expansive-soil-7-things-you-need-to-know-to-protect-your-investment/?slreturn=20190426090515, accessed May 26, 2019.

Mokhtari, M., and M. Dehghani. 2012. Swell-Shrink Behavior of Expansive Soils, Damage and Control. *Electronic Journal of Geotechnical Engineering* 17: 2673–82.

NRCS (Natural Resources Conservation Service). n.d. Soil Orders Map of the United States. https://www.nrcs.usda.gov/Internet/FSE_MEDIA/stelprdb1237749.pdf, accessed May 26, 2019.

Schaetzl, R., and S. Anderson. 2005. *Soils Genesis and Geomorphology.* New York: Cambridge University Press.

Srinivas, K., D.S.V. Prasad, and V.K.L. Rao. 2016. A Study on Improvement of Expansive Soil by Using CNS (Cohesive Non Swelling) Layer. *International Journal of Innovative Research in Technology* 3(3): 54–60.

Extinction

Extinction, a natural phenomenon, refers to the die-off or elimination of a species because of natural calamities. Scientists believe that up to 98 percent of all the plant and animal species that have ever lived on planet earth are now extinct. Most of these species disappeared during one of five times in the last 500 million years or so before the arrival of humans on the planet. These five different mass extinctions are often referred to as the Big 5; they were truly global events. These extinctions were caused by natural events like asteroid strikes, massive volcanic eruptions, global climate change, or a combination of these forces. Worth mentioning is that at onset, one destructive event leads to others. For example, a massive volcanic eruption triggers a sequence of events: it increases greenhouse gases in the atmosphere, which in turn causes global warming and ultimately climate change. Many land and sea species cannot cope with these changes and become extinct (Abbott 2008).

Some species become extinct because of biological reasons, such as reduced geographic area, low population size, competition, and predators or epidemic diseases. But biological reasons are not generally associated with mass extinctions: rather, these reasons lead indirectly to extinction. For example, predators do not have to eradicate a whole species. Instead, they can drive a population to such a low level that it has a high probability of extinction. Species extinction is an ongoing process that contributes to background extinction, which occurs at a fairly steady rate over geologic time. Background extinction is the result of normal evolutionary processes and affects only a few species in an ecosystem at any one time. Mass extinction is different from background extinction in the sense that the former involves a large number of species within a relatively short period of geologic time. Mass extinction is generally caused by two or more catastrophic global events that occur either simultaneously or closely spaced in time so that most species have no time to adapt. Among the five mass extinctions, volcanic activity seems to have wreaked the most havoc on earth's biota. It is implicated in at least four mass extinctions, while an asteroid is a suspect in just one. Past mass extinctions caused major changes in earth systems (e.g., ecology, atmosphere, surface, and waters) at rapid rates. These events also brought evolutionary changes as new species developed to take the place of those lost. Scientists maintain that the average "life span" of any species is about four million years (Abbott 2008).

At least five mass extinctions have been identified in the fossil record, coming at or toward the end of the Ordovician, Devonian, Permian, Triassic, and Cretaceous Periods. The Permian extinction, which took place 245 million years ago, is the largest known mass extinction in earth's history, resulting in the extinction of an estimated 90 percent of marine species. In the Cretaceous extinction, 65 million years ago, an estimated 75 percent of species, including dinosaurs, became extinct, possibly as the result of an asteroid colliding with the earth.

Biologists, ecologists, and geologists suspect that the earth is now in the midst of its sixth mass extinction of plants and animals. They believe that between 30 and 50 percent of all existing species will be extinct by the middle of the twenty-first century. Unlike past mass extinctions, the current crisis is almost entirely caused by human actions. In fact, 99 percent of currently threatened species are at risk because of loss of habitat, introduction of exotic species, and global warming. Although natural disasters, such as earthquakes, volcanic eruptions, floods, and hurricanes, cause local mortality of people and animals, the magnitude of such killing is almost insignificant compared to the powerful events that lead large numbers of species into extinction within geologically short lengths of time. For this reason, mass extinctions are often termed the biggest natural disasters (Abbott 2008).

Scientists believe that currently between 40 and 80 million species live in the world. This number represents less than 0.1 percent of the species in earth's history. This means that many more species that have ever lived on earth are now extinct. Biologically, a species is a group of individuals having common attributes and that are able to breed among themselves, but are not able to breed with members of another species. Correspondingly, members of a species share a common pool of genetic material. "A species is the smallest biologically real and distinct unit of individuals that share a common ancestor not shared with other organisms. Each species is unique; each new species is never entirely the same as any previous one" (Abbott 2008, 448).

EARLY AND CONTEMPORARY UNDERSTANDING OF MASS EXTINCTIONS

The French zoologist Georges Cuvier, founding father of paleontology, was probably the first scientist who in 1786 proved that mass extinction of species had occurred in geologically short lengths of time. He extensively studied fossils, evidence of extinct plants and animals from thousands of years prior, which he excavated. Scientists search for them in sedimentary rocks that are found across the continents. Cuvier reconstructed complete skeletons of unknown fossil quadrupeds and came to the conclusion that whole species of animals, such as giant salamanders and flying reptiles, had become extinct. These animals were far less similar to animals now living than those found in the more recent strata. Cuvier also noticed a profound change in the sedimentary rock record. Fossils of both animals and plants were found in abundance in overlying rock layers, but not in younger layers, which are deposited on top of old layers (Abbott 2008).

Cuvier was critical of theories of evolution, which involved the gradual transmutation of one form into another. He repeatedly emphasized that his extensive experience with fossil material indicated that one fossil form does not gradually change into a succeeding, distinct fossil form. His theory on extinction has met strong opposition from other notable natural scientists like Darwin and Charles Lyell. These scientists maintain that extinction was a sudden process, and thus animals and plants collectively undergo gradual change as a species. However, Cuvier's theory of extinction is widely accepted now by the scientific community. Proponents believe in mass extinctions that occurred in the last 500 million years or so, due in part to volcanic eruptions, asteroids, and rapid fluctuations in sea level and cooling and warming of seawater. During this vast period, many species fell and new species arose, precipitating the arrival of human beings.

Among the five waves of mass extinctions, the last one that occurred 65 million years ago led to the extinction of dinosaurs. This resulted from the combination of massive volcanic eruptions and the fall of a big asteroid. Scientists call the fifth mass extinction the K-T event after the German spellings of "Cretaceous" and "Tertiary." The K-T event had an enormous effect on life on earth. Geologically, this event marked the end of the Mesozoic Era. The evidence of the impact of K-T was first suggested by geologist Walter Alvarez and his father, a Nobel Prize winning physicist, Luis Alvarez, in 1980. They hypothesized that the mass extinction of the dinosaurs and many other living things during the Cretaceous–Paleogene extinction event was caused by the impact of a large asteroid on the earth. Evidence indicates that an asteroid of 6.2 miles (10 km) diameter plunged through the atmosphere and fell in the Yucatán Peninsula, at Chicxulub, Mexico, and created a large crater. The impact created by the asteroid, known as the Chicxulub Impact, caused an earthquake with a magnitude greater than 11 and tsunami 1.2–2 miles (2–3 km) high. The impact also sent rock materials and a plume of vaporized water into the stratosphere. Some of the gases and water vapor fell as acid rain, which destroyed many species in the oceans (Abbott 2008).

In 1980, Dutch paleontologist Jan Smit provided support for the hypothesis of asteroid impact posited by Alvarez. A 2016 borehole drilling project into the peak ring of the crater strongly supported the hypothesis, and researchers on the project reported that the peak ring comprised granite (a rock found deep within the earth) rather than typical sea floor rock, which had been shocked, melted, and ejected to the surface. Alvarez and colleagues found that rocks laid down previously at the K-T boundary contain extraordinary amounts of an iron-loving metal, iridium (a very hard, brittle, and silvery-white transitional metal of the platinum group), which is common in asteroids but very uncommon on earth. It is much rarer than gold, but iridium is twice as abundant as gold along the K-T boundary. Iridium is found in many places around the world and fell to the earth's surface from a cloud of debris that formed as an asteroid that struck Mexico. Thus, they firmly believe the iridium layer was created by the impact of a large asteroid with the earth (Cowen 1994).

It is worth noting that an asteroid strike is likely to cause acid rain, wildfires, tsunami, and a huge dust cloud. This dust blocks sunlight and creates weeks and months of dark winter, leading to cooled atmosphere and oceans, which are lethal to many species used to warmer temperatures. When the dust settles, voluminous gases may remain high up in the air, causing climate change via greenhouse gases and global warming.

PAST FIVE MASS EXTINCTIONS

Early life forms began to flourish during the Cambrian explosion, 540 million years ago. This event was accompanied by great diversification and abundance of life in the sea. Before the explosion, most organisms were simple, composed of individual cells occasionally organized into colonies. As the rate of diversification subsequently accelerated, the variety of life began to resemble that of today. From this time of diversification onward, the fossil record strengthened because many of the new groups of organisms began creating hard parts, such as shells, bones, and teeth, which preserve well as fossils. Two common requirements for organisms to become fossilized are possession of hard parts and rapid burial, which protects deceased organisms from being eaten by wild animals and birds such as vultures. Moving on through time, the earth has continued to fill with new forms of life, which has led to greater diversification. This diversification has been accompanied by setbacks and reorganization (Abbott 2008).

Extinctions of species are an ongoing fact of life on earth. This is because species are subjected to many physical, chemical, and biological changes both locally and globally. Oceans recede during glacier period and thus expose land. Climates change from hot to cold, and natural disasters occur from time to time. Some species can adjust to these environmental changes; others become extinct singly, or in groups.

First Mass Extinction

The first mass extinction occurred around 440 million years ago at the end of Ordovician Period, which lasted almost 45 million years. This period began 488.3 million years ago and ended 443.7 million years ago, and it is best known for its diverse marine invertebrates, including graptolites, trilobites, brachiopods, and conodonts. A typical marine community consisted of these animals, along with red and green algae, primitive fish, cephalopods, corals, crinoids, and gastropods. Ordovician extinction was a result of temperature fall (glaciation) that lowered sea level by at least 328 feet (100 m). Falling sea levels were possibly a result of the rise of the Appalachians mountain range. The newly exposed silicate rock sucked carbon dioxide out of the atmosphere, chilling the planet. The drop-off in sea level and cooling of the planet likely resulted in a major glaciation, which contributed to ecological disruption and mass extinctions of 60–70 percent of all marine invertebrate genera and 25 percent of all families. Then soon after, the ice melted, leaving the oceans starved of oxygen (Carrington 2017).

Second Mass Extinction

The second mass extinction wave took place during the late Devonian Period 364 million years ago. This extinction was caused again by temperature fall, which led to a messy prolonged climate change. This mass extinction again hit marine life, particularly in shallow seas, very hard and killed about 75 percent of species, including almost all corals. Scientists are divided over whether the late Devonian extinction was one single major event or a series spread over hundreds of thousands of years. Trilobites, which survived the Ordovician extinction due to their hard exoskeletons, were the most diverse and abundant of the animals that appeared in the Cambrian explosion 550 million years ago; however, they were nearly exterminated during the Devonian extinction. The likely culprit was the newly evolved giant land plants that emerged, covering the planet during the Devonian period. Their deep roots released nutrients into the oceans. The nutrient rich water resulted in massive algal blooms that decreased oxygen levels in the (deep) seas. Volcanic ash is also thought to be responsible for cooling earth's temperatures, which killed off the spiders and scorpion-type creatures that had made it onto land by this time (Worldatlas.com n.d.).

Third Mass Extinction

The third wave of extinction, at the end of the Permian, approximately 250 million years ago, is considered the largest mass extinction event in the history of the earth; it nearly ended life on this planet. Within less than 1 million years, around 95 percent of marine species and nearly 70 percent of terrestrial ones disappeared. On land, just one of four amphibian orders and only one of 50 reptile genera survived this extinction. Ancient coral species in the water were completely lost. Today's corals are an entirely different group. "The Great Dying," as this extinction is called, was probably caused by an enormous volcanic eruption in a large igneous province called the Siberia Traps. This eruption caused enormous amounts of lava flow on top of permafrost, creating immense sheets of igneous rock in the shallow crust. Direct heating of permafrost released tremendous amounts of methane that were frozen in hydrates. The total volume of eruptions covers an area the size of Western Europe and is nearly one mile deep. About two-thirds of this magma likely erupted prior to and during the period of mass extinction; the last third erupted in the 500,000 years following the end of the extinction event (Chu 2015).

The Siberian Traps eruptions caused a significant episode of global warming, filling the atmosphere with carbon dioxide, methane, and water vapor. Global temperatures surged while the oceans became acidic. During this geologic period, the formation of a super-continent (Pangaea) also affected climate and environment worldwide. The uniting of the continents eliminated most of the equatorial sea, but created a larger ocean basin, causing changes in ocean environment. Land environment also changed because the formation of a single super-continent reduced the shoreline. This means that greater percentages of land were located away from ocean water, which moderates the climate of landmasses. Thus the climate of the interior became arid, and thus lands were marketed by great deserts. Moreover, sea species were also affected because of the reduction in the extent of shallow water and because deep-ocean water became anoxic and carbon dioxide-rich. Meanwhile, the climate warmed because of flood-basalt volcanism. All of these factors set life back 300 million years (Abbott 2008).

Rocks after the Permian Period record no coral reefs or coal deposits. Moreover, life today has descended from the 4 percent of surviving species. In fact, after this extinction event, marine life experienced increased diversities and developed a complexity not seen before. Armored snails, urchins, crabs, and free-swimming predators, such as cephalopods and reptiles, emerged in abundant numbers (Worldatlas.com n.d.).

Fourth Mass Extinction

The fourth extinction occurred in the late Triassic Period between 199 million and 214 million years ago. Of all the mass extinctions, the one that ended the Triassic is the most ambiguous as no clear cause has been found. However, some scientists believe that this extinction resulted from a combination of events such as an asteroid impact, climate change, and huge outbursts of volcanism, which left the earth clear for dinosaurs to flourish. Scientists further claim that this extinction that precipitated occurred in several phases of species loss. During the beginning of the period, there were more mammals than dinosaurs. By the end of the Triassic period, their numbers were reduced, and thus the fourth wave of extinction allowed for the evolution of dinosaurs, which later existed for about 135 million years.

Fifth Mass Extinction

The fifth and last mass extinction occurred at the end of the Cretaceous Period, which brought on the extinction of nonavian dinosaurs 65 million years ago. Dinosaur refers to an especially large group of animals of different sizes that lived in vast areas of the earth over a long period beginning more than 230 million years ago. There are two types of dinosaurs: avian and nonavian, and they differ in several respects. One of the primary differences is that nonavian dinosaurs completely disappeared after the fifth mass extinction, while avian dinosaurs mostly survived and evolved into modern-day birds. They also differed with respect to bone structure and metabolic processes. Although both types of dinosaur had some similarities, they were vastly different. Avian dinosaurs had hollow bones while nonavian dinosaurs had dense bones. The nonavian dinosaur bones evolved to support their tremendous weight for which reason they could not fly. In contrast, light weight avian dinosaurs could fly, and therefore, their bones developed to allow them to fly.

They were also different in terms of metabolic processes. The nonavian dinosaurs were cold-blooded while the avian dinosaurs were warm-blooded. Interestingly, several theories exist to explain why avian dinosaurs were able to survive the last mass extinction. One is that avian dinosaurs had comparatively larger brains than nonavian dinosaurs, while another theory claims that avian dinosaurs were able to survive due to their diet. However, along with nonavian dinosaurs, 70 percent of all species then living on the earth disappeared within a very short period (Kiprop 2019). Most of the large and some of the smallest creatures of that time were casualties of the K-T event (Cowen 1994). The surviving species of this event were few in number, but numerous new species originated in the following millions of years.

Paleontologists speculated for many years about what could have caused this "mass extinction," particularly of dinosaurs. They claim that a combination of massive volcanic eruptions, a huge asteroid impact, and climate change were responsible for the last mass extinction. Scientists believe that a colossal eruption occurred in west-central India close to the plate boundary between India and Africa, which is known as the Deccan Traps Flood Basalt, at about the time of the fifth extinction. From this eruption, lava flowed on the other side of the plate boundary, much of which is now covered by the Indian Ocean. The Deccan Traps province, which lies across the K-T boundary, now covers an area of about 200,000 square miles (500,000 square km). Originally, it formed an area of 800,000 square miles (2.5 million square km). Its size has reduced because of erosion. Scientists maintain that the peak eruptions may have lasted only about one million years, beginning about 65.5 million years ago (Cowen 1994).

Immediately prior to the eruption, a large asteroid created the Chicxulub crater, which is an impact crater buried underneath the Yucatán Peninsula in Mexico. Its center is located near the town of Chicxulub, after which the crater is named. The crater is 120 miles (193 km) in diameter, 12 miles (20 km) deep, and about 6 miles (10 km) wide. The Chicxulub Impact event generated temperatures in excess of 10,000 degrees, which caused enormous fires, and indeed, the impact generated extremely high pressures and physically altered the form of the mineral quartz. The thermal pulse generated by the impact would have lasted for several minutes, and it would have been lethal for nearby life.

The Chicxulub Impact event produced a shock wave and air blast that radiated across the seas, over coastlines, and went deep into the continental interior. As a consequence, sea shoals suffered from tsunami and acid rains, the seabed was covered with enormous amounts of organic matter, and the climate worldwide was disrupted. Recent research suggests that the asteroid impact may have caused an increase in the intensity of Deccan volcanic activity. Scientists now believe that this asteroid carried iridium, and a number of other rare elements such as platinum, osmium, ruthenium, rhodium, and palladium, down into earth's core, along with much iron, when earth was largely molten. These materials were spread worldwide by the impact blast as the asteroid vaporized into a fireball and subsequently enhanced the clay layer worldwide. In fact, iridium has been identified in more than 100 places all over the world (Cowen 1994).

However, during the Cretaceous period, dinosaurs ruled the land, and squid-like ammonites ruled the oceans. But volcanic activity and a large asteroid-induced climate change had already placed the ammonites under stress. Thus, the asteroid impact that ended the dinosaurs' reign provided the final blow such that only a few dwindling species of ammonites survived. For instance, today the ammonites' oldest surviving relative is the nautilus. This extinction period ended 76 percent of life on earth and allowed the evolution of mammals on land and sharks in the oceans.

POSSIBLE SIXTH MASS EXTINCTION

Many scientists believe that the sixth mass extinction is imminent. Currently, the world is in the Holocene epoch—the time since the end of the last major glacial epoch, or "ice age." Thus, in general, the Holocene has been a relatively warm period in between two ice ages. Humanity has greatly influenced the Holocene environment by encroaching on and sometimes wiping out the habitats of other species primarily for three interrelated reasons: overpopulation, overconsumption, and overexploitation of natural resources. Besides the fact that all organisms influence their environments to some degree, few have ever changed the globe as much as, or as fast as, the current human species is doing. The vast majority of scientists agree that human activity, particularly burning of fossil fuels, is responsible for "global warming," an observed increase in mean global temperatures that is still going on. Additionally, habitat destruction, pollution, and other factors are causing an ongoing mass extinction of plant and animal species. It is no surprise then that extinction rates have risen substantially in the last 500 years. Available data suggest that 80 percent of land mammals, for instance, have lost their range in the last century. Scientists also maintain that at least 20 percent of all plant and animal species on earth will be extinct within the next 25–30 years. Thus, earth is currently losing species at a rate far higher than normal background extinction rates. This will lead to gradual loss of biodiversity similar to that seen during mass extinctions. Biologists predict that unless humans change course and begin to preserve more species, the earth's sixth mass extinction is inevitable.

Ecologists estimate that the present-day extinction rate is 1,000–10,000 times that of background extinction. They also claim that currently up to 70 plant and animal species become extinct every day. If this rate continues, it will take only 16,000 years for the disappearance of 96 percent of the existing flora and fauna of the world. The principal reason for the oncoming sixth mass extinction is destruction of ecological landscapes of plants and animal primarily by deforestation, overhunting, toxic pollution, invasion of alien species, and climate. About 1 percent of tropical rainforests disappear annually. At present, one-tenth of coral reefs perishes due to global climate change, reef fish catching, water contamination and warming, and hurricanes; about 30 percent more will be ruined in the next few decades. Paleontologists think that the species life span for contemporary mammals and birds has decreased up to 10,000 years, that is, it has become 100–1,000 times shorter than that of fossil forms. "If the habitat continues to be destroyed at the same pace, the life span of these species will soon make only 200–400 years. There are no such estimates for the invertebrates, but they are undoubtedly affected both by the global environment and climate change, and by disappearance of local biotopes" (Informnauka Agency 2004).

While modern advanced technology cannot ably deflect an unforeseen asteroid strike or put a plug in a volcanic eruption, human activity could significantly reduce the current extinction rate. For example, people could use alternative energy sources and reduce deforestation by changes in diet, which would help reduce emissions of carbon dioxide in the atmosphere and thus prevent global warming to a greater extent. Furthermore, people could reduce the extinction rate through policy changes to increase conservation efforts and curb production of greenhouse gases to slow climate change. People could become educated about the chain of events that led to past mass extinctions by studying the earth's history and by learning what traits make a species particularly vulnerable during a mass extinction. This would help people to be in a better position to break that chain today.

COMPLEXITY OF CAUSES OF MASS EXTINCTIONS

The forgoing discussion reveals causes of mass extinctions. In general, extinction occurs because of environmental forces, such as habitat loss and fragmentation, climate change, natural disaster, overexploitation of species for human use, and pollution, or because of evolutionary changes in their members, such as genetic inbreeding, poor reproduction, and decline in population numbers. However, some causes of extinction are complex and vary among events and species spatially. For example, during the last 100,000 years of the Pleistocene Epoch, some 40 percent of the existing genera of large mammals in Africa and more than 70 percent in North America, South America, and Australia became extinct (Gittleman 2019). This Epoch began about 2.6 million years ago and lasted until about 11,700 years ago. The most recent Ice Age occurred then, as glaciers covered huge parts of the earth.

In the context of causes of mass extinctions, several points need to be emphasized. First, not all massive asteroid impacts and powerful volcanic eruptions trigger mass extinctions. For example, the Manicouagan crater in central Quebec, Canada, is several miles wide and was formed following the impact of an asteroid with a diameter of three miles (five km) that struck the earth one and a half million years ago. The crater was originally about 62 miles (100 km) wide, although erosion and deposition of sediments have since reduced its size. But the fossil record indicates no major dip in diversity associated with this event. Similarly, the Karoo-Ferrar volcanic province, also called the Karoo and Ferrar Large Igneous Provinces (LIPs), covering what is now South Africa and Antarctica, indicates extensive volcanic activity around 180 million years ago. However, despite large-scale disruption, only a small rise in extinction rates occurred during that time period according to the fossil record.

Second, a particular catastrophic event can cause other events to unfold in a chain reaction, and the combined impacts trigger mass extinctions. For example, the asteroid that struck the earth at the end of the Cretaceous Period was responsible for extinction of dinosaurs. The catastrophic collision raised a dust cloud that obscured the Sun and lowered temperatures dramatically, resulting in environmental changes. Fallout from the impact covered the entire world with iridium and thus provided a key pointer to the impact event. The presence of this material at the K-T boundary has convinced many scientists that asteroidal impact was one of the principal reasons for the fifth wave of mass extinction. It also occurred because the collision happened to hit carbon-rich rocks, which probably led to ocean acidification, and hence the disruption of reef formation and the oceanic food web. However, the asteroid that caused the Manicouagan crater did not hit carbon-rich rocks and so did not set off this chain reaction or caused such a significant disruption of the earth's systems.

Third, mass extinctions seem to occur when multiple changes disrupt the earth's systems more rapidly than organisms can evolve and ecological connections can adjust. Although the Karoo-Ferrar volcanic activity was so large that it disrupted the earth's atmosphere and oceans, in this case, the changes came about very slowly. The volcanic activity was spread over millions of years. For comparison, the volcanic activity that may have caused the K-T event most likely occurred in less than 100,000 years, leaving no time for evolution to take place as habitats changed and widespread extinction ensued.

Fourth, mass extinctions may be caused by indirect impacts of events that disrupt the earth's system. For example, the Siberian Traps eruptions caused the end of the Permian mass extinction, probably occurring not through the eruption itself, but through secondary effects such as toxic gases that poisoned animals and plants and contributed to acid rain. Also, the rocks likely came in contact with lava that contained organic compounds and released large amounts of greenhouse and toxic gases. This leads to long-term global warming and hence widespread climate change, which may directly lead to the extinction of sensitive species and prompt others to shift their ranges, upsetting ecosystem dynamics and triggering additional extinctions. When land plants die, this increases erosion and damages delicate marine environments as sediments are carried into the ocean. Global warming also has the potential to reduce water circulation in the ocean, which ultimately could lead to drastically lower oxygen levels in the oceans. This, in turn, threatens some species and could severely interrupt the flow of nutrients through the marine food web (Understanding Evolution n.d.).

CONCLUSION

Geologically speaking, mass extinctions are rare and uncommon events, and not necessarily the worst thing to happen to the earth. Moreover, like many natural disasters, mass extinctions have beneficial impacts on the environment. For example, worldwide elimination of numerous old species provides new opportunities for different organisms to evolve new ways of life. "The constant elimination of old species and refilling of their vacated spaces in the environment by new species has created an incredible variety of life-forms during earth's history, increasing diversity fourfold since Cambrian time" (Abbott 2008, 448). Species diversity leads to biodiversity and ensures ecosystem resilience, giving ecological communities the scope they need to withstand stress. However, because of rapid population growth and remarkable scientific advancements, human activities are increasingly changing at an accelerated rate, creating prospects for losing species. Because every species' extinction potentially leads to the extinction of others bound to that species in a complex ecological web, the numbers of extinctions are likely to snowball in the coming decades as ecosystems unravel. Therefore, a comprehensive strategy for saving biodiversity is critical for attaining sustainable development of natural resources and conserving existing genetic diversity.

Bimal Kanti Paul

Further Reading

Abbott, P. L. 2008. *Natural Disasters*. Boston: McGraw Hill Higher Education.

Carrington, D. 2017. Earth's Sixth Mass Extinction Event Under Way, Scientists Warn. *The Guardian.* July 10, 2017. https://www.theguardian.com/environment/2017/jul/10/earthssixth-mass-extinction-event-already-underway-scientists-warn, accessed 25, 2019.

Chu, J. 2015. Siberian Traps Likely Culprit for End-Permian Extinction: New Study Finds Massive Eruptions Likely Triggered Mass Extinction. September 16, 2015. http://news.mit.edu/2015/siberian-traps-end-permian-extinction-0916, accessed September 26, 2019.

Cowen, R. 1994. *History of Life*. Cambridge, MA: Blackwell Scientific Publications.

Gittleman, J. L. 2019. Extinction: Biology. September 13, 2019. https://www.britannica.com/science/extinction-biology, accessed September 27, 2019.

Informnauka Agency. 2004. The Sixth Wave of Extinction. August 23, 2004. https://www.sciencedaily.com/releases/2004/08/040816001443.htm, accessed September 22, 2019.

Kiprop, J. 2019. What Are the Differences between Avian and Non-avian Dinosaurs? Dinosaurs Can Be Classified into Avian and Non-avian Types. May 23, 2019. https://www.worldatlas.com/articles/what-are-the-differences-between-avian-and-non-avian-dinosaurs.html, accessed September 23, 2019.

Understanding Evolution. n.d. Volcanic Activity and Mass Extinction. https://evolution.berkeley.edu/evolibrary/article/0_0_0/massextinct_09, accessed September 27, 2019.

Worldatlas.com. n.d. Timeline of Mass Extinction Events on Earth. https://www.worldatlas.com/articles/the-timeline-of-the-mass-extinction-events-on-earth.html, accessed September 22, 2019.

Floods

Floods are the most common natural disasters and affect nearly half the globe. They are the costliest natural disasters in the United States. Although there are many definitions of floods, most are restrictive in nature because they refer only to overflow of major rivers that spread water onto the floodplains (Smith and Ward 1998). Such definitions exclude floods occurring beyond floodplain areas. However, floods also occur in coastal areas due to daily tidal activity, storm surges, and tsunami waves. In fact, in Britain, coastal flooding dominates British flood history (Smith and Ward 1998). Thus, the flood is a hazard that causes water to overflow and submerge land that is normally dry. This resonates with Ward's definition, "A flood is a body of water which rises to overflow land which is not normally submerged" (cited in Smith and Ward 1998, 8). The Center for Research on the Epidemiology of Disaster (CRED) defines a flood as a significant rise of water level in a stream and includes lakes, reservoirs, and coastal regions (Jonkman and Kelman 2005).

The Federal Emergency Management Authority (FEMA) in the United States considers a flood to be a general or temporary condition where two or more acres of normally dry land are partially or completely inundated. The inundation can be derived from overflow of inland or tidal waters, unusual and rapid accumulation of runoff surface waters, or the collapse or subsidence of land along the shores of a water body (FEMA 2017). This definition is not applicable to all flood-prone countries of the world. For example, during the summer about one-third of land area in Bangladesh remains under water each year, which is considered a normal, beneficial, and welcome event. The event becomes a damaging phenomenon if it inundates more than 33 percent of Bangladesh.

TYPES AND CAUSES OF FLOODING

The causes of floods are numerous, multifaceted, and interrelated. These causes are differentiated not only by types of floods but also from one country to another. Although there are many types of floods, for convenience of analysis, here floods are broadly divided into three categories: river floods, flash floods, and coastal floods.

River Floods

The main source of flooding for this type of event is the bank overflow from rivers, which is generally caused by heavy rainfall in major river basins. If these basins expand over more than one country, excessive rainfall in the upstream basins can also cause floods in lower riparian areas. For example, the principal cause of most devastating floods in Bangladesh, a country known for flood events, is heavy rainfall in the drainage basins of the Ganges, Brahmaputra, and Meghna (GBM) rivers. The area encompassed by the GBM basins is disproportionately large compared to the relatively small area of Bangladesh through which they flow, eventually emptying into the Bay of Bengal. The combined drainage of the GBM rivers is 0.6 million square miles (1.6 million square km), only 8 percent of which lies in Bangladesh (Rasid and Paul 1987).

In North America, the immediate cause of floods is exceptionally heavy rainfall during the preceding months. Causes of heavy rainfall result from a series of severe summer storms that track across the upper Midwest. This is associated with the location of the jet stream across the Midwest instead of its usual summer location northward into Canada. River floods can also occur in North America when ice or floating debris causes a jam. Meanwhile, spring floods in North America are generally caused by snow melting in the mountains.

In South Asia, river or fluvial floods, which are also called monsoon floods, result from heavy rainfall in the Ganga-Brahmaputra-Meghna river basins and melting of snow in the Himalayas. A critical factor related to a flood occurrence is the coincidence of monsoon peak discharges in the major rivers with excessive summer rainfall within a specific country like Bangladesh and/or in the GBM basins located in other countries of the region. When intensive rainfall occurs simultaneously over the entire catchment area, the combined runoff from the tributary systems creates a tremendous volume of water draining into the Bay of Bengal. Consequently, river systems of South Asia cannot efficiently discharge the flow, which causes the riverbank overflow and flooding of land adjacent to the river. This type of flood can extend varying distances from the rivers.

Several hydraulic and topographic factors exacerbate the problem of discharging the enormous volume of floodwater rushing from upstream of the GBM basins. For instance, the very low gradient of almost all major rivers and the loss of channel capacity due to siltation results in rising water topping riverbanks and submerging the vast floodplains. Moreover, the inadequate dredging of riverbeds and the disruption of existing drainage systems by new road construction without adequate culverts and the indiscriminate building of houses has further contributed not only to the frequency of high floods in recent decades but also to prolonged flood duration. Unplanned urbanization, rapid population growth, and conversion of agricultural land to other uses have also caused floods.

Floods also occur because of construction of embankments, dikes, and other flood-related engineering schemes along the upstream reaches of large rivers. Most rivers (e.g., the Brahmaputra, the Danube, the Indus, the Limpopo, the Mekong, and the Rhine) associated with damaging flooding flow over several countries. The preceding construction devices reduce storage capacity of such rivers and thus induce higher flood peaks and faster rise in downstream water levels. For example, the government of Bangladesh, the media, and some flood experts have charged that opening of the Farakka barrage at the height of the monsoon is the prime reason for damaging floods in Bangladesh, particularly the flood of 1988. This barrage was constructed by India in 1975 on the Ganges River near the India-Bangladesh border.

Flash Floods

The flash flood is a subset of devastating floods caused by intense rainfall in a relatively short period of time. Flash floods are short-duration, localized, extreme events and occur suddenly, usually during the night, in a broad range of geographical (i.e., topography, soil conditions, and ground cover) and climatological conditions (i.e., intense localized thunderstorm activity). They are common mainly in desert and mountainous regions. Areas prone to this type of natural disaster are steep canyons, urbanized areas, arid and semiarid regions, and the small river courses that cannot handle large amounts of rainfall in short time periods. In addition to excessive rainfall, flash floods are also caused by a dam or levee failure, a sudden release of water held by an ice jam, and slow moving thunderstorms (American Red Cross 1992).

Flash floods are generally small in scale with regard to area of impact and population affected, but they are generally violent, resulting in a high threat to life and severe damage to property and infrastructure. For instance, a flash flood in dry and mountainous northern Iran killed nearly 2,000 pilgrims in 1954. Moreover, most flood deaths and damage in the United States are due to flash floods. A well-known and deadly flash flood occurred in the Big Thompson Canyon, Colorado, in 1976, killing 140 people. Flash floods can move boulders, tear out trees, and destroy bridges and buildings; also, they can trigger catastrophic mud slides in mountainous areas. Fast-moving water associated with flash flooding can even float heavy vehicles. Because of diminished vegetative cover and heavy sediment erosion, flash flooding in arid and semiarid regions can become severe. Also, because flash floods occur suddenly, they give little or no warning, which makes potential danger much greater.

Coastal Floods

Storm surges due to tropical cyclones/hurricanes and tsunami waves cause flooding especially in low-lying coastal areas of countries prone to such extreme events. Thus, not all coastal areas of a given country are subject to this type of flood. For example, the Gulf Coast of the United States often experiences flooding due to storm surges, but not the California coast. The reverse is true for floods caused by tsunamis. While tsunami-induced floods can occur any time of year, cyclones have a specific seasonal dimension. For example, hurricane seasons in the United States last from June through November. In South and Southeast Asian countries, there are two distinctive cyclone seasons: April to June and September to November. In such events, storm surge water or tsunami can flood a significant portion of coastal areas, differing in amount and intensity from one country to another. In fact, the overwhelming majority of deaths during cyclones/hurricanes are caused by storm surge flooding. In addition to storm surge and tsunami, coastal areas adjacent to estuaries generally flood twice a day due to astronomical tide, which is called tidal or estuarine flooding. The elevation of coastal areas determines the extent of inland flooding from the diurnal tides. All three subtypes of coastal flooding bring saline water inland and cause considerable damage to field crops.

In addition to the preceding three types of floods, some also consider rainfall floods, which are caused by high intensity and long duration rainfall in a river basin, affecting floodplains away from the main river. They are localized events as the local rivers cannot drain water quickly. In such locations, construction of embankment without adequate provision for unimpeded drainage of rainwater also causes this type of flooding during heavy rainfall. Rainfall floods also occur in arid and semiarid regions. Moreover, intense rainfall often causes drainage congestion in large urban areas, particularly in developing countries. Congestion is also caused because urban areas are largely paved. Thus, if urban centers are located near large rivers or coastal areas, such centers are liable to river and coastal flooding.

River, flash, and coastal floods are together called natural floods. Some floods are also a result of embankment/levee and dam failure and are called human-made floods. For example, at least 1,000 of the 1,576 levees were either breached or overtopped during the 1993 Great Midwestern Floods in the United States, and the flooding of New Orleans in 2005 was caused by levee failures. Conversely, a heavy rainfall-induced flood was the cause of an earthen dam collapsing in India in 1979, killing about 5,000 people. However, construction of dams often protects people who live nearby flood events, but they intensify flooding downstream.

Environmental degradation through deforestation and overgrazing also causes floods. These human activities intensify soil erosion, which ultimately causes rapid deposition of sediments on the riverbed, reducing the carrying capacity of floodwater. For example, environmental degradation in the form of deforestation in Nepal and the Indian state of Assam is often cited as the cause of devastating floods in Bangladesh. Such action results in rapid siltation of riverbeds in Bangladesh and is believed to increase the frequency of severe flooding in the country. Scientists also think that global warming will intensify frequency and intensity of future flooding across the globe.

PHYSICAL DIMENSIONS OF FLOODS

Like other natural disasters, severity of a given flood, to a large extent, depends on its physical characteristics such as magnitude, frequency, spatial extent, and duration. Flood magnitude is defined in several ways. In the United States, flood stage is widely used to measure flood magnitude. It is the elevation at which water overflows the natural banks of a river or other body of water in a given portion of the body of water, inundates adjacent areas, and damages flooded areas. Although the flood stage is different for different reaches of a water body, it is generally higher than or equal to bank-full stage. Flood magnitude is also often indirectly defined in terms of the number of people killed, the reported percentage of people affected, or estimated damage in dollar value for a particular event.

Flood frequency is defined as the recurrence interval (or return period), which is the time (in years) that elapses between two events of similar size. For example, a flood may be described as a 100-year event, meaning a flood of similar magnitude is expected to occur on average once every 100 years. The other two characteristics are spatial extent and duration. The former refers to the size of area inundated by floodwater, which differs from country to country. As noted, in a typical year, about one-third land of Bangladesh is inundated by floodwater. In contrast, the 1993 Great Midwestern Flood of the United States, which was the most devastating floods on record, inundated less than 1 percent of land in the country. Duration of flood means how long a flood persists in terms of days. This characteristic differs by flood types. Flash floods last from a few hours to a few days, while river and rain floods often last more than two months. As indicated, the extent of damage caused by floods does not depend on a single physical characteristic, but rather on a combination of several characteristics.

GEOGRAPHIC DISTRIBUTION

Floods are one of the most widespread natural disasters, accounting for 42 percent of all natural disasters that occurred between 1996 and 2015 (Table 1). This is because of the vast geographic distribution of floodplains and low-lying coastal areas across the globe. In terms of frequency, Asia was the most flood-affected continent during the preceding period, while Oceania was the least affected continent. However, care should be exercised in interpreting flood frequency in the global context. Among the continents, Asia has the largest and Oceania the lowest land area, and hence they are ranked the highest and the lowest, respectively, when flood frequency is considered for the preceding period.

Another reason for the highest frequency of flood events in Asia is that the continent has a large number of rivers historically known for devastating flood events. For example, the Huang Ho or Yellow River of China has flooded more than 1,000 times in the last 2,000 years; many of the floods were catastrophic, and for this reason, the river is referred to as "China's Sorrow." Over the centuries, this river has killed more people than any other river in the world. Deadly floods caused by the Huang Ho River occurred in 1887, 1931, and 1938, and these floods together killed nearly five million people. The Huang Ho is also called the Yellow River because its muddy water looks yellow from the air. The Yangtze, another Chinese river, is also known for destructive floods. Among the most recent major floods are those of 1870, 1931, 1954, 1998, and 2010.

Although floods occur on all continents, not all countries are equally prone to these events. For instance, among all countries of Asia, the most flood-prone country is China followed by India (15), Japan (7), Bangladesh (6), Pakistan (6), and South Korea (5). Since 1900, China has experienced 22 deadly floods that caused the deaths of at least 51 people. Other Asian countries have experienced at least one deadly flood during the preceding period, including Afghanistan (4),

Table 1 Frequency and impacts of floods by continents, 1996–2015

Continent	No. of Event (%)	No. of People Killed (%)	No. of People Affected (%)[a]	Total Damage in Dollars (%)[b]
Africa	728 (24)	16163 (11)	55821 (3)	6101 (1)
Americas	629 (21)	44214 (30)	33839 (3)	83897 (15)
Asia	1153 (38)	84186 (57)	2002174 (94)	341917 (62)
Europe	435 (14)	2485 (2)	9115 (<1)	103607 (19)
Oceania	84 (3)	216 (0.5)	992 (<1)	15615 (3)
Total[c]	3029 (42)	147264 (10)	2123940 (48)	551137 (26)
Grand total (all disasters)	7170	1533341	4417124	2115573

[a]In thousands.
[b]In millions of dollars. Damage from the 1996–2005 period is converted to 2005 prices, while for the 2006–2015 period it is converted to 2015 prices.
[c]Out of grand total.
Sources: IFRC (2006 and 2016).

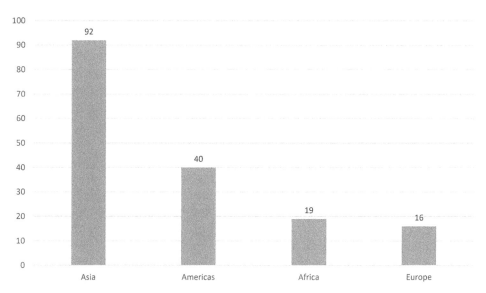

Number of deadly floods that killed at least 51 people, by continents, 1900–2017. (Compiled from EM-DAT)

North Korea (4), the Philippines (4), Iran (2), Indonesia (2), Nepal (2), Vietnam (2), Turkey (2), Lebanon (1), Jordan (1), Malaysia (1) , Saudi Arabia (1), Sri Lanka (1), Thailand (1), Yemen (1), Tajikistan (1), and Taiwan (1). In the Americas, the most flood-prone country is the United States, which has experienced 10 deadly floods since 1900 followed by Brazil (9), Peru (6), Colombia (3), Argentina (2), and Ecuador (2). Chile, Dominican Republic, El Salvador, Guatemala, Haiti, Mexico, Puerto Rico, and Venezuela each have experienced only one deadly flood.

Also, since 1900, the third-most deadly flood frequency has been in Africa followed by Europe. With four floods, South Africa is most prone to deadly floods. Meanwhile, Algeria, Kenya, and Nigeria, each have experienced two deadly floods, while Burkina Faso, Ethiopia, Ghana, Malawi, Mozambique, Morocco, Sudan, Somalia, and Tunisia each have experienced one deadly flood. Finally, the fewest deadly flood events have occurred in Europe, and none has occurred in Oceania. With three events, Italy and Spain are the European countries with the deadliest flooding followed by the United Kingdom and Russia, each having experienced two events. Other European countries that have experienced one deadly event are Belgium, Bosnia, Croatia, Czech Republic, France, Germany, the Netherlands, Portugal, Poland, and Servia. The number of deadly flood events is somewhat consistent with the number and percentage of flood events reported by the continent.

FLOOD IMPACTS

Direct impact of floods includes deaths, injuries, displacement, and destruction/damage to crops, property, businesses, livelihoods, environment, and critical infrastructure. Additionally, water-borne diseases emerging during and after floodwaters recede and chemical pollution are some of the indirect health impacts of floods. Worldwide, floods were responsible for nearly 10 percent of all deaths resulting from natural disasters during the 1996–2015 period. This means of the 1.5 million deaths, 147,264 were caused by floods. Flash floods, which are relatively more rapid onset disasters than river floods, have the highest average mortality rates per event and account for the majority of flood deaths in developed countries (Jonkman and Kelman 2005). River floods, on the other hand, are less likely to cause deaths because of sufficient time for warning and evacuation.

The immediate causes of flood death differ between developed and developing countries. Two-thirds of flood deaths in developed countries are caused by (vehicle-related) drowning, and one-third are from physical trauma, heart attack, electrocution, carbon monoxide poisoning, or fire. Moreover, drowning is a more significant problem in the United States than in Europe. The majority of flood deaths in developing countries are caused by diarrhea and other water-borne diseases, drowning, and snake bites. People in developing countries who die from drowning are usually children and the elderly. They also either die or suffer from infectious diseases like diarrhea, cholera, malaria, and respiratory ailments (e.g., cough, sneezing, wheezing, and sore throat) after floods recede. These diseases are caused by lack of nutrition and pure drinking water, poor hygiene, and deterioration of sewage and sanitation facilities. Also, like children and the elderly, the disabled are at greater risk of death due to flooding.

Deaths from floods, particularly from river floods, are influenced by several physical characteristics of the event, such as its depth (magnitude), spatial extent, duration, and frequency. Thus, the number of deaths is likely to increase with increasing flood depth, extent, and duration. However, these characteristics work in combination. For example, a flood that shallowly inundates a large area or a flood of short duration with low magnitude may not cause much death. In contrast, when flood depth and duration are substantial, deaths can increase significantly. Flood deaths are also influenced by the socioeconomic characteristics of individuals or households. Accordingly, poor people are at the highest risk of deaths from floods, and Jonkman and Kelman (2006) reported that the death rate from floods is higher for men than women in the United States and Europe.

Besides demographic characteristics, national characteristics such as political system, type of governance and government, level of economic development, rate of urbanization, population density, land area, and population size are important in explaining fatalities caused by floods. Several studies found that developed countries experience fewer deaths from floods than developing countries. This is primarily because governments of latter countries spend little or nothing on disaster mitigation and preparedness measures.

The last three columns of Table 1 present three selected impacts of flood disasters. In terms of number and percentage of people killed and affected by floods, Asia ranks at the top. Over the study periods, floods proved to be the costliest type of disaster for Asia, accounting for 62 percent of total reported damage caused by all natural disasters. Although floods caused 10 percent of all fatalities during the study period, flood events accounted for nearly half of the natural disaster victims.

Apart from human fatalities, floods also cause deaths of livestock, which can have serious consequences for the economy of rural households given the importance of these household assets and their role in the farming system of rural areas, particularly in developing countries. However, unlike many other natural disasters, flooding causes few injuries. Except flooding, the number of injuries caused by other natural disasters is reported to be higher than the number of deaths. For example, a death-to-injury ratio was 1:3 for Hurricane Katrina while Doocy et al. (2013) reported a death-to-injury ratio of 1:0.67 for flood events that occurred between 1980 and 2009.

Types of floods other than flash floods generally inundate large rural areas. Therefore, crop damage is the main economic loss in flood-prone countries. Additionally, flooding can cause isolation and disruption of everyday life due to closures of roads, bridges, schools, and businesses and due to destruction of dwellings. Both destruction and disruption can have severe effects on cash flows in flood-affected areas. Loss of income due to inability to work makes rural peoples' survival difficult. In addition to damaging roads, floods also damage sewage plants, water supply plants and lines, telecommunications, electricity, and gas installations. Such damage not only creates inconvenience but also poses serious threat to life, particularly from disease and electrocution.

Limited access to clean water and improper waste disposal, in particular, increase the incidence of water borne diseases. In the Great Midwestern Flood of 1993, many sewage treatment plants were flooded and forced to release untreated sewage into the floodwaters, causing serious health concerns for the flood-affected people (Coch 1995). Pools of standing floodwater also make an ideal breeding ground for mosquitos, and thus, malaria incidence increases during flooding. Destruction of public infrastructure also contributes to poor access to proper health care services at a time when the people need them the most. Likely as a result, both domestic and international migrations are believed to increase during floods and their immediate aftermath.

Available information reveals that almost all flood-prone countries by now have introduced flood forecasting and warning systems. Flood evacuation is also practiced in many countries so that people trapped by floodwater generally evacuate to flood shelters established on higher ground. Compared to other natural disasters, floods generate much small debris, which is generally removed by members of affected households. Type of debris generated by floods is also different from that of other natural disasters. Floods generate mud, sediment, appliances, tires, and hazardous waste. However, because of the typically relatively small volume, clearing flood debris usually takes little time, perhaps up to a week or so after floodwater recedes.

BENEFITS OF FLOODS

Floods produce a wide range of impacts, which are both positive and negative. Unfortunately, disaster literature emphasizes the negative impacts of any extreme event primarily because positive impacts are less understood and are more difficult to assess than the negative impacts. Moreover, the mass media also contributes in creating negative impressions of the impacts of natural disasters (Smith and Ward 1998). Therefore, little is known about any benefits of natural disasters. In the Western world, a flood is generally viewed negatively. But floods have beneficial ecological consequences. For instance, floodwater recharges groundwater and deposits fertile sediments on agricultural lands and thus increases soil fertility and crop production. These benefits tend to become apparent months or years after such events. For this reason, disaster researchers maintain that extreme natural events should be understood as ambivalent with respect to their effects on nature.

For the beneficial impacts, the people of Bangladesh, for example, perceive some floods as a resource. A normal flood, locally called *barsha*, resulting from predictable monsoon rainfall, is considered beneficial because it makes the land productive by providing necessary moisture and fresh fertile silt to the soil. For example, since standing water is required to grow rice, expected floods replace the need for artificial irrigation, which is time consuming and costly to build. Moreover, floodwater brings plenty of fish, which constitutes the main source of protein for many poor people. Millions of Bangladeshis depend on fisheries in the floodplains for their livelihood, and the life cycle of some species of fish depends on migration between the rivers and the seasonal floodplains.

Conversely, an abnormal flood (bonna), which occurs once every few years in Bangladesh, is regarded by its citizens as a damaging phenomenon. However, empirical evidence suggests that areas affected by abnormal floods provide bumper crop production in the following dry seasons. Such an increase in production is due to the availability of extra moisture in soil following a high flood and deposits of fertile silt on floodplains. This type of situation was observed in parts of Indonesia affected by the 2004 Indian Ocean Tsunami, which brought saline water from the ocean to coastal areas of the country. Farmers feared that soil contaminated with saline water would negatively impact agricultural production the following year. Fortunately, rains flushed out the salinity, and farmers received record yields of crops due to the nutritional value of silt deposited by the saline water. In some countries like Pakistan, salt is deposited on agricultural fields due to high rates of evaporation whereupon, during flood season, floodwaters prevent the land from becoming infertile.

FLOOD PREVENTION AND MITIGATION STRATEGIES

Negative impacts of flooding can be reduced using different measures and adjustments, which are undertaken before, during, and after such events. In the hazard literature, these measures are referred to as alternative adjustments to floods. Burton et al. (1968) classified these adjustments into four classes: (1) affect the cause, (2) modify the hazard, (3) modify loss potential, and (4) adjust to loss. The first two measures are similar, and they aim to control flooding by constructing dikes, levees, flood walls, dams, and channel improvement through dredging. The first three structures hold water off the land and prevent floodwaters from inundating the settled areas. All these structures provide the most direct means of flood protection. However, they also tend to increase the velocity of flood flow within the main channel and cause rise in flood peaks downstream.

After dikes, dams are the most popular and effective method of flood protection in the world. Today the United States has more than 50,000 public and private dams of different sizes. Although dams are not constructed with the sole purpose of flood control, they are intended to control the rate of streamflow (Coch 1995). Dams constructed with the sole purpose of flood protection are ideal in areas prone to flash floods. However, this type of dam is subject to heavy deposition of sediments in its basin. Finally, channel improvement and modification is another useful method of flood protection. In addition to dredging, increasing the discharge capacity of rivers and other water bodies can possibly straighten the channel and widen it by cleaning out vegetation along the river course.

Although channelization increases the cross-sectional area of the stream and thus enables it to carry more water, this method often causes destruction of aquatic life in adjacent water bodies. Another way to reduce flood damage is by creating floodways, which are areas adjacent to rivers, and existing structures in such areas are often relocated. This is similar to the buyout program in the United States, in which Federal Emergency Management Authority (FEMA) buys the land with a high risk of flooding. After the 1993 destructive Midwest flood in the United States, FEMA bought 25,000 flood-prone properties to reduce flood damage (Spence and Smith 2011). Once purchased, the properties were converted into wetland, open, or recreational space.

Third, modifying the flood loss potential includes a host of measures such as land-use change, flood-forecasting and flood-warning systems, evacuation, flood insurance, and flood proofing. The last measure refers to those activities taken up by communities or individuals before or during floods to reduce flood damage, and it is based on acceptance of the fact that flooding is an unavoidable event. Flood forecasting and warning, which appear to be universal now, are required to adequately prepare people of flood-prone areas for an imminent flood and hence enable them to reduce damage. By knowing the probability of impending flooding, people can store extra drinking water, keep canned and dry food on hand, move valuable items to higher ground, and remove animals from low-lying areas. Emergency services can also gather enough resources available ahead of time to respond to emergencies as they occur.

One common indigenous flood proofing practiced in developing countries is the construction of homesteads on natural levees that remain above the normal flood level. In many cases, such construction occurs on highlands and involves further raising by digging earth from nearby areas. Stilt houses are also common in flood-prone areas, particularly in developed countries. A stilt house is a raised structure that is most commonly built above floodwater level. During the time of flooding, people often put sandbags around their houses to block floodwater from entering. Sandbagging is also necessary to strengthen levees and protect other property. In the United States, local residents, National Guard members, and inmates generally participate in sandbagging along weak stretches of embankments, levees, and dikes. While measures taken to modify the loss potential include some private/individual constructions, these are usually considered nonstructural.

Finally, adjusting to losses is aimed at spreading, planning for, and bearing the losses. Through buying flood insurance and by receiving emergency relief goods from various private and public humanitarian organizations, flood survivors tend to distribute their losses. As emergency aid, what they receive either in the form of cash and/or in kind cannot fully compensate their losses. Thus, they must bear some flood-caused losses. While floods are not unexpected natural events for residents who live in flood plains, such residents often need to tap into their reserve funds or borrow money from friends and relatives to cope with the impacts of flooding. Although insurance is a common component of both spread and bear losses, flood insurance is not widely available even in developed countries. For example, in the United States, where flood insurance was introduced in 1968, only slightly over 20 percent of homes located in flood-prone areas are covered by flood insurance.

CONCLUSION

Flood plains are not immune to flooding. In spite of knowing this clearly, people are often forced to live in such plains because all available lands have already been occupied. Moreover, with rapid increase in population and expansion of urban centers, people will have no alternative but to live in flood-prone areas in the future. Therefore, they have no option but to live with floods. As noted, the negative impacts of flooding are numerous, and a wide range of responses is necessary to minimize the damage and inconvenience floods can cause. Finally, improved land-use planning, preparedness, response, and increased coping capacity may reduce flood impacts in the future.

Bimal Kanti Paul

Further Reading
American Red Cross. 1992. *Flash Floods and Floods . . . The Awesome Power!* Washington, DC: U.S. Department of Commerce.
Burton, I., R. W. Kates, and G. F. White. 1968. *The Human Ecology of Extreme Events.* Toronto, ON, Canada: University of Toronto.
Coch, N. K. 1995. *Geohazards: Natural and Human.* Englewood Cliffs, NJ: Prentice Hall.
Doocy, S., A. Daniels, S. Murray, and T. D. Kirsch. 2013. The Human Impact of Floods: A Historical Review of Events 1980–2009 as Systematic Literature Review. *PLOS: Currents Disasters*, April 18. https://doi.org/10.1371/currents.dis.f4deb457 904936b07c09daa98ee8171a, accessed March 18, 2018.
FEMA (Federal Emergency Management Authority). 2017. Definitions. https://www.fema.gov/national-flood-insurance-program/definitions, accessed March 20, 2018.
IFRC (International Federation of Red Cross and Red Crescent Society). 2006. *World Disaster Report 2006—Focus on Neglected Crises.* Geneva, Switzerland: IFRC.
IFRC (International Federation of Red Cross and Red Crescent Society). 2016. *World Disaster Report—Resilience: Saving Lives Today, Investing for Tomorrow.* Geneva, Switzerland: IFRC.

Jonkman, S. N., and I. Kelman. 2005. An Analysis of the Causes and Circumstances of Flood Disaster Deaths. *Disasters* 29(1): 75–97.

Rasid, H., and B. K. Paul. 1987. Flood Problems in Bangladesh: Is There an Indigenous Solution? *Environmental Management* 11(2): 155–173.

Smith, K., and R. Ward. 1998. *Floods: Physical Processes and Human Impacts.* New York: John Wiley & Sons.

Spence, P. R., and I. S. Smith. 2011. Floods. In *Encyclopedia of Disaster Relief,* edited by K. B. Penuel and M. Statler, 216–221. Newbury Park, CA: SAGE Publications.

Hail

Hail or hailstone is a weather phenomenon born of some thunderstorms, and it is also associated with tornadoes and, more specifically, with northwestern storms in South Asia. According to Allianz Global Corporate and Specialty (AGCS) (2013), every 10th thunderstorm is followed by hail. Hail is a layered ice ball and is most common in the late spring and summer when more energy is available in the atmosphere. Within the United States, most hailstorms occur between April and October. In North America, the timing of hail is associated with the position of the polar-front jet stream. However, within a cumulonimbus cloud, water droplets that freeze are the origin of hail (not to be confused with "ice pellets"). This type of cloud is a dense, towering vertical cloud that extends to elevations of 40,000–60,000 feet (12,200–18,300 m) and may persist for at least one hour. It forms from water vapor carried by powerful upward air currents called updrafts.

Hail is usually formed at the top of the cumulonimbus cloud. Similar to the opening stages of cloud formation, condensation nuclei serve as focal point in hail formation, providing surface area on which supercooled water will freeze upon contact (a process referred to as accretion). This occurs as the water droplets rise upward and encounter air temperatures below freezing. Strong convection is typically associated with thunderstorms, and when hail formation has occurred, these updrafts work to suspend the frozen droplets in the cloud structure and further work to cycle the nascent hailstone vertically, collecting more surface area and adding to the layered structure of the object; this is termed the wet-growth stage.

As the hail falls through the cloud, it may again encounter strong updrafts that again move the particle upward, collecting even more frozen water on its surface. This process can occur in a cycle several times until eventually the hailstone becomes heavy enough that the force of gravity pulling it toward earth cannot be overcome by the vertical air current. Cumulonimbus clouds containing hail can sometimes be easy to spot, as the thunderstorm clouds may take on an odd greenish coloring due to light reflecting off the blanket of hail contained within. Hail can also be detected using weather satellites and weather radar imagery. Hail has occurred throughout history in most parts of the world, but most often in middle latitudes.

SIZE AND LAYERING

Hail is usually pea-sized to marble-sized. Meteorologists typically compare the size of hailstones to everyday objects like pea (0.25 inch diameter), marble/

mothball (0.50 inch diameter), dime/penny (0.75 inch diameter), nickel (0.88 inch diameter), quarter (one inch diameter), ping-pong ball (1.5 inches diameter), golf ball (1.75 inches diameter), tennis ball (2.5 inches diameter), baseball (2.75 inches diameter), tea cup (3 inches diameter), grapefruit (4 inches diameter), and softball (4.5 inches diameter). This method of comparison, however, is not representative of officially sanctioned metrics by the National Oceanic and Atmospheric Administration (NOAA), although reliable comparisons can be drawn (NWS 2003).

In order to be considered a hailstone, the frozen water formation must be at least 0.04 inches (1 mm) in thickness, although hail in the United States typically measures between 1 and 1.75 inches (25 and 45 mm), with 1 inch (25 mm) being the lower bound for "large" hail. Damage most often occurs when hail has reached a diameter of 1 inch (25 mm). If hail measuring larger than 0.75 inches (20 mm) in diameter falls during a thunderstorm, it is classified as severe weather in the United States. The largest hailstone, by circumference, ever measured in the United States fell in Aurora, Nebraska, on June 22, 2003, and measured 18.74 inches (47.6 cm) and weighed 1.67 lb (0.76 kg), while the largest recorded by diameter fell in Vivian, South Dakota, on July 23, 2010, measuring 7.9 inches (20 cm) across and weighing 1.94 lb (0.88 kg).

How large a hailstone becomes as a result of vertical recycling is a function of the available water in the cloud, humidity (water vapor), and the strength of the updraft (given that gravity is a constant). Size can also be governed by the collision of two or more hailstones. During the wet-growth stage, a thin outer layer remains in a liquid state due to the release of latent heat as the hailstone adds mass. Should a hailstone in this state collide with another, they are likely to bond, forming a larger hailstone. Large thunderstorms typically produce large hailstones, and the stronger the updraft within the system, the larger a hailstone can grow (AGCS 2013).

The considerations of available supercooled water and water vapor also determine the appearance of the hailstone's layered structure. When the hailstone passes through portions of the cloud that contain more water droplets, the resulting layer is thicker and clearer (more translucent), and when it passes through sections with more water vapor, the layer is much thinner and less clear (opaque). During what is termed the dry-growth stage, the outer layer does not remain in a sticky and adhesive liquid state; when this occurs, pockets of air can become trapped during periods of rapid freezing, and upon entering the wet-growth stage, escaped air bubbles will produce a more translucent layer. The layering of hailstones suggests that they fall from the sky through parts of the thunderstorm cloud with greater and lesser amounts of supercooled liquid-water content.

Hail is considered one of four forms of precipitations; the others are snow (precipitation in the form of ice crystals or, more often, aggregates of ice crystals), drizzle (smaller droplets of rain, yet larger than mist), and sleet (falling small particles of ice that are clear to translucent). However, when hail falls from the sky, it can cover areas of varying sizes, ranging from a few acres (less than 1 ha) to an area of 1,000 square miles (2,590 square km). The spatial extent of an area affected by hail, referred to as hailstreak, is usually smaller than 193 square miles (500 square km) in central Europe, and the median affected area was 8 square miles

(20.5 square km) in the United States. However, piles of hail have often been so deep that a snow plow is required to remove hailstones, and occasionally, hail drifts have been reported. When viewed from the air, the path of hail, known as the hail swath, is usually very visible. One distinctive feature of hailstorms is the so-called hail shaft, which indicates hail falling at a distance in a sharply defined swath.

It is estimated that a hailstone of 0.4 inch (1 cm) in diameter falls at 20 mph (9 m/s). A large stone, however, can reach up to 134 mph (60 m/s). Damage is generally caused by hailstones larger than 0.6 inch (1.5 cm) in diameter. The potential damage caused by hailstones largely depends on their sizes and the wind speed in which they fall. Clearly, a 1.2-inch (30 mm) diameter hailstone falling in high winds has a greater damage potential than a stone of the same size falling in light winds. To a lesser extent, shape and hardness of hailstones and fall orientation can also affect the degree of damage (AGCS 2013). Hailstorms generally last for a few minutes, but 15–30 minute durations have also been frequently observed (AGCS 2013).

GLOBAL DISTRIBUTION

Only a few studies have undertaken efforts to estimate hail frequency distributions on a global scale. Despite this lack of information, hail is a common occurrence in any mid-latitude or continental interior location that regularly sees thunderstorm activity. Mountainous regions also experience a great deal of hail activity due to the presence of orographic lifting (the vertical motion of air) coupled with cooler temperatures at higher altitudes. Across the globe, many countries experience hail with seasonal regularity. India, especially in the northern Himalayan region, frequently sees widespread hailstorms producing large stones. Other Asian countries like Bangladesh and China also experience hailstorms each year.

Although hailstorms can occur in any part of Europe, particular hazard zones exist throughout the continent, and orography plays the most important role for the location of these zones. Most of mid-latitude Europe experiences significant and frequent hail (e.g., Bulgaria, Italy, France, the United Kingdom, and Germany). The number of hail events is the highest in mountainous areas and Alpine regions. Compared to the United States, hail in Europe tends to be less frequent and severe, mainly due to a different orientation of large-scale orography (i.e., the Alps) and related circulation patterns (Punge and Kunz 2016). Around a few thousand hailstorms occur annually in Europe, giving an average frequency of much less than one hail event per year at any location.

The number of hailstorms differs among the regions of Europe. Central Europe (e.g., Germany, Switzerland, and Austria) is highly exposed to hail hazard. Hail frequency in this region decreases from west to east and from south to north. The main reason for this decrease is the region's continental isolation increases in these directions, which leads to lower moisture content and weaker frontal systems (Punge and Kunz 2016). Hail frequency across Western Europe (e.g., France, the United Kingdom, Ireland, Spain, and Portugal) varies substantially and is mainly influenced by the proximity to the Atlantic Ocean. Parts of other regions like Southern Europe and Eastern Europe also experience hailstorms. In Northern Europe, hail is less common, mainly due to the prevailing colder climate. For the overwhelming majority of the countries in Europe, the hail season starts in April/May and lasts until August/September. The exception to this norm is near the coast, where hail is reported year-round. This is because the coastal climate is strongly influenced by the Atlantic Ocean or the Mediterranean (Punge and Kunz 2016).

Countries on other continents, such as Australia, Russia, and South Africa, also experience considerable hailstorms. Hailstorms possess low risk in South and Middle America, including the Caribbean. Among the countries of Africa, South Africa is most prone to hail. However, Kericho in Kenya experiences the greatest average hail precipitation, on average, 50 days annually. This location is close to the equator, but has an elevation of 7,200 feet (2,195 m). These two factors contribute to it being a hot spot for hail. Kericho reached the world record for 132 days of hail in one year (Glenday 2013). Weighing at 2.25 lb (1.02 kg), the heaviest hailstone was recorded in Gopalganj, Bangladesh, on April 14, 1986. Hailstones of 7.5 lb (3.4 kg) were reported in Hyderabad, India, in 1939. Hail is largely absent from tropical latitudes due to temperatures aloft that are usually too warm to allow for the formation process to occur.

UNITED STATES

Contrary to the widespread belief among insurance companies, the analysis of 30–100 years of hail events in the United States suggests no systematic increase in countrywide frequency of hail events and/or their severity. On average, the United States experiences around 6,000 hailstorms each year. In an average year, hail occurs on 158 days. Occurrences of damaging hailstorms differ by regions as well as by states. These events occur most often between the months of April and October. According to NOAA, May accounts for the largest number of severe hailstorms, followed by July. Texas and Oklahoma experience a great deal of hail in April, while by June, most of the hailstorms will have migrated northward into Montana, Wyoming, and South Dakota. In July, they move farther north to Canada, especially in the province of Calgary (Abbott 2008).

Among the states, Texas, Nebraska, Colorado, and Wyoming usually have the most hail storms (see Table 1). Except Texas, the remaining three states are generally identified as "hail alley." Other states prone to hailstorms are California, Indiana, Illinois, Kansas, Minnesota, Missouri, New Mexico, and South Dakota. Hailstorms are rare in Arizona, Idaho, Maryland, New Jersey, Nevada, Oregon, Utah, Washington, West Virginia, and New England states.

The states in hail alley experience a great deal of hail owing to their higher elevation. The freezing levels (the area of the atmosphere at 32 degrees Fahrenheit or less) in the high plains are much closer to the ground than they are at sea level, where hail has plenty of time to melt before reaching the ground. Allen et al.

Table 1 Top five states for major hail events, 2016

Rank	State	Number (%)
1	Texas	830 (15)
2	Kansas	569 (10)
3	Nebraska	376 (7)
4	South Dakota	324 (6)
5	Oklahoma	315 (6)
	United States	5,601

Note: Major events are defined as hail size one inch (2.54 cm) in diameter or larger

Source: Insurance Information Institute (https://www.iii.org/fact-statistic/facts-statistics-hail, last accessed May 7, 2018)

(2017) examined the recurrence of hail in the United States using data covering the period 1973–2013 and found that west of the Rocky Mountains, smaller hail is more prevalent, while east of the Rockies, larger hail is more common.

Several large metropolitan areas located in the Rocky Mountain region and moving east through the continental interior are routinely threatened by hailstorms. The presence of an enormous number of buildings, vehicles, and other types of property creates a massive inventory of objects that can be damaged by hail. Outside of metro areas, vast agricultural fields blanket the Central Plains, Midwest, and Eastern United States, providing millions of additional acres of area that could potentially be damaged by hail-producing thunderstorms.

DAMAGE AND DESTRUCTION

Property

Hail causes considerable damage to property (e.g., homes, cars and vehicles, equipment, and trees) and crops and occasionally causes death to farm animals, but rarely to humans. It can bring down electricity lines, causing blackouts. In developed countries, roofs are the most commonly damaged property resulting from hailstorms. Hail can cause holes in roofs, but roofs constructed from metal can be resistant to hail damage. Roofs on most of the houses in developing countries are made of either tin or thatch, and for this reason, they are rarely damaged by hailstones. In developed countries, hail can cause damage to roof-mounted equipment such as air conditioners, antennas, vents, cooling towers, heating units, and company signs.

Rooftop solar panel systems associated with residential and commercial buildings are also susceptible to hail damage. This technology has been rapidly growing across developed countries, particularly in the United States. Further, hail often causes damage to the glass or translucent plastic in skylights or the seal around the outside edge (AGCS 2013). Other than roofs, windows and siding can also be damaged by hail if it falls at an angle. Such hail can crack windows and siding on buildings. Additionally, Exterior Insulating and Finish System (EIFS) wall coverings can also be damaged if hail has any significant horizontal force (AGCS 2013).

Beyond residential buildings, property such as equipment, cars, and furniture lying in the yard can be significantly damaged by hail. It can also cause damage to trees planted around the house. Trees or their branches can fall on the house, cars parked in a driveway, or equipment because of the weight of hail and the winds that accompany hailstorms. Parked cars in open parking lots are extremely susceptible to hail damage. Hailstorms are very troublesome to drivers of vehicles as they can not only impair visibility but also fall onto them, causing damage to roofs, hoods, and sunroofs (even crashing through them, increasing the risk of injury or death). Windshields are often the target of hailstones and can crack or shatter them completely. Aircraft such as planes and helicopters are also susceptible to hail damage, and while in the air, significant risk is posed to the occupants of airborne craft.

Crops, especially long-stemmed ones such as corn, can be damaged by hailstones. In the United States, hailstorms coincide with the planting, growing, and harvesting season of crops, vegetables, and fruits. Thus, they can damage young to mature crops. In developing countries, which are mostly located in the tropical region, many different crops at various growth states are damaged by hail. In these countries, crop damage accounts for the largest damage caused by hailstorms. Farmers of developing countries often face the danger of starvation if the hailstorms occur during the time of crop harvesting. If the storms damage young crops, they tend to adjust loss by sowing crops in the field again.

Death and Injury

Occasionally, the death toll from a hail event can reach into the hundreds, but those incidents are extremely rare. In 1932, more than 200 people were killed during a hailstorm in China. In May 1986, a major hailstorm killed 100 people and injured 9,000 in China. In 1988, hailstones took the lives of 246 people in Moradabad, India. Hailstones caused the deaths of 92 people in central Bangladesh in 1986. In 2014, 16 people died during a hail event in China's Guangxi province. The largest number of deaths caused by a hailstorm was reported on April 13, 1360, in Britain, killing about 1,000 soldiers (Shaw 2016). Deaths from hailstorms are rare in the United States, but a pizza deliveryman was killed by hail in Fort Worth, Texas, in 2000; his was the most recently confirmed hail fatality in the United States (Shaw 2016). Before 2000, on average, only one death occurred per year in the country due to hail.

Characteristically similar to death tolls, the incidence of injury is relatively low for hail events when compared to other types of natural disasters. The number of injuries can reach several hundred in developing countries, but the number is much less in developed countries, particularly in the United States, where less than 50 people were injured annually from hail before 2000. Available data indicate that only 102 people were injured in the United States between 2012 and 2016, or 20 persons per year on average. In 2015, no hail-related injuries were reported in the United States. Staying outside during hailstorms is the main reason for injuries; seeking shelter immediately greatly reduces the chance of sustaining injury during a hailstorm.

The damage and destruction aspects of hailstones, including deaths and injuries, depend on the sizes (average and maximum) of the hailstones; the number of hailstones per unit area; the associated wind speed, duration, and the frequency with which the stones fall. Deaths often occur from strikes to the head or other vulnerable parts of the body. However, hail does not have to be large to be damaging. Small, pea-sized hail can completely wipe out a young crop growing in fields and be a costly loss for the farmer, but large, softball-sized hail can be fatal for livestock and humans.

In monetary terms, damage caused by hail has been increasing due to more frequent extreme weather conditions. According to Swiss RE, a reinsurance company based in Zurich, Switzerland, five out of the top 20 most costly insurance losses of 2011 were hail related (AGCS 2013). A hail storm causing the largest loss in Europe occurred in 1984 in Munich, Germany, resulting in a loss of approximately $1.9 billion. According to the National Oceanic and Atmospheric Administration (NOAA), hail causes over $1 billion in damage annually in the United States, mainly to houses, buildings, cars, and crops (Changnon 2009) (see Table 2). Events causing damage in excess of one billion dollars have occurred in the United States in recent years. The costliest hail year in the United States occurred in 2016, when 5,601 major hailstorms resulted in $3.5 billion in property and crop damage (see Table 2). In 2001, 2012, and 2017, hailstorms caused more than $2 billion in damage (AGCS 2013). Regional variation in hail damage is observed in the United States. For example, the crop damage in Midwest accounts for about 1–2 percent of the crop value, 5–6 percent of the crops produced in the High Plains, and much less in other regions. Hail damages, on average, account for up to 2 percent of the total crop value each year in the United States.

Table 2 Hail damage ($ million) in the United States, 2012–2016

Year	Property Damage	Crop Damage	Total
2012	2414.2	93.9	2,508.3
2013	1,245.5	75.0	1,320.5
2014	1,416.9	293.2	1,710.1
2015	586.0	133.0	719.0
2016	3,512.7	23.7	3,536.4

Note: Major events are defined as hail size one inch (2.54 cm) in diameter or larger

Source: Insurance Information Institute (https://www.iii.org/fact-statistic/facts-statistics-hail, last accessed May 7, 2018)

PREPAREDNESS AND RESPONSE

Hailstorms generally develop quickly, leaving little time to react. Moreover, unlike all other natural disasters, there is no provision for emergency relief for the people affected by hailstorms. Therefore, the people of hail-prone areas need to minimize hailstorm-induced damage by taking adequate preparation before the event. Allianz Risk Consulting recommends preparing a comprehensive pre-hailstorm planning, which should include a host of activities (AGSC 2013). Notable among them are (1) assembling necessary emergency supplies and equipment, such as plastic tarps, portable hail covers, mops, battery-operated radio and torchlight, tape for windows, and lumber and nails; (2) planning for salvage and recovery, including making a list of key vendors, contractors, and salvage services; (3) having a plan for restoring operations after the event; (4) designating a person to monitor the status and location of the hailstorm; and 5) inspecting roofs and if necessary addressing problems such as blocked or loose drains, gutters or downspouts, and uneven ballast distribution.

Allianz Risk Consulting further recommends either bringing outdoor equipment and machinery inside the house or installing protective shields, including protective shields for rooftop equipment, and installing protective screens over skylights or covering window glasses with translucent plastic (AGSC 2013). Hail damage to cars parked in open spaces can be reduced in several ways. One such way is to cover the car with several layers of mats. One should remember that such a mat will not provide complete protection if hail falls more vertically and the mat does not cover entire body of the car. Another way is to use a hail mat with a roll-up functionality. To ensure effectiveness, the mats should be extended to the side of the structure.

Most automobile dealers in the United States keep cars outside without any protection. Although they have hail damage insurance, they should construct low-cost protective roofs above parked cars. One roof can accommodate several cars, and it can be constructed with steel posts, beams, supports, and braces in the roof as well as frames that can hold elements made of hot-dip galvanized metal. Ideally, automotive companies should relocate their new cars from unprotected yard areas into shielded areas as well as place hail protection mats onto cars (AGCS 2013).

Forecasts and warnings of hailstorms exist in developed countries. When weather information networks and services warn residents of a potential hailstorm, people in the areas at risk need to take precautionary measures. During a hailstorm, they should take shelter inside houses and stay away from windows. At this stage, they should continue to monitor weather reports for information on potential hail damage and utility outages. They should not go out in the storm to try to protect property. Because hail can damage cars, after receiving a hail warning, people should move cars to a garage or other protective shelters. As noted, hail also damages roofs of houses. Therefore, when building a new home or replacing an existing roof, people should consider using hail-resistant roofing materials. It is worthwhile to mention that smooth roof coverings, including single-ply and built-up coverings, are more susceptible to hail damage compared to those with gravel, stone, or paving block ballast. For built-up roof coverings, hail damage can be minimized greatly by providing slag or gravel surfacing adhered with a flood coat of hot asphalt (AGCS 2013).

In countries where hail insurance is available, people owning such insurance should check trees, shrubs, and plants around their houses after each hailstorm to assess the extent and nature of damage. They should also check roofs, windows, patio covers, and screens for damage along with cars for dents and broken or cracked glass. After assessing the damage, victims of hailstorms need to protect individual property from further damage. If they find signs of hail damage, they should take immediate actions to guard against further damage. For example, if any window glass breaks, it should be covered, or if hail creates holes in roof, those should be covered so that no water can enter and damage the interior of the home. Similarly, people should cover any broken windows in their cars to prevent damage of the interior from rain.

As indicated, after the hailstorm, the victims must inform their insurance company. The longer one leaves hail damage un-repaired, the more damage can occur. They also need to notify utility companies of any outages or damage to equipment. At the same time, they should make phone calls to contractors to immediately begin repair work. Before workers arrive, victims of the hailstorm should promptly clear any debris and hailstone blockages from roof and yard drains, gutters, drain pipes, and catch basins. They should also move damaged stock and equipment to dry areas, review the effectiveness of the hailstorm emergency, and revise it as needed (AGCS 2013).

In the United States, there are about 10 policies that cover hail damage. Table 3 lists the top five policies identified in hail damage claims from January 1, 2013, through December 31, 2015. These five policies accounted for 96 percent of the total hail loss claims during the preceding time period. Personal property homeowner's policies ranked at the top with 1,118,469 hail loss claims, or 53 percent of all hail damage claims between 2013 and 2015, followed by personal auto policies,

Table 3. Hail loss claims 2013–2015 by top five hail loss policy types (in number)

Policy Type	2013	2014	2015	Total No. (%)*
Personal property—homeowners	376,076	418,410	323,983	1,118,489 (53)
Personal auto	270,519	288,342	164,882	723,743 (34)
Personal property—farm	14,122	59,892	34,463	108,477 (5)
Commercial multi-peril	17,996	18,403	12,772	49,171 (2)
Personal property—fire	8,340	10,449	10,808	29,597 (1)
Yearly top five policy totals	687,053	795,496	546,908	2,029,457
Top five: Percentage of yearly total	95	97	96	96

*Percentages have been rounded to the nearest whole number.
Source: Fenning (2016).

which accounted for 34 percent of the total number of hail loss claims during the preceding three years.

According to a 2016 National Housing Insurance Crime Bureau report, among all states, Texas ranked first in hail loss claims from 2013 through 2015 with 394,572 claims, or 19 percent of the total claims during the preceding time period. With 182,591 claims, Colorado had the second highest number of hail claims during the preceding period (Fenning 2016). The top 10 states (Texas, Colorado, Nebraska, Kansas, Illinois, Oklahoma, Missouri, Minnesota, South Dakota, and Indiana) accounted for 66 percent of the total number of hail loss claims during the preceding period. Although all states of the United States claimed hail loss between 2013 and 2015, the number of claims was very low in several states, including Alaska, Delaware, Main, New Hampshire, Hawaii, New Jersey, Rhode Island, and Vermont. However, the report further states that hail loss claims decreased by 21 percent from 720,473 in 2013 to 572,182 in 2015, with an annual average of 70,660 claims (Fenning 2016). The majority of the claims occurred in the spring and early summer months, between March and July.

CONCLUSION

Hailstorms are a major disaster type and a significant cause of insured and economic losses in the United States and, to a lesser extent, other parts of the world. This phenomenon typically occurs in conjunction with severe thunderstorm formations and can produce heavy and damaging hailstones. The lack of data, both in developed and developing countries, restricts analyzing the frequency of hailstorms and other important aspects of the hazard, including their impacts. This is particularly important due to the near absence of the local-scale extent of hail-affected areas and a lack of appropriate observing systems in most regions of the world. For this reason, hailstorms are not captured accurately and comprehensively, which makes statistical analysis of their frequency or climatology more difficult. However, in light of climate change, which makes the atmosphere more unstable, intensity and severity of hailstorms will increase across the globe in near future.

Mitchel Stimers and Bimal Kanti Paul

Further Reading

Abbott, P. L. 2008. *Natural Disasters*. Boston: McGraw Hill.

AGCS (Allianz Global Corporate and Specialty). 2013. Hailstorm Checklist. www.agcs .allianz.com/assets/PDFs/white%20papers/ARC%20Hailstorm%20Checklist %2014-Nov-2013.pdf, accessed May 6, 2018.

Allen, J. T., M. K. Tippett, Y. Kaheil, A. H. Sobel, C. Lepore, S. Nong, and A. Muehlbauer. 2017. An Extreme Value Model for U.S. Hail Size. *Monthly Weather Review*. 145(11): 4501–4519.

Changnon, S. 2009. Increasing Major Hail Losses in the U.S. *Climatic Change*. 96(1–2): 161. https://doi.org/10.1007/s10584-009-9597-z

Fenning, D. 2016. *Regarding: 2013–2015 United States Hail Loss Claims*. Des Plains, IL: National Insurance Crime Bureau (NICB).

Glenday, C. 2013. *Guinness World Records 2014*. New York: Guinness World Records Limited. ISBN 9781908843159.

NWS (National Weather Service). 2003. *National Weather Service Instruction 10—1605*. Silver Spring, MD: U.S. Department of Commerce, NOAA.

Punge, H. J., and M. Kunz. 2016. Hail Observations and Hailstorm Characteristics in Europe: A Review. *Atmospheric Research* 176–177: 159–184.

Shaw, J. 2016. 8 Deadliest Hail Storms in History. *Newsmax*. July 8, 2016. https://www.newsmax.com/FastFeatures/hail-deadliest-storms/2016/07/08/id/737837/, accessed May 4, 2018.

Hurricanes

Hurricanes are very large storms that originate in large water bodies in tropical regions. They have a very important spatial component as these events are not subject to all places of the world. They generally occur between 5 and 25 degrees latitude north and south of the equator. On the basis of their geographic location of origin, they are variously termed. If they originate in the east Pacific Ocean or the Atlantic Basin, which includes the Atlantic Ocean, Caribbean Sea, and Gulf of Mexico, they are called hurricanes. If they originate in the west Pacific, they are called typhoons. Although both hurricanes and typhoons are tropical cyclones, the term *tropical cyclone*, or simply *cyclone*, is used for similar events originating in the Indian Ocean, including the Bay of Bengal and the Arabian Sea. Thus, typhoons, hurricanes, and tropical cyclones are three different region-specific names for the same kind of storm system.

FORMATION

Hurricanes are rotating storms consisting of alternating bands of wind and housing a massive thunderstorm system, with characteristics similar to those found in land-based supercell systems. The center of a hurricane, termed the *eye*, is a center of low pressure and acts as the main convection engine driving the development of a hurricane. Beginning their life cycle as a tropical depression, the initial formation of this type of storm requires five key ingredients to strengthen into a tropical storm and ultimately a hurricane. The first is warm water, typically around or above 80 degrees Fahrenheit. This warm water provides the initial energy required to begin formation and also provides a mechanism for evaporation, which produces humidity and clouds within the system. The requirement of warm (tropical) water is the primary reason hurricanes do not form outside of the band of latitude from 5 to 25 degrees. Similar to how a tornado "ropes out" due to the conservation of angular momentum, hurricanes very rarely form within five degrees of latitude of the equator, and only in a few extremely rare instances has a Southern Hemisphere hurricane crossed into the Northern Hemisphere, or vice versa.

The second key ingredient hurricanes require is converging winds at the water surface. The process of convection (air moving vertically) is driven by warm air rising as well as these converging winds helping to force the air upward. Humid air rising in a column is the third ingredient and is closely tied to the process of convergence. This process is also responsible for producing the dense cloud network characteristic of hurricanes. The fourth necessary element is an outward flow of air at the top of the center of the structure, termed the *convective chimney*. Outward bound air acts as a sort of exhaust pipe in the system, moving the rising air out of the chimney, making room for the process of warm rising air to continue. Finally, the fifth ingredient is just the right level of winds to force the system to begin moving. This element is crucial, as winds that are too strong will essentially blow the system apart, disrupting the convection engine enough to cause the system to stop developing, and winds that are too light will not push the system into motion. Once these elements are in place, and a tropical depression reaches sustained winds of at least 74 mph (119 kph), a hurricane is born.

THE EYE AND EYEWALL

The center, or eye, of the hurricane is the calmest section of the storm. Due to the extremely low pressure in the center and the convergence of water-level winds, there is little activity in this section. However, immediately adjacent to the eye is the inner eye wall, which houses the strongest winds in the system. Similar again to the physics that assist in spinning a tornado, momentum is conversed as it begins to move cyclonically (in a circle), neither losing nor gaining energy unless some outside force acts on the air particle. This conservation of angular momentum results in the smallest (in radius) columns of air, located in the eye wall, spinning the fastest, while the far outer bands of the storm spin the slowest.

As the system spins, the shape of the eyewall expands with height, becoming stretched in radius as it reaches the top of the system. And, as an interesting self-regulating mechanism, the strength of mature storms can actually act as a check on the intensity of the storm. As strong thunderstorms migrate toward the eyewall, the added moisture can slow the rotation of the system temporarily, disrupting the convection occurring in the chimney. When this occurs, the eye structure becomes replaced with the former outer eye wall, and the system may then regain strength. This natural system of equilibrium is not well understood, but it's being widely studied due to the onset of climate change.

While the convective chimney is responsible for moving air upward in the system, in the exact center of a fully developed (termed *mature*) storm, cooler air descends, causing the center to be extremely calm in terms of atmospheric conditions; the water, though, may be roiling violently as the storm moves along its track. In fact, hurricane hunters, planes staffed with crews whose mission is to fly above hurricanes and collect data, can see directly down the center of a mature hurricane, all the way down to the water. However, in weaker and less-mature systems, the eye may be obscured by clouds.

ROTATION

In the Northern Hemisphere, hurricanes rotate counterclockwise, and in the Southern Hemisphere, they rotate clockwise, and this is due to what is known as the Coriolis Effect (also referred to as Coriolis Acceleration). As air moves into the low-pressure center, it must be subjected to some further action (i.e., it cannot simply "collect" in the center). As the earth spins on its axis, it drags air in the atmosphere with it, with the effect especially pronounced in the troposphere (the lowest region of the atmosphere). But air is deflected by the air moving into the center, and it is deflected to the right in the Northern Hemisphere and to the left in the Southern Hemisphere.

THE CORIOLIS EFFECT AND GLOBAL WIND PATTERNS

Established global wind patterns, created by what is called the Tri-Cell Model of Global Air Circulation, force hurricanes to move east to west. The Tri-Cell Model can be summarized by stating that energy input from the sun is greatest between 23.5 degrees of south latitude and 23.5 degrees of north latitude, as this band marks the southern- and northernmost extent of direct sunlight as the sun migrates throughout the year. As the surface is heated, warm air rises upward until it reaches the tropopause, approximately 10 miles (16 km) above the surface at the equator, at which point it diverges, deposits the majority of its moisture, and in a cooler and drier state, descends back to the surface at approximately 30 degrees of south and north latitude. This portion of the cell is called the Hadley Cell, named after George Hadley, the meteorologist who first described it. This action sets up air patterns in the other two cells, the Farrell and Polar Cells, which in turn produce the consistent global wind patterns seen on the earth. The band of latitude (5–20 degrees from the equator in either direction) in which hurricanes develop contains the wind pattern known as the "trades," the Northeast Trades in the Northern Hemisphere and the Southern Trades in the Southern Hemisphere, and these winds blow from the western coast of the African continent in a westerly direction toward the Americas. As such, the trade winds are responsible for pushing hurricanes on westerly tracks.

LANDFALL AND EFFECTS

Hurricanes spend the majority of their life cycle feeding off energy provided by warm tropical waters, and not all hurricanes make landfall, with some dissipating over the ocean if the system sufficiently weakens. While the figure varies by season, on average, one or two hurricanes out of typically a dozen will make landfall somewhere on the U.S. East Coast. While the outer bands of the system will make contact with land first, a hurricane is not considered to have made landfall when that occurs. Rather, the eye of the storm must move over land to have the event classified as having made landfall. An important distinction is made between what meteorologists term a *direct hit* and landfall. For the latter, this occurs when the eyewall, containing the strongest winds, comes ashore, but the eye may remain over water.

Once a hurricane makes landfall, it will immediately begin to weaken, having been robbed of the energy input provided by warm water. It is a common misconception that friction between the surface and the storm is the primary reason a hurricane will slow over land; while it is true that friction increases once the storm makes landfall, it is not the primary driver of the storm's demise. It is less important than the immediate absence of the main energy input, that is, warm water. Further adding to the weakening of the storm is the collapse of the eye. The storm will have enough rotational energy to continue tracking and spinning, but the same effect that imparts the counterclockwise spin to the system, the Coriolis Effect, will now act to deflect the storm westward. As the storm dissipates, it will continue to bring rain, as there is no longer an uplift mechanism to hold the warm moisture-laden air aloft. Depending on the size and track of the hurricane, rain can last for several days, extending deep inland, and is frequently responsible for inland flooding. Rain resulting from hurricanes has been shown to have one positive side effect: it can work to alleviate drought (Maxwell et al. 2012).

Damage is heaviest along the coastal regions where hurricane landfall occurs. The most immediate threat is the high winds that accompany the eyewall, which can be in excess of 74 mph (119 kph) and have been measured as high as 215 mph (346 kph), the result of Hurricane Patricia, which struck the Mexican state of Jalisco in 2015. (For comparison, those wind speeds are equivalent to an EF5 tornado.) High winds can tear down weaker structures, creating a deadly debris field, which can cause further damage to property, as well as claim lives. Prior to the 1990s, it was common to secure a roof to a home by simply nailing the support trusses to the walls. The trusses themselves were also constructed by a method known as "toe nailing," or driving a nail through two pieces of wood at an angle. This was shown to provide a discontinuous transfer load path from the roof to the foundation, which resulted in the roof being easily lifted off the foundation by high winds, aided by lift due to the pressure differential caused by those winds. Each state has different codes governing the use of hurricane ties in new construction, but the use of the device originated in Florida after Hurricane Andrew (1992). A hurricane tie works to securely connect the roof to the walls, thus providing a continuous transfer load path throughout the structure. With the roof securely fastened to the home, the occupants are much safer, as the structure stands a far better chance of withstanding the event.

Rainfall is also a major concern for coastal areas as hurricanes make landfall. Severe damage can occur to structures that lose their roofs or otherwise become exposed to heavy rains. Additionally, heavy rains can bring flooding, although the major driver of inland flooding resulting from hurricanes results from what is termed *storm surge*. This phenomenon occurs as water levels rise with approaching hurricanes due the low pressure associated with a hurricane, which causes the water to expand in reaction to the pressure change at a rate of approximately four-tenths of an inch per millibar. The high winds associated with hurricanes also act as a driver of the event, pushing the water up onto the shore. The threat of storm surge is another major contributor to the death and damage tolls resulting from hurricane landfalls, with one such event killing 500,000 people in Bangladesh in 1970 (Cyclone Bhola). Storm surge is responsible for the deadliest natural disaster in U.S. history, the 1935 Galveston Hurricane; while the exact death toll will never be known, it is estimated to have taken the lives of 6,000 to as many as 12,000 people (Larson 2000).

HURRICANE SEASON

In the Atlantic Basin, hurricane season officially begins on June 1 and lasts through November 30, although peak activity occurs in August and September. The Eastern Pacific basin also runs through the end of November, but begins earlier in the year, on May 15. In the United States, the agency responsible for predicting, tracking, and cataloging hurricane data and information is the National Hurricane Center (NHC), part of the National Oceanic and Atmospheric Administration (NOAA). The NHC maintains the North Atlantic Hurricane Database, called HURDAT for short. This system contains records of hurricanes going back to 1851. As the North Atlantic season gets underway, a different agency, the Climate Prediction Center (CPC), also a branch of NOAA, issues the Tropical Weather Outlook for the upcoming months. This report is based on long-term synoptic climatologies that allow CPC meteorologists to determine approximately how many hurricanes can be expected, how many could make landfall, and what areas are potentially at greater risk than others. The NHC issues shorter-term forecasts, extending out several days. According to the American Meteorological Society, the Atlantic Basin sees about twelve tropical storms, six hurricanes, and three major hurricanes each year (Avila and Stewart 2011).

WARNINGS

Hurricanes are relatively slow-onset events, forming days before any potential landfall will be made. Further, they move slowly, with an average forward speed of about 10 mph (16 kph). The NHC (Atlantic Basin) and Central Pacific Hurricane Center (CPHC Pacific Basin) both monitor tropical storm activity constantly and issue bulletins to warn the general public. When a tropical depression or storm develops, the appropriate agency issues an advisory, as well as a forecast cone, to help define possible paths the storm might follow. In the hours and days that follow, the advisory may expire if the storm diminishes, or it may be upgraded to a watch. If the depression shows the possibility of reaching hurricane status within 48 hours, a hurricane watch will be issued, which indicates that hurricane conditions are possible within the forecast cone. Should hurricane conditions be expected within 36 hours, a hurricane warning is issued.

While large-scale evacuations are typically not undertaken as a result of an advisory or watch, those conditions signal the need to be prepared to move out of harm's way should the situation worsen. A warning, however, is a good indication that further action will likely be necessary, including securing (boarding up) windows and other entrances to structures; ensuring adequate supplies such as food, water, batteries, and medications are at hand; and perhaps even evacuating the area. While evacuation orders are not legally enforceable unless martial law should be declared, following such advice from government agencies is strongly suggested. Failure or inability to evacuate sections of New Orleans, most notably, the Ninth Ward, led to the death of hundreds of people when Hurricane Katrina struck in September 2005 (Harrington et al. 2006).

THE SAFFIR-SIMPSON SCALE

Developed in 1971 by a civil engineer named Herbert Saffir and a meteorologist named Robert Simpson, the Saffir-Simpson Scale was put into operation in 1973 by the NHC. Following a category scheme common in many scales, the Saffir-Simpson Scale employs a one to five rating system to describe the characteristics of a hurricane. Similar in design to the Fujita Scale developed by Dr. Tetsuya (Ted) Fujita in 1971, the Saffir-Simpson Scale rated hurricanes based on wind speed, except that whereas the F-Scale rated damage postevent the Saffir-Simpson Scale showed damage that could be expected based on wind speed ranges. The authors also included information on flooding and storm surge, but those considerations were removed in 2010 by the NHC due to their perceived inability to account for those phenomena accurately. The current iteration of the scale is called the Saffir-Simpson Hurricane Wind Scale, although it is still commonly referred to by its shorter original title.

The scale classifies hurricanes based on wind speed, with categories one and two holding no special additional classification, but categories three through five describe what are called *major hurricanes*. Each category has a short descriptive label and the wind speeds that define the range. Category 1 (CAT1) hurricanes are described as very dangerous winds that will produce some damage. The range for CAT1 events is 74–95 mph (119–153 kph). Category 2 (CAT2) events are characterized by extremely dangerous winds that will cause extensive damage, with a wind speed range of 96–110 mph (154–177 kph). Category 3 (CAT3) is the lowest and first level to be labeled as "major" and comes with the warning that devastating damage will occur; wind speeds associated with CAT3 events range from 111 to 129 mph (178–208 kph). Category 4 (CAT4) events bring a great deal of destruction, with wind speeds ranging from 130 to 156 mph (209–251 kph) and foretelling that catastrophic damage will occur. The final category, 5 (CAT5), contains some of the most well-known events in history and can have wind speeds greater than 157 mph (252 kph). There is no CAT6 on the scale; however, recent large events during the 2017 season have sparked conversation among scientists that the scale may need to be adjusted to account for higher-end storms, such as the aforementioned Hurricane Patricia, 2015 (215 mph, 346 kph), and Hurricane Wilma, 2005 (185 mph, 298 kph) (Lin and Emanuel 2015).

HOW HURRICANES ARE NAMED

The origins of the current naming convention for hurricanes date back to the late 1800s, owing its inception to an Australian meteorologist named Clement Wragge, who began the practice of assigning Greek letters to the storms. The U.S. military adopted the practice of naming storms during World War II (1939–1945), initially using the names of forecasters' wives, but changed that practice to use the military phonetic alphabet in 1947. By 1953, the U.S. Weather Bureau was using female names to describe storms, and in 1978, it added a list of male names, with assignments alternating between male and female. Today, all hurricanes are named; however, there is some variation in practice depending on the agency and country, but the intention of reducing confusion is a common thread. It is interesting to note, though, that the naming conventions used are not created or controlled by the NHC; the lists are generated by a special committee at the World Meteorological Organization (WMO).

The list covers a period of six years, with a separate list for each year. When a list's term is up, it is replaced with the list deactivated seven years prior. Names are not removed from a list except by committee decision supported by the need to remove the list due to that particular storm causing extensive damage—once a name is retired, it will never be used again. Since 1954, the WMO has retired 87 names, including some very recognizable events, such as Andrew, Katrina, Ike, Maria, and Ivan. The practice of naming storms has bled into other aspects of meteorology as well, including naming winter storms.

MAJOR ATLANTIC HURRICANES AND HURRICANE RECORDS

Measuring and rating a storm's impact can be a difficult exercise, but when the record category is based on a physical and empirical data element, ranking becomes simple. Typically, hurricanes are measured in a physical sense by either their lowest pressure or the maximum sustained winds—both of these are considered measures of intensity. (Note that wind speed here is required to be sustained, which may cause some confusion when comparing these events to previously mentioned hurricane such as Patricia and Wilma; the measurements in those cases were peak wind speed, not sustained.)

The most intense hurricane on record as measured by pressure was the Atlantic Basin storm Wilma (2005), with a low atmospheric pressure level of 882 mbar (millibars) (for reference, atmospheric pressure on a typical day, at or near seal level, will usually be around 1,010–1,015 mbar). The remaining four hurricanes in the top five all recorded pressure levels below 900 mbar, with Gilbert (1988) at second position at 888 mbar. Very low-pressure hurricanes are typically more powerful than higher pressure storms, as pressure relates directly to the ability to generate high winds. Notable is the fact that three of the top 10 hurricanes in terms of low pressure occurred in the same year, 2005, which witnessed the life cycle of hurricanes Wilma, Rita, and Katrina.

Additionally, 5 of the top 10 on this list have occurred since 2005; Dean (2007) and Maria (2017) are added to the preceding 3 to arrive at that list of 5. These pressure readings reflect the lowest point at any stage in their life, but pressure level at landfall creates a new category. Sitting at the top of that list is the Labor Day Hurricane (1935), which recorded a landfall pressure of 892 mbar. Camille (1969) and Gilbert (1988) are tied for second at 900 mbar. Similar to the previous list, a grouping of recent events can be found here: 4 of the top 10 events in terms of low pressure at landfall have occurred since 2005; they are Katrina (2005), Dean (2007), Irma (2017), and Maria (2017).

As measured by sustained wind speeds, the top spot is held by Hurricane Allen (1980), at 190 mph (305 kph). A three-way tie for second is a short list comprising the Labor Day Hurricane (1935), Gilbert (1988), and Wilma (2005), all recording sustained wind speeds of 185 mph (295 kph). There is another three-way tie for the fifth-place spot, with Mitch (1988), Rita (2005), and Irma (2017), each recording speeds of 180 mph (285 kph). Finally, there is an eleven-way tie for the eighth spot, with all hurricanes in that group, including Katrina (2005) and Maria (2017), recording speeds of 175 mph (280 kph). Of the 18 hurricanes on this list, 7 have occurred since 2005, with 3 of those being the Wilma/Rita/Katrina trio as identified in the preceding low-pressure list.

CATEGORY FIVE EVENTS

Since 1924, there have been a total of 33 Atlantic Basin hurricanes that have reached CAT5 strength, with exactly one-third of those having occurred since 2003. In an even more elite list, there have been only three hurricanes in U.S. history that have reached CAT5 strength and maintained that level upon making landfall; these are the 1935 Labor Day Hurricane, Camille (1969), and Andrew (1992). CAT5 events in the Pacific basin are rarer, with 16 occurring in that region, and none ever making landfall at that level. The 2005 hurricane season is also the record holder for the most CAT5 events in a year, with four; in addition to 2005, only 1933 and 2017 had more than one CAT5 event, with each of those years witnessing two events at the top tier of the scale.

COST AND DEATH RECORDS

Hurricanes have the power to do an incredible amount of damage due to the multiple threat types that come with them, their sheer size and power, as well as the duration of the event. The infamous Hurricanes Katrina (2005) and Harvey (2017) are the costliest hurricanes in history, issuing a bill for $125 billion. In terms of loss of life, the so-called Great Hurricane of 1780, which struck several islands in the Lesser Antilles, killed an estimated 22,000 people. Hurricane Mitch took 11,374 (estimated) lives in 1992, with the majority of those deaths occurring in Honduras (7,000 estimated) and another 3,800 (again, estimated) occurring in Nicaragua. Just five of the deaths occurred on U.S. soil. The 1935 Galveston Hurricane killed an estimated 6,000–12,000 people.

DEADLIEST STORMS IN SOUTH ASIA

Outside of the Atlantic Basin, two historical hurricanes (called cyclones or typhoons, depending on the region) absolutely dwarf the records for deadliest events in the Atlantic Basin. In 1970, Cyclone Bhola struck the country of Bangladesh (then known as East Pakistan). Forming on November 3, 1970, the cyclone struck the coast on November 12. Carrying sustained winds of 115 mph (185 kph) and recording a low pressure of 960 mbar, the Bhola cyclone was not particularly notable in terms of those physical characteristics; what made Bhola so deadly was the storm surge coupled with the terrain of the Ganges Delta. Approximately the lower half of Bangladesh is very close to the mean sea level (MSL), with an average height above MSL of just under four feet (1.2 m). The storm surge that Bhola brought with it reached peaks of 13 feet (4 m), inundating tens of thousands of square miles. The fishing industry was severely affected, an estimate 85 percent of the homes in the east coast were destroyed, and the agricultural industry suffered heavy losses. The grimmest statistic, however, is the final death toll: the commonly accepted estimate of the number of people Bhola killed is 500,000, making it the deadliest cyclone (typhoon, hurricane) in history and the fourth deadliest natural disaster in history. Two cyclones struck India that caused approximately 300,000 deaths each, one in 1737 and the other in 1839.

HURRICANES AND CLIMATE CHANGE

One of the major components in the life cycle of a hurricane is the input of warm water, which feeds energy into the system. Warm water holds more energy and thus carries with it the potential to produce more intense hurricanes. It is suspected that changing global climate is warming global temperatures overall, and ocean temperatures as well. One of the primary areas of study in connection to hurricanes and climate change is the amount of precipitation the storms are likely to cause. A hurricane's ability to expand and thus hold (and ultimately drop) more rain is directly related to the sea surface temperature (Lin et al. 2015). As sea surface temperatures rise due to a warming climate, more moisture will be available not only to energize the system but also to provide additional moisture, which will mean more rainfall overland as the hurricane dissipates. (This is derived from the Clausius-Clapeyron equation/relation, which states that for every one degree of warming in the atmosphere, the capacity to hold moisture is expected to increase by 7 percent [Lawrence 2005].) Larger and slower-moving systems may also stall over land longer, further increasing the risk to populations in the rain bands.

Modeling results suggest that a warming climate might actually produce fewer events (the relationship concerning this connection is still unclear and in need of continued study). However, the intensity of those events will likely increase, leading to more destructive hurricanes, with more widespread damage. Further, an indirect effect of climate change as it relates to hurricanes will likely be worsening episodes of storm surge. Warm water expands and takes up more space than cooler water, termed thermal expansion. With the addition of no new water, a heated ocean will become a fuller ocean and provide a larger column of water to inundate coastal areas during hurricanes. Hurricane Harvey (2017) is believed to have shown higher levels of rainfall as a result of climate change. In 2015, it is noted that four CAT4 hurricanes occurred at the same time, a first, and likely connected to the increased availability of energy in the ocean as a result of warmer sea surface temperatures.

CONCLUSION

Hurricanes are extremely large meteorological events that contain immense power and destructive capability. Primarily driven by warm equatorial water, other considerations such as westerly winds and convection are required to assemble a hurricane and send it on a path toward land. Generally, hurricanes allow for ample warning time of populations that may be in the path, but regardless of this, many events still take scores of lives upon making landfall. In addition to high winds, hurricanes can also bring intense and voluminous rains, which in turn can produce flooding. Coastal storm surge is a common characteristic of this type of event, and as friction between the storm and the ground becomes present when the system moves over land, tornadoes can also develop. While the connection between global climate change and hurricanes is still being widely studied, it is known that a warming world will likely produce stronger hurricanes, thus supporting the need to better understand weather and climate systems in general.

Mitchel Stimers

Further Reading

Avila, L., and S. R. Stewart. 2013. Atlantic Hurricane Season of 2011. *Monthly Weather Review* 141(8): 2577–2596.

Climate Prediction Center, NOAA. 2018. 2018 Atlantic Hurricane Season Outlook. http://www.cpc.ncep.noaa.gov/products/outlooks/index.shtml, accessed September 1, 2018.

Harrington, L.M.B., J. Gordon, and B. K. Paul. 2006. Southeastern Louisiana Evacuation/Nonevacuation for Hurricane Katrina. In *Learning from Catastrophe: Quick Response Research in the Wake of Hurricane Katrina*, edited by Hazards Research Center, 327–352. Boulder: University of Colorado.

Maxwell, J. T., P. T. Soule, J. T. Ortegren, and P. A. Knapp. 2012. Drought-Busting Tropical Cyclones in the Southeastern Atlantic United States: 1950–2008. *Annals of the Association of American Geographers* 102(2): 259–275.

National Hurricane Center. 2018. Atlantic 2-Day Graphical Tropical Weather Outlook. https://www.nhc.noaa.gov/gtwo.php?basin=atlc&fdays=2, accessed September 1, 2018.

Larson, E. 2000. *Isaac's Storm*. New York: Vintage Publishers.

Lawrence, M. G. 2005. The Relationship between Relative Humidity and the Dewpoint Temperature in Moist Air: A Simple Conversion and Applications. *Bulletin of the American Meteorological Society* 86(2): 225–233.

Lin, N., and K. Emanuel. 2015. Grey Swam Tropical Cyclones. *Nature Climate Change* 31: 1–7.

Lin, Y., M. Zhao, and M. Zhang. 2015. Tropical Cyclone Rainfall Area Controlled by Relative Sea Surface Temperature. *Nature Communications* 6: 6591.

Ice Storms

An ice storm is a winter weather event characterized by freezing rain, which is precipitation that falls and freezes on contact with cold surfaces close to the ground. This creates a coating of ice on surfaces such as trees, power lines, and roads or walkways. Freezing rain is also known as glaze. The National Weather Service (NWS), which issues warnings for hazardous weather in the United States, distinguishes any freezing rain or freezing drizzle event from an ice storm by the amount of ice accumulation. Ice storms are those freezing rain events that cause significant accumulation of ice and are expected to have a significant impact. The NWS issues an ice storm warning when forecasted ice accumulation exceeds a locally defined criterion, typically one-quarter inch or more. This is the accumulation amount expected to cause a significant impact. Areas may change their predetermined threshold that will trigger an ice storm warning based on how rare or common the event is. Regions that experience ice storms more regularly use a higher threshold. For example, some NWS offices in New York and Maine use a one-half inch threshold.

ATMOSPHERIC CONDITIONS

The type of precipitation that falls in the winter time is dependent on the temperature profile of the atmosphere and specifically the thickness and location of layers that are above and below freezing. Because warm air is less dense than cold air, warm air often moves into an area at higher and higher levels until a thick

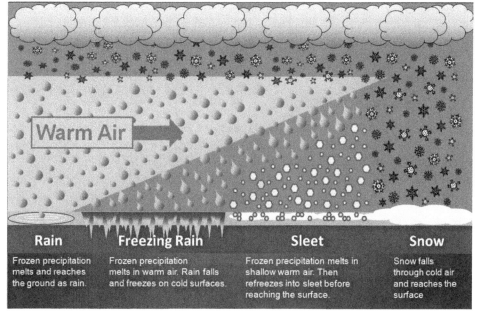

Typical winter precipitation progression as warm air enters an area. (National Weather Service)

warm layer overlies a cold layer. As the warm air grows deeper, the type of precipitation the atmosphere can support changes from snow, when the whole atmosphere is below freezing, to all rain, when the whole atmosphere is above freezing. Sleet and freezing rain can form in the transition between these two uniform zones depending on the thickness of the layer that is below freezing. While sleet will often occur mixed with snow or rain, freezing rain typically does not mix with these other precipitation types.

The icy glazing of surfaces occurs when raindrops, which have been cooled to below the freezing point, turn to ice on contact with a surface below freezing. When rain remains in liquid form even though its temperature is below freezing, it is called supercooled. In some cases, raindrops become supercooled and collide with each other, growing larger. When this occurs, it is often referred to as freezing drizzle. Freezing drizzle can be especially dangerous for the aviation industry because the ice can accumulate quickly (Kovacik and Kloesel 2014).

Forecasters will look at the temperature profile of the atmosphere to determine which type of winter precipitation will occur. If the atmosphere is below freezing at all levels, snow is the most likely precipitation type. When there is a deep layer of above freezing air near the surface, any snow that forms in the upper levels will melt before reaching the ground and fall as rain. In between those conditions, a warm layer can allow snow to partially melt as it is falling and then refreeze when it encounters a thick layer of below freezing air near the surface (and fall as sleet) or melt and refreeze only as it hits a surface that is below freezing (freezing rain). Forecasters have guidelines based on the specific temperatures of the layers as well as the thickness of the above-freezing layer that help them to tell the type of precipitation. For instance, while the whole atmosphere has to be cold enough that snow will not melt, it actually has to be several degrees below freezing for snow crystals to form. Also, the temperatures of the ground and the air above it are important. These can help determine how quickly or whether freezing rain can accumulate.

FORECASTING ICE STORMS

A combination of different weather features must come together to produce freezing rain. Atmospheric temperature is not the only factor that is important in determining winter precipitation. An ice storm will not happen unless there is enough moisture for the accumulation of ice to be significant. It does not take much ice for significant impacts, however. Often, the conditions occur in predictable patterns.

There are several scenarios that are associated with the likelihood of freezing rain. Researchers studied over 400 storms east of the Rocky Mountains that produced freezing precipitation and determined that there were seven large-scale weather patterns that are associated with it (Rauber et al. 2001). They found that the most common weather pattern is an arctic cold front ahead of a large high-pressure system. In this pattern, warm air rises above the cold air, and freezing rain occurs just beyond where the warm air meets the freezing line. Another common pattern involves a low-pressure system with both cold and warm fronts. The freezing precipitation occurs to the north of the warm front and wraps around the back end of the surface low-pressure center. The first two weather patterns also occasionally combine to produce severe freezing rain events. In this weather setup, a high-pressure area with an arctic cold front is positioned northward of the low-pressure system. Freezing rain occurs in the areas described previously, where these favorable areas intersect. Freezing rain can also occur on the western edge of an arctic high-pressure area, where warm humid air from the south intersects with the arctic air. Whereas freezing rain usually occurs in bands, this last setup often causes it to occur in compact circular areas.

Finally, there are weather patterns associated with the Appalachian Mountains that lead to freezing precipitation events. In each of these weather patterns, the mountains help trap cold air near the surface, while warmer air from either side of the mountains is allowed to rise above it. Freezing rain is always associated with colder air at the surface and warmer air above it, so if the specific conditions are right, the combination of the trapped or "dammed" cold air and the warm moist air can lead to a band of freezing precipitation parallel to the Appalachians.

A weather forecast map may depict several of the conditions described here. For example, a low-pressure system (shown as an "L") with cold and warm fronts is often found in the states bordering the northern Gulf of Mexico. North of the

Weather forecast map showing common conditions associated with winter storms. (National Oceanic and Atmospheric Administration)

low-pressure system is high pressure. In the warm part of the system ahead of the low-pressure system, and south of the warm front, thunderstorms are possible. Just ahead of the warm front, rain is expected. In this scenario, freezing rain is possible in two locations: in a narrow band between the high- and low-pressure areas and over the western Carolinas. Both areas are north of the warm front, where warmer air would be present higher up in the atmosphere. However, in the Carolinas, the Appalachian Mountains trap cold air at the surface, creating the perfect setup for freezing rain. Where the warm air has not yet reached north of the warm front and also behind the cold front, wrapping around the low-pressure area, snow would be likely because the whole atmosphere would be below freezing.

REGIONAL FREQUENCY

Climatologist Stanley Changnon and his colleagues performed several studies about the frequency of freezing rain. In one 50-year history of freezing rain events, they noted that the average annual occurrence of freezing rain events is not the same every year; some multi-year periods experienced particularly high or low numbers of events (Changnon and Karl 2003). Five high incidence areas were also identified based on the average number of days per year that have freezing rain. The area with the highest annual number of freezing rain days is eastern upstate New York, with seven freezing rain days per year, on average. From this peak, the average number of freezing rain days follows an inverted v-shaped pattern with arms of higher frequency stretching toward the Ohio Valley and also down the spine of the Appalachians. One region of higher incidence includes an elongated area from eastern Missouri into Pennsylvania; another region (the other "arm" of the inverted V) stretches from Maine to North Carolina. An isolated area of higher incidence is located in western Iowa and Minnesota.

A final isolated area of higher incidence occurs in the Columbia River valley in Washington and Oregon. Ice storms are most common east of the Rocky Mountains because of the conditions that must be present. However, the topography of the Columbia River Valley in southern Washington and northern Oregon can provide a favorable setup for freezing rain similar to events in the Appalachians because warmer Pacific air can be transported over the top of cold air being funneled down the river valley.

Researchers from the Southern Climate Impacts Planning Program conducted a study that examined whether ice storms were changing in their frequency and how these events changed over time (Kovacik and Kloessel 2014). These researchers looked at colder than average and warmer than average time periods and found that during colder time periods the area with the most ice storm events is in New England (including New Hampshire, southern Maine, and Massachusetts). During warmer periods, the area with the greatest number of events shifts further south to over eastern New York and eastern Pennsylvania/western New Jersey. In warmer years, ice storms are also more common in the Midwest and southern Appalachians.

The researchers also examined changing climate patterns called oscillations that might help explain the frequency of ice storms. There are many climate oscillations that can influence weather around the world through regional changes in temperature and/or pressure. The data analyzed in this study indicated that two climatological patterns were influential in the distribution of ice storms. One familiar climatological oscillation on the weather in the United States is El Niño/La Niña, also known as the El Niño Southern Oscillation (ENSO). ENSO characterizes the variation of sea level pressure and sea surface temperatures in the equatorial Pacific Ocean. El Niño years were associated with more ice storms in northern northeast (including eastern New York, Vermont, southern New Hampshire, and southern Maine). La Nina years were linked with more events in eastern New York, eastern Pennsylvania, and western New Jersey, that is, a southward shift in ice storms. The Atlantic multidecadal oscillation (AMO), which explains variation in sea surface temperatures in the North Atlantic, which occurs on the order of decades, is another oscillation that may influence ice storms in the United States. The researchers believed that the phase of the AMO worked along with El Niño/La Niña such that negative AMO and El Niño years were alike and positive AMO years had effects similar to La Nina years. When similar oscillations occur together, their influence can be enhanced.

SOCIETAL IMPACTS

There can be many impacts from ice storms, but the most significant societal impact is to the electrical power system. This includes damage to power lines and poles and the costs involved to restore electricity as well as for the secondary impacts that result from loss of power and downed power lines. The amount of damage depends on many factors, including the amount of icing, wind speed, and whether the area hit is urban or rural. Any amount of ice can be a nuisance to drivers as it can make roads and bridges slippery and require extra time to clear a windshield. A quarter to a half inch of ice can damage trees and power lines. A half inch (12.7 mm) of ice can add 500 pounds (226.8 kg) of extra weight to power lines. Wind is important because it can influence how much ice is allowed to accumulate on power lines and tree branches. It can also play a role after the storm. Higher winds can blow against lines or branches and cause them to fail. Urban areas often have a greater density of power lines, but rural areas tend to take longer to be brought back to full capacity after an event. Communications lines can be brought down the same way as power lines. Downed lines and branches may also cause disruptions to transportation if roads are blocked. Ice on road surfaces is also a major hazard of ice storms.

Vehicle accidents are a common cause of injury in ice storms. Other common types of injuries that occur include slips and falls, burns from fires, and hypothermia, which is when the body temperature becomes too low. Another significant impact resulting from loss of power is carbon monoxide poisoning. In a study of the 2002 North Carolina ice storm, researchers from the University of North Carolina found that the most common injuries were from falling debris and carbon

This image shows how disruptive ice storms can be to transportation and how destructive they can be to trees. (National Oceanic and Atmospheric Administration)

monoxide exposure (Broder et al. 2005). Other studies agree that carbon monoxide poisoning can be a dangerous side effect of ice storms and their resulting power outages. Carbon monoxide exposure results when heaters are used inside the home that are not meant to be used (such as charcoal or generators) or heaters that are not properly ventilated. The use of portable heaters and candles is also a common source of fires during power outages.

Another study examined data on insured losses maintained by the property insurance industry and determined that there had been 87 catastrophic freezing rain events from 1949 to 2000 (Changnon 2003). A catastrophe was defined as an event that caused $1 million or more in losses to insured property. Changnon estimated that property losses due to freezing rain amounted to about $18 billion (in 2000 dollars) when he added the amount resulting from catastrophic losses ($16.3 billion) to the amount expected to result from noncatastrophic events. From 1988 to 1995, the yearly average loss from catastrophes was $226 million in 2000 dollars. The climate region with the highest amount of insured losses was the Northeast (which included Connecticut, Delaware, Massachusetts, Maryland, Maine, New Hampshire, New Jersey, New York, Pennsylvania, Rhode Island, Vermont), followed by the Southeast (Alabama, Florida, Georgia, North Carolina, South Carolina, and Virginia) and the Central region (Illinois, Indiana, Kentucky, Missouri, Ohio, Tennessee, and West Virginia). The South (Arkansas, Kansas, Louisiana, Mississippi, Oklahoma, and Texas) reported about half as much insured losses as the Northeast. The rest of the United States (primarily those areas west of the Rockies, plus the northern Great Plains states) reported relatively little loss.

A similar pattern existed for the number of catastrophic events. The states with the highest numbers of catastrophes were located in the Northeast, and the area roughly corresponds to those areas where ice storms are most frequent. Despite not experiencing ice storms as frequently as other regions, the South and Southeast experienced a relatively high number of ice storm catastrophes over the 1949–2000 study period. Once again, the bulk of the events were east of the Rocky Mountains, with Washington and Oregon reporting the greatest number of events on the West Coast.

The same study also examined the average ice thickness values on telegraph wires. While the information used in the study was quite old (1928 through 1937), it was found that the ice accumulations were greatest on telegraph lines in the South and Southeast, possibly due to greater amounts of moisture in the air. The American Society of Civil Engineers develops standards for designing structures, including wires, in areas that experience icing (along with other weather phenomena such as snow or wind). Programs can determine the amount of ice that can be supported by structures of different materials and shapes before damage would occur. This information is incorporated into building codes.

Other economic impacts of ice storms are felt in the forest industry. Freezing rain events can affect the composition, structure, and condition of forest stands (Irland 2000). Irland described the impact of ice storms on individual trees as a range of severity from minor branch breakage to split trunks. Similarly, he reported that damage to larger areas can also vary in severity. One indirect effect of more large-scale damage from ice storms is the increased likelihood of infestation by pests. Insects are often attracted to weakened trees and can hasten their demise. In addition to the loss of timber, other products can be affected. An especially severe ice storm in 1998 caused huge losses to maple production in New York, Vermont, and New Hampshire, with New York reporting $6.4 million in lost production and hundreds of acres of lost production in New Hampshire (NEFA 1998).

NOTABLE ICE STORMS

Detailed records about ice storms have not always been kept. This is especially true regarding the impacts of such events. Social scientists have not studied ice storms to the same extent that other hazards such as hurricanes or tornadoes have been studied. The limits to both social and weather data make it difficult to truly identify the worst ice storms. However, some storms are notable due to their level of damage, deaths, or other unusual characteristics. NOAA maintains a list of all the weather events that cause over $1 billion in damage. There are 16 winter storms listed among the over 200 billion-dollar weather disasters that have occurred between 1980 and 2018. Two were ice storms: the

January 1998 northeastern ice storm and the February 1994 southeastern ice storm.

February 1994 Southeastern Ice Storm

The ice storm that occurred during February 9–13, 1994, primarily in Alabama, Mississippi, and Tennessee, was one of the costliest ice storms to affect the United States with over $3 billion in damages ($5.2 billion in 2018 CPI-adjusted dollars) (Lott and Sittel 1996). It also caused nine deaths. Lott and Sittel, who were employed by the National Climatic Data Center, described this event as being unusual in its aerial extent and also in the high amount of precipitation associated with it. For example, 43 stations reported 5 inches (2 cm) of precipitation or more, melted down. Two stations in Tennessee recorded over 7 inches (2.7 cm). The storm also affected Arkansas, Louisiana, Kentucky, North Carolina, South Carolina, Texas, and West Virginia. There were reports of ice accumulation of over 5 inches in northern Mississippi, with nearly 3 inches (1.2 cm) common in other affected states. The National Climatic Data Center described the damage in Mississippi as catastrophic, with some locations losing power for a month. Over 2 million customers were without power at some point throughout the storm's path through the South.

January 1998 Canada and Northeastern United States

The ice storm that occurred the first week of 1998 caused an estimated $1.4 billion in damage ($2.2 billion in 2018 CPI-adjusted dollars) and at least 16 deaths (Lott et al. 1998). The storm also caused nearly $3 billion damage in Canada,

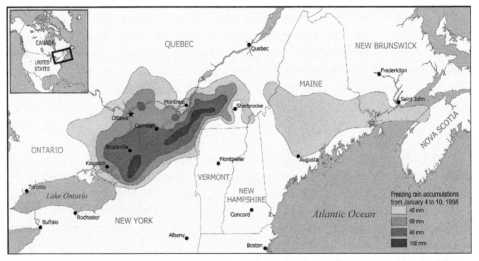

Freezing rain accumulation from 1998 storm. The map shows the levels of ice accumulation in millimeters. The accumulations range from approximately 1½ inches (40 mm) to almost 4 inches (100 mm). (National Oceanic and Atmospheric Administration)

making it one of the worst in Canadian history. NOAA reported that some locations received over three inches of freezing rain, with more than an inch accumulating on objects. Additionally, over 3 million people lost their power in Canada. Over 500,000 lost power in New England, including 80 percent of Maine.

During this event, an area of arctic high pressure continued to build eastward from Quebec toward Maine and Nova Scotia. According to the National Weather Service, a low-pressure area developed along the Gulf of Mexico, and along with a pronounced trough, it helped to provide moisture to supply the system. The low-pressure area tracked northward along a stationary boundary, ending up near New England on the main day of the event. The weather system brought several inches of precipitation, and along with surface temperatures for the most part in the 20s, it led to ice accumulations of 2–4 inches (5–10 cm) in the St Lawrence Valley and 1–2 inches (2.5–5 cm) in northeastern New York and northwestern Vermont.

January 2009

This ice storm was called "the biggest disaster in modern Kentucky history" by Kentucky's governor at the time, Steve Beshear (NWS Paducah 2009). Ice accumulations of 1 inch (2.5 cm) to over 1.5 inches (4 cm) were reported. Power outages in parts of Kentucky averaged for a week in duration, and parts of rural Missouri were without power for three weeks. Causes of deaths in Kentucky attributed to the ice storm included hypothermia, fires, and carbon monoxide poisoning. Missouri also reported deaths from carbon monoxide. Injuries resulted from carbon monoxide poisoning, heart stress, and gastrointestinal problems from eating spoiled food. Over 30,000 utility poles were downed or snapped (Dolce and Erdman 2017), and over 145 miles (218 km) of high voltage transmission lines were down (NWS Paducah 2009). Federal Emergency Management Authority (FEMA) reported that National Guard soldiers went door to door in some locations checking on residences and also helped with traffic control and debris removal, which was extensive. The event was the result of a low-pressure system that developed along the Gulf Coast and tracked northeast, while an area of high pressure stayed to the north, allowing cold air from the northeast to filter into the low levels of the atmosphere. An abundance of moisture was allowed to ride up over the cold air, causing days of precipitation over the region.

December 2002

The first week of December 2002 saw a winter storm system bring snow, sleet, and freezing rain to a large portion of the United States, from Oklahoma through the mid-Atlantic and continuing to the Northeast. The Carolinas were the hardest hit with accumulating freezing rain. December 4–5, 2002, marked one of the worst ice storms to hit North Carolina. The conditions came into place starting with a high-pressure system building into the eastern United States, with a cold front passing through the area, allowing for cold and dry conditions at the surface. Later, a series of surface low-pressure areas formed in the Gulf of Mexico and then off the East Coast, providing moisture. A ridge of high pressure caused air to get "wedged" in by the Appalachian Mountains, providing cold air at the surface. Climatologists at the Southeast Regional Climate Center reported that the ice storm was the worst in the 100-year history for Duke Energy, the largest provider of utilities in the Carolinas (Fuhrmann et al. 2009). Over 1.5 million people were without power due to the storm, and 10 percent of Duke Energy customers in one North Carolina county were still without power 10 days after the storm.

December 2000

In December 2000, multiple winter storms damaged Arkansas and Oklahoma and left many without power in both states. The storm affected people all the way from New Mexico to Arkansas, but Arkansas and Oklahoma were the hardest hit. Both systems involved an arctic high-pressure area to the north with a low-pressure system to the southwest, responsible for ushering in warm, moist air. The NWS office in Little Rock reported all forms of winter precipitation in their area in the first event, which occurred on December 12–13, with mainly freezing rain and sleet during the second event, December 25–28. Meteorologists believed the event was the worst icing event in the state since 1932, when another event occurred of similar magnitude (NWS Little Rock

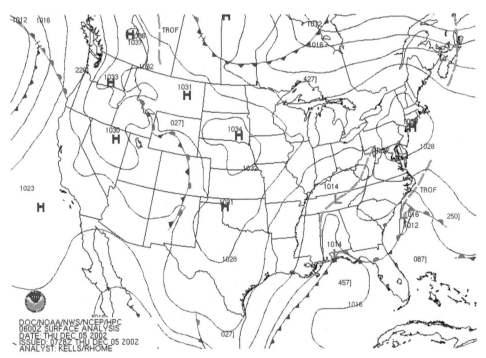

Surface weather map from the December 2002 storm. (National Oceanic and Atmospheric Administration)

n.d.). An inch (2.54 mm) of ice or more was reported across most of Arkansas during both events.

While many people lost power during both events, there were a couple of differences that made the latter event worse. Earlier in the month, surface temperatures were not quite as cold as later in the month when the soil was at or below freezing. A shallow layer of very cold temperatures combined with a deep layer of moisture to fuel long-lasting and heavy precipitation in Arkansas and Oklahoma. Over two dozen people died in the region, and 500,000–600,000 people were without power in each event.

Northern Idaho January 1961

One of the costliest storms on record was the one that hit northern Idaho on January 1–3, 1961. One NWS report stated this storm had the deepest ice accumulations ever recorded in the United States at the time with 8 inches (20 cm) in northern Idaho. The NWS report describes the event as a combination of freezing rain and dense fog. The addition of fog means the accumulation would not have been solely the result of freezing rain. The fog contributed to it.

Other Historic Events

The storm that hit Massachusetts during November 26–29, 1921, was the worst ice storm in New England history, with 4 inches (10 cm) of ice reported in eastern and central Massachusetts. Occurring at roughly the same time was one of the worst Pacific winter storms (though not technically an ice storm) in Dalles, Oregon, during November 18–22, 1921. In this event, a mixture of snow, sleet, and freezing rain combined into a 54-inch (1.4 m) deep mass of snow and ice. Six inches (14 cm) of ice was reported during January 22–24, 1940, in Northwest Texas, and December 29–30, 1942, in Upstate New York (Burt 2007).

SAFETY TIPS DURING AN ICE STORM

The best practices for staying safe during an ice storm are fairly similar to other hazards that are associated with power outages and dangerous travel conditions. The first thing to prepare for is the loss of electricity, which can mean the loss of heat. Layered clothing and blankets can help keep warm. Portable generators, grills (both gas and charcoal), and camp stoves should not be used indoors. Portable heaters designed to be used indoors should be used according to the manufacturer's instructions and properly vented. A battery-powered carbon monoxide detector can be used to ensure that air is safe. Candles and other sources of open fire can be another source of danger. Cell phones should be charged ahead of time and flashlights with extra batteries kept nearby. If the power does go out, refrigerators and freezers will keep food cold for a while (about 4 hours for a refrigerator or up to 2 days for a freezer) if unopened. It is also best to have extra water and food that does not require refrigeration.

Because a thin layer of ice can make driving difficult or dangerous, it is best to avoid travel. It might not be possible to travel right after the storm; it is a good idea to have enough essential supplies such as prescription medicine and baby items on hand ahead of the storm. Pets should be kept safe inside. Additionally, it is best to stay away from branches or other items that might fall due to the weight of the ice. Local news channels and National Weather Service are among the reputable sources for weather forecast information.

Kathleen Sherman-Morris

Further Reading

Broder, J., A. Mehrotra, and J. Tintinalli. 2005. Injuries from the 2002 North Carolina Ice Storm, and Strategies for Prevention. *Injury* 36(1): 21–26.

Burt, C. C. 2007. *Extreme Weather: A Guide & Record Book*. London, UK: W.W. Norton & Company.

Changnon, S. A. 2003. Characteristics of Ice Storms in the United States. *Journal of Applied Meteorology* 42(5): 630–639.

Changnon, S. A., and T. R. Karl. 2003. Temporal and Spatial Variations of Freezing Rain in the Contiguous United States: 1948–2000. *Journal of Applied Meteorology* 42(9): 1302–1315.

Dolce, C., and J. Erdman. 2017. The Nation's Worst Ice Storms. https://weather.com/storms/winter/news/top-10-worst-ice-storms-20131205, accessed January 15, 2019.

FEMA (Federal Emergency Management Agency). 2009. Kentucky Ice Storm Moves from Response to Recovery. https://www.fema.gov/news-release/2009/02/20/kentucky-ice-storm-moves-response-recovery, accessed January 15, 2019.

Fuhrmann, C., R. P. Connolly, and C. E. Konrad. 2009. Winter Storms: An Overlooked Source of Death, Destruction, and Inconvenience in the Carolina Piedmont Region. Paper given at the 66th Eastern Snow Conference. http://www.sercc.com/projects/WinterImpactsSERCC.pdf, accessed January 14, 2019.

Irland, L. C. 2000. Ice Storms and Forest Impacts. *Science of the Total Environment* 262(3): 231–242.

Kovacik, C., and K. Kloesel. 2014. *Changes in Ice Storm Frequency across the United States*. Southern Climate Impacts Planning Program. http://www.southernclimate.org/publications/Ice_Storm_Frequency.pdf, accessed January 15, 2019.

Lott, J. N., and M. Sittel. 1996. The February 1994 Ice Storm in the Southeastern US. In *Proceedings of the Seventh International Workshop on Atmospheric Icing of Structures*, 259–264. https://www1.ncdc.noaa.gov/pub/data/special/iwais96.pdf, accessed January 16, 2019.

Lott, N., D. Ross, and A. Graumann. 1998. *Eastern U.S. Flooding and Ice Storm January 1998*. Asheville, NC: NOAA's National Climatic Data Center.

NEFA (North East State Foresters Association). 1998. The Northeastern Ice Storm 1998: Effects on the Forests and People of Maine, New Hampshire, New York and Vermont. https://www.dec.ny.gov/docs/lands_forests_pdf/ice98.pdf, accessed January 17, 2019.

NWS (National Weather Service) Little Rock. n.d. Public Information Statement: Historic Ice Storms in Arkansas. https://www.weather.gov/lzk/pnsicestormtxt.htm, accessed January 15, 2019.

NWS (National Weather Service) Paducah. 2009. Ice Storm 2009: Beauty and Destruction. https://www.weather.gov/media/pah/Top10Events/2009/Ice%20Storm%20Jan%2026-28%202009.pdf, accessed January 15, 2019.

Rauber, R. M., L. S. Olthoff, M. K. Ramamurthy, D. Miller, and E. E. Kunkel. 2001. A Synoptic Weather Pattern and Sounding-based Climatology of Freezing Precipitation in the United States East of the Rocky Mountains. *Journal of Applied Meteorology* 40(10): 1724–1747.

Landslides

Landslides are gravity-controlled events where a mass of soil or rock moves downslope. Landslides are thus a form of mass movement—just like soil creep. Such mass movement differs from erosion caused by water in the sense that water-caused erosion picks up and transports individual particles of soil or rock rather than large amounts in one go. Landslides are dangerous. Thousands of people die per year because of landslides, and many more become homeless or lose property. Hundreds of scientists globally study landslides to help minimize these losses.

Cruden and Varnes' 1996 classification of landslide types by material and movement type. (Adapted from https://www.bgs.ac.uk/landslides/how_does_bgs_classify_landslides.html)

There are three main products of this work: first, landslide susceptibility assessments, which indicate where the probability of landsliding is largest; second, landslide hazard assessments, which indicate where and when which size of landslide is likely to occur; and third, landslide risk assessments, which add the aspect of expected damages to hazard assessments.

TYPES

The most used classification of landslides is based on the work by David Varnes (1978). In the most recent version, landslides are classified based on the kind of material that is moving (rock, debris, or earth) and the type of movement (fall, topple, slide, spread, flow, or a combination). Even though some of these combinations (for instance, a rockfall) do not immediately evoke what is thought of as a landslide, they are nonetheless all formally part of the landslide definition (Cruden and Varnes 1996). On occasion, the various types of landslide movements occur during the same event.

FALLS AND TOPPLES

Rock-, debris-, or earth-falls occur on the steepest slopes. The "fall" category implies vertical movement through the air or bouncing along a very steep slope. The material that is falling is (much) smaller than the distance that is covered during the fall, and the momentum that is gained often leads to continued movement down lower slopes that are less steep. With topples, the slope is still steep, but it's about the same size as the material that is toppling. Therefore, there is often less deformation of the falling material during toppling and less continued movement after the original topple.

SLIDES

Sliding as a way of moving often leaves the sliding material more intact. There are many examples of undisturbed surfaces, including vegetated surfaces with trees that have moved with the sliding material and that continue growing after the slide stops. During a slide, material moves along a sliding plane with the subsurface. In rotational slides, the distance of movement is on the same order as the size of the landslide, whereas in translational slides, the distance of movement is greater than the size of the landslide. Generally, the steeper and less interrupted a slope, the more likely a slide is to be translational.

SPREADS

Spreads involve almost horizontal movement of material due to the material behaving like a fluid. This so-called liquefaction can occur when an earthquake shakes up all soil material and forces soil water to mix with the material, creating a muddy fluid. This liquefaction happens some distance under the soil surface, and the muddy liquid can move down very flat slopes, carrying the less disturbed soil on top of it.

FLOWS

When more water is involved, landslides can more completely behave like fluids. The resulting "flow" type of movement completely deforms the material that is moving and covers the largest distances. Especially when rocks or debris are involved, these are among the most dangerous forms of landslides. Particularly damaging debris flows can happen after forest fires, when the lack of root cohesion from trees and other plants allows coarse material to be eroded and mixed into debris flows that can move downhill at speeds exceeding 100 miles (150 km) per hour. In 2018, debris flows that occurred one month after a series of forest fires in California in the United States killed 21 people and caused more than $100 million dollars in damages.

CAUSES

All landslides need a few ingredients: first, they need a slope to slide down. Second, they need material that can slide. Third, they need to overcome frictional forces with the subsurface (usually hard rock) and the surrounding hillslope. The first and second ingredients change very slowly, over geologic timescales. For instance, slopes can become steeper when a river is eroding the bottom of a slope, and slidable material can be created from rock by the process of weathering.

The third ingredient, which makes the slide overcome friction with the materials under it and to the side, often changes much more rapidly. That is why it should be a focus in discussions about landslide causes. If there is a sufficiently steep slope, and enough material to slide, all that is needed to trigger a landslide is overcoming friction. There are many ways that this can happen, but the main ones are through earthquakes, intensive or long-duration rainfall, or saturation from snowmelt.

Earthquakes can cause landslides by shaking the ground and reducing the friction, particularly the friction between the potential slide and the subsurface. This can be enough to set off a landslide. Most mountainous, steeply sloping areas do experience earthquakes, so earthquaking is a common cause for landslides. Large earthquakes can cause large numbers of landslides, and these landslides can in fact cause more damage than the earthquakes themselves. For instance, the 7.8-magnitude Gorkha earthquake that hit Nepal in 2015 caused more than 10,000 landslides that destroyed or damaged villages and roads.

Rainfall is another common cause for landslides. When intensive or long-duration rain infiltrates into the soil, it increases soil wetness. In some cases, deep infiltration becomes limited, and water pressure in the soil builds up. The increased wetness and the increased water pressure reduce the friction of a slide with the subsurface, increase its mass, and can therefore cause sliding. When rainfall is too intense to infiltrate the soil, it flows downslope over the soil surface and causes erosion without landsliding.

Snowmelt at the end of winter can take several days or weeks. In both cases, the melted water can infiltrate the soil slowly, and as long as there is still a snowpack, evaporation is limited as well. Therefore, similarly to rainfall, the soil can get saturated, which reduces friction and increases mass. This can lead to a landslide. Snowmelt can be particularly effective in causing landslides when part of the soil is frozen (e.g., in permafrost regions). Infiltration into the frozen part of the soil is often limited because the pores in the soil are blocked by ice, and the remaining, unfrozen, part of the soil becomes wetter and heavier quickly as a result.

Recently, there has been much attention for interaction between the various causing factors. Specifically, it has been recognized that the loss of friction with the subsurface that an earthquake causes persists for several years even if no landslide happens immediately. This means that after the earthquake, many hillslopes may be primed for landsliding in the next rainy season or the next rainstorm. This effect was clearly visible after the 2015 Gorkha earthquake that we discussed earlier: the monsoon season after the earthquake caused landslides that affected a 10-times larger area than monsoon seasons usually do.

SIZE

Landslides can be enormously large, or very small, but they cannot be infinitely large, or infinitely small. The maximum size of a landslide is the size of the slope on which it occurs, and the minimum size is determined by how much material is needed to overcome frictional forces on a slope (such as roots holding soil together).

The largest known landslides happened in prehistory. The largest landslide that happened on land may have happened near Saidmarreh in Iran about 12,000 years ago, where tilted geologic layers in a large mountain range facilitated an enormous rockslide that had a volume of 20 cubic kilometers—about 5 cubic miles. That is a volume equal to 1,500 times the largest building in the world. The landslide was so enormous that it dammed off two large rivers and created two lakes with a total surface area of almost 250 square kilometers. That is about as large as Lake Mead, the lake created by the famous Hoover Dam in the United States. The landslide was first recognized in 1938, but probably happened about 10,000 years ago (Harrison and Falcon 1938).

Even larger "landslides" happen under water, at the edges of the continental shelf, where coastal waters become oceans. These locations experience a lot of sedimentation over geologic timescales. This provides a lot of soft, unconsolidated material that can slide when disrupted, for instance, by an earthquake. This may be exactly what happened to the famous Storegga slides on the Norwegian coast. These three slides displaced a total volume of 840 cubic miles (3,500 cubic km)—more than 100 times as much as the Iranian slide. The sudden movement of all this material down the continental shelf to the ocean floor displaced an enormous amount of water, which in turn led to a large tsunami. These landslides happened more recently than the Iranian slide, about 6,000 years ago, and the tsunamis may have affected Stone Age populations living along the Scottish, English, and Dutch coasts, among others (Bondevik et al. 2003).

Of course, almost all landslides are much, much smaller than these record holders. Nonetheless, landslide size is as hard to predict as the moment that a landslide happens. But experts can predict how likely very large landslides are in a certain area or region. This is because of the so-called power-law behavior of landslide sizes. Power-law behavior means that every increase in landslide size, say from 1,076 square feet (100 square m) to 10,764 square feet (1,000 square m), or from 10,764 square feet (1,000 square m) to 107,639 square feet (10,000 square m), causes the same decrease in probability of occurrence. For instance, in a certain area, observations of previous landslides may indicate that there is an 80 percent probability that when a landslide happens, it is 1,076 square feet (100 square m) or larger, an 8 percent probability that a slide is 10,764 square feet (1,000 square m) or larger, and a 0.8 percent probability that a slide is 107,639 square feet (10,000 square m) or larger. In this example, every tenfold increase in landslide size becomes 10 times less likely to happen. This can be used to calculate how likely very large landslides are, even if they have never been observed in the area. In the example, there would be a 0.08 percent probability of a landslide having a size of 1,076,391 square feet (100,000 square m). Power-law behavior of this kind has been found for landslide sizes as well as for landslide volumes. As a result, experts can say how likely (or unlikely) very large slides are but are not yet able to pinpoint where exactly they will occur.

GEOGRAPHIC DISTRIBUTION

Many places experience landsliding, and known landslides often cover more than 10 percent of landslide-prone areas. This is because many places have the main ingredients for landslides: steep slopes, loose material, and water to help slides move. Global maps of landslide susceptibility show the likelihood of landslide occurrence at very large scales. Such maps show that landslide susceptibility is largest in the world's large mountain ranges, including the Himalayas, the Andes, the Rocky Mountains, and the European and Southern Alps.

Within these ranges, the wettest places are often the most dangerous ones. Many mountain ranges have a relatively wet and a relatively dry side, and so landslide susceptibility is often higher on one side of a mountain range than on another. A good example is the Himalayas, where landslides are more prevalent in wet India and Nepal than in dry Tibet. Zooming in further to individual mountain valleys, the steepest locations are the most susceptible, especially when they are made of soft rock or soil. Therefore, the factor that is most important in determining where slides occur changes with scale: from whether there are mountain ranges at global scale, to climate at continental scale, to the soil type and the shape of the landscape (geomorphology) at regional scale.

In such steep mountain landscapes, most of the area can be susceptible, with only few safe locations available. For example, the landscape in a region near the city of Perugia, Italy, is less steep than in many mountain ranges, but its rocks are a bit softer than is common in mountain ranges. As a result, it is prone to landsliding. Records from the last 100 years show that up to 50 percent of the area has experienced a landslide (Samia et al. 2017b). Yet this landscape has been intensively used and inhabited at least since Roman times, about 2,400 years ago, and landslides must have happened on even more of the area over that long period. Small medieval villages and old farmhouses are all situated on the few safe and stable locations on top of broad ridges, illustrating that the danger of landsliding has been recognized for a long time and that societies have adapted to them.

HISTORICAL EXAMPLES

The 563 CE Rockfall into Lake Geneva

Written reports mention that in 563 CE, a large tsunami hit towns on the shore of Lake Geneva. This tsunami arrived from the upstream end of the lake and was very high at the very downstream end, where the lake narrows before the Rhone river enters its gorge. The city of Geneva, which already existed at that time, was situated exactly in the wrong place. The tsunami, which has been calculated to be about 8 m high, overtopped the city walls, killed many, and destroyed much of the city. The reason? Probably a massive rockfall that hit the lake at its upstream end, possibly displacing part of an unstable delta. Seismic profiles of sediment in Lake Geneva indeed indicate such rockfall deposits. The risk of a repeated rockfall-with-tsunami is apparently not negligible.

Rockslide of Mont Granier, France

Possibly caused by an earthquake during a rainstorm, on the night of November 24, 1248, a large part of Mont Granier in France collapsed and buried five villages along with the town of St. André. There may have been up to 1,000 casualties. The geologic composition of Mont Granier (hard limestone on soft shale) facilitated the rockslide. The danger is not completely gone: a smaller, but substantial collapse happened in 2016 and came close to damaging houses again.

Earthquake and Landslide Near Villach, Austria

In the early afternoon of January 25, 1348, a large earthquake hit the Austrian province of Carinthia and the North of Italy. The quake lasted about two minutes and among many other impacts caused a large rockslide that blocked the river Gail upstream of the town of Villach in Austria. The dam created by the rockslide ultimately broke, which caused extensive flooding in Villach. The earthquake and landslide were very extensively described and commented on, and they were often linked to the arrival of the Black Death plague in Italy in the same year.

The Bridge of the Gods, Oregon

A large landslide called the Bonneville debris flow, which probably occurred around 1450, moved large amounts of debris into the Columbia Gorge near what is now Cascade Locks in Oregon in the United States. The debris blocked the Columbia river with a natural dam that was about 200 feet (61 m) high and 3.5 miles (5.6 km) long. As often the case, the blocked river filled up a lake behind the natural dam that ultimately washed away most of the debris. Native American legends explain that this "Bridge of the Gods" was made to allow two sons of the God Tyhee Saghalie to meet in the middle of their dominions.

The Haiyuan Earthquake and Landslides Near Ningxia, China

On December 16, 1920, an 8.5-magnitude earthquake hit the rural Haiyuan district in China. As a result, 675 large earthflows or mudflows happened in the weak loess material (loosely compacted deposits of windblown sediment) in the district. Of course, these landslides caused large numbers of casualties and enormous losses. It was estimated that more than 100,000 people lost their lives in the event—more than 50 percent of the district's population.

IMPACTS

Landslide impact greatly depends on the kind and size of landslides. Some common impacts are those on roads and other infrastructure, on buildings, and on crops and livestock. Impacts of landslides on individual persons are rare, although rockfall (a type of landslide) is widely recognized as a danger to mountain climbers.

A Google Earth image of the enormous Saidmarreh landslide in Iran in 2009. The area is currently submerged under a reservoir. The slope from which the landslide originated is white with black outline. The landslide deposit is in the middle, without outline, and the two ancient lakes caused by blocking by the landslide deposit are dark with black outline. Over time, these lakes have filled in with sediment. For scale: the white landslide slope is 9.3 miles (15 km) wide, and 3.1 miles (5 km) long. The landslide origin is at 32°59' N and 47°37' E.

Roads and Other Infrastructure

Roads in hilly and mountainous terrain often must cover large distances across or under slopes that are partly susceptible to landsliding. This makes damage from landsliding likely, and it is therefore quite common that mountain roads are closed to traffic after landslides and sometimes during periods of increased landslide hazard. Other forms of infrastructure, such as gas or electricity lines and irrigation canals, are similarly at risk in mountainous and hilly terrain. Just like for roads, limited damage to such infrastructure is often enough to cause large problems. One localized break in an irrigation canal, for instance, can mean that the entire downstream service area is without water.

Road construction itself can contribute to landslide susceptibility. This is because roads that cross a slope often are partly cut into the slope, which makes the overlying part of the slope steeper and therefore more susceptible to landsliding. For this reason, important mountain roads include engineering structures that are designed to support the overlying slopes.

Buildings and Towns

Buildings and towns are sometimes built in areas susceptible for landslides, especially where population pressure is high and less desirable terrain needs to be used for housing. Two types of such areas are steep slopes and alluvial fans on the edge of safe(r) valleys. One example of landslide-related problems in these places is formed by poor neighborhoods on steep land in Rio de Janeiro and Sao Paulo in Brazil. These neighborhoods are frequently affected by landslides, with losses of lives and damages to homes as a result. Another example are small towns and neighborhoods built on alluvial fans in the narrow coastal strip in Vargas state in northern Venezuela, which were hard hit during unusually heavy rains in December 1999. Between 10,000 and 30,000 people lost their lives, and damages amounted to several billion dollars.

Older towns and neighborhoods are safer from landslides because of the simple fact that they have not been destroyed during their existence, despite the extreme events that must have occurred. Using this reasoning, newer buildings and towns in hilly land can be more at risk, unless detailed planning has preceded their construction. Of course, extreme events and changing climate can still damage older structures, such as in the town of Ponzano in central Italy, which was damaged by a slow-moving earthflow in 2017. The earthflow followed an extreme earthquake in the region half a year earlier.

Crops

Although this is less often reported, landslides frequently affect crops and livestock. Many hilly and mountainous areas have no safe acreage for agriculture and livestock grazing. This means that these activities are performed in the more dangerous parts of landscapes, often with the implicit expectation that crops or livestock can be lost to various types of landslides during extreme events.

Landscapes

In addition to their impacts on our societies, landslides can have large effects on the landscapes in which they occur. The main effect of landslides is to flatten slopes that are overly steep. For this reason, landslides are seen by many as a safety valve in landscapes: whenever rivers cut valleys that are too deep and steep, landslides will ultimately happen, which move material into the valleys and reduce the slope steepness. In many mountain ranges, this process is the dominant way that mountain ranges get broken down—normal erosion, where flowing water picks up dirt and transports it from a hillslope to a river channel, or other mass movement processes are simply not fast enough in these cases.

The local and regional effects of landslides can also be substantial; however, not all landslides move all the way down a slope into a valley. Often, original landslides that do make their way off the slope into a valley dam rivers, such as the Saidmarreh landslide in Iran. If the landslide is large relative to the size of the river that is dammed, the dam may exist for hundreds and thousands of years and create a semi-permanent lake upstream. This is called "upstream flooding," and it represents a natural hazard in landslide-prone areas. After the famous 2014 Oso mudslide in Washington State in the United States, for instance, upstream flooding blocked a highway that hampered recovery efforts. Over geologic time scales, long-lived lakes that are formed by a dam may fill up completely or almost completely with sediment provided from further upstream. As is the case with the Saidmarreh landslide and many others, this land created by landslide damming can be very fertile.

In cases where the landslide is small relative to the size of the river that is dammed, the force of the water that is being dammed, or the force of the water flowing over the dam, can result in a collapse of the dam. This can cause large floods downstream and is therefore called "downstream flooding." In some cases, these floods represent a larger natural hazard than the original landslide. Therefore, some landslide dams are tunneled to release water in a controlled way. An example of this is the dam of Lake Waikaremoana in New Zealand, where electricity is now generated by the water that is released from the dam.

Secondary effects of upstream and downstream flooding present additional natural hazards. First, as upstream flooding inundates upstream land, it also saturates upstream slopes. This can cause secondary landslides. If that is not enough to cause a landslide, when dams fail or breach after overtopping, the catastrophic drop in water level above the former dam leaves upstream hillslopes saturated with water, but unsupported by standing water at their toes. This is another cause for secondary landslides. Finally, downstream flooding can be so intense that it undercuts and erodes slopes downstream of the dam and causes secondary landslides in that way.

In other cases, the original landslides do not make their way completely down a hillslope. Instead, they stop before that point, for instance, because the slope steepness is lower near the bottom of a slope or because not enough material is sliding to overcome friction with the terrain and the vegetation on it. Many rotational slides and earthflows stay on slopes in this way and therefore do not cause the natural hazards associated with damming. However, they do cause secondary natural hazards by their effects on the subsurface. This is the case when the movement of material over the subsurface creates a smeared layer that inhibits infiltration of rainwater. When this happens, rain that falls after the landslide event has difficulty infiltrating into the deeper subsurface and remains in the landslide deposit. As a result, the landslide deposit remains wetter and heavier than the surrounding landscape and can slide again. This effect is called the "bathtub effect," with the edge of the landslide as the bathtub and the wet slide as the water. The bathtub effect is the reason behind the annual movement of slow earthflows such as the Slumgullion earthflow in Colorado in the United States. The Slumgullion earthflow moves up to 20 feet per year, especially in the wet season (Varnes and Savage 1996). The bathtub effect may also explain why the study area near Perugia, Italy, appears to experience new landslides, especially where old landslides have occurred before (Samia et al. 2017a).

SUSCEPTIBILITY, HAZARD, AND RISK ASSESSMENTS

Much time and energy has been spent on trying to predict landslide occurrence and impact. The general process consists of three steps. First, attempts are made to find locations that are susceptible to landsliding. This step is called the susceptibility assessment. Second, knowing where slides are likely to occur, their timing and size are estimated. This step is called the hazard assessment. Finally, knowing the where and when of slides, it can be estimated what the losses and damages are. This last step is called the risk assessment.

Susceptibility assessment almost always uses knowledge of previous landslides as a starting point. By studying where landslides used to occur, experts try to estimate where they will occur in the future. The most popular technique to do this is using statistics that relate known landslide occurrences to factors such as slope and vegetation cover. If successful, this process leads to maps that predict the probability of landslide occurrence. This probability is high in places that have similar slopes, vegetation cover, etc., as places where landslides have already occurred. Many statistical techniques have been used for this purpose, including logistic regression, decision tree methods, and machine learning techniques.

Hundreds of susceptibility maps have been produced, sometimes for the entire world but much more often for smaller areas where the processes of landsliding and the geologic conditions are similar. In some cases, maps produced with different statistical techniques have been compared with each other, to learn which technique is most suitable. The main use of susceptibility maps is in planning construction of houses and infrastructure. In those cases, there is no need to know when exactly a landslide might happen—suitable locations simply need to be safe all the time.

Using susceptibility maps, the next step is predicting when a landslide will happen and how large it will be. This so-called landslide hazard assessment uses information about the susceptibility, the past weather, and the expected weather. Ideally, the hazard assessment changes rapidly with changing weather forecasts and is then rapidly communicated to those at risk. Using large-scale, satellite-based rainfall intensity data and predictions, modern hazard assessments are increasingly accurate in pinpointing the expected time and place of landslides. The information from hazard assessments is used for temporary blockages during particularly dangerous periods, for instance of roads or rail lines and for evacuations of towns or structures that are most at risk.

An important part of hazard assessments is knowing how much rain has already fallen and how much soil wetness has increased as a result. More recent and more intense rainfall has a greater effect on wetness than rain from longer ago or rain that is less intense. However, direct measurements of wetness are scarce, and therefore, rainfall maps are used to calculate a measure called antecedent wetness, which combines all recent rainfall into one number.

Alternative approaches to hazard assessments take a longer-term view. These consider the frequency of various extreme weather events to calculate which size of landslide occurs with which frequency. Although such hazard assessments are not updated before and during storms, they are very useful for flagging areas that are particularly at risk.

Finally, results from hazard assessments can be used to produce landslide risk assessments. In risk assessments, the expected damage from landslides is calculated. For example, in two areas with the same landslide hazard but with different land use, one of the areas is used as arable land, and the other is a village; the village will have a higher landslide risk because of the higher expected damage, even though a landslide is equally likely in both areas. Using risk assessments, authorities can better plan advance deployment of civil engineers and rescue personnel—or in the case of longer-term risk assessments, they can target protective measures to those areas that are most at risk.

CONCLUSION

Landslides are common and occur in many parts of the world. They are dangerous to societies, although they occasionally produce beneficial side effects over long timescales. To combat these dangers, much attention has been spent on attempts to predict landslide susceptibility, hazard, and risk. Although these attempts are incomplete, they have already resulted in a reduction of losses in many parts of the world. From a landscape perspective, landslides are successful processes of mass movement that affect not only their own slopes but also often large parts of the landscapes around them.

Arnaud Temme and Jalal Samia

Further Reading

Bondevik, S., J. Mangerud, S. Dawson, A. Dawson, and Ø. Lohne. 2003. Record-breaking Height for 8000-year-old Tsunami in the North Atlantic. *EOS, Transaction of American Geophysical Union* 84: 289.

Cruden, D. M., and D. J. Varnes. 1996. Landslide Types and Processes. *Special Report, Transportation Research Board, National Academy of Sciences* 247: 36–75.

Harrison, J. V., and N. L. Falcon. 1938. An Ancient Landslip at Saidmarreh in Southwestern Iran. *Journal of Geology* 46: 296–309.

Samia, J., A. Temme, A. Bregt, J. Wallinga, F. Guzzetti, F. Ardizzone, and M. Rossi. 2017a. Characterization and Quantification of Path Dependency in Landslide Susceptibility. *Geomorphology* 292: 16–24.

Samia, J., A. Temme, A. Bregt, J. Wallinga, F. Guzzetti, F. Ardizzone, and M. Rossi. 2017b. Do Landslides Follow Landslides? Insights in Path Dependency from a Multi-temporal Landslide Inventory. *Landslides* 14: 547–558.

Varnes, D. J. 1978. *Slope Movement Types and Processes*. Transportation Research Board Special Report No. 176. Washington, DC: National Academy of Science.

Varnes, D. J., and W. Savage. 1996. *The Slumgullion Earth Flow: A Large-scale Natural Laboratory*. U.S. Geological Survey Bulletin 2130. Denver, CO: USGS Information Services.

Lightning

Lightning is one of the most significant atmospheric hazards, often called an associated precipitation hazard. It is caused by electrical imbalance between thunderstorm clouds and the ground or within the clouds themselves. It is a giant discharge of electricity that causes current to flow over a short period of time. It is accompanied by a bright flash of light and a sound like a loud crack and is of relatively short duration. More distant thunder is known for its rumbling noise and echo because of the sound waves being reflected by air layers of different density as well as from hills and large buildings. Lightning is a brilliant flash of electricity, no more than one inch in diameter, and seems to be clear or a white-yellow color. Depending on atmospheric conditions, the sound of thunder can be heard as far away as 20 miles (32 km) from its origination (Ebert 2000).

Moreover, lightning is relatively widespread, particularly in tropical areas where more than two-thirds of the world's population lives, and it poses a significant threat to human lives and property. In fact, lightning kills and injures more people each year than some natural disasters. It is estimated that worldwide it causes approximately 24,000 fatalities with 10 times as many injuries. According to National Weather Service (NWS) data over the last 30 years (1989–2018), the United States has averaged 43 reported lightning fatalities per year. More recently, in the last 10 years (2009–2018), the United States has averaged 27 lightning fatalities annually, yet most incidents are avoidable (Jensen and Vincent 2019).

With modern technology, it is possible to detect lightning frequency anywhere in the world since lightning discharges generate high-frequency radio noise that can be accurately detected and located by monitoring satellites. Worldwide, there are on average 1.6 billion lightning flashes in a year, and 20 percent of these directly strike ground. Lightning is one of the oldest observed natural phenomena on earth, and humanity has encountered it since times immemorial. In addition to thunderstorms, it can be seen in volcanic eruptions, extremely intense forest fires, surface nuclear detonations, heavy snowstorms, and in large hurricanes (NSSL n.d.). In some countries like Bangladesh, lightning deaths have been increasing due to the unpredictability and frequency of strikes, lack of safety knowledge and awareness, lack of shelters and lightning safety precaution training, and high density of population. Despite disastrous consequences, this hazard is widely ignored by hazard researchers across the world.

CAUSES OF LIGHTNING

The formation of lightning is a complex process. It occurs when two clouds with opposite charge content (plus, proton or positive ions, and minus, electron or negative ions) come near each other or when the lower portion of clouds contains a negative charge and the upper portion contains a positive charge. If both charges meet, these give rise to upward lightning at a very rapid speed. This upward lightning is called thunder and is produced by thunderstorms, which in turn are produced from thunderclouds or cumulonimbus clouds. Such clouds contain many small bits of ice that bump into each other, and these collisions create an electric charge. After a while, the whole cloud fills up with electrical charges, causing the formation of positive charges in the upper portion of the cloud and negative charges at the bottom of the cloud. The top of the cloud extends far into the freezing zone, and the lower part of the cloud with negative charges occupies the largest part of the cloud base.

Since opposite charges attract each other, a positive charge builds up on the ground beneath the cloud through induction by the strong negative potential of the cloud base. This inducted positive electrical charge concentrates around any tall objects on the ground, such as mountains, high-rise buildings, people, animals, towers, telephone poles, or isolated single trees. The charge coming up from these points eventually connects with a charge reaching down from the clouds, thus causing lightning strikes (NSSL n.d.). This way of forming lightning is known as charge separation or polarization theory or "collision" theory and is aided by updrafts and the presence of ice in the cloud. However, lightning also occurs outside the regions of strong updraft. Given the frequency and unpredictability of lighting, the possibility of concentration of positive charge around a tall and exposed object clearly demonstrates the usefulness of such an object having a metal lightning rod to attract the lightning bolt and conduct the electrical discharge to the ground. Each building should have a properly installed lightning rod to protect buildings from lightning damage (Ebert 2000).

However, as opposed to the collision theory, the "convective" theory maintains that lightning precedes heavy precipitation. This theory is based on convectional updrafts instead of falling precipitation particles, leading to the buildup of negative and positive charge fields. Thus, after lightning occurs, torrential rains fall. Regardless of which theory is correct, separate regions of opposite electrical charges must build before the birth of lightning (Ebert 2000).

FEATURES OF LIGHTNING

Lightning has not always struck only tall objects. It can strike the ground in an open field even if a tree line is close by: it all depends on where the charges accumulate. Another feature is that lightning is extremely hot with an average core temperature of approximately 54,000 degrees Fahrenheit (29,982.2 degrees Celsius), or about five to six times hotter than the surface of the sun. Lightning can be both negatively and positively charged and can take both direct and alternating forms of current. Moreover, only about 10 percent of people who are directly struck by lightning are killed, leaving 90 percent with various degrees of permanent disability. Lastly, the duration of a lightning strike is one-tenth of a second.

The wind around both thunder and lightning becomes five times more heated than the surface of the sun, and lightning strikes with huge sound. If anyone hears the sound of thunder, particularly from dark thunderclouds, then that person is in danger from lightning, whereupon the best thing to do is to go indoors or get inside a fully enclosed vehicle, but avoid staying outside. If a person is caught outside, the person needs to move to a low place to avoid tall objects. The person should not lie down because lightning can flow along the surface of the earth. Since the speed of sound is slower than the speed of light, the sound is heard after a thunder strike. Another trait is that lightning carries a tremendous amount of energy. A bolt of lightning can contain a voltage exceeding 10 million volts with approximately 50,000 amperes of light as opposed to household light, which contains 15,000 amperes only. Sometimes, thunder/lightning may strike with 30 million volts of electricity.

TYPES OF LIGHTNING

The type of lightning strikes depends on three factors: strength of thunderstorm, location, and presence of contrasting electrical charge fields. There are four common types of lightning: intra-cloud lightning, inter-cloud lightning, cloud-to-ground lightning (CG), and ground-to-cloud lightning (GC). Thus, lightning can strike the ground, air, or clouds. More than 60 percent of all lightning strikes are classified as intra-cloud lightning. Both intra-cloud and inter-cloud lightning combined are termed cloud flashes (CF). Cloud flashes never leave the clouds but travel between differently charged areas within or between clouds. These flashes have visible channels that extend out into the air around the storm, called cloud-to-air or CA, but do not strike the ground. There are roughly 5–10 times as many cloud flashes as there are CG flashes.

Intra-cloud discharge may lead to what is often called sheet lightning. This lightning is embedded within a cloud such that it lights up or illuminates entire clouds during the flash. This type of lightning occurs when there are both positive and negative charges within the same cloud and looks like a bright flash of light. This type of lightning is harmless and usually occurs at high elevation. Inter-cloud lightning or cloud-to-cloud lightning (CC) is less common, and its strikes go from one cloud to another. This type of lightning is not harmful to objects on the ground, but it can be hazardous to aircraft.

The best-known lightning is cloud-to-ground lightning, which can strike the earth's surface at a rate of about 100 per second, and each CG bolt can contain up to one billion volts of electricity. Most cloud-to-ground lightning strikes come from the negatively charged bottom of the cloud traveling to the positively charged ground below. CG lightning strikes are most dangerous for objects on the ground, and these can cause fire and property damage. The bolts of this type of lightning are so intense that such lightning can split trees and may turn a sandy surface into a glassy crust (Ebert 2000).

GC lightning, also called triggered lightning, is the reverse of CG flashes and travels from ground to cloud. GC lightning originates from metal high towers on the ground or top of buildings such as radio, television, or telephone towers. As with CG lightning, a typical triggered lightning starts with a step-like series of negative charges, called a stepped leader, or simply leader. In CG, the stepped leader is created by excess electrons in a thunderstorm cloud. It travels downward from the bottom of a thunderstorm cloud toward the ground, looking for positive charges. The stepped leader is not visible to the human eye and shoots to the ground in less time than it takes to blink. When a stepped leader created by CG flashes travels near the ground within 150 feet (46 m), the negatively charged stepped leader is attracted to positively charged objects (e.g., a tree, person, telephone pole, and house), called streamers, powerful electrical currents begin to flow toward earth. Thus, a streamer develops as the downward moving leader approaches the ground. Typically, only one of the streamers makes contact with the leader as it approaches the ground and provides the path for the bright return stroke, which releases tremendous energy and thunder. The return stroke follows the same path as the original lightning flash. It is much larger and brighter than the original flash (Ebert 2000). CG lightning flash contains one or more return strokes.

Another related term, heat lightning, is lightning or lightning-induced illumination that is too far away for thunder to be heard, so called because it appears most often on a hot summer day when the sky is clear overhead. Also, forked lightning appears as jagged lines of light, which can have several branches. Forked lightning can be seen shooting from the clouds to the ground, from one cloud to another cloud, or from a cloud out into the air. This lightning can strike up to 10 miles (15 km) away from a thunderstorm.

Spider lightning refers to long, horizontally traveling flashes in the sky, which can be up to 120 miles (180 km) long. A subset of spider lightning, high-altitude lightning, has been given several names such as "red sprites," "green elves," and "blue jets." This form of lightning appears as brightly colored flashes high above thunderstorms. When a bolt of lightning separates due to wind and appears as parallel lightning streaks, it is called ribbon lightning. Meanwhile, chain or bead lightning is when a lightning bolt is broken into dotted lines while fading. Next, ball lightning is a rare form that appears as a reddish, luminous ball, but can come in any color. Ball lightning is usually spherical, and its size varies from a few inches to over six feet (2 m) in diameter. Lastly, anvil lightning is referred to as "the bolt from the blue" because it often appears suddenly from a seemingly cloudless sky.

THE IMPACT OF A LIGHTNING STRIKE

Lightning is not only spectacular but also dangerous. However, unlike other natural hazards, lightning bolts kill only one person in 91 percent of events. For this reason, deaths by a lightning strike are relatively few. However, there are some areas in the world where deaths occur at a higher rate than in others. For example, Central Africa has a greater incidence of lightning strikes than any other large region. Naturally, the majority of lightning deaths and injuries are caused when people stay in unsheltered areas such as play grounds, golf courses, parks, or open agricultural fields.

Relatively more people die from lightning in developing countries than in developed countries. The estimated death rate is three per million people per year in developed nations, while the corresponding figure is six per million in developing nations. The Nkhata Bay District of Malawi has the highest known rate in the world—84 per 1,000,000 people (Spring 2013). This district lies near the eastern Congo Basin and is considered the global "hot spot" of lightning flash density and experiences high but not highest rates of lightning strikes (Salerno et al. 2012). In a given country, mortality rates are higher for people living in rural areas than in urban areas. For example, rural areas in China accounted for 51 percent of total lightning fatalities despite less population per square mile. Likely, this is because a majority of the injuries and deaths occur to people working in fields (Zhang et al. 2011).

Meanwhile, the number of injuries due to lightning is estimated at 240,000 a year. Not all lightning injuries occur the same way, so injuries are classified as direct strike, side splash, contact injury, upward leader, ground current, and blunt injury. These terms are associated with the ways lightning strikes people. If a person is struck directly by lightning, the person becomes a part of the main lightning discharge channel. Direct strikes mostly occur to victims who are in open areas. Although this type of strike is not as common as the other ways people are struck by lightning, but it is potentially the deadliest. "In most direct strikes, a portion of the current moves along and just over the skin surface (called flashover) and a portion of the current moves through the body—usually through the cardiovascular and/or nervous systems. The heat produced when lightning moves over the skin can produce burns, but the current moving through the body is of greatest concern" (NWS n.d.). Injuries caused by this type of strike account for 3–5 percent of such injuries.

A side flash (also called a side splash) injury occurs when lightning strikes a taller object near the victim and a portion of the current jumps from the object to the victim. The victim acts as a "short circuit" for some of the energy in the lightning discharge. Side flashes generally occur when the victim nears an object that is struck. Next, because ground current affects a much larger area than the other ways of lightning strikes, it causes the most deaths and injuries, accounting for 50 percent, followed by side splash, which accounts for 33 percent (Jensen and Vincent 2019). It occurs when lightning strikes an object on the ground and much of its energy spreads outward from the strike in and along the ground surface. Any person outside near a lightning strike can be injured from such a current. Thus the distance from the lightning strike to the person is an important determinant for both deaths and injuries. Ground current also kills many farm animals (NWS n.d.).

Contact or conduct injuries are caused by touching an object that is struck by lightning. This type of injury can take place both inside a building or outside. Anyone in contact with anything connected to metal wires, plumbing, or metal surfaces that extends outside such as anything that plugs into an electrical outlet, water faucet, shower, or corded phone is at risk of death and injury. Likewise, being on a porch or near a door or window poses at risk also. One other way lightning strikes people is by streamers, but injury by this type of strike is not as common as the other types of injuries; streamers cause 3–5 percent of all lightning-related injuries (NWS n.d.). Upward leaders generally cause 10–15 percent of all injuries (Holle 2016).

As indicated, many people have survived lightning strikes but suffered from a variety of lasting symptoms, including memory loss, neurological problems, dizziness and weakness, numbness, seizures, and other life-altering ailments. A study conducted in northern Malawi reported that of the 317 studied victims who had survived lightning strikes, 85 percent made a full recovery, while permanent symptoms included hearing damage (7 percent), lasting pain (3 percent), headaches (3 percent), and psychological damage (2 percent) (Salerno et al. 2012). Death from lightning strike can be caused by cardiac and respiratory arrest. Lightning strikes can also cause severe burns. Some lightning victims may die a few days after resuscitation because of irreversible brain damage. Moreover, injuries often grow worse as years pass.

Lightning can injure a person in several ways. These include the effect of electric current on/through body tissues, burns on different parts of the body due to conversion of electrical to thermal energy, and damage by flying debris from nearby objects due to direct strike or shockwave, which is called traumatic or blunt injury. Bodily burns range from superficial to full thickness with the former being much more common. Lightning strikes also damage the central and peripheral nervous system and lead to vision loss, which is called blast trauma or ocular injury (Jensen and Vincent 2019).

The intense heat from lightning can ignite fires that damage structures, forests, and cropland. Lightning's extreme heat can also vaporize the water inside a tree, creating steam that may blow the tree apart. Regarding safety, being inside a car is safe as car tires conduct current, as do metal frames that carry a charge harmlessly to the ground. Thus, many houses are grounded by rods and other protection that conduct a lightning bolt's electricity harmlessly to the ground. Homes may also be inadvertently grounded by plumbing, gutters, or other materials. Grounded buildings offer protection, but occupants who touch running water or use a landline phone may be shocked by conducted electricity (National Geographic n.d.).

In terms of dollar value, the total damage caused by lightning is not as great as that associated with some other types of disaster. However, the most recent data on the damage value caused by lightning are not readily available. Nevertheless, this type of disaster is believed to cost several billion dollars a year in the United States alone as evidenced by the nearly 500,000 yearly insurance claims. In fact, half of all western U.S. forest fires are caused by lightning.

GEOGRAPHIC DISTRIBUTION

Lightning does not occur at the same intensity across the globe. Its occurrence is strongly correlated with areas having higher occurrence of thunderstorms. Thus, tropical areas are more prone to lightning strikes and account for nearly 80 percent of global lightning flashes. Areas with mountains often experience more lightning strikes than surrounding areas of lower elevation. Economic modernization is another reason for local and regional variations in lightning risk. Generally, the risk is much lower in developed countries than in developing countries. This difference is due to many factors such as higher urbanization rate, predominance of nonfarm occupation, structural improvement of buildings, and an increase in use of metal vehicles that provide protection from lightning in the former countries (Salerno et al. 2012). Regarding strike distribution, the two poles and the areas over the oceans have the fewest lightning strikes.

Six regions/countries of the world experience an unusual amount of lightning: Central Africa, Indonesia, the Himalayan Mountains, the Andes, the Pampas in Argentina, and the United States. In Central Africa, the Democratic Republic of Congo (DRC) has the highest frequency of lightning on earth. In this tropical country, thunderstorms occur year-round because of the presence of local convection and moisture-laden air masses from the Atlantic Ocean that encounter mountains as they move east from water to land. Year-round thunderstorms are also common in Indonesia, where moist winds from the Indian Ocean push warm, moist air up the mountains of Java and Sumatra and cause thunderstorms.

The foothills of the Himalayan Mountains have also one of the highest incidences of lightning. This region extends east-west and encompasses Nepal, Bhutan, Bangladesh, northern India, and eastern Pakistan. Unlike the previous two regions, thunderstorms here are a seasonal phenomenon. In the summer season, monsoon winds carry warm and moist air from the Indian Ocean, and the air strikes the Himalayan Ranges and rises up, causing thunderstorms and heavy rainfall in the Himalayan Forelands.

The remaining three regions are located in the Americas: two in South America and one in North America. First, thunderstorm activity is intense in northwestern South America, where warm winds from the Pacific Ocean carry moisture-laden air masses and strike the Andes Mountains in Colombia, Ecuador, and Peru, causing heavy thunderstorms. Second, the Pampas of southeastern Argentina is another region where intense thunderstorm activity is common in summer and spring because of moist seasonal winds coming off the Atlantic Ocean and causing violent thunderstorms. Finally, in the United States, Central Florida, between Tampa and Orlando, is known as "lightning alley." Here thunderstorm activity is common where warm, rising air pulls in sea breezes from the Atlantic Ocean and the Gulf of Mexico.

In the United States, lightning is one of the leading weather-related causes of deaths and injuries. However, the longitudinal data shows that despite increased population, lightning deaths have decreased over the past 50 years as the country has become more urban. About 300 people died per year before the 1940s, but that number has dropped to only 16 in 2017. More than 80 percent of lightning victims are males with most deaths occurring in individuals 20–45 years old. Most deaths occur within one hour of injury and are due to fatal arrhythmia or respiratory failure. Ninety percent of lightning deaths occur in the countryside between May and September with nearly 70 percent of incidents occurring in the afternoon or early evening. The Fourth of July is historically one of the deadliest times of year for lightning in the United States because so many people are outside. Specifically, two states (Florida and Texas) account for 25 percent of U.S. deaths. The remaining deaths are generally concentrated in the Great Lakes region, the South, and the southern Great Plains.

But the highest incidence of lightning strike occurs in states like Arizona, Arkansas, Mississippi, and New Mexico. During a life span of 80 years, a person in the United States has a one in 10,000 chance of being struck by lightning (Jensen and Vincent 2019). Additionally, as opposed to a declining trend in lightning deaths, lightning injuries in the United States have increased over the past decade. Currently, about 400 people are injured annually, and men account for 80 percent of such injuries.

LIGHTNING SAFETY TIPS

In many instances, most fatalities can be avoided with a better understanding of the nature of lightning hazard. Insurance companies, particularly in the United States, recommend to properly ground lightning rods in all buildings. Otherwise, the company may not insure or may charge higher rates of insurance for unprotected buildings. If one hears the sound of thunder, the person likely would be within 10 miles (16 km) of a storm, and that person has a high probability of being struck by lightning. As soon as one hears thunder, the best action is to seek shelter immediately to avoid being struck. The ideal place to go is a sturdy building or a car, making sure the windows in the car are rolled up. People should stay in a safe place until at least 30 minutes after the last clap of thunder, and if inside a house, they should be away from windows and doors. Additionally, people should avoid sheds, picnic areas, house porches, or any open field and high ground. If there is no shelter around, people should stay away from trees or keep twice as far away from a tree as it is tall. Standing under a tree is one of the most dangerous places to take shelter during a lightning thunderstorm (Holle 2016).

Other tips are to put feet together to minimize the chance of lightning strike and to place hands over ears to minimize hearing damage from thunder. If any person is with a group of people, they all should stay about 15 feet (4.6 m) from each other. Water is dangerous during lightning, and thus, people should avoid taking a shower or bath, washing hands and dishes, or laundering and drying clothes. People should also avoid water bodies and beaches because water is a great conductor of electricity. Thus, swimming, wading, snorkeling, scuba diving, standing in puddles, and, of course, being near any metal would be off limits. Further, people should stay away from clotheslines, fences, and drop backpacks because they often have metal on them. All in all, people would be wise to avoid outdoor activities before, during, and after thunderstorms.

People are warned against using a corded telephone during a storm because it is the leading cause of indoor lightning injury, particularly in the United States. Most people who get hurt by lightning while inside their homes happen to be talking on the telephone at the time. This is because lightning may strike exterior phone lines and can travel long distances, particularly in rural areas. Outdoors, lightning can travel along the outer shell of the building or may follow metal gutters and downspouts to the ground. Thus, inside a structure, lightning can follow conductors such as the electrical wiring, plumbing, and telephone lines to the ground. Thus, using electrical equipment such as computers and appliances during a thunderstorm is also unsafe as is being near windows and doors or being on a porch. In general, basements are a safe place to take shelter during thunderstorms. However, concrete walls may contain metal reinforcing bars, which would make those walls unsafe.

There are three main ways lightning enters homes and buildings: by a direct strike, through wires or pipes that extend outside the structure, and into the ground. Regardless of the method of entrance, once in a structure, lightning can travel through the electrical, phone, plumbing, and radio or television reception systems. Lightning can also travel through any metal wires or bars in concrete walls or flooring. Another consideration is that an umbrella can increase an individual's chances of being struck by lightning if it makes the person the tallest object in the area. Thus, one should always avoid being the highest object anywhere or taking shelter near or under the highest object and avoid being near a lightning rod or standing near metal objects such as a fence or underground pipes.

If a person is struck by lightning, medical care may be needed immediately to save the person's life. With proper treatment, including CPR if necessary, most victims survive a lightning strike, although the long-term effects on their lives and the lives of family members can be devastating. Since victims of lightning strike do not carry electrical charge and are not "electrified," it is safe to touch and help them. However, rubber shoes will not give any meaningful protection from lightning, but wearing dry clothes would offer some protection because such clothes serve as an insulator.

Public safety education is needed in many lightning-prone countries to increase awareness and knowledge about safety measures related to lightning risk. For instance, four lightning safety myths are false, and these myths can be avoided through proper education, particularly in developed countries and rural areas: (1) lightning is attracted to metal; (2) direct strike is the most common cause of fatalities; (3) person is safe outside of rain; and (4) rubber tires provide protection. There is also a need to initiate behavioral changes in people such as avoiding open space and bodies of water and seeking shelter in current-resistance buildings. Fatalistic attitudes that lightning strike is a matter of luck must be got rid of through safety education. People need to be informed of the safety rules, and awareness should be increased by disseminating proper and accurate information. As most lightning fatalities occur in agricultural fields, properly designed lightning safety structures near such fields are necessary to reduce death and injuries caused by this hazard.

CONCLUSION

Most lightning results from a strong separation of electric charge that builds up between lower and upper parts of cumulonimbus clouds. This separation helps lighter, positively charged water droplets and ice particles to rise to the top of the cloud while the heavier, negatively charged particles sink to the cloud's base, and this charge jumps to join the positive charge on the ground. This causes cloud-to-ground lightning, which is the most common type of lightning. In fewer cases, lightning will strike from the ground to the base of the cloud, which is called an upwardly forking lightning strike. Spatial and seasonal variations in lightning frequency and its impact are evident. Worldwide, low latitude has higher flash rates than high latitudes, and summer season has more lightning than the other season. Age and gender variations among victims are also reported. Mostly men who are outdoors are killed and injured by lightning strikes.

Spatial and temporal studies on various aspects of lightning hazard and its impacts are scanty and therefore should be studied particularly in developing countries. In addition to safety education, such a study can help experts to increase awareness among vulnerable people, which in turn could help minimize death tolls and injuries from lightning strikes around the world. As indicated, significant education and awareness programs have reduced the death rates and injuries in many developed countries like the United States. For example, studies show that the literacy rate among people is positively associated with lightning safety awareness, particularly in rural areas of developing countries. Thus, education is a powerful key that helps in communicating effectively to mitigate the impacts of lightning. Countries that do not have proper weather forecasting and warning systems need to introduce these two components of hazard mitigation. Lastly, many developing countries also need a comprehensive plan and policy to reduce deaths and injuries associated with lightning strikes. Further research should be conducted as a start to policies and plans for vulnerable people. With such an approach, public training needs to be introduced to make people more conscious and aware of lightning hazard.

Bimal Kanti Paul

Further Reading

Ebert, C. H. 2000. *Disasters: An Analysis of Natural and Human Induced Hazards*. Dubuque, IA: Kendall/Hunt Publishing Company.

Holle, R. L. 2016. A Summary of Recent National-Scale Lightning Fatality Studies. *American Meteorological Society*. https//doi.org/10.1175/WCAS-D-15-0032.1, accessed July 2, 2019.

Jensen, J. D., and A. L. Vincent. 2019. Lightning Injuries. StatPears (Internet). https://www.ncbi.nlm.nih.gov/books/NBK441920/, accessed June 28, 2019.

King, H. M. n.d. World Lightning Map. https://geology.com/articles/lightning-map.shtml, accessed June 30, 2019.

National Geographic. n.d. Lightning. https://www.nationalgeographic.com/environment/natural-disasters/lightning/, accessed June 28, 2019.

NSSL (National Severe Storms Laboratory). n.d. Severe Weather 101—Lightning. https://www.nssl.noaa.gov/education/svrwx101/lightning/, accessed June 28, 2019.

NWS (National Weather Service). n.d. Lightning Science: Five Ways Lightning Strikes People. https://www.weather.gov/safety/lightning-struck, accessed June 29, 2019.

Salerno, J., L. Msalu, T. Caro, and M. B. Mulder. 2012. Risk of Injury and Death from Lightning in Norther Malawi. *Natural Hazards* 62(3): 853–862.

Spring, J. 2013. Lightning Deaths and Injuries, by the Numbers. https://www.outsideonline.com/1912401/lightning-deaths-and-injuries-numbers, accessed June 28, 2019.

Zhang, W., Q. Meng, M. Ma, and Y. Zhang. 2011. Lightning Casualties and Damages in China from 1997 to 2009. *Natural Hazards* 57(2): 465–476.

Salinization

Salinization is the process by which the level of salinity—or the content of water-soluble salts of sodium, potassium, magnesium, and calcium chloride—changes in a body of water, soil, or another substance. Salinity is the saltiness or the amount of salt that is dissolved in a body of water. The measurement of salinity is traditionally in parts per thousand or grams per thousand grams. For example, normal seawater's salinity is 33 parts per thousand. The Red Sea has a high salinity of 40 parts per thousand. The opposite of salinization is desalinization, or the decrease in water-soluble salts in a substance.

Salinization can occur because of natural causes (primary salinity) or anthropogenic (or human-based) causes (secondary salinity). Natural long-term causes include less precipitation (infiltration), sea-level change, geology, and high salt content of the parent material or ground water. An example of geologic change are the near-shore karst formations located in Australia, Thailand, Vietnam, and the United Kingdom (UK), among several others—a topography formed from the dissolution of soluble rocks such as limestone, gypsum, and dolomite.

Anthropogenic causes of salinization include human soil management practices, over-pumping (usually in arid areas), illegal pumping, usage of salt-rich irrigation water, insufficient drainage due to water scarcity, soil sealing (reduced infiltration), and water retention infrastructure (e.g., dams). Because it affects both primary and secondary salinity, climate change is a powerful causative element regarding salinization.

Many salinization forms can be very harmful to humans, including factors affecting health. For example, the increased salinization of ground water and aquifers is having significant negative impacts on crop irrigation, drinking water, etc. In the world today, increased salinization is changing the chemical composition of natural water resources, such as rivers, lakes, and groundwater; lowering the quality of water for domestic and agriculture usage; quickening the loss of biodiversity; increasing taxonomic replacement by halotolerant species; diminishing the amount of fertile soil; causing a partial collapse of the fishery and agricultural sectors; changing local climatic conditions; and creating severe health problems (such as in the Aral Basin between Kazakhstan and Uzbekistan). Salinization is expected to increase in the future due to increasing water demands, rising sea levels, and changing climate. Once salinization has occurred, it is hard to remediate, making prevention a better strategy.

Different types, levels, and causes of salinization occur in the soil; rivers, lakes, aquifer, and geothermal waters; and oceans.

SOIL SALINIZATION

Soil salinity is the measure of the soil's salt content. Soil salinization is the process that leads to the excessive accumulation of water-soluble salts found in the topsoil. When soil salinization exceeds a certain point, it severely reduces the fertility of the soil and affects the soil structure. Increased soil salinization is a serious resource concern because the resultant excessive salts limit and shorten crop growth and yield by limiting the plants' ability to take up or absorb water. Any process affecting the soil-water balance possibly can change the movement and accumulation of salts found in the soil. If the negative process continues, the soil will be unable to grow crops permanently.

Salts occur naturally within soils and water. Natural processes can cause salinization, such as mineral weathering or by the gradual withdrawal of an ocean. The ions involved in salinization are sodium, potassium, calcium, magnesium, and chloride. However, when the amount sodium becomes substantial it can make soils sodic. Sodic soils usually have a very poor soil structure, which tends to prevent or significantly limit water infiltration and drainage. In areas with sufficient (but not overwhelming) precipitation, as soil minerals are weathered by the natural elements and release salts, these salts then are rinsed or leached from the soil by drainage water.

Dust and precipitation also deposit salts in soils. In the dry land environment, soils are naturally salty. This is caused by the long-term accumulation ground buildup of salts and the corresponding lack of adequate flushing. Salt accumulation and formation of efflorescent crusts have been found in the upper unsaturated zone and fracture surfaces in many arid regions. The salts remain in the upper soil areas. Such soils, for example, can be found in the Murray Basin of Australia and the southwestern United States. In Australia, ongoing soil salinization is now a serious environmental and economic disaster that has prompted resource management changes. The chemistry of the residual saline groundwater is controlled by characteristics of the original freshwater and the subsequent saturation of such typical minerals like calcite, gypsum, spiolite, and halite. It has been found that evaporation and mineral precipitation control the salinity of groundwater that flows to Deep Spring Lake, Death Valley, and the Sierra Nevada Basin, among others.

Excessive soil salinization mainly has been caused and stimulated by humans. Human-driven processes that have stimulated increased salinization include hydrology, irrigation, drainage, road salt, plant cover and rooting characteristics, farming practices, and climate change. Serious salinization most frequently occurs because of either harmful irrigation methods or by the overuse of coastal aquifers, which then yields seawater intrusion.

Rain or irrigation, without adequate leaching, brings salts to the surface by capillary action. Irrigation-caused salinity can occur wherever irrigation is utilized, because virtually all water—including natural rainfall—is composed of at least some dissolved salts. Salination from irrigation water is also greatly increased by poor drainage and use of saline water for irrigating agricultural crops. Soil salinity in urban areas usually comes from the combination of irrigation and groundwater processes. This has become more prevalent because irrigation is now common in the gardens and recreation areas of and around cities.

Over-irrigation can increase soil salinity because it transports the dissolved salts that are found in the groundwater. After the plants' capillary system uses the water, the salts are left behind in the soil, and eventually under normal conditions, they begin to accumulate. Since increased soil salinity makes it much more difficult for plants to absorb soil moisture, they tend to dry out. To remedy this, these salts must be leached out of the plant's root zone by applying additional water and thus raising the water table. This added water in excess of plant needs is known as the leaching fraction. Hence, irrigation salinity occurs due to increased rates of leakage and groundwater recharge, causing the water table to rise. This seriously affects both the plant's rate and characteristics of growth and soil structure. The salt remains in the soil when water is absorbed by plants or evaporates.

Normally, proper irrigation management mostly prevents excessive salt accumulation by supplying sufficient drainage water capable of leaching the added salts from the soil so that they will be discharged. However, a significant disruption of the existent soil drainage patterns that normally produce leaching also can result in disastrous salt accumulations. For example, when the Aswan High Dam was built in Egypt in 1970, the added level of ground water before the construction generated substantial soil erosion and a high concentration of salts in the water table. After the dam construction was completed, the resultant artificial high water table level produced deleterious salinization of the nearby arable land. The nearby land was rendered virtually unusable by agriculture.

Excessive salinization in drylands and grasslands also mainly occurs because of human actions. The main culprits are changes in land use, such as land clearing and the resultant replacement of the area's natural vegetation with agricultural crops and pastures. These land-use practices allow more rainwater to enter the aquifer than it can hold. The involved significant replacement of natural vegetation with crops and pastures with comparatively short roots sharply spikes the amount of water that percolates below the root zone, therefore upping the recharge rate and the rate of salt leaching into the groundwater. The salts naturally stored in the root zone thus travel through the unsaturated to the saturated zone, and this causes increased salinization of the groundwater.

Another type of anthropogenic-induced salinization in dry land and grassland environments is afforestation, the process of planting trees or sowing seeds in a land without trees to create artificially a forest. This happens because the deep tree roots draw water from the below groundwater, thus reversing the natural downward groundwater flow from the soil to the soil's saturated zone. Transpiration (the carrying of moisture through plants from roots to the small pores of leaves' underside, where it converts to vapor and joins the atmosphere) of water by deep roots settles its salt in the unsaturated zone, significantly reducing the natural groundwater recharge. The result is excessive soil salinization and shallow groundwater. Soil and groundwater salinization because of grassland tree planning is found, for instance, in the northern Caspian Sea area, Australia, and the Argentine Pampas. Excessive salinity in drylands also can occur because of the clearing of trees for agriculture as the deep rooting of the trees gets replaced by the annual crops' shallow rooting structures.

Efforts to combat soil salinization include growing more salt-tolerant plants, financial support for farming affected by high salinization, and switching to other crops less affected by high saline water. Plants can be grown in areas with high levels of soil salinity if salt-tolerant plants are utilized. Crops that are very sensitive to salt influx lose their vigor in even slightly saline soils, and most types of crops are negatively affected by moderately saline soils. Only salinity-resistant crops can thrive in excessively salinized soils.

Financial assistance currently is only available for large farms in the more developed countries. These nations tend to have much stronger safety nets composed of crop insurance and government supports, as well as opportunities and options for dealing with salty soils. The many millions of subsistence farmers across the globe meanwhile must desperately look for ways to make ends meet despite the much lower soil productivity. A number of farmers are switching to alternative crops.

Climate change already is severely affecting the soil and human life, exacerbating soil salinization through rising sea levels and heat stress that depletes groundwater resources and increases the saline contamination of soils. Sections of Australia, sub-Saharan Africa, and California already are affected by this.

The abnormal heating of the seas produces warmer water that is supplemented by water generated by melting ice sheets and glaciers. This water occupies more space and has led to increased sea levels that inundate low-lying coastal areas with saltwater, which gradually contaminates the soil. The net result is that global mean sea levels are projected to rise by at least 10–20 inches (254–508 mm) by 2100, pushing salty water further inland.

Some salts can be dissipated by rainfall. However, since climate change also is increasing the frequency and severity of extreme weather such as droughts and heat waves, there is more intensive usage of groundwater for drinking and irrigation. This further depletes the water table and eases the process of additional salt leaching into the soil. Rising soil salinity is negatively affecting agricultural production and internal migration in other areas. Bangladesh particularly now is suffering from this soil salinization. In coastal Bangladesh, studies have discovered that cultivated land with moderate saline contamination annually earns about 20 percent less in crop revenue than those farms with mild soil salinization (Alam et al. 2017).

Other studies have documented that rice farmers in India lose between 7 and 89 percent of their crop depending on the season and the level of saline contamination. This probably also will occur in many other coastal areas with nearby farming, including Asia and the U.S. Pacific and Gulf coasts. This is a severe problem especially for farming because it inherently involves very thin profit margins, even for large-scale farmers, so that if yields are lowered it can lead to less income and bankruptcy. It is currently estimated that 20 percent of global cultivated land is already experiencing lessened and uneven plant growth due to salt contamination. This translates into higher food prices and more food shortages (Shrivastava and Kumar 2015).

RIVER, LAKE, AQUIFER, AND GEOTHERMAL SALINIZATION

Over half of the world's major rivers are being seriously depleted and polluted, which causes degradation of the surrounding ecosystems and threatens the health and livelihood of people who depend upon these rivers for drinking water and irrigation. This is primarily happening because of the surging global demand for water and the consequent human actions to utilize existing water supplies. Normally, salts are deposited and stored in the unsaturated zone of land. They are ultimately taken by groundwater as it discharges into nearby rivers. Studies have estimated that half of the salinity comes from natural saline discharge, 37 percent from irrigation, and 16 percent from reservoir-storage effects and municipal and industrial practices (Vengosh 2013).

There are five major sources of soluble salts in river basins: (1) meteoric salts, (2) salts derived from water-rock interaction (e.g., dissolution of evaporitic rocks), (3) salts derived from remnants of formation water entrapped in the basin, (4) geothermal waters, and (5) anthropogenic salts because of human action. Meteoric salts are concentrated by in-stream net evaporation and evapotranspiration. In addition, meteoric salts can be recycled through irrigation and development of saline agricultural drainage water that flows to the river.

Excessive salinization of rivers also is caused by human action. This includes the purposeful diversion of the streams' natural flow pattern or via dam construction. The result is a significant reduction of the affected river's natural flow discharge. For example, the Amu Darya and Syr Darya rivers in Central Asia were practically dried out because water was diverted for cotton irrigation in the former Soviet Union. The annual flow in the rivers to the Aral Sea declined from an estimated 4,308 billion cubic feet (122 billion cubic m) to virtually nothing by the mid-1980s.

Increased salinization in aquifers mainly happens because of human action: excessive pumping to supply the water needs of agriculture, homes, and recreational areas. In coastal areas, the resultant decreasing groundwater levels can produce saltwater intrusion from oceans. Other causes that can impact salinity in rivers are rises in the sea level rise and the construction of dams. Both of these are influenced by the variability of the climate, including climate change.

Deep groundwater aquifers provide an alternative source of fresh and saline water that can be useable with desalination and/or treatment. Some of these deep groundwater resources are vulnerable to contamination from oil/gas and other human activities. Moderately saline groundwater aquifers, containing lower total dissolved solid (TDS) concentrations than seawater, require less desalination and are useable for drinking water.

Regarding geothermal water salinization, groundwater salinity typically increases with depth. Geothermal water, due to its high temperature, may contain a variety of dissolved chemicals that can be corrosive to various metals used in the heating system. Geothermal water composition varies by the exposed elements of its reservoir rock and the subsurface environment. Those ions with high concentrations in these waters are sodium, potassium, calcium, magnesium, hydrogen carbonate, carbonate, and sulfate. Other pollutants include heavy metals like mercury, copper, lead, silver, iron, zinc, arsenic, manganese, chromium, beryllium, selenium, vanadium, cadmium, nickel, strontium, uranium, cobalt, gallium, and antimony, as well as possibly boron and silica, which cause serious problems if found in high concentrations. Geothermal waters usually contain a high level of the ions from dissolved salts—TDS. Sodium chloride is the most prevalent salt in dissolved form that adds salinity. Across the United States, TDS concentrations vary widely.

Excessive salinization in geothermal waters is a problem because of its increasingly wide use for various applications. The geothermal heating of buildings and water, which has been used since the Roman era, involves utilizing hot water sources near the earth's surface. Sometimes this hot water that is produced is utilized in multiple buildings and locations. Fish farms utilize geothermal water to encourage the growth of aquatic animals such as shellfish, prawns, many amphibians, catfish, and trout. To prevent winter freezing, Iceland and Holland and other Northern Hemisphere nations use geothermal water to heat sidewalks, pavements, and smaller pedestrian roadways.

OCEAN SALINIZATION

The salinity of the oceans is affected by a mix of ongoing processes and counter-processes. Salinity has a major effect on ocean circulation, including affecting the flow of currents moving heat from the tropics to the north and south poles. The weathering of rocks delivers minerals, including salt, into the ocean. Evaporation of ocean water and formation of sea ice both increase the salinity of the ocean. However, these "salinity raising" factors are continually counterbalanced by processes that decrease salinity such as the continuous input of freshwater from rivers, precipitation of rain and snow, and melting of ice.

Salinity differently affects the ocean's three physical states: liquid, vapor, and ice. As a liquid, water dissolves rocks and sediments and reacts with emissions from volcanoes and hydrothermal vents. This creates a complex solution of mineral salts in our oceans. Like vapor and ice, water and salt are incompatible, rendering water vapor and ice virtually free of salt.

Evaporation and precipitation are the two main components of the global water cycle. Over 85 percent of global evaporation and more than 75 percent of global precipitation occur over oceans. Only 10 percent of moisture is transported to the land and returned via river runoff. Ocean surface salinity, temperature, and density vary by geography and with time. While earth surface winds drive the upper ocean's currents, changes in water density and buoyancy—determined by salinity and temperature—drive circulation in the deepest parts of the ocean waters. More heat is stored in the ocean's shallowest 10 feet (3.05 m) than in all of the atmosphere. Evaporation removes freshwater from the surface of the ocean, rendering the surface waters saltier. Precipitation and river runoff freshen the surface of ocean and lower sea surface salinity.

The global patterns of evaporation and precipitation are very different. The evaporation level is highest over the tropical warm pool regions and throughout the western boundary currents because of the increased air-sea temperature contrast in these regions. On the other hand, precipitation usually is greater in high latitude areas and below the Intertropical Convergence Zone (ITCZ), the meeting place of the northerly and southerly trade winds.

The density-temperature-salinity balance determines circulation, which transports heat in the oceans and maintains the earth's climate. The many ocean currents and wind systems that move heat from the equator north to the poles and then send the cold water back toward the equator make up the thermohaline circulation. In the circulation, thermo refers to temperature and haline indicates salt content because both are influential factors in the level of density of ocean water. This is also called the Great Ocean Conveyor, a term invented in 1987 by the American geophysicist Wallace Smith Broecker (1931–2019)—who also popularized the term *global warming*.

Scientists have discovered that even small variations in ocean surface salinity (i.e., the concentration of dissolved salts) can significantly affect ocean circulation and the water cycle. The thermohaline circulation depends upon the process where the cold dense waters sink deep into the ocean because the circulation is primarily driven by changing differences in the water's density. This process can be seriously affected by global warming because it warms the ocean surface waters and melts ice, which pumps freshwater into the circulation and thus renders it less saline. At a certain point, this fresher water can stop the cold water from sinking and subsequently alters the critical ocean current circulation patterns.

The circulation renders the cold ocean surface water saltier in the high latitudes, as some water evaporates and/or as salt is emitted as sea ice forms. Because this comparatively saltier, colder water is denser and heavier than the upper ocean waters, it soon falls deep into the ocean depths. Through the prevailing currents, it is transported at the lower levels of the ocean until it rises to the water's surface near the equator. This usually occurs in the Indian and the Pacific Oceans. Once on the ocean surface, the cold water is heated by the sun, and its subsequent evaporation increases the salinity of the waters. Circulation currents then carry this warm salty water northward, and it joins the Gulf Stream, which is a large, normally occurring, and powerful ocean current that is similarly affected by prevailing winds. The Gulf Stream takes the warm salty water north up the U.S. eastern coast and then crosses the North Atlantic Ocean area, where it releases heat and warms the United Kingdom and Western Europe. Once the water has released its heat and arrives at the northern North Atlantic, it becomes very cold and highly dense again. It sinks to the ocean depths, and the cycle continues. The thermohaline circulation plays an important role in determining the climate of the different parts of the globe.

Part of the thermohaline circulation is the Atlantic Meridional Overturning Circulation, which includes the Gulf Stream. This is an ocean circulation system that similarly transports heat northward from the tropics and Southern Hemisphere to the northern North Atlantic, Nordic and Labrador Seas. This also culminates in the colder waters sinking into the water's depths.

It is true that not every scientist believes that the Atlantic Meridional Overturning Circulation has been slowing or that, if it actually is, it is caused by global warming spurred by human action. Analysts still debate exactly why the Atlantic Meridional Overturning Circulation has weakened and what is the exact role of human action in causing variations in the climate.

Since 2011, there have been major scientific advances in measuring and analyzing global ocean salinity. Past measurements primarily were limited to summer observations made by ships traveling in the ocean shipping lanes. However, a European mission recently has started making ocean surface salinity measurements utilizing advanced technology. With the 2015 launch of NASA's satellite instrument Aquarius aboard the Argentine spacecraft Aquarius/Satélite de Aplicaciones Científicas (SAC)-D, scientists have collected more data in months than had been obtained in the previous 125 years. Aquarius/SAC-D was a collaboration between NASA and Argentina's space agency, Comision Nacional de Actividades Espaciales (CONAE), with participation from Brazil, Canada, France, and Italy. In addition, the Soil Moisture Active Passive (SMAP) satellite and the European Space Agency's Soil Moisture and Ocean Salinity (SMOS) platform can measure the salinity of sea surfaces.

The Aquarius data has shown that there are a multitude of global ocean salinity patterns. For example, the Arabian Sea, bordering the very dry Middle East, is much more saline than the nearby Bay of Bengal, whose sources include the normally intensive monsoon rains and significant freshwater flowing from the Ganges and a few other large rivers. The powerful Amazon River of South America, depending upon the prevailing seasonal currents, releases a huge amount of freshwater that then flows eastward across the Atlantic toward Africa or northward to the Caribbean Sea. The very heavy rainfall sections of the central Pacific Ocean push significant freshwater via currents to the coast of Panama. The data has reiterated that Mediterranean Sea is a very salty sea indeed. It also has been found that there currently is a large area of very saline water stretching across the North Atlantic Ocean that is its saltiest in the open ocean. Scientists have likened this to a land desert that has little rain and evaporation.

It also was found that the impact of the spreading abnormally saline blooms is vast and growing, especially in Oman, where locals have depended on artisanal fishing for centuries as their major source of protein. Thick blooms reduce visibility, making it difficult for divers to repair ships underwater. They clog the intake pipes of the desalinization plants that produce up to 90 percent of the country's freshwater, affect the intake waters for refineries, and are a deterrent to tourists who come to Oman for rest and recreation.

The NASA-funded expedition SPURS-1, the first Salinity Processes in the Upper Ocean Regional Study, investigated processes within the subtropical North Atlantic area known as an oceanic desert, where evaporation exceeds precipitation because the very dry air from the Sahara and Sahel regions absorbs water vapor as it travels west. It also studied other similar occurrences to analyze the causes of this high salinity and to validate Aquarius measurements. The SPURS-1 program spurred the development of new technologies to analyze salinity at the sea surface and in the depths.

SPURS-2 examined the precipitation-dominated region located in the eastern equatorial Pacific Ocean. The goal was to better understand the global oceanic freshwater cycle by analyzing the physical processes that control the upper-ocean salinity balance: air-sea interactions, transport, and mixing. It found that the observed patterns of moisture convergence and divergence agree with the satellite-derived data concerning evaporation, precipitation, and sea surface salinity features.

Recent worldwide temperature changes are beginning to disrupt the oceans' balance and the water cycle. Significant global temperature increases in the past 100 years have generated excessive heat that the oceans are absorbing and transporting in some cases to abnormal areas. Along with the temperature increases, the earth lately has seen a reduction in the normal temperature difference between the polar and temperate regions because the Arctic is warming twice as fast as other parts of the globe. Record ice melting has occurred because the jet stream air currents have been pulled northward, where its warm moist air has lingered over Greenland.

Studies have discovered that over the past 50 years in high latitudes and tropical areas, very high levels of rainfall have made the seawater fresher. In the subtropical areas that have high evaporation areas, the waters are becoming saltier. This is logical because a warmer world has an atmosphere that holds more water vapor. The high-latitude regions, where precipitation dominates over evaporation, usually were fresher than the average. The dry subtropics now are saltier than its longitudinal norm, with the exceptions of the South Pacific Ocean and the area off Australia in the southern Indian Ocean. The surge in freshwater in the latter may partly have been caused by the 2010–2012 La Niña that dumped significant rain over. Similarly, very heavy precipitation in the eastern Pacific Ocean may be linked to its increase in unusually freshwaters. From 2004 to 2013, the western Indian Ocean and areas near the equator in the western and central tropical Pacific, as well as in the high evaporation areas of the eastern subtropical Pacific in both hemispheres, became increasingly saline. In contrast, freshwater is a higher proportion of a growing part of the North Atlantic and eastern tropical South Pacific Oceans.

Scientific studies based on observational data and climate model simulations of global warming have predicted that precipitation will continue to surge in rainy regions of the globe while evaporation will increase in dry areas. The net result will be increasingly fresh ocean waters in those areas where it is already fresher and more salty areas of the parts of the ocean that are now saltier. These predicted changes to the water cycle are expected to continue and possible accelerate. Should the trends continue, the quickly melting ice sheets could spur alterations in ocean salinity that possibly could change global circulation patterns, weather, and the conditions for human life.

William P. Kladky

Further Reading

Alam, M. Z., L. Carpenter-Boggs, S. Mitra, M. M. Haque, J. Halsey, M. Rokonuzzaman, B. Saha, and M. Maniruzzaman. 2017. Effect of Salinity Intrusion on Food Crops, Livestock, and Fish Species at Kalapara Coastal Belt in Bangladesh. *Journal of Food Quality* 2017: 1–23.

Cho, R. 2017. *Could Climate Change Shut Down the Gulf Stream?* New York: Earth Institute, Columbia University. https://blogs.ei.columbia.edu/2017/06/06/could-climate-change-shut-down-the-gulf-stream/, accessed June 6, 2019.

Climate.gov. 2014. 2013 State of the Climate: Ocean Salinity. July 12, 2014. https://www.climate.gov/news-features/understanding-climate/2013-state-climate-ocean-salinity, accessed September 20, 2019.

Gupta, R. K., and I. P. Abrol. 1990. Salt-Affected Soils: Their Reclamation and Management for Crop Production. In *Advances in Soil Science*, edited by R. Lal and B. A. Stewart, 223–288. New York: Springer.

IAEA (International Atomic Energy Agency), Water Resources Programme. 2006. *Origin of Salinity and Impacts on Fresh Groundwater Resources: Optimisation of Isotopic Techniques Results of a 2000–2004 Coordinated Research Project.* Vienna, Austria: IAEA.

Lindstrom, E. Jo, J. B. Edson, J. J. Schanze, and A. Y. Shcherbina. 2019. SPURS-2: Salinity Processes in the Upper-Ocean Regional Study 2. The Eastern Equatorial Pacific Experiment. *Oceanography* 32 (2): 15–19.

Podmore, C. 2009. Irrigation Salinity—Causes and Impacts. *PrimeFacts*. PrimeFact 937. State of New South Wales. October 2009. http://www.dpi.nsw.gov.au/__data/assets/pdf_file/0018/310365/Irrigation-salinity-causes-and-impacts.pdf, accessed September 24, 2019.

Ridley, R. T. 1986. To Be Taken with a Pinch of Salt: The Destruction of Carthage. *Classical Philology* 81(2): 140–146.

Shrivastava, P., and R. Kumar. 2015. Soil Salinity: A Serious Environmental Issue and Plant Growth Promoting Bacteria as One of the Tools for Its Alleviation. *Saudi Journal of Biological Science* 22(2): 123–131.

Vengosh, A. 2013. Salinization and Saline Environments. In *Treatise on Geochemistry*. 2nd ed., edited by K. Turekian and H. Holland, 325–378. Amsterdam, Netherlands: Elsevier Science.

Storm Surges

Storms are powerful forces of nature accompanied by strong winds, heavy rain, and potentially hails and tornados. When storms reach the coast, the combination of wind and ocean water leads to additional hazards such as storm surge. According to the American Meteorological Society (AMS), storm surge is the abnormal rise and onshore push of seawater as a result primarily of the winds of the storm and secondarily of the surface pressure drop near the storm center (AMS 2019). The storm surge is measured as the height of the water above the normal predicted astronomical tides. When tides are included, storm surge is called storm tide. Tidal extremes, like spring tides and neap tides, have the greatest effect on water levels. The actual amount of seawater on land is further complicated by a run up of energetic waves onto the beach and local winds and currents.

Storm surge can be caused by both tropical cyclones, like hurricanes, and extratropical cyclones, like northeasters on the East Coast of the United States. In either case, these storms are centers of low pressure in the atmosphere. Low pressure means that the atmosphere exerts less force (or push) on the surface of the earth. When over the ocean, like hurricanes often are, the relaxation of this force allows the ocean to rise slightly higher, leading to a dome of water below the storm's center. However, this only contributes about 5 percent of the total storm surge, whereas the real culprit is the wind field. Surface winds are generated by the gradient of atmospheric pressure in a storm, between low pressure at the center and high pressure at the edge. The stronger the gradient, the faster the winds will blow, similar to how a ball rolls down a hill: the greater the slope of the hill, the faster the movement of the ball.

However, the speed of the wind is insufficient to predict the magnitude of storm surge at a given location. The Saffir-Simpson (SS) scale uses a hurricane's maximum sustained wind speed to classify it into one of five categories. For some time, it was normal practice to associate these classes with surge magnitudes (weak Category 1 storms having the lowest surges and intense Category 5 storms having the highest), but this has gone out of favor, because of the large variability of storm surge observations within and between storms. In addition to wind speed, the size of the storm is very important. Larger storms generally have greater surge, as they will have bigger areas of the ocean over which the wind blows in a consistent direction, known as fetch (Irish et al. 2008). Cyclonic winds generally blow over small fetches as the winds are constantly turning, circulating counterclockwise in the Northern Hemisphere about the low-pressure center.

Fetches will grow on the right and left sides of the storm if the storm moves in a constant direction. This is known as dynamic fetch. If a hurricane moves toward the coast, the right side of the system will have winds blowing landward and the left side of the system will have winds blowing seaward, due to the cyclonic rotation. Therefore, if a hurricane makes landfall perpendicular to the shoreline, its storm surge (and wave energy) will tend to be most destructive on the right side, whereas the left side of the storm may experience a drop in water level (Hall and Sobel 2013). Finally, the winds on the right side are also aided by the movement of the storm so that the total push of seawater is the addition of the wind speed and translational speed, which can be significant.

Things are slightly different for northeasters or other extratropical storms. Extratropical cyclones tend to have weaker wind speeds, but they are generally larger than tropical cyclones, and even though the highest surge values globally are related to tropical cyclones, arctic lows can produce significant surges. For example, Alaska has seen storm surge values as high as 13 feet (4 m). In the continental United States, extratropical storms travel from land to sea, and once they reach the ocean, it is the winds on the back (northwest) side of the storm that affect the magnitude of the surge. The winds often blow out of the northeast (hence the name northeaster) and can cover long distances, having large fetches. These storms also generally travel more slowly than tropical cyclones, especially if they run into a high-pressure system (blocking high). The pressure gradient increases, and the northeasterly winds strengthen between the two systems, blowing for long distances across the ocean and affecting a coastline over several tidal cycles. In fact, it has been known for extratropical storms to continue to produce surge when the skies are clear at the coast and the storm is long gone. The time over which winds blow in a consistent direction is known as duration. Winds with higher duration will lead to stronger and longer-lasting storm surge events (Munroe and Curtis 2017). Finally, tropical cyclones undergoing extratropical transition can lead to storms that retrograde westward, resulting in large and unexpected storm surges. Hurricane Sandy was an extreme case of this, where New York City saw unprecedented water levels (Hall and Sobel 2013).

While the meteorology is very important (wind, size of the storm, and forward speed), it is only one piece of the puzzle. Storm surge is also related to the geography of the coastal environs. This begins with the width and depth of the continental shelf and near shore bathymetry (the measure of depth of water in oceans, seas, or lakes). Waters that are shallow can amplify storm surge. The geographic features of the shoreline can also impact the magnitude of surge. For example, barrier islands can deflect storm surge from the mainland. Concave coastlines are more likely to concentrate and enhance surge as compared to convex coastlines. Bays and inlets can further focus and raise water levels much the same way a funnel constricts and intensifies flow through it. The interaction between storm surge and river flow from the mainland could also compound the flooding hazard.

FORECASTING AND MEASUREMENT

All of these complexities make the actual above-ground water height difficult to forecast, which is important for determining the extent of flooding. There are several storm surge models in operation around the world. The National Hurricane Center (NHC) uses the Sea, Land, and Overland Surges from Hurricanes (SLOSH) model to create a wind field. The model also incorporates the built and natural features of the shoreline, which can impact surge magnitude. Other models developed in the United States are the ADvanced CIRCulation (ADCIRC) coastal circulation and storm surge model and the Storm Surge Modeling System with Curvilinear-grid Hydrodynamics in 3D (SSMS-CH3D). A single model run is considered "deterministic." While it can be useful in some cases, it represents only one possible solution, and a single model run can have large uncertainty. Errors come from not only unknowns in the meteorological conditions—it is very difficult to predict the storm's winds—but also the inaccurate characterization of the constantly changing bathymetry. Thus, probabilistic forecasts are more common, where a distribution of possible surges is forecasted and the probability of exceedance of certain magnitudes can be estimated.

An example of this is the NHC's P-Surge product based on the SLOSH model. Another popular way to use numerical models is for planning purposes. Models can be run thousands of times to generate a distribution of storm surge magnitudes for different scenarios such as tide and Saffir-Simpson category, track and forward speed of the storm. The worst-case scenario for any given location is known as the maximum envelope of water (MEOW)—for example, the highest surge possible for a Category 3 storm that is moving toward the northwest at 20 knots and makes landfall at high tide. The maximum of MEOWs (MOM) can also be assessed for a particular category—for example, the highest surge possible from a Category 2 storm.

This type of information is very important in the design of coastal infrastructure and to make advance preparations. Storm surge causes immense destruction to whatever it encounters, but it shouldn't be confused with a wall of water or tsunami. On average, storm surge rises and ebbs over the course of two days, with storm surge generated by extratropical systems lasting somewhat longer (Munroe and Curtis 2017). For a given storm, the timeframe and spatial extent of flooding depends on elevation and drainage or connectivity to lower elevations. If the land gradually slopes upward, a 6 feet (1.8 m) storm surge at the coast would be measured as 2 feet (0.6 m) of flooding for a location that is at 4 feet (1.2 m) elevation. Depressions in the landscape (low areas surrounded by higher ground) can flood from rainfall, but they are relatively safe from storm surge. Areas that are flooded over long periods of time hinder recovery efforts, and the standing saline water can be particularly corrosive and harmful to crops. In fact, it may be difficult to grow crops in the subsequent agricultural season due to the contamination of the soil.

Storm surge is measured in two primary ways: tide gauge stations and high water marks. Tide, or water level, can be measured visually, mechanically (through floats), and acoustically in real time. An acoustic sensor sends an audio signal through a tube to the surface of the water. The time it takes for the sound wave to reflect back to the sensor is measured and converted to water depth. Recent improvements in tide gauges include the deployment of cost-effective microwave emission instruments that sit above the water. Of course, the primary purpose of tide gauges is to measure tides, and not storm surge, and thus the numbers of stations are insufficient for a good understanding of the complex spatial extent of surge, especially into estuarine environments. Thus, projects are underway to install mesonets of cheap tide gauges on canals, bridges, and docks in coastal communities. Direct evidence of surge extent can also be determined from high water marks after the storm has passed. Debris lines found on trees, buildings, or other structures demark the peak water surface. Extensive surveys are conducted after a storm, and high-water marks are GPS located and mapped.

The NHC and other government agencies are taking storm surge seriously because of the intersection between this hazard and society. In 2010, about 39 percent of Americans were living in counties on the shoreline according to the U.S. Population Census Bureau. Cities like New York and Miami are particularly vulnerable. New Orleans is built on the Mississippi River delta and is below sea level. Not surprisingly, it experienced the deadliest storm surge disaster in 2005 when an extremely destructive Hurricane Katrina made second landfall south of the city. Many deltas around the world are densely populated, such as Bangladesh, and are susceptible to storm surge. The NHC has recently invested in storm surge education and warnings.

Currently, the NHC issues storm surge watches and warnings; a storm surge watch indicates when life-threatening inundation is expected within 48 hours, and a warning reduces that to 36 hours. Emergency managers and other emergency support personnel are looking for extended range forecasts from four days to two weeks of lead time, even with the understanding of reduced skill. They find this useful for "additional time for planning and staging, thereby allowing for quicker recovery or return to normal operations" (Munroe et al. 2018). Recent studies have investigated using climate teleconnections to better characterize storm surge over a season. For example, in Duck, North Carolina storm surges generated from November to July tend to be more frequent, more intense, and longer lasting if an El Niño climate event is also occurring (Munroe and Curtis 2017).

IMPACTS AND MITIGATION

Storm surges cause more damage in the coastal environment than other storm-related hazards and can be quite deadly despite improvements in warnings. Most of the tropical cyclone associated deaths are, in fact, caused by drowning from storm surge flooding. According to the Centers for Disease Control and Prevention in the United States, most of the drownings are avoidable. Over half occur when a vehicle is driven into floodwater. Unfortunately, people are often unaware of the power of storm surge. Even six inches (152.4 mm) of rushing water can be difficult to stand in and can cause cars to become unresponsive or stall, and a couple of feet of water will float many large objects, including cars. The debris in the water inflicts additional damages. The U.S. National Weather Service (NWS) uses signs, videos, and other resources to encourage the public to "Turn Around Don't Drown" or avoid driving through flooded roadways.

Wetlands and mangrove forests are natural defenses against storm surge as they absorb some of the energy. However, loss of these ecosystems to development and aquaculture increases the economic losses from storm surge. Beach dunes also provide a natural barrier to storm surge, and so it is critical to ensure that they are not degraded through human activity. However, storm surge can reshape the coastal environment quite easily, moving tremendous amounts of sand. Dunes are overtopped and eroded with sand deposited landward. The United States Geologic Survey (USGS) has developed probabilities of collision (dune erosion), overwash, and inundation for sandy beaches along the east coast for both hurricane and northeaster storm surge and wave hazards.

There are several ways to mitigate storm surge damages and to reduce harm. As for all hazards, it is important to be prepared and have a plan. Minor surges may just require securing items or using sand bags to reduce the possibility of flooding. Hurricane evacuations will be issued if flooding is likely; thus, it is important to have an overnight bag, emergency kit, and extra tank of gas handy. However, a successful evacuation is limited by access to transportation, wayfinding, and traffic. People living in poverty may not have a car at their disposal, and public transportation may be inadequate. Even those who have a car may not know their evacuation zone or safe routes to higher ground. Finally, devastating consequences can occur if too many people evacuate at the same time, overwhelming the roadways. In terms of coastal construction, buildings should be elevated and set back from the dune lines. Hardened structures, such as seawalls and levees, can provide protection, but they can also be undermined by the erosive power of water and fail (such as during the Great Hurricane of 1900 and Hurricane Katrina in 2005). These structures are designed to withstand a maximum force of water and require upkeep. This static protection is in stark contrast to new developments in living shorelines, where natural buffers are established or renewed, making the coast more resilient over time due to accretion and land building.

NOTABLE EXAMPLES

In 1900, there was little understanding of tropical cyclones and few meteorological or oceanographic observations over the Gulf of Mexico. At the turn of the century, Galveston was a modest town of almost 38,000 residents living on a narrow sandy barrier island that protects Galveston Bay. The island's maximum height is 8.7 feet (2.7 m) above sea level, higher than the mainland, which is very low and marshy and frequently inundated by high tides. Issac M. Cline, a meteorologist and chief of the Galveston weather bureau, provided an eyewitness report of the Great Hurricane of 1900.

Cline wrote that waves and swells of water started to enter Galveston the morning of September 8, 1900, even though meteorological conditions did not indicate an approaching hurricane: the winds were out of the north, the clouds were broken stratus and stratocumulus, and there was little change in barometric pressure. However, the storm surge came in suddenly during the day, at one point rising 5 feet (1.5 m) in one hour, and by the morning of September 9, the seas were back to normal. The maximum surge was 15–20 feet (5–6 m), which drowned all of Galveston. Most buildings were either destroyed or sustained major damage, and some estimates put total fatalities at 12,000. After the storm, Cline surmised that a seawall would have minimized the loss of life and property from the hurricane (estimated to be Category 4 in strength). Today Galveston is known for its seawall, which has withstood many storm surges, but none like the Great Hurricane of 1900.

Several powerful northeasters have affected the United States, including the "perfect" storm of 1991 and the storm of the century of 1993. However, the northeaster that set several records for storm surge in the United States was the 1962 Ash Wednesday storm. This storm first struck the Outer Banks, without warning, on March 7 and lasted through five successive high tides. A massive low-pressure center extending 100 miles (150 km) out to sea and a strong blocking high in the Canadian arctic led to an enormous fetch of winds directed toward the East Coast. The Ash Wednesday storm caused flooding and coastal erosion from North Carolina to New Jersey. The storm produced Norfolk's, Virginia, highest recorded storm surge at 9 feet (2.7 m) (Stick 1987).

Bangladesh is no stranger to storm surge. Eight of the 10 highest storm tides were recorded in this country. An unnamed cyclone in 1876 set the global record at 45 feet (13.7 m). The world's deadliest tropical cyclone was the 1970 Bhola cyclone. The majority of the 500,000 fatalities in Bangladesh (which was then East Pakistan) were due to storm surge. On November 8, a tropical depression formed in the Bay of Bengal. It moved northward and strengthened, reaching Category 4 status. On November 12, it struck the delta near local high tide, causing a 33 feet (10 m) rise in water. Most of Bangladesh is less than 40 feet (12.2 m) above sea level. Villages, crops, livestock, and fleets of ships were completely annihilated, and the loss of life was incredible due to the dense population at the coast. International conflicts made the situation worse. Data from Indian ships in the Bay of Bengal were not shared with Pakistan, and the Pakistan government was initially reluctant to receive disaster assistance. Thus, the widespread devastation, lack of warning, and slow recovery were partially responsible for the Bangladesh Liberation War, which led to the country's independence a year later. This is why Bangladesh is sometimes known as "a nation born of a cyclone" (Anon 2008).

Finally, prior to 2005, most hurricane fatalities in the United States were caused by freshwater flooding. However, Hurricane Katrina changed the equation and set into motion much of the advances in storm surge forecasting and warning that continues today. Katrina was the twelfth named storm in a record-setting season that saw 26 named storms in the Atlantic. It was the costliest storm in U.S. history, estimated at $75 billion, and claimed over 1,800 lives, mostly due to storm surge. Katrina first struck southern Florida as a Category 1 hurricane on August 25. It then turned into the Gulf of Mexico and headed for the Mississippi delta. It reached Category 5 strength on August 28, but it weakened just before landfall during an eyewall replacement cycle. The exact storm surge from Katrina is complicated by the failure of tide gauges and lack of surviving structures with which to identify high water marks. Estimates of storm tide range from 24 to 28 feet (7.3–8.5 m) along a 20-mile (30 km) stretch of the Mississippi coast. The 28-feet (8.5-m) storm tide was recorded in Hancock County in Mississippi and is the largest on record for the United States (Needham and Keim 2012).

The surge penetrated as much as 12 miles (18 km) inland via bays and rivers. In Louisiana, winds from Katrina pushed water from Lake Pontchartrain into New Orleans. Surge values from 10 to 19 feet (3–5.8 m) breached, overtopped, or undermined the levee system that was built to protect the city from floodwaters. Water was able to rush through the canal system, flooding 80 percent of the city. It was not until October 11, 2005, 43 days later, that the city was completely drained. The NHC believes it was the immense size of the storm that was responsible for the extreme surge, but it was certainly the socioeconomic vulnerability that was responsible for the profound loss of life in New Orleans (Knabb et al. 2005).

CONCLUSION

It is likely that storm surge magnitudes will trend upward globally. Evidence from climate change theory and models points to tropical and extratropical storms becoming more powerful. A warming ocean will provide more fuel for growing tropical cyclones. Scientists have found evidence that this will lead to larger, longer-lasting, and more intense hurricanes in the Atlantic—all conducive to greater storm surge. Outside the tropics, the extreme warming of the arctic will lead to a reduction in poleward temperature gradients and slower jet stream. If the jet slows, it will be more likely to meander, meaning extended upper-air troughs for generating surface extratropical cyclones and extended upper-air ridges for generating blocking highs.

The combination could result in more frequent cases of long fetches, which could build surges similar to the 1962 Ash Wednesday storm. Furthermore, when the extratropical storms move over coastal waters, they will be able to tap into the enhanced oceanic heat content just like tropical systems. Warming will also accelerate current trends of increasing sea level. Sea level rise, caused by the thermal expansion of the oceans and the melting of glaciers and ice sheets, will add to storm surge magnitudes. A study for New York City found that a 3.3 feet (1 m) sea level rise could change New York City's 100-year surge event into a three-year event (Lin et al. 2012). Even small increases from sea level rise can affect whether storm surge overtops dunes or seawalls and levees.

Finally, with a thriving global economy and disregard for risk, people will continue to settle down and vacation in coastal areas. Therefore, even without any change in the atmosphere or ocean, more people will be vulnerable to storm surge in the future. If storms like the 1900 Galveston Hurricane or 1962 Ash Wednesday storm were to hit at the same locations today with the same surge, the loss of lives and economic devastation would be even more incredible (Hondula and Dolan 2010). Hence, if the lure of the coast cannot be abated, then new creative and sustainable mitigation solutions to storm surge must be imagined.

Scott Curtis

Further Reading

AMS (American Meteorological Society). 2019. Meteorology Glossary. glossary.ametsoc.org/wiki/Main_Page, accessed April 17, 2019.

Anon. 2008. Killer Cyclones: The World's Worst Disasters. Films Media Group. https://fod.infobase.com/PortalPlaylists.aspx?wID=103534&xtid=95244, accessed April 17, 2019.

Hall, T. M., and A. H. Sobel. 2013. On the Impact Angle of Hurricane Sandy's New Jersey Landfall. *Geophysical Research Letters* 40(10): 2312–2315.

Hondula, D. M., and R. Dolan. 2010. Predicting Severe Winter Coastal Storm Damage. *Environmental Research Letters* 5(3): 034004.

Irish, J. L., D. T. Resio, and J. J. Ratcliff. 2008. The Influence of Storm Size on Hurricane Surge. *Journal of Physical Oceanography* 38: 2003–2013.

Knabb, R. D., J. R. Rhome, and D. P. Brown. 2005: *Tropical Cyclone Report: Hurricane Katrina 23–30 August, 2005*. Miami, FL: The National Hurricane Center.

Lin, N., K. Emanuel, M. Oppenheimer, and E. Vanmarcke, 2012. Physically Based Assessment of Hurricane Surge Threat under Climate Change. *Nature Climate Change* 2(6): 462–467.

Munroe, R., and S. Curtis. 2017. Storm Surge Evolution and Its Relationship to Climate Oscillations at Duck, NC. *Theoretical and Applied Climatology* 129: 185–200.

Munroe, R., B. E. Montz, and S. Curtis. 2018. Getting More out of Storm Surge Forecasts: Emergency Support Personnel Needs in North Carolina. *Weather, Climate and Society* 10: 813–820.

Needham, H. F., and B. D. Keim. 2012. A Storm Surge Database for the US Gulf Coast. *International Journal of Climatology* 32(14): 2108–2123.

Stick, D. 1987. *The Ash Wednesday Storm: March 7, 1962*. Kill Devil Hills, NC: Gresham Publications.

Subsidence

Subsidence refers to the sinking of the land surface. The cause of such sinking falls into two broad categories: natural and human-induced changes. Natural causes of subsidence, for instance, include tectonic motions or dissolution of subsurface geologic layers, which causes the surface to subside, or collapse. Any human activity that extracts materials from below the surface such as pumping groundwater or natural gas, for instance, also may result in subsidence. The United States Geological Survey (USGS) estimates that, in the United States, more than 17,000 square miles (44,030 square km) of land in 45 states have experienced direct consequences of subsidence (USGS 2000).

Subsidence causes changes to the land surface ranging from slight to dramatic modifications to topography. For example, subsidence along coastal areas leads to a local decrease in relative sea level, resulting in differences in the shape of the coastline. Subsidence is more likely in certain types of geologic structures, such as areas dominated by karst, which comprise rocks such as limestone that gradually dissolve, forming caverns, sinkholes, and other erosion features.

Soil serves as another structure that can shrink to cause subsidence, a process ranging from slow to quite rapid. The collapse of a cavity would likely occur rapidly, while the settling of soil as it dries out, for example, can occur quite slowly.

PLATE TECTONICS

Tectonic forces, those which cause the crustal plates of the earth to shift, may result in subsidence. Tectonic subsidence can occur near or away from plate boundaries. In essence, tectonic subsidence occurs through earthquakes, crustal extension, cooling, and loading.

Where crustal plates converge, or come together, the denser plate dives beneath the other plate to form a subduction zone. As the subducting plate dives downward, friction between the plates may cause the surface of the upper plate to bulge upward (i.e., up warping). The frictional forces build until the plates slip, creating an earthquake. As the compression is allayed by the earthquake, the force creating the bulge in the upper plate is released, often leaving in its place an area of subsidence near the plate boundary.

Earthquakes may also cause subsidence where shaking consolidates sediments, packing them more closely together, causing a drop in the land surface. Earthquakes' moving faults can cause one side at the fault to rise and the other to fall.

At strike-slip faults, where the motion is lateral, a small-scale subsidence occurs at the fault line.

In places where the crust experiences a tension or pulling-apart force, the crust stretches until rifts form from a series of faults. Some areas remain at the original level (horst), while some areas subside (graben). Where the stretched crust has become thinned, the next, hotter, interior layer of the earth pushes up into the thinned space. Eventually, the tension force decreases and the crust cools. Crust expands as it heats up and becomes denser and sinks as it cools, a process called thermal subsidence.

Finally, the constant changing of earth's crust involves many other processes such as mountain building and erosion. Loading is the addition of weight by sediments or ice, for instance. The increased weight causes depression of the crust (i.e., isostatic adjustment), another form of subsidence.

KARST AND PSEUDOKARST TOPOGRAPHY

Landscapes dominated by karst are typically underlain with rocks made from minerals that easily dissolve in water, such as limestone, dolomite, or gypsum. Karst landscapes are characterized by caverns, underground streams, and sinkholes. A sinkhole may also be known as a cenote, swallet, doline, or swallow hole, all terms that are used interchangeably (Kohl 2001). There are two main pathways for sinkholes to form, but a sinkhole may form from a combination of these processes.

The first process that may form a sinkhole is dissolution. Rain, which is naturally slightly acidic, falls on a karst region and trickles down through the soil, picking up additional acids from plant roots. This water percolates through joints in the limestone. Acid in the water reacts with minerals in karst such as calcium carbonate ($CaCO_3$), which acts to break down the rocks. The dissolution of karst rocks leads to formation of a small depression, where drainage is focused. Water collected in the depression accelerates the dissolution processes.

The second process that results in sinkhole formation often starts with a subsurface void. Over time, the surface material can make its way into the void so that the unsupported ceiling of the void becomes increasingly closer to the surface. Eventually, the ceiling collapses, leaving behind an area of subsidence that resembles a pit or a hole at the surface from a few centimeters to several meters in depth and diameter. The area of low relief may have either vertical or overhanging sides that, if left undisturbed, will further weather to form a conical hole, or sinkhole. These cover-subsidence sinkholes are more common in areas with coarser sediments, so they are uncommon where clay soils dominate.

Sinkholes may form gradually, without the dramatic collapse phase, when the surface gradually subsides as subsurface material is removed by human activity. Many of these slow-forming sinkholes form due to removal of water or fossil fuels from the subsurface. The removal of material gradually creates a new void into which the surface subsides.

In the western United States, particularly in the states of Arizona, California, New Mexico, and Colorado, many sinkholes form from dissolution of evaporite minerals such as gypsum or halite (rock salt). Salt and gypsum underlie about 40 percent of the contiguous United States, while carbonate karst landscapes account for about another 40 percent of the country east of Tulsa, Oklahoma (White et al. 1995). In certain areas of Arizona, California, New Mexico, and Colorado, there have been large swaths of subsidence of hundreds or thousands of feet downward. The presence of such geologic structures that have the tendency to collapse has led to Colorado's mapping of a subsidence hazard report (Colorado Geological Survey 2019).

Evaporite rocks occur below the surface of about 35–40 percent of the United States, although they are not always close to the surface. Since salt and gypsum have high solubilities, solution-related subsidence can occur quite rapidly, from days to years. On the other hand, formation of subsurface voids in carbonate bedrock occurs much more slowly, from centuries to millennia. Of course, human activities, such as mining or pumping of groundwater, can speed up these processes.

One of the most spectacular examples of a large sinkhole transformed by human endeavors (in a good way) describes the Arecibo Observatory in Puerto Rico. The 1,000 feet (305 m) diameter radio telescope was constructed inside the depression left by a karst sinkhole. From its construction in 1963 until 2016, it was the largest single-aperture telescope in the world. Some geologic areas that don't form karst sinkholes are called pseudo-karst, which is still prone to collapse and subsidence. However, pseudokarst differs from karst because it comprises rocks that are not susceptible to dissolution. Instead, subsidence in pseudokarst occurs from other natural processes such as collapse of volcanic tubes, melting of permafrost, or melting of ice wedges, large buried sections of ice.

The island chains of Hawaii contain very little limestone due to volcanic origins, but it has areas of pseudokarst where volcanic fields full of subsurface lava tubes have collapsed. Other volcanic areas, such as those found in the Aleutian Islands of Alaska, contain similar features.

Alaska has also experienced rapid melting of permafrost in recent decades, which creates irregular subsidence. Permafrost is permanently frozen ground, which often contains ice. Since ice decreases in volume as it melts, permafrost subsides as it melts. Where permafrost has melted under homes or other infrastructure, it has caused and continues to cause great destruction (Nelson et al. 2001). As the Arctic warms, projections show acceleration of changes to permafrost in the future.

Permafrost that is ice rich subsides more than average permafrost. Known as thermokarst, when it melts, it creates a very irregular land surface full of small hummocks and marshy hollows. Thermokarst appears very much like a karst landscape at first glance, with its pitted surfaces. This ice-rich permafrost occurs in Arctic areas as well as on a smaller scale in high altitudes of the Himalayas and Swiss Alps. In Siberia, a large thermokarst depression known as the Batagaika crater provides an example of a large area of subsidence 328 feet (100 m) deep and about 0.6 mile (1 km) long, associated with thermokarst.

HUMAN-INDUCED SUBSIDENCE

Collapse of karst landforms occurs due to continued weakening of the subsurface rock ceiling of a void, by either natural or human-induced changes. Human causes of subsidence in karst landscapes are dominated by extraction of fossil fuels and ground water. Fossil fuel extraction is common in karst landscapes since about half of karst regions contain hydrocarbons such as coal and natural gas.

By far the most dramatic human cause of subsidence is groundwater extraction (USGS n.d.). According to the USGS, more than 80 percent of land subsidence in the United States is caused by extraction of ground water. Ground water is stored in natural reservoirs called aquifers. Some aquifers maintain their capacity and can be filled back up (recharged) by precipitation events. Other aquifers (i.e., unconsolidated aquifers) collapse once the water is withdrawn, decreasing its capacity to store water and causing subsidence at the land surface.

Ground water extracted from aquifers that cannot be replaced is considered to be mined water (USGS n.d.). Consequences of subsidence due to agricultural, industrial, and municipal mining of ground water are severe. For example, in the Santa Clara Valley in northern California, early agricultural extraction of groundwater for irrigation significantly contributed to subsidence in the area, resulting in increased flood risks in the region. Recent drought in the state has caused more irrigation so that parts of the region are subsiding by up to two feet/year (60 cm/year). In general, rural areas are less likely to experience land subsidence due to groundwater pumping since the rate of withdrawal often matches the level of replenishment (Cooke and Doornkamp 1977). However, where surface irrigation is extensive in an arid or semiarid region, such as the Central Valley in California, the rate of replenishment does not come close to matching withdrawals. With the compounding effects of large population centers surrounded by irrigated agriculture, the Central Valley has experienced dramatic levels of subsidence.

While California has experienced significant impacts from excessive pumping of groundwater, it has also had to address subsidence associated with pumping oil. The Los Angeles area has experienced widespread, subtle subsidence associated with groundwater pumping and localized, pronounced subsidence associated with extraction of oil, especially from the Wilmington oil field. In the 1970s, it was likely the most dramatic case of human-induced subsidence in the world. Between 1928 and 1971, the oil field's point of maximum subsidence reached 30.5 feet (9.3 m) (Cooke and Doornkamp 1977). Scientists tracked a clear association between oil production rates and subsidence rates.

Los Angeles city officials responded to the subsidence, enacting protections for coastal areas that had subsided, working to repair damaged oilfield structures such as oil wells, and to increase fluid pressure in the Wilmington field by injecting water into the ground. While the motivation for water injection was to increase oil production, it also significantly slowed or ceased subsidence in the area. These experiences paved the way for approval of further petroleum resource extractions with a better understanding of how to reduce impacts associated with subsidence (Cooke and Doornkamp 1977).

In the Las Vegas Valley, Nevada, the fast-growing metropolis of Las Vegas in a desert climate required extraction of a significant amount of water to transform it into a green oasis. As an unintended cost, the ground water depletion has caused significant subsidence. Finally, water-intensive agriculture in Southern Arizona caused such dramatic subsidence that it has resulted in the cracking open of earth's surface. In each of these examples, however, decreased extraction has either halted or slowed subsidence for now.

Mining of underground natural resources leaves void spaces that may collapse and cause subsidence. In states with abandoned coal mines such as Colorado and Virginia, the hazards associated with collapsing mines cause concern. Essentially, coal removed leaves a void where the weight of the overlying rock can shift and sink into the void. At the surface, it may appear as cracks, holes, or simply sagging of the land surface. According to the Colorado Geological Survey, 1 inch or about 3 cm of subsidence beneath a home can cause thousands of dollars in damage. Since many old mines are located near present-day urban areas, many homes have been unknowingly built on abandoned mines.

Not only do mine collapses have the potential to cause great property damage, but they can also lead to loss of life and devastation of other resources. In the case of the salt mine collapse in Genesee Valley, New York, the initial collapse led to a series of other events. Besides land subsidence, the mine was completely flooded, there were large declines in local ground water levels and quality, natural gases were released, and there were other detrimental effects to infrastructure and cultural aspects of the valley. The event was initiated by a magnitude 3.6 earthquake, which caused the collapse of 500-feet-by-500-feet (152.4-m-by-152.4-m) section of ceiling rock. Fortunately, active mining had been suspended that day for maintenance.

Texas is known for its rich petroleum and natural gas resources and also has large karst areas. Extraction of oil and natural gas resources coupled with ground water pumping has created some major problems for the state. Since World War II (1939–1945), there has been an increased demand for pumping water to fuel industrial development of the Gulf Coast of Texas, particularly in the Houston-Galveston area, where growth has been associated with increased port activities along with oil and aerospace industries. These developing industries also brought increased demand for water for urban population and rice irrigation. Surface waters provided by Lake Houston after 1954 could not keep up with demand, so by 1964 some areas in the Houston-Galveston region had subsided up to 4 feet (1.5 m) (Cooke and Doornkamp 1977).

While much of the subsidence was attributable to ground water pumping, there are clear areas where extraction of oil is the primary culprit in subsiding land, such as in the Goose Creek oil field north of Galveston (Cooke and Doornkamp 1977). Subsidence along the Gulf Coast has created severe and expensive coastal flooding hazards in the Houston-Galveston metropolitan area. Unfortunately, the Galveston Bay estuary has also been detrimentally affected by these practices.

SOIL AND SUBSIDENCE

Another human activity that leads to land subsidence is the drainage of soils that are rich in organic carbon. Typically, the motivation for draining such lands involves agricultural uses, but can be some other interest, such as building.

Naturally wet soils, such as those found in southern Louisiana (which would be commonly found in swamps or bogs), have rich quantities of organic matter because it's slow to decompose. Slow decomposition is due to the high moisture content that restricts ready access to oxygen, which depresses the activity of the microbes that do the work of decomposition.

In places where peat moss dominates, such as in Alaska, colder temperatures and acidic conditions depress decomposition. To prepare an area such as this for agriculture, the pH is adjusted to be more neutral, which causes rapid decomposition of the organic soil. So, in either case, once rapid decomposition occurs, organic carbon is readily converted into carbon dioxide gas and water. When the process is complete, the soil can be easily compacted, eroded by wind or water, or even burned—all leading to further subsidence of the land surface.

On the other hand, there are natural conditions that can lead to subsidence related to soils. Expansive soils (also known as swelling clays) readily absorb water into their clay minerals when wet, causing them to expand significantly up to 20–50 percent (Kohl 2001). By contrast, when they dry out, the water leaves the mineral structure, causing major shrinkage and subsequent cracking and subsidence of the land surface.

Subsidence related to organic soil compaction has occurred in two of the United States' most important wetland ecosystems (the Sacramento-San Joaquin Delta and the Everglades). In both cases, agricultural production and water supply infrastructure for large urban populations are affected, as well as complicating ecosystem restoration efforts by state and federal groups. In the Sacramento-San Joaquin Delta of California, levees weakened by subsidence increase the flood risk to Delta islands, which has the potential to disrupt the north-south water distribution system for the state. In the case of the Florida Everglades, drained soils subside not only with drying and natural compaction but also because of compaction due to agricultural practices, wind erosion, burning, and oxidation of organic matter (Cooke and Doornkamp 1977). In this region, subsidence has caused an acknowledgement of a looming end point for agriculture worth an annual $750 million (USGS n.d.).

Other generally smaller-scale incidences of subsidence are quite common, but they are typically less troublesome than the types listed earlier. For examples, some landowners dig holes of varying sizes to bury rubbish in—trash pits. Once they are covered with soil, the buried material becomes more compacted and begins to decompose, leading to subsidence at the surface.

Another organic-rich soil occurs in tropical and subtropical river deltas. As the organic carbon slowly decomposes and the sediments become compacted, natural subsidence occurs. However, this natural subsidence is typically offset by sedimentation from natural flooding. When people work to control the natural flooding, the renewal of sediments is virtually halted, which removes the offset to natural subsidence.

The Mississippi River delta provides an example of this situation. The causes for subsidence in the region are many, both natural and human-induced: sediment compaction, sinking of earth's crust from the weight of the sediments, faulting, and withdrawal of natural resources from below the surface (water, oil, and gas)

(NASA 2016). The primary contributors were groundwater pumping and dewatering, which is a process used to dry out soggy ground.

As the delta subsides and sea level rises, it has lost some of its natural coastal barriers, which increases flood risk. The regional response has been to add more flood control measures to protect valuable infrastructure and large population centers. After Hurricane Katrina, significant efforts at a cost of $14 billion were made to create a flood control system, which is unfortunately also sinking as the ground sinks. Because subsidence and sea level rise are occurring at faster rates than predicted, the flood control system needs constant uplifting in order to maintain the required safety level to maintain federal funding (Bowen 2019). Further, coastal communities throughout the world suffer from land loss due to coastal erosion and subsidence, which are exacerbated by global sea level rise.

SUBSIDENCE OUTSIDE THE UNITED STATES

In areas where subsidence is occurring along coastlines (such as in Galveston Bay), rising sea levels exacerbate flooding. Jakarta, the capital of Indonesia, is subsiding up to 6.7 inches (17 cm) per year. In response, city officials have built sea walls to protect the city from sea level rise and increased flooding. However, since the city is subsiding so rapidly, water comes over the sea walls at high tide (Ruggeri 2017).

On the other side of the world, in Mexico City, Mexico, the problem is not sea level rise but rapid subsidence from extraction of ground water. The overuse of groundwater has caused the aquifer to be overdrawn. Parts of the city are sinking at a rate of 12 inches (30 cm) per year. A large proportion of the water use comes from leaky pipes. Unfortunately, subsidence damages the water pipe infrastructure, causing more water leaks (Ruggeri 2017).

Some places have addressed the problem and reduced ground water extraction and, therefore, reduced subsidence as well. In Tokyo, Japan, in 1968, at the peak of ground water pumping, the land was sinking about 9 inches (24 cm) per year (Sato et al. 2006). To stop the subsidence, Tokyo's government passed laws to limit pumping. By the early 2000s, Tokyo had almost completely stopped subsiding, slowing it to 0.4 inch (1 cm) per year. Shanghai, China, took this idea one step further and, in addition to limiting pumping, actively worked to recharge their aquifers (Shen and Xu 2011).

Past mining activities in the United Kingdom (UK) has led to subsidence above the mines, which also sometimes leads to disruption of surface hydrology. For example, in the floodplain of the River Stour, near Canterbury, England, a lake began forming in 1933 as a result of the mining of a seam, which has been packed with sediment. The sediment had been placed in the mine with the goal to reduce subsidence, but it disrupted the surface waters, which caused a lake to form (Cooke and Doornkamp 1977).

In 2018, the United Kingdom took advantage of new subsidence mapping techniques that use satellite to measure and monitor land height changes, in both areas of rising and those subsiding (Amos 2018). The UK policy makers requested the information in order to better address current problems and to hopefully prevent more issues from occurring in the future. For instance, the United Kingdom is currently planning to build a high-speed railway that crosses old coal mine pits. Some areas above the pits are subsiding, while other areas are lifting up as the mines fill with water. The construction of the rails will continue, but it will take into consideration the current and potential future adjustments of the land.

MONITORING LAND SUBSIDENCE

Before satellites and Global Positioning System (GPS) technology, detection of regional-scale subsidence required that someone notice a benchmark had moved. Official land surveys establish benchmarks for setting property lines and locating roadways, pipelines, engineering infrastructure, and more. The officials identified which benchmarks had moved and which ones were stable. The stable benchmarks were then used to map regional-scale subsidence.

Today, geographers have the tools of GPS, which has increased in accuracy tremendously in the past decade. In addition, recently a new tool has emerged, interferometric synthetic aperture radar (InSAR), which uses satellites to generate radar images from repeat passes above the earth. Once an area of subsidence is identified, it can be monitored carefully to improve understanding of both subsidence processes and what practices can reduce subsidence rates.

These improvements to tools to measure subsidence were developed in response to a need for information on subsidence by the Panel on Land Subsidence of the U.S. National Research Council (NRC) in 1991. They determined that information and data on subsidence were needed to (1) recognize and assess future problems regarding the magnitude and distribution of subsidence, (2) find engineering solutions through research on subsidence processes, and (3) study cost-effectiveness of mitigation methods.

While the new technological tools such as the use of InSAR address the first item by measuring the magnitude and distribution of subsidence, there is more work to be done on finding engineering solutions and cost-effective mitigation methods. The USGS states that "to assess the total impact [of potential solutions], we would need to inventory the total costs to society of overdrafting susceptible aquifer systems" (USGS n.d.). Now that several other countries have nearly halted subsidence with water management solutions, perhaps the United States can also study these successful cases to determine their potential there.

CONCLUSION

Subsidence is caused by both natural and human impacts on the land. Natural subsidence occurs in tectonic settings, such as those caused by earthquakes or in the collapsing of old lava tubes. Another common natural setting for subsidence is in karst areas, where rocks are prone to dissolution by precipitation. The rocks slowly yet continually weaken, which leads to collapsing of surface layers into weathered voids below. These processes may occur rapidly, as in the sudden collapse of a cavern ceiling or quite slowly with a gently sinking of the surface. In addition, the vertical range of subsidence can range from centimeters to meters (inches to yards).

Sinkholes comprise a category of subsidence that form from both natural and human impacts. Any process that removes material below the land surface in a karst area can lead to a sinkhole. Human impacts in karst areas that can lead to sinkholes include removal of hydrocarbons such as natural gas and coal or the pumping of water from aquifers. A large proportion of subsidence in the United States is the direct result of pumping ground water. The USGS has identified overpumping of aquifers that cannot be recharged (unconsolidated aquifers) equivalent to mining water, since it cannot be replaced. Water mining in unconsolidated aquifers is common in dry climates with large population centers, such as in Las Vegas, Nevada, and the central valley of California, where extensive irrigation also plays a role.

Subsidence from human activities typically have the most rapid and dramatic impacts on the environment. Extraction of solids from under the surface, such as through mining activities, can cause surface subsidence, disrupt hydrology, and create hazardous situations. For illustration, Colorado is littered with abandoned mines, many of which have unknown locations until the land surface above a mine subsides. With expanding metropolises, homes built on abandoned mines can suddenly shift and collapse, a hazardous situation.

Drying organic soils, either from natural processes or drained artificially by humans, also causes land to subside. Soil that once had pore spaces filled with water shrinks. Pores now filled with oxygen accelerate decomposition of organic carbon in the soil, furthering subsidence. When the new pore spaces become compressed, the soil subsides. When organic soils dry, they become attractive to agriculture, which can also cause compaction. Organic soils are common in swampy, boggy, or marshy areas, including river delta areas such as the Mississippi River delta.

The Mississippi River delta continues to subside due to a combination of factors, but primarily from draining soils and pumping groundwater. In places like this, where flood control measures have been enacted to protect valuable infrastructure and population centers from flooding, the natural sediment deposits do not offset natural subsidence. Compounded by sea level rise, many coastal areas such as this and throughout the world are experiencing land loss.

Outside the United States, similar issues with subsidence exist related to natural and human-caused changes, often enhanced by sea level rise. Jakarta in Indonesia has enacted flood control measures, but they fail at high tide due to continued subsidence. Damage to water infrastructure in Mexico City from subsidence associated with pumping ground water is worsened by leaky pipes, a vicious circle. In Tokyo, Japan, dramatic subsidence has been virtually halted by dramatic water pumping policy changes enacted decades ago. Finally, the United Kingdom has used new technology to map and monitor land changes (both rising land and subsiding areas) with the goal of using this information in planning and policymaking to decrease costly impacts. The new technology used by the United Kingdom, called InSAR, was developed over the last decade. In concert with GPS, we now have the tools to continually monitor land elevation changes, including subsidence, with great accuracy. In part, these tools were developed to address the need for more and better information about subsidence in the United States by the National Research Council in 1991. The council also called for development of better engineering tools to address challenges of subsidence and ways to stop or recover from subsidence. The report stressed the need for a cost-to-benefit ratio since many solutions to subsidence in the United States will involve slowing the pumping of mined water, which supports agriculture, industry, and municipal needs. Clearly, subsidence is a complex issue that directly impacts millions in the United States and many more millions of people in the world who live and make their living on the land.

Trisha Jackson

Further Reading

Amos. 2018. Images for Amos 2018, UK Subsidence Map. https://www.google.com/search?q=Amos+2018%2C+UK+Subsidence+map&rlz=1C1CHMO_enUS523US523&oq=Amos+2018&aqs=chrome.0.69i59j0l7.8015j0j7&sourceid=chrome&ie=UTF-8, accessed May 20, 2019.

Bowen, B. S. 2019. Notice of Intent to Prepare a Draft Environmental Impact Statement for the Lake Pontchartrain and Vicinity General Re-evaluation Report, Louisiana. *Federal Register* 84(63): 12598–12599.

Colorado Geological Survey. 2019. Natural Subsidence. http://coloradogeologicalsurvey.org/geologic-hazards/subsidence-natural, accessed May 20, 2019.

Cooke, R., and J. C. Doornkamp. 1977. *Geomorphology in Environmental Management*. Bath, England: Oxford University Press.

Kohl, M. 2001. *Subsidence and Sinkholes in East Tennessee: A Field Guide to Holes in the Ground*. Nashville: State of Tennessee.

NASA (National Aeronautics and Space Administrative). 2016. New NASA Maps Show Just How Fast New Orleans Is Sinking. https://weather.com/news/climate/news/nasa-maps-new-orleans-sinking, accessed August 16, 2019.

Nelson, F. E., O. A. Anisimov, and N. I. Shiklomanov. 2001. Subsidence Risk from Thawing Permafrost. *Nature* 410: 889–890.

Ruggeri, A. 2017. The Ambitious Plan to Stop the Ground from Sinking. *BBC Future*. December 1, 2017. http://www.bbc.com/future/story/20171130-the-ambitious-plan-to-stop-the-ground-from-sinking, accessed May 20, 2019.

Sato, C., H. Michiko, and J. Nishino. 2006. Land Subsidence and Groundwater Management in Tokyo. *International Review for Environmental Strategies* 6(2): 403–424.

Shen, S. L., and Y. S. Xu. 2011. Numerical Evaluation of Land Subsidence Induced by Ground Water Pumping in Shanghai. *Canadian Geotechnology Journal* 48: 1378–1392.

USGS (United States Geological Survey). n.d. Land Subsidence. https://www.usgs.gov/special-topic/water-science-school/science/land-subsidence?qt-science_center_objects=0#qt-science_center_objects, accessed August 12, 2019.

USGS (United States Geological Survey). 2000. Land Subsidence in the United States. https://water.usgs.gov/ogw/pubs/fs00165/, accessed August 12, 2019.

White, W. B., D. C. Culver, J. S. Herman, T. C. Kane, and J. E. Mylroie. 1995. Karst Lands. *American Scientist* 83: 450–459.

Temperature Extremes

Temperature is among the most important weather and climate variables. Temperature stress or extreme temperature has become a major concern for human, animal, and plant health worldwide due to climate change. Extreme heat and cold have a far-reaching set of impacts including illness, significant loss of life, and economic costs in agriculture, transportation, energy, and infrastructure. Although both heat and cold can damage organisms in innumerable ways, extreme heat is currently a greater concern because the number of cold days and nights has dropped while the number of hot days and warm/tropical nights has increased globally since 1950 (Stocker 2014).

Summers are expected to be warm, but sometimes it can be severely hot. High temperatures and extreme heat are linked but not identical concepts. "Hot" is a relative term, locally defined based on the range of temperatures at a specific place. Extreme heat is considerably hotter than average weather conditions that may persist over a prolonged period. Extreme heat has many different names including a "heatwave," "hot spell," and an "excessive heat event," with these extremes generally referring to a condition that could be harmful to human health and infrastructure. A heatwave occurs when a system of higher atmospheric pressure and warmer temperatures extend from the surface to upper levels of the atmosphere. Clear skies and dry soils help to reinforce the warm conditions. Once formed, the dome of higher pressure and warmth tends to persist in the region. Extreme heat events are classified among the 10 deadliest disasters (Guha-Sapir et al. 2012), and they often result in an increase in hospital admissions for the heat-related illness. The effects of heatwaves on human beings and other organisms differ from one geographical region to another depending on the preexisting medical condition of human beings, their behavior, and acclimatization. In the United States, the National Weather Service (NWS) can issue a heat advisory when warranted.

Extreme cold or a cold wave, characterized by a number of cold days in succession, impacts industry, agriculture, transportation, commerce, and social activities. Characterization of extreme cold includes both the rate of temperature decrease and the low temperature. These factors differ based on the time of year and geographical region. Cold waves are mainly driven by a combination of cold air mass formation (a thermodynamic property) and large-scale circulation and movement of the air mass (atmospheric dynamics). Cold events have become less frequent and less severe over the past few decades in most land areas of the world (Stocker 2014). However, year-to-year variability occurs and extreme cold events occurred recently in North America (2014) and Europe (2012) (NASEM 2016).

The increase in extreme temperature events is having an impact on the environment, society, and economy. In the United States alone, the annual expenditure to address extreme heat and the associated droughts now exceed several billion dollars. During the spring and summer of 2012, most of the Central United States experienced a prolonged, catastrophic extreme heat event that resulted in $33 billion economic losses, a 10 percent drop of corn production, 123 human deaths, and considerable losses in the livestock industry. Extreme heat and drought together can set the stage for the costly and destructive wildfires that have burned in many areas throughout the country. Severe cold events in California in 2007 and in the Southern United States in 2017 produced economic damage with a price tag of several billion dollars (NOAA and NCEI 2019).

EXTREME TEMPERATURE TERMINOLOGY

Air Temperature and Trends

Air temperature, usually expressed in degrees Celsius or Fahrenheit, is a measure of atmospheric heat content and indicates how cold or hot the air is. Temperature is typically measured by a thermometer in the shade at a height of 4.1–6.6 feet (1.25–2.0 m) above the surface. Thermistors, thermocouples, and liquid-in-glass instruments are used to make temperature readings. Extreme high temperatures are most common in the late afternoon and extreme cold temperatures frequently occur just before sunrise on days with clear skies, snow cover, and weak winds.

Research summarized by the Intergovernmental Panel on Climate Change (IPCC) documents that the planet has warmed since 1901, with more pronounced upward trends during 1910–1940 and from 1970 onward. In a stable climate, the ratio of hot days and cold days is approximately even. However, in a warming climate, record numbers of hot days and nights have begun to outpace cold days and nights, with the imbalance increasing for the past three decades. An increase in extreme hot temperature events adds to the growing cost of climate change to the global economy.

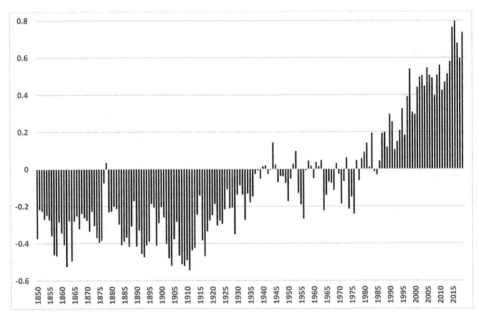

Worldwide annual average surface temperature anomalies since 1850, relative to 1961–1990. (Retrieved from the latest versions of the HadCRUT4 air temperature and sea surface temperature data set. Available from: https://crudata.uea.ac.uk/cru/data/temperature/HadCRUT4-gl.dat)

Heatwaves

Extreme heat events have been defined in a number of ways. Heatwave events are generally understood as a period of extremely hot days and warm nights, although there is no standard, single definition for these extremely hot temperature periods. The World Meteorological Organization (WMO) defines a heatwave as "a marked unusual hot weather (Max, Min, and daily average) over a region persisting at least two consecutive days during the hot period of the year based on local climatological conditions, with thermal conditions recorded above given thresholds" (WMO 2016). Other definitions consider a duration of two consecutive days or longer and may consider just the maximum temperature or both maximum and minimum temperatures together. Most definitions involve local climate data and identify the warmest 10 percent of days. Hazards researchers are interested in how often heatwave events occur (frequency), how hot they get (intensity), how long they last (duration), the impacts of the event, and whether or not society found ways to adapt to the extreme temperatures (Tavakol et al. 2020b).

Apparent Temperature

In humid regions, increases in temperature (maximum, minimum, and daily average) alone cannot precisely gauge the impact of extreme heat on organisms. An index of discomfort, the temperature humidity index (THI), was established in the 1950s and has been extensively used in livestock studies. When high THI values are forecast, livestock producers use a number of strategies to minimize the impact of the heat and humidity on their animals. A widely used improvement to THI and designed for human susceptibility to the combination of heat and humidity is the apparent temperature or heat index (Tavakol et al. 2020a). Other measures include the humidex, a wet-bulb temperature, and the dew point temperature.

Accompanying the planetary warming associated with global climate change has been an increase in the cumulative impact of temperature and humidity since the late twentieth century. The human body attempts to maintain a relatively constant, stable internal temperature (core body temperature) by losing excess heat through evaporation. Extremely warm temperatures and/or physical exertion create conditions that exceed the ability of the body to cool itself with evaporation. High humidity makes it even more challenging for the human body to cool off using evaporation during hot days. Some heatwave definitions use apparent temperature rather than just air temperature in their criteria.

Cold Waves

The WMO defines a cold wave, also known as a cold snap or cold spell, as a "marked cooling of the air, or the invasion of very cold air, over a large area" (WMO 2016). The U.S. NWS defines a cold wave as "a rapid fall in temperature within 24 hours to temperatures requiring substantially increased protection to agriculture, industry, commerce, and social activities" (AMS n.d.). Similar to heatwaves, the criteria used to define a cold wave are different for locations across the world. However, definitions generally considered the severity, duration, and perhaps the geographic extent of the cold wave. The criteria can be considered as either temperature values dropping below an absolute temperature threshold (e.g., below 32 degrees Fahrenheit (0 degrees Celsius) or below −4 degrees Fahrenheit (−20 degrees Celsius)) or a percentile value (e.g., the lowest 1 percent or lowest 10 percent of temperatures at that specific location).

Extreme Marine Temperatures

With more than 90 percent of the excess energy from global warming being absorbed in the ocean waters, researchers are now examining extreme marine temperatures and marine heatwaves. A marine heatwave is defined as a prolonged period of abnormally warm sea surface temperatures that persists for days to months. The above normal temperatures can extend to depths of thousands of kilometers. Marine heatwaves are caused by the combination of oceanographic and atmospheric processes and can produce severe impacts on marine ecosystem function and structure (Hobday et al. 2016). Coral reefs are especially susceptible to warming, and the number of coral bleaching events continues to increase.

As with land-based heatwaves, research on marine heat events is designed to characterize the duration, frequency, and intensity of these events to better understand how marine heatwaves form and evolve. Once extremely warm marine temperatures occur, the "blob" of above normal sea surface temperature may produce a reinforcing feedback with the atmosphere. This air-sea interaction can help maintain or even intensify the relevant oceanic and atmospheric circulation patterns that helped create the area of warmer waters. Marine heatwaves have doubled between 1982 and 2016 (Frölicher et al. 2018). One important difference between land surface and marine heatwaves is that heatwaves for land areas are ultimately doomed by the arrival of the low sun period (autumn in areas outside the tropics). However, due to the conservative properties of ocean waters, marine heatwaves may recur for more than one year as a result of a reemergence mechanism, whereby the warm deeper waters are mixed back into the near surface waters.

ACCLIMATIZATION

Acclimatization or acclimation is the process of adjusting to an environmental change in temperature. Humans are generally better able to cope with higher temperatures later in the warm season of the year. Acclimatization is a physiological adaptation that a person or an individual organism makes because of the repeated exposure to the changing conditions. Heat acclimatization produces an improvement in heat tolerance. The human body has two different reactions related to heat extreme: a quick response such as its ability to sweat in an attempt to cool off and the long-term response that includes hormonal and metabolic shifts that influence the body's temperature-regulation system.

EXAMPLES OF THE EXTREME TEMPERATURE HAZARD

St. Louis, Missouri, is known for a long history of hot summers and heat-related mortality (Smoyer 1998). In July 1980, St. Louis and Kansas City, Missouri, experienced a severe heatwave that, at the time, was one of the worst heatwaves on record and resulted in many deaths. Researchers who study heatwaves and human mortality usually calculate the number of heat-related deaths by looking at the number of deaths that exceed the average number of deaths per day at that location. Afternoon temperatures topped above 100 degrees Fahrenheit (38 degrees Celsius) for 17 days in a row and increased the number of deaths by 57 percent in St. Louis and 64 percent in Kansas City. Cities are typically warmer than surrounding rural areas, a phenomenon called an urban heat island. In rural areas, the increase in heat-related deaths was approximately 10 percent.

In mid-July 1995, a record-breaking heatwave influenced the U.S. Midwest with a devastating impact on the residents of Chicago. A deadly combination of temperature and humidity contributed to extremely high heat index values of greater than 100 degrees Fahrenheit (38 degrees Celsius) for five days in a row. The heatwave started on July 12, and the maximum heat index was greater than 115 degrees Fahrenheit (46 degrees Celsius) on July 13 and July 14. An unseasonably humid and hot air mass associated with a high-pressure system moved into the area and consistently produced maximum daily temperatures of higher than 90 degrees Fahrenheit (32 degrees Celsius) and minimum temperatures higher than 70 degrees Fahrenheit (21 degrees Celsius) at night. High humidity made a huge difference in the extreme nature of the event. The 1995 heatwave resulted in more than 3,300 excess emergency visits, at least 600 excess deaths, and a substantial number of near-fatal heatstroke victims (Dematte et al. 1998).

In August 2003, Europe experienced a severe heatwave with the highest recorded temperature since at least 1540. The extreme event caused almost 40,000 fatalities. Temperatures exceeding 105 degrees Fahrenheit (40.6 degrees Celsius) characterized the heatwave was concentrated in France, Germany, and northern Italy. A persistent anticyclone of high pressure over Western Europe, warm air movement in from the Sahara Desert, and soil moisture deficits were all associated with this European mega-heatwave.

A prolonged heatwave, which extended from July to August in 2010, affected western Russia. The summer of 2010 was the hottest recorded summer in Russia, and the magnitude and duration of the heatwave was so great that the hot and dry conditions enabled an increase in extensive wildfires. In Russia, the daily temperature exceeded 95 degrees Fahrenheit (35 degrees Celsius) on a majority of the summer days and Moscow experienced temperature of 32 degrees Fahrenheit (18 degrees Celsius) above normal for July. The combination of heat, fire, and smoke resulted in approximately 55,000 fatalities. A stagnant atmospheric flow pattern with persistent atmospheric blocking along with very dry soils was identified as distinctive aspects of the heatwave.

Large parts of North Africa and the Middle East experienced a heatwave in July 2016 that lasted for almost a week. On July 21, 2016, an intense heat event gripped Kuwait with a temperature of 129 degrees Fahrenheit (54 degrees Celsius) recorded by a weather station in Mitrabah, which is located in a remote and sparsely populated area in the northwest of Kuwait. This temperature in Kuwait is possibly the highest temperature ever recorded in Asia and Eastern Hemisphere (Voiland 2016). At the same time, a temperature of 129 degrees Fahrenheit (54 degrees Celsius) was recorded in Basra in Iraq. According to the WMO's World Weather and Climate Extremes, the hottest temperature ever recorded has been in Death Valley, California. On July 10, 1913, the temperature in Furnace Creek, surrounded by the vast and arid desert of Death Valley, reached 134 degrees Fahrenheit (56.7 degrees Celsius).

California experienced a multi-year period of exceptional warmth and drought from 2012 to 2015. The three-year period from 2012 to 2014 was the hottest and driest on record. During this period, water temperatures in the eastern Pacific were well above normal for areas from the Gulf of Alaska (in 2013) extending south to Baja California (2014 and 2015). These exceptional warm waters were nicknamed, "The Blob." In the atmosphere, high pressure was common at the surface, and the jet stream flow arched northward in a persistent pattern called "the ridiculously resilient ridge." This multi-year event is an example of compound hazards, where two (or more) hazards occur together.

Traditionally, dry and tolerable heat events are expected in semiarid Mediterranean climate regions. In California, temperatures normally increase during the day and cool off significantly at night. This mechanism allows animals and plants to recuperate from the afternoon heat during cooler nighttime hours. Since the 1980s, heat events have been associated with increased humidity and less cooling recovery at night. The 2006 California heatwave during the second half of July is an example of a period of severely hot and humid conditions that hospitalized 1,182 people, and caused 16,166 emergency visits (Knowlton et al. 2008). Negative impacts on agriculture, ranching, the energy sector, water supplies, and ecosystems were associated with these extreme temperatures.

Not all extreme temperature events are associated with hot conditions. Unprecedented freeze events were reported in Florida in December 2010. Extremely cold temperatures severely damaged the sugarcane crop with temperatures below 30 degrees Fahrenheit (−1 degree Celsius) on December 7 and 8 and temperatures below 28 degrees Fahrenheit (−2 degrees Celsius) on December 14, 15, 27, and 28. Unusual conditions in other areas can impact the weather in downstream areas in a process that atmospheric scientists call a teleconnection. A strongly negative Arctic Oscillation (AO) with warmth over the Arctic Ocean and a negative character to the North Atlantic Oscillation (NAO) with warm conditions near Iceland were associated with the cold air mass outbreaks that impacted Florida (NOAA 2010).

An early 2012 European cold wave occurred in January and February. Heavy snowfall accompanied the protracted severe freeze event. Arctic air moved west and south from Siberia and caused the temperature to fall below −22 degrees Fahrenheit (−30 degrees Celsius). This deadly cold wave affected southeastern and eastern Europe and caused above 550 deaths.

EXTREME TEMPERATURE HAZARDS

Human Health/Death

A major heatwave or cold wave can cause human health issues, including excess mortality and morbidity. During extreme hot conditions, human bodies dissipate the heat, attempting to keep the core body temperature constant, by losing water through the sweat glands and skin, varying the depth and rate of blood circulation, and by panting. Energy required to evaporate the sweat is extracted from the body, thereby cooling it. The stress that extreme heat places on human and animal bodies is not a simple function of temperature alone. Under conditions of high humidity, a combination of humidity and temperature limits the cooling effect of evaporation and leads to increased heat stress. Recognizable human health issues that result from extreme heat include sunburn, dehydration, heatstroke, heat exhaustion, heat syncope, and heat cramps. The population groups most vulnerable to extreme temperatures include the elderly, children, those with chronic health conditions, and the economically disadvantaged groups who are without shelter or live in a poorly insulated house. As the extreme heat events become more severe, more frequent, and longer lasting, scientists expect to see a growth in illnesses and deaths from heat.

Exposure to extreme cold can also result in life-threatening health emergencies. Wind chill, which is the combination of extreme cold and wind, may exacerbate the impacts of extremely cold temperatures. Hypothermia, frostbite, or unconsciousness are dangerous health conditions observed frequently during extremely cold periods.

In the United States, almost 10,000 deaths were attributed to excessive heat exposure between 1979 and 2001. This total is higher than the mortality from flood, hurricane, tornadoes, earthquake, and lightning. The mega heatwave of Europe in 2003 killed over 40,000 people in France and nearby areas. Hot summers were not common in that region, and most homes were not equipped with air-conditioning systems (Coppola 2006).

Drought

An important hazard that is frequently associated with extremely hot temperatures and possibly a heatwave is drought. Droughts are periods of abnormally dry conditions that continue long enough to produce impacts related to a decline in available water resources. During a heatwave, high temperatures increase evaporation and plant transpiration (evapotranspiration), resulting in a decline in soil moisture. Warmer temperatures associated with global climate change increase evapotranspiration with impacts for plant health and soil moisture supplies. During a drought, with less water available for evaporation and/or transpiration, most of the solar energy goes into raising the soil and air temperatures and amplifying the warming process. A reinforcing feedback between extremely hot temperatures and drought raises concerns regarding extreme event frequency and intensity as the global climate continues to warm.

Wildfire

The combination of extreme heat, vegetation with less moisture in plant tissues and drier soils, and plant litter results in a situation where wildfires are more likely. With hot and dry conditions, fires can grow larger and spread more quickly. The Russian wildfires of 2010 are an example of how a heatwave can reinforce environmental conditions that favor wildfire. In the summer of 2010, the unusual heatwave broke existing temperature records several times. Wildfires raged over European Russia from late July into early September and burned about 740,000 acres (299,467 ha). In late January and early February 2009, a major heatwave in Southeastern Australia broke all-time temperature records in several locations. The heatwave combined with dry days with high wind speeds reinforced a wildfire that burned about 1.1 million acres (445,154 ha).

Transportation

Periods of extreme temperatures during summer heatwaves can lead to severe road damage. Conditions associated with temperatures higher than 90 degrees Fahrenheit (32 degrees Celsius) are able to soften, expand, and even melt asphalt, which leads to asphalt rutting and potholes from heavy traffic. In addition, high temperatures increase tire wear and can cause vehicle engines to overheat. Expansion of bridge joints is also a risk when the temperatures increase. Expansion and contraction of bridge joints may lead to increased maintenance costs.

The steel rails used for railroads are at risk of overheating and warping. Deformities in rail tracks can result in a speed restriction if identified and maybe even a train derailment. Extreme heat impacts airline flights because the heat influences the physics of how aircraft fly. Hotter air is less dense, and the result is that an airplane needs to travel longer distances on the runway to generate enough lift for takeoff. Extreme heat events have produced a recent increase in canceled flights around the world. Numerous cancelations during the July 2017 extreme temperature event in Arizona provide an example.

Extreme cold and heavy snowfall can cause damage to structures and immobilize an entire region. Even areas that normally experience mild winters can be negatively affected by extreme cold or a major snowstorm. Just as with extreme heat, all types of transportations including roads, railroads, and airplanes are affected by extreme cold. With global climate change, there has been a decrease in the challenging impacts of extreme cold on transportation. Based on a report from National Research Council (NRC) on the potential impacts of climate change on the U.S. transportation, warmer winters have led to less frequent major winter storms.

ADAPTATION

Adaptation involves human adjustment in behavior in response to an extreme weather hazard. Whether dressing appropriately for a cold and windy day or heading to the local swimming pool when hot temperatures occur, humans have found ways to accommodate changing environmental conditions. Construction of different kinds of dwelling structures and development of different clothing options are examples of human adaptation. The substantial impacts of more frequent and more severe extreme temperature events serve as a warning that the extreme temperature events are likely to cause more damage to human and animal health, agriculture, infrastructure, and economic performance.

Adaptation, as a process of adjustment to changing conditions, provides an option to moderate damage while exploiting opportunities. There are numerous ways to categorize adaptation options, and IPCC scientists place options into structural (e.g., engineered environment, technological, and ecosystem services), social (e.g., educational, informational, and behavioral), and institutional (e.g., government policies, laws, and economic regulations) categories.

Advisories, Watches, and Warnings

Monitoring and early-warning systems, categorized as an informational adaptation, have played a crucial role in adjustment at the local scale. Early-warning systems apply diverse approaches and technologies, including earth system modeling and satellite information to help mitigate the potential damage. In the United States, the NWS issues extreme weather watches, advisories, and warnings to help citizens prepare for changing and dangerous weather condition.

Heat alerts are defined locally based on current and forecast heat index information. Notifications provided by NWS forecasters include a heat advisory, an excessive heat watch, and an excessive heat warning. When the heat index is expected to exceed 106 degrees Fahrenheit (41 degrees Celsius) for several hours to a few days, messaging changes from a heat advisory to a heat watch and then to excessive heat warning when stressful conditions last longer (including a lack of overnight cooling). The NWS issues extreme heat-related warnings and recommends preparation actions.

Cold season warnings from the U.S. NWS include forecasts of extremely cold temperatures. Forecasts of extreme wind chills, a combination of cold and biting winds, prompt advisories and warnings. Due to different levels of acclimation to the cold in different parts of the country, local NWS forecast offices use different criteria to determine when to issue relevant statements.

Cooling Centers

During an extreme heat event, cooling centers or cooling shelters (with air-conditioning) provide a safe and cool environment for vulnerable people. The many deaths that occurred in association with the 1995 Chicago heatwave helped document the need for cooling centers. Cooling centers are now an adaptation strategy commonly used in many cities across the United States. Examples of structures that are used as cooling centers include government-owned buildings, religious centers, private businesses, and recreation and community centers.

EXTREME TEMPERATURES AND CLIMATE CHANGE

Changes in the frequency and intensity of extreme temperatures are happening as the global climate changes. The increase in carbon dioxide in the atmosphere has resulted in extra energy being stored in the earth system and in generally warmer ocean and air temperatures. The ongoing changes pose risks to ecosystems, communities, and infrastructures. One way that climate scientists document the changing conditions is with descriptive statistics on the character of the extreme temperature events.

Observed Changes in Number of Record Highs versus Record Lows

The IPCC claims that the risk of extreme temperature is linked to global warming. A shift in the average temperature changes the likelihood occurrence of extreme temperature events. With global warming, there has been a shift toward more hot extremes and fewer cold extremes.

Arctic Amplification

Since 1880, the global average temperature has increased by 1.4 degrees Fahrenheit (0.8 degrees Celsius). However, the warming has not been distributed equally everywhere. Polar or Arctic amplification is the phenomenon where the temperature in the Arctic increases almost twice as fast as in mid-latitudes. Arctic amplification is related to reduced snow cover and the loss of sea ice in the Arctic Ocean. When the snow or ice melts, the highly reflective surface changes into darker land or open ocean. This shift causes more solar radiation to be absorbed and a local increase in warming. With Arctic amplification, numerous extreme high temperature records are being set across northern latitude areas. The existing influence of the Arctic amplification on ice sheet stability, glaciers, global sea level, and carbon cycle feedbacks makes this phenomenon a growing concern for the future.

IPCC Models and Future Scenarios

A rising rate of fossil fuel burning causes the addition of carbon dioxide to the atmosphere at a rate that exceeds removal or sequestration. The increase in the atmospheric carbon dioxide results in a change in the earth's energy budget with a net result of the warming of lower atmospheric and ocean surface temperatures. Reports from the IPCC scientists project additional changes in air and ocean temperatures and extreme temperature events. Projections suggest that the increases in global mean air and ocean temperatures will likely continue. Regionally, the temperature change may not be uniform. For example, land areas are projected to warm more than the oceans by the end of twenty-first century, and Arctic regions are highly likely to warm the most. The changes in average temperature will trigger more the extreme temperature events. More extreme heat and fewer extreme cold events are expected in the future.

CONCLUSION

Extreme climate events are startling, unusual, and severe climate events that may cause significant economic losses. Temperature extremes, including extreme heat and extreme cold events, have large impact on human safety, agriculture, water supplies, and infrastructure. The IPCC reported an increase in the intensity, frequency, and duration of extreme heat events as the climate system has warmed. Extreme heat events of Europe (2003) and Russia (2010) are examples of recent mega heatwaves. However, the extreme cold events are becoming less frequent because of global climate change. Both extreme heat and cold events are leading causes of morbidity and mortality in the world, with highest impact on vulnerable people (e.g., infants, the elderly). Extreme temperature early action warning systems, such as those developed by NWS, are helpful and can save lives all around the globe.

Ameneh Tavakol, Vahid Rahmani, and John A. Harrington Jr.

Further Reading

AMS (American Meteorological Society). n.d. Glossary of Meteorology. http://glossary.ametsoc.org/wiki/Cold_wave, accessed September 23, 2019.

Coppola, D. P. 2006. *Introduction to International Disaster Management*. Boston: Elsevier.

Dematte, J. E., K. O'mara, J. Buescher, C. G. Whitney, S. Forsythe, T. McNamee, R. B. Adjida, and I. M. Ndukwu. 1998. Near-fatal Heat Stroke during the 1995 Heat Wave in Chicago. *Annals of Internal Medicine* 129(3): 173–181.

Frölicher, T. L., E. M. Fischer, and N. Gruber. 2018. Marine Heatwaves under Global Warming. *Nature* 560: 360.

Guha-Sapir, D., F. Vos, R. Below, and Rand S. Ponserre. 2012. *Annual Disaster Statistical Review 2011: The Numbers and Trends*. Brussels, Belgium: Centre for Research on the Epidemiology of Disasters.

Hobday, A. J., L. V. Alexander, S. E. Perkins, D. A. Smale, S. C. Straub, E. C. Oliver, J. A. Benthuyseng, M. T. Burrowsh, M. G. Donat, M. Feng, N. J. Holbrook, P. J. Moore, H. A. Scannell, A. S. Gupta, and T. Wernberge. 2016. A Hierarchical Approach to Defining Marine Heatwaves. *Progress in Oceanography* 141: 227–238.

Knowlton, K., M. Rotkin-Ellman, G. King, H. G. Margolis, D. Smith, G. Solomon, R. Trent, and P. English. 2008. The 2006 California Heat Wave: Impacts on Hospitalizations and Emergency Department Visits. *Environmental Health Perspectives* 117(1): 61–67.

NASEM (National Academies of Sciences, Engineering, and Medicine). 2016. *Attribution of Extreme Weather Events in the Context of Climate Change*. Washington, DC: National Academies Press.

NOAA (National Oceanic and Atmospheric Administration). 2010. 2010 South Florida Weather Year in Review. http://www.weather.gov/media/mfl/news/2010WxSummary.pdf, accessed July 22, 2019.

NOAA (National Oceanic and Atmospheric Administration) and NCEI (National Centers for Environmental Information). 2019. Billion-dollar Weather and Climate Disasters: Overview. https://www.ncdc.noaa.gov/billions, accessed March 9, 2019.

Smoyer, K. E. 1998. A Comparative Analysis of Heat Waves and Associated Mortality in St. Louis, Missouri—1980 and 1995. *International Journal of Biometeorology* 42: 44–50.

Stocker, T. F. 2014. *Climate Change 2013: The Physical Science Basis: Working Group I Contribution to the Fifth Assessment Report of the Intergovernmental Panel on Climate Change.* Cambridge, UK: Cambridge University Press.

Tavakol, A., V. Rahmani, and J. Harrington Jr. 2020a. Changes in the Frequency of Hot, Humid Days and Nights in the Mississippi River Basin. *International Journal of Climatology.* 1–16. https://doi.org/10.1002/joc.6484.

Tavakol, A., V. Rahmani, and J. Harrington Jr. 2020b. Evaluation of Hot Temperature Extremes and Heat Waves in the Mississippi River Basin. *Atmospheric Research* 239, 104907. https://doi.org/10.1016/j.atmosres.2020.104907.

Voiland, A. 2016. NASA Earth Observatory: Extreme Heat for an Extreme Year. https://www.giss.nasa.gov/research/features/201608_extremeheat, accessed July 31, 2019.

WMO (World Health Organization). 2016. Guidelines on the Definition and Monitoring of Extreme Weather and Climate Events. https://www.wmo.int/pages/prog/wcp/ccl/opace/opace2/documents/DraftversionoftheGuidelinesontheDefinitionandMonitoringofExtremeWeatherandClimateEvents.pdf, accessed July 31, 2019.

Tornadoes

Tornadoes have been recorded in history as far back as 77 CE in a work called *Naturalis Historia* by Roman natural philosopher Pliny the Elder (23–79). In the Western Hemisphere, and more specifically, the Americas as they were called after 1504, records of tornadoes date back as far as 1521, occurring in the Aztec cities of Tlatelolco and Tenochtitlan, and are described in a work known as the *Florentine Codex*. The famous first governor of the Massachusetts Bay Colony, John Winthrop, described on July 5, 1643, what is probably the first recorded tornado in Colonial America, with reports of dark swirling clouds that removed trees from the ground. Another well-known Colonial American, and one of the Founding Fathers, Benjamin Franklin, wrote about what is now known as "waterspouts," a specific nonsupercell type of tornadic structure that forms over water, resulting from spanning at the ground, not the base of the cloud.

The vast majority of tornadoes in the world occur in the United States, and most of the research on this event has been conducted in the United States. However, tornadoes do occur on every continent except Antarctica, with Europe seeing the most of any region besides the United States/North America. Tornadoes can also form as a result of wind drag against the surface when hurricanes (cyclone, typhoons) make landfall, and as such, tornadoes can form anywhere tropical cyclones are prone to make landfall, including East, South, and Southeast Asia, in countries such as China, India, Bangladesh, Indonesia, and the Philippines.

TORNADO RESEARCH IN THE UNITED STATES

The formal study of tornadoes, however, went unattended through the 1700s, although events were recorded and described. It was not until the 1830s that the study of tornadoes began to move out of its infancy. James Espy, William Redfield, and Robert Hare examined in some detail a major event that occurred on May 31, 1830, in Shelby, Tennessee (Grazulis 1993). Toward the middle of the nineteenth century, famed physicist Joseph Henry (1797–1878), working though the Smithsonian Institute in the late 1940s, developed a network of storm observers who were charged with telegraphing their reports to him in Washington, DC, for analysis. By the 1860s, Henry was compiling data on characteristics such as location, speed, direction, path length, shape, and width in order to advance the understanding and possibly classification of tornadoes, but it would still be decades before any predictive models would be attempted. By the 1870s, the U.S. military was collecting information and data on weather conditions, with the intention of using such data to predict upcoming weather and provide forecasted warnings to naval vessels as well as private vessels. This endeavor is the precursor to the modern-day National Weather Service. This directive by the government also gave rise to a man widely considered the father of tornado forecasting, John Park Finley (1854–1943). A lieutenant in the U.S. Army Signal Corps, Finley was charged with collecting tornado data from 1884 to 1886 as part of a test program to determine if the storms could be accurately predicted. Finley did not uncover the ability to predict tornadoes, but he did have considerable success in predicting the onset of thunderstorms, a first in the field. Due to the perceived failure to create any precise predictive models, Finley's career as a meteorologist was essentially ended, and at the behest of his superiors, he was assigned to a desk job within the signal corps, and the tornado forecasting program was canceled (Bradford 1999). Also, born of the failure to predict tornadoes, it was decided that the use of the very word *tornado* would be barred from official weather forecasts, for fear that panic resulting from the possibility of a tornado would cause more damage than the tornado itself.

In 1891, the Weather Bureau came under control of the Department of Agriculture, and although it did very little research on tornadoes, even downplaying their significance, they did continue collecting weather data, which became important in the years to come. However, since no research or attempt at understanding, much less predicting, tornadoes occurred in the late 1800s and the early twentieth century, tornadoes occurred with no warning and with no definitive plan of action to respond to them. This lack of foresight is certainly to blame, at least in part, for the massive loss of life that occurred on March 18, 1925, in an event known as the Tri-State Tornado. The Tri-State Tornado struck parts of Missouri, Illinois, and Indiana and killed 695 people, the single deadliest tornado in U.S. history (although not the deadliest in the world).

Tornado research became more prominent in the 1940s and 1950s, with the first tornado forecast to successfully predict an event occurring on March 25, 1948. The forecast resulted when U.S. Air Force meteorologists Major Ernest Fawbush and Captain Robert Miller at Tinker Air Force base in Oklahoma City, Oklahoma, noticed that the synoptic conditions present on that day were remarkably similar to conditions seen just five days prior, a day which produced a deadly tornado that caused $10 million in damage to several aircraft (1948 dollars, or approximately $106 million in 2018 dollars). On March 21, Fawbush and Miller began to analyze the upper-air and surface weather data of the storms that had occurred on March 20, with the intention of assembling a tool to assist in future predictive efforts. Their work was validated on May 25 when they made their now-famous, and accurate, prediction of the March 25, 1948, tornado (Wagner 1999).

Arguably the most influential figure in tornado research is Dr. Tetsuya (Ted) Fujita (1920–1998), a name easily recognizable by the scale he developed, the F-Scale, or Fujita Scale (since revised, it is now known as the Enhanced Fujita Scale, or EF-Scale). Dr. Fujita's work began at the University of Chicago in the 1950s, and throughout his career, he discovered and described meteorological phenomena such as the bow echo and the downburst, which is a major threat to aircraft. Fujita's work on the downburst has been incorporated into pilot training programs around the world. He also led the way in developing uses for satellite imagery in describing the motions of clouds, work that can be found in our ability to track storms today.

Overlapping Fujita's time, the 1950s and 1960s saw great advancements in the use of instrumentation in aircraft as well as the development of Doppler radar. These advancements helped advance our understanding of what conditions were more likely than others to produce tornadic supercells. The 1960s brought the first realizations that the large storms called supercells were "prolific tornado producers" (Doswell et al. 1993). The 1970s is called the golden age of tornado research, as it was during this era that a nationwide Doppler network was established, more advanced modeling of mesoscale systems—large systems of unstable air capable of producing severe weather including thunderstorms and tornadoes—was being done with the help of advances in computer technology, and Dr. Fujita introduced the F-Scale to aid in estimating wind speeds based on postevent on-ground analysis. Advancements steadily continued, essentially improving on existing technologies, leading to today's era of advanced NEXRAD Doppler radar, encompassing 159 stations across the United States, and allowing for relatively accurate predictions of conditions that lead to severe weather and possibly tornadoes up to three days in advance (although warming time for specific tornado events is still far less than that, generally about 14 minutes, up from about 5 minutes in the 1980s) (Stimers 2011).

TORNADO FORMATION AND PHYSICAL CHARACTERISTICS

Tornadoes are defined as rotating masses of air that come into contact with the ground. In the United States, tornado genesis is the result of the intersection of several factors, but broken down to their most basic form, they are the product of warm and moist unstable rising air and the addition of air moving into a thunderstorm system from three different levels. The initial component of a thundercloud that may lead to the formation of the tornado is the column of rising air. As an air particle warms, it becomes more buoyant and thus moves vertically, much the same way a life jacket floats on the surface of water because of its buoyancy. Central to this is the low-pressure system at the ground level. With higher pressure air surrounding the low-pressure center, air rushes in, a process termed *convergence*, and is redirected upward. This can be observed as a cold front moves into the area, which pushes the warm air up and out of the way, lowering temperatures as a storm approaches.

The warming air mixes with other air particles as it moves upward in the system, creating instability and releasing heat as a result of its expansion. (Downward-moving air may also play a part in tornado genesis, but its role is not well understood at this time.) This latent heat release is what greatly assists in producing strong supercell thunderstorms. In addition, rainfall within the system will provide additional energy in the form of latent heat. As the thunderstorm matures, it may produce a tornado, but the mechanisms of heat release and vertical mixing coupled with wind shear will produce a tornado within a supercell about 20 percent of the time (NSSL 2018).

Additional components, namely wind shear, are necessary. Shear as it applies to wind (also called wind gradient) is defined here as the convergence of air masses at either (1) differing levels above the ground, termed *directional shear*, or (2) air colliding at differing speeds, termed *speed shear*. There are three levels of air that need to be added to a supercell in order to impart spin to the system, which greatly increases the chance of tornado genesis. The first of these is low-level moist air moving generally north to northeast, coming from the Gulf of Mexico. A second input of colder higher altitude, called mid-level air, comes from Canada, generally moving in from the northwest, or from the Rocky Mountain Range, also moving in an easterly direction, but its direction can vary substantially. Finally, the persistent jet stream (or Polar Jet) adds the third level at high altitudes, moving in an easterly direction.

When these three air formations meet a supercell, they may impart one or both of the types of shear, directional shear or speed shear. Directional shear occurs when the wind direction shifts with the height above the ground. In relation to the three air masses mentioned earlier, this type of shear will be caused by the varying heights of the streams. Likewise, speed shear is imparted from the convergence of these three air masses as they all move at different speeds, with the jet stream typically traveling upward of 150 mph (240 kph), the mid-level air moving at speeds reaching 50 mph (80 kph), and the maritime tropical air from the Gulf of Mexico moving the slowest, usually around 25–30 mph (40–48 kph).

As speed shear is added, the rotation will be along an axis parallel to the ground surface (think of a barrel rolling along the ground). Under circumstances outside of those found in a supercell, this would simply result in wind moving across an area. However, as this occurs along with a thunderstorm featuring uplift of warm air, this uplift will lift the column of rotating air, pushing it into a vertical position, and a tornado is born. This area of the supercell is termed a *mesocyclone* and can be identified on modern radar systems as a hook-like feature, appropriately called a "hook echo" (first identified by the famous meteorologist Ted Fujita).

Once the conditions are in place for a tornado to form, it is common to see a feature termed a *wall cloud* to develop at the leading edge of the system. This flat, broad downward extension of the cloud is the beginning of the tornado, which is characterized by moist descending air condensing into liquid; it is this liquid formation that people will first see as the funnel dropping out of the wall cloud, although the formation is not considered a tornado until it comes into contact with the ground. As the funnel moves out of the system, it will generally maintain the color of the cloud, but once in contact with the ground, it will change color rapidly as it picks up soil and other debris, sweeping it up into the rotating air mass. This color change can serve as an indicator that the funnel had made contact with the ground and can now be considered a tornado.

According to the National Weather Service, the duration of most tornadoes is approximately 10 minutes, but exceptionally strong and persistent systems can produce tornadoes lasting several hours. In the Northern Hemisphere, about 99 percent of tornadoes will rotate counterclockwise due to the Coriolis Effect (rotation of air particles due to the spin of the earth, the effect on tornadoes that was first described in the 1860s by a mathematician and teacher named William Ferrel), but 1 percent will rotate clockwise due to local conditions altering the direction of spin. The path-length, or track, of a particular tornado will be highly variable, with the majority of tornadoes classified as "short track" and very few classified as "long-track." (This is consistent with what hazards and disasters researchers identify as the Frequency-Magnitude relationship: there are many small events and few large events, distributed by what is called the power law.) An individual tornado can change direction and speed rapidly, and as such will move in unpredictable patterns, but tornadoes in general can be described as displaying southwest-to-northwest movement.

Tornadoes are also known to display what at first glance may seem like a random occurrence, wherein an area, a neighborhood, for example, may see several homes badly damaged or completely destroyed and others with minimal or no damage. These lesser damaged structures are actually the result of wind speeds slowing greatly as smaller vortices (funnels) along the edges of the main path work to temporarily cancel high winds from the main funnel. A structure that is passed over by the main funnel while the winds from a smaller vortex meet the main funnel head-on may experience diminished wind damage, but once the smaller vortex and the larger funnel show wind moving in the same direction, the speed can be amplified, completely destroying a structure just feet away from one that was barely damaged. Tornadoes may also "skip" along the ground, which can result in several structures in the path being spared major damage while structures in near proximity could be obliterated. As damage assessment occurs during postevent, this skipping tendency can be masked by the overall debris field, making positive identification of the occurrence difficult.

There are a few other types of structures that take on the appearance of a tornado, but are considered to be nonsupercell tornadoes. A gustnado is a spiraling dust storm that appears along the "gust front" of a storm. Waterspouts form over water, but if a waterspout moves onto land, which is not a rare occurrence, it is then classified as a tornado. Waterspouts also have three subvarieties: the nontornadic waterspout, the tornadic waterspout (which is the term given to a land-based supercell-derived tornado that makes its way over water), and the snowspout. Landspouts are similar to waterspouts in that they comprise a condensation funnel, but since the rotation is mainly due to drag at the surface, the structure is also a nonsupercell tornado. Firenados (also called fire whirls or fire devils) have the common characteristic of warm rising air as well as spin imparted from shear, but since firenados are never connected to a cloud base, they are not considered tornadoes at all, not even in the context of nonsupercell tornadoes.

TYPES OF TORNADOES

There are several different formations tornadoes can take on, each being born out of different conditions within the supercell. Visual identification of the types is fairly easy, as their characteristics are generally obvious. The "stovepipe" shape can be identified by the cylindrical shape of the funnel. Tornadoes that take on a V-shape are termed *wedge tornadoes*—these are typically very violent and strong tornadoes. As a tornado loses energy, the funnel will become smaller due to the conservation of angular momentum, and the funnel will take on a thin rope-like appearance; these are called "rope tornadoes," and at this stage, the dying tornado is said to be "roping out." Similar to rope tornadoes are the "drill bit" variety, which can be thought of as smaller than a large stovepipe funnel, but not yet at the rope stage. Multiple-vortex tornadoes are also possible, featuring more than one funnel protruding from the same cloud base. These are formed usually after the main vortex has reached the ground and occur as a result of the change in balance between the rate of rotation and rising air. Multiple vortex events may be difficult to identify if the secondary vortices are obscured by debris or the main vortex. Perhaps the most classic version of a tornado, though, is the "cone" variety. This type is wide at the top where it is connected to the cloud base, and it gets gradually narrower as the height from the cloud decreases. Although this may sound similar to the wedge variety, the cone does not take on a pronounced V-shape.

DISTRIBUTION

According to the National Severe Storms Laboratory (NSSL), a part of the National Oceanic and Atmospheric Administration (NOAA), the United States experienced an average of 1,141 tornadoes per year, based on data from 1985 to 2014. Texas has the highest average per year over that time frame, at 140, followed by Kansas (80), Florida (59), Oklahoma (56), and Nebraska (54) (SPC 2018).With the exception of Florida, the other four states in the top five are situated in what is known as "Tornado Alley," a loosely defined region extending from Texas north into Iowa and Minnesota and encompassing an east-to-west range from eastern Colorado to Kentucky and Tennessee. The term, first appearing in the 1950s, does not have an official definition by any government agency and is used largely in an unofficial capacity by news outlets and the general public. The area can more generally be thought of as the "Central United States," consisting of generally flatter, nonmountainous topography that allows for the convergence of conditions conducive to tornado genesis. A second "alley" is also loosely defined as "Dixie Alley," stretching through the lower Mississippi Valley and upper Tennessee Valley. The term was first used in 1971 by a meteorologist named Allen Pearson.

Tornadoes can occur during any month of the year in the United States, but a predictable seasonal pattern emerges when long-term data trends are examined. As moist air from the Gulf of Mexico moves northward during the spring months, pushing the Polar Jet farther north, the conditions for tornado genesis become more pronounced. March is the peak of tornado frequency as a whole, but it is important to note that due to the variable and continuous movement of conditions northward, the peak for any given location on a south-to-north spectrum will generally be later and later in the spring and summer months. As such, "tornado season" for the United States is not recognized as a whole but, rather, a range of timeframes for peak onset based on geography. Similarly, tornadoes follow a diurnal (day-to-night) cycle that can be for the most part relied upon to be fairly predictable. As solar radiation heats the planet during the daytime hours, it reaches a peak around mid-afternoon, which coincides with the largest concentrations of warm air rising from the heated ground, a condition crucial to the formation of supercells and potential tornadoes. For this reason, the majority of tornadoes will occur during the afternoon hours.

Since the United States is not the only country that has the key ingredients and topography to create these life-threatening events, other countries can also endure tornadoes. Italy, France, Spain, India, and Brazil are among the countries that experience tornadoes, typically on a yearly basis. Argentina, Uruguay, Australia, Japan, China, Russia, Ukraine, Poland, and Germany can be added to the list of countries that have reported tornadoes. According to Dr. Harold Brooks, senior research scientist with the NOAA National Severe Storms Laboratory, the location that gets the most tornadoes outside of the United States, other than the part of the southern Canadian plains, is an area in South America. The region includes southern Brazil, northeastern Argentina, Uruguay, and southeastern Paraguay. According to Brooks, "Bangladesh is another area that can get violent tornadoes, and there's a large, lower-grade threat in a large area of Europe from northern France through into Russia" (Quoted in Mitchell 2018).

Experts say the deadliest tornado in world history appears to be the Bangladeshi tornado of April 26, 1989, due to extremely high population density and poor construction. Bangladesh and nearby Northeast India are notorious for occasional big, powerful, and very deadly tornadoes, especially in April. An April 26, 1989, tornado was reported to have killed about 1,300 people in the Manikganj district of Bangladesh according to AccuWeather Meteorologist Jim Andrews (Mitchell 2018). While rare, tornadoes have even occurred in Middle Eastern countries like the United Arab Emirates, according to Andrews. "European Russia had a big, deadly outbreak of tornadoes back the 1980s, and Argentina had a big, destructive outbreak around that time," Andrews said.

Geographically large countries like Russia, China, or Australia have more land for tornadoes to touch down, which is why those locations are likely to have one in any given year. However, each country has regions where tornado development is more common, similar to the United States. According to Brooks, "Italy has a large number of non-super-cellular tornadoes in the peninsular region. They do get supercell tornadoes in the Po Valley" (Mitchell 2018). Tornadoes can spin in opposite directions. According to Andrews, most Southern Hemisphere tornadoes spin clockwise, whereas they normally spin counterclockwise on the northern side of the equator.

TORNADO OUTBREAKS

A supercell can produce a single tornado, as is often the case, but large systems can produce several tornadoes in a day before dissipating, usually at least six, a phenomenon called a tornado outbreak. If sufficient time exists between one outbreak and the next, typically six hours, the collection of events will be referred to as a separate outbreak. If weather patterns conducive to thunderstorm development continue over the course of several days, a tornado outbreak sequence can occur, commonly consisting of dozens of individual tornadoes over the span of the event. There have been six years in which major outbreak sequences have occurred since 1917, with the two most recent occurring in 2003 and 2011.

THE ENHANCED FUJITA SCALE

Fujita, in collaboration with Pearson, released the F-Scale in 1971 and designed it to allow for an estimate of the wind speed of a tornado based on a damage assessment of structures after the fact. Although designed to incorporate levels from F0 to F12, only F0 to F5 were used in practice, as an F6 was labeled to be "inconceivable damage," though still possible and probably likely to occur at some point (there has never been an F6/EF6 tornado, although it is possible that the 1999 Bridge Creek-Moore, Oklahoma, tornado exceeded the 318 mph upper limit of the original F-Scale in place at the time). The operational ranges of the F-Scale were 40–72 mph (64–116 kph) for an F0, 73–112 mph (117–180 kph) for an F1, 113–157 mph (181–253 kph) for an F2, 158–206 mph (254–332 kph) for an F3, 207–260 mph (333–418 kph) for an F4, and 261–318 mph (419–512 kph) for an F5. It is important to note that the wind speeds represent what the National Weather Service called "educated guesses," in that they were intended as a guide, not an explicit statement, about the actual measured speed of the wind, as no such measurements were a part of the construction of the F-Scale. It was known early on that the F-Scale posed some difficulties in accurately conveying information about the power or impact of a tornadic event.

On February 1, 2007, the Enhanced Fujita Scale, or EF-Scale, replaced the F-Scale. The new EF-Scale, developed at the Wind Science and Engineering Research Center at Texas Tech University, in Lubbock, Texas, incorporates more detailed and more widely studied and cataloged damage indicators, as well as additional structure and vegetation types, into the derivation of the final EF value. The major difference between the old and new scales is that the EF-Scale explicitly takes into account the quality of construction of a structure, which has been found to be a major factor in the likelihood of the structure surviving a tornado (and consequently, the avoidance of injury and/or death by any occupants). Still set on the range of zero to six, the scale uses the letters "EF" before the integer rather than just "F". The impetus to redesign and improve the F-Scale came from the realization that the original scale tended to overestimate wind speeds based on the limited information from which to draw a conclusion, and as such, tornadoes at F3–F5 values were likely lower in many cases. The modified scale employs the following categories for wind speed ranges and EF values: 65–85 mph (105–137 kph) for an EF0, 86 to 110 mph (138 to 177 kph) for an EF1, 111 to 135 mph (178 to 217 kph) for an EF2, 136 to 165 mph (218 to 266 kph) for an EF3, 166 to 200 mph (267 to 322 kph) for an EF4, and 200 mph (322 kph) or greater for an EF5.

FATALITIES AND DAMAGE

Fatalities from tornadoes have steadily been decreasing as forecasting and warning systems continue to improve, with about 60 deaths occurring annually. However, due to the violent nature and inherent unpredictability of tornadoes, as well as the economic conditions present in areas where the storms are more common, it is generally held by disaster researchers that the average annual death toll will probably not reach a much lower point than where it presently resides. According to research by economists Kevin Simmons and Daniel Sutter, 62 percent of tornado fatalities occur from strong and violent tornadoes, those rated at EF4 and EF5 strength (yet those tornado ratings make up just 1.2 percent of all occurrences) (Simmons and Sutter 2011). At the opposite end of this spectrum, Simmons and Sutter found that the largest percentage of tornado occurrences (78 percent) were weak events, rated EF0 or EF1, and account for just 5 percent of all fatalities (Simmons and Sutter 2011). There also follows a general power-law distribution, or the Frequency-Magnitude relationship, in which many small events occur, and a few large events, but here with the additional consideration that those few large events cause the majority of deaths.

Examined from another statistical perspective, EF5 (previously F5) tornadoes are responsible for 16.27 fatalities per event and 174.6 injuries per event, whereas the weakest tornadoes (EF0/F0) are responsible for just 0.0011 fatalities per event and 0.035 injuries per event. Fatalities and injuries as a function of time of day steadily increase throughout the day, morning to evening, but the largest proportion of deaths and injuries occurs as a result of tornadoes that strike overnight (for the obvious reason that people are typically sleeping and may be far less able to rapidly respond to the threat, if they are warned at all). Mobile or manufactured homes see the largest percentage of deaths resulting from tornadoes (43 percent) due to their weak construction and inability to protect the occupants. When a tornado strikes overnight or in the mid-afternoon, mobile home fatalities are approximately 30 percentage points higher than for occupants of well-constructed homes (e.g., wood-framed, brick). The states of Kansas, Arkansas, Mississippi, Alabama, and Oklahoma have the highest rates of fatalities per tornado event.

TORNADO RECORDS

In examining the official tornado record maintained by the National Weather Service, several notable tornadoes stand out as dubious record holders. Considering power of a tornado as a broad category, the highest winds ever observed in a tornado resulted from the Bridge Creek-Moore, Oklahoma, tornado on May 3, 1999. A mobile Doppler station measured a brief period of winds at 302 mph (486 kph), with plus/minus 20 mph (32 kph). Assuming the high end of that error range, then it is possible that this was the first ever F6 tornado; however, that rating has never been confirmed or entered into the official record. Physical characteristics of tornadoes also include path length, width, and duration. The longest path known is the 1925 Tri-State Tornado, at 219 miles (352 km). The Tri-State Tornado also holds the record for the fastest forward speed at 73 mph (117 kph). The widest path record was previously held by the Hallam, Nebraska, tornado of May 22, 2004, at 2.5 miles (4 km) wide, but it was just barely edged out of that top spot by the El Reno, Oklahoma, tornado of May 31, 2013, which reached 2.6 miles (4.2 km) in width. The El Reno tornado killed expert and long-time tornado chasers Tim Samaras and his partner Carl Young; Samaras's son, Paul, also died in the vehicle that they were using to chase the event when they were caught in the tornado's inflow.

The Tri-State Tornado holds yet another record—most persons killed by a single tornado event in U.S. history, with the official death toll standing at 695. This is not the record holder worldwide, however, as a tornado on April 26, 1989, in Bangladesh killed an estimated 1,300 people, with the death toll likely higher than that figure, but unknown due to poor reporting and record keeping. The costliest tornado in U.S. history is currently the May 22, 2011, Joplin, Missouri, tornado, which caused $2.8 billion (in 2011 dollars; adjusted to 2018 dollars, that figure reaches approximately $3.3 billion).

In 2011, a major and historic outbreak sequence occurred, spawning 360 tornadoes in total from April 25 through the 28, and has come to be known as the 2011 Super Outbreak. Of the 360 tornadoes in total, 216 occurred on just one day, April 27, which included an incredible four EF5 tornadoes (from 1950 to 2007, there were an average of 1.12 F/EF5 tornadoes per year). The 2011 Super Outbreak was also the costliest in U.S. history, inflicting $11 billion in damage (2011 dollars; $12.5 billion adjusted to 2018).

An interesting and not-likely-to-be-repeated record belongs to the small town of Codell, Kansas, in the northcentral portion of the state. In 1916, again in 1917, and yet again in 1918, it was hit by a tornado on the same day of the year, May 20. Moore, Oklahoma, also an unfortunate victim of several tornadoes, has seen major events strike the community in no fewer than seven years (not consecutive) and has been hit by an F/EF5 twice, once in 1999 (the aforementioned Bridge Creek-Moore tornado) and again in 2003. Moore is not the only community to be struck by an F/EF5 more than once; Tanner, Alabama, was virtually demolished by an F5 on April 3, 1974, and was struck again just 45 minutes later by a second F5 event. Almost four decades later, as part of the 2011 Super Outbreak, Tanner was yet again ravaged by an EF5 on April 27. Tanner is the only community in the United States known to have been hit by an F/EF5 tornado three times.

Tornadoes are deadly storms that can occur almost anywhere in the world, but they are particularly prevalent in the Central and Southern United States. Our current knowledge of the systems that produce these rotating storms is robust enough to provide upward of 15 minutes of lead time in most cases, although tornadoes can form with little to no warning as well. Research into this field continues, and as it does, more will be learned about the physicality of these events, with the ultimate goal of creating predictive tools that are accurate and precise enough to more fully predict the circumstances that will lead to tornado genesis and, hopefully, save lives in the process.

Mitchel Stimers

Further Reading
Bradford, M. 1999. Historical Roots of Modern Tornado Forecasts and Warnings. *Weather and Forecasting* 14: 484–491.
Doswell, C. A., S. J. Weiss, and R. H. Johns. 1993. Tornado Forecasting: A Review. In *The Tornado: Its Structure, Dynamics, Prediction, and Hazards*, edited by C. Church, D. Burgess, C. Doswell, and R. Davies-Jones. *Geophysical Monograph* 79: 557–571. Washington, DC: American Geophysical Union.
Grazulis, T. P. 1993. *Significant Tornadoes, 1680–1991*. St. Johnsbury, VT: Environmental Films.
Mitchell, C. 2018. Which Areas around the World Are Most Prone to Tornadoes? https://www.accuweather.com/en/weather-news/where-are-hotspots-for-tornadoes-outside-of-the-united-states/70001183, accessed October 1, 2018.
NSSL (National Severe Storms Laboratory) 2018. Severe Weather 101—Tornadoes. https://www.nssl.noaa.gov/education/svrwx101/tornadoes/types/, accessed September 1, 2018.
Simmons, K. M., and D. Sutter. 2011. *Economic and Societal Impacts of Tornadoes*. Boston: American Meteorological Society.
SPC (Storm Prediction Center). 2018. Warning Coordination Meteorologis's Introduction. https://www.spc.noaa.gov/wcm/, accessed September 1, 2018.
Stimers, M. J. 2011. A Categorization Scheme for Understanding Tornadoes from the Human Perspective. Dissertation. Kansas State University, Manhattan.
Wagner, K. K. 1999. The Golden Anniversary Celebration of the First Tornado Forecast. *Bulletin of the American Meteorological Society* 80(7): 1341–1348.

Tsunamis

Tsunamis are large-scale ocean waves triggered by seismic activity. Although sometimes called *tidal waves* in popular culture or casual conversations, scientists and researchers avoid this term because these waves are not triggered by tides. Tsunamis can move extremely quickly, reaching coastal areas in minutes in some cases, leaving many people with potentially little warning time for a hazard that, once seen, is often too close to outrun and that even when small in scale can be extremely powerful. The effects they have on an area are the product of a range of factors, including what triggered the tsunami, how far it traveled, its size, and the depth of the water it moves through as it approaches the coast. Increasingly, research has turned to the question of tsunami warnings issued by state agencies and weather centers, as well as investigation into long existing indigenous and local knowledge of tsunami warning signs, in an effort to give people warning even in the fastest-moving tsunamis.

In recent years, interest in tsunamis has increased, rooted in the large-scale destruction caused by the 2004 Boxing Day tsunami in the Indian Ocean. This 2004 tsunami, when combined with the triggering earthquake, caused widespread damage and left millions displaced in Bangladesh, India, Indonesia, Kenya, Malaysia, the Maldives, Myanmar, Somalia, South Africa, Sri Lanka, Tanzania, and Thailand. With the combination of people traveling and relocating for work and enjoyment, the tsunami killed more than 230,000 people from 73 countries and injured at least 280,000 people from 60 countries (Joseph 2011). While not the first recorded large-scale tsunami, the widespread impact of the event playing out on people around the world on news cycles pushed the event and the discussion of tsunamis squarely into the public eye and continued to pop up periodically there, even later becoming the subject of more popular media like the film *The Impossible* (2012). This pattern of large-scale tsunamis prompting an increasing rate of research or transformation in research of and responses to tsunamis is not new. In fact, the 1960 Chilean tsunami prompted the creation of new international-scale monitoring and warning systems.

Beyond such popular and public discussions of one specific event, research into and understanding of tsunamis has continued to grow over time. Such work has led to efforts to better warn people of potential tsunamis as quickly as possible, encouraging their relocation to higher ground or inland in case of a tsunami, sometimes even before one is spotted and confirmed. Learning about tsunamis is critical not only to general understanding of such hazards and environmental events but also to preparedness for them and the ability to save lives and minimize destruction in potentially affected areas.

HOW TSUNAMIS HAPPEN

Generally, tsunamis are started by seismic activity and disruptions tied to a range of naturally occurring and human-caused hazards, such as large-scale earthquakes, volcanic eruptions, landslides that occur in or near the ocean, meteorological disturbances, asteroid impacts, or even underwater explosions, although earthquakes are the most common cause (this includes "seaquakes" or earthquakes that occur underwater in the ocean). Of all forms of earthquakes and seismic activity, megathrust earthquakes, where one plate of the earth's crust slips suddenly and violently beneath another, are the most likely to cause extreme earthquakes and, related to this, severe tsunamis. Other types of earthquakes can cause tsunamis of varying sizes, but they are often both less likely to cause a tsunami and generally cause more moderate tsunamis. Triggering events like earthquakes can have epicenters under the ocean or simply in nearby coastal areas, or similarly, volcanic eruptions can occur under or near the ocean. In the case of other events like landslides, they can be either submarine or can occur onshore by dumping large amounts of debris into the water, causing a similar reaction. Differences in the earthquake itself are not the only factor that can determine whether or not a tsunami is triggered or the size and scope of the tsunami. Differences in the energy involved, the area impacted, and the movement of the waves themselves may determine whether there is a tsunami that strikes land or not.

Tsunamis do not consist of one solitary wave, as may sometimes be depicted in popular movies or other media, but instead, they include a series of multiple fast-rising and fast-moving waves that bring powerful currents with them, last for hours, and "do not 'break' like the curling, wind-generated waves popular with surfers" (USGS 2019). Tsunamis range in size from micro or small to mega or large or catastrophic. Micro or small tsunamis can only be detected by sensitive instrumentation, but they happen more commonly than larger tsunamis, often far from land or on a very small scale, potentially even mistaken as an extreme tide. In sharp contrast, mega or large tsunamis can affect entire oceans and multiple countries and continents, such as the 2004 Boxing Day tsunami in the Indian Ocean. These tsunamis are rarer but cause far more extensive and severe damage, as well as higher rates of injuries and deaths.

Tsunamis propagate much more quickly in deeper water. Despite this, they often also appear much smaller in deeper water. While a growing tsunami may produce wave heights of just short of one foot (approximately 30 cm) in deep water, this may escalate to nearly 100 feet (over 30 m) as it reaches the coastline (SMS Tsunami Warning 2019).

Tsunamis may also be categorized based on how long it takes for it to reach the coastline or a specific location: near-source-generated tsunamis, mid-source-generated tsunamis, and far-source-generated tsunamis. A near-source-generated tsunami takes 30 minutes or less to reach the site, which itself may also be affected by the earthquake or other tsunami-triggering event. A mid-source-generated tsunami takes between 30 minutes and two hours to reach a site. And a far-source-generated tsunami takes two hours or more to reach that site (Heintz and Mahoney 2008).

Tsunamis are most common in the Pacific Ocean, with approximately 70–75 percent of recorded tsunamis starting there. Tsunamis in the Indian Ocean in particular are rarer; they are also often extremely destructive. Since 2004, there have been 20 deadly tsunamis, and 12 have occurred in the Pacific Ocean and marginal seas (60 percent) (UNESCO 2019). However, two of the three most destructive global-level tsunamis in recorded history have originated in the Indian Ocean (Joseph 2011).

Understanding potential tsunami risks is not just a matter of knowing which ocean a person or community is located on. Instead, the risk is a result of a range of issues, including aspects of the hazards itself and the community potentially affected. In terms of the hazard itself, the potential risk can be shaped by the source of the tsunami, exposure to the eventual tsunami, or the depth and potential reach of the eventual tsunami. On the side of the community, there are questions related to population density, vulnerability, and the built environment, including potential shelter spaces.

TRACKING AND PREDICTING

Measuring, tracking, and predicting tsunamis can be complex work, but it's necessary for the protection and survival of people around the world. Several data sources may be used for predicting and monitoring tsunamis, including (1) seismic data, (2) coastal tide gage data, and (3) deep-ocean data. Each of these forms of data is important to measuring, tracking, and predicting tsunamis, but they also have both positives and negatives, often also related to how each data form relates to the tsunami and to the other data sets.

Seismic data comes from various sources, and it helps researchers and responders pin down the source of an earthquake or other activity that may trigger a tsunami, as well as potential parameters for the potential event. The initial warning sign of a potential tsunami is often rooted in detecting seismic activity. In some ways, this is useful because seismic data networks are denser than other forms of monitoring networks, providing a lot of information very quickly. However, time is needed to interpret this data, particularly regarding its potential for triggering a tsunami.

Coastal tide gauge data can be used to confirm tsunamis and for later research into tracking the course and progress of the tsunami from its source. Deep-ocean data from tsunameters (pieces of equipment kept in deep-water parts of the ocean to track wave and water data and potential tsunamis) can be available more quickly than coastal tide gage data because they are specifically placed in deep-water areas, where tsunamis propagate more quickly, and in areas that avoid contamination of measurements related to water interactions with the coastline. Since tsunamis propagate more quickly in deeper water, information sources like deep-ocean gages and tsunameters can be helpful in observations, often more so than monitoring directly on the coastline.

This tracking data is also helped by increasing understanding from modeling based on research on tsunamis throughout history. Many tsunami models fall into two larger categories: a data assimilation model or a forecast model. Data assimilation models adjust to real-time data and develop a solution to best fit that data, a system that was tested against data from historical tsunamis. In contrast, forecast models extend a simulation that best fits what data is available initially to fill in data gaps and create a forecast for potential impacts on other locations (Titov et al. 2005).

MONITORING AGENCIES, INDIGENOUS KNOWLEDGE, AND WARNINGS

There are two large overarching problems that cover the complications of tsunami warnings: (1) the speed at which they need to happen, since tsunamis can strike within minutes or hours, and (2) the fact that warnings must often be issued based on incomplete or unclear data, in part tied to the speed with which they must be issued. This can also sometimes lead to situations where a tsunami alert is issued because conditions are right for a tsunami, but no tsunami occurs, which has the potential to lead to ineffective local response. However, the reverse side of this is that confirmation of a tsunami may take too long, resulting in delayed warnings that may lead to deaths and injuries. Thus, many areas err on the side of caution and issue warnings in conditions ripe for tsunamis versus only when one is confirmed.

Given that tsunamis can often cover a large area or travel long distances, it makes sense that measuring, tracking, and predicting them, as well as issuing related warnings, are an international effort. The United Nations Educational, Scientific and Cultural Organization's (UNESCO) Intergovernmental Oceanographic Commission (IOC) coordinates the Intergovernmental Coordination Group for the Pacific Tsunami Warning and Mitigation System (ICG/PTWS) (formerly the International Coordination Group for the Tsunami Warning System in the Pacific or ICG/ITSU). The ICG/PTWS includes 46 IOC member states around and within the Pacific Ocean and other member states that are interested in helping in research work, monitoring, and issuing warnings for tsunamis.

Currently, the International Tsunami Information Center (ITIC), also coordinated by UNESCO and the IOC, maintains an operational headquarters in the United States (Hawaii) and subregional centers in Japan and the United States (Alaska). Beyond this, there are other tsunami information centers that serve other areas, including the Caribbean Tsunami Information Center (serving the Caribbean region), the Jakarta Tsunami Information Center (serving Indonesia and the Indian Ocean), and the Northeastern Atlantic and Mediterranean Tsunami Information Center (serving mostly Europe). While these systems are generally newer and are not directly under the umbrella of ITIC or PTWS, both organizations have used their own history and knowledge to help support these newer and growing tsunami centers around the world, particularly in the Indian Ocean, and to ensure consistency across the systems to allow them to communicate with one another and share data and information more easily.

Many warning signs are associated with tsunamis, rooted in both scientific research and local cultural traditions. Understanding not only the risks involved with various hazards but also people's understandings of them within and outside official emergency management structures, including indigenous knowledge, shared cultural knowledge, oral history, etc., is critical to effective preparedness and disaster risk reduction. In many cases, like those of the people of the islands of Simeulue in Indonesia, this is also rooted in historical contexts (Rahman et al. 2017).

EVACUATION QUESTIONS

In tsunamis, evacuation is often a matter of moving both inland and vertically. Given the speed with which they can strike and the potential for delays in warning, moving people up into higher floors of buildings or up hills or even trees may be the most effective way of getting people to some degree of safety rather than relying on them to only be able to run inland quickly enough. Being able to rely on hills or mounds (naturally occurring or human-made or altered) or high buildings requires them to (1) be available and accessible in high tsunami-risk areas, (2) be high enough to move people out of potential tsunami waters, and (3) be strong enough to resist the impact of tsunami strikes and waters.

As tsunamis make landfall, they can cause significant damage to the human-made and natural environment. This damage can come in a range of forms.

Tsunamis themselves and the force of the water in them can cause damage through impact and inundation. Debris being carried by the force of the water can cause damage by hitting into objects, buildings, or people and when combined with combustible liquids can cause and spread fire. Debris can be carried up to several miles inland.

The devastation caused by tsunamis can be especially extensive in poorer nations where infrastructure, construction regulations, and buildings may be less resilient to impacts from hazards such as tsunamis. Due to this potential damage, in any affected area buildings that are used for vertical evacuation for people in tsunami-affected areas must be built with this risk of impact in mind and be able to remain both upright and relatively intact, even if they are damaged or have missing parts. This also puts limits on potential evacuation locations when combined with how fast a tsunami can move, particularly in relation to the amount of warning time.

POTENTIAL TSUNAMI IMPACTS

The impact of this debris and the water itself can be extensive, ranging from damage to plants and wildlife to damage to and destruction of buildings and vehicles to severe injuries and deaths. Wide-scale environmental impacts are possible in tsunamis and often last long beyond the initial moment of impact. Trees and vegetation may be uprooted and destroyed, including the animal habitats in them, and animals themselves may be killed or injured. Beyond this, the chemicals and debris that are dumped into the environment, along with the saltwater from the tsunami itself, may cause long-term contamination and problems. Some of these may be difficult to clean up or dispose of, especially in initial and fast-moving attempts to clean up, which may include burning or dumping materials.

Tsunamis, both via the water itself and the debris they move and damage they case, often also cause extensive injuries and deaths. This is especially true with near-source-generated tsunamis and mega- or large-scale tsunamis, which may have less or not enough warning time for people in nearby countries. Injuries may be caused by debris strikes, but there may also be situations of circumstances like "near drowning." Complicating these injuries further may be the destruction of local medical centers and hospital, as well as the potential destruction of, damage to, or contamination of needed medical supplies. When medical supplies from abroad arrive, there may be additional confusion or delays tied to language use, dosage measurement, or other complexities with converting what is used in one system into another system. In addition to these complications, the scale of the death toll in tsunamis may overwhelm local systems, both for finding the dead and storing their remains until they can receive proper traditional burials as per local cultural and religious customs. These delays and the scale of damage, destruction, and potential confusion related to family separation may also lead to problems with and delays in identifying the dead.

Since 1850, tsunamis have been responsible for at least 430,000 deaths, many of whom died in the especially deadly and devastating 2004 Boxing Day tsunami. These deaths are the result of not only being hit by moving debris, the impact of water, drowning but also from fires and explosions tied to gas leaks and damaged gas tanks. This is further complicated by the multiple waves that may come with a tsunami. People who are injured in the initial wave may be more severely injured or killed in subsequent waves. In fact, the rate of deaths in some areas, like Nagapattinam, India, in the 2004 Boxing Day tsunami, far outweighed the number of injuries (there 6,000 dead to just 2,000 injured) (Yamada et al. 2006).

The scale of tsunami impacts may lead to an equally large-scale attempted response, particularly in the modern era where news about and images of the tsunami's effects may reach a worldwide audience almost immediately. This can help when local and national governments and communities lack the capacity to respond to such events alone. However, it can also raise complications like problems with cultural differences and knowledge of cultural beliefs and problems with communication in different languages at work, which can also require additional resources, skills, and personnel to overcome.

In addition to deaths and injuries from water and debris, tsunamis and their aftermath can result in other complications. The influx of seawater can contaminate soil and vegetation. Infrastructural systems such as those for sewage or freshwater may be contaminated or damaged. Such contaminations may in turn cause problems with drinking water supplies, which can also lead to the rise of diseases made even more complicated by the destruction of infrastructural systems such as hospitals and sanitation. This combination, particularly for people who lack access to clean water, sanitation, and health care under the best of circumstances, may lead to extensive disease spread and additional illness, injury, and deaths. Moreover, such contaminations may cause secondary and long-term effects. Saltwater itself and related salination can contaminate soil and freshwater sources, impacts that can last for years and affect potential agricultural yields over time. Contamination from gas or chemicals moved by the water into soil, food sources, water, homes, or other spaces may result in people becoming ill not just in the short term but also over time, especially if people are unable to fully clean the contamination or are unaware of the specifics or extent of it.

Tsunami impacts, response, recovery, and cleanup, both in the short term and long term, can be not only extensive but also extremely expensive. The scale of damage caused by a tsunami when combined with the potential for long-term effects through issues like contamination is problematic for affected areas. The people affected may need financial assistance for cleanup efforts, rebuilding, and medical issues, as well as the potential financial impacts from lost work. Beyond this, they may need sources of food and clean water. Beyond this, large-scale replacements are also needed, like repairs to or reconstruction of infrastructure and long-term cleanup of contamination. In many cases, such aid efforts come from a variety of sources, including governments, international organizations such as the United Nations and other governments, community and local organizations, and large- and small-scale nongovernmental organizations (NGOs).

Beyond the physical effects of a tsunami, there are potential mental health effects as well. Psychological problems in general and post-traumatic stress disorder (PTSD) can manifest among many survivors in such events. Some people may start to recover from them in days; others may find themselves facing these issues for the rest of their lives. For example, following the Boxing Day tsunami in 2004, survivors in Sri Lanka were examined and interviewed by the World Health Organization (WHO). Among those they spoke with, 40 percent of adolescents in particular had PTSD symptoms four months after the tsunami (SMS Tsunami Warning 2019).

The Sri Lanka example in the 2004 Boxing Day tsunami also demonstrates a number of other problems with tsunami impact, evacuation, and response. As an island, Sri Lanka found itself especially hard hit by the tsunami with more than two-thirds of the nation's coastline being affected. Moreover, it demonstrates the multiple-wave impact often found with tsunamis in contrast to popular media or common associations with a single wave; in parts of Sri Lanka, a wave over 3 feet (1 m) high struck initially, followed 10 minutes later by a second wave that was over 30 feet (10 m) high. On an island with 19 million people, 1–2 million were directly affected (5–10 percent), including over 31,000 thousand deaths and 440,000 displacements (Yamada et al. 2006). Many of these deaths came with subsequent waves in the tsunami, largely from drowning.

RESPONSE AND RECOVERY EFFORTS

The significant scale of tsunami destruction also creates problems with recovery, as entire areas may have been wiped out in their entirety, leaving people needing a large range of services and facilities to meet basic needs like food, water, and shelter. As international response begins, there can be problems between different nongovernmental organizations, government agencies, and other groups and with potentially overlapping responses or gaps left between responses. Donated items may not meet local needs or may not be compatible with the local environment, no matter how well intended by donors.

The 2004 Boxing Day tsunami also demonstrates the response to tsunamis and how it can work or fail. During and after disasters, people often display pro-social behavior, moments where they take care of and assist others. Regular people who could came out immediately to offer food, shelter, and other things people needed to those who were affected as they searched for survivors and tried to ensure they received medical treatment at the scene or in hospitals. Religious leaders reached out to help others and organize their spaces and organizations to help more people than they could alone. However, the Sri Lankan government's response also revealed some of the problems that can emerge in tsunami response. Their lack of a set incident command system to respond to large-scale disaster caused delays, and preexisting ethnic conflicts between different groups in Sri Lanka led to a reported lack of aid to some of the affected and conflicts over those delayed or blocked aid efforts, particularly for members of the Tamil ethnic group (Yamada et al. 2006).

HOW THE IMPACTS AND AFTERMATH OF TSUNAMIS CAN CAUSE CHANGE

In 1907, a 7.8-magnitude earthquake and subsequent tsunami struck Indonesia, including Simeulue, killing 70 percent of the people there. From this experience emerged the *Smong* tradition, whereby survivors shared stories and experiences via oral traditions and passed down to younger generations, a process that continued over time through community-level interactions, family discussions, conversations with elders, etc. The shared oral stories are direct in their instructions: "If the strong earthquake is followed by the lowering of seawater, please find in a hurry a higher place" (Rahman et al. 2017). Such shared knowledge in this cultural context, where it was made clear that these stories were not just stories but warnings rooted in lived experiences, saved lives in 2004.

In 1946, an earthquake of 8.6 magnitude struck near the Aleutian Islands in Alaska in the United States, triggering a tsunami that affected not only Alaska but also other areas in the Pacific Ocean. This resulted, in 1949, in the establishment of the PTWS. The PTWS issues tsunami threat information for earthquakes in the Pacific or nearby with a magnitude of 6.5 or higher. Currently, this results in messages being sent to designated national authorities in areas that may be affected, which in turn issue more specific warnings for their residents and in the PTWS, issuing tsunami wave forecasts where needed.

Then, in 1960, a 9.5-magnitude earthquake struck Chile and triggered a tsunami throughout the Pacific Ocean, including effects on and deaths in Hawaii, Japan, and the Philippines. The aftermath of this tsunami led to the creation of an international tsunami early-warning system in 1965 that was, and remains, reliant on freely shared seismic and sea level data and timely communications. The ICG/PTWS disseminates information about tsunamis or tsunami threats to National Tsunami Warning Centers (NTWS), who then issue tsunami warnings to citizens working with emergency management agencies at the national, regional or state, and local levels.

While such elaborate efforts are critical to our understanding of and warnings for tsunamis in the Pacific Ocean and attached marginal seas, such systems do not exist or are not as well organized in other areas or have not been well organized historically. For example, following the 2004 Boxing Day tsunami in the Indian Ocean, from 2005 to 2012, the PTWC and the Japan Meteorological Agency cooperated to provide information regarding potential tsunamis in the Indian Ocean. At the time, the Indian Ocean lacked a comparable tsunami warning system. The lack of a consistent and solid tsunami tracking and warning system in the Indian Ocean in 2004 was part of what led to the relative lack of warning about the tsunami, especially for countries and people nearest to the epicenter of the earthquake near Sumatra. This gap, and the resulting deaths from the tsunami in nearby nations like Indonesia, Sri Lanka, and India, reinforced the importance of data gathering and sharing and the critical need for communications with national and local governments, as well as people in the area.

CONCLUSION

Tsunamis may be small- or large-scale events, but the majority of damage and deaths comes from the larger and rarer tsunamis. Triggered by a range of events, but most commonly by earthquakes, tracking, monitoring, and issuing warnings for tsunamis are a complex process that relies on early and rapid warnings, especially in near-source-generated tsunamis and larger-scale tsunamis like the 2004 Boxing Day tsunami. Since tsunamis can move rapidly, in some cases striking in minutes or hours depending on the site of the triggering event, this sort of rapid warning effort is crucial, even if it results in some so-called false alarms where a potential tsunami does not manifest.

This is also, along with the impacts of historical tsunamis, what has led to the creation of national and international organizations dedicated to monitoring and tracking tsunami development and movement and larger tsunami research in an effort to develop better warning, mitigation, preparedness, and response capabilities in all potentially affected areas. Tsunamis are devastating events that can cause extensive damage and may trigger long-term recovery processes that involve cleaning up contamination, restoring access to clean drinking water and food sources (including food growth), and construction of new buildings in the aftermath shaped by a consideration of potential future events.

Jennifer Trivedi

Further Reading

Heintz, J. A., and M. Mahoney. 2008. *Guidelines for Design of Structures for Vertical Evacuation from Tsunamis*. Washington, DC: FEMA. ftp://jetty.ecn.purdue.edu/spujol/Andres/files/15-0021.pdf, accessed April 14, 2019.

Joseph, A. 2011. *Tsunamis: Detection, Monitoring, and Early-Warning Technologies*. Burlington, MA: Academic Press.

Rahman, A., A. Sakurai, and K. Munadi. 2017. Indigenous Knowledge Management to Enhance Community Resilience to Tsunami Risk: Lessons Learned from *Smong* Traditions in Simeulue Island, Indonesia. IOP Conference Series: Earth and Environmental Science 56. https://iopscience.iop.org/article/10.1088/1755-1315/56/1/012018/pdf, accessed April 14, 2019.

SMS Tsunami Warning. 2019. Tsunamis: The Effects. https://www.sms-tsunami-warning.com/pages/tsunami-effects#.XLVC—hKjIU, accessed April 14, 2019.

Titov, V. V., F. I. González, E. N. Bernard, M. C. Eble, H. O. Mofjeld, J. C. Newman, and A. J. Venturato. 2005. Real-Time Tsunami Forecasting: Challenges and Solutions. *Natural Hazards*. http://citeseerx.ist.psu.edu/viewdoc/download?doi=10.1.1.1032.3208&rep=rep1&type=pdf, accessed April 14, 2019.

UNESCO (United Nations Educational, Scientific and Cultural Organization). 2019. Intergovernmental Coordination Group for the Pacific Tsunami Warning and Mitigation System (ICG/PTWS). http://itic.ioc-unesco.org/images/stories/ptws/PTWSinfo_feb19.pdf, accessed April 14, 2019.

USGS (United States Geological Survey). 2019. Natural Hazards: What Are Tsunamis? https://www.usgs.gov/faqs/what-are-tsunamis?qt-news_science_products=0#qt-news_science_products, accessed April 14, 2019.

Yamada, S., R. P. Gunatilake, T. M. Roytman, S. Gunatilake, T. Fernando, and L. Fernando. 2006. The Sri Lanka Tsunami Experience. *Disaster Management and Response* 4(2): 38–48.

Volcanic Activity

As a geophysical disaster, volcanic activity has remained far less significant than earthquakes in terms of frequency, number of deaths, and people affected annually. This activity accounted for just 1.7 percent of all disasters from 1996 to 2015. In particular, 55 volcanic eruptions caused the deaths of 460 people from 2006 to 2015. Moreover, such activity has increased in the past 20 years (CRED and UNISDR 2016). A recent study reported that more than 278,000 people died from volcanic activities from 1500 to 2017, amounting to 540 deaths per year on average (Fosco 2017). Unlike many natural disasters, volcanoes provide advance warning signs of imminent eruptions, which partially explains relatively low volcanic-induced fatalities. However, some volcanic eruptions have killed thousands of people in the past. For example, the 1815 Tambora volcanic eruption in Indonesia killed more than 92,000 people. Even as recently as 1985, the volcano Nevado del Ruis ravaged the Colombian town of Armero, killing nearly 25,000 people. Yet, often due to remote location, large volcanic eruptions do not kill many people. For example, no one died from the 1912 Alaska eruption because no one lived nearby.

VOLCANIC MATERIALS AND ERUPTIONS

A volcano is an opening in the earth's crust through which lava reaches the surface and buries everything in its flow path. In addition to lava, ash and gases also escape from volcanoes to the atmosphere. Thus, volcanoes are vents for magma with great variations in shape, height, and structure. Magma is composed of molten rocks that have been heated because of very high internal temperature and intense pressure beneath the earth's crust. It is located beneath or within the earth's crust, from which igneous rock is formed. It can be found within 60 miles (100 km) of the earth's surface and 15 miles (25 km) underneath the ocean floor (Ebert 2000).

The most basic requisite to form a volcano is the presence of magma, which forms as follows. Mantle, another term associated with volcanoes, is the mostly solid bulk of earth's interior, which lies between earth's dense, super-heated core and its thin outer layer, the crust. It is about 1,802 miles (2,900 km) thick and makes up to 84 percent of earth's total volume. Whenever the mantle is able to melt (becoming magma) and is forced to reach the earth's surface through a volcano vent, zones of structural weakness, or fissures due to sufficiently high pressure, it is called lava. The pressure mostly occurs where the tectonic plates meet and one plate moves under another (subduction). Volcanoes and volcanic activity tend to occur along subduction or convergence as well as divergent plate boundaries. They are also found within lithospheric plates away from tectonic plate boundaries. The plates are huge pieces of earth's crust that constantly and slowly move across the surface over millions of years (Ebert 2000).

The rising of magma is partly driven by pressure from dissolved gas. As indicated, liquid magma contains dissolved gases. When the magma rises, pressure decreases, allowing the gases to form bubbles. Both its gas content and chemical composition control how the lava behaves when it reaches the surface. For example, lavas that contain low silica have low viscosities and flow freely, allowing any gas bubbles to escape readily, while lavas with high silica contents are more resistant to flow so that any trapped gases cannot escape gradually. The most abundant gas is water vapor, which accounts for about 70 percent of the bulk of volcanic gases. The water vapor is generally released in steam form. The remaining 30 percent of the gases contains carbon dioxide; carbon monoxide; sulfur trioxide and sulfide; chloride; hydrogen fluoride; and small amounts of ammonia, methane, hydrogen thiocyanate, and some rare gases (Rittman 1962).

However, the structure of magmatic materials differs considerably, and this difference largely determines the shape of volcanoes and their activities. One such activity is the eruptions, which are frequently deadly, killing and injuring people and animals and damaging or destroying localities and properties over wide geographic areas. Volcanic eruptions have wiped out many cities across the globe in the distant and recent past. However, currently more than 800 million people or 10 percent of the world's population live within 62 miles (93 km) of an active volcano (Fosco 2017).

TYPES OF VOLCANO

Volcanoes can be classified in different ways: by their eruptive activities, shapes or structures, and type of eruption. On the basis of eruptive activities, volcanoes are classified into three types: active, dormant, and extinct. Active volcanoes are those that have erupted recently or are most likely to erupt in the near future. Excluding those submerged under oceans, there are approximately 500 active volcanoes in the world. Between 50 and 70 volcanoes erupt every year, and about 160 have erupted in the last decade. Next, a dormant volcano has not erupted recently, but geologists think that it may still be capable of erupting. The line between active and dormant volcanoes is sometimes blurred; some volcanoes can remain dormant for thousands of years between eruptions. For this reason, a dormant volcano is often the most dangerous as people living within the vicinity of such a volcano typically are not prepared for eruption. Finally, an extinct volcano is one that will never erupt again. This type of volcano has not had an eruption for at least 10,000 years, and therefore, extinct volcanoes are considered dead.

The primary reason for volcanoes to become dormant is constant shifting of the earth's plates above the volcanic hot spots. This movement eventually shuts off the magma chamber beneath dormant volcanoes. A "hot spot," named for a high concentration of heat in a small area, is a stationary source of basaltic magma in the mantle from which hot plumes rise upward to the earth's surface, creating new active volcanoes on the overlying crust. Over time, the continual outpouring of lava can form a sea mount or island volcano if the hot spot is under the ocean floor, as in the case of the Hawaiian Islands. However, extinct volcanoes are cut off from their supply of lava because the tectonic plate slowly moves over the hot spot underneath, effectively stifling any volcano.

On the basis of shapes, volcanoes are classified into four groups: (cinder) cone, shield, composite, and lava dome, which are largely determined by the composition of lava. The cinder cone volcano is the most common type and is built from particles and blobs of congealed lava ejected from a single opening. It is a steep-sided and cone-shaped volcano composed of layers of pyroclastic materials. This type of volcano is also called a stratovolcano. When the lava and gas are blown furiously into the sky, the material splits into small fragments that solidify and fall as clinkers a short distance from the vent to form a circular or oval cone. Most cone volcanoes have a bowl-shaped crater at the summit and rise up to 1,200 feet (370 m) or so above their surroundings. Cone volcanoes are less dangerous than other types and are found near larger volcanoes as well as away from other volcanoes. The exact composition of a cinder cone depends on the composition of the lava ejected from the vent. They are common in western North America as well as throughout other volcanic terrains of the world.

Shield volcanoes are dome-shaped volcanoes with a wide base and are flat on the top. They are formed by successive lava flows without interbedded pyroclastic materials. The lava is composed almost entirely of basalt, which is very fluid when it erupts. Shield volcanoes are not steep and form due to continuous flow of lava from a central summit vent or group of vents. This is the largest and tallest type of volcano in size and height, respectively. Shield volcanoes can be up to 30,000 feet (9,000 m) in height, and their slopes are roughly 10 degrees near the base and five degrees near the top. They are formed slowly by the accretion of thousands of highly fluid lava flows that spread widely over great distances and then cool as thin, gently dipping sheets. They are the common product of hot spot volcanism and are also found along subduction-related volcanic arcs or all by themselves. In northern California and Oregon, many shield volcanoes have diameters of 3 or 4 miles (4 or 6 km) and heights of 1,500–2,000 feet (457–762 m).

In some cases, basaltic lava erupts quietly from long fissures instead of central vents and floods the surrounding countryside with continuous lava flows, forming broad plateaus. Examples of such plateaus can be found in Iceland, southeastern Washington, eastern Oregon, and southern Idaho. Along the Snake River in the western United States, these lava flows are beautifully exposed and can measure more than a mile in total thickness.

Composite volcanoes are also called strato volcanoes. They are steep-sided and composed of many layers of volcanic rocks, usually made from thick sticky lava, ash, and rock debris. Composite volcanoes are reasonably large and can rise up 8,000–10,000 feet (2,400–3,048 m). Moreover, they can range anywhere from 1 to 7 miles (1–10 km) in diameter. Composite volcanoes have a broad base and steeper sides than cone or shield volcanoes. Their eruptions are dangerous and explosive in nature because of the many layers of lava and pyroclastic materials. For this reason, among all types of volcanoes, they kill the most people. Moreover, slow-moving lava often is accompanied by deadly mudflows, commonly known as "lahars." Mount Fuji in Japan and Mount St. Helens and Mount Rainier in the United States are famous composite volcanoes.

Lava domes are the fourth type of volcano and are significantly smaller than the other three types. They are formed when the lava is too viscous to flow a great distance. As the lava dome slowly grows, the outer surface cools and hardens as the lava continues to pile within, and the dome height can reach up to 330 feet (100 m). Such lava domes are found on the flanks of larger composite volcanoes (Meer 2018).

Two basic classifications of volcanoes are based upon the type of eruption they produce: explosive (or central) and quiet (or fissure). Explosive eruptions are caused by the buildup of gases under highly viscous, thick, and slow-flowing magma trapped deep within the volcano. When an explosive eruption occurs, the ejecta is widely spread due to the force of the blast and by the wind. Explosive eruptions are rapid and violent, often spewing lava, ash, and volcanic material high into the air. They can form pyroclastic flows that reach valleys, destroying everything in their path. Explosive lavas move as fast as 40 miles (60 km) per hour. On the other hand, quiet eruptions, also known as effusive eruptions, usually emit great volumes of thin and liquid-like lava slowly moving along a long fissure or fracture. Lavas typically have low viscosities, so the gases are not prevented from readily escaping.

TYPES OF LAVA

The most common way to divide lava flows is into four distinct types: Aa lava flow, Pahoehoe lava flow, Blocky lava flow, and Pillow lava flow, all named after traits associated with the volcanic nature of the Hawaiian Islands. Sometimes Turbulent lava flow is also added, but because of its presence in the distant past, primarily, this type of lava flow is not currently seen. Not surprisingly, the type of lava flow is intimately associated with the type of lava that forms from a volcano, and therefore, these two terms are used interchangeably. However, more accurately, the type of lava flowing from a volcano depends on its mineral content. Some lava is very thin and thus can outflow a large distance. Other lava is very thick and therefore can flow a short distance before cooling and hardening. Yet some lava is so thick that it cannot flow at all and therefore stops an opening of a volcano.

Aa lava is a basaltic lava that moves very slowly. The uppermost part of this lava is composed of loose, clinker, unstable blocks, and when it hardens, the sharp spiny surface of this lava is extremely difficult to walk across. This type of lava erupts at temperatures from 1,832 to 2,012 degrees Fahrenheit (1,000–1,100 degrees Celsius). Aa lava can be very destructive to all types of structures because of its massive impact (Ebert 2000).

Pahoehoe lava is a smooth and continuous lava crust that forms when the effusion rate is low, and consequently, the velocity of lava flow is at least 10 times slower than that of typical Aa lava. This type of lava also is much thinner, less viscous, and less common than Aa lava and can flow down the slopes of a volcano.

The glassy surface of the lava congeals into a thin crust that looks like coils of rope. Pahoehoe and Aa lava are strikingly different in appearance, but their composition may be very similar or even identical. Lava flow that was originally Pahoehoe may transform into Aa lava, but it never reverts back to smooth Pathoehoe form (Francis and Oppenheimer 2003). This type of lava erupts at temperatures of 2,012–2,192 degrees Fahrenheit (1,100–1,200 degrees Celsius). Moreover, only low-viscosity (usually basaltic) lava can form Pahoehoe. Aa lava is much more common than pahoehoe and the best-known examples of Pahoehoe are from the Big Island of Hawaii.

Blocky lava is composed of a large block that forms when silica content generally is over 55 percent. This lava has a much smoother surface and more angular edges than Aa flows and has common dimensions from a few decimeters to several meters. Also, Blocky lava moves much more slowly and is thicker than Aa flows.

Pillow lava is typically formed when fluid lava has erupted or is flowing under water. Thus, it is found erupting from underwater volcano vents, subglacial volcanoes, or is a lava flow that enters the ocean. When it contacts the water, it cools down and forms a hardened shell. As more lava issues from the vent, the shell of the lava cracks, and more "pillows" come out of these cracks. The term *pillow* comes from its pillow-shaped cross-section. This lava ranges from less than a foot to several feet in diameter, each pillow having a glassy outer skin formed by the rapid cooling by water. The bulk of the submarine part of a Hawaiian volcano is composed of pillow lavas.

GEOGRAPHIC DISTRIBUTION

Although no reliable estimate is available as to how many volcanoes are in the world, scientists believe that there could be more than 20,000 volcanoes on the earth's ocean floor alone, the majority of which are dormant or extinct. Additionally, there could be 1,000 active volcanoes around the globe both in the ocean and on land. Two notable zones of volcanoes in the ocean are called the Pacific Ring of Fire and the Mid-Atlantic Ridge. On land, volcanoes are found in continental rift valleys and mountainous areas. Rift valleys are linear-shaped lowland regions that form both on land and the sea floor, where earth's tectonic plates move apart. However, while the total number of volcanoes in the world is not accurately known, their basic distribution pattern is clear.

Over 75 percent of the world's volcanoes are found in an area of intense seismic and volcanic activity known as the "Pacific Ring of Fire." The area is located around the edges of the Pacific Ocean running from the southern tip of South America; up along the west coast of North America; across the Bering Strait; down through Japan, Taiwan, the Philippines, and Indonesia; and into New Zealand. The Pacific Ring of Fire has a shape more like a 25,000-mile (40,000 km) horseshoe. However, several active and dormant volcanoes are found in Antarctica "close" to the ring. This zone is the home of an estimated 452 volcanoes.

The Ring of Fire is made up of a long convergent plate boundary, which is formed by tectonic plates crashing into each other. Convergent boundaries are

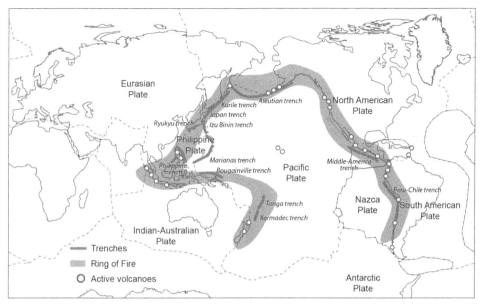

The Ring of Fire is a band of active volcanoes that loops around the Pacific Rim. This region is home to the world's most active volcanoes. (U.S. Geological Survey)

often subduction zones, where the heavier plate slips under the lighter plate, creating a deep trench. Along these boundaries, a huge number of earthquakes as well as volcanic eruptions occur. Most of the active volcanoes on the Ring of Fire are found on its western edge, from the Kamchatka Peninsula in Russia, through the islands of Japan and Southeast Asia, to New Zealand.

There are 169 active volcanoes in the United States, chiefly in Alaska, California, Hawaii, Oregon, and Washington. Except Hawaiian volcanoes, all other U.S. volcanoes are located in the Ring of Fire. In fact, the Cascade Range, which runs through southwestern Canada and the Pacific Northwest of the United States, contains most of the volcanoes in North America. These volcanoes are the result of subduction of the Juan de Fuca plate beneath the North American plate.

The Mid-Atlantic Ridge (MAR) includes a deep rift valley that runs along the north-south axis on the floor of the Atlantic Ocean. The MAR is a divergent tectonic plate boundary that separates the Eurasian and North American plates in the North Atlantic, and in the South Atlantic, it separates the African and South American plates equidistantly. The section of the ridge that includes Iceland is known as the Reykjanes Ridge. The ridge and some portion of the MAR are the sites of volcanic activities.

Active and other volcanoes are also found in the East-African Rift Valley, the Baikal Rift Valley, the West Antarctic Rift Valley, and the Rio Grande Rift Valley. The East African Rift Valley is a contiguous geographic trench, approximately 3,700 miles (6,000 km) in length, that runs from the Middle East in the north to Mozambique in the south. This is also known as the "Great Rift Valley." Throughout the East African Rift, the continent of Africa is split into two. The African plate, sometimes called the Nubian plate, carries most of the continent, while the smaller Somali plate carries the Horn of Africa. The valley is geologically active and features volcanoes, hot springs, geysers, and frequent earthquakes. A hot spring, also called a thermal spring, is a spring with water at temperatures substantially higher than those of the surrounding region. It is produced by the emergence of geothermally heated groundwater that rises from the earth's crust.

The Baikal Rift Valley (sometimes called the Baikal Rift Zone), a series of continental rifts centered beneath Lake Baikal, cuts through 1,200 miles (2,000 km) of Siberia, in southeastern Russia. It is a valley of extinct volcanoes, but hot springs are present here. The West Antarctic Rift, which contains at least 138 volcanoes, is a series of smaller rifts that roughly separates the two regions of earth's southernmost continent: West Antarctica and East Antarctica. The former is one of the most difficult rift valleys to study, because it lies beneath the massive 1.2 miles (2 km) thick Antarctic Ice Sheet. Geologists believe that volcanic eruptions here may not reach the surface but could melt the ice from beneath and drastically destabilize it (McKie 2017).

Mount Erebus, the most southern active volcano on earth, sits over the eruptive zone of the Erebus hot spot in the West Antarctic Rift Zone. This glacier-covered volcano has a lava lake at its summit and has been consistently erupting since it was first discovered in 1841. The intense heat causes the mantle to melt, and the ensuing molten magma often pushes through cracks in the crust to form volcanoes. Finally, the Rio Grande Rift, which stretches from central Colorado to the Mexican state of Chihuahua, is a series of rift valleys along faults in the Southwestern United States. Geologists consider the volcanic features in this rift region to be dormant, not extinct.

VOLCANIC HAZARDS

Many different kinds of hazards are associated with volcanic eruptions. For example, volcanic earthquakes are triggered by stress created in solid rock due to the injection of magma. They typically are much smaller than earthquakes caused by nonvolcanic sources, but they can produce cracks in the ground, deform ground, and damage human-made structures. Volcanic earthquakes are of two types: volcano-tectonic (VT) earthquakes and long-period earthquakes. The former type is caused by slip on a fault near a volcano. The second type of volcanic earthquake is caused by movement of magma or other fluids within the volcano. Pressure within the system increases, and the surrounding rock fails, creating small earthquakes. Volcanically caused long-period earthquakes are an indication of magmatic activity and may be a precursor to an eruption.

Directed blasts are powerful volcanic explosions that can generate devastating high-energy pyroclastic density currents. Over the past century, there have been 31,000 fatalities from blast eruptions. Directed blasts are not uncommon; several people were killed by the directed blast of Mount St. Helens on May 18, 1980 (Salleh 2001).

Tephra is fragmental material of any size and origin produced by a volcanic eruption. It is also called "pyroclastic material." Tephra ranges in size from ash to blocks and bombs (the latter two kinds being greater than 2.5 inches or 64 mm). Tephra is thrown high up into the air from the volcano. Large size blocks and bombs fall out near their source while smaller pieces of tephra can travel as far away as 53 miles (80 km). Smaller ejecta are convected upward by the heat of the eruption and will fall out farther from the volcano. Most small particles will fall out within 30 minutes of the time they are erupted, and the smallest particles can stay in the atmosphere for long periods of time after a volcanic eruption. Some climate scientists believe that these particles may contribute to global warming.

The distance that ejecta travel away from a volcano depends on several factors: the height of the eruption column, temperature of the air, wind direction, and wind speed. A very high-erupting column that reaches into the stratosphere will be sheared by strong winds. This causes the eruption cloud to spread out over a larger area, resulting in the sky becoming dark even during the middle of day. However, the temperature of the air during an eruption will increase due to the hot material ejected into the atmosphere, causing a buoyant force that carries tephra higher into the atmosphere, which allows it to be deposited over a larger area.

Tephra produces a wide range of hazards. When the ejected material is in the atmosphere, it often produces lightning. Other hazards are produced when the ash is deposited on the ground. Ash can disrupt electricity, television, radio, and telephone communication lines; bury roads and other structures; damage machinery; cause fires; and clog drainage and sewage systems. In the atmosphere, it can produce poor visibility and respiratory problems as well as traffic accidents and hazards to airplanes; in fact, volcanic eruption clouds cause rerouting, cancelations, and flight delays. Moreover, tephra may bury entire towns, such as the ancient Roman city of Pompeii in Italy in the eruption of Mount Vesuvius in 79 CE.

If volcanic ash builds up on rooftops, it often causes collapse of houses. Most of the 350 deaths caused by the eruption of Mount Pinatubo in 1991, which also affected weather around the globe, were due to collapsing roofs. Tephra can also destroy vegetation and crops, which can result in famine, the largest indirect hazard produced by volcanic eruptions. In 1815, after the eruption of Mount Tambora, Indonesia, 80,000 people died due to famine (Riley n.d.). Many hazards caused by tephra falls can be mitigated with proper preparation and mitigation. For example, designing roofs with steep slopes or strengthening roofs and walls would be helpful measures in areas with volcano eruption.

In addition to other materials, molten rock released into the atmosphere from volcanoes contains dissolved gases, some of which are deadly. Although the largest portion of these gases is water vapor, some of the gases (e.g., hydrogen sulfide, carbon dioxide, carbon monoxide, and sulfur dioxide) are toxic, and these gases often kill animals and people. For example, a disastrous outburst of built-up carbon dioxide from the bottom of Lake Nyos in Cameroon killed 1,746 people and burned many more. This toxic gas release also killed 3,500 livestock on August 21, 1986 (Ebert 2000). Some of the volcanic gases are transported away from vents as acid aerosols while other gases such as sulfur compounds, chlorine, and fluorine react with water to form poisonous acids. These acids damage the skin, eye, and respiratory systems of animals even in very small concentrations, and they also burn vegetation. Most volcanic gases are noxious and smell bad and can even produce acid rain when concentrations are high enough.

Lava flows are the least hazardous of all processes in volcanic eruptions because they generally move slowly and thus rarely threaten human life. However, lava flows are very hot and can cause injuries by burning skin, charring eyebrows, and melting the soles of boots. Lava flows also destroy property by burying cars, burning homes and vegetation, and crushing everything in their path. Sometimes lava flows can melt ice and snow to cause floods and lahars. Melting of ice beneath a glacier may produce very large floods called glacier bursts. Additionally, lava flows can dam rivers to form lakes that might overflow and break their dams, causing floods.

Volcanic structural collapse can cause several hazards such as avalanches, mudflows, rock fall, or landslides. These events can be almost any size ranging from a few loose rocks falling from the crater rim of a volcano to large avalanches. Large-scale debris avalanches and landslide deposits are found in many volcanic areas in the world. A debris avalanche is formed when an unstable slope collapses, and debris is transported away from the slope. Large-scale avalanches generally occur on very steep volcanoes and move particularly fast because they have stored energy in the material within them. Some large avalanches can carry large blocks and travel several miles from their source. There are two general types of debris avalanches: cold and hot. A cold debris avalanche usually results from a slope becoming unstable, while a hot debris avalanche is the result of volcanic activity such as volcanic earthquakes or the injection of magma.

A landslide is a gradual movement of an avalanche while mudflows occur near volcanoes if the surrounding areas receive heavy rainfall or melt snow and glacial ice near the crater or even if a crater lake bursts. All these events can produce a considerable number of hazards. They can dam rivers and lakes and cause flooding and tsunamis. They can also decrease in pressure and cause a volcanic explosion. The mixture of debris from a landslide or avalanche with water may produce lahars, which can affect people living far away from the volcano's summit.

Pyroclastic flows and surges, also known as pyroclastic density currents, are fluidized masses of rock fragments and gases that move rapidly away from the volcano in response to gravity. They can move up to 430 mph (700 km/h) and can reach temperatures of about 1,830 degrees Fahrenheit (1,000 degrees Celsius). Pyroclastic flows contain as much as 80 percent unconsolidated materials. The flow is fluidized because it contains water and gas from the eruption, water vapor from melted snow and ice, and air from the flow overriding air as it moves downslope. Pyroclastic flows are often accompanied by a glowing cloud of ash separated from the flow.

Pyroclastic surges are low concentration density flows of pyroclastic material, which can expand over hill and valley and move fast. The reason they are low density is because they lack a high concentration of particles and contain a lot of gases. Both pyroclastic flows and surges are potentially highly devastating because of their mass, high temperature and velocity, great mobility, and deadly gases that can asphyxiate people. They can kill, bury, burn, and destroy things upon impact. Probably the greatest number of people killed by pyroclastic surges was in 1902 near Mount Pelee in the town of St. Pierre, Martinique, when 30,000 people lost their lives.

Since pyroclastic surges are fast and not constrained by topography, it is hard to find a safe place to be when they move. The best way to reduce risk of death is to evacuate people prior to such eruptions from areas likely to be affected by such flows and surges. One of the greatest volcanic hazards is lahars. Lahars are similar to pyroclastic flows, but they contain more water and usually travel down valleys. They can travel long distances and transport very large boulders. They can float boulders, cars, buildings, and bridges.

POSITIVE EFFECTS OF VOLCANIC ERUPTIONS

Volcanic eruptions have some positive consequences. For example, the Plain of Catania is one of the most fertile plains in Sicily as well as in Italy and has been created by ashes from nearby Etna. The plain produces the best citrus and pistachios in the country, and wines produced in the plain are of high and unique value among local products. These are important sources of income for local residents, and nearby towns in higher elevations attract tourists—another source of income. In addition, rocks of different types created by volcanic lava are used for building construction, decoration, housing flooring, facades, and typical dry stone walls. Lava also has health benefits via soaps with exfoliating properties or homeopathic products based on fine volcanic ash for various diseases.

Volcanoes provide stunning landscapes with high-rise peaks, which attract domestic as well as international tourists who come for sightseeing as well as climbing and skiing. Also, hot springs associated with volcanoes are used to produce geothermal power, which provides energy security for Japan. In addition, many mineral resources, including gold, are found in the vicinity of volcanoes.

CONCLUSION

Volcanoes provide early-warning signs of imminent eruptions. These signs include tremors and gas, stream, and ash emissions from volcanic vents. Also, a variety of scientific methods (e.g., laser beams, seismometers, and special microphones) are employed to monitor volcanic eruptions. Some gases can be monitored extensively and consistently using satellite technology. Satellites also monitor temperature readings and deformation. However, these methods have proven to be nearly as unreliable as those for predicting an earthquake. Many pieces of evidence, such as the history of previous volcanic activity, earthquakes, and slope deformation, can mean that a volcano is about to erupt, but the time, duration, and magnitude of the eruption are difficult to pin down. However, scientists continue to work to improve the accuracy of their predictions.

Since volcanologists are usually uncertain about an eruption, the best way for people to be safe is to evacuate the danger area after observing signs of volcanic eruption. If a volcano erupts, people need to stay clear of lava, mud flows, and flying rocks and debris; use goggles or eyeglasses, wear an emergency mask, or hold a damp cloth over the face; stay away from areas downwind from the volcano to avoid volcanic ash; and stay indoors until the ash has settled (Meer 2018).

Those people who do not evacuate should store adequate food and water, stay indoors, avoid going outside without wearing a wet cloth or some sort of filter over their mouth and nose, stay in areas that are unlikely to receive large amounts of tephra, avoid staying in flat-roofed buildings that can collapse due to heavy ash deposit, and be prepared to be without electronic communication.

Bimal Kanti Paul

Further Reading

CRED (Center for Research on the Epidemiology of Disasters) and UNISDR (The United Nations Office for Disaster Risk Reduction). 2016. *Poverty & Death: Disaster Mortality 1996–2015.* Brussels, Belgium: CRED.

Ebert, C.H.V. 2000. *Disasters: An Analysis of Natural and Human-Induced Hazards.* 4th ed. Dubuque, IA: Kendall Hunt Publishing Company.

Fosco, M. 2017. 500 Years of Volcano Deaths Could Help Save the 800M People at Risk Today. June 10, 2017. https://www.seeker.com/earth/500-years-of-volcano-casualty-data-could-help-improve-safety-around-eruptions, accessed February 9, 2019.

Francis, P., and C. Oppenheimer. 2003. *Volcanoes.* 2nd ed. New York: Oxford University Press.

McKie, Robin. 2017. Scientists Discover 91 Volcanoes below Antarctic Ice Sheet. *The Guardian.* August 12, 2017. https://www.theguardian.com/world/2017/aug/12/scientists-discover-91-volcanos-antarctica, accessed February 9, 2019.

Meer, A. H. 2018. 4 Different Types of Volcanoes According to Shape. https://owlcation.com/stem/4-Different-Types-of-Volcanoes-Cinder-Cones-Lava-Domes-Shield-and-Composite-Volcanoes, accessed June 18, 2019.

Riley, M. n.d. Types of Volcanic Hazards. http://www.geo.mtu.edu/volcanoes/hazards/primer/, accessed June 17, 2019.

Rittman, A. 1962. *Volcanoes and Their Activities.* New York: John Wiley & Sons.

Salleh, A. 2001. How Volcanoes Kill. January 17, 2001. http://www.abc.net.au/science/articles/2001/01/17/234135.htm, accessed May 2, 2019.

Waterspouts

The most common waterspout is a small, relatively weak rotating column of misty air beneath a towering *cumulus congestus* cloud occurring over tropical or subtropical waters. A waterspout descends from a cumulus cloud to an ocean or a lake. They form from winds close to the water's surface, usually do not suck up water, get their mist from sea spray, and rise toward the clouds above. These are "fair-weather waterspouts."

Usually appearing as a funnel-shaped cloud, waterspouts look like tornadoes. Some waterspouts have a more intensive columnar vortex, while some others form connected to a cumuliform cloud and some to a cumulonimbus cloud. Only a relatively few waterspouts actually are tornadoes (tornadic waterspouts) that form during a thunderstorm over a body of water. Waterspouts generally dissipate before coming on shore or just as they come on shore. Both tornadic and fair-weather waterspouts require high levels of humidity and a relatively warm water temperature relative to the overlying air.

There is some scientific debate about the exact definition of waterspout. The term mainly is reserved for small vortices over water that are not associated with storm-scale rotation. In other words, they are the water-based equal of landspouts. But most call almost any rotating column of air a waterspout if it contacts the surface of water.

Waterspouts form mostly in the tropics and subtropical areas of the globe. There also are some in Europe; Australia; New Zealand; Antarctica; and in the United States, in the Great Lakes, Great Salt Lake, Florida and the East Coast, and the California coast. Although rare, waterspouts have been observed in connection with lake-effect snow precipitation bands. They can also happen during blizzards.

In the United States, if a waterspout moves onshore, the National Weather Service (NWS) issues a tornado warning. Typically, fair-weather waterspouts decay quickly if they make landfall and rarely go very far inland (NWS 2019).

HISTORY OF WATERSPOUTS

Waterspouts have been observed and talked about since ancient times. For most of history, they have been feared and considered mysterious, accompanied by fanciful speculation. Originally, the word *waterspout* was a nonmeteorological term used at least since the fourteenth century to describe a spout through which water is discharged, such as a drainpipe for removing water from a roof. It was considered this until very recently. The atmospheric phenomenon called a waterspout probably was seen by sailors from the earliest sea voyages. The 1611 edition of the Hebrew Bible's Psalm 42 mentions "Deepe calleth unto deepe at the noyse of thy waterspouts," though this probably is not referring to atmospheric waterspouts.

One of the earliest records of a waterspout that caused damage is the powerful Malta tornado, which occurred on September 23, 1551, at the Grand Harbor of Valletta. This waterspout-turned-tornado killed over 600 and sank several boats and at least four galleys of the Order of Saint John, known as the Maltese Navy after 1530.

The initial print references in English date from no earlier than the eighteenth century. Many thought that the waterspout's column reaching from cloud to sea was really water that was pouring from the cloud, or it was water in the process of being drawn upward from the sea to the cloud.

American scientist, inventor, and statesman Benjamin Franklin (1705–1790), in his book *Physical and Meteorological Observations: Conjectures and Suppositions* compiled in 1751 and published in the Royal Society's *Philosophical Transactions*, LV (1765), opined that whirlwinds and waterspouts were about the same. He reasoned that waterspouts differed only in their location over water. Nonscientists continued to believe that a waterspout was a column of water. When a funnel cloud was witnessed over the surface of land, most thought it was—like a waterspout at sea—water pouring from the cloud but somehow suspended in midair without reaching the ground. People watched the funnel nervously, thinking that it might burst and flood the area.

Until recent years, most waterspouts were under-recorded because they caused no damage. The earliest record of a waterspout was of two spouts that were seen off England's southern coast in June 1233. The monk Roger of Wendover, in his world history chronicle *Flores Historiarum* (published in Latin in 1567), described a waterspout as two huge snakes fighting in the air. English historian C. E. Britton, in his *A Meteorological Chronology to AD 1450* (published in 1937), thought that Wendover's account was probably a description of an aurora because the terminology was commonly used by old writers in describing auroral appearances. The earliest record of a waterspout on sea was in 1456 when a whirlwind of water was seen on the sea near Ancona, Italy.

Sailors in the eighteenth and nineteenth centuries commonly thought that shooting a broadside cannon volley would disperse a waterspout. Russian Admiral Vladimir Bogdanovich Bronevskiy (1784–1835) agreed that this was successful, having seen one dissipate in the Adriatic Sea while serving as a midshipman on the frigate *Venus* during 1806, under Russian Admiral Dmitriy Senyavin (1763–1831). The 1869 shipwreck of the kerosene-carrying *Trovatore*—en route from Trieste to Palermo—off Sicily, Italy, is blamed on a waterspout, though one firsthand account falsely states that a waterspout sucks up water. Also, a waterspout strike has been proposed as a possible reason for the inexplicable 1872 abandonment of the American merchant ship *Mary Celeste* in the Atlantic Ocean off the Azores Islands.

On Cape Cod in the United States, the *Barnstable Patriot* (Massachusetts) reported that on August 23, 1870, a waterspout entered Barnstable Harbor, struck land, scattered some wood, and then moved beyond Sandy Neck into Cape Cod Bay. In June 1871 and August 1880, two waterspouts were reported to have occurred southwest of Hyannis and off Bourne.

Two waterspouts ultimately were blamed for the December 28, 1879, plunge of a passenger train off the Tay Bridge (Tayside) into the Tay Estuary, Scotland, when the middle section of the bridge collapsed. While the bridge was poorly constructed and had been weakened by previous gales, the waterspouts were identified as causing the final collapse.

FORMATION OF WATERSPOUTS

A typical waterspout is only a few feet in width, but can be several hundred feet tall. The average spout is around 165 feet (50.3 m) in diameter, with 50 miles (75 km) per hour winds, which is roughly about as strong as the weakest types of tornadoes on land. The largest waterspouts can be 330 feet (91.44 m) in diameter.

Most are caused by different atmospheric dynamics. They often develop in moisture-heavy areas as their parent clouds also are developing. It is theorized they spin as they move up the surface boundary from the horizontal shear near the surface and then stretch upward toward the cloud once the low level shear vortex is aligned with a developing cumulus cloud or thunderstorm. This formation process also is typical of some weak tornadoes (i.e., landspouts).

More than one waterspout can occur at the same time in a limited location. For example, up to nine have been reported on Lake Michigan. The International Centre for Waterspout Research reported 16 waterspouts on the Great Lakes in one week in 2019. The record for waterspouts reported over a seven-day period is over 66 in 2003.

Waterspouts most commonly happen in the late summer, with September as the leading month in the Northern Hemisphere. Waterspouts are rare occurrences on the western coasts of the world's continents. Waterspouts can occur in bodies of water other than saltwater. For example, there have been many on freshwater lakes and rivers including the Great Lakes and the St. Lawrence River, as well as many other locations across the globe. Normally, waterspouts are more frequent within 60 miles (90 km) of the coast than out at sea. The fair-weather waterspout occurs mainly in tropical and subtropical climates, with some 400 annually in the Florida Keys most commonly from late spring to early fall. More are reported in the lower Florida Keys than in any other place in the world, sometimes causing damage like the one near Oldsmar, Florida, that happened on July 9, 2013. National Oceanic and Atmospheric Administration (NOAA) scientist Joseph H. Golden holds that the United States waterspout capital is Key West, where 100–500 form annually from about mid-May to late September within a 35-nautical-mile (53 km) radius of the city. The next highest frequency is on the southeast coast of Florida and the northern Gulf of Mexico.

Other parts of the United States sometimes have waterspouts. On the Great Lakes, waterspouts form when a convergence zone moves over a Great Lake at the same time as cooler air aloft is moving in. In other words, a cold front crosses the water, bringing with it a large amount of overriding cold air that has a great temperature differential from the warm water and high humidity in the lowest several thousand feet of the atmosphere. The waterspout forms at the sea surface as the nearby warm air is light and rises higher in the atmosphere. The cold air from the front does the opposite and sinks toward the sea floor. At the same time, the front-driven winds shift from southwest to northwest. Along that line where the air swings in direction to the northwest, spinning of the air occurs and a waterspout forms.

Waterspouts can occur at any place over the world's coastal regions. Though the tropics host the majority of waterspouts, they also can occur in the world's temperate zones and across the western coast of Europe, the British Isles, and some locations in the Mediterranean and Baltic Seas. About 160 waterspouts annually are reported in Europe. The most in that continent occur in the Netherlands at 60, followed by Spain and Italy at 25, and the United Kingdom at 15. Waterspouts are frequently observed off the east coast of Australia, with several being described by Joseph Banks during the voyage of the Endeavour in 1770.

TYPES OF WATERSPOUTS

Waterspouts are similar to tornadoes over water. Waterspouts are much less dangerous than tornadoes, whose wind speeds normally are over 113 mph (182 km/h) and can last up to three hours. Waterspouts are generally broken into two categories: fair-weather waterspouts and tornadic waterspouts.

Nontornadic or fair-weather waterspouts are the most frequently occurring and develop on the surface of the water and move upward. These are waterspouts that are not associated with a rotating updraft of a supercell thunderstorm. They usually move very slowly if at all because the cloud to which they are attached is horizontally static. By the time the funnel is visible, a fair-weather waterspout is beginning to mature. It is formed by vertical convective action instead of the normal subduction/adduction interaction between colliding weather fronts. Its visible cloud column is formed by water vapor condensation. Such waterspouts look like and act like landspouts and often behave like them if they go ashore.

Fair-weather waterspouts mostly occur during light wind conditions in coastal waters and are connected with dark, flat-bottomed, developing convective cumulus towers. Fair-weather waterspouts are associated with developing storm systems but not a storm. This type of waterspout rapidly develops and dissipates, with a total life cycle varying from 5 to 10 minutes on average but up to one hour often during early to mid-morning and occasionally in the late afternoon. Fair-weather waterspouts are fragile and can be easily broken apart by too much wind, a shift in wind direction, or a burst of cold air.

A tornadic waterspout commonly is a nonsupercell tornado that occurs over water. This most powerful type of waterspout can be large and are capable of causing significant destruction. They are very similar to a land-based tornado in that they form from severe weather called mesocyclones that are connected with severe thunderstorms. Mesocyclones form when strong changes of wind speed and/or wind direction with height (wind shear) stimulate parts of the lower atmosphere to spin in tube-like rolls. The thunderstorm's convective updraft next draws up this spinning air, tilting the rolls from parallel to the ground to perpendicular. This finally causes the whole updraft to rotate as a vertical column. The air thus rises and rotates on a vertical axis that extends from the cumulonimbus or severe thunderstorm clouds to the ground. Like tornadoes, tornadic waterspouts are often accompanied by high winds and rough seas, large hail, and dangerous lightning that strikes frequently.

In the United States, tornadic waterspouts are much less frequent than fair-weather waterspouts because most mesocyclone thunderstorms occur in land-locked areas. In some other parts of the planet—like the Adriatic, Aegean, and Ionian seas—tornadic waterspouts occur much more often and can constitute half of the total number of waterspouts.

The longest-traveling European tornado also was the earliest-known occurrence in France during September 1669. It went almost 249 miles (401 km) from La Rochelle (Charente-Maritime) to Paris. It was thought to have begun as a waterspout over the Bay of Biscay, though official records are sketchy at best. On May 21, 1950, a tornado touched down at Little London (Buckinghamshire), England, and traveled over 66 miles (106 km) Coveney (Cambridgeshire), as a funnel cloud another 32 miles (42 km) to Shipham (Norfolk) and then out across the North Sea. It traveled the longest distance of any English tornado.

In January 2002, twin tornadic waterspouts swept across the Mediterranean Sea between Rhodes, Greece, and southwestern Turkey. During this winter season, the strong southerly flow (winds from the southern quadrant) in the eastern Mediterranean was blowing very hard (9 on Beaufort Scale). Right before the two funnels descended, the winds suddenly shifted to the north and rain started. Despite this tornadic waterspout with the powerful thunderstorms, fortunately there were no reports of related deaths or injuries.

LANDSPOUTS, GUSTNADOS, AND SNOWNADOS

A fair-weather waterspout is similar to a dust devil (or landspouts), which usually develop on sunny days where the heating of the sun creates a swirling updraft and a well-formed whirlwind of dust. A landspout is a nonsupercell tornado with a narrow, rope-like condensation funnel that forms while the thunderstorm cloud is growing with no rotating updraft. The spinning motion originates near the ground. In 1985, American research meteorologist Howard B. Bluestein coined the name "landspouts." Bluestein also co-invented with Al Bedard and Carl Ramzy of NOAA a tornado measuring device (TOTO) that inspired the fictional "Dorothy" in the American epic disaster film *Twister* (1996) about a severe outbreak of tornadoes in Oklahoma. Landspouts emerge near the land or surface where the winds are converging and have flat bases and a fluffy, almost cotton-like appearance. Unlike a true tornado, lands spouts do not form from a mature mesocyclone. Like a waterspout, a landspout forms with a slow-moving (or not-moving), developing cloud system. Damage tends to be minimal.

Another nonsupercell tornado like a waterspout is the gustnado, a swirling of dust or debris at or near the ground with no condensation funnel, which forms along the leading edge front of a thunderstorm. The name derives from "gust front of a tornado," and a gustnado looks like a tornado. It begins to form as very strong thunderstorms produce a vigorous downward push of air called a downdraft. These downdraft winds go outward when they hit the ground and cause a rush of wind on the land's surface. Rotation may develop if there is enough instability, and thus, gustnado is formed. Once it forms, the gustnado spins upward from the surface to about 30–300 feet (9.1–91.4 m) high. Since the rotating air column in a gustnado is not fused to the cloud, it is different from a tornado. The average gustnado is short-lived, from a few seconds to a couple minutes, and is relatively weak. Rain may also occur. Some gustnados reach wind speeds of 60–80 mph (90–120 km) and cause damage like that of an EF-0 or EF-1 tornado.

A snownado—or winter waterspout, a snow devil, a snow spout, an ice spout, or an ice devil—is a very rare waterspout that has formed under the base of a snow squall, usually over frozen lakes or other snow-covered areas. It often forms in air temperatures of −0.4 degrees Fahrenheit (−18 degrees Celsius) or colder. Some snownados have been 30 feet (9.14 m) wide, 45 feet (13.7 m) high, and capable of lifting objects over 1,500 pounds. They mostly happen under or before a snow squall—a short, intense period of heavy snow often with strong winds—and frequently indicate that more snow is going to fall soon. The snownado starts to form when a warmer surface causes the snow or ice to emit fog or steam. If there is a column of colder, low-pressure air above this fog, it will rise. The wind shear or currents then will make it rotate counterclockwise (the usual, but not always, direction in the Northern Hemisphere) and pick up loose snow to approximate the funnel shape.

Severe snownados are very rare because its necessary causal elements are relatively odd. More likely, a rather weak snownado like looks like a landspout on snow. Very cold temperatures have to be present over a body of water to produce a temperature differential that is warm enough to produce fog resembling steam above the water's surface. This usually needs temperatures of −0.4 degree Fahrenheit (−18 degree Celsius) or colder if the water temperature is no warmer than 41 degrees Fahrenheit (5 degrees Celsius). Usually, strong lake-effect snowing in a small or banded form also must be occurring. The wind speed has to be slow, usually less than about 6 mph (or 5 knots).

Until recently, few pictures of snownados existed, including four in Ontario, Canada, near Whitby on January 26, 1994, and one at 2,800 feet (853.4 m) during a snow squall over Flathead Lake near Missoula, Montana, on November 4, 2013. Thanks to continuing human interest and online video sharing, there are a number of videos of snownado activity around the world. One was taken on the southern portion of Lake Champlain in Vermont on January 15, 2009. The video shows that a steamy fog is faintly evident near the water surface, and the air temperature was about −5 degrees Fahrenheit (−20.56 degrees Celsius) with the water temperature near freezing. Five waterspouts also were observed over the lake during that mid-morning. Smaller low-level whirls (sometimes called steam devils) were also around, having come from the arctic sea smoke or steam fog near the water surface. Other snownado videos include one in the Alps near Lessach, Austria, in April 2015; an occurrence in upstate New York during a lake effect snowstorm on December 8, 2016; one at Val di Sole in the Folgarida-Marilleva resort in Folgarida-Marilleva, Italy, in March 2016; and another on February 20, 2019, during a snowstorm north of Albuquerque, New Mexico.

LIFE CYCLE OF A WATERSPOUT

In 1974, noted NOAA scientist Joseph H. Golden was the first to identify the traditional five stages of all waterspouts' life cycles. The stages are discrete but overlapping. Not every waterspout that has been observed since it began goes through all of the five stages. Some seem to skip a stage or merely die out after attaining one of the lesser stages. Stages one and four are the longest in duration. It also should be noted that the stages are approximate.

First, termed the *Dark Spot Stage*, a prominent circular, light-colored disk appears on the surface of the water, surrounded by a larger dark area of varying shape. The dark spot may or may not have a funnel cloud over it at first. The vast majority of dark spots appearing over the water do not evolve into waterspouts. A parent cumulus cloud above must be present.

Second, there is a Spiral Pattern Stage, as after the formation of the colored disks, an alternating pattern of light- and dark-colored spiral bands emerge and develop from the dark spot on the water surface. There often will be one major dark band coming from a nearby rain shower. This stage is the primary growth period.

The third stage is the Spray Ring, or Incipient Spray Vortex. This is indicated when a dense annulus (a ring-shaped object) of sea spray called a cascade appears around the dark spot with what seems to be an eye. The eye is like those seen in hurricanes. The funnel cloud above is lengthening. This is the shortest stage of the cycle, often going for less than two minutes.

Fourth, signifying the Mature Vortex Stage, the waterspout becomes visible from the surface of the water to the overhead cloud. Its funnel frequently appears to be hollow, with a surrounding shell of turbulent condensate (liquid formed by condensation). This spray vortex attains its maximum funnel cloud length and diameter. It can rise to several hundred feet or more and often creates a visible wake and a wave train as it is moving. The spiral pattern gradually weakens.

Finally, the fifth stage is Decay. As the amount of inflow of warm air into the waterspout's vortex weakens because of the cool down air drafts of a nearby rain shower, the funnel and spray vortex begin to dissipate. The waterspout slows, the spray vortex weakens, and the funnel becomes shorter and diffuse. The rain shower in effect has overtaken the waterspout. This can sometimes happen abruptly. This stage normally varies from one to three minutes, but can persist to seven minutes.

WATERSPOUTS AS MARINE HAZARDS

Waterspouts long have been recognized as serious marine hazards. Stronger waterspouts pose threats to watercraft, aircraft, and people. With winds possibly gusting to 70 mph (105 km/h), a waterspout can flip a boat and injure its passengers as well as other people by its hurling debris onshore.

People are advised never to move closer to investigate a waterspout because some may be as dangerous as tornadoes. At sea, experts advise mariners to listen for special marine warnings about waterspout sightings that are broadcast on NOAA Weather Radio or via the NWS Mobile app and to watch the sky for certain types of clouds. If a waterspout is sighted, one is urged to head immediately at a 90-degree angle from the apparent motion of the waterspout. Mariners are advised never to try to navigate through a waterspout and to go below to avoid any flying objects. Although waterspouts are usually weaker than tornadoes, they can produce significant damage to mariners and their boat.

There have been very few incidents of waterspouts causing severe damage and casualties. In 2015, a waterspout came onshore at a beach in Fort Lauderdale, Florida, and lifted an inflatable bounce house with three children higher than the palm tree line. Fortunately, the children were tossed out of the house and only slightly injured as the waterspout first flipped over the beach. The waterspout had moved from the ocean onto the beach sand, tossed a canopy aside, and then rolled the bounce house before lifting it up. The waterspout also snapped a concrete pole holding a basketball hoop. An American flooring installer from Fort Lauderdale, Florida, and his family were on the sand and watched the waterspout pick up the bounce house. Reportedly, they barely felt anything when the waterspout passed over them, noting that tablecloths on nearby picnic tables were not disturbed. They solely complained about getting stung by the spinning sand.

Animals living in the water can be severely affected by waterspouts. Depending on how fast the waterspout winds are going, anything within about 3 feet (0.9 m) of the water surface—including fish, frogs, and turtles—can be lifted into the air. Sometimes, a waterspout can suck smaller animals like fish completely out of the water and up into the cloud. Even if the waterspout stops spinning, the fish in the cloud can be carried over land and pushed around with the cloud's winds until the currents dissipate. Depending on how far they travel and how high they are taken into the atmosphere, the unfortunate fish are sometimes dead by the time they reach the earth. Fish from such a strong waterspout can be rained on neighborhoods as far away as 100 miles (150 km) inland. Fish also can be sucked up from rivers. Raining fish though is not a common weather phenomenon, with fewer than 10 annually in most waterspout-prone areas.

The United States National Weather Service often will issue special marine warnings about when waterspouts are likely or have been sighted over coastal waters or about tornado warnings when waterspouts are expected to move onshore.

RESEARCHING AND FORECASTING WATERSPOUTS

Modern scientific interest in waterspouts started when a very large, long-lasting spout occurred on August 19, 1896, off the northeastern Martha's Vineyard town of Cottage City (now Oak Bluffs), Massachusetts. It was observed by literally thousands of vacationers and several scientists. It was estimated to be 3,593 feet (1,095 m) high and 840 feet (256 m) wide at the crest, 141 feet (43 m) in the center, and 240 feet (73 m) at the base. The shroud of spray surrounding the central funnel was about 656 feet (200 m) wide near the water surface and projected 394 feet (120 m) in the air. The spout's circulation lasted for over 35 minutes. This was the highest reliably measured waterspout as of 2017. Most waterspouts are smaller than this one and have much shorter lives. The Cottage City spout is an untypical example of one that was spawned by thunderstorm-squall conditions similar to those that produce tornadoes over land.

Aside from Golden and Fujita, early waterspout researchers included F. M. Maury, M. Reye, Alfred Wehener, Verne H. Leverson, Peter C. Sinclair, Frank Kieltyka, Alexander Keul, and Michalis Sioutas, among others.

Until recently, meteorologists reviewed available forecast graphics to attempt to determine when and where were the necessary simultaneously occurring weather conditions for a waterspout. Most weather centers previously issued a Special Marine Warning (in the United States) or a Waterspout Advisory (Canada) only after a waterspout was sighted. Their predictive specificity and accuracy were greatly improved with the development of the Szilagyi Waterspout

Index (SWI), completed in 2005 by Canadian meteorologist Wade Szilagyi. Now widely utilized, the SWI ranges from −10 to +10, where values greater than or equal to zero indicate whether the conditions are favorable for waterspout development.

To develop the SWI, Szilagyi first gathered meteorological data during waterspouts to develop a forecast technique he called the Nomogram. This Nomogram was initially used at the Weather Center in Toronto and refined and improved with additional data. The first forecasting model was completed in 2012. Szilagyi's Experimental Waterspout Forecast System (EWFS) produces forecast values of SWI. Ongoing improvements to the EWFS include a higher model resolution, simplification of the display, and the incorporation of surface convergence. Surface convergence is essential for waterspout formation.

The International Centre for Waterspout Research (ICWR)—formed in 2008—is a nongovernmental international organization of individuals interested in waterspouts from a research, operational, and safety perspective. First founded as a forum for researchers and meteorologists, the ICWR has expanded. To date, the ICWR has jointly produced waterspout research papers and increased public awareness of waterspouts, including its "Live Waterspout Watch" on its website. It is working on the development of a Global Waterspout Forecast System, Global Waterspout Database, and Global Waterspout Watch Network. These will improve waterspout forecasting and preparation considerably.

Today, the United States' National Weather Service includes waterspout warnings as part of its severe local storm warning. Various parameters are considered in a forecast for an alert, including air temperature, water temperature, wind speed, and moisture near the water surface. National forecasting services around the world normally issue bulletins and warnings of a waterspout tornado well before the 12–24 hour window of their possible occurrence. Work of the ICWR and numerous scientists will help improve future forecasts.

William P. Kladky

Further Reading

Brown, Paul R., and G. Terence Meaden. 2016. Historical Tornadoes in the British Isles. In *Extreme Weather: Forty Years of the Tornado and Storm Research Organisation (TORRO)*, edited by R. K. Doe, 17–30. Chichester, UK: John Wiley & Sons, Ltd.

MI News Network. 2019. Facts about Waterspouts at Sea. *Marine Insight*. September 11, 2019. https://www.marineinsight.com/know-more/8-facts-about-water-spouts-at-sea, accessed September 26, 2019.

National Geographic. 2019. Waterspout. *National Geographic, Resource Library*. Encyclopedic Entry. https://www.nationalgeographic.org/encyclopedia/waterspout/, accessed October 2, 2019.

NWS (National Weather Service). 2019. About Waterspouts. https://www.weather.gov/mfl/waterspouts, accessed October 3, 2019.

Rafferty, John P., ed. 2010. *Storms, Violent Winds, and Earth's Atmosphere*. New York: Britannica Educational Publishing.

Revkin, Andrew, and Lisa Mechaley. 2017. *Weather: An Illustrated History*. New York: Sterling Publishing Co.

Rowe, Mike. 2016. Tornado Extremes in the United Kingdom: The Earliest, Longest, Widest, Severest and Deadliest. In *Extreme Weather: Forty Years of the Tornado and Storm Research Organisation (TORRO)*, edited by R. K. Doe, 77–90. Chichester, UK: John Wiley & Sons, Ltd.

Sioutas, Michalis, Wade Szilagyi, and Alexander Keul. 2009. *The International Centre for Waterspout Research*. Preprints, 5th European Conference on Severe Storms, Landshut, Germany, October 12–16, 319–320.

Teague, Kevin Anthony, and Nicole Gallicchio. 2017. *The Evolution of Meteorology: A Look into the Past, Present, and Future of Weather Forecasting*. Hoboken, NJ: Wiley Blackwell.

Williams, Jack. 2017. *Pocket Guide to the Weather of North America*. Washington, DC: National Geographic.

Wildfires

Wildfires are dramatic hazards that can be either natural or anthropogenic in origin and begin in rural and wildland areas. Wildfires are the out-of-control burning of forest, brush, or grasses that may also threaten or consume human structures. Sometimes wildfires will spread into more built-up suburban and urban places from rural areas. Once initiated, wildfires can burn rapidly if supplied with a considerable volume of dry fuels like grasses, dead leaves, and branches, especially if driven by strong winds. Wildfires can modify the local weather, resulting in lightning and intensifying the fire by increasing airflow into it.

Wildfires are becoming bigger, hotter, and more frequent. They can be deadly, with dozens killed in several of the worst recent wildfire disasters. In recent decades, the economic cost, number of human lives lost, and the ecological impact of fires have been increasing. In the United States alone, the annual expenditure for addressing wildfires exceeds several billion dollars. With increase in very large fires, some people have begun calling fires of over 100,000 acres (40,468 ha) "megafires" and those reaching over 1 million acres (404,685 ha) "gigafires" (*Wildfire Today* 2019). The most intense of modern fires defy traditional modeling and firefighting; some can release 100,000 kW of heat energy per square meter, while "[e]ven at 4,000 kilowatts, firefighters cannot go near the flames and require aerial support" (*Horizon* 2019). These are the types of wildfire that can generate thunderstorm clouds known as pyrocumulus.

Considerable human effort is expended to understand, control, and extinguish wildfires, especially when the fires threaten the built environment. In the United States, land management agencies use computer models to estimate available fuels, to characterize the hazard, and to identify the risk of a wildfire based on the likelihood of natural ignition, especially by lightning when surface fuels are dry. Ignition refers to fire-starting, which happens with the application of sufficient heat to flammable, or burnable, materials. Wildfires can be categorized based on the type of fuel, such as grassfires, forest fires, brushfires, and peat fires.

Drought, which helps prepare the vegetation for more rapid burning by drying out the fuel, is frequently a precursor to wildfire. This was the case for the

Yellowstone fires in 1988. When two or more hazards occur together, such as drought and wildfire, scientists refer to these as compound hazards or events. Drought and associated wildfires have been more destructive in various parts of the world in recent decades, but parts of the United States and Australia have seen some of the worst wildfire disasters in the twenty-first century. In Australia, wildfires are commonly called "bushfires." The worst Australian wildfire outbreak on record is known as "Black Saturday." Numerous fires broke out in the state of Victoria during a major heat wave in February 2009 (midsummer for the Southern Hemisphere). The Black Saturday fires were ignited by a variety of sources (natural, accidental, and intentional), burning over 1 million acres (404,685 ha), completely destroying several towns, and killing 173 people. One firestorm killed 120 people in the affected area.

HUMAN-RELATED WILDFIRE IGNITION

Although fire scientists often focus on modeling conditions for natural ignition and fire behavior, it is normally much more difficult to model fire ignition through human action. People may accidentally start wildfires when, for example, they lose control of a campfire or a waste-burning pile, when a cigarette or match is dropped, or when a hot tailpipe of a car comes into contact with dry grass. Sometimes arsonists intentionally set fires.

The destructive and deadly 2018 Camp Fire in California provides an example of a wildfire from a human-caused, technological hazard occurring in a wildland-urban interface zone (WUI). Strong Diablo winds interacted with an electrical power line that had become detached early in the morning on

Landsat 8 satellite image of the smoke plume from the Camp Fire in northern California on November 8, 2018. (NASA Earth Observatory, https://earthobservatory.nasa.gov/images/144225/camp-fire-rages-in-california)

November 8, 2018. Diablo (devil) winds are the local name in northern California for strong and gusty winds blowing toward the ocean. A large local fuel load and northeast winds approaching 50 mph (75 km/h) allowed a spark from the downed power line to grow rapidly into a wildfire that devastated the town of Paradise. The powerful winds pushed the fire and its accompanying smoke plume toward the southwest. The wildfire, which burned over 150,000 acres (60,702 ha), resulted in 86 deaths and economic losses estimated to exceed $16 billion. It took over two weeks for the Camp Fire to be completely contained. The active outer edge of a wildfire is its perimeter, and containment occurs when all the perimeters are under control. Once a fire is contained, burning can still continue inside the perimeter.

An intentional human ignition starting a fire along a popular hiking trail on November 23, 2016, in combination with other fires likely sparked by wind-downed trees hitting power lines, is thought to have been a part of the Great Smoky Mountains fire of 2016. Fire management officials thought they had the fire contained until November 28, when the weather situation changed and strong winds increased the fire severity. Winds that fanned the flames reached the equivalent of hurricane force and caused multiple patches of fire to spread. When two or more fires are burning at the same time in the same area (sometimes merging), the result is referred to as a wildfire complex. The complex of fires left the park boundaries and impacted the tourist-oriented communities of Pigeon Forge and Gatlinburg. Regional drought and a large number of consecutive dry days, which were categorized as exceptional, played a major role in setting the stage for the wildfires, which burned into December. In the end, this wildfire complex cost 14 lives. The 2016 wildfires in the southern Great Smoky Mountains were the deadliest in the eastern United States since fires in Maine resulted in 16 deaths in 1947.

NATURAL WILDFIRE IGNITION AND SUPPORT

A fire needs three things: heat, fuel, and oxygen. Wildfire occurs when these three elements of the "fire triangle" occur together. Since the atmosphere contains 21 percent oxygen and a fire only needs about 16 percent oxygen, that element of the fire triangle is usually not a limiting factor.

Lightning produced by thunderstorms is the common natural source for wildfire. Once started, a fire can continue burning even at lower surrounding temperatures, as long as the other ingredients are sufficient. Thunderstorms that produce wildfires are more common in the warm season when higher temperatures at the earth's surface help create the atmospheric instability that leads to thunderstorm development. In areas with relatively dry air, the base or bottom of the thunderstorms forms higher up in the atmosphere. These high-base thunderstorms (also called elevated or dry thunderstorms) produce rain that falls into a very dry atmosphere beneath the storm cloud. As the rain falls, it evaporates. In situations where little to no rainfall reaches the surface, lightning from the thunderstorm can spark a wildfire. This is more common in mountainous areas of western North America, especially in the summer.

Humans alter the character of wildfires in three general ways: (1) shifting the location and frequency of ignitions, (2) altering the timing of when fires start, and (3) changing the character of the fuel available for the wildfire (Balch et al. 2017). Human-caused fire ignitions, whether due to carelessness or due to arson, are becoming an increasing problem. Human ignitions expand fires into places where lightning tends not to occur when fuels are dry or in places where lightning is rare. Humans have accounted for the vast majority of wildfire ignitions within the conterminous United States in recent decades, and these anthropogenic fire starts have greatly expanded the fire season. Human-initiated fires now account for approximately half the area burned each year by wildfire (Balch et al. 2017). Data for the period 1992 to 2012 indicate that human-caused ignitions are more evenly distributed throughout the year, with a seasonal maximum in spring and a single-day maximum associated with fireworks linked to patriotic celebrations on the Fourth of July (see Figures 1 and 2a in Balch et al. 2017).

Lightning-induced fires are most frequent in summer, with more than 75 percent of all natural fire starts in the United States happening in June, July, and August. Lightning-caused fire starts occur when warmer atmospheric conditions have dried the potential fuel for a wildfire and when thunderstorms are more likely across the elevated terrain (when a warm surface can contribute to instability and thunderstorm formation) in the western United States. There is a daily cycle to summer thunderstorm development in the western United States. Warming that occurs following sunrise produces a daytime local wind flow pattern (called valley winds) that moves upslope, helping to create the rising buoyant air that produces afternoon thunderstorms.

Conditions supporting wildfire strengthening and spread include wind, air mass characteristics, and topography. In southern California, winds flowing away from western United States' interior desert areas toward the Pacific Ocean are labeled Santa Ana winds. As with the Diablo winds in northern California, these northeasterly winds (blowing from northeast to southwest) cross the mountains and flow down from higher elevations; the air warms and the relative humidity of the air mass decreases. These atmospheric changes increase the risk of an out-of-control, extreme wildfire: not only are conditions dryer and hotter, but the wind can also help spread fires. Topography affects fire behavior by channeling wind in valleys or canyons. Fire also naturally tends to burn upward toward the fuel on hill and mountainsides. Of course, rugged topography also makes fire-fighting much more difficult than in flat and gently rolling landscapes.

Some fires burn for lengthy periods, perhaps smoldering at lower intensities after the high-intensity fire has burned through. The major 1988 United States wildfire complex in Yellowstone National Park began in June as a number of small fires. In total, approximately 250 different fires eventually converged into a single large conflagration that resulted in the park being closed to visitors in September. Fires in the Yellowstone complex burned for nearly six months and were not completely extinguished until the cold season snows arrived in November.

FIRE IN THE ENVIRONMENT

Fire ecology is the study of fire and its impact on vegetation. In a number of ecosystems, surface fires are a natural occurrence and paleoenvironmental evidence documents that fires have occurred frequently in the past. Some tree species actually need periodic fires to remove excess "litter" (tree needles or leaves and dead branches) from the surface and to release natural seed protection so they can germinate and grow new trees. In more humid forested areas, wildfires tend to be rare and are frequently linked to a lengthy drought.

A surface fire, or ground fire, burns through the underbrush beneath the tree canopy at a slow rate. If the burning occurs without an active flame, it is called smoldering. Trees can be well adapted to ground fires, with thick, protective bark, so little damage is done. If surface fires are suppressed, the underbrush can accumulate. Accumulation of fuel at middle levels of the forest can allow a fire to climb upward and burn the fuel in the treetops (the canopy). Fuels that allow a fire to grow vertically are called ladder fuels. Crown fires occur when fire reaches the upper parts of trees in the forest canopy. Crown fires burn much hotter, spread much faster, and are far more dangerous and destructive than ground fires. Crown fires are referred to as stand replacement fires, with an entire stand of new trees getting started in the decades following the fire. Suppression of surface fires has contributed to the buildup of ladder fuels and an increase in the wildfire hazard.

Airflow associated with winds can fan a fire and increase the rate at which oxygen enters the conflagration as the combustible material burns. As mentioned earlier, the fire itself can affect airflow and pull more oxygen in, keeping the fire burning.

Warmer temperatures from an advancing fire front helps evaporate moisture from the combustible (burnable) material so that the fire can continue to move ahead rapidly into new fuel. The perimeter typically is where the hottest parts of a wildfire occur. Super-heated air can be a characteristic of a rapidly advancing fire perimeter. When fire temperatures reach approximately 1,000 degrees Fahrenheit (537 degrees Celsius), "flashover" occurs as the cells of the plant material undergo a thermally induced breakdown and the process releases highly flammable gases. After the active edge of a wildfire has raced through an area, some of the remaining fuel will smolder and burn at a much slower rate. This happened with the 1988 complex of fires in Yellowstone National Park. Fires in areas with organic soils or peat can smolder for a considerable amount of time. Fire suppression crews work to put out areas that are smoldering because a future increase in wind speed can provide more oxygen to reactive the fire.

Following a wildfire, the removal of the vegetation cover can set the stage for an additional hazard: a mudflow, debris flow, or landslide. Heavy rains impacting areas that have recently burned are susceptible to the downslope overland flow of water along with soil and debris. Especially in hilly and mountainous areas, waters from flash flooding can pick up ash and other debris from the fire along with loose soil and carry the material rapidly downslope.

WILDFIRE SUPPRESSION AND FIREFIGHTING

Efforts to control and eventually put out a wildfire are called wildfire suppression. Prior to the middle of the twentieth century, it was considered appropriate to suppress all wildfires. This policy resulted in a buildup of fuels in a number of forests and has resulted in an increase in the number of crown or stand replacement fires in recent years. Today, a coordinated fire suppression effort for active fires involves firefighters on the ground and specialized aircraft. The combined aerial- and ground-based efforts help control locations where a fire is the hottest (hot spots). Ground crews create firebreaks or fire lines designed to stop an advancing fire in order to protect the natural forest resource and human structures. Firebreaks are areas ahead of a fire perimeter where the vegetation has been removed in an effort to halt the fire's spread. Sometimes existing roads serve as firebreaks, but firefighters may also use bulldozers to remove plant material (when terrain and accessibility allow) or may work to create narrower firebreaks with hand tools like a shovel, hoe, and Pulaski (a specialized firefighting tool). Rivers are natural firebreaks. Unfortunately, embers carried in windy conditions may mean that even wider firebreaks can be jumped by a wildfire.

Smokejumpers are specially trained firefighters who parachute into the area ahead of a fire, especially in more inaccessible areas, and use a variety of techniques to either stop or slow the advance of the fire perimeter. Aircraft assist by dropping water on hot spots or releasing fire retardant chemicals. The National Interagency Fire Center in Boise, Idaho, coordinates wildfire suppression efforts in the United States. Wildfire suppression costs have grown to over four billion dollars annually in the United States.

FIRE WEATHER

Weather forecasting services provide information that is helpful in identifying and mitigating the wildfire hazard. Central to providing useful information are forecasts related to humidity, wind, lightning strikes, and temperature. Low relative humidity, which tends to occur with warmer temperatures, is a critical metric used in determining the level of wildfire risk. A fire weather watch can be issued to warn individuals in fire suppression and land management agencies when problematic conditions are likely to develop in the next 12–72 hours. When conditions are forecast to reach above a critical local threshold, a red flag warning is issued.

Warming of the earth's surface during the day can produce locally warmer areas where the air becomes more buoyant and rises. The upward air movement from surface heating is called convection. Other areas nearby will see slightly denser air sinking to take the place of the rising pocket of buoyant, warmer air. This turbulence—up and down movement—in the lowest layers of the atmosphere occurs most frequently in the afternoon. Since the air at slightly higher levels in the atmosphere tends to be of lower humidity, turbulent convection can bring both windier conditions and drier air to increase the severity of a wildfire. Wildfires can produce significant local heating and convective turbulence in the atmosphere.

A special cloud name, pyrocumulus or flammagenitus, is used for the tall cloud associated with an active hot spot along the perimeter of a fire. In extreme situations, pyrocumulus clouds can produce enough convection to generate lightning that might start another fire. The largest, most extreme fire-related clouds may be referred to as "pyroCbs," or pyrocumulonimbus clouds—the components of the name refer to fire (pyro), vertically-oriented tall clouds (cumulo) with rain (nimbus). A pyroCb has been described as "an explosive storm cloud actually created by the smoke and heat from fire, and which can ravage tens of thousands of acres. And in the process, 'pyroCb' storms funnel their smoke like a chimney into Earth's stratosphere" (Finneran 2010).

WINTER WET, SUMMER DRY CLIMATES AND WILDFIRE

Climate can play a major role in setting the stage for wildfire. Normal seasonal climate patterns that promote vegetation growth during the cooler wet season and then dry out that vegetation during a lengthy dry summer help establish the fuels for a wildfire. Across the planet, five regions of "Mediterranean climate," with

Ground-based view of a large pyrocumulus cloud produced by the Elk Complex fire in Idaho on August 10, 2013. (NASA Earth Observatory, https://earthobservatory.nasa.gov/images/81841/lightning-fires-in-central-idaho)

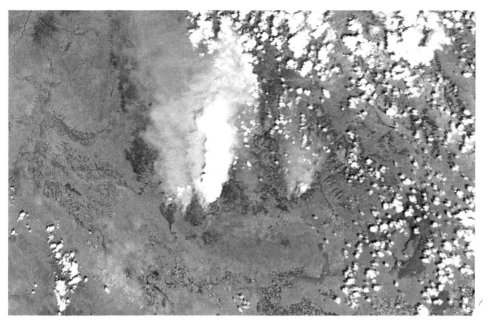

MODIS Aqua satellite-based view of a large pyrocumulus cloud produced by the Elk Complex fire in Idaho on August 10, 2013. (NASA Earth Observatory, https://earthobservatory.nasa.gov/images/81841/lightning-fires-in-central-idaho)

moist winters and warm, dry summers, exist in subtropical latitudes on the western side of continents. The countries with this climate type include South Africa, Portugal and Spain, Chile, Australia, and the United States (especially in the State of California). Unfortunately, ongoing climate change is likely to make these areas even more susceptible to wildfire as summer temperatures increase.

The Mediterranean climate type actually sets up the environment for two differing hazards, with one following the other. The vegetation is generally a shrubland or dry forest, moist in the rainy season, and dry and flammable in the summer season. In the Mediterranean climate region of California, wildfires can remove most of the vegetation during dry months, with a loss of the surface soil stabilization that plants provide on slopes. The loss of plant cover leads to water moving quickly downhill when the rains return, creating mudflows and landslides. In the actual Mediterranean region and adjoining areas of southern Europe that can see seasonally dry conditions, wildfires can be an important hazard. Greece had deadly fires in 2007, killing 77 (IAFRS 2018). Portugal has had several severe wildfire events in recent years, including 2017, 2018, and 2019. The June 2017 Pedrógão wildfire killed 64 and injured at least 135 (Valente et al. 2017), with later estimates rising to over 200 injured. The five-day burn occurred during a heat wave. The early August 2018 Algarve wildfires likewise occurred with a severe heatwave, with record-breaking temperatures and drought across most of Europe. More than 1,100 firefighters were mobilized to battle this complex. Other fires burned in Spain during the same period. Chile has likewise seen severe wildfires in recent years. In 2017, fires that started in January (summer) burned for weeks, killing several people, destroying homes, and covering a very large area.

The fire season began early for Portugal in 2019. By the end of March, hundreds of wildfires had already ignited with the early hot and dry conditions that would not normally cause extensive fires until at least May.

EXPOSURE, SENSITIVITY, AND ADAPTIVE CAPACITY

Humans are very vulnerable to the wildfire hazard, where vulnerability includes aspects of exposure, sensitivity, and adaptive capacity. The exposure component addresses the frequency and intensity of fires. Unfortunately, human actions are causing more ignitions and increasing exposure. Climate change has been implicated in increasing exposure by changing moisture and temperature conditions that contribute to fire spread. Recently, California has seen a shift from past climatic conditions, with a moist winter and dry summer, to having dry enough conditions for even winter wildfires. Climate change also promotes more widespread insect outbreaks in forests, contributing to more tree deaths and an increased availability of dry fuel.

Exposure also is increased by where human populations are spreading and the types of environments where homes are being built. The growth of cities and their suburbs into surrounding forest or other areas of natural fuels can put humans and their homes in danger in a zone referred to as the wildland urban interface, or WUI (Radeloff et al. 2018). Wildland-urban interface areas are especially problematic for wildfire occurrence in subhumid areas and in the Mediterranean climate type. The October 1991 Oakland firestorm is a tragic example of a wildfire impacting the WUI. What started as a small grassfire that firefighters thought they had under control quickly raged out of control the next day, when Diablo winds gusting to over 60 mph pushed the fire into areas to the south and west. The environmental conditions created a firestorm, where the fire helped create its own local weather. Burning embers were carried by turbulent motion in the wind to places ahead of the main fire area, creating new ignitions. Losses of 25 lives and more than 3,000 houses, apartments, and condominiums prompted a major research effort and the development of new communication tools to help educate people and hopefully to reduce the wildfire hazard for those living in the wildland urban interface.

Sensitivity identifies the losses associated with a fire, including deaths, injuries, and economic losses. Health problems and reduced visibility related to smoke from a fire impacting areas downwind, or even regionally, are included in the sensitivity category. An increase in "smoke waves," or the amount and length of time that people have to deal with smoke and poor air quality from wildfires, is an important health issue (Schoennagel et al. 2017).

Adaptive capacity involves the ability of people to take actions that can reduce, or mitigate, the wildfire hazard. These actions include weather forecasting to identify conditions that are problematic, such as red flag days. Red flag warnings are issued in the United States by National Weather Service (NWS) offices when very low humidity levels, warm temperatures, and windy conditions combine to make the rapid spread of a wildfire more likely. Other adaptive actions include human efforts to put out fires, building and having individuals occupy fire lookout towers, creating firebreaks to help contain an existing fire, and having controlled burns to reduce the fuel load and try to prevent out-of-control crown fires. Scientific investigations to better understand all aspects of wildfires, including modeling the character of the spread of individual events, also adds to adaptive capacity.

HUMAN RESPONSE TO WILDFIRE

Humans have a long history of using fire for beneficial purposes. Observations of the impacts of fire on the environment led early humans to use fire in food preparation, in hunting, or in encouraging the growth of plant species that would attract game (Roos et al. 2018). In modern land use, farmers or ranchers use fire to assist with resource management. For example, the ranchers of the Flint Hills in Kansas burn their grasslands in early spring to help with nutrient cycling and to speed the green-up of the forage grasses for their cattle herds.

In the early twentieth century in the United States and as a response to human carelessness with fire, there was an effort to extinguish all wildfires, including the beneficial ground fires that helped clear out fuels on the forest floor. This policy of fire suppression was implemented in response to the damage caused by uncontrolled wildfires in the recent past and because fire was thought to be a threat to the valuable timber resource. Historic fires from the late 1800s and early 1900s that had an impact on fire suppression policy include the 1871 Peshtigo Fire in Wisconsin and Michigan and the Great Fire or "Big Blowup" of 1910 in Idaho and Montana. The Peshtigo fire was the worst in U.S. history, burning an area of about 10 miles (15 km) wide by 40 miles (60 km) long (over 1.5 million acres or 0.61 million ha) [Rosenwald 2017]) and destroying entire communities on the night of October 8. Death toll estimates are over 1,200, and perhaps as many as 2,500, people. Strong winds related to an advancing cold front and dry conditions set the stage; small fires that had been set for clearing and removing woody debris leapt out of control, merging into the huge fire complex.

The Big Blowup of August 1910 involved about 3 million acres (1.2 million ha), destroyed several towns, and killed nearly 90 people. Dry conditions, exceptionally high temperatures, numerous small fires (often started by cinders from trains), and hurricane-force winds came together to create a monstrous fire. Although not the deadliest wildfire in the United States, the Big Blowup covered the largest area on record and influenced the developing U.S. Forest Service firefighting and fire suppression efforts.

Fire suppression has not been a good idea in those forested ecosystems where ground fires were a natural component of the healthy ecosystem. Suppression resulted in building up fuels, greatly increasing the wildfire hazard—with more forest litter and thus more fuel, crown fires became more likely. Efforts are now underway to reverse the situation. In actively managed forests, thinning out the area by removing the underbrush and/or the use of controlled burns can help restore more natural plant communities and ecosystems. In some wildland forested areas, like relatively inaccessible wilderness areas, land management agencies often follow a "let burn" policy unless conditions indicate that a fire may grow to a disastrous size or threaten people and their property.

As various locales face increasing problems of drought and fuel accumulation, as well as the increased exposure related to more population growth and building in the wildland-urban interface, there are concerns and pressures to continue fire suppression efforts and to avoid controlled burns (which have rarely blown up into uncontrolled wildfires). When a wildfire is advancing toward a populated area, evacuation orders may be given. It is important for citizens to receive the orders to leave an area, but contacting people who do not have landline telephone service is an increasing problem. An additional issue with an evacuation is that many citizens wait till what is seemingly the last minute and traffic flow is impacted with congestion.

CONCLUSION

Wildfire can be a devastating and deadly hazard. Unfortunately, human actions have increased the hazard over time through increases in the number of ignitions, changes to the natural environment, and unwise forest management activities. In the United States and other countries (such as Australia, South Africa, Chile, and Portugal), the number of fires, the amount of area burned on an annual basis, and the human and economic cost of wildfires have been increasing in recent decades. It is thought that the impacts of global climate change will continue to increase the risk of wildfire as temperatures warm and dry climate regions become even drier (Schoennagel et al. 2017).

John A. Harrington Jr. and Lisa M. B. Harrington

Further Reading

Balch, J. K., B. A. Bradley, J. T. Abatzogloue, R. C. Nagya, E. J. Fusco, and Adam L. Mahood. 2017. Human-started Wildfires Expand the Fire Niche across the United States. *Proceedings of the National Academy of Sciences (PNAS)* 114(11): 2946–2951.

Bowman, D.M.J.S., J. Balch, P. Artaxo, W. J. Bond, M. A. Cochrane, C. M. D'Antonio, R. DeFries, F. H. Johnston, J. E. Keeley, M. A. Krawchuk, C. A. Kull, M. Mack, M. A. Moritz, S. Pyne, C. I. Roos, A. C. Scott, N. S. Sodhi, and T. W. Swetnam. 2011. The Human Dimension of Fire Regimes on Earth. *Journal of Biogeography* 38: 2223–2236.

Finneran, M. 2010. Fire-Breathing Storm Systems. https://www.nasa.gov/topics/earth/features/pyrocb.html, last updated October 2010, accessed May 27, 2019.

Higuera, P. E. 2015. Taking Time to Consider the Causes and Consequences of Large Wildfires. *Proceedings of the National Academy of Sciences (PNAS)* 112(43): 13137–13138.

Horizon: The EU Research & Innovation Magazine. 2019. "It Eats Everything"—The New Breed of Wildfire That's Impossible to Predict. *Physical Organization.* February 21, 2019. https://phys.org/news/2019-02-everythingthe-wildfire-impossible.html, accessed May 25, 2019.

IAFRS (International Association of Fire and Rescue Services) (CTIF; Comité Technique International de Prevention et d'extinction de Feu). 2018. An Overview of the Deadliest Wildfires in Europe. https://www.ctif.org/news/overview-deadliest-wildfires-europe, accessed May 25, 2019.

NOAA (National Oceanic and Atmospheric Administration), and National Centers for Environmental Information (NCEI). 2019. U.S. Billion-Dollar Weather and Climate Disasters. https://www.ncdc.noaa.gov/billions/, accessed March 28, 2019.

Radeloff, V. C., D. P. Helmers, H. Anu Kramer, M. H. Mockrin, P. M. Alexandrea, A. Bar-Massada, V. Butsic, T. J. Hawbaker, S. Martinuzzi, A. D. Syphard, and S. I. Stewart. 2018. Rapid Growth of the US Wildland-urban Interface Raises Wildfire Risk. *Proceedings of the National Academy of Sciences (PNAS)* 115(13): 3314–3319.

Roos, C. I., M. N. Zedeño, K. L. Hollenback, and M.M.H. Erlick. 2018. Indigenous Impacts on North American Great Plains Fire Regimes of the Past Millennium. *PNAS* 115(32): 8143–8148.

Rosenwald, M. S. 2017. "The Night America Burned": The Deadliest—and Most Overlooked—Fire in U.S. History. *The Washington Post*. December 6, 2017. https://www.washingtonpost.com/news/retropolis/wp/2017/10/10/the-night-america-burned-the-deadliest-and-most-overlooked-fire-in-u-s-history/, accessed July 4, 2020.

Schoennagel, T., J. K. Balch, H. Brenkert-Smith, P. E. Dennison, B. J. Harvey, M. A. Krawchuk, N. Mietkiewicz, P. Morgan, M. A. Moritz, R. Rasker, M. G. Turner, and C. Whitlock. 2017. Adapt to More Wildfire in Western North American Forests as Climate Changes. *PNAS* 114(18): 4582–4590.

Valente, L., C. B. Ribeiro, H. Torres, A. Campos, D. Dinis, L. Alvarez, A. M. Photos, C. Carvalho Silva, V. Ferreira, A. Sanches, P. Guerreiro, L. Botelho, D. Queiroz de Andrade, H. D. Sousa, C. Lamelas Moura, S. Rodrigues, L. Borges, and F. D. Real. 2017. To the Minute: To the Minute: "We Can Be Watching a Favorable Evolution," Says Marcelo. 64 Confirmed Dead. [Translated from Portuguese.] *P.* June 19, 2017. https://www.publico.pt/2017/06/18/sociedade/noticia/ao-minuto-o-que-se-passou-para-morrerem-19-pessoas-num-incendio-1776047#16879, accessed May 25, 2019.

Wildfire Today. https://wildfiretoday.com/, accessed March 29, 2019.

Aid Organizations

ActionAid International

ActionAid is an international development and humanitarian aid organization working with about 15 million people in 47 countries with over 400 program areas. The organization was started in Johannesburg, South Africa, where its headquarters is located. It has international hubs in Rio de Janeiro, Bangkok, Brussels, London, and Nairobi. The organization envisions a world that is impartial and sustainable, where all its citizens have a right to a life of dignity and are free from poverty and oppression. Through its works, ActionAid International wants to achieve social, economic, environmental justice, and gender equality and to eradicate poverty. It also wants to accomplish its works by involving people.

Typically, humanitarian aid organizations save and protect millions of people across the world who are impacted by crisis. But humanitarian actions are not always fast, and needs are not generally evenly met. According to a recent report by United Nation's Office for the Coordination of Humanitarian Affairs (OCHA), even if a record sum of money is raised, increasing vulnerability to natural and man-made disasters worldwide is increasing the gap between need and response. ActionAid believes that communities should be empowered to be self-sufficient to be more resilient. This means that rather than using the conventional global core and periphery division to facilitate relief aid, communities will work alongside agencies in a spirit of equality, democracy, and accountability, which is more efficient for community empowerment.

ActionAid is a people's organization, meaning it believes the people it serves deserve to run the organization. Its goal is to use community expertise, knowledge, resources, and power. In other words, the organization believes that community empowerment comes from within community and aims to work with the underprivileged and those discriminated against to fight poverty and injustice. ActionAid works with the people who live in poverty and/or have been excluded from societal structures. Unlike other relief aid organizations, ActionAid has its headquarters in the developing world, where it works the most, having offices across Asia, America, and Europe.

AREAS OF WORK

ActionAid International focuses on four areas: climate change and land, emergencies, women's rights, and politics and economics. First, climate change is the most persistent threat to global stability, and its associated risks are spatially uncertain and uneven. However, human-induced climate change is happening, and all societies need to learn to cope with the changes predicted such as warmer temperatures, drier soils, and changes in weather extremes (Adger et.al 2003). The uncertainty arises based on where people live and how they generate their economic support and whether that is influenced by ambient climate. As climate is spatially variable, societies have developed different coping mechanisms to suit their needs. All societies adapt to climate change and other risks, but there is a difference between how the developed world and the developing world cope with these challenges. Some of the coping mechanisms or strategies are technologically dependent, better resourced, and more robust in developed countries than in developing countries. Frequently in the developing world, some sectors are more sensitive, and some groups are more vulnerable to risks posed by climate change. ActionAid has recognized that the burden of climate change impacts unfairly falls on people whose actions have least contributed to the problem: poor and marginalized individuals.

In developing countries, socioeconomic challenges can aggravate climate change associated risks, particularly as these societies are dependent on natural resources, which are sensitive to climate variability. Moreover, people living below the poverty line are more susceptible to effects of climate change, meaning that this population is vulnerable not only to extremes of normal climatic variability but also to the increase in frequency and magnitude of extreme weather events and disasters from climate change. Therefore, under its climate change and land theme, ActionAid works for climate justice, climate-induced migration, food security, and natural resource management and land rights.

First, climate justice ensures that people who suffer the impacts of climate change receive support and compensation. In this sector, the ActionAid team encourages adaptation and building of resilient communities by strengthening national plans and processes. Additionally, the organization protects and financially supports people facing climate change-induced losses and migration, advocating for climate action and adaptation in poorer countries by coercing wealthy nations to support initiatives such as the Green Climate Fund. Also, it ensures that land-based solutions advocate for system change and local solutions to deal with climate change impacts. Moreover, ActionAid ensures that land-based solutions respect human rights and food security and promote solutions such as agro-ecology. For example, it works closely with coastal communities in Bangladesh to devise strategies for people to survive and mitigate for climate change and climate change-induced migration.

Second, climate change-induced migration impacts 1,300 people daily in Dhaka, Bangladesh's capital city. The city is already among the most densely populated places on the earth at 18,301 sq. miles (47,400 people per sq. km), and with 500,000 new arrivals each year. Moreover, Bangladesh receives about 40 percent of the impact of total storm surges in the world (Dasgupta et al. 2010). In particular, ActionAid works closely in cyclone-prone and disaster-impacted Lalua union (a union is the lowest administrative unit) in this country. Devastation from cyclones and flooding are challenges this area faces frequently, and since 2012, the union has lost 40 percent of its population. Accordingly, the resilience and climate justice lead for ActionAid Bangladesh said that ActionAid works with women in the community and trains them to be prepared for disasters. For instance, when warning signs of a major disaster are given by authorities, women are trained to pack emergency bags with documents, food, and drinking water and move to a higher elevation to protect their children. Additionally, communities have raised their home foundations and animal pens to prevent flooding.

Third, food security and therefore farming are the next area under climate change and land where ActionAid International concentrates its work. Agriculture in developed countries is largely industrial, but while a large proportion of farmers in the developing world still practice subsistence agriculture, some are leaning toward more industrial agriculture for economic benefits and financial stability. ActionAid believes that the industrial agriculture model is environmentally damaging with its heavy reliance on chemical usage, improved seeds, and selling of produce to corporate supply chains. In this process, the efforts of millions of women food producers, farmers, indigenous communities, and other food producers are not recognized as most of industrial agriculture is driven by governments and transnational companies. Yet women in rural areas play an important role in environment and natural resource management with their active involvement in food production and related activities. Nevertheless, it is widely understood that women in developing countries constitute one of the poorest and most disadvantaged groups in society (Denton 2002).

Therefore, ActionAid and its global partners support sustainable agriculture. Their aim is to help farmers with small holdings improve their resilience against climate change impacts, improve food security, and better the livelihoods of under-represented demographics, such as women, indigenous youth, other young people, and others; to challenge industrial and corporate agriculture by working with social movements; and finally to improve women's access to markets. To achieve their aims, they challenge industrial and corporate agriculture solutions and policies. They suggest agro-ecology as an alternative to industrial agriculture, as this approach enables food to remain where it is most needed. Agro-ecology is defined as "the science of sustainable agroecosystems . . . ; it is a set of farming practices, and a social movement" (Wijeratna 2018, 6). Additionally, ActionAid uses evidence-driven advocacy to create support access to national and international markets for women.

For example, currently, over 500 million family farms produce 80 percent of the world's food. These include smallholding farmers, pastoralists, fisherfolk, and indigenous communities, most of whom are women. ActionAid argues that peasant-based agro-ecological systems are superior to industrial farming methods as they are more productive, highly sustainable, resilient to climate change, and offer benefits beyond economics. In fact, the agro-ecological approach brings together social, biological, and agricultural sciences that integrate traditional, indigenous, and local farmer knowledge and culture. ActionAid recommends scaling out agro-ecology so that multiple Sustainable Development Goals (SDGs) can be achieved.

The last area in ActionAid's climate change and land protocol is natural resources and land rights. With advancement in technology, the demand for natural resources is expanding. Across the globe, we are witnessing unprecedented and violent scrambles for natural resources. However, like the climate change impact burden falling on people who least contribute to it, extraction-based ecological crises affect the most who are least responsible for causing it. To satisfy this increasing need to extract resources, many governments tend to focus on unsustainable resource exploitation to fuel growth. Such commodity-dependent communities or nations often face unstable economies and unsustainable growth. At the same time, often community livelihoods directly depend on such ecosystems. Through their program on natural resources and land management, ActionAid tries to achieve a balance between biodiversity conservation, sustainable use, and equitable sharing of the benefits arising from natural resources.

For example, ActionAid's research is focused on understanding mining supply chains and on how to support communities where this extraction industry is active. It advocates for community in a way that communities can support the claim to their natural resource rights, and it lobbies for people-centric approaches. In case of any human rights violations, they support communities by holding corporations accountable for their actions. In particular, through regional and international forums and initiatives, ActionAid supports women's land right claims.

For instance, ActionAid Zambia works closely with communities in the Chingola area through its partner, the Catholic Diocese of Ndola. Chingola is a city in Zambia's Copperbelt Province, the country's copper-mining region. The ActionAid Zambia chapter provides a platform and resources to support the villagers' claims to their rights to the resources on their land and advocates for regulation of actions of corporate companies. For instance, they fought for 13 years for compensation for damage, remediation, and an end to continual pollution that has impacted the life of the community members caused by a mining giant from the United Kingdom. After more than a decade of fighting for compensation, villagers in Chingola won the case.

CONCLUSION

To efficiently implement its aims and objectives, ActionAid International uses a federal model of governance for the organization. It means that the administration of the organization comprises several self-governing affiliates and associates unified by an international "federal" structure. Its leadership is embedded in various stages and levels within the organizational structure. Shared leadership is an important component for ActionAid in its quest for good governance. In 2015, ActionAid reorganized its administrative structure to ensure country members play a stronger leadership role, where member countries lead in making management decisions and setting the priorities for the organization. The Chief Executive is the top administrator of the organization and is appointed by a General Assembly and supported by a team of International Directors. Meanwhile, ActionAid International Affiliate Members and Associate Members have National Boards and some have General Assemblies. Ultimately, the administrative structure may vary in each country with respect to its size.

Finally, ActionAid is now a part of a new campaign that calls for effective law to require companies and investors to act to prevent human rights abuses, worker exploitation, and environmental harm in their global operations, activities, products, services, investments, and supply chains.

Avantika Ramekar

Further Reading
Adger, N., H. Saleemul, B. Katrina, D. Conway, and M. Hulme. 2003. Adaptation to Climate Change in the Developing World. *Progress in Development Studies* 3(3): 179–195.
Dasgupta, S., M. Huq, Z. H. Khan, M.M.Z. Ahmed, N. Mukherjee, M. F. Khan, and K. Pandey. 2010. *Vulnerability of Bangladesh to Cyclones in a Changing Climate Potential Damages and Adaptation Cost.* Washington, DC: The World Bank.
Denton, F. 2002. Climate Change Vulnerability and Adaptation Assessments. *Sustainable Gender and Development* 10(2): 10–20.
Lunda, J. 2019. Villagers in Zambia Win Right to Take Legal Action against Mining Giant Vedanta in the UK. https://actionaid.org/opinions/2019/villagers-zambia-win-right-take-legal-action-against-mining-giant-vedanta-uk, accessed June 10, 2019.
Wijeratna, A. 2018. Agroecology: Scaling-up, Scaling-out: More Information on ActionAid International. https://actionaid.org/land-and-climate/resources-and-land-rights, accessed June 11, 2019.

American Red Cross

Humanitarian Clara Barton and her acquaintances in Washington, DC, founded the American Cross on May 21, 1881. The organization received its first congressional charter in 1900 and a second modified charter in 1905. It is the only congressionally mandated organization to provide services to members of the American armed forces and their families and offer disaster relief and mitigation in the United States and across the globe through other Red Cross networks. The Red Cross is responsible for fulfilling the mandates of the Geneva Convention within the United States, and it joined the International Federation of Red Cross and Red Crescent Societies (IFRC) in 1919. The organization developed the first nationwide civilian blood donation program in 1948 and now supplies more than 40 percent of the blood and blood products in the United States.

The American Red Cross has historically been known under other names, American Association of the Red Cross (1881–1892) and American National Red Cross (1893–1978). Although the Red Cross has been given mandate by Congress, it is not a consistently federally funded organization. Largely, it is a nonprofit, charitable organization that depends on generous contributions of time, blood, and money from the American public to carry out its lifesaving services and programs. However, it receives federal funding on an irregular basis.

The primary components of the American Red Cross consist of five services: (1) Biomedical Services, (2) Health and Safety Services, (3) Disaster Services, (4) Armed Forces Emergency Services, and (5) International Services. Biomedical Services provide collection, testing, and distribution of the nation's blood supply. Health and Safety Services helps save lives and strengthen communities through training of volunteers in CPR, first aid, and water safety. Disaster Services focus on meeting immediate needs of survivors of extreme natural events, including health and mental health services. Armed Forces Emergency Services (AFES) include among other services connecting with family members overseas in the military and have chapters throughout the United States and abroad.

ORGANIZATIONAL STRUCTURE

The governing body of the American Red Cross is a 50-member Board of Governors. The members have all powers of governing and directing and of overseeing the management of the business and affairs of the organization. The president of the United States is the honorary chairperson of the Board of Governors and also appoints eight of the board members. The remaining 42 members are elected at the annual national convention. The board elects 12 of them and delegates elect the remaining 30 from various Red Cross chapters whose members attend the convention. The organization is headed by a president, who is elected by the members of the Board of Governors. The members are volunteers, while the president draws a salary from the organization. The rest of the Red Cross's roster is made up of approximately 1.2 million volunteers and about 40,000 paid employees, mostly nurses. The American Red Cross is headquartered in Washington, DC.

According to the organizations website, currently, the American Red Cross has over 800 regional or city-based chapters throughout the United States and its territories, which are organized into seven divisions and 60 blood service regions. Each chapter is officially chartered by the national Board of Governors and headed by a director who has some degree of autonomy in selecting programs and services that are most vital in that area. The American Red Cross volunteers respond to over 70,000 disasters and train nearly 12 million people annually.

CLARA BARTON AND THE AMERICAN RED CROSS

Barton was born in 1821, and during the American Civil War (1861–1865), she worked with the sick and wounded soldiers, earning the nickname "Angel of the Battlefield" for her dedication to the frontline soldiers. After the Civil War, Barton tirelessly worked to establish an American version of the International Red Cross (which had been founded in Switzerland in 1863) as well as for the United States to sign the Geneva Convention, which the country ratified one year after establishing the American Red Cross. She became the first president of the Red Cross and led the organization for the next 23 years, dying in 1912 at the age of 91 (Rosenberg 2018).

Barton went to Europe in 1870 as a volunteer for the International Red Cross and returned to the United States in 1873. Four years later, she organized an American branch of the International Red Cross. Moreover, just days after the first local chapter of the American Red Cross was established in Dansville, New York, on August 22, 1881, the organization participated in its first disaster relief operation when it responded to the devastation caused by major forest fires in Michigan. The fire killed 125 people and left thousands homeless. The organization's role grew during the 1889 Johnstown flood in Pennsylvania, when the American Red Cross set up large shelters to temporarily house those displaced by the disaster, which killed over 2,000 people. The flood was the first great test for the newly organized American Red Cross. Barton herself arrived in Johnstown with 50 doctors and nurses five days after the disaster to lead the relief efforts, and she stayed for five months. The American Red Cross also distributed new and used emergency supplies valued at $211,000, and some 25,000 people were helped (McCullough 1968).

When Barton arrived in Johnstown, she was 67. Before she came to Johnstown, she had already been in Ohio during the floods of 1884 and in Texas with emergency aid during the famine of 1887. She participated in distributing emergency relief items among the tornado victims in Illinois in 1888, and later that year, she went to Florida during a yellow-fever epidemic (McCullough 1968). In 1898, during the Spanish-American War, Barton provided food and medical supplies for wounded American soldiers and civilians. In 1907, the American Red Cross began working to combat consumption (tuberculosis) by selling Christmas Seals to raise money for the National Tuberculosis Association. A year before, it had participated in distributing emergency relief items for the victims of the San Francisco earthquake in 1906.

RELIEF AND OTHER ACTIVITIES

The American Red Cross experienced phenomenal growth during the First World War (1914–1918). Its chapters, volunteers, and funds exponentially expanded, and the organization sent "thousands of nurses overseas, helped organize the home front, established veterans hospitals, delivered care packages, organized ambulances, and even trained dogs to search for [the] wounded" (Rosenberg 2018). Its nurses also participated in combating the worldwide influenza epidemic of 1918. After the war, the Red Cross focused on serving veterans and enhanced its programs in safety training; classes on first aid, nursing, accident prevention, and home care for the sick; and nutrition education. It provided emergency relief for survivors of the Mississippi River floods in 1927 and severe drought and the Depression during the 1930s.

In the Second World War (1939–1945), the American Red Cross provided extensive services to the U.S. military, Allies, and civilian war victims. According to an official website of the Red Cross, the organization enrolled more than 104,000 nurses for military service and prepared 27 million packages for American and Allied prisoners of war. At the request of the military, it also initiated a national blood program that collected 13.3 million pints of blood for use by the armed forces (Rosenberg 2018). In 1990, the Red Cross added a Holocaust & War Victims Tracing and Information Center. During this decade, it initiated a massive modernization of its blood services operations to improve the safety of blood products.

The American Red Cross provided services to members of the armed forces and their families, including during the Korean (1950–1953), Vietnam (1955–1975), and Gulf Wars (1990–1991, 2003–2011). The organization also expanded its services into such fields as civil defense, cardiopulmonary resuscitation (CPR) and automated-external defibrillation (AED) training, HIV/AIDS education, and the provision of emotional care and support in the wake of disasters. Since 2006, the Federal Emergency Management Authority (FEMA) and American Red Cross have started to work together following natural disasters in the United States in helping coordinate relief efforts and providing emergency assistance to victims of extreme events. It is a member of National Voluntary Organizations Active in Disaster (VOAD) and works closely with other agencies like the Salvation Army and the Amateur Radio Emergency Service with each of which it has a memorandum of understanding.

The American Red Cross participated in several recent high-profile disasters, such as the September 11, 2001, terrorist attacks in New York City and Washington, DC, Hurricane Katrina in New Orleans in 2005, the 2010 Haiti earthquake, and Hurricane Sandy in 2012. Within a month of Katrina, the organization had raised almost half a billion dollars. It deployed 74,000 volunteers in the first two weeks after the storm, and the volunteers provided shelter to 16,000 evacuees in its 700 temporary shelters and distributed more than 7.5 million hot meals (Paul 2019). It also responded to other two hurricanes (Wilma and Rita) that occurred after Katrina in the same year. For these disasters, the American Red Cross distributed 346,980 comfort kits (which contained toothpaste, soap, washcloths, and toys for children) and 205,360 cleanup kits (e.g., brooms, mops, and bleach).

The American Red Cross collected donations for the survivors of the Haiti earthquake and distributed food, water, medical care, emergency shelter, cash grants, and other essentials to people in need. Less than 10 months after the earthquake, a severe cholera outbreak occurred in the earthquake-affected areas in Haiti, whereupon the American Red Cross helped save lives again by distributing soap and water purification tablets, educating people about how to prevent and treat cholera, and providing 70 percent of the funds needed for the country's first cholera vaccine. The Red Cross served 17 million meals to the 2012 Hurricane Sandy survivors and provided millions of supplies and housed thousands of people in its shelters. More recently, it supported short- and long-term recovery efforts following Hurricanes Harvey and Irma in 2017, as well as California wildfires.

However, according to national and international print media, the distribution of donations by the American Red Cross for several large natural and man-made disasters has been widely criticized by public officials, disaster survivors, and even its own volunteers. After the 9/11 terrorist attacks, the American Red Cross raised $530 million in donations, but the organization decided to put over $200 million of that money into accounts for long-term goals or administrative costs. In fact, only $40 million of the $530 million collected has been distributed to families of victims, yet donors thought the money donated was going straight to the affected families (AP 2005).

In the context of the 2005 Hurricane Katrina, the *Associated Press* reported that the Red Cross operations were chaotic in some places, inequitable in others. Additionally, it raised about $1.1 billion—its record so far for a single disaster—but the organization was criticized when it earmarked $200 million for future crises rather than to help victims of Katrina. Red Cross President Bernadine Healy resigned, the money was shifted back to the Liberty Fund, and the organization promised greater accountability in future fundraising campaigns (AP 2005). The poor management of the Red Cross during 9/11 and Katrina was once again repeated in the 2010 Haiti earthquake, which killed 222,750 people. National Public Radio (NPR) and ProPublica investigated the donations appropriations and found a string of poorly managed projects, questionable spending, and dubious claims of success. The Red Cross says it has provided homes to more than 130,000 people, but the number of documented permanent homes the charity has built is six (Elliot and Sullivan 2015).

The American Red Cross was also criticized by for its blood donation program. For example, the advocacy organizations representing lesbian, gay, bisexual, and transgender (LGBT) opposed its policy of not receiving blood donations from gay men. In fact, this restriction is imposed by the Federal Drug and Administration (FDA) for all blood collection companies and organizations in the United States. The FDA has instructed collection agencies to defer blood collection from donors for one year from the most recent such sexual contact. As a result, Red Cross was legally unable to collect blood from gay men.

CONCLUSION

Despite criticism, the American Red Cross has continued to be an important humanitarian organization, offering aid to millions affected by wars and disasters and supporting military service personnel through its Armed Forces Emergency Services program. It provides financial assistance to service members or their families and offers a wide range of resources. The American Red Cross enjoys a long tradition of humanitarian service and remains the nation's premier emergency response organization.

Bimal Kanti Paul

Further Reading

AP (Associated Press). 2005. Despite Huge Katrina Relief, Red Cross Criticized. *NBC News*. September 28, 2005. http://www.nbcnews.com/id/9518677/ns/us_news

-katrina_the_long_road_back/t/despite-huge-katrina-relief-red-cross-criticized/#.XNNI9xRKhph, accessed May 30, 2018.

Elliot, J., and L. Sullivan. 2015. How the Red Cross Raised Half a Billion Dollars for Haiti and Built Six Homes. *ProPublica*. June 3, 2015. https://www.propublica.org/article/how-the-red-cross-raised-half-a-billion-dollars-for-haiti-and-built-6-homes, accessed June 21, 2018.

McCullough, D. G. 1968. *The Johnstown Flood*. New York: Simon and Schuster.

Paul, B. K. 2019. *Disaster Relief Aid + Changes & Challenges*. Gewerbestrasse, Switzerland: Palgrave Macmillan.

Rosenberg, J. 2018. American Red Cross. ThoughtCo. https://www.thoughtco.com/american-red-cross-1779784, last updated July 30, 2018, accessed May 30, 2018.

Catholic Relief Services (CRS)

Catholic Relief Services (CRS), an international humanitarian agency of the Catholic community in the United States, aims at alleviating suffering of the poorest of the poor in more than 100 countries worldwide. With more than 75 years of experience, it also provides assistance to people in need in more than 100 countries, without regard to race, religion, or nationality. Although CRS is a Catholic charity, it employs both Catholics and non-Catholics. Initially, CRS was founded to help and stand beside the people wounded during World War II (1939–1945). It also provided aid to the refugees of Europe displaced by the war. After the war, it expanded its vision to promote human development by responding to major emergencies, fighting disease, alleviating poverty, and thus nurturing a peaceful and just society. In 2017, CRS worldwide helped more than 136 million poor and vulnerable people. It is ranked as one of the most efficient humanitarian organizations in the world. In that one year, it sent 94 percent of its donations directly to the programs that benefited poor men, women, and children overseas. CRS meets all 20 of the strict Charity Standards set by the Better Business Bureau's Wise Giving Alliance and has an "A+" rating from CharityWatch (CRS 2018a).

ORIGIN AND BACKGROUND

Founded in 1943 by Catholic Bishops of USA, Catholic Relief Services is a humanitarian organization funded by donations of the U.S. government and various other sources, and it has worked tirelessly to provide help to communities in need, especially poor and disadvantaged people outside of the United States. Currently, all the activities of the organization are motivated by the fundamentals of the Gospel of Jesus Christ since it contains methods for mitigating human misery, developing mankind, and nurturing charity along with social justice throughout the world (CRS 2018b). CRS directly helps poor people and assists them in staying on course regarding personal development and realization of potentiality.

Inside the United States, CRS helps people learn to develop their moral accountability toward siblings of the world by helping them in times of need, working toward alleviation of poverty in general, and in promoting justice. CRS has a lot of programs with very specific goals for targeted communities. It helps people to stand on their feet by enabling them to become self-reliant and self-sufficient. Among its programs, some worth mentioning are its environment program, agricultural program, health and education programs, emergency humanitarian response, AIDS program, and safety, microfinance, water, community banking, and peace-building programs.

In the environmental and agricultural programs, CRS provides support with tools and seed for harvesting necessary crops (CRS 2018b). The health program targets child and adult health needs to improve public health situations, while the AIDS program provides support to people who are affected already and educates others how to take preventive measures. Meanwhile, access to help, equity among people, and quality education for everyone are the backbone of the education program. The organization provides assistance to almost 130 million people throughout the world in around 100 countries and on four continents. During its expansion in the 1990s, CRS was committed to working for development of civil society in Central America, Kosovo, and many other regions in order to address emergency situations whether man-made or natural (CRS 2018b).

The theory of change that CRS currently embraces is that of Integral Human Development known or IHD. This theory promotes the betterment of each person. The idea refers to the sustainable growth of everyone with the appropriate perspective to enjoy and embody everyone's cultural, natural, social, economic, political, physical, and spiritual wholeness. It also includes gifts of learning and dignified living through a rich civic life. CRS believes IHD is a long-term and highly dynamic process that correlates with human rights and dignity. It works using various key role players and the structure of society to transform the lives of people and closes gaps in the structure with active participation and engagement activities. With more than 75 years of experience globally, it understands the necessity of global solidarity and continuously working toward it. Its work draws totally from the foundation and principles of Catholic Social Teaching (CRS 2018b).

MISSION

Catholic Relief Services carries out the universal mission of the Bishops of the United States to assist the poor and vulnerable people of the world on the basis of need, not creed, race, or nationality. It works with local, national, and international Catholic institutions and structures to promote human development on the basis of Jesus Christ's Gospel that upholds the sacred and holy dignity of life, justice, and Catholic teaching. As noted, CRS partners with Catholic and other organizations to help domestic and overseas communities. The three principles that built the mission of CRS are faith, action, and results (CRS 2018c).

Faith is the foundation of CRS. The organization promotes faith in those people whom they serve by their activities and shared ability to create just societies and a peaceful world. However, CRS believes that a wish just to serve is not enough to uphold faith, human dignity, and development. The mission needs collaborative actions to facilitate real development of people's quality of life and creation of that desired peaceful society. Consequently, faith and resultant actions can be justified only by outcomes that are measurable and effective in alleviating human miseries, reducing causes of suffering, and empowering people to realize their full potential.

GUIDING PRINCIPLE

CRS follows the principles that are built on the rich tradition and culture of Catholic scripture and social teaching. It guides people to understand what a socially just world looks like through sharing values across religious, geographical, and cultural boundaries in search for common ground for those who seek a long-lasting, peaceful, and justice-based world (CRS 2018c).

Sacredness and Dignity of Life: Since created by God, every human life is celestial and sacred and worthy of dignity directly through creation but not by human action.

Rights and Responsibilities: Regardless of social or political identity, each and every person has basic rights and accountabilities inherent to the dignity of human life by virtue of being a human being. The rights can be innumerable but must include those that truly make a person human. This principle also addresses our accountabilities as human beings to respect others' rights and to work toward the common good for everyone.

The Common Good: Specific social fabric is needed to fully grow and develop as a human being, and this fabric draws from political, economic, material, and cultural conditions with the power to impact human dignity. This is commonly known as the Common Good.

Social Nature of Humanity: Through a natural process, we are social and live in a community. We cannot realize our full potentiality in isolation, but we can within the community. This principle refers to the building of families and societies that have direct effects on dignity as well as human potentiality.

Solidarity: Regardless of national, religious, ideological, economic, racial, and cultural differences, the human being is one member of the family on an interconnected planet. Therefore, solidarity with people and loving and caring for them is a global dimension.

Subsidiarity: Subsidiarity is the principle that a higher level of government or organization has no right to perform any action that can be handled successfully by the lower level community that is directly involved in the problem and knows it better.

Option for the Poor: While working for the betterment of societies through financial, political, and social decision-making processes, we should give great consideration to the people who are in need and vulnerable. To do so provides strength to the entire community and empowers the powerless, thereby reducing need and detriment for all society.

Stewardship: Careful stewardship of resources is vital to maintain integrity and equitable allocation and includes planning for future generations too.

PROGRAMS AND ACTIVITIES

CRS use policy analysis as one of its tools for humanitarian development to address the main causes of poverty, struggle, and marginalization. It analyzes issues that fuel the well-being of people all around the world. The main focuses of policy analysis are global hunger, preservation of and improvement in public health, timely emergency response, funding, and general assistance including addressing issues based on geographical location.

Through the agriculture program, CRS has served 6.1 million people in 51 countries. In 2017 alone, it operated 124 agriculture and livelihood projects. In particular, it provides agricultural support to farmers through the Pathway to Prosperity approach. The organization promotes saving and lending groups among the farmers, which help them to save and borrow within their group community. It also has a farmer's exchange program that helps locals to meet U.S. farmers in their country to share their expertise for better productivity (CRS 2018c).

CRS has an education program in 38 countries, and so far, 5 million people have been served through its program. In 2017, this program operated 76 education projects. Education resides at the heart of CRS mission, and so its projects contain a vast range of programs such as early childhood education (ECD); school feeding programs; and support programs for primary, secondary, and higher education in formal and nonformal settings (CRS 2018c).

In 2017, CRS ran 119 health projects, hosted by 42 countries and serving 93 million people. CRS operates in many areas where other organizations do not. Moreover, it tailors projects to meet the demand of the communities in which it works. Major health programs address issues such as malaria, nutrition, HIV, tuberculosis, the well-being of children, and the strengthening of the local health system (CRS 2018c).

In 2017, CRS's water and sanitation programs served 8 million people with 62 projects in 32 countries. It works for access to clean water and improved sanitation. The Water, Sanitation, and Hygiene (WASH) projects of CRS mostly focus on the poorest communities, and many of its projects are being operated in sub-Saharan Africa. These programs work in connection with priority areas of CRS such as agriculture, emergencies, and health (CRS 2018c).

CRS microfinance programs work with local and international organizations as well as with community-based groups that lend money. In this capacity, CRS focuses on poor households and encourages savings and investment for income generation by establishing microenterprise. In 2017 alone, it ran 89 projects in 41 countries and served 5 million people (CRS 2018c).

CRS believes in solidarity, social justice, and compassion through response in emergency situations and recovery that meets the needs of the most vulnerable people through a market-based approach, using shelter and settlement, food security, disaster resilience, and technology. It served 12.4 million people in 52 countries through 201 projects in 2017 (CRS 2018c).

CRS engages civil society along with private and public institutions to ensure equity, accountability, and inclusion. It works toward conflict resolution and peace building. In 2017, it invested $31.4 million in 76 projects that operated in 33 countries (CRS 2018c).

Among its other programs, the CRS has some other important programs. Its capacity strengthening and partnership program served 1.4 million people through 202 projects in 57 countries in 2017. It also worked in the leadership development of youth who are growing up around conflict and violence zones. In order to achieve sustainable development, CRS believes in empowerment of young people.

CONCLUSION

In addition to its numerous global projects, CRS has contributed funding after major natural disasters. In 2004, it allocated $190 million to 6,000 survivors of the Indian Ocean Tsunami as an emergency grant and for reconstruction efforts. It provided a grant worth $200 million to help the needy in the 2010 Haiti earthquake and $23.7 million to help those in the 2013 Typhoon Haiyan in the Philippines. CRS also supported earthquake victims in Nepal, where nearly 9,000 people lost their lives and more than 22,000 people were injured (CRS 2016). Moreover, since the Syrian Civil War began in 2011, CRS has been coordinating with its church partners in Lebanon, Jordan, and Egypt to provide urgent assistance, including necessary medical care, to nearly 1 million Syrian refugees.

Catholic Relief Services has been working with organizations around the world to help impoverished and vulnerable groups overcome emergencies and earn a decent living through agriculture and access to affordable health care. It provides development assistance and fosters charity and justice. It has been working to motivate people to live for a reason and run for a cause. By collaborating with each other, one day, CRS, along with other humanitarian agencies, could well establish a just and peaceful human society without poverty and injustice.

Md. Nadiruzzaman

Further Reading
CRS (Catholic Relief Services). 2016. Top 5 Humanitarian Stories to Watch in 2017. *Huff-Post*. December 21, 2016. https://www.huffpost.com/entry/top-5-humanitarian-stories-to-watch-in-2017_b_585a9a94e4b04d7df167cc16?guccounter=1, accessed September 27, 2018.
CRS (Catholic Relief Services). 2018a. *CRS Annual Report: Fiscal Year 2017*. Washington, DC: Library of Congress.
CRS (Catholic Relief Services). 2018b. About Catholic Relief Services. https://www.crs.org/about, accessed September 27, 2018.
CRS (Catholic Relief Services). 2018c. Our Work Overseas: Working in Solidarity with Our Global Neighbors to Achieve Lasting Results. https://www.crs.org/our-work-overseas, accessed September 27, 2018.

Concern Worldwide

Concern Worldwide, often referred to as Concern, is a nongovernmental, international humanitarian, and development organization focused on transforming the lives of the world's most vulnerable people. It specializes in emergency response, recovery of the most vulnerable after natural disasters, fighting malnutrition and deadly diseases, combating hunger, enhancing health, promoting climate resilience, and helping the extremely poor in their livelihoods to improve their standard of living. In essence, the organization is dedicated to the reduction of suffering and working toward the ultimate elimination of extreme poverty in the world's poorest countries. In accomplishing its mission, Concern's programs include both short-term disaster relief and long-term projects designed to promote a greater degree of self-sufficiency. Projects are typically small scale and grass roots in their orientation. Concern's visions are as follows:

- No one lives in poverty, fear, or oppression.
- All have access to a decent standard of living and the opportunities and choices essential to a long, healthy, and creative life.
- People are equal in rights, and everyone must be treated with dignity and respect.

ORIGIN

Irish couple John and Kay O'Loughlin-Kennedy founded Concern in 1968 in response to the famine in Biafra, a province of Nigeria, which was fighting for independence against the government forces of the country. This caused the displacement of millions of people, and the subsequent consequence of a blockade of food, medicine, and basic necessities by the Nigerian authorities created a severe famine in Biafra. At the height of the famine in the summer of 1968, nearly 6,000 children were dying every week because of lack of food. This famine reminded the Irish about their famine just a few generations earlier. Besides the emotional connection, some 700 Irish church missionaries had been working in Biafra before the famine. But their number was not sufficient to serve the people in need of food and other assistance. This necessity inspired John and Kay O'Loughlin Kennedy to form Concern. It is the largest aid and humanitarian agency in Ireland. The headquarters are located in Dublin.

After raising the equivalent of more than $6 million from Irish people, on September 6, 1968, Concern sent a ship with vital supplies of powdered food, medicine, and batteries from Ireland to Sao Tome, a Portuguese island off the coast of Biafra. African Concern volunteers distributed the supplies among the famine-stricken people of Biafra at night to avoid the Nigerian authorities and bypass the blockade. These volunteers provided necessary supplies over a period of 11 months and saved countless lives.

Later, in 1970, a deadly tropical cyclone hit the coast of East Pakistan, now Bangladesh, and the public was asked to respond to this natural disaster. In the following year, an independent war broke out in East Pakistan to secede from West Pakistan, now Pakistan. As a consequence of the war, more than a million people were displaced from their homes and became refugees in the neighboring Indian state of West Bengal. Thus, the organization Concern Africa became Concern Worldwide. Even later, in the 1990s, Concern Worldwide U.S. was founded with offices in New York and Chicago.

Over time, it was not just collection and distribution of emergency assistance to disaster survivors that was the focus, but also Concern volunteers—responding quickly with practical solutions—whose work provided incentive to take the organization in a new direction. Later, major countries established their own Concern organization. For example, Concern Worldwide U.S. is an affiliate of Concern Worldwide. It has offices in New York City and Chicago. However, Concern works in partnership with small community groups as well as governments and large global organizations.

FUNDRAISING

Fundraising is necessary to ensure that sufficient funds are available to conduct projects all over the world. Although Concern strives to diversify its income base, the primary sources of its funding are the governments of Ireland, the United Kingdom, the United States (along with Concern U.S.), and the European Union. Thousands of individuals from these countries and South Korea also contribute money to this organization. The funds are raised through a variety of efforts such as special public appeals during Christmastime, particularly to grandparents, who donate cash and other items to Concern's children programs. Funds are also sought from corporations, large banks, and companies within and beyond Ireland. Donations can be in any form: cash and gifts, such as hygiene kits and school supplies. In 2017, Concern raised a total of $217 million of which 57 percent came from government sources, and the remaining contribution came from largely individual contributions (Concern Worldwide 2018).

ACTIVITIES

Concern's effective responses in the early 1970s were not only confined to Bangladesh, but it also responded in a timely way to the famines in Ethiopia in 1973 and again in 1984. The organization also provided aid to Cambodian refugees who fled their country to take shelter in Thailand close to the Thai-Cambodian border during the communist rule (1975–1979) of the Khmer Rouge. As a consequence, many Cambodians escaped the brutality and atrocities of the Khmer Rouge, who killed about 2 million of the country's 8 million people. In 1979, the neighboring country of Vietnam invaded Cambodia and overthrew the Khmer Rouge, but along with other groups, the Khmer Rouge fought a guerrilla war against the Vietnamese occupiers. This war also threatened the lives of many Cambodians and brought the prospect of famine, leading to the second Cambodian refugee flow to Thailand. For these refugees, Thailand opened many refugee camps inside the country, and sometime after, a large number of these refugees ultimately gained entry into the United States. Concern provided food and other emergency assistance to Cambodian refugees both in Thailand and in the United States.

In the 1990s, a considerable number of African countries confronted famines, genocide, and epidemics, which displaced many people within or beyond the various countries' national borders. Concern provided support to these internal and external refugees as well as famine-stricken Africans. It further supported the people of the war-torn country of the Democratic Republic of Congo, formerly known as Zaire. At the end of the last millennium and the beginning of this millennium, the country faced two wars. In 2010, Concern was involved in providing emergency assistance to the earthquake victims in Haiti, helping flood survivors in Pakistan, and assisting famine-stricken Somalians.

In recent years, Concern has responded to conflict-driven emergencies in Syria, Iraq, Lebanon, Turkey, and South Sudan. The Syrian Civil War (2011–) displaced many of its citizens, most of whom have taken refuge in neighboring Lebanon. Considering the overcrowded and poor conditions in the Lebanese refugee camps, Concern sponsored accommodation for 21,950 Syrian refugees either rent-free or in reduced rent homes for one year and longer, along with support to landlords to rehabilitate the properties. The organization also actively encouraged over 1,100 Syrian refugee children to attend the Lebanese school system in 2017. Other activities in Lebanon included establishing vegetable gardens around the compound of Informal Tented Settlements (ITS) and training 900 low-skilled Syrian refugees to grow food and providing them with gardening kits (Concern Worldwide 2018).

In Syria, Concern hired almost 3,000 Internally Displaced People (IDP) through cash-for-work projects to help maintain living and hygiene standards within two camps. It also provided work for women who lost their husbands in the war both in 2016 and 2017. The organization installed 22 water and 13 sewage system, which reached 256,839 Syrians, along with distributing food baskets, essential household and hygiene kits, and kerosene stoves (Concern Worldwide 2018).

In 2017, successful emergency operations were conducted in drought-affected and/or conflict-afflicted countries in East Africa such as Somalia, Kenya, Ethiopia, the Republic of Sudan, and South Sudan. In Somalia, Concern distributed cash to the drought victims and provided food and water to them, and it also supported over 2,100 vulnerable households in building their shelters. In Kenya, Concern installed water infrastructure during the emergency drought period to ensure ongoing supply of drinking water to over 21,000 people. In Ethiopia, Concern was involved in building the resilience of drought-stricken communities through multi-sectoral programs, focusing on livelihoods, food security, natural resource management, and health and nutrition (Concern Worldwide 2018).

In South Sudan, the organization opened 40 Farmer Field Schools to teach farmers about climate-smart techniques, such as seed preservation, pest management, and general agronomic skills. It also constructed dykes and ponds and distributed fishery and farming kits. In the Republic of Sudan, nearly 20,000 mothers were provided with maternal and child health (MCH) education services including antenatal care, safe delivery, postnatal care, nutrition and health counseling, and immunization and health services to approximately 48,270 children. In addition, Sierra Leone, which experienced a massive landslide in 2017, received Concern support for over 30,000 people with hygiene kits and clean water within a few days of the emergency. In these countries, nearly 23 million people needed humanitarian assistance, but Concern was able to directly assist 1.5 million people (Concern Worldwide 2018).

In Bangladesh, Concern provided emergency assistance to Rohingya refugees, who were forced to flee their homeland Myanmar to Bangladesh. Nearly 671,000 refugees came to Bangladesh to escape military brutality in Myanmar, and most of them temporarily sheltered in Cox's Bazar district, Bangladesh. Concern screened over 61,000 refugee children under five for malnutrition and provided therapeutic feeding for around 2,700 severely malnourished children. It also provided health, nutrition, and counseling services to over 13,000 women (Concern 2018).

Concern was also involved in various development programs and emergencies in other Asian countries. For example, in Asia, it has such programs in Afghanistan, Bangladesh, Nepal, Pakistan, and North Korea. In Afghanistan, Concern built a suspension bridge, which provided easy access to medical facilities and allowed children to attend primary schools. In addition to providing support to Rohingya refugees, with support from Sajida Foundation and Water Aid, Concern Worldwide established a three-Pavement Dweller Center (PDC) in Dhaka, Bangladesh, to provide temporary night shelter, safe drinking water, sanitation facilities, and other essential services for homeless people and the extremely poor. These centers regularly serve about 4,000 poor and homeless people (Concern Worldwide 2018).

In summer 2014, Nepal experienced a massive landslide-induced flood that displaced 5,000 households. Concern Worldwide took part in disaster relief efforts and provided shelter, drinking water, food, and other essential items to over 76,000 flood victims. In Pakistan, the organization provides safe drinking water and sanitation facilities to 52 schools in the Balochistan province. Moreover, in North Korea, Concern has water projects that regularly serve 133,000 people living in remote rural areas (Concern Worldwide 2018).

Currently, Concern Worldwide employs nearly 4,000 workers of some 50 nationalities. More than 90 percent of these workers work in their home countries. Although Concern will provide its services wherever needed, so far it has worked in 50 different countries. For example, Concern worked in 27 of the world's poorest and most vulnerable countries and helped approximately 27 million people in 2017 through its long-term development projects. In that year, it responded to 65 emergencies. One year earlier, the agency also served 27 countries and helped 22 million people. In 2017, it spent over $188 million on its overseas programs consisting of five main categories: emergencies, livelihoods, health, education, and integrated (Concern Worldwide 2018).

Concern's emergency relief efforts and other humanitarian activities are generally limited to countries in Africa, Asia, and the Caribbean. In emergencies, both natural and man-made, Concern acts quickly to save lives and remains on the ground to rebuild livelihoods and infrastructure so that the affected communities are better prepared for future crises. The primary objective of the emergency effort is to fulfill its humanitarian mandate and effectively respond to, and mitigate, natural and human-influenced disasters. The specific objectives are to (1) respond in a timely way to save lives and reduce suffering; (2) improve access to food, water, and health care; and (3) prevent and reduce the impact of emergencies. In 2017, the organization participated in 65 emergencies in 24 countries, reaching approximately 12.9 million people, 6 million of whom were direct beneficiaries (Concern Worldwide 2018).

CONCLUSION

Although Concern has responded to both major and minor natural and man-made disasters, it is known as an innovational force in many other areas such as nutrition, poverty alleviation, maternal and child health (MCH) and child education, empowerment of women and activities related to improvement of their status, and climate-smart agriculture. This humanitarian organization has almost always followed a continuum from response to recovery to long-term development. One distinctive characteristic of this organization is that it takes a "results-based management" approach to development programs. This means it monitors and evaluates all of its programs to ensure timely accomplishment of the project objectives.

Bimal Kanti Paul

Further Reading
Concern Worldwide. n.d. Concern Worldwide was Born in 1968, a Year of Global Upheaval. https://www.concernusa.org/about/our-history/, accessed February 4, 2019.
Concern Worldwide. 2018. Annual Report & Financial Statements 2018. https://admin.concern.net/sites/default/files/documents/2019-05/Concern_AR18_WEB.pdf, accessed July 5, 2020.

Cooperative for Assistance and Relief Everywhere

Nongovernmental organizations, commonly referred to as NGOs, have become an important partner of human development—by improving communities and promoting citizen participation in both developed and developing countries but, more importantly, in developing countries. Cooperative for Assistance and Relief Everywhere (CARE) (formerly Cooperative for American Remittance to Europe) is one such major international NGO that is committed to providing emergency relief and humanitarian assistance to several development projects. As per the annual reports of the organization, CARE is one of the largest and oldest aid organizations dedicated to saving lives and ending poverty in most developing countries. CARE works in more than 90 countries, reaching over 50 million people with aid programs (CARE Annual Report 2018).

CARE has two goals: (1) long-term help to fight poverty and other negative consequences of underdevelopment and (2) emergency relief, including food and health care for disaster victims. Such disasters, for example, have included civil wars in both Syria and Yemen, earthquakes in Nepal and the Southern Coast of Mexico, droughts in the Horn of Africa (Ethiopia and Kenya), and 1-million-plus Rohingya refugees fleeing to Bangladesh after Myanmar's armed forces launched a crackdown in August of 2017.

Like many other NGOs, CARE is dedicated to eradicating poverty and achieving better health care for the poor. In a broader sense, it works to promote the role of social justice and good governance that are indispensable for doing away with poverty and deprivation. CARE also supports the United Nations' Agenda for Sustainable Development Goals (SDGs), which is an example of this goal.

FOUNDING

Originally intended as a temporary organization, CARE International was founded on November 27, 1945, with its headquarters in Geneva, Switzerland. It is a global confederation of 14 member organizations, each of which is registered as an autonomous, nonprofit, and nongovernmental organization; they are also nonsectarian and impartial in nature—and work together to improve the lives of the poor. CARE has also four affiliate members in Sri Lanka, Indonesia, Egypt, and Morocco.

Following the devastation of World War II (1939–1945), then President Harry Truman ordered to ship out millions of tons of food, medicines, and other basic supplies for individuals and families in a war-ravaged Europe. He also decided to let private organizations provide relief to the people who suffered from starvation due to war. A large number of NGOs and other compassionate organizations—including Red Cross, Oxfam, UNICEF, Save the Children, Doctors Without Borders, Catholic Relief Organization, World Vision, and CARE, to name a few—came forward to alleviate the suffering of these societies.

CARE started, along with 22 other American organizations, to distribute food packages to the survivors of World War II. On May 11, 1946, CARE delivered the first 15,000 packages, containing milk powder, cheese, rice, and beans, to Le Havre, France. In the same year, CARE also opened a new office in Canada, and in the following decades, it established CARE offices in Germany, France, England, Austria, Norway, Australia, Denmark, and Japan. An additional umbrella organization (CARE International) was also created to coordinate and avoid duplication among the various CARE organizations.

Later, the size of supplies contained more goods, including canned meats, dried fruits, and few comfort items such as chocolates, coffee, and cigarettes. By 1949, CARE had shipped out more than 2.5 million packages with 12 different items of nutritious food to war victims and low-income people, including schoolteachers. Free supplies of cigarettes were withdrawn, and coffee had been substituted for tea, both for health benefits.

DEVELOPMENT

The United States Agriculture Act of 1949 made a surplus of food available to many countries, particularly to Europe and Japan in large quantities. From 1950 onward, CARE shifted its focus to Latin America, followed by Africa and Asia. Between 1950 and 1975, CARE broadened its geographic focus and extended its functions beyond food distribution. For example, the medical aid organization

MEDICO merged with CARE and thus began to deliver more products, as well as offer more variety. The organization introduced a school feeding program and health-care education, followed by clean water supply and sanitation programs in Chile, Columbia, and Honduras, in addition to extending its inventory of supply for more general relief goods.

CARE also advocates at the national level for human rights, especially rights for women and rights for children. Concurrently, CARE denounces the inhumanity of cutting budgets and foreign aid—by rich countries, including the United States and Canada—to countries in need. Foreign aid is an important factor in reducing extreme poverty and is crucial in recovering from emergency situations, such as the famine in South Sudan and the earthquake in Bohol, Philippines. CARE has also worked in many other places to help provide relief for similar disastrous situations.

CARE has been continuously active in strengthening the resilience of communities in order to cope with the effects of current humanitarian crises in Kenya, Syria, Yemen—and Rohingya refugee camps in Cox's Bazar, Bangladesh. Toward the end of the 1960s, CARE moved from "Food Packages" to "Virtual CARE Packages" with food aid, plus services including civil engineering services for the construction of schools, local roads, and health-care centers.

Then President John F. Kennedy established the Peace Corps, which is a volunteer program run by the U.S. government. At the time, CARE was in charge of selecting and training Peace Corps volunteers who served all over the world. CARE continued to provide the training up until 1967. While the number of Peace Corps volunteers has declined in recent years, CARE's domain of work has increased, including providing emergency relief for millions of the world's most vulnerable people and basic education plus primary healthcare to children living in refugee camps or on the streets. CARE advocates human rights and gender equality and an inclusive national development. However, the main function of the organization is to deliver emergency relief.

With the introduction of the Marshall Plan—which channeled more than $12 billion in 1947 (nearly $100 billion dollars in 2018), Europe was put to fast recovery. At this time, CARE found itself getting more involved in the humanitarian problems that the developing nations of Asia were facing. The organization first moved to Japan, followed by the Philippines, Korea, India, and Pakistan and then in a number of Latin American countries in the 1960s and 1970s, while CARE became an organization whose work ranged from "sending remittances in Europe to relief everywhere."

Since 1980, CARE has broadened its developmental work and changed the nature of its projects, for example, addressing the issues of the 1983–1985 famine in Ethiopia and the 1991–1992 famine in Somalia. During this same period, CARE also focused on agroforestry, reforestation, sand soil conservation, and to improve drought tolerance crop variety in East Africa. Today CARE—and other humanitarian aid organizations—takes a multidimensional perspective of poverty, including how climate change impacts vulnerability—and thereby addresses the root cause of poverty.

FUNDAMENTAL PRINCIPLES

History is full of examples of humanitarian assistance and action in many different forms to promote welfare of others, not only from religious activities but also from philanthropic or charitable groups. However, at the end of World War I (1914–1918), humanitarian aid emerged in a more organized fashion. The Treaty of Versailles is one such example, having accepted these principles of humanitarian aid. In the aftermath of World War II, a record number (close to 200) NGOs came into being; aid became more global. From the 1990s onward, NGOs have formed various partnerships with private sectors. They remain more active today and keep moving well.

CARE is one of such NGO that is based on core principles and values. CARE functions independently of political, commercial, and religious objectives. It delivers assistance on the basis of need, regardless of race or nationality. In other words, their method of operation is impartial. These principles are, however, similar to other humanitarian agencies such as Oxfam, World Vision, and Save the Children International. A set of principles followed by CARE includes:

- Is nonprofit, nonsectarian, nongovernmental, and nondiscrimination organization
- Works in partnership with others to promote development
- Ensures accountability, financial transparency, and work responsibility
- Opposes discrimination and violence
- Promotes nonviolent resolution of conflict for sustainable results
- Promotes empowerment of women, children, and the downtrodden
- Seeks sustainable results of their works

CARE strives for achieving the highest standards of ethical practice. Its members abide by the confederation's statutes and code of conduct and works with truth, integrity, and good faith—at all times and in all circumstances. According to CARE's vision statement, the organization recognizes that it is committed before God "to uphold these core values individually and as corporate entities" (World Vision n.d.)

The organization is careful to avoid bad decisions when it is not moral by their judgment. For example, in 2007, CARE walked away from some $45 million per year in federal financing because American food aid was not only plagued with inefficiencies but also could hurt some of the poorest people it aimed to help.

CARE's commitment to victims of civil war, natural disasters, and extreme poverty has increased over time in coverage and quality of service. Cox's Bazar, Bangladesh, has been a scene of mass exodus of Rohingya population from Myanmar. Between 1978 and 2017, they faced recurring military crackdowns and fled Myanmar in significant numbers. In 2019, they constituted a devastated population of 1.3 million. Women, children, and infants make up over 70 percent of the refugees. They are in dire need of humanitarian assistance.

CARE has directly reached nearly 250,000 people through the distribution of food, nonfood items (NFIs), health support, shelter, protection, and water and sanitation services. Additionally, more than 180,000 children have been reached indirectly by CARE through the community-based management of acute malnutrition (CMAM) activity, where CARE is the technical partner.

As another example, in its response to war-torn, displaced people inside northeast Syria and in Turkey, Egypt, Iraq, and Jordan, CARE has reached close to 5 million in those countries since the beginning of the Syrian Civil War in 2011. The organization has provided support for—in addition to humanitarian assistance—job training, valuable life skills, and entrepreneurial incentives among the refugee men and women so that they can earn a living.

Since the escalation of civil war in Yemen in 2015, there has been the worst outbreak of cholera and famine in Yemen's history. The economy has suffered a collapse, and health services do not work. More than half of the population has been displaced, living in extremely vulnerable conditions. CARE is very much on the ground, working toward minimizing the suffering of these people from displacement and disease by "providing food and supplies to those who have no alternatives."

ORGANIZATION AND SOURCE OF FUNDING

CARE receives contributions from individuals, organizations, corporations, and governments. Sometimes it enters into agreements with government agencies and receives partial financing for development projects. Moreover, while working with various social organizations, members of concerned organizations contribute in the way of subscriptions. MEDICO, an organization of health professionals, advises CARE on its primary health projects.

CARE's website provides a list of donors and shows great appreciation and gratitude for their kindness. These donors are divided into three groups: (1) Multilateral funding partners, (2) Bilateral partners, and (3) Individual funding.

Multilateral funding partners include the European Union (EU) and major organs of the United Nations, such as Food and Agricultural Organization (FAO); International Labor Organization (ILO); United Nations Children's Fund (UNICEF); and United Nations Development Programs (UNDP), to name a few partners.

Bilateral partners are the governments of Australia, Austria, United Kingdom, United States, Japan, and Taiwan. The donor agencies who extend support to CARE include nearly a dozen donor agencies from the United States, the United Kingdom, France, Japan, Taiwan, and Scandinavian countries.

Individual funding comes from "over one million individuals, corporations and private foundations worldwide whose financial gifts make our work possible" (CARE n.d.b).

CONCLUSION

CARE is a global leader for humanitarian relief, dedicated to saving lives of millions. The organization has an international network of 14 Members, three

Candidate Members, and one Affiliate Member with a common vision and mission to defeat global poverty, help economic development, and promote social justice and the environment.

Each CARE member is an independent organization that leads programs, raises funds, advocates on key issues, and communicates to the public in their respective countries. These efforts provide lifesaving and life-changing work in more than 90 countries presently. Around the world, CARE works with a broad network of partners and allies to help rebuild and improve the lives of the most disadvantaged with a particular focus on displaced women and children.

Despite all the accolades used to describe the role of CARE, critics say that many of their lofty goals are out of reach—that a more coordinated effort between government and nongovernmental organizations is required to achieve that goal. Nevertheless, CARE—during an emergency response—is among the first to respond and the last to leave during a humanitarian crisis such as war and conflict.

Syed A. Hasnath

Further Reading
CARE. n.d.a. What We Do: Emergency Response. https://www.care-international.org/what-we-do/emergency-response-1, accessed October 15, 2019.
CARE. n.d.b. Where We Work. https://www.care-international.org/, accessed October 15, 2019.
CARE. 2018. *Annual Report.* Atlanta, GA: CARE.
World Vision. n.d. Life in All Its Fullness Our Vision Statement. https://www.wvi.org/about-us/our-vision-and-values, accessed October 16, 2019.

Doctors Without Borders

Doctors Without Borders (Medecins Sans Frontieres in French, MSF) is a highly respected international and independent medical humanitarian organization, whose international headquarters is based in Geneva, Switzerland. It is also known for being the most active and most outspoken humanitarian organization. Its volunteers are mostly in medical professions from all over the world, and they work in war-torn countries, countries devastated by natural disasters and affected by epidemics, and the organization helps people of any country with little or no access to healthcare. Doctors Without Borders also assists refugees, asylum seekers, migrants, and people who have been internally displaced due to civil unrests, wars, and conflicts. Volunteers of this organization treat those in need, ignoring religion, politics, and national boundaries. Treatment is guided by medical ethics and the principles of impartiality and neutrality.

Doctors Without Borders has two different missions: respond to an immediate humanitarian crisis and provide emergency medical assistance, long-term health care, and medical training in countries where health structures are insufficient or even nonexistent. MSF has initiated efforts to fight infectious diseases, and its volunteers also help provide vaccines, safe drinking water, and sanitation to people of remote areas. The organization has staff in over 71 countries at all times, and it had nearly 40,000 personnel—mostly local doctors, surgeons, psychiatrists, nurses, mid-wives, and other medical professionals—in 2016. The organization is open to people from all other professions who might help it achieve its goals. In 2016, private doctors from all over the world provided about 95 percent of its funding, and the remaining 5 percent came as donations from governments, corporations, or large institutions for a total income of $1.7 billion (MSF 2016).

ESTABLISHMENT AND ORGANIZATIONAL STRUCTURE

While formally founded in 1971, a young French physician, Bernard Kouchner, sowed the seeds for MSF in Nigeria in 1968. Along with a number of French doctors and medical journalists, Kouchner volunteered to work for the French Red Cross in the Biafra region of Nigeria. The Biafra War, or Nigerian Civil War, broke out in the region on July 6, 1967, and ended on January 15, 1970. The Nigerian government forces fought against the forces that supported a separate independent state for the ethnic Igbo people of Biafra, an oil-rich, southeastern coastal state of Nigeria. The war caused massive civilian deaths and property destruction largely when Nigerian government forces encircled and blockaded the Biafrans, causing widespread starvation, malnutrition, and displacement of Igbos.

While major world powers like the United Kingdom, the United States, and the Soviet Union directly or indirectly supported the Nigerian government's position, only France sided with the Biafrans. After entering Biafra, French doctors and journalists were subjected to attacks by the Nigerian army, and they witnessed the army killing civilians and violating human rights. Also, the French were not allowed to provide external emergency assistance to the rebelling population. The French doctors stationed in Biafra publicly criticized the Nigerian government's actions and the long silence of the Red Cross. This situation frustrated the French volunteers, who concluded that a new humanitarian organization was urgently needed to serve victims of war and natural disasters. These volunteers came to Nigeria after signing an agreement, in which they agreed to maintain neutrality (Rostis 2011).

After returning to France, the doctors formed an organization called the Groupe d'Intervention Médicale et Chirurgicale en Urgence ("Emergency Medical and Surgical Intervention Group") in 1971 to provide aid to populations in distress and to emphasize the importance of victims' rights. At about the same time, Raymond Borel, the editor of a French medical journal, established another organization called Secours Médical Français ("French Medical Relief") in response to the deadly 1970 cyclone that struck Bhola, East Pakistan (what is now called Bangladesh), which killed up to 500,000 people. On December 22, 1971, the two organizations merged into one and formed the Médecins Sans Frontières or Doctors Without Borders (Bortolotti 2004).

For almost for two decades, Doctors Without Borders operated from Europe, and then in 1990, the organization opened its offices in the United States. This marked the expansion of offices beyond Europe, and in the following two years, MSF opened offices in Canada and Japan and subsequently in the UK, Australia, Germany, Austria, Denmark, Norway, and in other countries. Now, Doctors

Without Borders has offices all over the world, and it consists of specialized organizations, called satellites, that are in charge of specific activities like providing humanitarian relief supplies, conducting medical and epidemiological research, and researching humanitarian and social action. Although each national organization of Doctors Without Borders operates independently, they are all held together by their common principles of medical ethics and the principles of independence and impartiality. The international body of the organization is made up of representatives from each national organization, but the organizations from developing countries have no decision-making power, which lies exclusively with the organizations of Western countries (Rostis 2011).

In 2011, MSF's International Governance Body (IGB) structure changed, and it was registered in Switzerland. In the same year, it organized the first annual International General Assembly (IGA). The IGA comprises two representatives of each national organization, two representatives elected by the individual members of its International wing, and the International President. The IGA is the highest authority of Doctors Without Borders and is responsible for protecting its humanitarian mission and providing strategic orientation to all national organizations. The IGB acts on behalf of and is accountable to the IGA (MSF 2016).

PRINCIPLES

The humanitarian actions of Doctors Without Borders are guided by medical ethics and the principles of independence and impartiality. According to its medical ethics, the organization offers neutral and impartial medical support solely on the basis of health needs of people experiencing diverse crises. Volunteers of this organization treat patients with dignity and with respect for their cultural and religious beliefs. They respect patients' confidentiality, autonomy, and their right to informed consent. MSF is different from other humanitarian organizations in the sense that it assesses the needs of the people affected by natural and man-made disasters and accordingly offers emergency assistance to the affected people. Also, the organization's independence is reflected in its marginal dependence on funding from governments and intergovernmental organizations.

Doctors Without Borders gives priority to people who need urgent assistance and does so irrespective of race, religion, gender, and political affiliation. The organization is extremely vocal when medical facilities come under threat, when the assistance is inadequate and unfairly distributed among the vulnerable people, or when emergencies are neglected. It is transparent and acts sensibly. It also carries out evaluations of each emergency in which it participates. Thus, MSF has a reputation for its neutrality and its work under fire. Moreover, its volunteers arrive at the disaster or war sites faster than any other relief agencies.

DISASTER RELIEF

The first mission of Doctors Without Borders was in Nicaragua after the country was hit by a devastating 6.3-magnitude earthquake in 1972. The event killed between 4,000 and 11,000 people, injured 20,000, and displaced 300,000 people. It almost destroyed Managua, the country's capital, and it completely destroyed four hospitals in the capital. MSF has also solicited funds for and provided emergency aid to victims of almost all major natural disasters and civil wars since then. However, it did not seek funding for the 2011 Japan earthquake, but it instead sent an emergency response team to the country.

Because of numerous ethnic conflicts, MSF has been active in Africa, particularly in the Central African Republic, since the mid-1990s. More than 60 percent of its activities are conducted in Africa, 25 percent to Asian, and the remaining 15 percent occur on other continents. Because Africa suffers from numerous ethnic conflicts and consequently is the principal source and destination of large numbers of refugees and internally displaced persons (IDPs), MSF has a heavy presence on this continent. Also, being tropical makes the continent more prone to many diseases such as malaria, yellow fever, river blindness, and HIV/AIDS.

After the Vietnam War (1955–1993) and during the Cambodian Genocide (1975–1993), MSF performed its first large-scale and long-term relief missions in several war-torn countries of Southeast Asia to provide aid to survivors of the mass killings. The organization also helped in reconstructing the health care systems of these countries. A similar long-term mission was also administered during the Lebanese Civil War (1976–1984) to perform surgeries on both Muslim and Christian victims in the hospitals across the country (de Deus and Reis 2017). This mission established its international reputation as an organization with a strict policy of neutrality.

Doctors Without Borders also worked with refugees displaced by wars in Southeast Asia, Central Africa, and the Middle East. It monitored the Cambodian refugees who fled the Khmer Rouge (1975–1978) in Thailand and assisted victims of Rwandan genocide (1994) in the Democratic Republic of the Congo. The organization continued extending its support and assistance to Somali refugees taking shelter in Kenya and Ethiopia for more than one decade, until they were forced to leave the country to avoid conflict and hunger. MSF also launched its relief operation in Afghanistan after the Soviet Union invaded the country in 1979. Five years later, in 1984, volunteers of this organization assisted the famine-affected people of Ethiopia.

Moreover, MSF has been helping Palestinians who were displaced after the formation of Israel, along with those who were displaced in the Latin American countries of Colombia and Mexico. After the 2004 Indian Ocean Tsunami, Doctors Without Borders deployed more than 160 volunteers and 200 tons of supplies to those affected by the disaster. More recently, since August 2017, the physicians of this organization have been providing health care services to over 500,000 Rohingya refugees, who were forced to leave their home country of Myanmar for Bangladesh. MSF denounced the inadequate international response to the West African Ebola virus epidemic (2013–2016) and also put tremendous pressure on pharmaceutical companies to lower the price of expensive medicines for the people of developing countries.

CONCLUSION

In terms of disaster funding and relief operations, Doctors Without Borders has unique characteristics that makes the organization distinctive compared to most other international relief agencies. It follows a decentralized policy and strongly supports participatory democracy. Once members think that the necessary funding for a disaster has already been secured, this organization stops appealing for such funding. For example, it closed an appeal after realizing that the 2004 Indian Ocean Tsunami was overfunded. Additionally, MSF does not always accept funding from external sources. For instance, it rejected funding from the European Union and its member countries in 2016 because of their unacceptable policies with regards to the Syrian and other Middle Eastern refugees and migrants to European countries. It also refuses donations from countries involved in military intervention. For example, it refused funding from American government agencies in 2004 due to the Iraq War (2003–2011). This policy is pursued by the organization to guarantee the neutral and impartial provision of disaster and medical aid (de Deus and Reis 2017).

With numerous projects worldwide, Doctors Without Borders is the world's leading independent international relief organization. For nearly five decades, the organization has provided assistance to survivors of natural disasters, armed conflicts, disease epidemics, and other emergencies. Its volunteers are highly professional and known for their rapid response to natural and man-made disasters. Starting with the development of emergency kits during the 1980s, MSF also developed a variety of kits (e.g., shelter kits, water purification kits, and cholera kits) adapted to different needs. For its contribution to humanity and underserved people, the organization was awarded the Nobel Peace Prize in 1999. Despite abduction and killing of its volunteers in Afghanistan and Iraq, the organization remains committed to pursuing its objectives throughout the world.

Bimal Kanti Paul

Further Reading

Bortolotti, D. 2004. *Hope in Hell: Inside the World of Doctors Without Borders.* Richmond Hill, ON, Canada: Firefly Books.

de Deus, S., and R. Reis. 2017. Doctors Without Borders: Coherent Principles. *International Journal on Human Rights* 14(25): 259–264.

MSF. 2016. *International Financial Report* 2016. Geneva, Switzerland: MSF.

Rostis, A. 2011. Doctors Without Borders. In *Encyclopedia of Disaster Relief*, edited by B. Penuel and M. Staller, 123–124. Los Angeles, CA: Sage Publications, Inc.

International Federation of Red Cross and Red Crescent Societies (IFRC)

The International Federation of Red Cross and Red Crescent Societies (IFRC) is the world's largest humanitarian network. Along with the International Committee of the Red Cross (ICRC), it is part of the International Red Cross and Red Crescent Movement. Currently, the IFRC comprises 190 National Societies and more than 160,000 local units/branches worldwide. All National Societies and units are coordinated by the IFRC. The IFRC has over 17 million volunteers, and 33.9 million people donate blood to its National Societies each year. It provides disaster and other emergency assistance to 150 million people each year through its National Societies and local units (IFRC 2017).

As a partner in development, and in response to both natural and man-made disasters, the IFRC is dedicated to improving the lives of vulnerable people through the goals of reducing their vulnerabilities and increasing their resiliency against extreme events. It does so with impartiality to nationality, race, gender, religious beliefs, class, and political opinions. Its work focuses on three key areas: disaster response and recovery, development, and social inclusion. It also works toward a codified, worldwide ban on the use of land mines and offers medical, psychological, and social support for people injured by land mines. The IFRC promotes a culture of peace across the world, advocates humanitarian principles and values, and has the ability to provide a global voice to vulnerable people.

FOUNDING

Representatives from the National Red Cross Societies of Britain, France, Italy, Japan, and the United States came to Paris in 1919 to found the League of Red Cross Societies, later renamed the International Federation of Red Cross and Red Crescent Society. The representatives from these five countries sought to improve the health of people in countries that had suffered greatly during World War I (1914–1918). The stated goals of the League were to strengthen and unite Red Cross Societies that existed at that time and to promote the creation of new societies. Henry Davison, president of the American Red Cross War Committee, envisioned a worldwide federation of all Red Cross Societies. An international medical conference initiated by Davison resulted in the birth of the League of Red Cross Societies in 1919.

The idea to form an international voluntary relief organization originated with Swiss businessman and activist Henry Dunant in 1859. Dunant had personally witnessed the aftermath of the bloody battlefield between the armies of imperial Austria and the Franco-Sardinian alliance in Solferino in modern-day Italy. Some 40,000 soldiers had either already died or were about to die on the battlefield, and the wounded lacked immediate medical attention. With help from local civilians, particularly women and girls, Dunant provided necessary medical assistance to wounded soldiers and fed and comforted them. After returning to his birthplace of Geneva from the battleground, he founded the International Committee of the Red Cross (ICRC) in 1863 to aid and protect the sick and the wounded in combat situations.

At that time, however, the ICRC was not truly an international humanitarian organization; it was a European relief agency. World War I showed a need for close cooperation between Red Cross Societies, which, through their humanitarian activities on behalf of prisoners of war and combatants, had attracted millions of volunteers and built a large body of expertise. A devastated Europe could not afford to lose such a resource. In 1983, the League of Red Cross Societies was renamed to the League of Red Cross and Red Crescent Societies to reflect the growing number of National Societies operating under the Red Crescent symbol.

The name of the League was changed again in 1991 to the International Federation of Red Cross and Red Crescent Societies (Davey et al. 2013).

Although only five countries formed the League of Red Cross Societies originally, the number of National Societies increased markedly after World War II (1939–1945) and again in the 1960s in response to decolonization. By the end of the 1960s, more than 100 societies had formed around the world. In 1963, the IFRC (still known at that time as the League of Red Cross Societies) and the ICRC jointly received the Nobel Peace Prize.

The ICRC initially opposed the formation of the League of Red Cross Societies. Like the IFRC, the ICRC operates worldwide, helping people affected by conflict and armed violence and promoting the laws that protect victims of war. It is an independent and neutral organization, and its mandates stem from the 1949 Geneva Convention. The ICRC is funded by voluntary donations from governments and from National Red Cross and Red Crescent Societies. It employs some 16,000 people in more than 80 countries (IFRC 2017).

FUNDAMENTAL PRINCIPLES

There are seven fundamental principles that guide the work of the IFRC and its members: humanity, impartiality, neutrality, independence, voluntary service, unity, and universality (IFRC 2013b). These principles were adopted in 1965 but incorporated into IFRC's statutes in 1986. The first three fundamental principles are similar to UN Resolution A/RES/46/182, which states that all humanitarian actions should be guided by these three principles.

Humanity refers to the centrality of saving human lives and alleviating suffering wherever it is found. Its goal is to protect life and health and to ensure respect for all. "It promotes mutual understanding, friendship, cooperation and lasting peace among all people" (IFRC 2013b, 262). It is consistent with the key objective of its origin in 1863. The IFRC desired to provide assistance without discrimination to the wounded soldiers on the battlefield and to prevent and alleviate human suffering all over the world.

Impartiality consists of the implementation of all humanitarian actions solely on the basis of need, without any discrimination pertaining to nationality, race, religious beliefs, social class, caste, or political opinions. It endeavors to relieve the suffering of individuals and give priority to the most urgent cases of distress.

The third principle of IFRC is *neutrality*, which means that humanitarian action must not favor any side in an armed conflict, complex emergencies created by humans, or other dispute where such action is carried out. Neutrality is necessary to secure the confidence of all, and it means the IFRC "may not take sides in hostilities or engage at any time in controversies of a political, racial, religious or ideological nature" (IFRC 2013b, 262).

The International Red Cross and Red Crescent Society's principle of *independence* means that all the national societies of IFRC must always maintain their autonomy so that they may be able at all times to act in accordance with fundamental principles of the organization.

Voluntary service means relief efforts must not be prompted by any desire for gain.

Under *unity*, there can be only one Red Cross and or Red Crescent Society in any one country, and it must be open to all. The organization must carry its relief operation throughout its territory.

Universality stipulates that all Red Cross and Red Crescent Societies in the world have equal status and share equal responsibilities and duties in helping each other (IFRC 2013b).

ORGANIZATION AND SOURCES OF FUNDING

Like the ICRC, the IFRC is based in Geneva, Switzerland. It has five regional offices and numerous country and multi-country cluster offices around the world. The IFRC along with the ICRC supports the formation of new National Societies in countries where no such society exists. Each new Society is admitted as a member to the IFRC, provided the society is recognized by the ICRC. The IFRC works closely with the National Societies of the countries that experience extreme events; these countries are called Host National Societies (HNS). The National Societies of nonaffected countries willing to offer assistance are called Partner National Societies (PNS). About 25–30 PNS regularly work in other countries. The most active are the American Red Cross, the British Red Cross, the German Red Cross, and the Red Cross Societies of Sweden and Norway.

The IFRC is governed by a Governing Board consisting of a president, four vice-presidents, the chairman of the finance commission, and 20 National Society representatives. The highest body of the IFRC is the General Assembly, which convenes every two years with delegates from all 190 National Societies. Among other tasks, the General Assembly elects the President.

The IFRC receives funding from various sources, such as National Societies, governments of both developed and developing countries, the delivery of field services to program partners, and voluntary contributions from corporations and individuals. The criteria for the contributions of each National Society are established by the Finance Commission and approved by the General Assembly. Additional funding, especially for unforeseen relief assistance missions, is raised by emergency appeals. Donating to the local National Societies or the IFRC rather than the National Societies is often preferred because the National Societies may take a substantial cut of the donations for themselves.

The IFRC provides financial support to National Societies from its Disaster Relief Emergency Fund (DREF) to ensure immediate relief needs. The DREF offers two possible forms of funding to National Societies: the "loan facility" to respond to large-scale disasters and the "grant facility" for responses to small- and medium-scale disasters for which no international appeal will be launched or when there is no expected support from elsewhere. Generally, the IFRC makes international appeals for large-scale man-made or natural disasters (IFRC 2013a).

DISASTER REPORTS AND ACTIVITIES

The IFRC occasionally publishes reports on specific disaster events. One of the important and useful publications of this organization is its World Disasters Report–published annually since 1993. This report brings together the latest trends, facts, and analysis of contemporary crises and disasters. Each report has a specific focus. For example, the focus of the 2016 report was "Resilience: Saving Lives Today, Investing for Tomorrow." The IFRC evaluates activities of National Societies after their involvements in almost all emergencies. Its evaluations are comprehensive and relatively impressive. In most cases, National Societies' response and relief efforts are timely and adequate. In a few cases, such as the 2009 Zambia Floods, the response was delayed. In contrast, in the 2012 Hurricane Sandy Operation, the Jamaican Red Cross was praised for its excellent coordination with the government and other NGOs. The response to the Haiti earthquake was disappointing, as there were problems with management and cooperation, as well as insufficient engagement at the highest inter-agency or government levels (Giving What We Can n.d.).

The success of IFRC's involvement in conducting disaster relief and recovery efforts depends largely on the National Societies: their proximity to emergency sites means excellent response times, but nonproximity may create coordination problem and hence delays in providing assistance to the victims of emergency. The IFRC seems likely to be effective at responding to small, less well-publicized disasters due to having a permanent National Society in almost every country. The first large-scale postdisaster mission that the IFRC participated in came after the 1923 earthquake in Japan. It killed about 200,000 people and left countless more wounded and without shelter. Due to its excellent coordination, the Red Cross Society of Japan received goods from its sister societies, reaching a total value of about $100 million.

The IFRC's largest mission to date was in response to the 2004 Indian Ocean Tsunami. More than 40 National Societies worked with more than 22,000 volunteers to bring relief to the countless victims left without food and shelter and endangered by the risk of epidemics. Despite its unsatisfactory performance in a few emergencies, the IFRC has already saved and will save the lives of millions of people throughout the world. It has been providing emergency supplies to millions of disaster victims each year, and its humanitarian service must be appreciated by humanity.

Bimal Kanti Paul

Further Reading

Davey, E., J. Borton, and M. Foley. 2013. *A History of the Humanitarian System: Western Origins and Foundation*. HPG Working Paper. London, UK: Overseas Development Institute.

Giving What We Can. n.d. Emergency Aid. https://www.givingwhatwecan.org/research/other-causes/emergency-aid/, accessed December 10, 2017.

IFRC (International Federation of Red Cross and Red Crescent Societies). 2013a. *Plan and Budget 2014–2015*. Geneva, Switzerland: IFRC.

IFRC (International Federation of Red Cross and Red Crescent Societies). 2013b. *World Disasters Report: Focus on Technology and the Future of Humanitarian Action*. Geneva, Switzerland: IFRC.

IFRC (International Federation of Red Cross and Red Crescent Societies). 2017. Who We Are. https://www.ifrc.org/en/who-we-are/, accessed July 22, 2017.

International Organization for Migration (IOM)

The United Nations (UN) established the International Organization for Migration (IOM) in 1951; it is the premier inter-governmental organization regarding migration, both domestic and international, and development. Since 1991, the organization has operated within the Inter-Agency Standing Committee (IASC), which is the longest-standing and highest-level humanitarian coordination forum of the United Nations system. Moreover, the IOM works closely with migrants, governmental entities, nongovernmental entities, and other partners. With 166 member states, a further 8 states holding observer status, and offices in over 100 countries, the IOM promotes the orderly and humane management of migration challenges by providing services and advice to governments and all types of migrants, including those affected by natural disaster-induced migrations. Additionally, the IOM promotes international cooperation on migration issues, assists in the search for practical solutions to migration problems, and provides humanitarian emergency assistance to migrants in need, including people who are displaced by natural and man-made disasters.

The Constitution of the Organization emphasizes the association between migration and economic, social, and cultural development, as well as the right to freedom of movement. It works in four broad areas of migration management: migration and development, facilitated migration, regulated migration, and forced migration. The five activities of the IOM that intersect across these four areas are the promotion of law of international migration, policy debate and guidance, protection of migrants' rights, migration health, and issues related to the gender dimension of migration. The IOM also facilitates and regulates voluntary and forced migration. It provides information and supports migration policy to promote increased dialogue between migration stakeholders at national, regional, and global levels.

STRUCTURE

Regarding implementation of policies related to migration, the IOM is a key partner of the European Union (EU). It started to collaborate with the EU and its affiliate commissions to increase cooperation on migration, development, humanitarian response to emergencies, and human rights issues in 2012. The chief of the organization is the Director General who serves a five-year term, and oversees more than 10,000 staff all over the world, with an annual budget of more than $1 billion. As an international organization, the IOM maintains a policy of equitable geographical balance among its staff, and thus it recruits staff from both Member and Non-Member States. The organization seeks committed professionals with a wide variety of skills in migration, and its staff works in multi-cultural environments in which diversity and cultural sensitivity are valued.

DISASTER-RELATED ACTIVITIES AND PARTICIPATION

Hazard research consistently provides evidence that large and deadly natural disasters displace many survivors across the globe, and a considerable number of these survivors invariably migrate temporarily or permanently to areas not affected by such events. Furthermore, the IOM claims that between 2008 and 2015, disaster displaced 25.4 million people per year on average (IOM 2017). Realizing the role of migration in mitigating some of the impacts of extreme natural events, and the interplay between natural disasters, migration and development, the IOM has started to become involved actively in postdisaster humanitarian response. Another reason for the IOM's involvement in disaster relief is the support provided by the diaspora of affected countries in sending emergency assistance as well as funding for recovery (IOM 2007).

In 1998, the IOM for the first time responded to a major disaster when Hurricane Mitch hit several Central American countries. This and subsequent natural disasters displaced millions who fled internally as well as externally. As a response to Hurricane Mitch, the IOM initiated humanitarian measures as well as developed disaster risk management policies to support survivors, particularly those who were displaced by the event. Later, the organization enhanced its policies and extended them to include disaster victims of North America (PBS 2018). Furthermore, the IOM has become even more active in responding to natural disasters since the 2004 Indian Ocean Tsunami (IOT).

In addition to participating in relief efforts, the IOM examined the migratory patterns in post-tsunami Indonesia, Sri Lanka, and Thailand. These three countries were severely impacted by the tsunami, as well as represented "a range of preexisting internal and international migratory movements" (IOM 2007, 10). For example, all the selected countries had guest workers in the Gulf and other foreign countries prior to the 2004 IOT, which generated constant flow of remittances, and that formed an important part of the economy in each country. The IOM considered migration in tsunami-affected countries as well as migration from tsunami-affected countries. Further, it studied migration flows both out of and into affected areas and analyzed the extent of trafficking in persons from the affected areas to both domestic and international destinations. The IOM concluded that irrespective of the type of migrants, they face increased vulnerability in a postdisaster period primarily because they are often overlooked for humanitarian assistance and support. The tsunami case studies highlight that migrant communities must be included in disaster recovery planning so that they are treated in accordance with the principles of international human rights law (IOM 2007).

Experts largely agree that the international response to the 2004 IOT relief effort was delayed and inadequate because of the inability of involved humanitarian organizations to mobilize their capacity and resources. Observing this, the UN Office for the Coordination of Humanitarian Affairs (OCHA) introduced a Cluster Approach in 2005 to improve international relief response to emergencies. Clusters comprise at least 11 sectors of humanitarian actions (e.g., food security, shelter, health, and water, sanitation, and hygiene or WASH). Each of these sectors is assigned to either a UN or non-UN humanitarian organization. The Cluster Approach was applied for the first time following the 2005 earthquake in Pakistan. Nine clusters were established within 24 hours of the earthquake (Paul 2019). In 2006, the IOM assumed the lead role for Camp Coordination and Camp Management (CCCM) in major natural disasters, and it also participates actively in other clusters such as emergency shelter, logistics, health, protection, or early recovery,

In 2007–2008, the IOM participated in the 22 flash appeals related to natural disasters. In the same period, it also completed more than 60 projects, responding to such events in some 20 countries. Notable past participation of the IOM in natural disasters include Cyclone Nargis, the Pakistan flood, the Haiti earthquake, and the flood in Nepal. Cyclone Nargis hit Myanmar in 2008, and the IOM responded with medical care and shelter. In the same year, Nepal experienced a devastating flood in the Koshi River valley, and the IOM was assigned to CCCM.

In July 2010, the Indus valley of Pakistan experienced a catastrophic flood considered one of the largest in the country's history. That flood affected about 18 million people and destroyed or damaged over 1.7 million houses. The IOM responded through the "One Room Shelter Project" to provide 30,000 flood-resistant durable shelters. A component of the project was to empower flood victims to reconstruct their damaged houses by their own initiatives. For this purpose, the IOM provided a system of conditional cash payments to household heads through a "build back better" approach such that survivors reconstructed houses using locally available resources and salvageable materials and local labor. The keys to the approach were to provide training in flood-resilient building methods and to put emphasis on adoption of local designs. "IOM also supported a national communications initiative that raised awareness of assistance entitlements, disseminated key risk information and promoted a humanitarian hotline. The communications project led by the IOM was subsequently incorporated into the Government's national disaster risk management strategy in recognition of its important contribution to the strengthening resilience in recovery and reconstruction" (IOM 2017).

After the 2010 earthquake in Haiti, the IOM responded within 24 hours, and it provided and coordinated life-saving services. The organization eventually cooperated with the international humanitarian community and the government of Haiti to aid displaced people. Most of the 1.5 million homeless lived for more than one year in makeshift camps built in public squares. Even 14 months after the event, approximately 700,000 were still housed in camps. In this instance, the UN formed 12 clusters, and the IOM was in charge of CCCM (Paul 2018). It had 43 active projects and spent $191 million on these projects in areas such as: (1) Camp Management Operations, WASH, Site Planning, Protection; (2) Data Management, Census, Enumeration; (3) Health, Psychosocial; (4) Communications; (5) Temporary Shelter, (6) Non-Food Items; (7) Community Stabilization and Early Recovery; (8) Disaster Risk Reduction; (9) Counter-Trafficking; and 10) Capacity Building on Migration Management. The IOM is also committed to supporting reconstruction and relocating people from camps to communities (IOM 2012).

As indicated, the IOM participates in disaster rebuilding efforts. It believes that involvement in such activities will make disaster-affected communities more resilient and hence prevent further forced migration. The IOM subsequently has cultivated such resilience against natural disasters in Colombia, Cambodia, Mozambique, and the Philippines. In 2017, the IOM conducted a workshop in Papua New Guinea to train first responders of natural disasters about camp coordination and camp management. The workshop highlighted the importance of safe shelter in effective disaster risk management. It also conducted another workshop in the country on Participatory Approach on Safe Shelter Awareness (PASSA) for at-risk communities. Additionally, the organization works together with governments and disaster-affected communities to reduce the risk of extreme events, which is consistent with the objectives of the United Nations' International Strategy for Disaster Reduction (ISDR) programs.

Further, the IOM's projects on Disaster Risk Reduction (DRR) and mitigating environmental degradation also are consistent with the Sendai Framework for Disaster Risk Reduction (DRR) outlined for the 2015–2030 period. The Framework and the IOM are committed to reducing risk from natural disasters and increasing community resiliency by promoting mobility-based strategies. Both programs have paid particular attention to at-risk communities, migrants, and other vulnerable mobile populations. Their primary aims are to: (1) reduce disaster-induced displacement by mobilizing the dimensions of mobility in disaster prevention and preparedness; (2) alleviate the impacts of displacement through risk-informed response; (3) bolster resilience by building back better in recovery and reconstruction; and (4) enlarge partnerships with government and local authorities to support integration of mobility dimensions in global risk reduction efforts. The IOM considers these to be four thematic pillars of work (IOM 2017).

CLIMATE CHANGE

The involvement of the IOM seems positioned to be more intense in the future with the recognition that climate change is a global reality and will in all likelihood lead to frequent natural disasters and freak weather events. Many people in many countries with coasts likely will be displaced from coastal areas in the near future. For example, the First Assessment Report of the Intergovernmental Panel on Climate Change (IPCC) estimated that climate change-induced sea level rise along the coast of Bangladesh could range from 3.281 feet (1 m) by 2050 to 6.56 feet (2 m) by 2100. This means that by 2050, about one-fifth of the land area of Bangladesh could be lost due to sea level rise, which would displace 15 million people.

A report published in *New Scientists* on September 7, 2017, claims that at least eight low-lying islands in the Pacific Ocean have disappeared due to sea level rise. The report further claims that five of the Solomon Islands of western Pacific have been lost to climate change since the middle of the last century (Klein 2017). In this part of the Pacific Ocean, the sea level is believed to have risen by up to 0.4 inches (12 mm) over the last 50 years. Disappearance of other small islands in the Pacific was also reported by other climate change scientists in Australia.

Additionally, remote sensing images have revealed that another six low-lying islands in Micronesia became submerged between 2007 and 2014. As sea levels continue to rise, many residents of low-lying islands across the world will be forced to move to higher ground. This is already happening in several islands of Papua New Guinea, where a resettlement scheme is underway to move the population to a higher island 60 miles (90 km) away.

CONCLUSION

Since the last decade of the twentieth century, the IOM has been supporting people affected by the impact of natural disasters. It has been doing this by disbursing humanitarian assistance, reducing and preventing risk of natural disasters, increasing community resiliency, providing transitional shelters and settlements, and setting up the roadmap for the safe and orderly return and reintegration of the displaced population. Additionally, the organization has provided critical services to both governments and migrants. For instance, it helped Asian and African refugees during migrant crises in recent years and the refugees at the U.S.-Mexico border. Naturally then, the role of the IOM is predicted to increase in years to come because natural disasters will intensify with climate and environmental change, fast-paced urbanization, and population growth.

Bimal Kanti Paul

Further Reading

IOM (International Organization for Migration). 2007. *Migration, Development and Natural Disasters: Insights from the Indian Ocean Tsunami.* Geneva, Switzerland: IOM.

IOM (International Organization for Migration). 2012. *Haiti: From Emergency to Sustainable Recovery IOM Haiti Two-Year Report (2010–2012).* Port-au-Prince, Haiti: IOM.

IOM (International Organization for Migration). 2017. Disaster Risk Reduction and Environmental Degradation. https://www.iom.int/disaster-risk-reduction, accessed March 7, 2019.

Klein, A. 2017. Eight Low-lying Pacific Islands Swallowed Whole by Rising Seas. *New Scientist.* September 7, 2017. https://www.newscientist.com/article/2146594-eight-low-lying-pacific-islands-swallowed-whole-by-rising-seas/, accessed August 8, 2018.

Paul, B. K. 2019. *Disaster Relief Aid: Changes and Challenges.* Gewerbestrasse, Switzerland: Palgrave-Macmillan.

PBS (Policy Brief Series). 2018. *Central and North America: Migration and Displacement in the Context of Disasters and Environmental Change.* Geneva, Switzerland: IOM.

Islamic Relief Worldwide

Islamic Relief Worldwide (IRW) is an independent development and humanitarian organization. With an active presence in over 40 countries through national offices, affiliated partners, and field offices, this international aid agency focuses on providing poor people with access to vital services, such as healthcare, sanitation, education, and sustainable means out of poverty, protecting communities from disasters and delivering lifesaving emergency aid, and empowering vulnerable people to transform their lives and their communities. IRW works in partnership with other international aid organizations, church groups, and local relief agencies to help individuals and communities irrespective of race, religion, political affiliation, and gender.

Based on the values and teachings of the Koran, the central religious text of Islam, and the prophetic example (*sunnah*), IRW was established in 1984 in response to the famine in Africa. It was founded by Dr. Hany El-Banna and students from the University of Birmingham in the United Kingdom. A pathologist by education, Dr. El-Banna was attending a conference in Sudan in 1983 during a period of famine in the region; the poverty and distress he witnessed compelled him to return to the United Kingdom and set up IRW to help people in need. It is claimed by Mohammed Kroessein, associate in the Research and Policy Unit at the headquarters of IRW in Birmingham, that El-Banna wanted his yearly *zakaat*—the yearly alms that Muslims who can afford give to the poor, and the third pillar of Islam—to be donated to an institutional humanitarian cause in favor of Muslims in hungry Africa (Danckaers 2008). The organization launched an appeal, and through door-to-door collections raised by supporters, it provided food for people affected by the famine. In 1985, IRW inaugurated its first project to sponsor a chicken farm in Sudan; that same year, a modest office was established in Mosely, Birmingham, from where the organization raised money for the famine (IRW 2019).

In the next decade, IRW extended its work in Mozambique, Iran, Pakistan, Malawi, Iraq, and Afghanistan, among other countries. Starting with strong grassroots community support, today IRW raises funds across Europe, America, Australia, and with its international headquarters based in Digbeth in Birmingham, United Kingdom, it has offices in Bangladesh, United States, Germany, Iraq, Lebanon, Sweden, Australia, Malaysia, South Africa, and Mali. In 2017, IRW reported an annual income of $168 million through private donations and funding from UN agencies, USAID, and other institutions (Charity Commission 2019). In addition, it registered with the Charity Commission of England and Wales in 1989, and it remains an independent, nonpolitical, nongovernmental organization (NGO). IRW is also a signatory of the Red Cross Code of Conduct, an international standard on working with people and communities affected by emergencies in a nonbiased manner and providing aid without discrimination.

GOVERNANCE

IRW operates through numerous entities across the globe. The main task of IRW's independent national entities when first established was funds development for international humanitarian projects; today, their tasks have expanded to include advocacy and implementation of local projects. IRW's field offices in Africa, the Middle East, Asia, and Southern Europe are also involved in humanitarian projects, fundraising, and advocacy. IRW's headquarters in Birmingham, UK, oversees global standards, and a few of its functions are: (1) it identifies new sources for funds development; (2) it oversees efficiency of response to emergencies and disasters; (3) it supports IRW family members with marketing and other media and publicity material; (4) it protects the organization's reputation and intellectual property; (5) it coordinates multi-lateral institutional relations, such as with the UN Economic and Social Council and the UN; and (6) it develops and coordinates its global strategy.

The Board of Trustees of IRW is responsible for the governance of the organization. The day-to-day running of IRW is delegated to Executive Directors, headed by the Chief Executive Officer (CEO). Executive directors make sure that the policies laid down by the Board of Trustees are implemented and support the work of other staff and volunteers. Currently, the global governance structure of IRW has the following aims: (1) a tighter IRW family working together to reach the needy in every part of the world efficiently and quickly, (2) greater transparency and accountability at all levels, (3) full participation of members of the IRW family in its operations, and (4) operating to the highest standards globally.

IRW has also identified steps to sustain and bolster this new governance structure; some of the steps are: (1) IRW Board of Trustees are elected by representatives of eligible IRW family organizations and individual members; (2) eligible family organizations (and members) are those independent national Islamic Relief organizations that have signed a license agreement with IRW, including agreement to international operating policies, procedures, and standards; (3) representatives of eligible IRW family organizations collectively form the IRW International Assembly (IA); (4) the IRW Board members are elected for a term of four years and may be re-elected for only one further consecutive term; (5) there is a transitional period of four years, during which all necessary legal and policy requirements to realize the new governance arrangements are worked out and agreed on; and (6) during the transitional period, half of the members of the existing Board resign, allowing new members to be elected to the Board by the IA.

GLOBAL STRATEGY

IRW's global strategy emphasizes the role organizations can play in responding to humanitarian and development challenges in the world. By 2021, IRW aims to become one of the leading humanitarian International NGOs (INGO) and the leading humanitarian INGO in the Muslim world (IRW 2019). In line with that goal, IRW hopes to be one of the first to respond to international appeals for assistance and to mobilize resources and act in response to humanitarian needs. It aims to be recognized in the INGO sector for its expertise in delivering shelter during times of emergency and in helping individuals and communities become resilient. The aim is to also influence the policies and practices of governments regarding humanitarian priorities, the need for community-based preparedness, and climate change adaptation.

VISION, MISSION, AND VALUES

IRW envisions a world where individuals and communities are empowered and social obligations are fulfilled, with people responding as one to the suffering of others. With this as its vision, IRW's mission is to mobilize resources, build partnerships, and develop local capacity to: (1) enable communities to prepare for the occurrence of disasters; (2) support communities to lessen the effect of disasters and respond effectively by providing relief, protection, and recovery; (3) promote integrated development with the aim to create sustainable livelihoods; and (4) empower the marginalized and the vulnerable to voice their needs and address the root causes of poverty through IRW's programs and advocacy efforts.

IRW's vision and mission are guided by the Islamic principles of sincerity (*ikhlas*), excellence (*ihsan*), compassion (*rahma*), social justice (*adl*), and custodianship (*amana*). To elaborate, in responding to suffering, IFW's efforts are driven by sincerity to God and a need to fulfil the obligations to humanity; it's actions to tackle poverty are marked by excellence of operations and deportment; motivated and driven by compassion, IFW believes in the well-being of every life and thus comes together with other humanitarian actors to allay suffering brought on by disasters, poverty, and injustice to individuals and communities; with an eye toward social justice, IRW's work is founded on empowering the dispossessed and transforming communities worldwide; and finally, IRW works as a transparent and accountable custodian of the earth and its resources in a way that connects these resources to humanitarian and developmental endeavors for the poor and vulnerable. IRW is also a policy leader on Islamic humanitarianism, and its research programs focus on developing practical solutions to key problems. Thus, IRW's efforts are broadly categorized under two types of programs—humanitarian and developmental.

HUMANITARIAN EFFORTS

A few of IRW's major humanitarian interventions include providing lifesaving aid during the Bosnia and Kosovo wars in the 1990s and medical assistance during the Afghanistan and Iraq wars. It has also responded to natural disasters including the 2004 Asian tsunami and the Kashmir earthquake in 2005. IRW supported 2.5 million people at risk of famine and disease in Yemen (Bisset 2018), assisted violence-afflicted communities in Myanmar, and delivered aid to 1.4 million people in war-torn Syria (BBC n.d.). IRW runs an annual global food distribution program during the holy month of Ramadan.

DEVELOPMENTAL EFFORTS

The goal of IRW's development programs is to focus on long-term sustainability through climate adaptation and livelihood support, including Islamic microfinance and orphan sponsorship programs. In 2001, IRW set up Waqf programs supported by donors and run as an Islamic endowment scheme, a Waqf being a sustainable and ongoing charitable endowment widely used throughout Islamic history to develop and support communities. In 2006, IRW signed a Program Partnership Agreement with the UK government's Department for International Development, recognizing its capacity to contribute to the Millennium Development Goals. IRW's current development strategy is also aligned to support the UN Sustainable Development Goals.

ADVOCACY INITIATIVES

In August 2015, IRW launched an Islamic Climate Change Declaration with GreenFaith and the Islamic Foundation for Ecology and Environmental Sciences. The objective was to call on Muslims to actively participate in climate action at local, national, and international levels. The declaration made an Islamic faith-endorsed case for protecting the environment with support from global Muslim leaders. In 2017, IRW implemented 50 projects worldwide to reduce the impact of climate change (IRW 2019).

In 2018, as part of its gender justice work, at the UN Commission on the Status of Women conference in New York, IRW announced a forthcoming Islamic Declaration on Gender Justice. It is a call to action against gender inequality from an Islamic faith perspective and seeks to tackle discrimination and harmful practices, especially against women and girls in Muslim communities. This initiative is based on key Islamic principles of justice and balance to challenge cultural practices and social traditions that leave women and girls disadvantaged and vulnerable.

CONTROVERSIES

Muslims generally strive to be generous and humble in their charitable contributions, with *zakaat* being one of the five pillars of Islam, but it is getting increasingly harder for individuals and organizations to do so in today's climate of suspicion and fear. For instance, some Islamic organizations have been shut down on allegations or proof that they have rerouted funds to terrorist causes. IRW has also been mired in its share of controversies.

In June 2014, Israel added IRW to a list of organizations banned from operating in Israel for allegedly funding Hamas, and its West Bank offices were raided. In response, in late 2014, IRW asserted that an audit carried out by a global audit firm found no evidence of links to terrorism. The Israeli government in turn responded by claiming that its decision to declare IRW illegal was based on information accumulated over years. IRW has challenged the decision in the Israeli courts.

Consequently, on November 15, 2014, the United Arab Emirates (UAE) placed IRW on a list of proscribed organizations due to alleged links to the Islamist Muslim Brotherhood (MB), which the UAE considers to be a terrorist organization; IRW has denied links to the MB. In 2016, it was reported that the banking group HSBC decided to sever ties with IRW over concerns that cash meant for humanitarian aid could potentially end up with terrorist groups abroad. The government of Bangladesh barred IRW from aiding the Rohingya people in Cox's Bazar, alleging funds were used to preach Islam, construct mosques, encourage radicalism, and fund militants. IRW has refuted these allegations too and continues its humanitarian and developmental work across Bangladesh, where it has been operating for over 25 years.

CONCLUSION

Islamic Relief Worldwide is a professional international relief organization with its ideology based on the ethics of Islam. With its humanitarian, developmental, and advocacy initiatives, it is also considered to be the largest Islamic aid agency in Great Britain. Despite criticisms and controversies, IRW has provided aid to individuals and communities in a sustained and unbiased manner, thus showing that humanitarian and/or developmental initiatives need not be exclusive to organizations that have secular or Western/Christian ideals.

Soumia Bardhan

Further Reading

BBC. n.d. Syrian Refugees Tell Their Stories. https://www.bbc.co.uk/programmes/articles/1yk0lNKbDg3vY77HKtFwV4/syrian-refugees-tell-their-stories, accessed August 27, 2019.

Bisset, V. 2018. Yemen Conflict: Hudaydah's "Calm before the Storm." *BBC News*. https://www.bbc.com/news/world-middle-east-44482432, accessed August 29, 2019.

Charity Commission for England and Wales. 2019. Islamic Relief World. http://apps.charitycommission.gov.uk/Showcharity/RegisterOfCharities/CharityWithPartB.aspx?RegisteredCharityNumber=328158&SubsidiaryNumber=0, accessed August 28, 2019.

Danckaers, T. 2008. In the Name of Allah: Islamic Relief Organizations in Great Britain. Mondiaal Nieuws. https://www.mo.be/en/article/name-allah-islamic-relief-organizations-great-britain-last, accessed August 29, 2019.

IRW (Islamic Relief Worldwide). 2019. https://www.islamic-relief.org/, accessed August 29, 2019.

Lutheran World Federation

The Lutheran World Federation (LWF) is an international humanitarian organization that specializes in providing emergency relief to survivors of natural disasters. The organization also prepares people for such destructive events, works for improving their resiliency to natural disasters, and promotes principles of sustainable development. The vision, purpose, and values of the organization center on areas of interfaith relations, humanitarian assistance, international affairs and human rights, and mission and development. LWF is committed to addressing human needs with empowering and effective service worldwide, resulting in assistance for 2.7 million people annually. Organizational membership represents more than 75.5 million Christians in the Lutheran tradition in 99 countries. LWF is a global community of 148 member churches, including two associate member churches, 10 recognized churches and congregations, and two recognized councils. In addition to its membership in the Action by Churches Together (ACT) Alliance, a global network of Protestant and Orthodox churches and church-related nongovernmental organizations (NGOs) committed to working together for humanitarian aid and development, LWF continuously receives high rankings from Charity Navigator, Charity Watch, GreatNonprofits, and the Better Business Bureau. Additionally, LWF is the fifth largest partner of the United Nations High Commissioner for Refugees (UNHCR).

HISTORY

In 1947, LWF was founded in Lund, Sweden, as a federation of Lutheran churches. Initial attempts to establish a global association of Lutheran churches were proposed in the first and second Lutheran World Conventions held in Europe in 1923 and 1929, respectively. LWF was formed at the third convention in Lund. This is an umbrella organization of the various national Lutheran churches.

A primary objective for the formation of LWF was to end human suffering and provide aid for European people and others in the aftermath of World War II (1939–1945). Because not all Lutheran churches are members of this organization, however, LWF primarily serves as a forum for intra-Lutheran discussion and ecumenical consultation with other churches.

Since its founding, LWF has held world assemblies, called by the president of the organization every five to seven years. Although LWF has headquarters in Geneva, Switzerland, the organization's 500 World Service staff members are responsible for all programs and projects worldwide. The world assembly elects the president and other members of the executive committee. The executive committee meets annually and is responsible for electing the general secretary, who is a full-time employee of the organization. The committee provides guidance for the organization's work and oversees 7,548 local, regional, and international staff and volunteers.

CORE COMMITMENTS

The five core commitments of LWF include human rights protection and peace, impartiality, accountability, gender justice, and climate justice and environmental protection. The first commitment is associated with human rights education and advocacy, peace building, and conflict resolution. LWF considers respect for human rights to be fundamental to poverty reduction or elimination and asserts that every person has dignity and the inherent right to achieve sustainable livelihoods and food security. The organization is committed to ensuring that human rights are preeminent, thereby producing justice, peace, and sustainable development. LWF seeks to understand the causes, drivers, and triggers of violent conflict and accordingly initiates necessary long-term programs and interventions to resolve conflict. It also promotes and adheres to international humanitarian law.

The second commitment, impartiality, means that LWF assists people affected by natural and man-made disasters based on need without discrimination or favoritism and irrespective of race, religion, ethnicity, gender, political affiliation, and socioeconomic class. Thus, the organization secures equal access to food, water, and other necessary emergency items for disaster survivors. LWF strictly follows and promotes the Code of Conduct for the International Federation of Red Cross and Red Crescent Societies (IFRC) and nongovernmental organizations (NGOs) in participating in disaster relief efforts. That is, emergency aid is given neutrality regardless of the race, creed, or nationality of the survivors of natural disasters. The organization calculates aid priorities based on need alone (IFRC 2013).

The third core commitment, accountability, means LWF maintains a culture of accountability at all its activities. In 2012, for this reason, LWF received the 2010 Humanitarian Accountability Partnership (HAP) Standard in Accountability and Quality Management certification. The organization is committed to strengthening accountability practices, particularly in the areas of participation, information sharing, and complaint-handling mechanisms.

Gender justice, the fourth core commitment of LWF, pertains to the conviction that women can play a critical role in economic and social development and ensures the equal participation of women in all programs and projects. LWF integrates gender perspectives, undertakes specific advocacy to change attitudes and practices and institutionalize gender justice, and encourages female empowerment through creating specific opportunities and projects for women.

The fifth and final core commitment relates to climate justice and environmental protection. LWF considers climate change to be a climate emergency. It advocates for global climate justice but prioritizes local communities that are extremely vulnerable to the effects of climate change. Considering the escalating effects of climate change on developing countries, particularly poor and marginalized people groups, LWF recommends adaptation activities and promotes initiatives to mitigate greenhouse gases. In addition, the organization expresses its deep concern for environmental degradation by initiating development programs and supporting sustainable strategies. In addition, LWF integrates ecological considerations into agricultural and emergency programs and promotes natural resource protection and education. LWF partners with communities to reduce disaster risks, build sustainable livelihoods, and improve resiliency.

DISASTER RELIEF

LWF has proactive partnerships with communities to prepare for natural disasters, rapidly respond to the needs of survivors of extreme events and complex emergencies, and increase community resilience. The organization prepares communities to care for their most vulnerable populations, including the elderly, women, and children. Its emergency responses to natural disasters and complex emergencies include providing food, shelter, water, medicine, and other lifesaving and sustaining activities at the onset of a crisis. In addition to material aid, LWF provides psychosocial support to disaster survivors, particularly survivors who are most vulnerable and least able to exercise their human rights. Through its regional emergency response hubs in San Salvador, Nairobi, Kathmandu, and Lusaka, LWF quickly disburses necessary relief items to disaster survivors. If necessary, LWF can also rapidly deploy additional staff by coordinating with other humanitarian organizations at global, national, and community levels to reduce duplication and gaps in assistance. Further, it regularly collaborates with the ACT Alliance in respond to natural disasters.

Utilizing disaster risk reduction (DRR) and emergency preparedness, LWF builds resilient communities and reduces vulnerability before disaster strikes. When crises arise, the organization partners with communities to initiate measures that reduce disaster risk, including development of early-warning systems. In addition, LWF helps communities establish contingency plans, food security measures, and public shelters. It also helps disaster-displaced households return to homes within their home countries or resettle in other countries. LWF supports member churches as they work together for risk reduction, emergency preparedness, and emergency responses. Although most humanitarian organizations participate in disaster response and recovery efforts, LWF links relief and recovery to development programs.

LWF has a history of participating in response and recovery efforts in major natural disasters that occurred in the United States. For example, LWF assisted after Hurricane Katrina hit the Gulf Coast in 2005, as well as after Hurricane Sandy occurred in 2011 and Hurricane Matthew in 2016, which also severely hit Haiti. Immediately after these extreme events, LWF provided survivors with shelter, food, water, hygiene kits, cash, and psychosocial support to survivors. Five months after Hurricane Matthew affected Haiti, LWF supported people in recovering from the hurricane and rebuilding their lives (French 2017). In fact, even before these hurricanes had formed in the Atlantic Ocean, LWF began preparing people in the areas of potential impact.

Similarly, after the 2015 Nepal earthquake, LWF helped in rapidly providing emergency assistance to survivors, rebuilding homes and schools, restoring water and sanitation systems, improving livelihood, and providing job-training and income-generating projects such as beekeeping and farming. LWF also worked with local partners and international humanitarian organizations to help the migration crisis in the Middle East, particularly in Syria, Central America, and South Sudan, Africa.

CLIMATE CHANGE

As one of its five core commitments, LWF wholeheartedly supports adaptation and mitigation measures to reverse global climate change. It believes that the world has been facing a climate change crisis, and it supports a recently resealed report of the Intergovernmental Panel on Climate Change (IPCC) that states that global temperature has increased to 2.7 degrees Fahrenheit (1.5 degrees Celsius) above preindustrial levels. This increase has created a significant barrier to sustainable development and efforts to eradicate poverty (LWF 2018). The organization also believes that recent disasters, such Hurricane Dorian, Cyclone Idai, the

Amazon fires, flooding in India, Bangladesh and Myanmar, and drought in the Horn of Africa, are evidence of a global climate crisis.

Although LWF primarily blames industrialized development countries for climate change, most victims of this event are people of developing countries who have contributed least to climate change and are most significantly affected. These countries are still contributing about half of the total human-caused greenhouse gas emission. For this reason, LWF maintains that developed countries must take stringent steps to curb their share of emission, as well as they should also subsidize some of the costs of emission controls in developing countries.

At the Climate Action Summit on September 23, 2019, the ACT Alliance, LWF, and the World Council of Churches, which collectively represents 580 million Christians globally, unanimously called for climate justice and immediate action. LWF urged governments worldwide to take stronger measures and present concrete plans to immediately address climate change. Further, it urged developed countries to drastically decrease carbon emission, "increase their ambition in meeting emissions targets, in providing financing, and in focusing on adaptation and mitigation for those most affected by climate change" (LWF 2019). The existing levels of greenhouse gases in the atmosphere can be either stopped or significantly reduced by a number of ways: reduction of energy consumption, use of alternative energies, changes in behavior, and carbon trading, which is a market-based way to lower greenhouse gases in the atmosphere.

SOURCES OF FUNDING

In 2017, 34 percent of LWF funding came from Augusta Victoria Hospital, 31 percent came from Lutheran member churches and related agencies, 29 percent came from United Nations (UN) and government grants, and the remaining 6 percent came from non-project-related and other sources. In 2017, LWF received $187 (ERO 155.5) million; total expenditures for that year were $181 (EUR 151) million, with nearly half of those expenditures being DRR and emergency response. LWF regularly receives gifts, donations, and other contributions to strengthen its financial stability and assure a strong future. Nearly half of the expenditure in 2017 went into disaster risk reduction. Coordination overheads costs were maintained at 3 percent of the total yearly income. Other expenses included nonproject expenditures (1 percent), community-led action for justice and peace (4 percent), sustainable livelihoods (11 percent), and Augusta Victoria Hospital (32 percent) (LWF n.d.).

CONCLUSION

LWF has a reputation for quickly responding to natural and man-made disasters, particularly events in Asia, Africa, and Latin America. Although it primarily works behind the scenes, the organization employs its extensive network of local partnerships around the world to work effectively in postdisaster situations. The sustainability of LWF is intrinsic to the way in which it responds to disasters. LWF is generally committed to disaster-affected areas far beyond the response phase of the disaster management cycle. Whenever needed, the organization may be involved in monitoring the response. LWF field officials regularly meet with community members, leaders, and representatives from local partners to review progress and ensure that assistance is aiding the community.

Anjana Paul

Further Reading
French, M. 2017. The Slow Road to Recovery after Hurricane Mathew. March 3, 2017. https://www.lutheranworld.org/news/slow-road-recovery-after-hurricane-matthew, accessed October 4, 2019.
IFRC (International Federation of Red Cross and Red Crescent Societies). 2013. *World Disaster Report: Focus on Technology and the Future of Humanitarian Action*. Geneva, Switzerland: IFRC.
LWF (Lutheran World Federation). n.d. Where Our Funds Came From. https://www.lutheranworld.org/content/dws-finances, accessed October 4, 2019.
LWF (Lutheran World Federation). 2018. Climate Justice: Time to Ramp Up Efforts. October 10, 2018. https://www.lutheranworld.org/news/climate-justice-time-ramp-efforts, accessed 4 October 4, 2019.
LWF (Lutheran World Federation). 2019. As Climate Summit Begins, Churches Call for Action Now! September 23, 2019. https://www.lutheranworld.org/news/climate-summit-begins-churches-call-action-now, accessed October 4, 2019.

Mennonite Central Committee (MCC)

The Mennonite Central Committee is a global, nonprofit faith-based organization that strives to make peace across the globe through emergency relief and development programs. It accomplishes all its programs with local partners and churches. It is a worldwide ministry of Anabaptist churches, whose members include several Christian groups such as the Mennonites, the German Baptists, the Amish, and the Hutterites. Approximately 4 million Anabaptists live in the world; most of them are in North America. More than 50 percent of Anabaptists are Mennonite, and for this reason, MCC is popularly known as a Mennonite organization.

When responding to disasters, MCC works with local groups to distribute relief aid in ways that minimize conflict. In development, it partners with community and church groups to make sure the projects meet their needs. And the organizations advocates for policies that will lead to a more peaceful world. Anabaptism's peacemaking is rooted in a unique historical and theological heritage, and MCC is best known for working for peace by other international relief and development nongovernmental organizations. Its "peacemaking is evident in the choice of its partner organizations, the personal ethics and religious beliefs of its employees, and its particular ethos of humble, relationship-oriented work" (Welty 2016, 534). Since the 1980s, justice has become increasingly integrated into the definitions of MCC's peace.

VISION AND MISSION

The Mennonite Central Committee focuses its work in five strategic areas of service: (1) caring for the lives and futures of uprooted and other marginalized people; (2) providing food, water, and shelter first in times of hunger, disaster, and conflict and then education and ways to earn income; (3) working with churches and communities to prevent violence and promote peace and justice; (4) investing in opportunities for young people to serve in Canada, the United States, and around the world; and (5) serving with humility and in partnership to meet local needs with local solutions (MCC 2019b). The mission statement of MCC does provide a nuanced approach to peace, which is not achieved through economic sanctions, or armed intervention, but through building "right relationships."

A brochure of MCC outlines the 10 primary ways to engage in peacemaking: cultivating a personal spirit of peacefulness, providing a reconciling presence in the midst of tension, supporting local peacemaking efforts, explaining peacemaking as at the heart of Christian life, providing trainings in peacemaking, sponsoring seminars or meetings with elected officials in North America, building relationships with church and community leaders overseas, sharing peacemaking information with North American constituents, and sponsoring seminars on peace (Welty 2016). These 10 ways for peacemaking reveals MCC's belief that peacemaking is rooted in personal relationships. In fact, most of MCC's relief and development work is also driven by a peace ethic, particularly in Africa. Its volunteers demonstrate their commitment to nonviolence and peace both by incorporating peacemaking into their assignments and by exemplifying it in their daily lives. Similarly, its partners also exhibit personal as well as professional commitments to peace (Welty 2016).

STRUCTURE OF THE ORGANIZATION

MCC was founded on September 27, 1920, in Chicago. MCC has headquarters in both the United States and Canada. The U.S. headquarters are in Akron, Pennsylvania, and the Canadian headquarters in Winnipeg, Manitoba. In the United States, there are four regional offices: MCC Central States (headquarters at North Newton, Kansas), MCC Great Lakes (headquarters at Goshen, Indiana), MCC East Coast (headquarters in Philadelphia, Pennsylvania), and West Coast MCC (headquarters in Fresno, California). Currently, MCC has seven affiliated organizations: Beachy Amish Mennonite Churches, Brethren in Christ Church, Conservative Mennonite Conference, Mennonite Church Canada, Mennonite Church USA, Mennonite Disaster Service, and U.S. Conference of Mennonite Brethren Churches.

MCC has also an office in the United Nations (UN). This office keeps MCC staff and partners informed about UN policies and program that may affect them. It also brings the expertise and experience of MCC partners and personnel to the UN diplomats and employees who establish and carry out UN policies. MCC United Nations office has five specific efforts: (1) networking with

UN diplomats and ambassadors to advocate for international policies that reflect MCC's stances on peace and reconciliation, (2) leading and participating in NGO working groups to advocate for UN action together as members of the civil society, (3) connecting MCC staff or local partners with members of the UN community, (4) advocating for humanitarian and human rights issues in ways that reflect MCC's commitments and support its programming, and (5) articulating and pursuing a Christian pacifist vision for peace and justice in international affairs (MCC 2019a).

Both MCC United States and MCC Canada jointly employ administrators in both countries, manage international programs, and share financial responsibility for it. About 775 people from the United States and Canada, together with 230 people employed in-country, carry out MCC's work. As noted, overseas, MCC works in partnership with churches, communities, and faith-based and civil society organizations. Every year, more than 100 young adults from around the world leave their homes to serve in another country through MCC's Global Service Learning programs. However, MCC staff outside these two countries includes country representatives, area directors, service workers, and SALT (Serving & Learning Together)/YAMEN (Young Anabaptist Mennonite Exchange Network) volunteers, along with salaried national staff. Country representatives are the direct supervisors of local staff and the indirect supervisors of service workers and the volunteers. They report to area directors who are responsible for regional clusters. MCC selects its country staff including volunteers based on their commitment to peacemaking, and it trusts them to replicate Mennonite principles of peacemaking in the field.

However, MCC service workers typically commit to a three-year term during which they are provided with housing, food, medical insurance, and other necessities in addition to a small monthly stipend. Their work assignments vary, but many are technical advisors doing capacity-building work with the local organizations to which they are attached. The local organization is the primary supervisor for service workers. All global programs of MCC are led by representatives who serve for five years or longer during which time their basic needs and a small stipend are provided. Service workers typically commit to a three-year term during which they are provided with housing, food, medical insurance, and other necessities in addition to a small monthly stipend (Welty 2016).

In 2018, the organization supported 716 programs in 57 countries around the world and worked with 508 different partner organizations (MCC Annual Report 2018). In fiscal year 2017–2018, MCC spent nearly $65.8 million: $17.8 million on relief, $35.8 million on development, and $12.2 million on peacebuilding activities (MCC Annual Report 2018). The primary sources of MCC's financial support are individuals and congregations of the Mennonite and Brethren in Christ churches in Canada and the United States. The support also comes from a wide variety of generous individuals who identify with the mission and values of MCC. MCC thrift shops and relief sales supply about 20 percent of MCC's budget. Many supporters can donate material resources, such as comforters and relief kits full of supplies, to be sent to people in the midst of crisis.

ACTIVITIES

Over the past decade, MCC worked every country in the world, except Antarctica and Australia. It has offices in more than 50 countries and also offer relief, development, or peacebuilding work in at least 10 more. Currently, it serves the highest number of countries in Africa (20), followed by Asia (19). Its partners and the number of projects differ by year. In a typical year, MCC is involved in over 700 projects. From these projects, several recent projects are presented here.

Cyclone Response

Cyclone Fani made landfall on May 3, 2019, in Odisha, India. The cyclone affected 16 million people, killing 89 people, and it resulted in damages totaling $1.9 billion. In preparation for the advancing cyclone, the Odisha state government evacuated over 1.2 million residents from vulnerable coastal areas and moved them into 9,000 cyclone shelters built a few miles inland. The authorities opened about 7,000 kitchens to feed evacuees. In Bangladesh, the government opened shelter in 19 coastal districts and evacuated more than 1.2 million people. Luckily, the cyclone did not cause discernible damage in Bangladesh. However, with local partners, MCC participated in Odisha cyclone relief efforts and provided emergency food supply to cyclone survivors. It also monitored the storm as it approached the state of West Bengal and Bangladesh and assessed the situation with local partners. MCC has worked in India for more than 75 years and in Bangladesh since its independence in 1971.

Nearly two months earlier, Cyclone Idai made landfall near the coastal city of Beira in Mozambique. About 3 million people were affected by Cyclone Idai, killing more than 730 people and displacing 200,000 people. Additionally, it destroyed at least 988,421 acres (400,000 ha) of crops. Most of the displaced people were living in makeshift camps and temporary shelters such as schools. More than 1,000 cases of cholera, diarrhea, and malaria were reported during the postdisaster period. The cyclone also affected people of neighboring Malawi. MCC provided food assistance for the cyclone victims for two months. Additionally, the organization distributed tarps, pails, blankets, soaps, hygiene kits, and school supplies among the victims. In providing emergency assistance, MCC partnered with the Christian Council of Mozambique.

Refugee Response

Recent political instability and economic crisis in Venezuela caused 3 million Venezuelans to flee to neighboring countries in Latin America and the Caribbean. These forced migrants urgently needed food assistance, medical services, shelter, and legal support. Through ongoing projects in Colombia, which hosts more than 1 million Venezuelan refugees, and Ecuador, MCC appropriately responded to the needs of the displaced people from Venezuela. In Colombia, MCC partnered with the Mennonite Church in Riohacha to provide daily meals and worked with the Mennonite Brethren Church in Valle del Cauca to deliver food, household items, shelter, and small business support. In Ecuador, MCC partnered with the Quito Mennonite Church to provide food, medical support, hygiene items, stoves, blankets, and start-up funds for refugee families.

In addition to Venezuela, MCC has been working with refugees displaced by wars and other violence. For example, in war-torn Syria, MCC has been working with local churches to provide financial support to those who are internally displaced to buy food, water, heating fuel, or housing. In Iraq, Lebanon, and Jordan, MCC's partners also help displaced people with basic needs, trauma healing, and vocational training.

The Democratic Republic of Congo Emergency Response

The Democratic Republic of Congo (DRC) has been facing ongoing violence across the country, particularly in the Ituri, Kansai, and Kivu regions. Recent political violence centered around the national election that was held in late December 2018. More than 100 armed groups are believed to operate in these regions. Despite the presence of UN peacekeepers, these groups continue to terrorize communities and control weakly governed areas. Millions of civilians have been forced to flee the fighting: the United Nations estimates that there are currently 4.5 million internally displaced persons in the DRC and more than 800,000 DRC refugees in other nations. In Kansai region alone, more than 1.4 million people were forced from their homes due to violence. About half of the displaced people have not been able to return home because it was not safe or because their homes and villages no longer exist. Families who are displaced still need food, education, medicine, and shelters until they can return home. MCC and Mennonite Brethren churches have been continuing to respond to the needs of displaced families by providing tools, seeds, and pigs to get started farming. .

Other Projects

Besides, MCC also supports other programs. For example, in Afghanistan, the organization has a project to improve maternal, newborn, and child health. This project is a joint venture with the Government of Canada and is expected to benefit 121,600 individuals through health and nutrition training, support for health facilities, construction and rehabilitation of water sources, and hygiene and sanitation training. In Latin America, MCC has been supporting small coffee farmers since 2003 so that they can get a fair price for their product. The organization is also providing drinking water to school students in Jordan.

CONCLUSION

MCC is a global, faith-based organization that strives to share God's love and compassion for all through relief, development, and peace. In all its projects, it is committed to serving the vulnerable people by establishing relationships with local churches and partners. As an Anabaptist organization, MCC strives to make peace throughout the world, and its projects and policies will lead to a more peaceful world.

Bimal Kanti Paul

Further Reading

MCC (Mennonite Central Committee). 2018. MCC Annual Report 2018. https://mcc.org/sites/mcc.org/files/media/reports/us/2018-11-20usannualreport2018web.pdf, accessed June 25, 2019.

MCC (Mennonite Central Committee). 2019a. MCC United Nations Office. https://mcc.org/get-involved/advocacy/un, accessed June 25, 2019.

MCC (Mennonite Central Committee). 2019b. Vision and Mission. https://mcc.org/learn/about/mission, accessed June 25, 2019.

Welty, E. 2016. A Mennonite Peace? An Analysis of Mennonite Central Committee's Work in East Africa. *Mennonite Quarterly Review* 90: 533–566.

Oxfam International

Oxfam International was formed in 1995 by a group of independent nongovernmental organizations (NGOs) with the general aim to work together all over the world through a local accountable organization for greater impact on the international stage in fighting structural causes of poverty and related injustice. Its specific objectives are to (1) prevent and relieve poverty and to protect vulnerable people; (2) support sustainable development; and (3) promote human rights, equality, and diversity. This charitable and humanitarian organization is also a world leader in the delivery of emergency relief items to survivors of natural as well as man-made disasters. In November 2000, Oxfam adopted the five rights-based approaches as the guiding principles for all the work of the confederation and its partners: the right to a sustainable livelihood, the right to basic social services, the right to life and security, the right to be heard, and the right to an identity.

The name "Oxfam" comes from the Oxford Committee for Famine Relief, founded in Oxford, England, in 1942 by a group of quakers, social activists, and Oxford academics. The group founded Oxfam to supply food to starving women and children in enemy-occupied Greece during the Second World War (1939–1945). After the war, Oxfam continued its philanthropic activities by providing materials and financial aid to alleviate the plight of poor people throughout Europe. As the situation of the poor improved in Europe, Oxfam's attention shifted to the needs of the people in developing countries, and in the late 1960s, its officials began to conceive of Oxfam as a global organization rather than just a leading British NGO. To realize this vision, it supported the establishment of Oxfams in a number of potential donor countries. For example, the U.S. Oxfam office was opened in Washington, DC, in 1970. Initially, Oxfam America received funding and loans exclusively through Oxfam Great Britain (Oxfam GB), and its original headquarters was located in Washington, DC, and then relocated to Boston in 1973. By the end of the 1970s, Oxfam America became both financially and administratively independent of Oxfam GB.

Today, 20-member organizations comprise the Oxfam International confederation. They are based in Australia, Belgium, Brazil, Canada, Denmark, France, Germany, Great Britain, Hong Kong, Ireland, India, Italy, Japan, Mexico, The Netherlands, New Zealand, Quebec, South Africa, Spain, and the United States. In addition to these 20 affiliates, Oxfam is currently supported by public engagement offices in South Korea and Sweden. The main objective of these offices is to raise money and engage the public in Oxfam's activities. The Oxfam International Secretariat is based in Oxford, and the organization works in more than 90 countries. In 2014, the Oxfam confederation consisted of more than 10,000 paid staff members and nearly 50,000 volunteers (Oxfam Novib n.d.).

FOCUS AREAS

Oxfam has four main focuses for its resources. Economic justice is its first focus, which means the organization aims at working with small-scale producers and farmers in developing countries, especially women, in their efforts both to earn a sustainable livelihood by providing access to appropriate marketing outlets and to increase their business skill and trading capacity. This focus is reflected in its Fair Trade program. Essential service is its second focus, and the organization demands that national governments be responsible for equitable delivery of good-quality health, education, water, and sanitation as well as be accountable for the delivery of essential services to their citizens. Oxfam also maintains that governments should wholeheartedly support civil society organizations. Rights in crisis is its third focus, which means the organization centers its efforts on improving the ability to deliver better protection and greater assistance and on working within the framework of human security. Lastly, gender justice focuses on empowering women, improving their status, and eliminating all forms of gender-based violence.

As noted, Oxfam International has been a leading supporter of "Fair Trade," focusing on the elimination of four trade practices: commodity dumping, high tariffs, unbalanced labor rights for women, and irregular patent issues. The organization maintains that dumping of agricultural crops is when highly subsidized surplus commodities from developed countries are sold at low prices in developing countries. Farmers in the latter countries are unable to compete with cheaper foreign crops. Next, high tariffs imposed by governments of developed countries make products of developing countries costlier than domestic products and thus restrict their sales. Thirdly, unbalanced labor rights for women refer to lower wages for women than for males in the same kind of works. Finally, irregular patent issues result in the increase in the prices of some items, such as medication and software, and thus such items are often out of reach for people of developing nations.

ORGANIZATIONAL STRUCTURES

Oxfam International has a two-tier governance structure comprising a Board of Supervisors and an Executive Board. The members of the Board of Supervisors include the Chair and also the Treasurer, if independent, and the Affiliate Chairs. The duties of the Board of Supervisors are to: (1) appoint and dismiss the Executive Director; (2) supervise the work of the Executive Board and the Oxfam International Secretariat and approve their reports and programs; (3) supervise the work of the Board of Supervisors' Committees; (4) approve and adopt the annual financial accounts of the organization; (5) approve the Oxfam Strategic Plan, Code of Conduct, and the Rules of Procedure; (6) address any other supervisory task or duty delegated to the Board of Supervisors; and (7) address the referral of resolutions or other documents of any nature.

The members of the Executive Board are Executive Directors of the Oxfam Affiliates whose duties and responsibilities include the following: (1) manage and organize the jointly agreed common activities of the Affiliates; (2) prepare, implement, and update the Oxfam Strategic Plan through consultations with the Affiliates; (3) maintain and manage the risk register of the organization and the affiliates; (4) prepare an annual accountability report, the Code of Conduct and Rules of Procedure, and the Oxfam Strategic Plan; (5) make proposals to the Board of Supervisors concerning the admittance or expulsion of Affiliates; and (6) prepare any other policy and strategy documents and programs and delegate the development of more detailed plans, policies, and protocols to working groups. Additionally, the Executive Board regularly updates the Board of Supervisors on its activity and informs the Board of Supervisors of any matters relevant to the overall well-being of the organization and the common interests of the Affiliates (Oxfam International 2018).

Although each Affiliate is independent and retains its own Board of Supervisors and an Executive Board, all 20 Oxfam Affiliates share a single Strategic Plan: "The Power of People Against Poverty," and they work together to build a fairer and safer world for all people, particularly for the poor and the vulnerable. However, a secretariat coordinates the Oxfam confederation. "It provides leadership, coordination and facilitation to the Confederation as a whole, and support to individual members where needed" (Oxfam Novib n.d. 13).

FUNDRAISING

Oxfam International and its Affiliates collect funding from many sources. Each Affiliate donates a maximum of 1 percent of its annual income to Oxfam International. All Affiliates seek funds from public, governmental, corporate, and multilateral donors. For example, more than half a million people make a regular financial contribution to Oxfam Great Britain (Oxfam GB) in the United Kingdom, including many London Marathon participants. Funds are also received from gifts in people's wills. Also, Oxfam has shops all over the world that sell many fair-trade and donated items such as handicrafts, old books, music CDs and instruments, coffee, second-hand clothing, furniture, chocolate, toys, and food. Oxfam buys fair-trade products from developing countries in Africa, Asia, and Latin America, and selling such items is not only profitable to the organization but also helps producers in developing countries achieve better trading conditions, which creates equity in the international trading system. The proceeds from selling these items support Oxfam's mission and relief efforts around the globe.

RELIEF EFFORTS AND CRITICISMS

Oxfam participates in disaster relief in all major extreme events throughout the world where charitable and relief organizations often receive more than their set goals. One of the important tenets of Oxfam is that it discontinues seeking fund for a particular disaster event when the organization reaches its target. For example, Oxfam America adjourned raising money for the 2010 Haiti earthquake after reaching its target of approximately $100 million. It also stopped fundraising for the 2004 Indian Ocean Tsunami after raising its goal of $225 million (McGowan 2014).

However, Oxfam International and its Affiliates are frequently criticized for many reasons such as misuse of relief funds, lack of timeliness and quality control in relief distribution, and being less than accountable for its emergency activities, although the organization's annual review provides public information regarding its work. Certainly, Oxfam International publishes progress reports, and Oxfam GB evaluates all its operations. Apparently, although Oxfam GB's "effectiveness reviews" are good at measuring whether or not its individual missions reach certain quality controls, they do not assess cost-effectiveness (McGowan 2014).

Furthermore, Oxfam often violates political neutrality, which is one of the three guiding principles of all humanitarian actions set by the United Nations. For example, Oxfam GB was strongly criticized in 2005 for becoming too close to the Labor government in the United Kingdom. Similarly, the organization supported the Palestinians and endorsed the two-state solution to end the Israeli-Palestinian conflict in 2013. As recently as a year ago, Oxfam solely blamed Israel for poor economic development in the Palestinian territories. The organization also criticized the U.S. government's slow and inadequate response to the crisis in Puerto Rico in the aftermath of Hurricane Maria in 2017. The group specifically criticized President Donald Trump's administration for not protecting the most vulnerable in times of humanitarian crisis.

Journalists all over the world have often accused Oxfam of deliberately providing inaccurate information to the press. For example, prior to the Davos, Switzerland, economic meeting in 2015, Oxfam released a report, claiming that the richest 1 percent of people in the world owned 48 percent of the world's wealth. Some critics questioned the accuracy of these statistics (Vara 2015). They further accused the organization of a lack of attempt to seek the opinions of disaster beneficiaries on the assistance being provided. Moreover, in 2006, Oxfam had a conflict with Starbucks about Ethiopian coffee, where the organization urged Starbucks to boost prices paid to farmers. In the following year, an Australian think tank, the Institute of Public Affairs, lodged a complaint, accusing Oxfam of violating the Trade Practices Act in its promotion of Fairtrade coffee. In both cases, Oxfam was accused of misleading conduct under the Fair

Trade Practices Act in ways that would benefit the agency rather than the producers of coffee.

Additionally, Oxfam has been trying aggressively to expand its specialist bookshops, but the charity has been criticized for two reasons. First, it has been adopting the tactics of multi-national corporations, and the expansion has come at the expense of independent, second-hand book sellers and other charity shops in many areas of the United Kingdom. Critics also claim that Oxfam has a corporate-style, undemocratic internal structure. They further accuse the organization of addressing the symptoms rather than the causes of international poverty.

Oxfam was also severely criticized for use of prostitutes by its workers while providing relief in Haiti immediately after the 2010 earthquake. The agency spent four years in Haiti launching long-term development programs. However, in this instance, the Oxfam country director was at the center of a sex scandal. Sources claimed that a group of Oxfam male workers committed sexual misconduct and that some of the prostitutes were girls aged 14–16, below the age of consent. In response, the UK government threatened to cut off its funding to Oxfam until the organization made wide-ranging reforms in its relief practices. Paying for sex is banned under Oxfam's code of conduct and is against UN guidelines for aid workers.

Oxfam investigated the Haiti sexual misconduct claims and dismissed a number of senior aid workers (Hodge 2018). Similarly, Oxfam's director of relief operations in Chad in 2006 and other staff had repeatedly used prostitutes at the Oxfam team house (Ratcliffe and Quinn 2018). More recently, Oxfam has alleged that its shop volunteers in the United Kingdom were sexually harassed by the staff. There are about 650 Oxfam shops across the United Kingdom, staffed by 23,000 volunteers.

CONCLUSION

Oxfam is an international confederation of charitable organizations focused on the alleviation of global poverty. The organization is determined to reduce the extent of poverty by working together with partner organizations and alongside vulnerable people to end the injustices that cause poverty.

Bimal Kanti Paul

Further Reading

Hodge, M. 2018. Abuse of Power: How Many Oxfam Shops Are There in the UK, Where Are They and What Did Helen Evans Says? *The Sun*. February 14, 2018. https://www.thesun.co.uk/news/5564456/oxfam-shops-uk-helen-evans-claims-allegations-sex-scandal/, accessed July 5, 2018.

McGowan, G. 2014. The Giving What We Can Trust. https://www.givingwhatwecan.org/post/2014/02/giving-what-we-can-trust/, accessed June 29, 2018.

Oxfam International. 2018. Our Governance. https://www.oxfam.org/en/our-governance, accessed June 29, 2018.

Oxfam Novib. n.d. Oxfam Novib Annual Report: About Us in 2014–2015. https://www.oxfamnovib.nl/Redactie/Pdf/Oxfam%20Novib_Annual%20Review%202014-2015_About%20us.pdf, accessed July 5, 2018.

Ratcliffe, R., and B. Quinn. 2018. Oxfam: Fresh Claims That Staff Used Prostitutes in Chad. *The Guardian.* February 11, 2018. https://www.theguardian.com/world/2018/feb/10/oxfam-faces-allegations-staff-paid-prostitutes-in-chad, accessed July 5, 2018.

Vara, V. 2015. Critics of Oxfam's Poverty Statistics Are Missing the Point. *The New Yorker.* January 28, 2015. https://www.newyorker.com/business/currency/critics-oxfams-poverty-statistics-missing-point, accessed July 5, 2018.

Pan American Health Organization (PAHO)

The Pan American Health Organization, or, in short, PAHO, is an international public health agency established to serve the health needs of the population of both Latin and North America. It has been working to combat infectious and other diseases and improve the health and living conditions of the residents of the region since its formation in December 1902. It ensures timely health care access to the people of the Americas and emphasizes primary health care strategies. Members of this organization include 27 countries of the region, along with Puerto Rico, which is an associate member. France, the Netherlands, and the United Kingdom are participating states, and Portugal and Spain are observer states. The PAHO is the oldest international health agency in the world and was the first international organization to promote health research and education. It is led by a four-person executive management team and establishes policies through its three governing bodies, the Pan American Sanitary Conference, the Directing Council, and the Executive Committee. Scientific experts are stationed in PAHO offices throughout the hemisphere.

The PAHO works in technical cooperation with its member countries to identify causes and maintain control of both communicable and noncommunicable diseases and to strengthen health systems. It also responds to emergencies and natural disasters. It promotes evidence-based decision making to improve health as a driver of sustainable development and supports the right of everyone to good health. To accomplish its stated goals, the PAHO encourages technical cooperation between member countries and works in collaboration with health ministries and other government agencies, civil society organizations, other international agencies, universities, social security agencies, and other relevant partners. It promotes the inclusion of health in all public policies and the engagement of all sectors in efforts to ensure that people live longer and have healthier lives. It further sets regional health priorities and mobilizes action to address health problems.

HISTORY

The Pan American Health Organization, formerly the Pan-American Sanitary Bureau (PASB) and the Pan American Sanitary Organization (PASO), was founded in 1902. It is part of the United Nations system, serving as the Regional Office for the Americas of the World Health Organization (WHO) and as the health agency of the Inter-American System. Initially, the political leaders of North and South America wanted to form the PASO in order to control the outbreak of yellow fever from Brazil to Paraguay, Uruguay, Argentina, and ultimately to the United States. By the end of the nineteenth century, more than 100,000 people suffered from yellow fever, and it caused the deaths of 20,000 people. In 1947, the PASB was renamed the Pan American Sanitary Organization, while the name Pan American Sanitary Bureau was retained for PASO's executive committee. The PASO was integrated into the United Nations system as the regional office for the WHO in 1949. The organization's name was changed again in 1958 to Pan American Health Organization.

Between 1946 and 1958, the PASB created six zones to place its resources nearer to its sphere of action. With headquarters in Washington, DC, Zone I included the United States, Canada, and the non-self-governing territories and had field offices in Jamaica and El Paso, Texas. Zone II with headquarters in Mexico City covered Cuba, the Dominican Republic, Haiti, and Mexico. Zone III with headquarters in Guatemala City, included seven countries (Belize, Costa Rica, El Salvador, Guatemala, Honduras, Nicaragua, and Panama) of Middle America. The remaining three zones included countries of South America. With Lima as headquarters, Zone IV comprised five countries (Bolivia, Colombia, Ecuador, Peru, and Venezuela). Zone V covered only one country (Brazil) with headquarters in Rio de Janeiro, Brazil. With Buenos Aires as headquarters, Zone VI covered Argentina, Chile, Paraguay, and Uruguay. Each Zone office had one or more medical officers, a sanitary engineer, a public health nurse, and other medical staff.

The organization established three Pan American Centers during this period. It established the Institute of Nutrition of Central America and PaLatewrnama (INCAP) in Guatemala in 1946; in 1951, the Pan American Foot-and-Mouth Disease Center (PANAFTOSA) was established in Rio de Janeiro; and in 1956, the Pan American Zoonoses Center (CEPANZO) was established in Buenos Aires to promote and strengthen activities against zoonoses in the Americas. Later in 1972, the PAHO established a Latin American Center for Educational Technology in Rio de Janeiro and another one in Mexico City in 1973 to train physicians, dentists, and nurses.

The PAHO receives funding from three main sources. First, member governments provide annual funds to run its activities. Each member country is allocated contributions (quotas) to make to this organization. Second, it receives the portion of the WHO regular budget approved for the region. Lastly, individuals and others can contribute to the organization to help carry out its health programs and education projects through the PAHO Foundation. The contribution is tax-deductible. While funding received from the preceding first and second sources is considered un-earmarked, voluntary contributions can be categorized as either earmarked or unearmarked. Through its Strategic Fund, any member state can purchase medicines and supplies. For example, in 2015, Latin American and Caribbean countries made over 400 requests to purchase over $70 million worth of medical goods.

ACTIVITIES

The PAHO works in cooperation with both governmental and nongovernmental organizations to coordinate international health activities in the Western Hemisphere. Although initially founded to control spread of yellow fever from one country to another, it has been fighting to control other infectious diseases, such as malaria, cholera, and HIV/AIDS. Moreover, because of its intervention measures, the PAHO has been successful in eradicating smallpox and polio from the Americas. Smallpox was eradicated in 1973 through concerted efforts to produce a vaccine, provide training to medical staff, and also provide essential supplies. The organization set a goal in 1985 to eradicate polio, and within 10 years, it was able to do so. The last case of polio in the Americas was identified on August 23, 1991, in Peru.

Currently, the PAHO is working to eradicate measles from the Western Hemisphere, and it is introducing new vaccines to fight other deadly diseases. Also, the PAHO has achieved remarkable progress in reducing meningitis and respiratory infections and deaths from water-borne diseases. It has made considerable progress in improving life expectancy, reducing health gaps between the rich and poor, and improving water safety and sanitation to eliminate cholera in Hispaniola. In short, the PAHO promotes equity and improved living standards for impoverished communities. One of the primary goals of the Pan American Health Organization is to reduce infant mortality rates and improve maternal health in its member countries. Thus, it closely works with women of reproductive age and children by providing prenatal and perinatal health services, including immunization programs for the children, the same program that helped eradicate both smallpox and polio from the Americas. Furthermore, the PAHO has a program to make the Americas a tobacco-free region.

The PAHO also helps spread scientific and technical information through its publications program; its Internet site; and a network of academic libraries, documentation centers, and local health libraries. The organization publishes books, reports, and documents in English, Spanish, and Portuguese. It has already published dozens of clinical textbooks in Spanish and Portuguese and distributed them among students of the health professions in the region at prices much lower than commercial prices. The organization launched a quarterly journal on a wide range of health subjects in 1966. To disseminate essential health information, the PAHO established the Regional Library of Medicine and Health Sciences (BIREME) in Sao Paulo in 1967. It also established the Latin American Center for Perinatology and Human Development in 1970 in Montevideo. The center stresses breastfeeding and focuses on investigation and application of new methods of pregnancy delivery.

The PAHO not only provides services to refugees and other disadvantaged persons but also organizes emergency preparedness and participates in emergency relief efforts following major natural disasters in the Americas. Additionally, it has published several handbooks on how to manage humanitarian and medical supplies after natural disasters. The organization feels that the health sector faces exceptional challenges in emergencies and disasters. These challenges may have consequences in the form of interrupted flow of medical supplies and incompetent health professionals, and hence well-being of victims of such events. For instance, large quantities of expired and useless medicine being sent to disaster sites as well as acute shortage of necessary medicine at such sites is widely reported in disaster literature. Therefore, special attention must be paid to the management of these supplies.

The PAHO's response to health crises after the 2010 Haiti earthquake was particularly demonstrative. Four days after the disaster, the PAHO/WHO began holding daily coordinating meetings as a lead organization of the Health Cluster identified in the Post-Disaster Needs Assessment (PDNA) by the United Nations, the World Bank, the European Community, and the Inter-American Development Bank. The organization distributed 345,000 boxes of emergency medical supplies between January and March from its medical warehouse. It also participated in postdisaster vaccination programs and responded to cholera outbreaks. Along with other organizations like the Centers for Disease Control (CDC), the PAHO established 51 surveillance sites within and outside the earthquake-affected areas of Haiti. These sites offered general care for all injuries whether or not related to the earthquake. After the earthquake, people were injured in traffic accidents as well as due to violence. One year after the Haiti earthquake, the PAHO/WHO continued to support rebuilding the country's devastated health system and improving the health of the Haitian population (PAHO 2011).

After Hurricane Sandy in 2012, four PAHO response teams (including clinicians, logisticians, and water and sanitation specialists) were deployed in the affected countries (the United States, the Bahamas, Cuba, and Haiti). These agencies teamed up and supported the health authorities of the affected countries in the areas of response coordination, investigation of health alerts, and provision of supplies to cholera treatment centers, including decentralized warehouses. A similar response has been also extended after the major natural disasters that have occurred since 2012 in Latin and North America, including the 2016 Columbian earthquake and three monster hurricanes in 2017, namely, Hurricanes Irma, Maria, and Harvey.

In coordinating and providing emergency aid to victims of hurricanes and other disasters in the Caribbean and Central America, the PAHO realized that a significant proportion of incoming donations from domestic and international sources were low priority (LP) and/or nonpriority (NP) items or, simply, items that were useless or irrelevant. Both types of supplies increase workload and occupy storage space. Therefore, the organization classified donated items in three categories: urgent or high priority (HP), nonurgent or low priority (LP), and nonpriority (NP). The HP items are those required for immediate distribution and consumption (PAHO 2001). It is worth mentioning that some supplies that arrive at disaster sites may not be a priority but might prove to be useful later. Therefore, these items need be classified, labeled, and stored until they are needed. Those items that are considered useless due to their condition (damaged, expired, and entirely inappropriate culturally and/or weather wise) should be discarded as soon as possible, to make room for useful supplies. Ultimately, the practice of sending useless emergency goods to survivors of extreme natural events led to the identification and development of the convergence phenomenon in disaster literature.

CONCLUSION

Since its formation, the PAHO has adopted the Pan American approach and has continued to cooperate among countries of the Americas in controlling and eliminating infectious and communicable diseases and improving health and well-being of the population of the region. It has been helping its member countries to accomplish common goals and has been initiating multi-country health ventures. The PAHO assists vested countries in mobilizing the necessary resources to provide immunization and treatment services for all vaccine-preventable diseases. Finally, it has achieved success in eradicating several diseases from Americas, and its health services have tremendously benefited millions of people in the region.

Bimal Kanti Paul

Further Reading
PAHO (Pan American Health Organization). 2001. *Humanitarian Supply Management and Logistics in the Health Sector.* Washington, DC: PAHO.
PAHO (Pan American Health Organization). 2011. *Earthquake in Haiti—One Year Later, PAHO/WHO Report on the Health Situation.* Washington, DC: PAHO.

Refugees International

In 1979, Sue Morton founded Refugees International in Washington, DC, as a response to the Indochinese refugee crisis in Vietnam, Laos, and Kampuchea (formerly known as Cambodia). The Vietnam War (1955–1975), the subsequent political turmoil during the Khmer Rouge regime in Kampuchea, and the Vietnamese invasion of Kampuchea all generated a massive flow of refugees, on a scale the world has not seen since World War II (1939–1945). The organization is an independent and international organization to protect and serve the needs of the people displaced from their ancestral homes by both natural and man-made emergencies. The organization does not accept funding from any government or United Nations agencies. This allows its advocacy to be independent and fearless.

Its principal mission is to provide lifesaving emergency assistance in a timely way to all displaced people, protect them, and find out ways to solve displacement crises globally. More specifically, displaced people receive food, medicine, and education from Refugees International. It also arranges for displaced families to return home, and it convinces political leaders to resolve refugee crises through negotiations. It is a leading advocacy organization in favor of vulnerable displaced people worldwide.

TARGET POPULATION

The main target of Refugees International are displaced people of three types: refugees, internally displaced persons (IDPs), and asylum seekers. A refugee is someone who has been forced to flee his/her home country for a neighboring or other foreign country to avoid a persecution, war, violence, or impact of natural disasters. Refugees generally live in camps. In 2017, there were about 27.5 million refugees in the world, nearly two-thirds of them having come from just five countries: Syria, Afghanistan, South Sudan, Myanmar, and Somalia. Developing countries host about 80 percent of all refugees, and about 50 percent of such people are under the age of 18. About 75 percent of the world's refugees reside in countries neighboring their country of origin. Under international law, refugees are not allowed to be forced back to the countries they have fled (HREA 2014).

An internally displaced person (IDP) is similar to a refugee, but he/she has not migrated across an international border. Unlike refugees, IDPs are not protected by international law or eligible to receive many types of aid because they are legally under the protection of their own government. However, the United Nations High Commissioner for Refugees (UNHCR) provides assistance to IDPs. In 2017, there were 40 million IDPs in the world, with some of the largest internally displaced populations being in countries such as Colombia, Syria, Democratic Republic of the Congo, and Somalia. Of these 40 million, 18.8 million IDPs in 135 nations were the product of sudden onset natural disasters (IDMC 2018). Finally, an asylum seeker is someone who has migrated to another country in hopes of being recognized as a refuge. Given the 3.1 million asylum seekers in 2017, there were 68.5 million displaced people in the world that year, which was roughly equivalent to the population of France. These numbers fluctuate year to year. For example, the total number of people displaced in 2010 came to more than 26 million, of which 10.55 million were refugees, 14.7 million were IDPs, and 0.8 million were asylum seekers (UNHCR 2011).

Refugees International provides necessary support to the first two groups. Moreover, the organization also supports more than 15 million stateless people living without citizenship. A stateless person is someone who is not a citizen of any country. Such people are often denied access to basic rights such as education, healthcare, employment, and freedom of movement. A large concentration of stateless people is in Asia (e.g., Nepal, Myanmar, Nepal, and Thailand), including those in the nomadic pastures (locally called Bidoon) in the Middle East. Stateless people are also found on all other continents. In Europe, for instance, statelessness is linked to the dissolution of the former Soviet Union and Yugoslavia. Similar to the number of other displaced people, the number of stateless people also fluctuates. For example, there were 12 million stateless people in 2010 in 65 countries (UNHCR 2011). Stateless people are able to obtain legal status from Refugees International.

ACTIVITIES

As indicated, Refugees International acts quickly for displaced people so that crises do not turn worse or spread across borders and the suffering of vulnerable people is minimized. In addition to providing help for displaced people who are subject to extreme weather such as floods, cyclones, and droughts, and conflicts and wars, the organization has programs for "climate refugees," who are forcefully driven from their low-lying coastal ancestral homes either inside their countries (i.e., IDPs) and/or across borders (i.e., refugees) due to extreme weather and changing climate patterns in the near future. However, researchers working in the field of environmental degradation and climate change do not know what the magnitude of this eventual displacement will be. Some estimate that by the end of the twenty-first century, between 200 million and one billion people are expected to be displaced by sea level rise (SLR) caused by global climate change (Christian Aid 2007; Myers 2005). According to the Intergovernmental Panel on Climate Change (IPCC), rising sea levels due to global warming will displace tens of millions of people, in particular in coastal areas and on island countries. People will also be displaced due to climate change-induced extreme weather events such as higher frequency and magnitude of storms, floods, riverbank erosion, and droughts. There is growing consensus among climate researchers and politicians that the negative impacts of climate change will affect an increasing number of people, leading to migration of affected people.

Climate and environmental scientists, including IPCC experts, maintain that the number of the displaced will increase during the coming years as rising sea levels devour coastal areas of tropical and other countries. In such a scenario, the governments of such countries would have to relocate huge numbers of internally displaced people. Because few resources are available to these governments, and because many of these countries are not responsible for climate change-induced displacement, the international community, particularly developed countries, should come forward with resources to relocate and rehabilitate the people likely to be displaced by sea level rise. As a result, Refugees International promotes a strong climate change program, which ensures that politicians and voluntary organizations at different levels listen to and prioritize the persistent needs of victims of flooding, drought, storms, and other climate-related events.

Those who are suspected to be displaced by climate-induced sea level rise are variously termed *environmental refugees*, *ecological refugees*, or *climate refugees*. However, the United Nations High Commissioner for Refugees (UNHCR), the International Organization for Migration (IOM), the UN Office for the Coordination of Humanitarian Affairs (OCHA), and other international organizations have expressed their view that using the term *refugee* in relation to environmental stressors is problematic. Refugee has a specific legal meaning in the context of the 1951 Geneva Convention relating to the Status of Refugees. This convention defined refugees as people with a well-founded fear of being persecuted because of their religion, race, nationality, membership in a particular social group or for political opinion. The grounds for persecution are important in this definition, and the factors such as "nature" or "the environment" or "the climate" do not seem to be grounds for persecution. People who will likely be displaced due to SLR or other environmental causes are not legally protected in any way by this convention. Therefore, members of these organizations argue the term *refugees* should not be used for those who are already or will be displaced due to climate change; indeed, the term *climate refugee* is not endorsed by UNHCR, but Refugees International has been trying to petition for "climate refugees" to be included in the Geneva Convention of 1951.

Refugees International began the Climate Displacement Program in 2009 to advocate for improved assistance, protection, and solutions for vulnerable communities and individuals uprooted by weather and climate change-related extreme natural events. After observing firsthand displacement crises caused by climate change in several countries, the organization has pleaded with national governments, UN agencies, donors, and others to address the increasing impacts of climate change on displacement and migration. The program was funded by several organizations and personnel, and it successfully initiated several lifesaving projects for refugees and displaced people caused by climate-related disasters and conflicts. In fact, since 2009, Refugees International has initiated over a dozen programs in countries with high risk of climate-related displacement. Additionally, the organization has played a key role in the development of a variety of initiatives aimed at promoting action to mitigate and reduce climate-related displacement (RI n.d.).

Another important program of Refugees International concerns women and girls who are most vulnerable as well as are victims of armed conflict and natural disasters, and thus, they are confronted by exceptional danger. Clearly, they are displaced more than their male counterparts and face more difficulties than men. In both types of circumstances, women and girls live with the constant threat of gender-based and domestic violence, such as rape and sexual assault. They are also victims of sex trafficking within and beyond national boundaries. Moreover, each year, more than 200 million women and girls are subjected to a traditional painful and harmful practice, female genital mutation (FGM). The procedure has no health benefit and is carried out in some 30 countries in Africa, the Middle East, and other Asian countries. Further, in many developing countries, girls are forced to marry at a young age. Thus, Refugees International has specific programs to elevate the well-being and safety of women and girls (RI n.d.). The organization has programs to prevent or minimize women's risks of gender-based violence, and it encourages policy makers and humanitarian actors to prioritize the needs of females.

Refugees International has 15 field missions across the globe to determine the needs of people who have been displaced by natural disasters and conflicts. Such fact-finding missions seek to provide emergency and basic services such as food, water, housing, health care, and education. Based on these missions, RI advocates for people of concern, soliciting government officials, policy makers, and international aid organizations to improve the lives of displaced people around the world. The organization currently focuses its attention on displacement crises in Colombia, Democratic Republic of Congo, Mali, Myanmar, Bangladesh, Somalia, South Sudan, and Syria.

One of the serious crises that recently attracted the widespread attention of international media is the Rohingya issue between the two neighboring countries of Myanmar and Bangladesh. Refugees International has been deeply involved in this refugee issue and has outlined five key priorities to address the Rohingya Crisis (Sullivan 2018). Although Muslim Rohingya refugees started to flee the Arakan state of Myanmar to Bangladesh in 1978, driven out by gross violations of human rights perpetrated by the Myanmar security forces, the influx in 2017 was the largest in volume as more than 700,000 Rohingya have fled to Bangladesh, joining at least 200,000 others who had fled previous bouts of violence and persecution. Many experts consider that the military actions in Myanmar have been marked by indiscriminate killing, widespread sexual violence, and wholesale burning of villages. They further consider that the military has committed crimes against humanity; hence, RI outlined the five key priorities to solve this crisis: (1) crimes against humanity have taken place and must be exposed, (2) ongoing persecution of Rohingya inside Myanmar must be stopped, (3) Rohingya in Bangladesh are still at risk and must be protected, (4) repatriation of Rohingya to Myanmar is a dim prospect but an essential part of any sustainable solution, and (5) several steps toward accountability are available but have not yet been taken. RI strongly maintains that accountability will be essential to any sustainable solution for the Rohingya crisis (Sullivan 2018).

CONCLUSION

Refugees International originated to identify the needs of displaced people all over the world through its field missions and to help these people in all possible ways. It advocates for their rights and urges the international community to meet their needs. It has two specific programs designed for climate refugees and for women and girls who have been forced to flee their homes due to violent persecution and/or natural disasters. It urges the UN, U.S. government, humanitarian actors, donor nations, and accountable host governments to meet the specific needs of displaced people. Efforts of the organization along with those of international community and aid agencies have undoubtedly improved the circumstances of refugees and displaced people and elevated their security and dignity.

Bimal Kanti Paul

Further Reading

Christian Aid. 2007. World Facing Worst Migration Crisis. https://www.business-humanrights.org/en/christian-aid-says-world-faces-forced-migration-crisis-caused-by-climate-change-conflict-development-projects#c30599, accessed April 16, 2019.

HREA (Human Rights Education Associates). 2014. The Rights of Refugees. http://www1.umn.edu/humanrts/edumat/studyguides/refugees.htm, accessed April 16, 2019.

IDMC (Internal Displacement Monitoring Center). 2018. *Global Report on Internal Displacement*. Geneva, Switzerland: IDMC.

Myers, N. 2005. *Environmental Refugees: An Emergent Security Issue*. Prague, Czech Republic: Economic Forum.

RI (Refugees International). n.d. What We Do. https://www.refugeesinternational.org/currentwork, accessed April 18, 2019.

Sullivan, D. 2018. 5 Key Priorities to Address the Rohingya Crisis, Refugees International. August 2018. https://www.refugeesinternational.org/reports/2018/8/22/5-key-priorities-to-address-the-rohingya-crisis, accessed April 30, 2019.

UNHCR (United Nations High Commissioner for Refugees). 2011. *60 Years and Still Counting: UNHCR Global Trends 2010*. Geneva, Switzerland: UNHCR.

Save the Children

Save the Children (SCF) is one of the world's largest child-focused nongovernment organizations (NGOs). As an international network, it raises and expends over $2 billion annually across more than 120 countries. SCF estimates that it directly impacts nearly 60 million children each year through both its development activities and its humanitarian work (SCF 2016b).

FORMATION

In addition to being one of the world's largest NGOs, SCF is also one of the world's oldest aid agencies, being established in 1919. The impetus for its establishment was the humanitarian crisis faced by 5 million starving and destitute children in Central and Eastern Europe following the cessation of the World War I (1914–1918) (Watson 2014). The driving force in the founding of SCF was Eglantyne Jebb.

Initially, Ms Jebb sought to raise public awareness of the plight of these children, expecting the British government to respond to what she hoped would be public demand for action and lift the ongoing blockade of defeated countries. To advocate for such a change, Jebb stood in public squares and handed out leaflets showing starving children and proclaiming that "all wars are waged against children." Jebb's early humanitarian advocacy actions though were largely unsuccessful, and she was arrested, convicted, and fined for her actions, which were considered in breach of the Defence of the Realm Act. Although undeterred, she faced ongoing opposition to her efforts to raise support for these children.

> Later, [Ms Jebb hired] . . . the Royal Albert Hall for a public meeting. Many come only to hurl insults, believing that Britain should help its own, rather than the enemy's children. In the face of nationalism and gender-stereotyping, Eglantyne calls out, "Surely it is impossible for us, as normal human beings, to watch children starve to death without making an effort to save them." (SCF 2016a)

However, from this public meeting, SCF was formed, and its fundraising began, resulting in its first grant being made in May 1919 (for Austria). True to its name, SCF raised funds that were specifically focused on addressing the needs and providing some relief to affected children across war-ravaged Europe with additional grants soon being made to Armenia and Germany. These early grants supported the provision of food and medical aid. However, SCF also utilized funds to address wider humanitarian needs including the provision of linen, food, and soap (Watson 2014). Within a very short period, SCF had organized itself to become quite an effective fundraiser, quickly identifying messages that would illicit financial responses from their supporters. While current fundraising techniques are noticeably more nuanced, it is not difficult to identify a direct link between humanitarian appeals launched approximately 100 years ago by SCF and those that are currently used today by all aid agencies:

> The most dreadful siege in history is taking place in Europe to-day—a short journey from where you are reading now. Famine, Cold and Disease are the besiegers. The daily casualties, numbering always hundreds and often thousands, are innocent

little children—not strong fighting men. And this siege had been going on for two years—two years during Millions of Children have Died! (SCF 1921, 64)

Such descriptive and emotional requests for financial support were also accompanied by graphic photos of emaciated children (and often their mothers) living in squalid conditions and clearly highlighting the desperate reality they faced. This approach to humanitarian fundraising was successful, and SCF soon become one of the larger NGOs responding to humanitarian needs within Europe.

CONTEMPORARY SCF ACTIVITIES

From these small but determined beginnings, SCF has evolved over the last century. While it remains very focused on the needs of children in terms of health, education, and safety, it does now invest considerable funds in longer-term development projects as well as continuing to respond to humanitarian crisis. Fundraising for development activities is often through to child sponsorship, while its humanitarian responses are generally funded through specific public appeals and donor grants. SCF member organizations raise nearly $1 billion a year to support children and communities affected by humanitarian emergencies.

SCF has approximately 30 member organizations as well as an overarching international coordination body. While each member organization is managed and governed independently, there is a strong level of coordination and partnership between members. This is particularly so in circumstances of complex humanitarian emergency where member organizations can cede some management and responsibility to colleagues working for SCF International. This structure supports investment in humanitarian specialization, allows greater opportunities for effective delivery of humanitarian aid, provides greater coordination of member organizations in regional humanitarian events, and ensures more efficiency in fundraising efforts in supporting member organizations. It is a model that SCF believes facilitates the optimal response to significant humanitarian need.

Currently, SCF responds to more than 100 humanitarian emergencies annually. Staying true to its origins, SCF prioritizes children in their responses. SCF has stated goals that:

Ensur[e] children caught up in a natural emergency have access to food and a place to live is essential. But beyond that, they need protection. During times of drought, or following floods or an earthquake, children are most vulnerable. With or without family, they are at risk of abuse and exploitation. They often stop going to school and resulting poverty leads to a rise in child marriage. When responding to natural emergencies, we ensure children survive, and that their right to learn, play and enjoy a childhood is protected. (SCF 2016a, 7)

Of course, best practice in humanitarian response requires SCF to work with whole communities and not to isolate or focus solely on children. This ensures optimal outcomes of such interventions. However, SCF continues to make their community responses child-centered so that planning, implementing, and evaluating humanitarian interventions all include preferences and the needs of children. It is evident therefore in large complex humanitarian emergencies that SCF will be involved in the provision and operation of child safe spaces, pop-up schools, and child and maternal health clinics. SCF has become a recognized global leader in the provision of humanitarian responses aimed at children.

EXAMPLES OF SCF HUMANITARIAN RESPONSES

Humanitarian events occur across the globe, in developing as well as developed countries. SCF responds to these events in more than 60 countries every year. As discussed, when planning, implementing, and evaluating their humanitarian responses, SCF remain focused on the needs of children.

In 2016, Ethiopia experienced its third year of failed "spring rains," resulting in a severe drought that also affected the neighboring countries of South Sudan, Somalia, and Kenya. In the face of this worsening drought and without access to affordable food or water and with increasing spread of disease, 30 million people became displaced as they left their homes in search of basic necessities. SCF estimated that at this time, 6 million children in Ethiopia did not have access to clean water and were at risk of starvation. In response, SCF established over 100 stabilization centers in which they were able to treat children experiencing malnutrition and other illnesses associated with the lack of clean water and adequate food. SCF also operated mobile clinics across the affected region to address a range of medical issues, including provision of peri-natal care to pregnant women. Conditional cash transfers were also made available to families so they could access food, water, and medical care. Across this response, SCF worked closely with other aid agencies and the Ethiopian Government.

The small island state of Haiti was devastated in 2010 by a significant earthquake. Some six years later, while long-term reconstruction efforts continued, Haiti was ravaged by Hurricane Matthew:

> Winds of up to 240km/h flattened homes and trees, while torrential rain caused enormous mudslides and the sea was driven inland. . . . It quickly became the worst humanitarian crisis since the 2010 Earthquake in Haiti: at least 500 people—and likely more—were killed, 175,000 people were displaced and more than 1.4 million Haitians—600,000 of them children—were left in need of humanitarian assistance. (SCF 2016a, 17)

In an environment with so much damaged infrastructure (from both the hurricane and the previous earthquake), it was necessary for SCF to establish portable, mobile health facilities. These facilities were set up across the small country, including in very remote and hard-to-access locations. SCF distributed food, clean water, and basic medical supplies. They also undertook sanitation programs in order to minimize the spread of diseases. Continuing the past efforts of reconstruction, SCF also worked with local communities in rebuilding homes, schools, and other community facilities—engaging local builders wherever possible. In addition to all of these community-based activities, SCF also ran a number of child-friendly spaces that provided opportunities for children to play and have some access to education.

SCF also responds to emergencies in developed countries. For example, SCF works in Italy, assisting refugees who are fleeing poverty and conflict by undertaking dangerous sea crossings across the Mediterranean Sea from northern Africa to Italy. SCF estimate that 90 percent of children who undertake such crossings do so as unaccompanied minors—that is, without parents. Children are at great risk not only during the dangerous sea crossing but also when they arrive without parental protections in Italy. In response, SCF operates a search and rescue ship on the Mediterranean to assist those at risk of drowning. Upon rescue, SCF staff provide immediate medical care and then provide longer term support, particularly to accompanied children, throughout the refugees' immigration process.

CONCLUSION

From its inception in 1919, SCF has constantly focused on children as the center of their activities. Over time, SCF has raised funds for and supported both longer-term development activities and response to humanitarian events. Just under half of the $2 billion that SCF raise and expend annually is used to support children and communities affected by humanitarian emergencies in more than 60 countries.

Children and communities affected by natural or human-caused disasters are vulnerable. This vulnerability is caused by the destruction or loss of infrastructure, access to clean water and sanitation, food, and medical care. Their vulnerability is further acerbated by the psychosocial dimensions that such destruction brings. Working with vulnerable children and communities requires the highest level of integrity of both individuals and organizations. Within aid agencies, zero tolerance of any behaviors that take advantage of power imbalances or vulnerability must therefore be enforced.

From very humble and controversial beginnings, SCF has over its 100-year history become one of the world's largest child-focused NGOs. It works across the globe and directly impacts nearly 60 million children each year. Just as its founder, Eglantyne Jebb, responded to the materially deficient circumstances experienced by children following WWI in Europe with advocacy, fund raising, and programming that prioritized children, SCF continues this work effectively and efficiently. SCF has grown and expanded beyond anything its founder may have possibly dreamed, but there is little doubt that the work SCF currently does across the globe to promote health, education, and protection of children in all its development and humanitarian program would be immediately recognizable a century later.

Matthew Clarke

Further Reading
SCF (Save the Children). 1921. The Record of Save the Children Fund 1(4): 49–64.
SCF (Save the Children). 2016a. Rising to the Challenge. Save the Children Annual Report 2016. https://www.savethechildren.net/annualreview/ui/docs/Save_the_Children_Annual_Report_2016.pdf, accessed May, 10, 2018.

SCF (Save the Children). 2016b. Save the Children International: Trustees' Report, Strategic Report and Financial Statements for 2016. https://www.savethechildren.net/sites/default/files/libraries/Save%20the%20Children%20International%20Trustees%20Report.pdf, accessed May 10, 2018.

Watson, B. 2014. Origins of Child Sponsorship: Save the Children Fund in the 1920s. In *Child Sponsorship: Exploring Pathways to a Brighter Future*, edited by B. Watson and M. Clarke, 18–40. London, UK: Palgrave.

United Nations Children's Fund (UNICEF)

When it was founded in 1946, UNICEF became the acronym of the United Nations International Children's Emergency Fund. More commonly, this humanitarian program, launched by the United Nations, is known currently as the United Nation's Children Fund. The mission of this special UN program is to provide humanitarian assistance to underprivileged children living mostly in developing countries. In cooperation with governments and nongovernment organizations (NGOs), UNICEF is devoted to improving the health, nutrition, education, and general welfare of children—and their mothers—across 193 countries and territories. In a broader sense, the organization works to save imperiled children's lives, defend their rights and those of their mothers, and ultimately help them to fulfill their potential.

UNICEF is the best-known organization for dedicated, bold work in this area. For example, with famine stalking in Sudan, more than 80 percent of the employees are engaged in improving the condition of children affected by poverty and social exclusion, especially those living in urban slums, such as the Dharavi slum in Mumbai. The list of humanitarian actions by UNICEF is long, and its history is rich. In recognition for its contribution to promoting health, primary education, and well-being, UNICEF was awarded the Nobel Peace Prize in 1965 and the Prince of Asturias Award of Concord in 2006. The organization's present work includes, among other things, delivering food, water, and medical supplies, along with providing temporary shelter, schooling, and family reunification services to those children whose need is ongoing and during emergencies.

A variety of emergencies that could occur anywhere may be grouped under two headings: natural disasters such as earthquakes, floods and drought, and man-made disasters, including civil war, famine, and forced migration. UNICEF spends 88 percent of its budget to help alleviate such suffering.

FOUNDING

In the aftermath of World War II (1939–1945), Europe was a devastated continent, and the most affected victims were European children. They suffered from famine and disease, starvation, and malnutrition. Thus, on December 11, 1946, the UN General Assembly created UNICEF by Resolution 57-I to address the plight of this population. The organization started distributing daily supplementary meals, including milk, and various articles of clothing and providing vaccinations against tuberculosis. UNICEF rebuilt milk processing and distribution centers and initiated training for workers for maternal and child care services.

Since 1946, many countries have won independence from European colonial rule. At the time of independence—and even today—most of those independent countries suffer from underdevelopment in pervasive ways, including poverty, disease, malnutrition, and lack of safe water and sanitation. A functional education is rarely provided, and violence against children and women looms large.

Since being established, UNICEF has broadened its scope of humanitarian work to Asia, Africa, and Latin America. For example, the organization became substantially committed to working with the government of India in 1949 "to ensure that each child born in this vast and complex country gets the best start in life, [and] thrives and develops to his or her full potential" (UNICEF 1996). Some noted contributions of UNICEF in India include providing equipment and technical assistance for India's first penicillin plant established in 1949, funding 13 milk processing centers, and building the first DDT plant in 1954 to eradicate malaria. Moreover, it supplied drilling rigs to provide safe drinking water for two drought-stricken Indian states (Bihar and Uttar Pradesh) in 1966.

DEVELOPMENT

In January 1961, the United Nations decided that the 1960s would be the Decade of Development. During this decade, UNICEF embarked on a series of humanitarian projects in 13 newly independent states in Africa, including Kenya, Ghana, and Tanzania. These welfare projects were associated with improving the food and nutrition status of residents of 13 countries and providing maternal and child health care, education, abundant and clean water supply, and sanitation. UNICEF also provided material assistance in the form of equipment, drugs, vehicles, and training stipends.

Considering UNICEF's substantial contribution to social development in developing countries, in 1972, the United Nations formally declared that "UNICEF was a development, rather than a welfare, organization and began to review its work under its economic and social, rather than humanitarian, machinery. And it was not until later still that the idea of investing in children would move away from the notion of philanthropy and into the development mainstream" (UNICEF 1996).

The early 1970s saw two important events of international development discourse. Many developing countries during the previous two decades had achieved high rates of economic growth, but little had "trickled down" to the poor. To remedy this problem in 1972, Robert McNamara (1916–2009), then President of the World Bank, recommended governments of developing countries for new projects and policies to improve the status and the standard of living of the poorest segment of their people. At that time, this segment constituted 40 percent of the population, on average, of developing countries (UNICEF 1996).

Under the changed, but more favorable, world development environment, UNICEF expanded the scope of its activities, including its successful agenda of "advancing children's rights to education, healthcare, and nutrition" (UNICEF

1996). In the 1980s, the organization launched its Women and Development Program (WDP) and introduced a Primary Health Care Program (PCHP) for children to improve their health by monitoring their growth, providing oral rehydration therapy, advocating breastfeeding, and providing immunization. From 1989 onward, UNICEF has used the UN Convention on the Rights of the Child as guidance for its programs (DeLargy 2016).

Under the Executive Directorship of James P. Grant (1980–2005), UNICEF promoted the Child Survival Revolution. Its message was simple: "Low-cost, practical methods for saving young lives and curtailing the spread of easily preventable diseases can quickly make a difference. He advocated treating diarrhea, a leading killer of children, with doses of water, sugar and salt that village people could make themselves" (Crossette 1995).

The first 30 years of the twenty-first century are known as the United Nations Development era: in two stages, this has involved Millennium Development Goals (MDGs), 2001–2015, and Sustainable Development Goals (SDGs), 2016–2030. Six of the eight goals and 13 of 48 indicators of MDGs are primarily about children. Furthermore, for the case of each of these goals, UNICEF assumes the central role in meeting them. In order to achieve the goals of SDGs, children must be put at the center of the agenda. UNICEF continuously achieves several goals, using indicators directly and indirectly relevant to child welfare, for example, its WASH (water, sanitation, and hygiene) program. This program is currently implemented in over 100 countries worldwide, and its purpose is to improve water and sanitation services as well as basic hygiene practices (UNICEF 2018).

According to the UNICEF Annual Report 2017, the organization put new emphasis on (1) global efforts to reduce child mortality, (2) drastically reducing the annual number of new HIV infections, (3) expanding the WASH program, (4) improving the nutrition program, and (5) curtailing gender-related barriers. This emphasis aligns with the 2030 Agenda for Sustainable Development across countries and emphasizes developing countries (UNICEF 2018).

FUNDAMENTAL PRINCIPLES

The fundamental principles of UNICEF are stated in its mission statement. The salient features of the statement read as follows:

- UNICEF is nonpartisan and its cooperation is free of discrimination. In everything it does, the most disadvantaged children and the countries in greatest need have priority.
- UNICEF aims, through its country programs, to promote the equal rights of women and girls and to support their full participation in the political, social, and economic development of their communities.
- UNICEF works with all its partners towards the attainment of the sustainable human development goals adopted by the world community and the realization of the vision of peace and social progress enshrined in the Charter of the United Nations (UNICEF 2003).

In addition to UNICEF's contributions to children's welfare, two unique contributions—disaster relief and refugee rehabilitation in times of emergency—deserve mention. For example, after the devastating earthquake in Haiti in 2010, UNICEF responded quickly and efficiently. The organization responded to the emergency needs to children and their parents and provided them with temporary shelter, pure water, sanitation, food, health care, and child protection. Recently, the organization helped more than a million Rohingya Muslims who were forced to leave Myanmar and who took shelter in refugee camps in southeastern Bangladesh. UNICEF has been working with the government of Bangladesh and several national and international organizations to alleviate the misery of refugees, particularly that of the children, living in camps. UNICEF is providing shelter and food and water in its efforts to protect children and women. It has plans to install water pumps and deep tube wells in the camp compounds. Moreover, it is distributing therapeutic food to malnourished and undernourished camp children and has a plan to vaccinate 150,000 children against measles, rubella, and polio.

UNICEF humanitarian actions for war-torn children in South Sudan and massive relief work in Nepal, after the country suffered two consecutives earthquake in 2015, are two more examples.

ORGANIZATION AND SOURCES OF FUNDING

Although the official headquarters of UNICEF is situated in New York, the core functions, providing structure, support, facilitate growth, and overall management, are carried out by the organization's seven regional offices, located in Panama City, Geneva, Bangkok, Nairobi, Amman, Kathmandu, and Dakar. UNICEF has also six other offices each with a unique function, such as Supply Division in Copenhagen, Innocewnti Research Center in Florence, Brussels Office, Office for Japan in Tokyo, and Global Shared Services Center, Budapest.

The functional areas in which UNICEF operates have been categorized into four broad areas:

- *Program and Policy*, including cross-cutting concern for development, such as Adolescent Development, Child Protection, HIV/AIDS, and Human Rights
- *External Relations*, including Alliance and Resource Mobilization, Communication/External Relations, Private Fundraising and Partnership, and Inventory Management
- *Emergency Program*, including programs and operations that are needed immediately to ensure the survival of a community isolated within the larger community. Examples are the victims of the 2018 earthquake in Indonesia and the ongoing wars in Syria and Yemen
- *Operations*, including Administration, Finance and Accounting, Internal Audit, and Supply and Logistics.

The preceding UNICEF works are guided and monitored by a 36-member Executive Board made up of government representatives.

UNICEF receives five types of funding: regular resources on which UNICEF builds programs, thematic funds that help meet basic needs of children and protect their rights, earmarked funds for a specific program purpose, humanitarian funds for emergency relief assistance, and pooled funds and trust funds that represent more than one donor's contributions, including funds held in trust, that is, not considered revenue to UNICEF.

UNICEF annual revenue has been increasing over the years. Its revenue in 2014 was $5,266 million; in 2017, it was $6,577 billion, demonstrating an increase of 24 percent over three years. Its sources of revenue include voluntary contributions from governments, NGOs and the private sector, inter-organizational arrangements, and other income. Since 1993, the United States has been the biggest donor. UNICEF maintains high standards of integrity and transparency in all expenditures (UNICEF 2018).

CONCLUSION

Since its inception in December 1946, UNICEF has been devoted to bringing about humanitarian change. Today UNICEF works in 193 countries and territories of the world to address the immediate postwar needs of Europe's children. It provides children with lifesaving aid and long-term assistance for functional education, adolescent care, gender equality, and child protection. The organization has achieved, in cooperation with governments and other organizations, a considerable amount of success to help alleviate childhood poverty and to improve mothers' health and well-being.

UNICEF's priorities are usually to help in emergencies—be it due to natural disaster or civil war. However, in addition to emergency response and regular welfare programs for underprivileged children in developing countries, the organization's present role seems to be moving toward overseeing the overall development of these children. The vision is defined as "UNICEF for every child a fair chance at life." But under the condition of glaring disparities and growing inequalities that persist in a great many developing, and some developed countries, where far too many children are left behind, working by itself UNICEF can achieve only a modest success.

Syed A. Hasnath

Further Reading

Crossette, B. 1995. James O. Grant, Unicef Chief and Aid Expert, Is Dead at 72. *New York Times*. January 30, 1995. https://www.nytimes.com/1995/01/30/obituaries/james-p-grant-unicef-chief-and-aid-expert-is-dead-at-72.html, accessed December 24, 2018.

DeLargy, P. 2016. Europe's Humanitarian Response to Refugee and Migration Flows: Volunteerism Thrives as the International System Falls Short. Humanitarian Practice Network, September 2016. https://odihpn.org/magazine/europes-humanitarian-response-to-refugee-and-migrant-flows/, accessed December 22, 2018.

UNICEF (United Nations Children's Fund). 1996. The State of the World's Children 1996. https://www.unicef.org/sowc96/1960s.htm, accessed December 12, 2018.

UNICEF (United Nations Children's Fund). 2003. UNICEF's Mission Statement. https://www.unicef.org/about/who/index_mission.html, accessed December 23, 2018.

UNICEF (United Nations Children's Fund). 2018. UNICEF Annual Report 2017. https://www.unicef.org/reports/unicef-annual-report-2017, accessed December 23, 2018.

World Food Program (WFP)

The World Food Program, one of the important agencies of the United Nations (UN), is one of the world's largest humanitarian organizations fighting hunger worldwide, delivering food assistance in emergencies, and working with communities to improve nutrition. It also participates in disaster rehabilitation efforts as well as developing programs. Two-thirds of its work is in conflict-affected countries where people are three times more likely to be undernourished. Sixty percent of residents of these countries do not know where their next meal is coming from. In 2018, the WFP allocated 80 percent of its resources to six countries: Bangladesh, the Democratic Republic of Congo (DRC), Nigeria, South Sudan, Syria, and Yemen. In these countries, 32.9 million people received lifesaving food assistance (WFP USA 2018).

The WFP has a reputation of responding quickly to emergencies even in difficult environments. In 2018, the agency assisted 86.7 million people across 83 countries. On any given day, the WFP delivers food and other necessary assistance to those in most need using 5,000 trucks, 20 ships, and 100 planes. Thus, every year, it distributes more than 15 billion rations at an estimated average cost per ration of $ 0.31 (WFP n.d.). In 2018, it delivered 3.9 million metric tons of food from 600 warehouses scattered all over the world (WFP USA 2018).

The WFP is the food assistance branch of the United Nations. The idea of establishing a multilateral food aid program was conceived initially at the 1961 Food and Agriculture Organization (FAO) Conference. In 1963, it was formally established by the FAO and the United Nations General Assembly on an experimental basis. Two years later, its activities were extended across the globe, particularly in the developing world, and since then, the UN has been supporting the WFP's activities on a continuing basis. Like the FAO, the WFP has its headquarters in Rome. It is a member of the United Nations Development Group and part of its executive committee. Currently, the WFP has a staff of 17,000 persons, with nearly 1,000 staffers at its Rome headquarters. Close to 92 percent of WFP employees are working at field offices in 82 different countries.

MISSION STATEMENT

The main goals of the WFP are as follows: achieve zero hunger in 2030 by eradicating hunger and malnutrition and gain worldwide food security by eliminating the need for food aid. In 1994, it adopted a mission statement that established its focus on the following: (1) use food assistance to support economic and social development; (2) save lives of refugees and survivors of both man-made and natural disasters; (3) improve the nutrition and quality of life of the most vulnerable people, promote the self-reliance of poor people and communities, and increase world food security in accordance with the recommendations of the UN and its FAO.

The activities of the WFP include not only to directly distribute food but also to purchase food items from rural areas in developing countries. Thus, the organization supports local farmers as well as the local economy. It provides food to children and mothers to reduce child mortality and improve maternal health, respectively. The WFP also distributes food to people combating diseases, including sufferers of HIV/AIDS and tuberculosis for whom proper nutrition is especially important. In addition, it also provides food vouchers to the needy, implements food for work programs (FWPs) for the rural poor in developing countries, and replants crops following disasters or other crises, which strengthens environmental and economic stability and increases agricultural production. Under FWPs, the participants work to rebuild damaged infrastructure primarily caused by natural disasters.

In 2018, 15.8 million women and children received malnutrition treatment or preventive services by the WFP. Additionally, 16.4 million children in 61 countries received school meals, snacks, or take-home rations. Sixty-two percent of the WFP beneficiaries were children, and 52 percent were women and girls. In that year, the organization bought food grains worth $31 million from small-scale farmers. Also, it constructed 3,000 ponds, wells, and reservoirs for livestock operations and production of fisheries. Through its FWPs, it constructed or repaired 6,000 miles (9,656 km) of roads and rehabilitated 470 square miles (756.3 km) of land in 2018 (WFP USA 2018).

ORGANIZATION AND ACTIVITIES

The WFP is governed by an Executive Board, which consists of 36 Member States. The organization is headed by an Executive Director, who is appointed jointly by the U.N. Secretary General and the Director-General of the FAO. The Executive Director is appointed for fixed five-year terms and has the responsibility to administer the organization and implement its programs and other activities. The WFP also has one Deputy Executive Director and three Assistant Executive Directors with specific tasks. The organization's strategic direction is mapped out in its Strategic Plan, which is renewed every four years. Moreover, the European Union (EU) is a permanent observer in the WFP and participates in the works of its Executive Board.

In 1992, the UN created the Office for the Coordination of Humanitarian Affairs (OCHA) to improve coordination among the many organizations and stakeholders involved in emergency responses. Given the coordination failures during the humanitarian response to both the Darfur crisis in Sudan in 2003 and the Indian Ocean Tsunami (IOT) in 2004, the OCHA introduced a Cluster Approach in 2005 with the primary goal of improving the effectiveness and timeliness of international response to emergencies. Depending on the nature of disasters, OCHA has identified 10–20 clusters (e.g., health, education, emergency shelter, and early recovery) for major disasters in developing countries since 2005.

The OCHA also has assigned a particular international humanitarian organization to lead one or more of these clusters.

Due to its expertise in food security, nutrition, and logistics, the WFP was chosen by OCHA as the lead agency for these clusters. For example, after the 2010 Haiti earthquake, the WFP led three of the 12 clusters created by OCHA: telecommunications, food, and logistics. Additionally, it opened an emergency relief distribution center only for women survivors (Paul 2019).

One of the crucial aspects of the WFP is that it works to partner with parallel organizations and agencies in mitigating starvation, whether caused by civil strife or natural disaster. As noted, it provides food assistance to refugees, and thus the organization closely works with UN High Commissioner for Refugees (UNHCR). It also coordinates and cooperates with government agencies such as the United Kingdom Department of International Development (DFID) and the United States Assistance for International Development (USAID) as well as with international nongovernmental organizations (NGOs) such as Save the Children, Catholic Relief Services, Educational Concerns for Hungry Children (ECHO), and Norwegian Refugee Council. In a given year, it generally partners with nearly 100 different corporations and private organizations and over 300 local, national, and international NGOs.

There is also a World Food Program (WFP) USA, which has the same mission to solve global hunger. Before 2010, it was known as Friends of the World Food Program. The WFP USA continues to lead the way in the fight against the food crisis. Its efforts have fed families and saved lives, but its members realize that in a world with increasing need, there is still much more to be done.

FUNDING

All operations of the WFP rely entirely on voluntary contributions for funding, accepting both cash and kind. Its principal donors are governments, but it also receives donations from corporations and individuals. Among the governments, the USA is the top donor followed by the EU and especially Germany. The organization has one of the best track records because of its cost-effectiveness with its administrative costs being around only 7 percent of its budget, one of the lowest and best among aid agencies.

In 2007, the WFP launched the first annual World Hunger Relief Week to increase awareness about hunger and raise funds needed to deliver food to the needy. Nearly 35,000 restaurants around the world participated in this program, which flared a global movement to end hunger and generated support for the WFP programs. Nearly 1 million food employees and their family members volunteered close to 4 million hours to aid hunger relief efforts. Since then, the initiative has been repeated every year.

In 2005, the WFP launched the "End Hunger: Walk the World" annual event to raise funds for and awareness of the organization's efforts to fight child hunger and malnutrition. It was a joint effort of the WFP and its corporate partners, their employees, and NGO partners to show their solidarity in the fight against hunger and malnutrition. More than 200,000 people walked in

296 locations. In 2010, an estimated 150,000 people took to the streets in 70 countries around the world and raised enough money to provide a whole year of meals for more than 10,000 school children. This event was also part of the campaign to achieve the Millennium Development Goals, which intended to halve the number of people who suffered from hunger and poverty by 2015 (New World Encyclopedia n.d.). The WFP also uses its website to inform potential partners and supporters and to provide opportunities for giving and participating in its initiatives.

CRITICISM

One serious criticism of WFP's fight against hunger is that it should not be addressed by providing ongoing food assistance, because such assistance is counterproductive, can increase dependency, and can contribute to the underdevelopment and marginalization of hunger victims in developing countries. Often it supplies too much food aid, leading to reduction in crop production by local farmers and creating a situation where farmers have no incentive to increase food production. To address this concern, the organization can support programs to increase local food production such as by distributing seeds to poor farmers and by providing modern agricultural inputs at a nominal price and training them to increase food production or by providing employment opportunities.

Because the WFP feeds only a small proportion of the estimated 800 million people who face starvation, the organization should deal with the root causes of shortage of food, which can be achieved by initiating sustainable development programs, controlling overpopulation, and improving food distribution. For instance, in many instances, the WFP has channeled food assistance to the national governments. However, in most cases, governments of developing countries are corrupt and impose strict control measures in distributing food, often unfairly and by nepotism.

Often, the WFP does not store food grains in its own warehouses or directly distribute food to eligible people; instead, it gives contracts to private agencies or middlemen to distribute food. Some corrupt agencies and middlemen take this opportunity to hijack food aid and make money by selling it in the open market. Often they sell the original food grains, buy inferior-quality grains at a lower price from local markets, distribute the inferior-quality grains to the people, and thus make money. For example, the WFP delivered food assistance to the 2015 earthquake survivors in Nepal. It used the stocks that WFP already had in country from its existing operation prior to the earthquake, but it distributed those stocks through middlemen. Those who received food aid complained that the quality of rice was not good. Some earthquake survivors did not use the rice, while others fermented it to produce domestic alcohol, or mixed it with better-quality rice and ate it, or gave it to cattle.

The authorities of WFP in Nepal initially denied distribution of spoiled and damaged rice to earthquake survivors. The organization claimed that it identified the problem before distribution began and removed the bags containing spoiled rice, replacing them with good rice. Yet, an investigation into damaged rice distribution among the earthquake survivors was undertaken by the WFP, and the agency later disposed of a large quantity of damaged foodstuffs (including rice, peas, and biscuits) that was stored in the warehouse at Deurali in the Gorkha district, Nepal (Kathmandu Post 2015).

The WFP was created as an experiment to provide food aid through the UN system, as the organization believes that food and food-related assistance lies at the heart of the struggle to break the cycle of hunger and poverty. Despite some criticisms, the WFP has been providing food aid not only to end world hunger but also to make the world more stable and peaceful.

Bimal Kanti Paul

Further Reading

Kathmandu Post. 2015. WFP Destroys Rotten Food Items. December 30, 2015.

Mohammed, F. n.d. The History of the World Food Programme. https://borgenproject.org/the-history-of-the-world-food-programme/, accessed July 3, 2019.

New World Encyclopedia. n.d. World Food Programme https://www.newworldencyclopedia.org/entry/World_Food_Programme, accessed July 3, 2019.

Paul, B. K. 2019. *Disaster Relief Aid: Changes & Challenges*. Gewerbestrasse, Switzerland: Palgrave Macmillan.

WFP (World Food Program). n.d. Overview. https://www1.wfp.org/overview, accessed July 3, 2019.

WFP (World Food Program) USA. 2018. *Annual Report 2018: Hunger on the Rise*. Washington, DC: WFP.

World Health Organization (WHO)

The World Health Organization (WHO) is a specialized agency of the United Nations that works on improving international public health. Its founding principle was that health is a human right and all people should enjoy the highest standard of health. The organization was established on April 7, 1948 (now celebrated as World Health Day), and is headquartered in Geneva, Switzerland. Its predecessor, the Health Organization, was an agency of the League of Nations. The WHO incorporated the Office International d'Hygiène Publique and the League of Nations Health Organization. It is controlled by delegates from its 194 member states, who set policies, elect the director-general, and run the six regional and 149 country offices. The WHO is responsible for the World Health Report, the worldwide World Health Survey, World Health Day, and many other initiatives. The current Director-General of the WHO is the Ethiopian Tedros Adhanom, who started his five-year term in 2017.

The WHO is funded by the members' mandatory dues (20 percent) and voluntary donations from governments and private partners (80 percent). The United States, the United Kingdom, and the Bill and Melinda Gates Foundation are the largest voluntary contributors. As the largest donor, the United States gives $845 million or 15 percent of the total budget. Because voluntary contributions often are specified for specific initiatives, the WHO cannot totally pursue its own agenda, and it has become increasingly dependent on voluntary contributions. Some donors have pressured the organization to change its goals. For example, in 2019 the United States reportedly threatened to slash its donations if the WHO enacted a resolution encouraging breastfeeding. Recent budget cuts have been identified as contributing to the WHO's slow response to the 2014 Ebola outbreak. Another looming budget problem may be the nearing complete eradication of polio, because anti-polio funds amount to almost 75 percent of all WHO employees' salaries.

The WHO's activities have expanded from its initial concentration on women's and children's health, nutrition, sanitation, and fighting malaria and tuberculosis. Now, the WHO is recognized as the directing and coordinating authority on international health. Its current priorities are communicable diseases, especially HIV/AIDS, Ebola, malaria, and tuberculosis; the alleviation of noncommunicable diseases (NCDs), such as sexual and reproductive health, community development, and aging, nutrition, food security, and healthy eating; occupational health; substance abuse; and enhancing improved reporting, publications, and networking. The WHO also monitors and coordinates ameliorative activities regarding several health-related issues, such as tobacco and drug use, genetically modified foods, and climate change. It valuably serves as an international arbiter of standards, norms, and best practices in several health and health-related areas. For example, the WHO has developed and maintained since 1977 an essential medicines list and diagnostic tests list it recommends to hospitals. A recommended list for devices, such as an ultrasound machine, is under development. As a result of this work, the WHO reputably has significant credibility, especially in low- and middle-income countries.

ORGANIZATIONAL RESPONSE TO EMERGENCIES

The WHO relies on its member countries for initial identification of emergencies, and countries historically have been slow to report outbreaks because of economic repercussions. In a severe crisis, the WHO can declare a public health emergency of international concern (PHEIC). This has been done five times: the 2009 swine flu (H1N1) epidemic, the decline in eradicating polio in 2014, the 2014 Ebola outbreak in West Africa, the 2016 Zika virus outbreak in the Americas, and the Ebola epidemic hitting Goma in the Democratic Republic of the Congo (Grady 2019). PHEIC declaration hastens international action and often research on the relevant disease. However, most PHEIC declarations are controversial with some holding that they can worsen an outbreak.

During a PHEIC, the WHO gives nonbinding information to the country about the best emergency response, including travel and trade restrictions. It spells out treatment guidelines in an effort to not provoke panics. The WHO acts as a global coordinator by getting scientific data and experts to the affected areas. The organization attempts to stop nearby states from overreacting and thus economically hurting the country in crisis. Despite the WHO's efforts, many countries continue putting onerous travel and trade restrictions, such as what happened during the

2014 Ebola PHEIC. For less severe emergencies that are not a PHEIC, the WHO also gives guidance and coordination.

WHO staff works alongside governments and health professionals on the ground, as well as with ministries of health, government agencies, and other government departments at the national level. WHO regularly tracks how it is perceived by its external stakeholders and uses the information for programming and budgeting.

HISTORY AND ACCOMPLISHMENTS

In partnership with countries, the WHO has many achievements in promoting health and well-being. It led the largely successful struggle against smallpox. The organization's work has played a great role in development of internationally recognized standards for air and water quality. The WHO's prequalification program has led to increasingly safe and effective vaccines and medicines, as well as widely utilized height and weight charts for children as health guides. In addition, it has popularized its professional guidelines and advice on the prevention and treatment of many health conditions, including asthma, hepatitis, malnutrition, and Zika. Its International Classification of Diseases, a common standard for reporting diseases and identifying health trends, is used in over 100 countries and is in more than 40 languages.

Earlier, the WHO focused on fighting infectious killers such as smallpox, polio, and diphtheria. Its Expanded Programme on Immunization (1974) has—in partnership with UNICEF, Gavi, the Vaccine Alliance, and others—vaccinated millions of children. After the discovery of polio vaccines during 1952–1957, the WHO-facilitated global campaign has nearly eradicated poliomyelitis. In 1969, the World Health Assembly issued the first International Health Regulations, stemming from an agreement by WHO member states to work communally to monitor and control cholera, plague, yellow fever, smallpox, relapsing fever, and typhus. In 1972, the WHO established the Special Programme of Research, Development and Research Training in Human Reproduction to research sexual and reproductive health and rights.

In 1975, the WHO began hosting the Special Programme for Research and Training in Tropical Diseases, which helped nearly eliminate five of the eight targeted diseases by 2016. The "Health for All" goal set by the 1978 International Conference on Primary Health Care, in Alma-Ata, Kazakhstan, was the basis for the WHO's recent advocacy of universal health coverage. In 1995, the WHO started the DOTS (directly observed treatment) strategy to lessen tuberculosis's suffering. By 2013, this strategy had saved more than 37 million lives.

The WHO's Millennium Development Goals, including halving extreme poverty, stopping the spread of HIV/AIDS, and providing universal primary education, were agreed to by global leaders in 2000. In 2003, the WHO Framework Convention on Tobacco Control, its first global public health treaty, was adopted. In 2005, the World Health Assembly revised and updated the International Health Regulations to prevent and respond to acute public health threats instead of just specific diseases.

The WHO created a new health emergencies program in 2007, focusing on early-warning sign identification and assistance provision for countries to prepare. After the new H1N1 influenza virus was detected in 2009, the WHO collaborated with its partners to develop effective influenza vaccines. In sum, WHO-recommended earlier, simpler treatment and improved access to cheaper generic medicines resulted in 21 million people getting lifesaving HIV treatment. Recently, outbreaks of plague and Marburg virus disease were contained, and there were major largely successful cholera and yellow fever vaccination campaigns.

In 2010, WHO started advocating universal health care to improve popular access to essential health services. Eight years later, the WHO recommitted to the fundamental goal of health for all through universal health coverage. In 2012, WHO member states set concrete goals for the prevention and control of heart disease, diabetes, cancer, and other NCDs.

When in 2014 West Africa had the largest outbreak of Ebola virus disease, the WHO sent several thousand technical experts, support staff, and medical equipment. In 2016, the WHO found that there were no cases of Ebola in West Africa, though Zika virus infections continued to be a PHEIC. The next year the WHO contributed $1.5 million from its Contingency Fund for Emergencies when Madagascar was hit by its worst outbreak of pneumonic plague.

In response to newly arisen health threats, the WHO also has fruitfully worked on global health security by switching its emphasis from biological warfare to bioterrorism and emergent and re-emergent infectious disease outbreaks (natural or intentional), human security, and risks from dual-use technologies and emerging life sciences research (Fidler and Gostin 2009). The WHO has endeavored to educate the world about such events' wider societal impacts, such as disruption of critical infrastructure and services; diminished law and order; and other negative social, economic, and military impacts. The WHO has played a critical role in identifying pathogens as threats and in coordinating intergovernmental efforts to retard their spread and effects.

In its history, the WHO's work has contributed to many countries' increased healthcare spending, capacity building, health system strengthening, and improved responses to public health emergencies. This has included in a number of nations the passage of various new laws and regulations designed to enhance surveillance; curtail civil liberties; regulate behavior; and/or ease technical cooperation on the national, regional, and international levels. The Global Health Security Initiative and the European Commission's Health Security Committee now serve as international health forums. The organization has significantly aided the passage of many bilateral and some multilateral agreements, such as the revised International Health Regulations (2005) and the 2011 Pandemic Influenza Preparedness (PIP) Framework.

CONCLUSION

In 1952, the WHO was almost closed over a controversy about chemical and physical contraceptives with the Roman Catholic Church. This stemmed from the WHO, at India's request, announcing in 1951 that it would support a birth control education program there. The WHO reasoned that the severe problems in too many children and infant deaths were best addressed with Western scientific medicine. The dispute led to the WHO not pursuing family planning for the next decade (Farley 2008).

Many have faulted the WHO for its slow and poorly coordinated response to the 2014 Ebola outbreak when it waited five months before declaring a PHEIC, despite urgings from Doctors Without Borders and others. As a result, the WHO created a reserve force of public health workers, a $100 million emergency fund, and an incident management system so that medical responders, equipment, and supplies could be immediately at the sites while the organization coordinated a broader response.

The organization has been criticized for inefficiencies from being too bureaucratic and centralized. Other critics have cited the political friction between the WHO headquarters and its very autonomous six regional offices. For instance, some hold that tension between its headquarters and the WHO's Africa office, in Brazzaville, Republic of Congo, led to its slowness during the 2014 Ebola outbreak.

Others have criticized the WHO for unsavory relationships within its institutes, particularly during global health threats. For example, in 2009 pharmaceutical companies (e.g., Glaxo and Novartis) had questionable connections with the countries that had signed agreements to purchase swine flu vaccines from them and the WHO.

Some claim that WHO ignores the realities of situations in exchange for funding. For example, the WHO supposedly underestimated the number of deaths caused by a nuclear explosion because it was working with the International Atomic Energy Agency (IAEA), which was initially founded as UN's "Atoms for Peace" agency (Al Jazeera 2018).

Finally, the WHO has been attacked for wading into areas beyond its remit, such as the 2003 World Health Assembly's unanimous approval of a treaty curbing tobacco-related disease and death. Critics declared this a violation of personal liberties because people have a right to engage in risky behaviors such as smoking. The same arguments have been made by opponents of vaccination. However, the WHO has been always working to improve the well-being of the people of the world.

William P. Kladky

Further Reading

Al Jazeera. 2018. Trust WHO: The Business of Global Health. December 15, 2018. https://www.aljazeera.com/programmes/specialseries/2018/12/trustwho-business-global-health-181205092342434.html, accessed August 25, 2019.

Farley, J. 2008. *Brock Chisholm, the World Health Organization, and the Cold War.* Vancouver, BC, Canada: UBC Press.

Fidler, D., and L. Gostin. 2009. *Biosecurity in the Global Age: Biological Weapons, Public Health, and the Rule of Law.* Chicago, IL: University of Chicago Press.

Grady, D. 2019. Ebola Outbreak in Congo Is Declared a Global Health Emergency. *New York Times.* July 17, 2019. https://www.nytimes.com/2019/07/17/health/ebola-outbreak.html, accessed August 25, 2019.

Kamradt-Scott, A. 2015. *Managing Global Health Security: The World Health Organization and Disease Outbreak Control.* Houndmills, UK: Palgrave Macmillan.

Lee, K. 2009. *The World Health Organization (WHO).* Oxford, UK: Taylor & Francis.

Prentzas, G. S. 2009. *The World Health Organization.* New York: Chelsea House.

WHO (World Health Organization). 2006. *The World Health Report 2006: Working Together for Health.* Geneva, Switzerland: WHO Press.

World Vision International (WVI)

World Vision International (WVI) is the world's largest nongovernmental organization working to improve the lives of the world's poorest people. Rev. Bob Pierce founded WVI in the United States in 1950, but its genesis dates back to 1947. At this time, Pierce traveled through China as a Christian missionary. Following one of his sermons, a young Chinese girl, named White Jade, converted to Christianity. This conversion led her family to abandon her and leave her without care or support. Upon being told of her circumstances, Pierce provided her a small amount of money and promised to continue providing financial support each month. Pierce continued this financial support and encouraged others to do likewise. Three years later, drawing on this model of direct financial assistance, WVI was created as a Christian organization focused on supporting abandoned or orphaned children resulting from the Korean War (1950–1953). Over time, WVI began working in many other countries where there was obvious need; yet it is these two cores and historical attributes of its Christian identity and child sponsorship that continue to characterize WVI nearly 70 years after its formation. World Vision International's values include "We Are Christian; We Are Committed to the Poor; We Value People; We Are Stewards; We Are Partners; We Are Responsive" (WVI 2017c, 3).

As the world's largest NGO (with more than 40,000 staff), WVI raises and expends nearly $3 billion annually across 100 countries (WVI 2017c). As a child sponsorship organization, WVI supports over 3 million sponsored children, and approximately half of its annual budget is raised through and expended on child sponsorship program. However, unlike the original model of direct financial support of White Jade by Pierce, child sponsorship has evolved so that whole communities form the basis of the organization's development program, and these projects provide optimal outcomes for children (and their communities). As a result, WVI reaches over 40 million children each year through its Area Development Programs—which is the current approach used by WVI to implement its child sponsorship activities, World Vision International strengthened advocacy efforts in the twenty-first century, especially those related to child survival. The organization increased activities related to child labor, children in armed conflict, and the sexual exploitation of women and children (WVI 2017b).

Given the size of its annual budget, the number of countries in which it works, the number of sponsored children, the outreach model it utilizes, WVI has a very significant "development footprint" in both developing and developed countries in terms of the millions of people who sponsor a child or regularly donate to support WVI's work.

HUMANITARIAN WORK

While Peirce was certainly responding to the humanitarian needs of a vulnerable and abandoned child in 1947, his response was to address her immediate welfare needs. This was right and appropriate in the circumstances as an individual forming a personal relationship with another individual in need. However, once WVI as an aid agency was formed, it began to seek long-term development outcomes over short-term welfare responses. At the same time, WVI began to respond to humanitarian events when the local communities with whom they were working were affected by such events, and in time (during the 1970s), it began to respond to more complex humanitarian emergencies in countries and regions where it did not have an existing presence. In this way, WVI expanded to become a humanitarian NGO as well as a development NGO. As a result, it aspires to 20 percent of annual budget being expended on humanitarian responses. Given the overall size of its budget, this equates to approximately $400 million annually.

In recent times, WVI has responded to more than 130 disasters in more than 60 countries and has assisted more than 15 million affected people each year. These disasters include natural events such as earthquakes, cyclones, floods, droughts, and so forth as well as human-induced events including war, famine, and climate-change. WVI remains a Christian organization however and is nondiscriminatory in terms of with whom it works and partners. WVI also remains committed to working with children and has a goal of reaching 20 percent of all affected children in any disasters or emergency to which they respond. WVI has set itself the following goals in relation to its humanitarian activities:

- Expand our global operational footprint in conflict and fragile contexts
- Target our programming, policy and resources agenda to end violence against children
- Continue to increase our local, regional and global technical capacities in humanitarian protection, child protection in emergencies, educations in emergencies, water, sanitation and hygiene, health, nutrition and livelihoods so that the humanitarian needs of children are met
- Scale-up our humanitarian cash programming through leveraging private partnerships and the latest technologies
- Inform and lead the global humanitarian dialogue around the role and relevance of faith actors in the humanitarian action (WVI 2017a, 5)

Cash Transfers: Extreme Hunger

The use of cash transfers within development projects has a history dating back over two decades. Perhaps the best-known development project utilizing cash transfers to improve development outcomes is the Prospera program that originated in Mexico, which has now been replicated in over 50 countries. The principles of this development project are straightforward: provide cash to targeted families to remove barriers to children attending school and accessing health care.

In Mexico, nearly 6 million families have been involved in this program with marked improvement in poverty and inequality across the country.

More recently, the idea of conditional cash transfers has been applied to humanitarian responses. In the past, aid agencies would typically "truck and chuck" relief supplies to affected communities. These supplies would normally be sourced from nonaffected communities or regions elsewhere and transported (often over long distances) to the area of need. One of the unintended consequences of this was that local economies in affected communities would be underutilized following a disaster, and there would be a reduction in local economic activity that would have a lasting impact on communities. Conditional cash transfers allow affected communities to source and purchase (ideally locally produced) goods, which stimulates the local economy, which in turn has wider recovery consequences. To this end, conditional cash transfers not only allow affected communities to have their immediate humanitarian needs met but also increase local economic activities, which in turn provides opportunities for positive development outcomes. WVI has set itself some very ambitious targets in this programming approach, seeking to "deliver 50 percent of its humanitarian aid through a multi-sectoral and multi-purpose cash first approach by 2020, where context appropriate, such as in urban settings. This will leverage digital payment and identification systems and shared value partnerships with others, such as with Mastercard Corp, to track the delivery of assistance from donor to beneficiary" (WVI 2017a, 6).

Already WVI is utilizing this approach when addressing humanitarian issues associated with extreme hunger. In 2016, WVI implemented 180 food assistance projects in nearly 40 countries, which positively impacted over 5 million children. One such program was in South Sudan, where WVI worked alongside a community of more than 40,000 people. Each eligible family received around $45 per month to purchase basic food necessities. In this program, participants used their thumbprint as proof they had received their allowance.

Regional Health Emergencies: The Zika Virus

In a short nine-month period in 2015 and 2016, more than 40 countries in Latin America and the Caribbean were affected by the rapid spread of the Zika virus. Within this short period, 1.4 million were infected, resulting in increased risks to unborn children of developing microcephaly as well as other health complications. Given the scope and rapid escalation of infection, this event was declared a health emergency by the World Health Organization (WHO), noting it was of international concern and calling on the global community to respond. Since WVI was already working within the region, the organization took measures to stop the spread of the virus. It focused on protecting children and pregnant women and cooperated with health authorities and other agencies to establish diagnostic centers. These efforts were initiated to ensure that the children of women infected with the virus could be referred to appropriate healthcare services. In addition, WVI disbursed mosquito nets to help prevent further infection and eliminated mosquito breeding grounds (WVI 2017c).

WVI provided detailed health information to the public regarding the spread of virus so that people were educated about its symptoms. Additionally, it conducted a study in the six Latin American countries on public knowledge, attitude, and practices about the virus. The study was partially funded by the Pan American Health Organization and the WHO. "The data collected and analysed from the study informed national programming in each of these countries and also was shared with the international community via peer-reviewed journals, international conferences, and webinars. By embedding our response in community-focussed programming and engaging local leaders and faith leaders, World Vision reached more than 3 million people in the region. We will continue to work with these communities through our ongoing programming and will continue to support public health efforts in relation to containing the Zika virus" (WVI 2017c, 15).

In a six-month response, over $1.5 million was expended by WVI across the activities described earlier, including nearly 1,000 cleaning campaigns, over 800 actions to eradicate mosquito breeding sites, more than 8,000 protection kits distributed to pregnant women, and behavior change campaigns that reached 2 million people across the region.

CONCLUSION

World Vison was established through a single act of humanity when Reverend Peirce responded immediately to the needs of a child abandoned by her family following her conversation to Christianity in China. Pierce was committed to provide for this child's needs on a monthly basis, and on his return to the United Sates, he encouraged others to make similar commitments. In time, and in response to the humanitarian needs resulting from the Korean War (1950–1953), WVI as a formal organization was formed to facilitate the sponsoring of children. From these small beginnings, WVI has grown to become the world's largest aid agency. Seven decades later, its approach to community development has evolved, yet its identity as a Christian organization sponsoring needy children has remained at its core.

For more than 40 years, WVI has also responded to humanitarian events beyond those that affect the children whom they sponsor. These events include both natural and human-induced disasters and involve responses in more than 60 countries each year with more than 15 million assisted annually. Like many other aid agencies, WVI is now incorporating conditional cash transfer into its standard responses. It has committed to using this model of cash transfers in half its humanitarian responses in the next few years. This is a relatively new and innovative approach to meeting the humanitarian needs of vulnerable children and their communities affected by humanitarian events, yet it was immediately familiar to Rev. Bob Pierce and White Jade.

Matthew Clarke

Further Reading

World Vision International. 2017a. FY16: End of Year Report. World Vision Humanitarian and Emergency Affairs. https://www.wvi.org/sites/default/files/WVI_HEA_EOFY16_Report_Final.pdf, accessed May 8, 2018.

World Vision International. 2017b. Our History. https://www.wvi.org/our-history, accessed May 8, 2018.

World Vision International. 2017c. World Vision Annual Review 2016. https://www.wvi.org/sites/default/files/WVI-Annual-Report-2016_final_0.pdf, accessed May 8, 2018.

Natural Hazards and Disasters

Natural Hazards and Disasters

From Avalanches and Climate Change to Water Spouts and Wildfires

VOLUME 2: Natural Disasters
Bimal Kanti Paul, Editor

An Imprint of ABC-CLIO, LLC
Santa Barbara, California • Denver, Colorado

Copyright © 2021 by ABC-CLIO, LLC

All rights reserved. No part of this publication may be reproduced, stored in a retrieval system, or transmitted, in any form or by any means, electronic, mechanical, photocopying, recording, or otherwise, except for the inclusion of brief quotations in a review, without prior permission in writing from the publisher.

Library of Congress Cataloging-in-Publication Data

Names: Paul, Bimal Kanti, editor.
Title: Natural hazards and disasters : from avalanches and climate change to water spouts and wildfires / Bimal Kanti Paul, editor.
Description: Santa Barbara : ABC-CLIO, 2021. | Includes bibliographical references and index. | Contents: v. 1. Natural hazards and aid organizations — v. 2. Natural disasters.
Identifiers: LCCN 2020019064 (print) | LCCN 2020019065 (ebook) | ISBN 9781440862151 (v. 1 ; hardcover) | ISBN 9781440862168 (v. 2 ; hardcover) | ISBN 9781440862137 (set : hardcover) | ISBN 9781440862144 (ebook)
Subjects: LCSH: Natural disasters—Encyclopedias.
Classification: LCC GB5014 .N389 2021 (print) | LCC GB5014 (ebook) | DDC 363.3403—dc23
LC record available at https://lccn.loc.gov/2020019064
LC ebook record available at https://lccn.loc.gov/2020019065

ISBN: 978-1-4408-6213-7 (set)
       978-1-4408-6215-1 (vol. 1)
       978-1-4408-6216-8 (vol. 2)
       978-1-4408-6214-4 (ebook)

25 24 23 22 21   1 2 3 4 5

This book is also available as an eBook.

ABC-CLIO
An Imprint of ABC-CLIO, LLC

ABC-CLIO, LLC
147 Castilian Drive
Santa Barbara, California 93117
www.abc-clio.com

This book is printed on acid-free paper ∞

Manufactured in the United States of America

Contents

VOLUME I

Guide to Related Topics xi

Preface xv

Introduction xvii

Natural Hazards 1

Avalanches 1

Blizzards 11

Climate Change 22

Coastal Erosion 33

Desertification 43

Droughts 53

Earthquakes 64

Erosion 75

Expansive Soils 87

Extinction 97

Floods 107

Hail 119

Hurricanes 129

Ice Storms 139

Landslides 151

Lightning 162

Salinization 172

Storm Surges   181

Subsidence   189

Temperature Extremes   199

Tornadoes   210

Tsunamis   220

Volcanic Activity   230

Waterspouts   240

Wildfires   250

Aid Organizations   263

ActionAid International   263

American Red Cross   267

Catholic Relief Services (CRS)   272

Concern Worldwide   276

Cooperative for Assistance and Relief Everywhere   281

Doctors Without Borders   286

International Federation of Red Cross and Red Crescent Societies (IFRC)   290

International Organization for Migration (IOM)   295

Islamic Relief Worldwide   299

Lutheran World Federation   304

Mennonite Central Committee (MCC)   309

Oxfam International   314

Pan American Health Organization (PAHO)   319

Refugees International   323

Save the Children   328

United Nations Children's Fund (UNICEF)   332

World Food Program (WFP)   337

World Health Organization (WHO)   341

World Vision International (WVI)   346

VOLUME 2

Guide to Related Topics xi

Natural Disasters 1

Bam Earthquake, Iran, 2003 1

Bangladesh Flood, 1998 6

Bengal Famine, 1943–1944 10

Bhola Cyclone, Bangladesh, 1970 14

Big Thompson Canyon Flash Flood, Colorado, 1976 18

Black Saturday Bushfires, Australia, 2009 23

BP Deepwater Horizon Oil Spill, United States, 2010 27

Brisbane and Queensland Flood, Australia, 2011 32

Buffalo Blizzard, New York, 1977 36

California Drought, 2012–2016 40

Chicago Heat Wave, Illinois, 1995 45

Chi-Chi Earthquake, Taiwan, 1999 50

Christchurch Earthquake, New Zealand, 2010–2011 54

Colombia Floods, 2010–2011 59

Colorado Flood, United States, 2013 64

Cyclone Gorky, Bangladesh, 1991 69

Cyclone Nargis, Myanmar, 2008 73

Cyclone Pam, Vanuatu, 2015 77

Cyclone Phailin, India, 2013 82

Cyclone Sidr, Bangladesh, 2007 87

The Dust Bowl, 1930s 91

East African Drought, 2011–2012 96

Edmonton Tornado, Canada, 1987 101

European Heat Wave, 2003 105

Eyjafjallajökull Eruption, Iceland, 2010 109

Grand Forks Flood, North Dakota, 1997 114

Great Ice Storm of 1998, Canada   118

Great Kanto Earthquake, Japan, 1923   122

Great Mississippi River Flood, United States, 1993   127

Gujarat Earthquake, India, 2001   132

Haiti Earthquake, 2010   137

Heat Wave and Wildfires, Russia and Eastern Europe, 2010   141

Hurricane Andrew, United States and the Bahamas, 1992   146

Hurricane Charley, United States, 2004   150

Hurricane Galveston, United States, 1900   155

Hurricane Harvey, Texas and Louisiana, 2017   159

Hurricane Ike, United States, 2008   164

Hurricane Irma, Florida, 2017   168

Hurricane Katrina, United States, 2005   173

Hurricane Maria, Puerto Rico, 2017   179

Hurricane Matthew, United States, 2016   183

Hurricane Mitch, Central America, 1998   188

Hurricane Stan, Guatemala, 2005   193

Indian Ocean Tsunami, 2004   197

Iowa Flood, United States, 2008   202

Izmit/Marmara Earthquake, Turkey, 1999   207

Johnstown Flood, Pennsylvania, 1889   211

Joplin Tornado, Missouri, 2011   216

Kashmir Earthquake, Pakistan, 2005   221

Kerala Floods, India, 2018   225

Kobe Earthquake, Japan, 1995   230

Loma Prieta Earthquake, California, 1989   235

Mexico City Earthquakes, Mexico, 1985   239

Millennium Drought, Australia, 2001–2012   244

Mozambique Flood, 2000   248

Nepal Earthquakes, 2015   253

Pakistan Flood, 2010   258

Sichuan Earthquake, China, 2008   262

Sulawesi Earthquake and Tsunami, Indonesia, 2018   267

Summer Floods, United Kingdom, 2007   272

Superstorm Sandy, United States, 2012   276

Tangshan Earthquake, China, 1976   281

Thomas Fire, California, 2017–2018   286

Tohoku Earthquake and Fukushima Tsunami, Japan, 2011   290

Tri-State Tornado, United States, 1925   295

Tropical Storm and Floods, Yemen, 2008   299

Typhoon Haiyan (Yolanda), Philippines, 2013   304

Valdivia Earthquake, Chile, 1960   309

Vietnam Flood, 1999   313

Yangtze River Flood, China, 1931   317

Bibliography   323

About the Editor and Contributors   331

Index   335

Guide to Related Topics

VOLUME 1

CLIMATOLOGICAL
Climate Change
Desertification
Droughts
Ice Storm
Temperature Extremes
Wildfires

GEOPHYSICAL
Avalanches
Coastal Erosion
Earthquakes
Erosion
Expansive Soils
Extinction
Landslides
Subsidence
Tsunamis
Volcanic Activity

HUMANITARIAN AGENCIES
ActionAid International
American Red Cross
Catholic Relief Services (CRS)
Concern Worldwide
Cooperative for Assistance and Relief Everywhere
Doctors Without Borders
International Federation of Red Cross and Red Crescent Societies (IFRC)
International Organization for Migration (IOM)
Islamic Relief Worldwide
Lutheran World Federation
Mennonite Central Committee (MCC)
Oxfam International
Pan American Health Organization (PAHO)
Refugees International
Save the Children
United Nations Children's Fund (UNICEF)
World Food Program (WFP)
World Health Organization (WHO)
World Vision International (WVI)

HYDROLOGICAL
Floods
Salinization
Storm Surges
Waterspouts

METEOROLOGICAL
Blizzards
Hail
Hurricanes
Lightning
Tornadoes

VOLUME 2

CLIMATOLOGICAL
Bengal Famine, 1943–1944
Black Saturday Bushfires, Australia, 2009
California Drought, 2012–2016
Chicago Heat Wave, Illinois, 1995
The Dust Bowl, 1930s
East African Drought, 2011–2012
European Heat Wave, 2003
Great Ice Storm of 1998, Canada
Heat Wave and Wildfires, Russia and Eastern Europe, 2010
Millennium Drought, Australia, 2001–2012
Thomas Fire, California, 2017–2018

GEOPHYSICAL
Bam Earthquake, Iran, 2003
Chi-Chi Earthquake, Taiwan, 1999
Christchurch Earthquake, New Zealand, 2010–2011
Eyjafjallajökull Eruption, Iceland, 2010
Great Kanto Earthquake, Japan, 1923
Gujarat Earthquake, India, 2001
Haiti Earthquake, 2010
Indian Ocean Tsunami, 2004
Izmit/Marmara Earthquake, Turkey, 1999
Kashmir Earthquake, Pakistan, 2005
Kobe Earthquake, Japan, 1995
Loma Prieta Earthquake, California, 1989
Mexico City Earthquakes, Mexico, 1985
Nepal Earthquakes, 2015
Sichuan Earthquake, China, 2008
Sulawesi Earthquake and Tsunami, Indonesia, 2018
Tangshan Earthquake, China, 1976
Tohoku Earthquake and Fukushima Tsunami, Japan, 2011
Valdivia Earthquake, Chile, 1960

HYDROLOGICAL
Bangladesh Flood, 1998
Big Thompson Canyon Flash Flood, Colorado, 1976
BP Deepwater Horizon Oil Spill, United States, 2010
Brisbane and Queensland Flood, Australia, 2011
Colombia Floods, 2010–2011
Colorado Flood, United States, 2013
Grand Forks Flood, North Dakota, 1997
Great Mississippi River Flood, United States, 1993
Iowa Flood, United States, 2008
Johnstown Flood, Pennsylvania, 1889
Kerala Floods, India, 2018
Mozambique Flood, 2000
Pakistan Flood, 2010

Summer Floods, United Kingdom, 2007

Vietnam Flood, 1999

Yangtze River Flood, China, 1931

METEOROLOGICAL

Bhola Cyclone, Bangladesh, 1970

Buffalo Blizzard, New York, 1977

Cyclone Gorky, Bangladesh, 1991

Cyclone Nargis, Myanmar, 2008

Cyclone Pam, Vanuatu, 2015

Cyclone Phailin, India, 2013

Cyclone Sidr, Bangladesh, 2007

Edmonton Tornado, Canada, 1987

Hurricane Andrew, United States and the Bahamas, 1992

Hurricane Charley, United States, 2004

Hurricane Galveston, United States, 1900

Hurricane Harvey, Texas and Louisiana, 2017

Hurricane Ike, United States, 2008

Hurricane Irma, Florida, 2017

Hurricane Katrina, United States, 2005

Hurricane Maria, Puerto Rico, 2017

Hurricane Mathew, United States, 2016

Hurricane Mitch, Central America, 1998

Hurricane Stan, Guatemala, 2005

Joplin Tornado, Missouri, 2011

Superstorm Sandy, United States, 2012

Tri-State Tornado, United States, 1925

Tropical Storm and Floods, Yemen, 2008

Typhoon Haiyan (Yolanda), Philippines, 2013

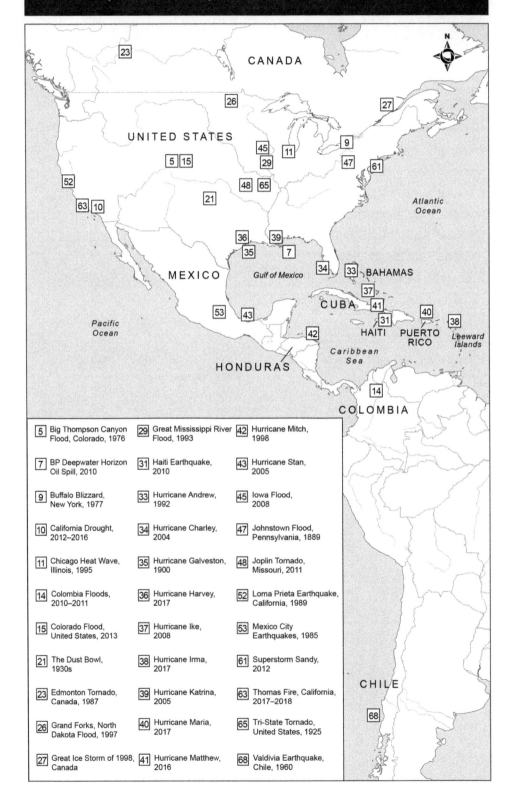

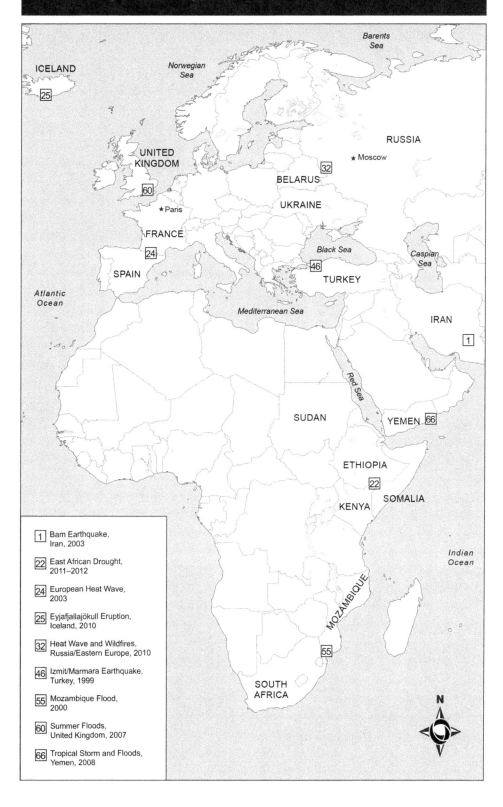

Asia and Oceania

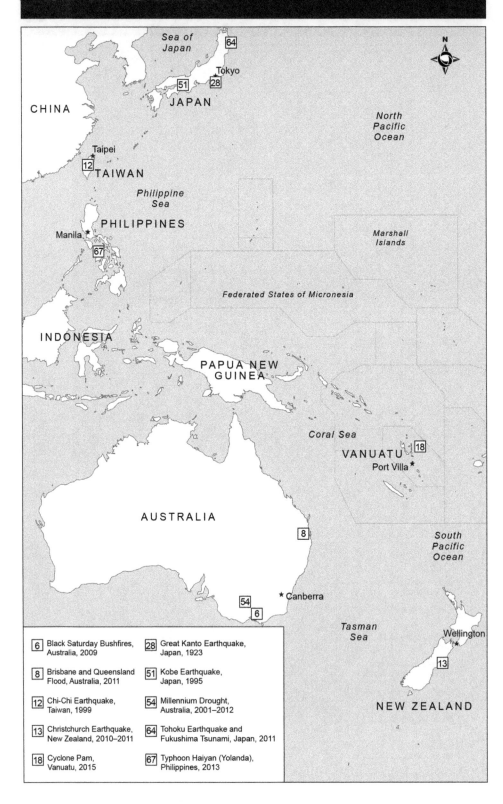

South Asia

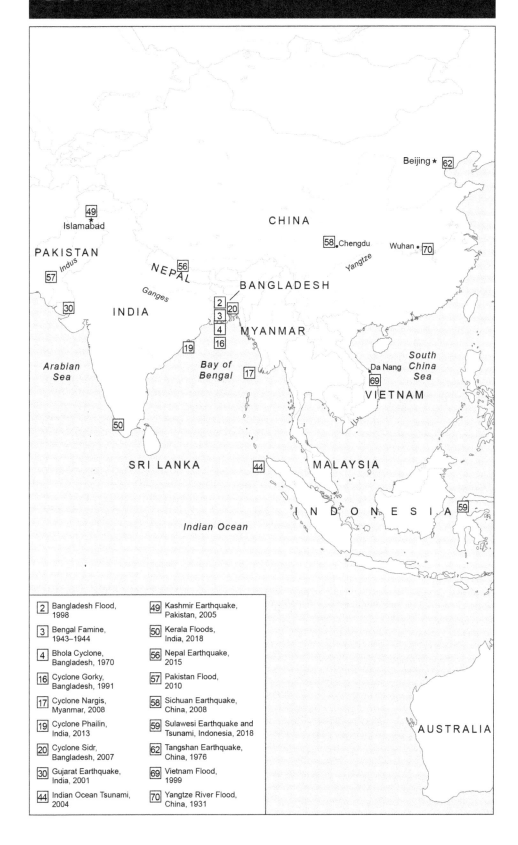

Natural Disasters

Bam Earthquake, Iran, 2003

The city of Bam is located in the southeastern part of the Kerman province of Iran. On December 26, 2003, a massive earthquake struck the city and some of the neighboring villages at 5:26 a.m. local time while most people slept. The magnitude of the earthquake was 6.5 on the Richter scale, and the epicenter was just outside the city. Approximately 30,000 people lost their lives, and 75,000 others were rendered homeless. In all, the event affected nearly 200,000 people. The earthquake destroyed the historical citadel of Arg-e-Bam, also considered the largest mud-brick construction in the world (prior to the earthquake). Although other parts of Iran experienced major earthquake events in the past, the regions around Bam had no historical earthquake records and were not considered the most seismically active zones of the country prior to the 2003 event (UN 2004; Ghafory-Ashtiany and Hosseini 2008).

CAUSE OF THE EARTHQUAKE

Iran is tectonically a part of the Alpine-Himalayan orogenic belt and is extremely vulnerable to earthquakes. The Central Iranian Block is one of the major tectonic features in this area. This block consists of several other blocks including the Lut Block. Several strike slip faults, or vertical fractures where the blocks move mostly horizontally, border these blocks, and the city of Bam is located in the southern part of the Lut Block. The Bam fault, a 33-mile (50 km) long right-lateral, hidden strike slip fault with a north-south trend, is another major tectonic feature of the area. This is a hidden fault that goes down to the surface of the bedrock and is very thickly covered with sediments. This fault's rupture caused the December 2003 earthquake in Bam and the surrounding areas (Nadim et al. 2004).

Damage from the 2003 Iran earthquake in the ancient town of Bam. (Mathess/iStockphoto.com)

DAMAGE AND LOSS

The Bam earthquake and several major aftershocks almost destroyed the city. Naturally, major damages were recorded in places that experienced the strongest shakings, while away from the epicenter, the severity of damage decreased. Notably, symmetry in the damage pattern was noticed within 2 miles (3 km) west of the Bam fault surface. The intensity of damage, however, decreased with increasing distance from the surface of the fault line. Residential and public buildings, including hospitals and schools, were severely damaged, the city's infrastructure collapsed, and rural agriculture suffered significant damage (Ghafory-Ashtiany and Hosseini 2008).

The city of Bam then consisted of different types of buildings made of a variety of materials and for this reason suffered different intensities of damage from the earthquake. Approximately, 25,000 houses in Bam and nearby villages were destroyed. This is because the structures made of earthen materials (adobe structures) and bricks were the least resistant to earthquakes and therefore the most vulnerable. The earthquake completely destroyed about 90–95 percent of the adobe structures. The remaining 5–10 percent of the structures were damaged beyond repair. Similarly, the brick structures failed almost completely during the earthquake. Approximately all of the one- or two-story brick structures either entirely collapsed or were damaged beyond restoration. However, the city's steel and concrete structures fared much better than the adobe and the brick structures (Hisada et al. 2004; Ghafory-Ashtiany and Hosseini 2008).

Health facilities also suffered significant damage. The earthquake destroyed three district hospitals, and 10 urban health centers, 14 rural health centers, and 95 health houses suffered severe damage. The disease surveillance system and primary healthcare, including postnatal and antenatal care, was disrupted. Following the earthquake, the education infrastructure of the city of Bam suffered a huge loss. All 131 school buildings were heavily damaged and rendered nonoperational. In addition to these major structural damages, significant damage to the city's roads, power distribution, communication network, sanitation facilities, and water supply system was also reported. Two of the eleven drill wells and main pipes and branch pipes supplying water to the city of Bam were damaged. About 20 percent of the wells supplying water to the city and six tanks along with the water quality monitoring laboratory collapsed.

Because the sanitation system in the city suffered considerable damage, human waste and garbage became one of the most severe issues following the earthquake. Additionally, as a huge number of the affected people relocated to other places, mostly temporarily, the receiving communities suffered shortage of necessary resources and facilities. Agricultural facilities also suffered varying degrees of damage mostly due to blockage and leakage of the underground canals, collapse of orchard walls, and landslides (UN 2004; Ghafory-Ashtiany and Hosseini 2008).

Food security for nearly 100,000 people was severely compromised after the earthquake, and this followed a severe food shortage resulting from a drought only a year prior. Post-earthquake loss of life; destruction of infrastructure; loss of employment, business, personal assets, and livelihood; and diminished economy directly impacted the purchasing power of the people, as simultaneously, access to food became difficult for the survivors (UN 2004).

Between 2,000 and 3,000 children lost their parents or family members in the earthquake, and nearly 5,000 children lost both parents. Thousands of children were wounded and taken to hospitals in other cities away from their families. It was believed that about 40 percent of Bam's population, including a substantial number of children, would suffer from posttraumatic stress disorder (PTSD) afterward and require psychological support (UN 2004).

RELIEF

The Iranian government and many nongovernmental organizations (NGOs) carried out immediate relief operations. Specifically, Iran Red Crescent Society, Iranian law enforcement, various branches of the United Nations, and private volunteers, both Iranian and internationals, participated in the rescue and relief measures, among other entities. A six-member committee, chaired by the Governor of Kerman province, coordinated the relief efforts. A massive evacuation operation was launched with 725 vehicles and 8,500 workers. Dead bodies were recovered from the rubble and buried with religious ceremony. Approximately 12,000 injured people were airlifted to hospitals in other Iranian cities for medical treatment. To track survivors and expedite assistance, the affected people were registered by the civil registration organization and issued registration cards for proper and systematic disbursement of relief.

Within two days of the earthquake, the United Nations Children's Fund (UNICEF) sent flights with emergency care, blankets, and tents. The UN also deployed doctors, logisticians, and information officers in Bam. Additionally, the World Health Organization (WHO) was active in rescuing and providing relief and worked in collaboration with the Ministry of Health and Medical Education, Iran. The WHO conducted a rapid health assessment survey of the affected area. Following this survey, a communicable disease surveillance system was formed in Bam. Meanwhile, the United Nations High Commissioner for Refugees (UNHCR), WHO, World Food Program (WFP), and United Nations Population Fund (UNFPA) each donated emergency cash grants of $100,000 and the United Nations Office for the Coordination of Humanitarian Affairs (OCHA) donated $50,000.

Relief also reached Bam from nearly 44 nations. The United Nations Disaster Assessment and Coordination (UNDAC) team assisted by the United Kingdom conducted an aerial survey of the affected area. Within a few days, the UNDAC team and International Humanitarian partnership established an on-site operation and coordination center to coordinate the international rescue and disbursement of relief. The WFP responded to the crisis by disbursing dry foods, mostly biscuits, to more than 100,000 people by January 1, 2004 (UN 2004; Ghafory-Ashtiany and Hosseini 2008).

The earthquake rendered a huge number of people homeless. For immediate relief, the UNHCR disbursed 1,000 tents, 3,000 mattresses, and 10,000 blankets. Household items, emergency services, and temporary and intermediate shelters were also provided by the UNHCR. The Iranian Red Crescent Society also provided temporary emergency shelters and distributed more than 50,000 tents within 24 hours of the earthquake. Intermediate shelters were also provided by the governor's office in Kerman and by the Interior Ministry. These shelters were equipped with water heating facilities, air-conditioning, sink, and sanitary facilities. In addition, about 18,000 showers were installed for sanitation needs. Most of the survivors were moved to intermediate shelters from temporary tents by March 2004 (UN 2004).

RECONSTRUCTION

Besides the immediate and short-term relief measures, a massive long-term reconstruction plan was endorsed by the Government of Iran for the affected areas. As a reconstruction effort, separate housing complexes were constructed for the affected people. Nearly 66 villages were almost completely destroyed by the earthquake, and a total of 25,022 rural houses were extensively damaged in 136 different villages. Therefore, rural reconstruction involved both rebuilding of totally damaged homes and carrying out repairs of the ones that were damaged. By the year 2008, nearly 93 percent of the rural reconstruction was complete, and life in the villages had almost normalized (Ghafory-Ashtiany and Hosseini 2008).

Besides monetary help to the affected, the government provided low-interest and long-duration (15 years) loans for reconstruction and repair of damaged commercial and residential units. These funds were made available to every household whose home was either completely or partially damaged by the earthquake. Banks were also encouraged to provide long-term loans at a reasonable interest rate to people rebuilding their homes (Ghafory-Ashtiany and Hosseini 2008). Reconstruction of schools, hospitals, government buildings, and sports complex was also needed. Under the school safety program, priority was given to the reconstruction of 131 schools and repair of 33 schools. This restoration and reconstruction that began in the summer of 2004 was funded by the government, private donors, foreign nations, and various international organizations. A preliminary assessment report suggested that the newly constructed schools in Bam were of better quality and better equipped for disaster management than were the previous schools. Later in 2006, a $4-million bill for the reconstruction of damaged classrooms and improvement to remaining ones was passed by the Iranian parliament (Ghafory-Ashtiany and Hosseini 2008).

As noted, the healthcare system in Bam nearly collapsed in the earthquake. Even though immediate healthcare services and emergency relief were provided, massive reconstruction of the healthcare units in Bam was required as a part of the long-term plan for the city. By 2005, a hospital in Bam and 18 other healthcare units became operational. By 2008, two additional new hospitals were constructed, and several other clinics and healthcare units were also under construction. A large-scale project to provide psychosocial support to the survivors to help cope with PTSD was funded by UNICEF and the Ministry of Health, Iran. In addition, several rehabilitation centers were constructed to cater to the post-earthquake mental health needs of the survivors.

Besides, cultural and recreational centers, other entities such as sports complexes, centers for intellectual development of children and youth, libraries, markets, government buildings, and television stations, were also rebuilt or reconstructed. The reconstruction of Arg-a Bam and other historical monuments began with the close cooperation of UNESCO and foreign nations. Water supply and electricity networks were also restored and repaired while construction of 20 new wells, a water reservoir, and main and branch pipelines were completed by mid-2006 (Ghafory-Ashtiany and Hosseini 2008).

CONCLUSION

The destructive earthquake in December 2003 in the city of Bam and surrounding areas resulted in a massive loss of life and property. However, the government of Iran responded to the event as an opportunity to create awareness among people about earthquake-risk reduction in the country. Following the earthquake, the government developed an integrated strategy entitled "Iran's Strategy of Earthquake Risk Reduction." This approach included policy changes, disaster risk mitigation, and spread of awareness and public preparedness as effective ways to reduce risk. The reconstruction of the affected areas was possible due to the joint endeavor of several organizations and very effective, planned implementation of reconstruction strategies. The lessons learned from the previous earthquakes and also from this event in 2003 entailed greater attention to people's demands in the wake of disaster along with a fast economic recovery, creation of jobs, and provision of loans. As a result of reconstruction, the city of Bam and the surrounding areas are safer at present than ever before.

Subarna Chatterjee

Further Reading

Ghafory-Ashtiany, M., and M. Hosseini. 2008. Post-Bam Earthquake: Recovery and Reconstruction. *Natural Hazards* 44(2): 229–241.

Hisada, Y., A. Shibayama, and M. R. Ghayamghamian. 2005. Building Damage and Seismic Intensity in Bam City from the 2003 Iran, Bam, Earthquake. *Bulletin of the Earthquake Research Institute* 79(3/4): 81–93.

Nadim, F., M. Moghtaderi-Zadeh, C. Lindholm, A. Andresen, S. Remseth, M. J. Bolourchi, and E. Tvedt. 2004. The Bam Earthquake of 26 December 2003. *Bulletin of Earthquake Engineering* 2(2): 119–153.

UN (United Nations). 2004. Flash Appeal Bam Earthquake of 26 December 2003, Islamic Republic of Iran Relief, Recovery and Immediate Rehabilitation. New York. https://reliefweb.int/sites/reliefweb.int/files/resources/6D743F4669054C41C1256E15005DE64E-ocha-irn-8jan.pdf, accessed May 20, 2018.

Bangladesh Flood, 1998

Bangladesh is located at the convergence of the Ganges-Brahmaputra-Meghna river system, the third largest river system in the world by freshwater discharge. The floodplain of these three rivers and their distributaries occupies around 80 percent of the country. Less than 10 percent of the 0.60 million square miles (1.55 million square km) catchment area of these rivers lies within the borders of Bangladesh. Thus, the rainfall in neighboring India, Nepal, Bhutan, and China and snowmelt in the Himalayas are major determinants of water flow through Bangladesh. This geographical location makes Bangladesh a vulnerable place for seasonal flooding, and as such, Bangladesh encounters flooding almost every year. Yet, the 1998 flood is remarkable in history because of its intense and extensive nature. Deemed the "flood of the century," it inundated more than two-thirds of the country, including parts of Dhaka, the nation's capital, that were 6.5 feet (2 m) under floodwater.

The country had previously suffered from a severe flood just 10 years earlier in 1988, which was considered the most disastrous flood event in the recorded history of Bangladesh. More than half of the country went under water, and an estimated 45 million people were directly affected. However, compared to the 1988 flood, in 1998, the water remained above the danger level for a longer time in the three main rivers, the Ganges, the Brahmaputra, and the Meghna. As such, the extent and magnitude of the 1998 flood were greater than those of the 1988 flood, but the 1998 flood was particularly remembered for its extraordinary length of stay: eleven weeks. In total, floodwater covered 53 of the 64 districts of Bangladesh (a district is the second largest administrative unit in the country) at the height of the disaster. Notably, the most severe flooding occurred along the main river courses, especially at the confluence of the three major rivers. In particular, along the Ganga, the flood surpassed all previous records, including those of the 1988 flood (Hofer and Messerli 2006).

CAUSES

A combination of several factors caused the 1998 flood event: excessive rainfall in the catchment area; simultaneous high peaks of the three major rivers between

September 7 and 11; high tides in the Bay of Bengal (particularly affecting the mouth of the Meghna estuary); a combination of flash floods, river floods, and tidal floods; a monsoon trough that was positioned further north than normal; and a La Niña situation that created favorable conditions for a large flood (Hofer and Messerli 2006). The 1998 flood primarily began with torrential rains in the month of July. Specifically, July 13–16 saw more than 58 inches (1,472 mm) of rainfall, around 300 percent more than the typical July average, which triggered the widespread flooding. In the following 12 days, 37 out of 64 districts went under water at varying heights. Initially, flooding was mainly confined to the southeastern hilly basin and the Meghna basin in the northeast of Bangladesh. During the third week of July, a heavy onrush of water in the Brahmaputra added water to rising levels in the Ganges basin in the western part of the country. By August 30, 1998, around 23,000 square miles (37,000 square km) (41 percent of the country) were affected by flooding (del Ninno et al. 2001). The peak water level climbed to 37.6 feet (11.45 m) above danger level. So that by September 7, around 51 percent of the country was completely inundated whereupon the water level started to fall, and by September 25, no monitoring stations reported flows above danger levels.

Unlike other major floods, the 1998 flood prevailed for an extraordinarily long time, lasting for nearly three months, while the regular monsoonal flood period in Bangladesh is about three weeks. Abnormally high tides in the Bay of Bengal, high glacial melt, and continuing monsoon rains throughout the three months (July–September) caused the flood's long duration. High tides created a barrier for floodwater to drain down to the sea, and the prolonged heavy rains led to all three major rivers reaching peak flow at the same time. Additionally, climate change has been attributed to increased snowmelt and higher sea levels. The reason for the extensive, erratic rainfall is not certain, but some experts have hypothesized that the dynamics of the El Niño Southern Oscillation (ENSO) played some role (Chowdhury 2000). ENSO is an irregularly periodic variation in winds and sea surface temperatures over the tropical eastern Pacific Ocean, affecting the climate of much of the tropics and subtropics. The effects of El Niño prevailed over the Indian Subcontinent at the beginning of 1998. Later in the year, excessive rainfall occurred in the northeastern part of India and in Bangladesh due to the effect of La Niña. Accordingly, from June to September 1998, this ENSO phenomenon created anomalies in monsoon activities over the Indian Subcontinent and thereby increased the total rainfall to 24 percent above the normal in Bangladesh.

LOSS OF LIFE AND EXTENT OF DAMAGE

The vast areal extent and unusually long duration of the 1998 flood makes it one of the most disastrous in the history of Bangladesh, affecting approximately 30 million people. Post-flood epidemics of water-borne diseases are usual in Bangladesh, but in the case of the 1998 flood, epidemics started before the water receded. Around 400,000 people were affected by diarrhea and other diseases (e.g., cholera and typhoid), which killed 500 people largely because of lack of clean water supply. Another 500 people died because of drowning and snake bites. The consequences of this flood were much more damaging in rural areas than in urban areas. For example, more than 300,000 tube wells, a basic source of clean water in rural areas, were destroyed. Also, due to severe flooding, 3.35 million tons of crops were lost, the equivalent to an economic loss of $1.35 billion. Particularly, rice production encountered a massive loss of $610 million with 3.7 million acres (1.5 million ha) of damaged production area (Mustafi and Azad 2003).

The 1998 flood also prevented the planting of *aman* rice (which is harvested in December). As a result, the households of the flood-affected districts experienced an average of 1,609 lbs. (730 kg) of food deficit that year; meanwhile, an estimated 25 million people faced malnutrition. Available information shows that the flood affected 7.8 million cattle/buffaloes, 4.2 million goats/sheep, and 352 million poultry/ducks. In all, 5,326 cattle died because of lack of fodder. Furthermore, the flood severely damaged an estimated 10,000 miles (15,000 km) of roads, 7,000 bridges and culverts, and 3,000 miles (4,500 km) of river embankments. Schools and other educational institutions are often used as flood shelters, but this flood damaged around 20,000 educational institutions, reducing the number of available locations for shelter. In addition to such public infrastructure, the flood also damaged over 500,000 houses, leaving 25 million people homeless. Overall, with an approximate total loss of $2 billion, the Gross Domestic Product (GDP) growth of Bangladesh dropped to 3.3 percent for 1998–1999, from 5.6 percent in 1997–1998 (Shah 1999; Hofer and Messerli 2006).

RELIEF WORKS

Considering its limited resources, the Bangladesh government sought external aid and managed to amass $700 million by February 1999 (Paul 2003). With this external (and internal) support, the Bangladesh government mobilized the local, national, and international nongovernment organizations (NGOs) in a massive relief and rehabilitation program. Based on a small sample, Paul (2003) found that because of this effort, around 75 percent of the flood affected people received relief assistance either from the government or from NGOs. The remaining people, who did not receive any assistance, were mostly economically well-off. The relief came primarily in the form of dry food, crop seeds, pure water, water purification tablets, clothing, and medicine: each household received around $15 in cash money (Paul 2003).

Furthermore, the government started a special Vulnerable Group Feeding (VGF) program that provided 4 million vulnerable households with 35 lbs. (16 kg) of wheat and rice per month. The government also substantially expanded food allocations for existing public works and transfer programs: Food-for-Work (43 percent increase), Test Relief (100 percent), Food for Education (6 percent), and Gratuitous Relief (12 percent). These expansions were to help contain the adverse impact of the floods on the poor (del Ninno et al. 2001). Around 750 NGOs, which operated micro-credit schemes in the rural areas, played a crucial role in relief and rehabilitation. No major NGO suspended repayments, but most eventually gave branch managers discretion to reschedule repayments wherever needed. Additionally, fresh loans were made available in some cases.

RECOVERY

To cope with the impacts of the flood, the rural people adopted a variety of livelihood options such as petit trade, boat transport, and fishing. Where employment alternatives were limited, some of the poor moved closer to urban areas to work in the nonfarm sector, though early reports suggest that the extent of migration to urban areas was not significant. After the floodwaters receded, employment prospects improved with the creation of jobs in public works, to repair damaged houses, and in fishing.

Fortunately, flooding in Bangladesh brings suspended sediment from upstream, which recharges soil nutrients to some degree and consequently provides good harvests. For example, because of the 1998 flood, Bangladesh experienced a bumper crop of *boro* rice in 1999 that exceeded even the government's pre-flood projections. Consequently, employment prospects improved appreciably as a result of good crop production. All of these positive outcomes were made possible due to a comprehensive program of agricultural rehabilitation, including the timely provision of credit and agricultural inputs. However, recovery in the manufacturing sector is a little more complicated. Growth in manufacturing, which was already weakening before the floods, dropped to a significantly negative rate afterward, which was reflected in a marked deceleration of export growth in 1998–1999. Private investment growth also appears to have decelerated in 1998–1999 despite the World Bank providing a $200 million loan for improving the manufacturing and garments industry. In sum, various efforts to recover from the flood losses were mostly successful, but the 1998 flood remains a major disaster because of its catastrophic impacts.

CONCLUSION

The 1998 flood in Bangladesh was the most severe of the twentieth century for that country in terms of both duration and extent. Fifty-three of the 64 districts were affected to different magnitudes. About half of the country was under water for up to 67 days, at depths of up to 10 feet (3 m). Flood conditions lasted from July to mid-September and became severe on three particular dates: July 28, August 30, and September 7 when 30 percent, 41 percent, and 51 percent of the country was under water, respectively. The Bangladesh government apparently provided flood relief to the affected people in a timely manner, and the distribution of emergency aid was more or less fair while domestic and international responses to the flood were noteworthy. The recovery period was also short. Lessons possibly learned from the 1988 flood helped facilitate the successful response to the 1998 flood.

Asif Ishtiaque

Further Reading

Chowdhury, M. R. 2000. An Assessment of Flood Forecasting in Bangladesh: The Experience of the 1998 Flood. *Natural Hazards* 22(2), 139–163.

del Ninno, C., P. Dorosh, L. Smith, and D. Roy. 2001. *The 1998 Floods in Bangladesh: Disaster Impacts, Household Coping Strategies, and Response*. Research Report 122. Washington, DC: International Food Policy Research Institute.

Hofer, T., and B. Messerli. 2006. *Floods in Bangladesh: History, Dynamics and Rethinking the Role of the Himalayas*. Tokyo, Japan: United Nations University Press.

Mustafi, B.A.A., and S. Azad. 2003. The 1998 Flood Losses and Damages of Agricultural Production in Bangladesh. *Journal of Biological Sciences*. https://doi.org/10.3923/jbs.2003.147.156.

Paul, B. K. 2003. Relief Assistance to 1998 Flood Victims: A Comparison of the Performance of the Government and NGOs. *Geographical Journal* 169(1): 75–89.

Shah, S. 1999. Coping with Natural Disasters: The 1998 Floods in Bangladesh. In *Seminar Paper Presented in June to the World Bank, Washington, DC*. July 1999.

Bengal Famine, 1943–1944

The Bengal famine of 1943–1944 was a preventable human-made disaster that took millions of lives. Some social, political, economic, and environmental factors caused this famine. The Bengal famine seemed to be an outcome of British colonial neglect toward the well-being of its subjects. European colonialists, including the British, French, and Dutch, voyaged to the South Asia region toward the end of the fifteenth century. Among these nations, the British were able to extend their trading stations rapidly throughout India. They established their first trading bases in Surat in 1612, in Madras (present Chennai) in 1640, and in Calcutta (present Kolkata) in 1690. Bengal was a historically wealthy and prosperous region located in the eastern part of India. The British East India Company took control of Bengal when they defeated Nawab Siraj-ud-Daulah, the provincial governor, at the Battle of Plassey in 1757.

A series of protest movements between 1757 and 1857 across the territory, organized by the Indian Nationalists against the company's rule, provided the opportunity for the British colonial rule to be established in this region. Subsequently, numerous famines occurred in rural India after 1857 due to the cash crop plantation system, tax burden, and unequal access to land, which affected the nationals. The colonial government introduced the Zamindari system—a permanent land tenure system where native landlords were contracted to pay revenue collected from the peasants. This system deprived the farmers of their fair shares and destroyed their independent ability to produce their own food and to have control over their production, distribution, and consumption systems.

In 1943–1944, Bengal's population was approximately 60 million; of them, 1–5 million people died from the famine. When the British administration left the Indian subcontinent in 1947, Pakistan and India appeared as two independent countries. Present-day West Bengal (a state of India) and Bangladesh (formerly known as East Bengal or East Pakistan and now a separate country) were the territories under Bengal until 1947. The Bengal famine of 1943–1944 affected more people in East Bengal than in West Bengal, more females and children than males, more rural people than the urban people, and more the poor than the rich. During the famine, the British administration provided inadequate sustenance to the famine-stricken people, though some humanitarian organizations provided food, shelters, and medicines.

CAUSES AND CONSEQUENCES

Scholars of famine and poverty studies have identified two dominant causal factors behind famine: (a) natural disasters that destroy food crops or other primary sources of sustenance and (b) human actions that determine unequal distribution of power and resources, (re)shape unjustifiable social structures, exacerbate discriminatory economic practices, and sustain anti-poor power structures. Both causes combined to create the Bengal famine in 1943–1944. For instance, the cyclone of 1942, flood, crop disease, crop failure, the cash crop plantation system, inflation, rising food prices, government negligence and corruption, and World War II (1939–1945) all played a part in causing the Bengal famine (Lohman and Thompson 2012).

Former Bengal was located in a natural basin formed by three major rivers: the Brahmaputra, the Ganges, and the Meghna. These rivers produced alluvial sediments, which made the soil fertile in this region. However, these rivers along with the Bay of Bengal made this region quite susceptible to flooding. Floods and tropical cyclones caused numerous famines in this region. Approximately 40,000 people died by the hurricane on October 16, 1942. Also, drought, salinity, and river erosion reduced agricultural land, which is also considered a prominent cause of crop failure and subsequent famine. Heavy rains during the monsoon season also destroyed crops in this region. Thus, any natural or human-induced disturbance in crop productions could cause a food shortage, starvation, and famine in Bengal.

In greater India, scholars have reported some devastating famines in 1876–1878, 1896–1897, and 1899–1901. These famines occurred due to droughts and the discriminatory economic policies implemented by the British administration (Dyson 1991). In Bengal, some scholars have reported three significant famines in 1770, 1943, and 1974. The 1770 famine, caused by a series of droughts and monopolistic trading practices by the British East India Company, killed more than 10 million people in Bengal. The 1974 famine also killed 1.5–1.8 million people in the newly independent country of Bangladesh. This famine occurred due to severe flooding of the Brahmaputra River. Also, some human actions such as unequal distribution of food, food hoarding, price speculation, and market fluctuations contributed to this famine (Lohman and Thompson 2012).

A large body of research, however, argues that the Bengal famine of 1943–1944 occurred mainly due to human activities. Natural disaster—such as the cyclone of 1942—played an insignificant role in this case. This famine thus has been identified by many as a "human-induced crisis." Population growth was an essential factor for reducing agricultural land because of land fragmentation among family members. The cash crop plantation system also reduced a vast amount of agricultural land on which millions of peasants used to live. Furthermore, construction of houses, dams, roads, railways, and bridges reduced a significant amount of agricultural land in Bengal. These infrastructure projects sometimes protected crops from floods, but in most cases, they were also the cause of flood, starvation, and famine between 1942 and 1944.

These researchers also found that food shortage and lack of access to food by the poor in many countries due to the devastating impact of the WWII contributed to the famine. This political factor linked with natural disasters according to such researchers brought numerous famines in many parts of the world, including the Dutch Famine of 1944–1945 and the Bengal Famine of 1943–1944. The threat of Japanese invasion of British India in 1942 and Japanese control over Burma (now Myanmar) contributed to rising food prices in Bengal, and the food supplies for the population were cut off. Those food supplies then were sent to feed Japanese soldiers. Also, some internal threats posed by the Indian Nationalists disrupted domestic transportation systems, including damaging the railroads, which eventually cut the supply of food to Bengal, Bihar, and Assam in British India.

Amartya Sen (1981), one of the leading scholars in famine and poverty studies, explains that the Bengal famine of 1943–1944 occurred due to lack of access to food. He argues that during the famine, adequate food was available, but the poor were not able to buy any due to rising food prices. The Famine Inquiry Commission (FIC), directed by the British government, identified two major causes of famine in 1943–1944: (1) a severe food shortage due to the cyclone of 1942 and Burma's limited rice export to Bengal and (2) a massive hike in food prices. However, Sen challenges the government's arguments about the famine and questions the validity of data regarding food production and availability at the rural district levels. He mentions that food shortage was not the primary cause of hunger; he instead argues that unequal distribution and inability of the poor to buy food were the primary causes of famine and millions of deaths.

Some scholars show that ineffective government policies and actions played an important role in expediting the mortality rate and increasing the extent of suffering. Also, there were some significant corruption cases among British government officials, who played an unpleasant role in killing millions of Bengalis by arbitrarily distributing food supplies to the region in a discriminating way, which resulted in raised food prices (Lohman and Thompson 2012). The lack of sufficient food and therefore inadequate nutrition made the famine-stricken people vulnerable to diseases and epidemics. Consequently, deaths and sufferings in the Bengal population continued for several years after the famine was over. People also were then suffering from malaria, cholera, smallpox, and dysentery. The famine killed mostly the rural Bengal poor, including many country people in cities who had left the villages for food. Famine also killed more women than men due to the traditional food distribution system among the family members, where men used to eat "first" and "more in volume" than women.

Sen (1981) also finds that demographically, the people who most affected by the famine were agricultural laborers, transporters, and fishermen, while peasants and sharecroppers who had some food stored at their home were less affected. Agrarian laborers had the least opportunity to sell their labor power because landowners were unable to pay for the labor force. Fishermen were significantly affected due to the government's "boat denial" policy, which had its roots in the British administration's fear of an invasion by the Japanese army, who could use boats to enter India from Burma. The administration thus prohibited fishermen from using boats to earn their living. This restriction also affected internal transport workers, who used boats on rivers and canals to carry people from one place to another.

PRIVATE AND PUBLIC RELIEF EFFORTS

Lance Brennan mentions that there was no attempt by the British administration to prevent the massive death toll through timely relief during the famine. He cites examples from three different regions of Bengal—Midnapore, Dacca, and Faridpur—to show how poor the responses were in addressing the crisis. Also, the "boat denial" policy of the British administration, which confiscated medium to large capacity boats, severely hampered the relief activities in most rural areas. In some cases, the administration was reluctant to send adequate relief to the distressed areas, blaming them as Nationalists (Brennan 1988).

One of the high concentrations of population in Bengal was found in Dacca region during the famine. These people lost their main earning source since the jute production was lowest at that time, which made them incapable of buying rice. The government did not supply adequate food to address this disaster. Also, the administration did not provide sufficient relief to Dacca to save lives. Indeed, as the government failed to feed the starving population, the Dacca Central Relief Committee, a voluntary citizen group, fed over 5,000 people per day.

The colonial administration did initiate "soup kitchens" to feed people during the famine in Dacca, Faridpur, and other areas. But that attempt did not save lives in Bengal because of its small-scale supply and insufficient food distribution. Most of those relief camps were "mismanaged, unorganized, and corrupt" (Lohman and Thompson 2012). Also, politics determined how relief programs were run. The British government denied much-needed relief assistance because the government thought that such assistance could undermine its imperial power. Also, the government introduced rationing and subsidized pricing for rice in Calcutta since most government employees, soldiers, and their families were living in that city, while simultaneously, they deprived the rural population of such rations and pricing.

British government officials disregarded the urgency of supplying food to Bengal during the first few months of the famine, which could have prevented the loss of millions of lives. In 1943, Lord Archibald Wavell, the newly appointed Viceroy in India, established some camps to provide shelters and foods to the famine victims. He even deployed the army to distribute food, but the anticipated threat of a Japanese invasion shifted the government's attention to the war. Thus, this relief effort was seen as too little too late. Some other private initiatives, however, including the British Friends Ambulance Group and the Indian Ram Krishna Mission, fed many famine victims.

Some studies show that the famine of 1943–1944 reshaped the demography and social structure of the region due to the loss of millions of people by starvation, malnutrition, and disease. Many researchers have called this famine a "holocaust." Some scholars report that approximately 50,000 famine-stricken people were dying of disease and starvation each week as they were trying to live on cattle fodder, tree leaves, water hyacinths, banana skins, and melon rinds (Lohman and Thompson 2012). They were described as "walking skeletons." This famine left the poor who were alive in poverty for generations.

The Bengal famine was a preventable human-induced disaster. The colonial administration, however, failed to save millions of lives due to their negligence and exploitative and oppressive socioeconomic and political practices. Consequently, this famine reshaped the demographic composition and sociopolitical landscapes in British India.

Lipon Mondal

Further Reading

Brennan, Lance. 1988. Government Famine Relief in Bengal, 1943. *Journal of Asian Studies* 47: 541–566.

Dyson, Tim. 1991. On the Demography of South Asian Famines Part I. *Population Studies* 45(5): 5–25.

Lohman, Andrew D., and Wiley C. Thompson. 2012. Bengal Famine: 1943–1944. In *Food and Famine in the 21st Century: Vol. 2: Classic Famines*, edited by William A. Dando. Santa Barbara, CA: ABC-CLIO.

Sen, Amartya. 1981. Ingredient of Famine Analysis: Availability and Entitlements. *Quarterly Journal of Economics* 96:433–464.

Bhola Cyclone, Bangladesh, 1970

On the evening of November 12, 1970, a severe cyclonic storm, equivalent to a Category 3 hurricane, made landfall on the southwest coast of East Pakistan (present-day Bangladesh). The cyclone's destructive path swept over coastal areas of the Sundarbans to 60 miles (95 km) north of Chittagong, passing directly over the islands of Manpura and Bhola. The storm, known by a multitude of names including The Killer Cyclone of 1970, unleashed torrential downpours and unrelenting winds. The precipitation and winds, combined with a high tide and relatively flat topography, resulted in 33 feet (10 m) storm surges that claimed the lives of anywhere from 300,000–600,000 people in the densely populated region. The massive death toll resulted in the Bhola Cyclone ranking as the deadliest tropical cyclone in recorded history (Smillie 2009; Penna and Rivers 2013).

PREPARATION AND MITIGATION

The Bhola Cyclone began as a tropical depression from the remnants of Cyclone Naru in the middle of the Bay of Bengal on November 8. The depression quickly gained energy from the Bay waters to become a cyclone the following day. Prior to modern predictive equipment, little was known about cyclones ahead of landfall, and most warnings were developed utilizing skills of indigenous peoples living in the area. Fisherfolk also played a crucial role in this system by alerting coastal villages of impending storms, when they would moor. This system was greatly hampered by restrictions in both time and location as most warnings merely spread by word-of-mouth.

Reliance on personal observations and word-of-mouth to predict the cyclone gave little warning to populations in (present-day) Bangladesh prior to landfall, although prior warning was received from the United States, who had captured remote-sensing satellite imagery of the storm's formation (Penna and Rivers 2013). Despite being present for decades, radio was a fairly new technology to the Bengali people, and much of the populace did not trust weather reporting due to grossly inaccurate reporting of the era (Frank and Husain 1971). An additional warning sign of the impending disaster came from the eyewitness testimony of the crew of the M. V. *Mahajagmitra*, which was enveloped by the cyclone. The ship reported the intense storm at midnight on November 12 and then went silent. All 50 crew members were lost at sea (Karmakar et al. 1998).

Historical land-use patterns also increased levels of risk and vulnerability. The delta of Bangladesh was formed by sediments from the Himalaya Mountain range, creating fertile low-lying farmlands that incentivized individuals to settle there for agriculture as well as fishing. The loss of mangroves during the colonial era also made soils and coastal areas unstable. This loss of binding vegetation increased erosion, although there were active reforesting programs aimed at reducing erosion (Tatham et al. 2009). Finally, annual monsoons unleash copious volumes of precipitation that further erode riverbanks and encroach upon human settlements, leading to enhanced vulnerability to precipitation events—especially mega disasters such as the Bhola Cyclone.

IMPACT, RESPONSE, AND AID

In the wake of the storm, 300,000 people were confirmed dead, although the actual number of fatalities is much higher considering a lack of vital records, proper death recording, and the thousands of bodies that were washed out to sea and/or unrecovered. In coastal islands such as Bhola, most of their population was completely wiped out. Massive numbers of farm animals also perished in the storm, totaling up to 280,000 livestock and 500,000 poultry. Coastal areas were inundated with saline water, causing the degradation of soils and the destruction of crops. Economic damages associated with crop loss totaled $63 million (Frank and Husain 1971), while freshwater wells were contaminated with saltwater and many buildings were washed away, fashioning a context rife with starvation, dehydration, and disease. These secondary health impacts made the post-disaster scenario as deadly as the initial disaster itself (Penna and Rivers 2013).

The Bhola Cyclone hit coastal areas the hardest. The elevation of these areas was no more than 20 feet (6 m), which were easily overwhelmed by the 20–33 feet (6–10 m) storm surges. Fisherfolk, who reside near the coast, were the hardest hit demographic. The loss of fisherfolk resulted in huge economic losses in the fishing industry at large and local economies at a smaller scale, not to mention the decrease in protein intake that ensued in already malnourished populations, who needed the calories provided by fisherfolk (Frank and Husain 1971).

Following the cyclone, a young oil baron by the name of Fazle Abed witnessed the grave inhumanities and jumped into action. Abed was living in a Shell Oil compound in Chittagong, where he opened his house to 50 relief workers. Five days after landfall, Abed and others went to the island of Manpura, where they observed the mass destruction that the cyclone had left behind. Only one-third of the population of Manpura survived the cyclone, and human and animal carcasses were scattered about. After visiting Manpura, Abed and Father Richard William

Timm, Marty Candy, Vikarul Chowdhury, and Akbar Kabir formed the Heartland Emergency Lifesaving Project (HELP), which over time evolved into the versatile, life-sustaining nonprofit known today as Building Resources Across Communities, or BRAC. BRAC remains the largest nonprofit in Bangladesh and is one of the largest and most comprehensive nonprofits in the world.

The United Kingdom deployed military troops to affected regions. However, they were not cleared to provide aid until a Pakistani official arrived on scene. While the British had materials on-site (e.g., bags of rice, boats, motors, and tents), they ended up digging latrine pits and tanning until the official arrived. Meanwhile, locals looked on at the soldiers without giving any assistance. Days later, the official arrived, and relief aid was finally allocated to affected populations.

Prior to the storm, political tensions were palatable between Bangladesh (i.e., East Pakistan) and Pakistan. These tensions were further strained in the aftermath of the cyclone. The border to Pakistan was completely closed within two days of the cyclone making landfall, and Pakistan's inability to show interest in aiding their noncontiguous state sparked discontent among local governments and the Bengali masses, which resulted in sweeping political changes in their first ever democratic election. Thus, during the first national election the following month, the people of Bangladesh made their voices heard by electing 160 of the 162 Awami League (AL) candidates in the National Assembly of Pakistan. This result gave political dominance of the AL, a Bengali Nationalist and pro-independence party (Siddiqi 2004; Smillie 2009).

Although East Pakistan was the eastern wing of Pakistan, it was never truly part of Pakistan. When the British finally relinquished control of India and the surrounding territories in mid-twentieth century, they split the territories based on the dominant religion of each area. This resulted in a piece of the Indian subcontinent, present-day Bangladesh, becoming part of Pakistan despite being separated by the country's adversary, India. From the outset, they were divided by ethnicity, geography, and especially language.

Conflict erupted along the border as Bengalis were enraged by the then Pakistani President Yahya Khan's invasion. In March 1971, Bangladesh officially announced secession and publicly released their Declaration of Independence. Beginning March 25, 1971, Khan initiated a series of attacks on villages and civilian targets in Bangladesh with the objective of curbing the actions of local militias operating under the banner of the Bengali Nationalist Movement. Within 24 hours of the beginning of Operation Searchlight, Bangladesh officially declared war on Pakistan.

Plagued by intense guerrilla warfare by local militias and a lack of supplies, Operation Searchlight was an abject failure. However, Bengalis were galvanized against Pakistan, and the number of active militias grew rapidly. Overwhelmed, the Pakistani military eventually retreated toward the border and began widespread bombing campaign of Dhaka. With the city in ruins, the Pakistani military pushed back into Bangladesh, making it all the way to the coast with the order to slaughter as many Bengalis in Dhaka as possible. Pakistani troops burned villages and destroyed buildings, and in Dhaka, they rounded up scholars working in the country and shot them outside a brick-making factory. Itching to seek revenge on Pakistan for the last war, a mere six years earlier, India jumped into the fray on the side of Bangladesh. The war lasted to the end of the year; however, a second war between the three countries began soon afterward accompanied by smaller border skirmishes in the interim. Peace between Bangladesh, Pakistan, and India was not achieved until late 1979. The result, which was initiated by the Bhola Cyclone, was the formation of the independent state of Bangladesh.

RECOVERY AND RECONSTRUCTION

Recovery and reconstruction are critical for the mitigation of future disasters, but due to civil strife, such processes were stymied and took much longer than was expected. BRAC, at that time HELP, provided clothing, food, blankets, pond desalination technology, crop seeds, and agricultural tools and offered support groups to the people on the island of Manpura. BRAC has done further work in Bangladesh in terms of disaster recovery and reconstruction, and their methods are lauded because they target long-term developmental projects rather than immediate efforts to provide aid (Smillie 2009). There is evidence that reconstruction activities, both from BRAC and other organizations, has led to greater mitigation and resilience from future disasters. Many cyclones following the Bhola Cyclone have been more powerful, but less deadly. This is due to better weather reporting, warning systems, communication media, and structural mitigation measures.

The construction of embankments along rivers has been essential for the prevention of flooding in many coastal areas, but sometimes these embankments fail because people decide to reside near them as regular weather events and storm surges are less impactful in such areas (Tatham et al. 2009). While embankments slow erosion and keep water out of low-lying areas, when large flooding or storm surges occur the areas are inundated. Thus, this practice of risk transference leads to lower vulnerability in normal times at the risk of disaster, drownings, disease, crop failure, and livelihood disruption in larger scale events, which, unfortunately, are common in Bangladesh.

A successful reconstruction element was the large-scale construction of cyclone shelters from 1970 to 1991, which can cumulatively house approximately 350,000 people. However, the shelters are sometimes unable to accommodate the affected populations of the regions they serve, not to mention that the structures are sometimes vulnerable themselves and often lack proper amenities for women as well as water storage capabilities (Tatham et al. 2009).

Infrastructure is essential for mitigation, but civil engineering can only do so much. Following the Bhola Cyclone and many others throughout recent history, there has been a call for the development of education, communication systems, local economies, social capital, and community capacity. More recently, there has been the establishment of disaster management programs with educational outreach components, which conduct research and dissemination knowledge on disasters and their effects. Nearly all international aid to Bangladesh is provided by nonprofit institutions.

CONCLUSION

The 1970 Bhola Cyclone is the most devastating weather event in recorded history. It created a realization of the need for comprehensive disaster mitigation plans including emergency drills, real-time forecasting, and local training of the public. Immediate response to the disaster was poor, but there is evidence that reconstruction has reduced both mortality and destruction in the region from more recent storms such as Cyclones Sidr and Gorky. Furthermore, the Bhola Cyclone triggered an independence movement that resulted in the formation of Bangladesh, one of the world's newest political states. The Bhola Cyclone provides an excellent case study of both what not to do in the case of a major disaster and how efforts to mitigate disasters have proven effective. Thus, the Bhola Cyclone serves as a focal point of many studies on disaster preparedness and mitigation.

Ben Clark, Ian MacNaughton, and Luke Juran

Further Reading

Frank, N. L., and S. A. Husain. 1971. The Deadliest Tropical Cyclone in History. *Bulletin of the American Meteorological Society* 52(6): 438–445.

Karmakar, S., M. L. Shrestha, M. N. Ferdousi, and M. S. Uddin. 1998. SMRC Research Report No 1. Dhaka, Bangladesh: SAARC. Meteorological Research Centre (SMRC).

Penna Anthony N., and Jennifer S. Rivers. 2013. *Natural Disasters in a Global Environment*. Hoboken, NJ: Wiley Blackwell.

Siddiqi, B.A.R. 2004. *East Pakistan—The Endgame*. Karachi: Oxford University Press.

Smillie, Ian. 2009. *Freedom from Want: The Remarkable Success Story of BRAC, the Global Grassroots Organization That's Winning the Fight against Poverty*. Sterling, VA: Kumarian Press.

Tatham Peter, Karen Spens, and Richard Oloruntoba. 2009. Cyclones in Bangladesh: A Case Study of a Whole Country Response to Rapid Onset Disasters. POMS 20th Annual Conference, Orlando, Florida, May 1–4, 2009. https://www.pomsmeetings.org/ConfProceedings/011/FullPapers/011-0029.pdf, accessed March 19, 2018.

Big Thompson Canyon Flash Flood, Colorado, 1976

On a Saturday evening, July 31, 1976, the Big Thompson Canyon in Larimer County, Colorado, experienced a devastating flash flood that swept down the steep and narrow canyon combining with surface materials of very low water-absorption capacity, which meant that the canyon could not hold the enormous runoff from a sudden series of torrential downpours. The event claimed the lives of 143 people and injured 150 others. Some bodies were carried away as far as 25 miles (37 km), while five bodies were never recovered. One man who left the area that day was presumed a victim until he was located decades later, living in Oklahoma. Bodies were pulled from debris piles, and not till the death toll surpassed 100 people did many realize just how bad this event was. It was the deadliest flash flood in Colorado's recorded history and was more than four times as powerful as any in the 112-year record available in 1976, with a discharge of 35,000 cubic feet (1,000 cubic m) per second. It claimed the lives of a relatively large number of elderly people. Of the victims, 45 percent were male and 55 percent were female (W-ME 1977). Among the deaths, one State Patrol officer lost his life while warning citizens of the flood danger.

Big Thompson Canyon is formed by the uppermost course of the Big Thompson River—a small tributary of the South Platte River, approximately 78 miles

(123 km) long. The river descends some 2,500 feet (760 m) through the 25 mile (40 km) long canyon. Three major communities are located along the canyon: Cedar Cove, Drake, and Glen.

CAUSES

A stationary thunderstorm over the upper part of the canyon, just east of Estes Park—a popular summer resort about 50 miles (80 km) northwest of Denver—triggered the flash flood. Estes Park marks the western end of the canyon, and it ends near Loveland, where the foothills Rockies meet the plains. On July 31, 1976, a cool Canadian air mass moved in to become stationary over the Big Thomson basin, while a strong low-level warm front from the east pushed a moist air mass upslope into the mountains. The unstable air continued to rise, powered by heat released from condensing water vapor. Late-afternoon cloud formation is a typical phenomenon in the eastern edge of the Rockies. On that particular day, the mid- and upper-level winds were weak, which was why the thunderstorm remained stationary (Abbott 2008).

The thunderstorm caused 12 inches (300 mm) of rain in less than 4 hours in the mountains around Estes Park. This rainfall amount accounted for more than 75 percent of the average annual rainfall for the area. More than four inches (100 mm) of rain fell over the lower section of the canyon, where many of the victims were found, compounded by unusual weather patterns that allowed the huge storm system to stall over the area. The eastern edge of the Rocky Mountains rises

Before (left) and after (right) images of the Big Thompson Canyon flash flood in Colorado, 1976. (NOAA)

dramatically above the Interior Plains of central North America, immediately west of Denver, Colorado. This terrain promotes moist summer air masses from the east often to rise into the Front Range of the Rockies, become cool, and dump moisture in thunderstorms (Hyndman and Hyndman 2006).

DAMAGE

On the evening of July 31, 1976, approximately 4,000 people were staying in or passing through the recreational area of Big Thompson Canyon, most of them from outside the area and one-third from out of state. A great majority of them were tourists, campers, climbers, and others enjoying the rugged beauty of the canyon. They came on the eve of the 100th anniversary of Colorado statehood, and July 31, 1976, was the start of a three-day centennial weekend. These visitors were in addition to the full-time canyon population of 600 and the part-time residents who numbered approximately twice that. Many tourists were attracted by the trout fishing, streamside motels, and campgrounds. The weather that day was pleasant. However, heavy rain started at around 6:30 p.m., and within an hour or so, the amount of downpour provided an early indicator of a major flood. Around 9 p.m. on July 31, a wall of water more than 20 feet (6 m) high roared down through the length of the canyon. The flood crest moved at 15 mph (25 km/h) through the entire canyon, which did not allow authorities to issue a timely flash flood warning.

The flood destroyed 438 vehicles, 316 homes, 45 mobile homes, and 52 businesses. It caused major damage to 73 homes and washed out most of U.S. Route 34, the highway that follows the Big Thompson River. Additional private damage or destruction included damage to land, fences, irrigation systems, crops, farm buildings and equipment, wells, private wastewater systems, cars, recreational vehicles, private bridges and roads, landscaping, propane tanks, and household goods. The 1976 flash flood destroyed telephone and power lines, bridges, and roads. Huge boulders, trees, houses, propane tanks, mobile homes, and everything else in the path of the wall of water were tossed around as if in a giant blender. The frightening roar of the churning debris was illuminated by frequent lightning strikes. The flood caused damages totaling $36 million, and the total funds either spent or committed as on December 31, 1977, for disaster relief and related expenses amounted to $58 million (W-ME 1977).

While emphasis was put on loss of life, infrastructure damage, and property loss to people, the flash flood also impacted the natural environment and wildlife population. Water-oriented species such as beaver, muskrat, and waterfowl suffered extensive habitat damage, and damage to nesting along the Big Thompson Canyon by bird predator species was widespread. However, nongame species such as reptiles, amphibians, and rodents were the most impacted by flood damage due to habitat loss. As a result, their number decreased immediately after the flood. However, most large mammal populations that used the river were not seriously affected. Meanwhile, access to the river way for recreationalists, fishermen, and hunters was either destroyed or damaged.

RESPONSE

The National Weather Service (NWS) in Denver issued a scattered showers and thunderstorms forecast for the Big Thompson Canyon area first at 4:00 a.m. and then at 10:00 a.m. on July 31, 1976. Then the forecast issued at 4:00 p.m. indicated a chance of afternoon and evening showers and thunderstorms, but there was no mention of heavy and abnormal rainfall at nighttime. However, a special weather statement issued at 9:00 p.m. mentioned a probability of heavy thunderstorms that could produce local flooding around the Big Thompson Canyon. Next, a flash flood warning was issued at 11:00 p.m. for the portion of the Big Thompson River east of the canyon. Also, the NOAA Weather Wire Service (NWWS) was used to disseminate thunderstorm and flood warnings at 7:35 p.m., 9:00 p.m., 11:00 p.m., and 11:15 p.m. At that time, NWWS, a teletypewriter system, was the principal means of distribution for weather information to the news media to disseminate to the public (U.S. Department of Commerce and NOAA 1976).

Suspecting a major potential flash flood event, one patrolman from Fort Collins arrived in Drake, an unincorporated community in the Canyon between Estes Park and Loveland, and began to warn campers and residents shortly after 8:00 p.m. to leave the area. Other patrolmen soon followed and began going door to door to warn residents. Because heavy summer rain is very common in the area, many visitors and residents did not believe they were in danger and did nothing. For many residents, it was difficult to realize that they were not safe where they were. Still others were openly defiant. Historical records suggest that floods have occurred about once in six years. But no one expected the 1976 flood would be the deadliest one. Even some law enforcement officers on the site early had difficulty recognizing the magnitude of the flood and the need for immediate lifesaving measures.

However, some of those who escaped the flood did so with barely time to climb, drive out to outrace the flood, or abandon their cars and run upslope. At least 139 who tried to drive out ahead of the storm or wall of water were trapped in their cars and swept to their deaths. Many of those visitors who did not leave the area faced difficulty in breathing in the moisture-laden air as the rain drove straight down with little wind, creating a heavy spray all around them. Meanwhile, water gathered speed as it washed over the steep rocky hillsides and flushed through the flatter meadows, all of it heading for the bottom of the V-shaped canyon.

Many rescuers were on the scene quickly, searching for flash flood victims from one stop to another. Family members, friends, and relatives converged on the disaster site to determine the status of their loved ones. Families also gathered at the old Loveland Memorial Hospital to identify bodies. After hearing the news of the deadly event, an ambulance crew rushed to the canyon to provide aid to the survivors. The ambulance slammed into a wedge on the canyon wall, and the crew survived by climbing out of the wrecked ambulance (Hyndman and Hyndman 2006). Along with public rescue teams, private rescue teams, comprising volunteers, also participated in the rescue operation. More than 800 people were evacuated by helicopter the following Sunday morning on August 1, 1976. The National Guard also participated in the rescue operation, which ended within a few days, but cleanup took long time. The terrain made damage assessment, reconnaissance, and rescue operations difficult and extremely hazardous.

On August 1, 1976, both state and federal governments declared the Big Thompson Canyon a disaster area. Subsequently, federal personnel actively helped the local and state efforts. On the same day, the American Red Cross (ARC) began establishing temporary shelters for the flood survivors. The private sector provided donations of food, clothing, and personnel assistance. Neighbor helped neighbor, and local churches initiated and coordinated private disaster responses. The church groups not only provided emotional support and personal counseling for flood victims but also served as advocates for them, and they gave physical and financial assistance where needed. The Army Corps of Engineers organized the debris cleanup program. The effort by the state continued for more than a year to help the victims and their families recover from the disaster and to plan for the future.

However, recovery started late and was delayed because of the lack of flood recovery experience at the local level, lack of adequate resources, absence of a recovery plan, and lack of coordination between local, state, and federal agencies. Had the floodplain been delineated before the disaster and had the residents purchased flood insurance, the recovery effort would have been easier and quicker. While recovery efforts focused on humans, little effort was directed toward the loss and devastation experienced by the wildlife populations of the Big Thompson Canyon. In fact, popular coverage of the impact of the flood on local fisheries and wildlife was negligible. However, the Colorado Division of Wildlife initiated fishery rehabilitation and terrestrial wildlife recovery efforts.

CONCLUSION

Although each major natural disaster is destructive, such an event provides opportunity to implement measures that reduce damage and risk of future disaster. Colorado is much better prepared today for flash flood than it was in 1976. Following the event, the authorities put up signs advising people to "Climb to Safety" in the event of a flash flood along U.S. Highway 34 and in canyons across the state. Also, flood plains were delineated and re-drawn. Prior to the flash flood, flood warning systems in Colorado were not well defined. Based on the experience of the flood, authorities have made progress in improving drainage basins-based warning as well as warning dissemination. For this purpose, new radar systems were purchased.

Each year, residents, friends, family, and survivors gather at the Big Thompson Canyon Association and Memorial Site, about one mile south Drake, 13 miles (19 km) west of the Kmart on U.S. Highway 34 in Loveland. In 2001, on the 25th anniversary, a stone memorial was placed near Drake, Colorado. The marker lists the names of those who perished in the 1976 flash flood.

Bimal Kanti Paul

Further Reading
Abbott, P. L. 2008. *Natural Disasters*. Boston: McGraw Hill Higher Education.
Hyndman, D., and D. Hyndman. 2006. *Natural Hazards and Disasters*. Belmont, CA: Thomson.

U.S. Department of Commerce and NOAA (National Oceanic and Atmospheric Administration). 1976. *Big Thompson Flash Flood of July 31–August 1, 1976: A Report to the Administration.* Natural Disaster Survey Report 76-1. Rockville, MD.

W-ME (Wright-McLaughun Engineer). 1997. *Big Thompson Flood Disaster: Final Report to the Governor of Colorado.* Denver, CO: W-ME.

Black Saturday Bushfires, Australia, 2009

Almost every year, Australia experiences bushfire. On average, over 50,000 bushfires of varying extent and intensity strike the country annually; however, the most extensive and severe bushfires occur generally after prolonged drought over several years. The bushfires of Black Saturday, February 7, 2009, became one of the worst natural calamities in Australian known history. The Black Saturday bushfires refer to a series of bushfires that started on the same day, Saturday, in different parts of the State of Victoria. The bushfires claimed 173 lives and injured over 400 people. More than 1 million wild and domesticated animals were killed, around 1.1 million acres of land area were burnt, and 3,500 buildings, including more than 2,000 houses, were destroyed. Some of the Black Saturday bushfires are believed to have started because of direct or indirect human activity (Cruz et al. 2012). Regardless, disaster preparedness and response strategies in the State of Victoria and in Australia, in general, have gone through significant modifications since the Black Saturday Bushfires.

CAUSE OF THE BUSHFIRE

The bushfires of Australia are linked to climatic variation and climate change. Specifically, the Black Saturday Bushfires of 2009 along with "Ash Wednesday" bushfires in 1989, and several other significant bushfires of Southern Australia, were preceded by exceptionally dry conditions, which are associated with the phases of El Niño-Southern Oscillation and Indian Ocean Dipole events (a situation when the eastern Indian Ocean becomes cooler than the western part, resulting in below average winter and spring rainfall in southeastern Australia). Prior to the Black Saturday bushfires, the southern part of Australia experienced a record-breaking decade-long drought. An unprecedented series of positive Indian Ocean Dipole events worsened the situation. Consequently, the lower rainfall and higher temperature influenced soil moisture and dry conditions and increased available summer load (or fuel for bushfire). From 2002 to 2008, five such dipole events in the Indian Ocean were observed, three consecutively between 2006 and 2008 (Cai et al. 2009).

The threat of bushfire in southeast Australia is the highest in summer and autumn (January to April) as grass and forests become dry during that time. Weeks before the bushfire event, a heatwave covered south-eastern Australia, fueled by a two-month long spell of hot and dry weather conditions. The Victorian State Capital, Melbourne, encountered a record high temperature of 116 degrees Fahrenheit (43 degrees Celsius) for three consecutive days in late January 2009. The hot and dry weather, combined with extremely low levels of humidity, resulted in "tinder-dry" conditions across the bush lands of the State of Victoria. Meanwhile, strong wind caused calamitous propagation of fire.

The daytime temperature in the State of Victoria on February 7, 2009, was around 110 degrees Fahrenheit (40 degrees Celsius) at 11 a.m., and in Melbourne, it increased later in the day to 122.8 degrees Fahrenheit (46.4 degrees Celsius). Around midday, a fallen power-line on a farm near Kilmore East started a blaze, which triggered the Black Saturday Bushfires. Eventually, fire broke out in Kinglake, Murrindindi, and other areas of the state. In combination with exceedingly dry fuel and weather conditions, strong wind intensified the fires, and within a short period of time, the fire had spread over a vast area at an estimated speed of between 223 and 502 feet (68 and 153 m) per minute. A thick black cloud at least 8.7 miles (13 km) high developed in the lower atmosphere from the large amount of smoke and other burning products.

According to the 2009 Victorian Bushfires Royal Commission, about half of the bushfires were started at the first location from direct and indirect human activities. Specific evidence shows that a significant number of Black Saturday bushfires were initiated because of failure of the state's age-old electricity assets. Several others began from fires that people deliberately or accidently created. Generally, one out of three bushfires in Victoria is believed to be lit by people acting with mischievous or criminal intent. Notably, however, if the weather condition had not been extreme, then either most of the fires would not have started or they would have had a less devastating impact (Cruz et al. 2012).

DAMAGE AND IMPACT

The Black Saturday Bushfires took a heavy toll on human and animal life and property; of the more than 300 fires across the State of Victoria that day, 13 became devastating. The deadliest and largest bushfire of Black Saturday occurred at Kilmore East and spread over 250,000 acres (101,171 ha) of land within 12 hours. The Kilmore East fire killed 119 people, and that accounts for 70 percent of the death toll of Black Saturday Bushfires. Additionally, 40 people in Murrindindi and 11 people in Churchill were killed. A majority of these people died on the day the bushfire started with only a few dying in the following days or weeks. Children, seniors, and people with disabilities were the hardest hit; around half of the casualties were younger than 12, older than 70, or had physical disabilities (Cruz et al. 2012).

The Black Saturday Bushfires consumed vast amounts of property and infrastructure located in the burnt area (approximately 1,737 square miles [4,500 square km]). Poor land use planning and weak building infrastructure could be blamed for some of the damage. The bushfires burned vast patches of forest and farmland and left behind devastating impacts on the biodiversity and ecosystem that will take a long time to recover from. Estimates point to the total cost of the bushfire disaster being about $3 billion (Cruz et al. 2012).

The Black Saturday Bushfires also have had a far-reaching impact on the communities living in fire-prone areas of Australia. People of Victoria and Australians in general have become more scared of bushfires. A study, led by the University of Melbourne with over 1,000 participants, identified at least one out of four people

Devastation near Mount Tassie caused by the Black Saturday Bushfires in South Gippsland, Victoria, Australia, 2009. (Stephen Denness/Dreamstime.com)

in high-impact communities experiencing symptoms of depression, post-traumatic stress disorder (PTSD), or severe psychological distress nearly six years later (Teague et al. 2010).

RESPONSE AND RECOVERY

The Black Saturday Bushfires required a massive coordinated response operation as attention was required in so many places at the same time. Over 4,000 fire service volunteers and professional firefighters responded to more than 300 fires across Victoria on February 7, 2009. Subsequently, more than 10,000 personnel were involved in coordinated emergency response and community recovery operations, which were the largest of their kind in the history of the State of Victoria. Moreover, many individuals began fighting fires with their own multipurpose farm utility equipment. Different organizations of the State launched recovery efforts for the impacted communities. However, continuing fire, inaccessible roads, and loss of power and telecommunications networks largely hindered the immediate relief activities. Naturally, it was difficult to deliver medical services to the communities living in remote areas.

The 2009 Victorian Bushfires Royal Commission was formed to investigate the cause and consequence of the fire. The commission was also directed to evaluate the disaster planning and preparedness and actual emergency response and recovery operation and other relevant issues. The commission acknowledged that while it might be possible to reduce the number of severe fires and to be better prepared for fire management, bushfire will never be eliminated from the Australian landscape. However, for better preparedness and response to future bushfires, the Royal Commission came up with over 60 recommendations.

The Royal Commission recommended revising and strengthening the state's community safety bushfire policy. The commission emphasized the improved bushfire warning system, timely dissemination of warnings, improvement of community education and awareness regarding bushfire, and better coordination among different agencies supporting emergency responses. Based on the commission's recommendations, the Forest Fire Danger Index (FFDI), a measure of severity of bushfire, was updated in Victoria. Also, a new category "catastrophic" or "Code Red" was added to the index, which indicates a situation where evacuation is mandatory.

The Royal Commission also recommended replacing the age-old electric power infrastructure of the state and improving electric power safety. Additional recommendations included initiating criminal investigation of arsonists associated with some of the bushfires, increasing scientific and technical understanding of bushfires, and implementing relevant regulatory reforms. The Royal Commission commented that the Black Saturday Bushfires should not be treated as a "one-off" event, but rather, Australia should expect increased risk of such events because of the growing population in rural-urban interfaces and the rising impacts of climate change.

In December 2014, Victoria's Supreme Court approved a $500 million payout to the victims of the Black Saturday Bushfires. The decision came from a case that was filed after the Royal Commission indicated that the Kilmore East-Kinglake bushfire was initiated by an aging power line of a utility company, named SP AusNet. Although the company did not admit any liability for the fire, they agreed to the settlement. It is considered the biggest class action suit in Australian legal history. Under the settlement, a total of 10,500 individual property claims and 1,750 personal injury claims have been made (Gibbs et al. 2016).

Another study conducted by the University of Melbourne a few years after the incident found that individuals and communities impacted by the Black Saturday Bushfires have recovered by and large from the disaster and subsequent interruptions (Teague et al. 2010). The impacted individuals and communities have adapted to a changed life and environments. However, community recovery was relatively faster than individual recovery as evidence of delayed impacts of the fire on individual mental health even five years later attests.

CONCLUSION

The Black Saturday Bushfires are a perfect example of a deadly combination of extreme environmental conditions and human activities. Australia had never experienced such severe bushfires, and as a result, the country had to redefine its preparation, assessment, and response mechanisms for bushfire incidents. The Black Saturday disaster has had a long-lasting impact on Australian society and created an increased level of awareness of the potential risk associated with bushfire.

Such risk is on the rise for a number of reasons including climate change, population growth, and community practices. Studies indicate that climate change is directly linked to major bushfire incidents in Australia, and clearly the Black Saturday Bushfires happened at a time when a perfect storm of dry conditions was created by a decade-long drought associated with climate change. However, while extreme weather conditions contributed to the Black Saturday Bushfires, some of the fires were initiated by direct and indirect human activities. Nevertheless, whatever the cause, strong wind increased the severity and extent of the bushfires.

The unprecedented level of response and immediate recovery actions in the long run helped to support the impacted communities. Over time, the community and the environment recovered; however, significant improvement in disaster preparedness and response planning could be made to manage such incidents and to reduce the exposure of people and properties to such devastation. The Black Saturday Bushfires tested the endurance of Australian society and could be considered a wake-up call for the country to minimize future risks of and exposure to bushfires.

Salim Momtaz and Palash Basak

Further Reading

Cai, W., T. Cowan, and M. Raupach. 2009. Positive Indian Ocean dipole events precondition southeast Australia bushfires. *Geophysical Research Letters* 36: L19710. https://doi.org/10.1029/2009GL039902, 2009.

Cruz, M. G., A. L. Sullivan, J. S. Gould, N. C. Sims, A. J. Bannister, J. J. Hollis, and R. J. Hurley. 2012. Anatomy of a Catastrophic Wildfire: The Black Saturday Kilmore East Fire in Victoria, Australia. *Forest Ecology and Managemen,* 284: 269–285.

Gibbs, L., R. Bryant, L. Harms, D. Forbes, K. Block, H. C. Gallagher, and D. Lusher. 2016. *Beyond Bushfires: Community Resilience and Recovery Final Report.* Melbourne, VIC, Australia: University of Melbourne.

Teague, B., R. McLeod, and S. Pascoe. 2010. *Final Report, 2009 Victorian Bushfires Royal Commission.* Melbourne VIC, Australia: Parliament of Victoria.

BP Deepwater Horizon Oil Spill, United States, 2010

The BP Deepwater Horizon Oil Spill was the largest accidental oil spill in history, occurring on April 20, 2010, in the Gulf of Mexico. An explosion on BP's Deep Water Horizon oil rig caused the spill, which left 11 dead. Cement designed to protect the outside of the well pipe failed, which caused the blowout. The oil leak was discovered two days later when a slick layer of oil, called a sheen, became visible. Live underwater cameras captured leaking oil and gas for much of the duration of the spill. The spill affected approximately 1,100 miles (1,770 km) of coastline ranging from Texas to Florida (NRDC 2015). Cleanup efforts included physical barriers and the release of chemical dispersants. However, the oil spill left lasting effects on wildlife, local fishing, and tourism economies, and people who came into contact with the oil and chemicals used to treat the spill.

Fireboat response crews battle the blazing remnants of the offshore drilling rig Deepwater Horizon on April 21, 2010. (U.S. Coast Guard)

PREPARATION AND MITIGATION

There have been large-scale oil spills in the past, notably the *Exxon-Valdez* spill that occurred when an oil tanker hit a reef off the coast of Alaska in 1989 and the Arabian Gulf spill in 1991, a result of deliberate dumping of oil by Iraqi soldiers. Marine blowouts have also occurred, such as the Ixtoc I oil spill in 1979 that occurred in a similar manner to the Deepwater Horizon spill. Both the Ixtoc I and Deepwater Horizon spills occurred due to a combination of pressure being too high on the oil well cement and the cement failing to hold up to the pressure. While the cement casing had a built-in failure mechanism, the blowout preventer failed and thus did not "plug" the oil well. When the drill pipe and casing exploded, the oil being extracted flowed freely into the Gulf of Mexico. The Department of Energy's (DOE) best estimate is that the well, which was located 5,000 feet (1,524 m) deep on the ocean floor, leaked approximately 53,000 barrels per day from April 20 through July 15, 2010.

Efforts to halt oil from spilling initially occurred via remote-controlled vehicles that attempted to trigger the blowout preventer that initially failed. After these attempts failed, BP placed a cone over the area from where oil was escaping in order to mitigate the uncontrolled flow of oil into the Gulf of Mexico. The ultimate solution to close the well was to achieve both "top kill" and "bottom kill." Top kill was achieved by pumping mud, sand, and other materials into the well opening to stop oil from escaping. Once enough of the mixture was pumped into the well, oil stopped flowing and the well was cemented. Bottom kill was achieved by drilling two relief wells into the initial well pipe below the seafloor. This was done by drilling at an angle by using magnetic sensors to locate the original well pipe. Once the relief pipes were functional, oil was safely extracted from the original well pipe, and it was eventually depleted and closed. United States President Barack Obama declared the well officially closed on September 19, 2010.

IMPACT AND RESPONSE

Impacts of the Deepwater Horizon oil spill were severe. After 87 days of leaking, roughly 3.2 million barrels of oil spilled directly into the Gulf of Mexico (Ocean Portal 2018). The spill seriously impacted marine and terrestrial ecosystems, local economies, and cleanup workers. Negative effects were observed immediately, and it is expected that the effects will continue to be observed in the future.

Two main cleanup methods were deployed to remove oil from the ocean: physical barriers and chemical dispersants. However, it should be noted that cleanup of areas near human settlements (e.g., harbors and coastlines) and important biological habitats (e.g., marshes and estuaries) took priority. It was not possible to clean up all of the oil, so the most critical and sensitive areas received elevated assistance.

Since oil floats to the surface of water, cleanup workers surrounded the slick with physical barriers called floating booms. The booms contained the oil slick and helped prevent it from further spreading. Next, skimmer boats were used to separate oil from the water, which then collected the segregated oil. Next, since the skimmers did not collect all of the oil, cleanup workers spread sorbents to help collect the remaining oil. This process utilized three main types of sorbents comprising organic materials, inorganic materials, and materials that are similar to plastics.

Cleanup crews also deployed dispersants to help clean up the oil spill. Dispersants are chemicals that break oil down into smaller particles that can more easily mix with the ocean water. About 1.4 million gallons of dispersants were released into the water using airplanes, helicopters, and even methods of injection into the wellhead (Ocean Portal 2018). The dispersants left a 22-mile (35 km) long plume, and there is concern as to how the chemicals adversely impacted water quality and marine life.

Wildlife faced devastating effects from the oil spill, and the repercussions were immediate. Birds, especially pelicans, were visibly covered with oil, which affected their health and also hindered their ability to fly. Dead, oil-covered fish and turtles were also seen washed up on beaches. Perhaps the most concerning aftermath was the presence of deformed fish and shrimp. Fish were caught with visible lesions, and, in one instance, a fisher found only eyeless shrimp inside his net. Dolphins, corals, and other marine flora and fauna were observed with noticeable health consequences. Furthermore, it is important to note that petrochemicals and chemicals used to clean up the oil spill invariably moved up the food chain to affect entire ecosystems, and the cumulative effects will take years to fully dissipate.

The oil spill severely impacted fishing economies throughout the Gulf of Mexico. Fishing grounds began to close immediately following the spill, and at one point fishing was not permitted in 37 percent of federal waters in the Gulf of

Mexico. It took an entire year to reopen all fishing grounds, and as a result, the commercial fishing and shrimping industry sustained a loss of around $247 million (NRDC 2015). Furthermore, the tourism industry (e.g., beaches, resorts, restaurants, watersports, and chartered fishing expeditions) lost approximately $22.7 billion, and the total cost of the spill is estimated at $61.6 billion (NRDC 2015).

Some workers faced serious injury from cleaning up the oil spill. Workers encountered a variety of symptoms including dizziness, skin rashes, nausea, cramping, fainting, and even women became infertile, to name a few. Many workers developed these symptoms as a result of breathing in the chemical dispersants they were handling. In one case, a worker became blind from exposure to the dispersants. In another case, a young boy swimming in a pool near a beach that was treated with dispersants experienced seizures. Lastly, a fisherman's wife developed a blood disorder from washing her husband's oil-stained clothing. Many workers and residents continue to confront such health issues (Marsa 2016).

RECOVERY

Recovery is ongoing and will continue for a long time given that the liable party, BP, has been ordered to make compensatory payments until 2031. The most recent plan for recovery is the 2016 update to the Comprehensive Plan originally issued in 2013 (GCERC 2016). The Comprehensive Plan intends to distribute the responsibilities among various stakeholders, including nongovernmental organizations, government agencies, and commercial and environmental parties. The plan charts a 10-year horizon for funding and strives to attain clear, measurable goals across all participants while simultaneously establishing a holistic agenda for future actions.

Included in the Comprehensive Plan is the RESTORE Act, signed into law by President Obama in 2012, which calls for 80 percent of fines incurred from the Clean Water Act to be placed in the Gulf Coast Ecosystem Restoration Trust Fund (GCERTF) to fund recovery efforts. The remaining 20 percent is to be allocated to a separate trust, the Oil Spill Liability Trust Fund, which seeks to leverage interest gained from the fund to supplement ongoing recovery and ecological restoration activities. Of the GCERTF, 35 percent is to be disbursed to ecosystem restoration, economic development, and tourism promotion in the five most affected states (i.e., Alabama, Florida, Louisiana, Mississippi, and Texas). The next 30 percent of the GCERTF is to be used to implement policies laid out in the Comprehensive Plan. Another 30 percent of the GCERTF is to be allotted to the five affected states to implement their own projects, while the remaining 5 percent is to be administered by the National Oceanic Atmospheric Administration (NOAA) and Centers of Excellence (COE) Research Grant Program to monitor ecosystems and recovery efforts as well as to conduct research in the Gulf Coast region.

On April 4, 2016, BP agreed to a settlement of $20.8 billion to resolve all government and civil claims. This is the largest penalty to be paid under an environmental statute in the United States. The $20.8 billion accounts for $5.5 billion for Clean Water Act violations, $8.8 billion for natural resource damages, and $600 million in other payments. The remaining $5.9 billion is to be paid to the five affected Gulf States and their local governments.

In addition to ecological and economic relief, in 2016, the Obama Administration enacted legislation to create stricter regulations on the construction, maintenance, and oversight of oil wells (BSEE and DOI 2012). In particular, 16 points were specified to address numerous mistakes that caused the Deepwater Horizon oil spill. The overarching measure was to ensure that blowout preventers are operational and successful in stemming the flow of oil in the event of future disasters. In addition to blowout preventers, the legislation calls for real-time monitoring of high-temperature and high-pressure drilling activities. This is important because it allows the Bureau of Safety and Environmental Enforcement (BSEE) to take action in the event of dangerous drilling activities, much like the ones drill operators ignored in the BP Deepwater Horizon incident.

These legislative changes are in jeopardy of being undermined by the current Trump administration as of December 29, 2017. In fact, changes to the regulations were introduced with the objective of decreasing the cost of monitoring and maintenance for oil companies. The proposition claims that oil companies have learned from the Deepwater Horizon spill and that the previously introduced legislation is now outdated. The Trump administration claims that fewer regulations will boost the oil economy, which proponents argue is important in this era of energy independence.

CONCLUSION

The BP Deepwater Horizon oil spill devastated marine life; coastal ecologies; and fishing, tourism, and natural resource-based economies throughout the Gulf of Mexico. The spill is one of the greatest man-made disasters of the twenty-first century due to its geographic scope and far-reaching social, economic, and environmental impacts. Moreover, the impacts are still cascading and will continue to unfold for years to come. Beyond the initial oil spill, the use of massive quantities of chemical dispersants will continue to be evaluated due to unforeseen impacts on human and marine health. While fishing grounds have been reopened for the past seven years, effects upon the wildlife are still noticeable. Regulations enacted after the spill may prevent similar disasters in the future; however, recent changes to such regulations may allow for such disasters to occur again.

Andrew Stevens, Kelsey Compton, and Luke Juran

Further Reading

BSEE (Bureau of Safety and Environmental Enforcement) and DOI (Department of the Interior). 2012. Well Control Final Rule Fact Sheet. https://www.doi.gov/sites/doi.gov/files/uploads/FACT%20SHEET-%20Well%20Control%20Final%20Rule%20-%20FINAL.pdf, accessed May 8, 2018.

GCERC (Gulf Coast Ecosystem Restoration Council). 2016. *Comprehensive Plan—Update 2016: Restoring the Gulf Coast's Ecosystem and Economy.* New Orleans, LA: GCERC.

Marsa, L. 2016. 6 Years after Deepwater Horizon Oil Spill, Thousands of People Are Still Sick. October 10, 2016.https://grist.org/article/6-years-after-deepwater-horizon-oil-spill-thousands-of-people-are-still-sick/, accessed May 8, 2018.
NRDC (Natural Resources Defense Council). 2015. Summary of Information Concerning the Ecological and Economic Impacts of the BP Deepwater Horizon Oil Spill Disaster. https://www.nrdc.org/sites/default/files/gulfspill-impacts-summary-IP.pdf, accessed May 8, 2018.
Ocean Portal. 2018. Gulf Oil Spill. Smithsonian's National Museum of Natural History, http://ocean.si.edu/gulf-oil-spill, accessed May 8, 2018.

Brisbane and Queensland Flood, Australia, 2011

Vast areas of the Australian State of Queensland and its capital city, Brisbane, are floodplains. Because of its tropical location, Queensland receives high amounts of rainfall, a sizeable portion of which is discharged into the Pacific Ocean through the Brisbane River and its tributaries. The Brisbane River runs through Brisbane and other towns and causes infrequent but extensive flooding. Queensland also experiences prolonged drought, especially before a major flood. Prior to 2011, Brisbane experienced major flood events in 1864, 1893, and 1974. Inhabitants of the area still remember the devastating flood of 1974, which killed 16 people, inundated 6,700 homes, and impacted around 13,000 buildings, including buildings in the central business district of Brisbane city (van den Honert and McAneney 2011).

Between November 2010 and January 2011, significant parts of Queensland, and Brisbane in particular, were inundated. It was one of the worst flood events in Australia's written history, and it is commonly referred to as the "Brisbane and Queensland flood 2011" or "Queensland floods 2010/2011." Isolated flood events started in different parts of the state in December. On January 10, 2011, major flash floods occurred in the city of Toowoomba and downstream Lockyer Valley, about 83 miles (125 km) west of Brisbane. The flash floods caused several deaths and severe damage to property. Brisbane experienced major flooding in January 12–14, 2011, which became severe on January 13, when the height of the Brisbane River at the City Gauge reached 14 feet (4.16 m). Enormous rescue, relief, and recovery efforts were required to respond to the flood. Direct and indirect economic loss from the flood totaled in billions of dollars.

CAUSES OF THE FLOOD

The Queensland Floods Commission of Inquiry (Flood Commission) stated that prolonged and extensive rainfall over large areas of Queensland, coupled with already saturated catchment areas, caused the severe flooding in 2011. Reportedly, the 2010 La Niña, an unusual weather pattern that brings wet weather to eastern Australia, was the strongest experienced in Australia since 1973. As a result, Queensland experienced huge rainfall in the last weeks of 2010 and early 2011. In fact, December 2010 was the wettest month on record for Queensland. During November 2010 and January 2011, most of the Queensland coastal areas received

Aerial view during the Brisbane and Queensland flood, Australia, 2011. (OnAir2/Dreamstime.com)

24–48 inches (600–1,200 mm) of rainfall. Some areas to the north and west of Brisbane experienced over 48 inches (1,200 mm) of rainfall. Tropical cyclone Tasha, which made landfall on December 25, 2010, in the northern part of Queensland, contributed a sizable portion of the downpour. In some areas, the cyclone caused more than 9.8 inches (250 mm) of rainfall (van den Honert and McAneney 2011).

Inefficient dam management was also partially responsible for the flood event. Operation of Wivenhoe Dam, located about 53 miles (80 km) northwest of Brisbane, caused an overflow of water in the Brisbane River catchment area. Constructed in 1984 to protect Brisbane and nearby areas from major floods, such as the one of 1974, Wivenhoe Dam also created the reservoir, Lake Wivenhoe. On January 11, 2011, Lake Wivenhoe filled to a level equivalent to 191 percent capacity. Based on dam operating procedure, storage over 100 percent was required to be released progressively, even though the dam can hold up to 225 percent of its supply capacity. Then the dam was operated, assuming there would be no additional rainfall. However, the modeling by the dam management was found to be incorrect. On January 13, 2011, most of the Brisbane River catchment area experienced major flooding when water released from the dam combined with the water from the local intensive rainfall. Toowoomba, Lockyer Creek, and Bremer River catchment areas and Brisbane consequently were severely impacted by the flood.

DAMAGE AND IMPACT

At least 33 people died and three people went missing because of the flood. The flood directly or indirectly impacted over 78 percent of the State of Queensland and was subsequently declared a disaster zone. The size of the disaster zone was bigger than the combined area of France and Germany. Approximately 29,000 homes and businesses were inundated. It is estimated that over 2 million people (nearly 10 percent of Australia's 2011 population) were impacted by the flood, and the total damage was estimated at about $3.7 billion. Moreover, there was evidence of looting in Queensland during the flooding period (Queensland Flood Commission 2012).

The flood caused the greatest possible amount of devastation as there was no apparent warning of the event. First, the Toowoomba/Lockyer Valley flash flooding occurred due to heavy rainfall from several intense storms in a brief period. Then, runoff occurred as the rainwater moved over the already saturated catchment area. This caused the banks to burst on the creeks, and the wall of muddy water of the runoff rose up to 36 feet (11 m). The maximum effect of the runoff was observed in the down gullies and streets of Toowoomba. People and animals sought shelter on the top of their houses, and cars were washed away. Most of the deaths (about 70 percent) occurred during the flash flood over Toowoomba and Lockyer Valley areas.

Following a warning given 24 hours prior, part of the Brisbane Central Business District (CBD) and several suburbs went under water as did several highways. Meanwhile, road and railway communication was severely impacted. Ultimately, approximately 300 roads, including nine major highways, were closed. More than 12,667 miles (19,000 km) of road and about 28 percent of Queensland's railway were damaged along with three major ports. Fortunately, Brisbane airport was still in operation, but business activity in the third largest city of Australia was halted for about a week. The pattern of inundation in the 2011 flood was almost identical to that of the 1974 flood.

The floodwater entered several coal and other mines so that over 80 percent of Queensland's coal mines were either closed or had restricted production for a prolonged period, resulting in economic loss estimated at $4.2 billion (about 2.2 percent of annual production). The Queensland government consequently lost billions of dollars in earning opportunity as royalties. Potentially harmful waste and substances from the mines mixed with the floodwater, which eventually was discharged into the Pacific Ocean. Scientists claimed that coral bleaching at the Great Barrier Reef increased after the 2011 flood. Floodwater discharge and contaminants of coal mine waste were partially responsible for that.

Increase in population and the construction of buildings along the flood path led to greater impact in 2011, despite floodwaters being lower than the 1974 flood. People who migrated to Brisbane after 1974 were unaware about the potential risk of flood, and many believed that a major flood would not happen in Brisbane after the construction of the Wivenhoe Dam. This view also played a role in the conservative assessment of the potential for flooding in land use and building planning, which led to controls being considered inadequate to restrict settlement in flood-prone areas.

RESPONSE AND RECOVERY

The immediate rescue and relief operations were challenging, but there was no breakdown in the conventional emergency management system. Helicopters and boats were used to recue people, while throughout Queensland, about 12,000 people took shelter in 34 evacuation centers managed by the Red Cross. Several other official and informal evacuation centers were also in operation to accommodate flood-affected people. The Premier of Queensland, Ms. Annastacia Palaszczuk, pledged for Flood Relief and raised about $22.5 million by January 9, 2011. The Australian federal government, other Australian states, and sports associations joined the relief effort.

After the flood, a massive recovery operation started. Over 55,000 registered volunteers along with thousands more unregistered volunteers helped people to clean up flood-affected houses. They also helped clean up business and commercial facilities and community facilities such as stadium, parks, and schools. Mobile and web technology was used to find where help was needed and who could help. Eventually, the army was mobilized to restore operations to road, rail, and power networks.

Immediately after the flood, the Queensland Floods Commission of Inquiry formed to evaluate the planning and preparation for the flood by the government agencies and communities. Other objectives of the commission were to assess response adequacy, arrangement for essential services, forecast and early-warning systems, insurance issues, dam operation strategy, and land-use planning for minimizing flood impact. The commission found the dam operating manual to be noncompliant and ambiguous. It also found that flood risk was not adequately addressed in land-use planning. The commission made over 160 recommendations for government and different agencies, some of which were implemented before the next wet season.

A sizable number of individuals and businesses made insurance claims for flood damage. Data from the Insurance Council of Australia showed that insurance companies received about 56,200 claims, with a total insured amount of $1.9 billion. However, insurance did not cover everyone. Additionally, many people thought that they were covered by insurance, but wording in such policies precluded coverage. Some insurance policies, for example, covered "flash flood" but not "flood."

In July 2014, law firm Maurice Blackburn lodged a class action suit before the state of New South Wales Supreme Court on behalf of 6,000 Brisbane and Ipswich flood victims. The class action was against the negligence and nuisance of two dam operators of Wivenhoe and Somerset dams: Seqwater and SunWater, and the State of Queensland in relation to the devastating Queensland flood of 2011. The suit alleged that the dam operators failed to use rainfall forecast information in formulating operating strategies. Additionally, the operators failed to preserve reasonable storage capacity for the maximum protection of urbanized areas from floodwater inundation. According to the lawsuit, the dam operators' negligence contributed significantly to the downstream flooding in Brisbane and Ipswich. It is one of the largest damage lawsuits in Australia and comes with a twist in that the class action suit was filed earlier in Queensland courts, but it was not heard there. The final verdict of the lawsuit is yet to be decided.

CONCLUSION

The Brisbane and Queensland flood of 2011 was an unprecedented event and completely unexpected in many places. Accumulation of water from several rain events in late 2010, huge rains in late December 2010 and early January 2011, and mismanagement of the flood protection dam all resulted in a catastrophic flood event. In other words, the devastation of the flood was a combined effect of extreme weather conditions, inadequate controls in land-use planning, and substandard dam management. The flood caused extensive human suffering, huge property and economic loss, and environmental damage. The third largest city of Australia, Brisbane, was almost nonfunctional for about a week.

The effects of the flood would have been less severe if land use and building planning controls had adequately accounted for and provided flood mitigation measures. Also, dam management strategy of Queensland should have been better, and insurance policies for both private and public sector could have been more inclusive. However, despite all the negativity, the flood was an excellent example of community resilience. People, on many occasions complete strangers, helped each other to a quick recovery from the flood damage.

As severe flooding is infrequent in this area, so documentation and record keeping of such events are critical for future generations. Adequate planning and preparation and awareness about such events are critical because there are growing indications of climate change in Australia, which may fuel similar hazards in the future. Several risk reduction and management measures have already been taken by Queensland Government as per the recommendation of the Flood Commission, but their effectiveness awaits testing.

Salim Momtaz and Palash Basak

Further Reading

Bohensky, E. L., and A. M. Leitch. 2014. Framing the Flood: A Media Analysis of Themes of Resilience in the 2011 Brisbane Flood. *Regional Environmental Change* 14(2): 475–488.

Box, P., D. Bird, K. Haynes, and D. King. 2016. Shared Responsibility and Social Vulnerability in the 2011 Brisbane Flood. *Natural Hazards* 81(3): 1549–1568.

Queensland Flood Commission. 2012. Queensland Floods Commission of Inquiry: Final Report. http://www.floodcommission.qld.gov.au, accessed July 30, 2018.

van den Honert, R. C., and J. McAneney. 2011. The 2011 Brisbane Floods: Causes, Impacts and Implications. *Water* 3(4): 1149–1173.

Buffalo Blizzard, New York, 1977

The Buffalo Blizzard besieged parts of western and northern New York state in the United States and Southern Ontario, Canada, from January 28 to February 1, 1977. This blizzard hit the Buffalo area on Friday, January 28, 1977, around 11:00 a.m., and it is considered the blizzard of the century and millennium. Located on the eastern side of Lake Erie, Buffalo received the brunt of the storm, which gained the city's name. The day started off like any other typical winter day, with temperatures in the twenties and sunny skies, until the weather took a turn for the worse. During the blizzard, temperatures plunged to 10 degrees Fahrenheit (−12 degrees Celsius) and remained there for several days. Hurricane force winds produced gusts ranging from 46 to 70 mph (74–113 km/h), with snowfall as high as 100 inches (254 cm) in some areas. Temperature and wind combined to create a wind chill of −60 degrees Fahrenheit (−51 degrees Celsius). The blizzard was characterized by heavy blowing of snow, and visibility was reduced to zero, which remained there from 11:30 a.m. on January 28 until 12:50 a.m. on January 29, 1977. The blowing snow and extreme cold paralyzed the Buffalo area for several days (Ebert 2000).

CAUSE

Like all other blizzards, the Buffalo Blizzard was the result of the collusion of warm air with very cold air. For a blizzard to form, the air temperature should be near or below the freezing point to make snow, winds need to pick up moisture or water vapor, visibility should be less than one-quarter mile (0.38 km), and warm air must collide with cold air; for example, an excellent source of water is available when air must blow across a large water body, such as a lake or an ocean. A blizzard has sustained wind speeds of 35 mph (53 km/h) or more and lasts for more than three hours. The climatic situation for development of the Buffalo Blizzard was ideal for producing this monster storm. A deep low-pressure system had moved just to the north of the Great Lakes from the polar region. In the immediate south, a stationary high-pressure system developed. The cold air from the low-pressure system collided with the warm air of the high-pressure system because the former moves counterclockwise and the latter moves clockwise. This led to a persistent west-to-northwest flow of extremely cold air into the western New York region and Southern Ontario (Ebert 2000).

Weather conditions in November and December 1976 also contributed to this blizzard. The month of December experienced record-breaking cold temperatures, bringing about 93 inches of (234 cm) of snow, 13 inches (33 cm) above the average annual snowfall in the Buffalo area. By December 14, 1976, Lake Erie was entirely frozen, which set an early freezing record. This facilitated wind to flow without any barrier. The surface area of Lake Erie is 9,910 square miles (25,667 square km), and in many places, the lake was covered with about 30 inches (76 cm) of ice. As a result, the wind could not pick up moisture from the surface of the lake. Instead, the blizzard winds picked up some of December's snowfall, and it also carried new snow into the affected areas, including the metropolitan area of Buffalo (Ebert 2000). Fierce winds blew this into drifts 30–40 feet (9–12 m) high. Meteorological reports from the National Weather Service (NWS) in Buffalo indicated that up until the blizzard began, it had snowed every day since Christmas of 1976.

IMPACTS

The Buffalo Blizzard struck suddenly with little warning, stranding over 1 million people in their cars, streets, office buildings, schools, bars, factories, or their own homes or the homes of strangers, and isolated farms across the affected areas. Effectively, Buffalo was cut off from the rest of the world. This blizzard made travel impossible while snowdrifts buried houses up to the second story.

Friday night, 250 people were stranded in the International Nickel Company plant in Port Colborne, Canada. Although schools closed after receiving warning of the impending blizzard, the rapid onset of the storm resulted in about 1,000 students being stranded overnight in Port Colborne and Wainfleet schools (about 2,000 students were stranded in the Niagara region alone). By Saturday on January 29, 800 students were still trapped, with 600 of them in the above two cities. On Sunday, the remaining students were taken from the schools by buses, several of which became stranded on the streets, and so some students were housed in nearby homes. Snow blocked train lines and closed airports in the affected region.

In western New York, 29 deaths were reported due to the storm with causes ranging from frost bite to breathing difficulties and hypothermia. Nine persons died in the Buffalo area in automobiles from hypothermia, carbon monoxide poisoning, or a heart attack. Many cars were stalled on expressways in and around Buffalo city because of brake failure, which, in turn, was caused by a combination of three factors: bitter cold temperatures, very high winds, and blowing snow. Thus, people were not able to walk to homes without freezing to death. The blizzard-induced snow was so densely packed that snow plows and other snow removal equipment were virtually useless on most of the roads, and earth-moving equipment had to be used.

Sixteen out of 25 towns in Erie County, as well as the City of Buffalo, declared state of emergency and banned all nonessential traffic at some time during the storm. The parts of Southern Ontario lying near the northern border of Lake Erie were also hit hard by the blizzard. Given the geographical proximity to western New York and sharing a location on the edge of the frozen and snow-covered Lake Erie, parts of Southern Ontario experienced similar conditions. Reports seemed to indicate the worst conditions in Southern Ontario were limited to closer to the lakeshore than to western New York. Here too, zero visibility and blowing and drifting snow made roads impassable and stranded many vehicles. In response to the blizzard event, the greater Regional Municipality of Niagara declared a state of emergency on January 29, and this remained in effect until February 2, 1977 (Rossi 2005).

The direct economic losses were very large, estimated at about $300 million. However, indirect loss was also very high. Loss of production, wages, and sales amounted to at least $297 million for Erie and Niagara Counties. There was extensive physical damage to property, including homes, trees, and cars, and products looted from abandoned vehicles. Additionally, the storm severely damaged orchards in Southern Ontario, and traffic and communication were severely disrupted. Power lines collapsed under the weight and power of the wind and snow. Dairy farmers in Wainfleet, Canada, dumped milk because it could not be transported out, and they also had trouble getting feed to their animals. One interesting effect of the prolonged confinement at home during the blizzard period was a marked increase in births at local hospitals 10 months after the event. In particular, an almost 18 percent increase in births was reported in areas affected by the blizzard in Southern Ontario (Rossi 2005). This consequence was also reported in subsequent blizzards in the United States.

RESPONSE

For the first time in U.S. history, the federal government declared the blizzard-affected areas a disaster. On Saturday, January 29, 1977, President Jimmy Carter issued a declaration of emergency for four counties in western New York. Later, five additional counties, including two northern New York State counties, were added. With international attention focused on this disaster, President Carter declared the nine counties a "major disaster area" on February 5, 1977. This was the first such declaration ever made for a blizzard, and it allowed extensive federal relief, including direct reimbursement to local governments to cover their snow removal costs (Rossi 2005).

The U.S. military deployed 2,000 troops to the Buffalo area to clean the roads and help those stranded. Snow removal equipment from Colorado, New York City, and Toronto were used to clean up snow, while abandoned cars were towed; nevertheless, it took two weeks to clear the streets. Snow was transported to other parts of the country by train at a cost of more than $200 million, five times the annual budgeted amount, to reduce the risk of local spring flooding. Hamilton, another blizzard-affected city of Southern Ontario, helped Buffalo by loaning snow removal equipment as part of an international rescue operation (Bryant 1991).

Personnel from offices of the United States Army Corps of Engineers (USACE) participated in response operations using $6.8 million to pay 216 private contractors to plow 3,186 miles (4,779 km) of road in nine counties in western New York. In addition, the USACE employed its work force at a cost of $700,000. Other military assistance included 500 National Guard troops, 320 U.S. Army Airborne troops, the 20th Engineer Brigade from Fort Bragg, 65–70 U.S. Marines, and assistance from the U.S. Air Force. The National Guard operation was titled Task Force Western, with headquarters at the Connecticut Street Armory in Buffalo, and it provided 9–10 army-type ambulances along with operators for use in the city of Buffalo when almost no traditional ambulances were able to operate (Rossi 2005).

In northern New York, people of blizzard-affected counties faced a serious shortage of food and other essential goods. Ultimately, the authorities in these counties lifted the travel ban on the last day of the storm on February 1, 1977. As a result, nearly 2,000 stranded travelers could leave the area. Several of the counties are known for the dairy industry, which was the hardest hit because the farmers were not able to get milk to the market. In Jefferson County alone, about 85 percent of dairy farmers were forced to dump milk because tank trucks could not reach farms. This contributed to $8 million in agricultural losses. Other problems included barns collapsing from the weight of snow, feed and grain shortages, inability to dispose of manure, and farmers being unable to reach barns to feed cattle, which resulted in loss of cattle, resulting in livestock deaths. Five deaths were reported in northern New York, all due to heart attacks, which occurred while four persons were shoveling snow and one was stranded in his car.

In Southern Ontario, the Canadian military forces assisted under police direction. Specifically, Canadian Army Reserve Battalion troops were sent to blizzard-affected Southern Ontario. Their initial priorities were to save lives and remove snow from main roads around the communities of Port Colborne and Fort Erie. In addition, 156 reserve militia and nine regular force soldiers helped distribute disaster relief items among the survivors. The military was also involved in areas that were less severely impacted by the blizzard such as the London, Ontario, area (Rossi 2005).

Throughout the blizzard-affected areas, people who were not stranded in their homes helped their trapped neighbors, area radio stations relayed messages and provided crucial information, and many individuals donated their time and loaned equipment to remove snow. People also helped to bring the needed food or medicine to their neighbors. Buffalo city police as well as Niagara Regional Police extended their help to transport medical personnel to city hospitals by snowmobile. Police also received many calls to deliver important prescriptions and medicines to stranded people. More than 1,000 volunteers of the Salvation Army provided hot meals for 67,000–176,000 people. They also donated clothing to about 4,500 people and housed 851 people for a cost of $75,000–$150,000. Also, the Salvation Army donated over 400 snowmobiles. In addition, the American Red Cross distributed 5 tons (4500 kg) of food at 84–92 locations, feeding about 50,000 people (Rossi 2005).

CONCLUSION

The Buffalo Blizzard of 1977 was unique in the sense that the instigating storm system did not by itself produce even larger amounts of snow because the cold temperatures completely froze Lake Erie. One important ingredient a blizzard needs is the wind to pick up moisture, and the lake could not supply this ingredient. Indeed, the blizzard would likely not have happened at all if it were not for the anomalously cold conditions in the last two months of 1976; the air temperature during the months of November and December was 42.8 degrees Fahrenheit (6 degrees Celsius) colder than the climatological normal, breaking records that went back to the 1880s. Meteorologists maintain the Buffalo Blizzard largely occurred because the lake froze over so early.

Bimal Kanti Paul

Further Reading
Bryant, E. A. 1991. *Natural Hazards*. New York: Cambridge University Press.
Ebert, C.H.V. 2000. *Disasters: An Analysis of Natural and Human-Induced Hazards*. Dubuque, IA: Kendall/Hunt Publishing Company.
Rossi, E. 2005. *White Death—The Blizzard of '77: A Canadian-American Survival Classic*. Southfield, MI: Seventy Seven Media Publishing Company.

California Drought, 2012–2016

Starting in 2012, California experienced a prolonged period of extremely low precipitation and high temperatures that lasted through 2016. While drought conditions are common in California (one lasting from 2007 to 2009, another lasting from 1987 to 1992, and a third from 1987 to 1992, which are the three most recent droughts preceding the 2012–2016 drought), the extended drought of 2012–2016 was particularly severe. Precipitation in California set a new record for the driest period from 2012 to 2015. A contributing factor to the persistent low precipitation was a zone of atmospheric high pressure over the Pacific Ocean, which altered normal wind patterns that typically push moisture-rich air over California in the winter, pushing it northward instead of west (Jones 2015). In addition, the years 2012–2016 also experienced above average temperatures (He et al. 2017).

The combination of both low precipitation and high temperatures created conditions that led to low soil moisture, tree canopy water loss, record low snowpack, and dangerously low river and reservoir levels. Snowpack hit a record low in April 2015 with the total snowpack estimated to be at only 5 percent of normal levels (He et al. 2017). As snowmelt is a major contributor to freshwater supplies in California, the loss in snowpack led to continually declining reservoir and river levels. Moreover, due to the above average temperatures seasonal snowmelt patterns were shifted, with more snowmelt occurring during winter months instead of during late spring and summer, resulting in lower summer streamflow and increased water temperatures. From 2013 to 2015, the annual average stream flows were at less than 50 percent of the average, and reservoir storage steadily decreased, approaching 50 percent of the average by 2015 (He et al. 2017).

The drought had significant and wide-ranging impacts on many aspects of life across the state of California. Most notably, the drought significantly affected agricultural production. California's Central Valley is one of the most productive agricultural systems on the planet. California ranks first among all U.S. states in terms of the value of agricultural products sold, accounting for more than 10 percent of cash farm receipts in the United States. This huge agricultural system accounts for more than 70 percent of the state's water use (Jones 2015). In addition, the drought led to a loss of domestic drinking water supply, reduced water bird habitat and native fish populations, death of forest trees, and increased wildfire risk (He et al. 2017).

DAMAGES

The total estimated economic impact of the 2012–2016 drought is about $10 billion dollars (Jones 2015). This accounts for agricultural losses, hydropower generation losses, cost of groundwater well drilling and pumping, and emergency response costs. However, it does not account for wildfires and increased wildfire risks or ecological impacts.

The largest source of economic damages was to the agricultural sector. This sector lost nearly $2.7 billion in 2015 alone. The majority of these losses were the result of fallowing (not cultivating a piece of land) of hundreds of thousands of acres of agricultural land. The cost of the loss of crop revenue due to fallowing in 2015 was estimated to be more than $900 million. Loss of milk production from dairy cows and loss of beef cattle due to lack of grazing fodder on nonirrigated pasture land also contributed significantly to agricultural sector losses with estimated losses of $350 million in 2015. In addition, while increased groundwater pumping was able to offset a large amount of the shortages in surface water supply caused by the drought, pumping costs increased during the drought from drilling of new wells and pumping from increasingly low water tables. The cost of groundwater pumping in 2015 alone was estimated to be $590 million (Howitt et al. 2015).

When considering the impact of direct agricultural losses on other sectors of the economy, such as reductions in fertilizer and other farm service purchases, the estimated cost of drought-related damages to the agricultural sector increased by 150 percent. These economic damages do not account for the loss of individual income due to losses of tens of thousands of seasonal farm jobs each year of the drought. Losses of individual income from farm jobs across the state of California were estimated to be $720 million just in 2015 (Howitt et al. 2015).

Streamflow reductions during the drought led to a reduction in hydropower production of up to 50 percent during 2014 and 2015. In order to replace this loss in electricity production, more expensive gas-turbine electricity generation was increased at an estimated cost of $2 billion. In addition, costs associated with maintaining domestic water supply, particularly in rural areas, increased during the drought. In some communities, new groundwater wells were drilled, or the cost of using existing wells increased as groundwater levels declined. In others, bottled water was used or potable water was trucked in and held in large storage tanks. An additional monetized cost of the drought was emergency community assistance. More than $500 million from the funds given by the state and federal governments was spent on emergency assistance for drought-impacted communities including funding for food, housing, and new drinking water systems (Lund et al. 2018).

One of the most significant unquantified costs of the drought was the death of forest trees. More than 102 million trees in California's forests were believed to have died due to stress caused by the drought (Lund et al. 2018). In addition to increasing soil erosion rates, tree death leads to the accumulation of wildfire fuel, creating conditions that can result in more severe wildfires. The dry and warm conditions of the drought likely contributed to wildfires that occurred between 2012 and 2015. The Rush Fire in 2012 and Rim Fire in 2013 were some of the largest wildfires in California's history, burning more than 500,000 acres (202,347 ha). In addition, the Valley Fire and Butte Fire in 2015 were among the most damaging wildfires in California's history, burning nearly 3,000 buildings (He et al. 2017).

Damages to ecosystems, particularly those to critical habitats that support native fish and water bird populations, were also particularly severe. Although the State Water Resources Control Board prioritizes water allocations to ecosystems in order to maintain the minimum required flows to support natural habitats, the severity of the drought was such that even water flows to protected ecosystems were cut back. Water flows to protected water bird habitats in the Sacramento-San Joaquin River Delta region of California were reduced by about 25 percent during the drought, reducing the amount of habitat, increasing salinity levels, and reducing food supply. These ecosystem impacts have negative impacts on water bird mortality. Reduced stream flows and subsequent increased water temperatures and decreased water quality across the state had a negative impact on native fish populations, including many endangered species. Fish populations in the Delta region during the drought reached record lows. Salmon and trout, which are highly sensitive to water temperatures, were also negatively impacted through loss of eggs and newly hatched fish to warm river water temperatures, placing some local species at high risk of extinction. While impacts to the damaged ecosystems and wildlife populations have not been quantified in monetary terms, the cost associated with emergency support of ecosystems during the drought was roughly $67 million (Lund et al. 2018).

RESPONSE

The decreased surface water supply during the drought of 2012–2016 led to increased pumping of groundwater, irrigation water curtailments, and mandatory water conservations measures. During a typical year, groundwater comprises about 30 percent of the total irrigation water for agriculture; however, during the drought this percentage increased to more than 50 percent (Howitt et al. 2015). Groundwater levels in many locations across the state dropped by about 10 feet (3 m) between 2013 and 2014. Within the San Joaquin valley, an area of intensive agricultural activity in California's Central Valley, groundwater levels during the drought dropped to about 100 feet (30.5 m) below historic levels, resulting in land subsidence rates of two inches (5.1 cm) per month (Jones 2015).

In response to the drought conditions, allocations of water via the State Water Project and Central Valley Project by the California State Water Resources Control Board were severely reduced. In 2014, agricultural irrigation water delivered via the Central Valley Project was completely curtailed, while allocations to both agricultural irrigation and urban water supply via the State Water Project were reduced to 5 percent of their full allocation (He et al. 2017; Jones 2015). The reductions in irrigation water supply resulted in not only water shortages, which were partially offset by groundwater pumping and water transfers, but also in crop loss and a reduction in cultivated land. In addition to increased groundwater pumping, shortages in domestic water supply were in some cases offset by increased water purchases from other sources and increased water recycling.

In addition to the responses above, Governor Jerry Brown declared a statewide drought emergency in 2014 and put in place mandatory water conservation measures. Domestic water suppliers were required to reduce urban water use by 25 percent. Suggested conservation measures included requests to residents to reduce lawn and landscape watering, car washing, cleaning of sidewalks and driveways, and decorative water fountain use. In addition, parks and golf courses were asked to reduce irrigation water use, hotels were asked to reduce washing of towels and sheets, and restaurants were asked to provide water to customers only when requested (Jones 2015).

PREPARATION FOR THE FUTURE

Although not conclusive, the results of many recent studies suggest that future changes in climate in California will continue to negatively impact surface water supply quantities and seasonal availability, leading to more frequent and severe drought conditions. When considered along with rapid population growth and increasing agricultural demand, these changes to the climate will place a significant strain on water resources in the state. During the 2012–2016 drought, increased pumping of groundwater was used to offset much of the shortage in surface freshwater supplies; however, many climate and groundwater studies suggest that increasing groundwater extraction may not be a sustainable response for future drought events. In addition to groundwater extraction rates currently exceeding groundwater recharge rates, land subsidence has reduced the total amount of storage capacity of groundwater aquifers.

Prior to the drought, groundwater use was not regulated at a state-wide level. Unlike surface water, groundwater was not subject to a state's permitting and regulation process, and most regulation of groundwater was limited to localized programs in urban areas where participants were charged for pumping. However, in 2014, the Sustainable Groundwater Management Act was signed into law, in part in response to the continued increase in groundwater pumping rates and associated land subsidence in agricultural regions. The act requires that groundwater basins subject to critical conditions of overdraft develop a sustainable groundwater management plan by 2020 and reach sustainable operations by 2040 (Jones 2015).

In response to the drought, California's state government, as well as the federal government, provided funding for investments in California's water system. About $2.7 billion was set aside during the drought for projects that would repair, maintain, and improve the existing water transport and water supply system within the state, with a focus on increasing the resilience of the state to future droughts. In addition, new techniques for managing saltwater intrusion in the Delta were tested during the drought, leading to an increased capacity to implement these and similar measures for protection of critical ecosystems (Lund et al. 2018).

CONCLUSION

The California drought of 2012–2016 was characterized by extremely low precipitation accompanied by extremely high temperatures. These conditions led to record low snowpack, low soil moisture levels, low streamflow, and low reservoir storage. The reductions in surface water supply associated with these drought conditions led to curtailments in water allocations for agriculture and unprecedented efforts to reduce urban water consumption. The impacts of water shortages were not only particularly severe for the agricultural sector but also impacted domestic drinking water supplies, hydropower production, and ecosystem health. While the estimated economic impact of the drought is about $10 billion, this is a relatively small percentage of California's annual economy, suggesting that water management and emergency responses enacted during the drought helped to mitigate much of the potential damages. However, while the drought was severe, the impacts to precipitation, streamflow, and reservoir storage were not unprecedented. Furthermore, future climate changes in the state are expected to lead to even more severe drought conditions, signifying an increasing need to proactively address, manage, and prepare for water shortages in the future. The enactment of the Sustainable Groundwater Management Act is a positive first step in this direction with the potential to help regulate the use of water in California to the benefit of future generations and the environment.

Katherine S. Nelson

Further Reading

He, Minxue, Mitchel Russo, and Michael Anderson. 2017. Hydroclimatic Characteristics of the 2012–2015 California Drought from an Operational Perspective. *Climate* 5(1): 5.

Howitt, Richard, Duncan MacEwan, Josue Medellín-Azuara, Jay Lund, and Daniel Sumner. 2015. *Economic Analysis of the 2015 Drought for California Agriculture*. Davis, CA: University of California, Davis.

Jones, Jeanine (prepared by). 2015. California's Most Significant Droughts: Comparing Historical and Recent Conditions. California Department of Water Resources. https://water.ca.gov/-/media/DWR-Website/Web-Pages/Water-Basics/Drought/Files/Publications-And-Reports/California_Signficant_Droughts_Comparing-Historical-and-Recent-Conditions.pdf, accessed January 20, 2019.

Lund, Jay R., Josue Medellin-Azuara, John Durand, and Kathleen Stone. 2018. Lessons from California's 2012–2016 Drought. *Journal of Water Resources Planning and Management* 144(10): 04018067.

Chicago Heat Wave, Illinois, 1995

During July 12–15, 1995, Chicago experienced a deadly heat wave that claimed 830 lives nationally, with 525 deaths in Chicago, mostly the poor, elderly, and others on society's margins. Because nearly 63 percent of all heat wave-related deaths occurred in Chicago, this is called the Chicago Heat Wave, which caused the highest number of heat wave-related deaths in the history of the United States. Over the past 30 years (1988–2017), 134 people have died annually from heat waves. Among all weather disasters such as flood, lightning, tornado, hurricane, wind, and winter storms, heat waves have killed more people cumulatively in the United States. There are many definitions of heat waves. It is generally defined as the persistence of extreme high temperatures at least for several consecutive days. According to the World Meteorological Organization (WMO), the duration of high temperatures should be for five consecutive days or more. The definition of heat wave is region specific because the average temperatures in one region may be contemplated as a heat wave in another region. Heat waves in a given region generally last for several days to several weeks (Rafferty 2018).

On July 12, 1995, the third day of an intensifying heat wave reached 97 degrees Fahrenheit (36 degrees Celsius) during daytime, and the night temperature did not

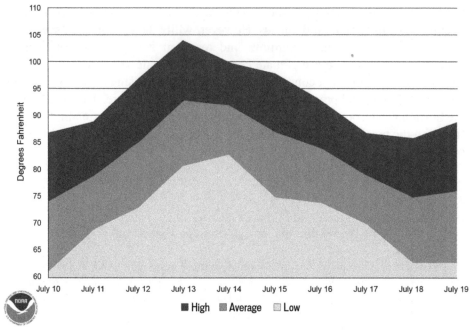

Graph showing the daily temperatures during the Chicago heat wave, 1995. (NOAA)

cool much because of high humidity. The combined heat and humidity made it feel as if the air temperature were 100 degrees Fahrenheit (37.8 degrees Celsius). During the next three days, heat records were broken at many locations in the central and northern Great Plains. On July 13, 1995, the temperature hit 106 degrees Fahrenheit (41 degrees Celsius), and the heat index, driven by extraordinarily high humidity, climbed to 126 degrees Fahrenheit (52.2 degrees Celsius). On a single day, July 15, the number of heat-related deaths reached its highest daily tally of 215, and refrigerated trucks were summoned in Chicago to handle the overflow of corpses. The city emergency officials were not prepared for this heat wave. Weeks later, when a second heat wave arrived, the city and its inhabitants responded quickly. Neighbors checked on neighbors, and police knocked on doors. This action saved many lives.

This high concentration of pressure makes it difficult for other weather systems to move into the area, which is why a heat wave can last for several days or weeks. The longer the system stays in an area, the hotter the area becomes. The high pressure inhibits winds, making them faint to nonexistent. Because the high-pressure system also prevents clouds from entering the region, sunlight can become punishing, heating up the system even more. The combination of all of these factors come together to create the exceptionally hot temperatures we call a heat wave.

CAUSES

Similar to many weather hazards, the origin of a heat wave is directly associated with the movement of a high-pressure system into an area. The area experiences the heat wave until a new pressure system is strong enough to push the high-pressure system away. In July 1995, a strong atmospheric high pressure sat on top of a slow-moving, hot humid air mass on the surface in the central United States and Canada, and it caused a deadly heat wave (Abbott 2008). This system consistently produced maximum temperature during the day with high minimum temperatures at night. The heat wave also brought extremely low wind speeds and high humidity.

In fact, an interesting climatic anomaly developed in the Great Plains and the East Coast of the United States in April 1995 and lasted through August 1995. These regions experienced heavy rainfall from early April through late June 1995, causing flooding in several states. As noted, high temperatures began developing across the regions during early July 1995, resulting largely from the high record dew point values, which, in turn, was the result of high evaporation from the wet soil. A five-day period of exceptionally high air pressure and high dew points occurred during July 11–15, 1995 (Changnon et al. 2007).

IMPACT

A heat wave creates several impacts, but the most critical impact is on human health. When human body temperature reaches 105 degrees Fahrenheit (41.5 degrees Celsius), dehydration and hyperthermia set in, which is likely to increase the probability of death from heat strokes. In 1995, mostly from July 13 to 16, while deaths occurred in several regions of the United States, they were primarily concentrated in urban areas in the Illinois and in Wisconsin. In addition to Chicago, the heat wave killed 60 people in Milwaukee. Urban heat islands, particularly in Chicago, greatly exacerbated the heat problems. Because of the high concentration of buildings, and paved surface, urban centers absorb more heat in the day and radiate more heat at night than comparable rural sites. As a result, large cities generally experience a lot less cooling at night than do rural sites (Angel n.d.).

Illinois State Climatologist, Jim Angel (n.d.), claimed that the high number of heat wave-related deaths in 1995 was exacerbated by several other factors such as an inadequate local heat wave warning system, power failures, inadequate ambulance service and hospital facilities, and a relatively large aging population in Chicago. Additionally, such emergency measures as Chicago's five cooling centers were not fully utilized. The inability of many residents of Chicago to properly ventilate their residences due to fear of crime or a lack of resources for fans or air-conditioning also contributed to the high number of deaths (Changnon et al. 1996). "The medical system of Chicago was severely taxed as thousands were taken to local hospitals with heat-related problems. In some cases, fire trucks were used as substitute ambulances" (Angel n.d.).

In addition to Illinois and Wisconsin, seven states (Indiana, Iowa, Kentucky, Michigan, Minnesota, Missouri, and Ohio) also reported heat wave-related deaths. As in previous heat waves such as in 1936, 1955, 1966, and 1988, most deaths began two days after the heat wave started as bodies weakened and failed. The 1995 heat wave mostly affected people who had no access to air-conditioning and those in Chicago who lived isolated from others. Patrick Abbot (2008) reported that more than 50 percent of the deaths in the city occurred among those who lived on the top floor of their buildings, where heat buildup was the greatest. However, previous heat waves caused a much lower number of deaths than the 1995 event. For example, the 11-day heat wave in 1936 killed 297 people. Even the summer of 1955 had been hotter, but it caused far fewer deaths than that of 1995.

Stanley Changnon and his colleagues (1996) provided two reasons for more deaths in 1995. Before the 1995 event, residents of Chicago lived a lifestyle that seems exceptional today. People were not afraid to go outside at night to sleep in parks, on beaches, or simply on their front porches, leaving windows and doors open. During the heat wave in 1995, it was not possible to sleep outdoors because of the remarkable increase in violent crimes. Sociologist Eric Klinenberg wrote, "Fewer than 10 percent of Chicago homes at the time had air-conditioning, but the simple strategy of sleeping outside helped to keep the mortality rate during the 1955 crisis down to roughly half the level of the 1995 disaster" (Klinenberg 2015, 23). As noted earlier, older people are most likely to die from heat waves. In 1995, the number of elderly people had grown rapidly since the early or mid-twentieth century, resulting in a greater number of vulnerable persons in the city in 1995 (Changnon et al. 2007).

Additionally, increased use of energy for air-conditioning contributes to heat waves, and the 1995 heat wave was no exception. Usage of electricity reached all-time record highs by mid-July in Chicago and Milwaukee. Compounding the problem, power outages are also common during heat waves. A densely populated section of Chicago experienced a major power outage on July 13–14, which affected 40,000 people, a factor associated with many deaths. One by-product of increased energy production and higher energy prices was it generated additional revenue in July, amounting to $510 million for regional power companies (Changnon et al. 2007).

Further, repairing and maintaining of streets, highways, and railway tracks cost more than $200 million to address heat-induced heaving of pavement and bucking of tracks. Retail sales also decreased in the heat wave-affected city and rural areas, while the region experienced lowered food production and hence increased prices of crops and forages. It also affected livestock, killing 850 dairy cows in Wisconsin and also a large number of poultry. "Milk production declined 35 percent and agricultural losses were $38 million" (Changnon et al. 2007, 50). The heat waves contributed to the record wildfire season because the heat dried out vegetation, creating tinder for fires. However, no wildfire directly contributed to the 1995 heat wave.

RESPONSE

The 1995 heat wave somehow caught Chicago by surprise. Thus, City Hall failed to declare a weather emergency until it was too late, failing to release a heat emergency warning until July 15, the final day of the heat wave. For this reason, city officials received extensive criticism. Although the U.S. National Weather Service (NWS) generally provides warning for upcoming heat waves, the organization did not issue either a timely heat advisory or a heat warning. A heat advisory triggers public safety regulations, such as a ban on evictions and electricity shutoffs. A heat warning, on the other hand, alerts hospitals to prepare for an increase in emergency calls, and some cities open public air-conditioned centers. However, the governments of Illinois and Wisconsin did ultimately declare Chicago and Milwaukee disaster areas and requested federal government assistance. Accordingly, President Bill Clinton allocated $100 million in emergency energy funds to low-income homeowners in the affected areas (Changnon et al. 2007).

LESSONS LEARNED

The heat wave event of 1995 changed the way Chicago responds to and prepares for all emergencies. In that year, the Chicago Office of Emergency Management and Communications was just getting underway with fire and police dispatch operations in one location. Today, the office includes emergency responders, as well as the 311 Call Center to reach City Hall, the Traffic Management Authority, and a public infrastructure center to effectively handle any emergency, including heat waves. The office also developed heat advisories for the city. The city hospitals and Cook County hospitals, where Chicago is located, are all now equipped to handle many heatstroke patients. In 1995, such a facility was limited. One of the important changes after the 1995 event was improved technology and software to produce data and pick up any abnormal trends, making it easier to address crises before they happened.

Like many other natural disasters, heat waves benefit some people, regions, and sectors. For instance, power companies benefited from the 1995 Chicago heat wave because the event increased the demand for electricity and hence the price. Tourism flourished in cooler areas of Wisconsin and Minnesota, where people went to escape the heat. The number of visitors in these two states increased 20 percent over the previous year. Manufacture and sales of air-conditioning units also increased dramatically in the affected states.

CONCLUSION

The 1995 Chicago heat wave was an extremely sad and deadly event, but it triggered improved forecasting of extreme heat conditions in the United States. As a result of heavy criticism for the large number of deaths in Chicago, Mayor Richard Daley appointed a Commission on Extreme Weather Conditions to determine what went wrong and what should be done in the future. This led to a new and more comprehensive "heat warning plan" on July 20. The city also established a "weather center" so that the NWS can quickly feed heat wave information. Many positive policies and actions were implemented after the 1995 heat wave, which were successful in reducing the deaths from heat waves after 1995. No subsequent heat wave has killed so many people in an event that exceeded all other weather events in the United States combined.

Bimal Kanti Paul

Further Reading
Abbott, P. L. 2008. *Natural Disasters*. Boston: McGraw Hill Higher Education.
Angel, J. n.d. The 1995 Heat Wave in Chicago, Illinois. https://www.isws.illinois.edu/statecli/General/1995Chicago.htmm, accessed February 1, 2019.

Changnon, S. A., K. E. Kunkel, and D. Changnon. 2007. *Impacts of Recent Climate Anomalies: Losers and Winners*. Champaign: Illinois State Water Survey, Illinois Department of Natural Resources, and University of Illinois at Urbana-Champaign.

Changnon, S. A., K. E. Kunkel, and B. C. Reinke. 1996. Impacts and Responses to the 1995 Heat Wave: A Call to Action. *Bulletin of the American Meteorological Society* 77: 1497–1506.

Klinenberg, E. 2015. *Heat Wave: A Social Autopsy of Disaster in Chicago*. Chicago, IL: University of Chicago Press.

Rafferty, J. P. 2018. Heat Wave Meteorology. https://www.britannica.com/science/heat-wave-meteorology, accessed January 28, 2019.

Chi-Chi Earthquake, Taiwan, 1999

Taiwan experienced the most powerful earthquake of the twentieth century on September 21, 1999, at 1:47 a.m. local time. This 7.6-magnitude earthquake, known as the Taiwan earthquake of 1999, is also known as the 921 earthquake because it occurred on September 21 or the Chi-Chi earthquake because it occurred near the town of Chi-Chi in Nanton County, occurring at a depth of 4 miles (7 km). The earthquake was produced by a south-to-north rupture on a well-mapped fault called the Chelungpu thrust in central Taiwan. The hanging wall of the thrust fault moved westward and upward 3–7 feet (1–2 m) along the entire length of the rupture. The earthquake was followed by 10,252 aftershocks until October 10, 1999, with six aftershocks measuring at magnitudes greater than 6.5 (Moh et al. n.d.). This earthquake was unexpected because the central and western regions of the country were considered less vulnerable to damaging earthquakes than the eastern region of the country, where the Eurasian and Philippines Sea plates collide. On a geologic time scale, Taiwan, an island country, has virtually erupted from the ocean floor as the Philippines Sea plate pushes northwestward into the Eurasian plate (RMS n.d.).

The Chi-Chi earthquake provided abundant opportunities for a wealth of observations and earthquake data from modern instruments, such as global positioning system (GPS) devices, consequently advanced seismological research and engineering design as well as allowed mapping of ground change, crustal movement, and structural damage. Because Taiwan is in a seismically active zone of the Pacific basin, the country installed more than 600 digital instruments, including GPS, and a dense network of seismographic stations prior to the 1999 earthquake under the Taiwan Strong-Motion Instrumentation Program (TSMIP). In addition, the Taiwan Weather Bureau Seismic Network (CWBSN) established digital telemetered stations and a dense network of seismographic stations that electronically issued earthquake information (hypocenter, magnitude, and isoseismal map) to all pertinent emergency management agencies within minutes of the main shock (Shin and Teng 2001).

DAMAGE

The Chi-Chi earthquake, which lasted for more than 40 seconds, destroyed approximately 10,000 buildings and partially damaged 7,500 others in central western Taiwan, leaving more than 100,000 people homeless. Less than 1 percent of those displaced had no earthquake insurance. Although approximately 9,000 industrial plants in 53 industrial parks reported some form of damage, the overall impact on Taiwan's industry was not considerable and did not significantly interfere with the country's export volume. However, if the epicenter had been further south or north, the industrial sector would have sustained heavy damage.

Earthquake-induced shaking was felt as far as 100 miles (150 km) from the earthquake's epicenter, including the most densely populated northern region of Taiwan. Several tall buildings in Taipei, the capital of Taiwan and located 90 miles (135 km) from the epicenter, were also affected. Many buildings in the affected areas were destroyed or damaged because they were not constructed under modern seismic design during the country's building boom in the early 1990s. Although Taiwan had adequate building codes, enforcement was lenient, meaning constructors often ignored construction regulations to reduce building costs (Prater and Wu 2002).

Cracked roads and sidewalks in Taichung, Taiwan, after the Chi-Chi earthquake in 1999. (Aaron Siirila)

The Chi-Chi earthquake caused more than 2,400 fatalities, and more than 10,700 people were injured. Nantou County had the highest mortality rate followed by neighboring Taichung County. Mortality was higher in localities that were nearer to the epicenter. Liang et al. (2001) reported that about 77 percent of the victims died in their own houses, 16 percent died outdoors, and the remaining 7 percent died in hospitals. A majority of deaths resulted from the collapse of low- and mid-rise buildings less than eight stories high. In both high- and low-casualty areas, women had a slightly higher mortality than men. Unlike most earthquakes, however, minimal fire damage was reported because buildings in Taiwan typically have tile or stone facades and the use of natural gas is limited (RMS n.d.).

The Chi-Chi earthquake ruptured along the 67-mile (100 km) Chelungpu fault, with vertical displacements ranging up to 23 feet (7 m) along the northern part of the rupture and 16.4 feet (5 m) in the southern part. Areas that experienced the strongest shaking reported many damaged roads, and at least 40 were damaged that had to be demolished and rebuilt. Ground subsidence caused some of the bridge damages, but most structures along the fault were nonengineered structures constructed prior to implementation of modern design codes. One significant aspect of the Chi-Chi earthquake was its large lateral and vertical displacements at the surface; structures constructed across faults typically cannot withstand such displacements.

The earthquake also generated thousands of landslides throughout the epicentral region. These landslides diverted rivers, causing the formation of new lakes, which consequently caused people to abandon several villages when water was impounded behind the new lakes. Nearly all the landslides were located to the right of the Chelungpu fault (on the hanging wall), and majority of the landslides were relatively shallow slips in residual soils, typically 3.2–16.4 feet (1–5 m) deep. One massive landslide, however, killed 32 people (Moh et al. n.d.). Approximately 786 schools were damaged, and other critical facilities, including hospitals, police and fire stations, power houses and transmission, were also severely damaged. The earthquake also caused significant damage to high-voltage transmission lines, nearly destroying the Chunglia substation. Estimated earthquake damages totaled $14 billion, an estimate that represented 3.3 percent of the gross national product (GDP) of Taiwan, and it was expected to decrease the country's GDP by 0.5 percent in 1999 and 6.05 percent in 2000 (Bolt 2004; RMS n.d.).

RESPONSE

The Taiwanese government quickly engaged in response efforts following the earthquake, and the success of these efforts was due in large part to the high capacity of the central government and the prompt arrival of foreign aid (Prater and Wu 2002). Prior to the earthquake, the government had installed seismic instruments across the country, and these instruments were connected by telephone lines. This provided rapid earthquake information, which significantly improved timely emergency response and effective dispatching of rescue missions. Immediately following the earthquake, the government of Taiwan established the Council for Disaster Response (CDR), and on September 25, 1999, the president declared the central part of the country to be disaster districts. Subsequently, an emergency decree was issued to empower the local government to override laws and statutes to quickly mobilize recovery. Ten medical teams were dispatched to disaster areas, 12 helicopters were used to deliver supplies, and six temporary rescue centers were established.

Numerous international and national humanitarian organizations, charity groups, and individuals provided emergency assistance to earthquake survivors. Twenty countries, including Japan, South Korea, United States, and Thailand, sent more than 700 nurses, physicians, paramedics, engineers, emergency managers, aid workers, and relief experts to Taiwan to participate rescue and relief operations. The Taiwan Red Cross participated in both relief and recovery phases. The private sector also contributed in many ways to the disaster response. For example, local members of one Buddhist religious organization distributed tents immediately after the earthquake, quickly built temporary shelters, and participated in rescue operations. Within three hours after the earthquake, the organization's members had set up a vegetarian kitchen in cooperation with other religious groups and offered emotional and psychological support to earthquake survivors.

The organization also provided monetary assistance of $625 to families of the deceased and $156 to the injured (Prater and Wu 2002). In addition, 15 foreign rescue teams with 530 members and 38 dogs helped search for survivors (Moh et al. n.d.). These teams, along with local teams, pulled out more than 5,000 people from earthquake wreckage. The teams formed by the International Association of Search and Rescue of the Republic of China played an important role in rescue operations.

In addition to local and international aid, the Taiwanese government dispatched 5,000 military personnel to provide emergency services. They opened shower stations and shelter centers, provided security in hard-hit areas, deployed mobile medical teams, and removed debris and human remains. They also distributed tents, blankets, vehicles, and food and water. By September 28, one week after the earthquake, they had completed 90,000 supply and evacuation trips and more than 1,000 flights (Prater and Wu 2002). Military forces were also deployed to demolish severely damaged buildings and repair bridges and roads. The National Center for Research on Earthquake Engineering (NCREE) of Taiwan formed several reconnaissance teams to collect firsthand information about the earthquake. Most of these teams arrived in the affected areas within hours after the event.

On September 23, 1999, the Taiwanese government appealed to international and domestic communities for donations and received more than $431 million within the first week after the government made appeal for donations. Ultimately, the government received donations from all domestic and foreign sources to a total of approximately $1 billion. These funds were used primarily for emergency relief supplies and the construction of temporary shelters and schools. A portion of the donations were also used for disaster recovery (RMS n.d.). Many survivors were living in tents in open fields and parks because they were afraid to live in damaged or un-damaged buildings. So the government constructed temporary shelters for earthquake survivors on public land. As of early January 2000, however, over 1,000 people were still living in tents. The government planned to move earthquake survivors off unstable lands while implementing large urban renewal projects into recovery efforts. The Taiwanese government identified 116,000 acres (47,000 ha) of land for resettlement in the central region of the country (RMS n.d.).

RECOVERY

One week after the disaster, the Taiwan government formed the 921 Earthquake Reconstruction Committee, which established 13 task forces, including finance, education, and environmental preservation, to expedite recovery. Electrical power was restored within two weeks after the earthquake, and all damaged bridges were repaired or replaced within three months of the event occurrence. The Council for Economic Planning and Development (CEPD) prepared a five-year Post-Earthquake Reconstruction Plan (PERP) with $7 billion allocated for reconstruction over three years, including $6,500 for survivors whose houses had been completely destroyed and half of that amount for homeowners with partially damaged homes. Seven hundred experts were appointed to verify damage status and build damage appraisals (RMS n.d.).

In addition, the Central Bank of Taiwan allocated $3 billion for low-interest, long-term loans to earthquake survivors for reconstruction (RMS n.d.). The economy of Taiwan had been experiencing a boom prior to the Chi-Chi earthquake, so the government had minimal debt burden and was able to fund recovery efforts.

CONCLUSION

The Chi-Chi earthquake increased awareness of potential seismic threats, particularly in the western and central regions of Taiwan, and provided an opportunity to improve earthquake preparedness. Several mitigation measures were initiated after the 1999 earthquake. For example, the government of Taiwan revised the criteria for seismic design of structures and initiated the expansion and enhancement of the hazard prevention system, which included earthquakes and other types of natural hazards. One of the most comprehensive programs developed a software to simulate earthquake scenarios. In addition, earthquake data gathered during the post-disaster period helped seismic experts model ground motion worldwide. However, the Chi-Chi earthquake also had drastic political implications because the political party in power at the time of the disaster lost in the national election held in 2000. Some critics blamed this loss on inadequate response to the event. However, most of the citizens of Taiwan were generally satisfied with the performance of the government in responding to the disaster.

Bimal Kanti Paul

Further Reading

Bolt, B. A. 2004. *Earthquake*. New York: W.H. Freeman and Company.

Liang, N.-J., Y.-T. Shih, H.-M. Wu, H.-J. Wang, S.-F. Shi, M.-Y. Liu, and B. B. Wang. 2001. Disaster Epidemiology and Medical Response in the Chi-Chi Earthquake in Taiwan. *Annals of Emergency Medicine* 38(5): 549–555.

Moh, Z.-C., R. N. Hwang, T.-S. Ueng, and M.-L. Lin. n.d. 1999 Chi Earthquake of Taiwan. https://pdfs.semanticscholar.org/a86b/403ff71602fb6358046ffe9680bed9ba31ba.pdf, accessed September 7, 2019.

Prater, C., and J.-Y. Wu. 2002. *The Politics of Emergency Response and Recovery: Preliminary Observations on Taiwan's 921 Earthquake*. College Station, TX: Hazard Reduction and Recovery Center, Texas A&M University.

RMS (Risk Management Solutions, Inc.). n.d. *Event Report: Chi-Chi, Taiwan Earthquake*. Menlo Park, CA: RMS.

Shin, T.-C., and T. Teng. 2001. An Overview of the 1999 Chi-Chi, Taiwan, Earthquake. *Bulletin of Seismological Society of America* 91(5): 895–913.

Christchurch Earthquake, New Zealand, 2010–2011

The 2010–2011 Christchurch sequence (twin) earthquakes, as they are called, which occurred in New Zealand's Canterbury Plains of South Island, struck within five months of each other. The first one, with a magnitude of 7.1, happened on September 4, 2010, in the early morning hours, with its epicenter near the rural town of Darfield. The second one, with a magnitude of 6.3, occurred on February 22,

Devastation in the southern central business district caused by the Christchurch earthquake in New Zealand, 2011. (Nigel Spiers/Dreamstime.com)

2011, during lunch hours, with its epicenter at 6 miles (10 km) southeast of the center of Christchurch.

CAUSE

New Zealand, located along the Australian and Pacific convergent boundary, consists of two islands, the North Island and the South Island. The Canterbury Plains, where the Christchurch sequence earthquakes occurred, are situated on the Southern Island's eastern part. These plains are formed by the deposition of silt carried by the rivers from the Southern Alps. Of the two islands, the tectonics of the South Island especially is fascinating. The northern half of the South Island lies on the zone of convergence where the Pacific plate in the east is sub-ducting beneath the Australian Plate in the west, while in its southern half, this movement is reversed (Potter et al. 2015).

Home to over 500,000 people, the Canterbury region's main city, Christchurch, has a population of about 370,000. Agriculture-based industry supported by alluvial plains and tourism comprises its main economic activities. Before the twin earthquakes, the Christchurch area was known to have suffered only a few moderate intensity earthquakes (Downes and Yetton 2012). Of them, only two stand out: a shallow earthquake in New Brighton in 1869 and a deeper one called the Lake Ellesmere earthquake in 1870. These two particular events provide information about historic earthquake faults beneath the alluvial plains where Christchurch is situated (Downes and Yetton 2012), but until 2010, this region's people had little idea about earthquake damage. After the 2010 event, however, their perception of earthquake risks increased considerably.

The September 4, 2010, earthquake occurred at a depth of about 7 miles (11 km) along a previously unidentified fault, later named the Greendale fault located about 25 miles (40 km) west of Christchurch. This earthquake caused a 20-mile (30 km) surface rupture trending east-west across agricultural land, a rare phenomenon linked to earthquakes. After this, the February 22, 2011, earthquake unleashed one of the largest aftershocks in New Zealand history.

IMPACT

The consequences of the twin earthquakes' impact on Canterbury's natural and built environments were evident as location and intensity coupled with population density and the nature of the built environment determined the lethal outcomes of each earthquake. For instance, the 2010 earthquake, occurring in a rural area, recorded 100 injuries but no related deaths. In contrast, the 2011 earthquake toppled buildings and registered 185 deaths. The other major impact was liquefaction, which dispersed laterally and adversely affected natural and built environments, water and air quality, ecology, and biodiversity and caused land contamination, flooding, and earth movement (Potter et al. 2015).

As noted, the Canterbury plains are formed by the deposition of alluvium carried by the rivers flowing to the east coast. This soil's top 66–82 feet (20–25 m) layer is geologically young and weakly consolidated, and it mostly contains a mixture of sand, silt, and peat. In the event of the ground quaking, the alluvial sediment foundation renders the area vulnerable to liquefaction, whereby soils rapidly transform into a liquefied state. Liquefaction can render the ground unsteady, causing buildings to sink or collapse. This process occurs more readily if the soil is loose, as happened in the Christchurch area. There, liquefaction ejected large volumes of silt onto the surface, triggering massive alteration in land levels in parts of the Canterbury plains (CERC 2012).

An offshoot of liquefaction is lateral spreading, which results in large amounts of ground displacement ranging from a few inches to several feet (CERC 2012). Lateral spreading has a cascading effect: it destabilizes riverbanks, it increases erosion by waterways, and it narrows river channels. In this case, liquefaction and lateral spreading also damaged underground infrastructure. Damaged pipes heightened the threat of untreated sewage waste mixing with waterways, further jeopardizing water quality. Fortunately, the quality of ground water remained largely unaffected, but as pipes and wells were damaged, a threat of contamination loomed, prompting the city council to chlorinate the water until 2011.

On the surface, silt washed into waterways, increasing the concentration of suspended particles in water and so affecting its quality. Moreover, surface water had lower dissolved oxygen, which affected the organisms living in the water, especially sensitive fish species. This became evident from the high level of bacteria in lower reaches, according to Christchurch's City Council. In some areas, owing to the altered ground conditions, coastal forests were disturbed. Furthermore, liquefaction mounds covered the estuarine region of Avon Heathcote Estuary. In other estuarine areas, reduced organic material resulting from sedimentation impacted spawning areas of certain aquatic species. Additionally, sediment discharge in rivers, liquefaction of riverbeds, and lateral spreading altered the rivers' capacities. Elsewhere, subsidence of land, especially closer to the coasts, increased inundation and erosion of riverbeds by seawater. This land subsidence also increased the risk of flooding and surges in water table levels. Moreover, the earthquake impacted slope integrity in hilly areas, triggering mass movements, such as rock falls, rock rolls, cliff collapse, landslides, and slumps.

The 2010–2011 earthquakes severely damaged several unreinforced masonry buildings, mostly because of ground quaking and deformation by way of liquefaction, uplift, subsidence, and tilting. Among those affected buildings were schools, healthcare facilities, central business districts, historical landmarks, and heritage buildings.

In an earthquake event, factors such as location, intensity, building materials, and time of day determine the number of fatalities. As noted, no death occurred in the 2010 earthquake, but the event in 2011 killed 185 people, the majority occurring in the collapse of two multi-story office buildings. Also, the 2011 earthquake alone injured about 7,000 people, three times more than the 2010 earthquake did.

The earthquakes' impact on the communities continues to shape everyday life, in terms of not only physical effects but also the social environment that supports the recovery process. After the 2010–2011 events, the quality of life was reported to be lower in Christchurch as compared to neighboring Selwyn and Waimakariri. As time passed, secondary stresses related to loss of facilities, ongoing aftershocks, insurance issues, relocation, and decision-making about damage, repairs, and relocation increased. Stress in workplaces also increased. On the other hand, though, positive impacts on the mental and emotional well-being of people manifested as an increase in pride, a heightened sense of community, resilience, ability to cope with tough situations, and a renewed appreciation of life. For instance, spontaneous volunteering, making new connections, and forging bonds with neighbors to help one another through issues unleashed by the earthquake were evident.

RESPONSE

In the earthquake's aftermath, Christchurch witnessed typical earthquake havoc: damaged and collapsed buildings, power outages, and broken communication lines. Still, several relief and rescue teams worked together to bring the situation under control.

In 2010, the Territorial Authorities (TA), responsible for local emergencies and disaster management, set up Emergency Operations Centers and declared local emergency for three cities: Christchurch City, Waimakariri, and cities in the Selwyn Districts. But no regional emergency was declared. A couple of days later, the Canterbury regional authority teamed up with the TA's recovery managers, infrastructure providers, and social welfare agencies to help with impact assessment and recovery planning. Politically, Prime Minister John Key established a new Cabinet (Senate) position along with an ad hoc Cabinet Committee (Senate

Review Committee) on the Canterbury Earthquake Recovery. This new minister (Senator) removed bureaucracy to expedite recovery and ensure better coordination between central and local governments.

The 2011 earthquake response evolved out of the 2010 experience. Moreover, the 2011 earthquake's impact was so severe that it almost immediately mobilized people and organizations within the Canterbury region, across New Zealand, and throughout the world. It was the first ever state of national emergency declared by the Ministry of Civil Defense. It aimed at ensuring maximum cooperation and coordination between local and central resources and demonstrating the government's commitment to helping the Canterbury people. Help began pouring in immediately. Two companies of soldiers joined two Urban Search and Rescue (USAR) teams from Auckland and New Plymouth and worked together with the local USAR team, fire officers, paramedics, police officers, and ambulance staff. Then, a 73-member search and rescue team arrived from New South Wales, Australia, as did international rescue teams from Japan, Taiwan, Singapore, the United Kingdom, and the United States. A week later, about 330 new Australian police personnel were inducted, adding to the 1,200 New Zealand police on patrol and active duty (Lambert and Mark-Shadbolt 2012).

As a long-term response to the twin Canterbury earthquakes, the New Zealand government updated its national and local risk and emergency management strategies. Specifically, it introduced a revised framework to ensure a more "integrated, comprehensive and decentralized approach to risk management and societal resilience" (Mamula-Seadon and McLean 2015, 83).

Post-earthquake reports pointed at relief-related challenges faced by New Zealand's indigenous Maori population and leadership (Lambert and Mark-Shadbolt 2011; Potter et al. 2015). This community received disproportional relief in terms of reduced financial resources, less access to necessities, proper sanitation, and less support from responders. Several nominated leaders themselves lived in damaged homes and suffered stress from the earthquakes' aftershocks. Further, the governmental and nongovernmental leadership often received preferential treatment over the Maori leadership.

RECOVERY

After the twin earthquakes, Environment Canterbury, under the name Natural Environment Recovery Program, spearheaded the recovery program and strategy for the Greater Christchurch region. Through this program, 17 projects aimed at managing and repairing the ecological damage and improving the ecosystems' sustainability are being undertaken. The New Zealand Treasury estimated the capital cost of the Canterbury earthquakes to be around NZ $40 billion, about 20 percent of country's gross domestic product (GDP). After the 2010 event, the estimated repair and rebuilding cost was gauged to be around NZ $5 billion. Meanwhile, financial markets remained largely unaffected. Subsequent to the aftershock, the cost of repairs was considerably higher, and in the aftermath, the New Zealand dollar's value dropped immediately, though it recovered soon after. Overall, the earthquake's long-term impact was not pronounced on either the regional or the national economy.

CONCLUSION

Short- and long-term relief operations included establishment of the Canterbury Earthquake Temporary Accommodation (CETA) service with the help of the Building and Housing Group (Ministry of Business, Innovation and Employment) and the Ministry of Social Development of the Government of New Zealand. Services included community connectors, earthquake counselors, earthquake support coordination services, residential advisory and Whanau Ora navigators (which address community-based Maori collaboration), financial assistance, and matching and placement services for people whose homes were being repaired or rebuilt after the twin earthquakes. Finally, the city of Christchurch was divided into zones depending on the severity of the impact and for residents and businesses to be eligible for assistance. This service was active from 2010 until 2018, and later support lines were established for assistance.

Avantika Ramekar

Further Reading
CERC (Canterbury Earthquake Royal Commission). 2012. *Summary and Recommendations in Volumes 1–3. Seismicity, Soils and the Seismic Design of Buildings.* Wellington, New Zealand: CERC.
Downes, G., and M. Yetton. 2012. Pre-2010 Historical Seismicity Near Christchurch, New Zealand: The 1869 Mw4.7-4.9 Christchurch and 1870 Mw5.6-5.8 Lake Ellesmere Earthquakes. *New Zealand Journal of Geology and Geophysics* 55(3): 199–205.
Lambert, S., and M. Mark-Shadbolt. 2011. Maori Experiences and Expressions of Leadership through the Christchurch Otautahi Earthquakes. *International Indigenous Development Research Conference*, 242–247. Auckland, New Zealand.
Mamula-Seadon, L., and I. McLean. 2015. Response and Early Recovery Following 4 September 2010 and 22 February 2011 Canterbury Earthquakes: Societal Resilience and the Role of Governance. *International Journal of Disaster Risk Reduction* 14: 82–95.
Potter, S. H., J. S. Becker, D. M. Johnston, and K. P. Rossiter. 2015. An Overview of the Impacts of the 2010–2011 Canterbury Earthquakes. *International Journal of Disaster Risk Reduction* 14: 6–14.

Colombia Floods, 2010–2011

The 2010–2011 cold phase El Niño-Southern Oscillation (ENSO) event is considered one of the worst disasters in Colombia's history, based on the area and number of people affected, the impacts on the population, and the total costs of economic damages. Months of rain, starting around September 2010 and petering off by February 2011, culminated in December 2010. The elevated precipitation resulted in significant flooding and landslides throughout over half of the country.

Hundreds of people lost their lives in the event, and the damages to buildings, infrastructure, agriculture, and other sectors topped are estimated to have topped $5 billion, making it the costliest disaster in Colombia's history (World Bank 2012). Recovery efforts were largely financed by loans and donated resources from international aid organizations and other countries.

ENVIRONMENTAL CONTEXT

The flooding in Colombia was directly caused by the cold phase of the El Niño-Southern Oscillation phenomenon (ENSO), called La Niña. El Niño is an episodic atmospheric and oceanic phenomenon that occurs every three to seven years in the equatorial Pacific Ocean near Christmastime. During typical years, strong westward winds blow warm oceanic surface water off the equatorial north-west coast of South America toward Southeast Asia. Cool water from the ocean's depths is then drawn upward toward the surface. This pattern is broken during El Niño (Spanish for "the boy") years, when global pressure systems shift and cause winds to blow eastward across the equatorial Pacific near Christmastime. As a result, warm surface waters converge off the coast of the northwestern portion of South America. In Colombia, El Niño often causes abnormally dry conditions (Hoyos et al. 2013). Central America and the western coast of the United States, in contrast, receive unusually high amounts of precipitation.

La Niña (Spanish for "the girl") is a variation of the ENSO cycle. This phenomenon occurs less predictably and causes the waters of coastal South America to become unusually cool rather than warm as global pressure systems shift the direction and strength of winds across the Pacific. La Niña has the opposite effect on precipitation patterns than El Niño, causing conditions in Colombia to have become unusually wet in the last 40 years. During the 2010–2011 La Niña event in Colombia, elevated precipitation levels began roughly in September, with the heaviest amounts of rain falling in December 2010. The increased precipitation caused the inundation of rivers in the western, central, and northern regions of the country. Roughly 60 percent of the country was affected, including 28 out of 32 Colombian departments (IFRC 2011).

DAMAGES

The economic impacts of the event in Colombia were tremendous. Estimates place economic damage as exceeding $5.1 billion, or roughly 2 percent of the country's Gross Domestic Product (GDP). Only roughly 7 percent of the flood victims possessed insurance (World Bank 2012). Some of the worst areas of flooding during the event occurred in the Magdalena and lower Sinu River basins, where over 3.7 million acres (1.5 million hectares) of land were inundated beyond typical rainy season flooding (Hoyos et al. 2013). In addition to the extreme flooding of rivers, landslides from the saturated and destabilized soils posed a significant threat to human populations by causing property damage and many deaths. Windstorms and lightning associated with the extreme weather event were also reported to have caused major damages to buildings and infrastructure. The departments that sustained the most damage were Bolivar, Magdalena, Sucre, Cordoba, and Choco.

Over 600 cities, towns, and municipalities were affected by the extreme weather event, with severe damages occurring to residential and agricultural properties (World Bank 2012). More than 16,000 homes were destroyed in the flooding and associated landslides, while over 550,000 homes were damaged (IFRC 2011). Poor infrastructural planning by the Colombian government exacerbated the flooding. Extensive networks of dykes and levees exist on many of the rivers in Colombia. Their purpose is to direct the flow of water for agricultural and sanitation uses as well as to serve as protection for populations and agricultural lands against annual small-scale flooding during the rainy season. During the events, drainage systems overflowed, and several of the dykes and levees broke, allowing the floodwaters to travel into towns and agricultural areas. Poor zoning regulations meant that homes and apartment buildings were constructed in flood plains in many locations, and they were particularly susceptible to inundation. Damage further occurred in several areas as looters broke into properties to steal items after residents evacuated from their homes and apartments (Otis 2010).

The death toll from these events was high. In total, 486 reported deaths are attributed to the 2010–2011 events. An additional 595 people suffered injuries and 44 people went missing (IFRC 2011). During one particularly tragic landslide event in December 2010, mud engulfed 30 homes in the city of Medellin. As a result, dozens of people were killed. By the time rains diminished, approximately 4 million people, or about 9 percent of Colombia's total population, were rendered homeless by the damages from flooding and landslides (Hoyos et al. 2013). People already displaced by violence from Colombia's ongoing internal conflict were reported to have been particularly vulnerable to the effects of the event (IFRC 2011).

IMPACTS TO PUBLIC INFRASTRUCTURE

Infrastructure and public services were also extensively damaged. Over $1 million in damage is estimated to have accrued to transportation networks that include roads, railways, airports, and river and maritime ports (World Bank 2012). Many roads were washed out or flooded during episodes of heavy rainfall, hindering the evacuation of residents from their homes and the movement of disaster response workers. More than 1,067 miles (1,600 km) of roadways in Colombia were affected and at least 90 bridges were destroyed (World Bank 2012). Additionally, over 400 health facilities as well as 3,000 schools were affected. Several schools were closed temporarily due to damages (IFRC 2011).

IMPACTS TO AGRICULTURE

Farmland abuts many of the major rivers in Colombia and thus was heavily affected as dykes and levees were topped or broken. Over 3,000 square miles (7,770 square km) of farmland were inundated by floodwaters (BBC 2010). One of the agricultural industries hardest hit by the events was the coffee industry. Over 220,000 coffee producers experienced crop damage, reducing the output of coffee producers in Colombia by an estimated 1 million sacks in 2010 (World Bank 2011). The weather conditions also allowed crop diseases, such as coffee rust, root wilt disease in coconut palms, and moniliasis in cacao, to rapidly spread. Overall, the coffee industry is estimated to have lost over $285 million from the destruction of crops (World Bank 2011).

Livestock were also lost. The beef and dairy industries particularly suffered, as an estimated 130,000 cattle died from the flooding and landslides. Many that survived suffered nutrient deficiencies because adequate forage could not be obtained for them. Much agricultural infrastructure was also damaged, as landslides and floods swept away greenhouses, barns, stables, and other buildings (World Bank 2012). The closure of transportation routes further affected the agricultural sector as farmers could not reach much of their livestock to move them to safety. The washed-out and flooded roads furthermore hindered the ability of farmers to transport their products to market, leading to greater financial losses.

HEALTH AFFECTS

As was the case in Colombia, El Niño events frequently involve human infectious disease transmission. Excessive rainfall can particularly increase the risk of disease transmission, as standing water makes for an ideal habitat for breeding mosquitoes and damage to flooded sanitation systems often results in increased exposure to pathogens. Outbreaks of food-borne illnesses, leptospirosis, diarrheal disease, malaria, and respiratory and skin infections thus accompanied Colombia's flooding events. Typically vulnerable populations, including pregnant women, children, adolescents, the elderly, and people with disabilities, were most affected by the outbreaks (PAHO 2011). Flooding and landslide damages to hospitals, clinics, and other health facilities aggravated the outbreaks, as many closed temporarily, and the infected people were unable to access care. This issue was compounded by damage to transportation systems, which furthermore impeded access to health facilities that remained operational during the events.

GOVERNMENT RESPONSE

Colombia's government declared the event an economic, social, and ecological national emergency on December 7, 2010. On the same day, the Colombian President Juan Manuel Santos made direct appeals to the international community for providing assistance to flood and landslide victims, claiming that the capacity of the government was overwhelmed. The government also unveiled a new framework for flood relief called "Colombia Humanitaria" intended to efficiently delegate money and personnel to address aid efforts (IFRC 2011). Colombia's then-president Santos petitioned the World Bank for $150 million to aid victims and recovery (World Bank 2012). The United Nations allocated nearly $6 million to be used for the communities that were most severely affected by the flooding and mudslides events. International assistance furthermore came from a $1.3 million donation from the United States; a donation of equipment, foodstuffs, and other supplies from Israel; a donation of food and supplies from Ecuador; and a donation of medical and relief supplies from Russia.

To acquire money within the country for relief and reconstruction efforts, the Colombian government lowered the threshold of the net wealth tax to raise approximately $1.6 billion. The government also sold shares from the state oil company. The Colombian government furthermore utilized at least $25 million in money and properties seized by police from the drug trade to provide relief and shelter to those affected by the flooding. After the floods and mudslides subsided for the year of 2010, the Colombian government set aside approximately $850 million for public works to try and mitigate future flooding events (*The Economist* 2011). Although there were reported positive outcomes of "Colombia Humanitaria" for disaster response and recovery, the framework rolled out during the disaster was widely criticized as being slow to respond to the thousands of victims from the events.

CONCLUSION

The 2010–2011 flooding and landslide events associated with sustained and heavy precipitation deeply impacted populations throughout Colombia. In addition to the losses accrued from homes and infrastructure, the agricultural industry suffered devastating losses of crops and livestock. The flooding further intensified as levees broke and drainage systems overflowed. Health of populations was further impacted as sanitation systems malfunctioned and standing water supported an increase in mosquito populations. The disaster further served as the platform to introduce the Colombia Humanitaria framework for disaster response and recovery in Colombia.

Audrey Joslin and Rose G. Micke

Further Reading

BBC. 2010. Colombia Flooding Continues with Thousands Homeless. December 16, 2010. https://www.bbc.com/news/world-latin-america-12006568, accessed February 22, 2019.

CBS. 2010. Colombia Landslide Leaves 100 Missing, 20 Dead. December 7, 2010. https://www.cbsnews.com/news/colombia-landslide-leaves-100-missing-20-dead/, accessed February 24, 2019.

The Economist. 2011. The Americas: That Damned Nina; Colombia's Floods. December 10, 2011. *The Economist* 402: 42. https://www.economist.com/the-americas/2011/12/10/that-damned-nina, accessed February 22, 2019.

Hoyos, N., J. Escobar, J. C. Restrepo, A. M. Arango, and J. C. Ortiz. 2013. Impact of the 2010–2011 La Niña Phenomenon in Colombia, South America: The Human Toll of an Extreme Weather Event. *Applied Geography* 39: 16–35.

IFRC (International Federation of Red Cross and Red Crescent Societies). 2011. Emergency Appeal Operation Update. http://www.ifrc.org/docs/appeals/10/MDRCO008eu3.pdf, accessed February 24, 2019.

Otis, J. 2010. Heavy Rains and Flooding: Colombia's Katrina? *GlobalPost*. https://www.pri.org/stories/2010-12-02/heavy-rains-and-flooding-colombias-katrina, accessed February 24, 2019.

PAHO (Pan American Health Organization). 2011. La Niña Ravages Colombia. https://reliefweb.int/report/bolivia/la-ni%C3%B1a-ravages-colombia, accessed February 24, 2019.

World Bank. 2012. Analysis of Disaster Risk Management in Colombia: A Contribution to the Creation of Public Policies. Bogota, Colombia/Washington, DC: World Bank.

Colorado Flood, United States, 2013

People in the Denver, Colorado metropolitan area, and beyond watched in disbelief as an urban area along the Rocky Mountains known as the Front Range experienced a deluge of record rain starting September 9, 2013, and lasting about a week. The region was experiencing drought when prolonged, heavy rainfalls caused massive flooding that destroyed homes and businesses, washed out infrastructure, displaced people, and caused fatalities. Some parts of the region received more rainfall during this week than in an average year, breaking many precipitation records. Water levels rose over the top of riverbanks causing severe flooding, flash floods, and landslides. Over 18,000 people were forced to leave their homes because of the flooding. The heavy rainfall in the foothills and mountains resulted in over 1,100 landslides and led to flood damage that destroyed at least 1,882 structures and closed nearly 485 miles (781 km) of roads. The flooding ultimately claimed 10 lives and caused over $2–3 billion of damage. The level of destruction was the worst the region had experienced in over 20 years (Draper 2013/2016).

EVENT SUMMARY

The Front Range runs north-south along the Rocky Mountains from southern Wyoming to southern Colorado and includes Denver, Boulder, Fort Collins, and other smaller towns. Extreme heat dominated the Front Range region prior to the flood event. Colorado had been exceptionally warm and dry, breaking or tying three high temperature records in Denver from September 2 to 8. As a slow-moving cold front collided with tropical moisture, temperatures dropped and intense rains occurred in areas across the Front Range region, most forcefully in Boulder County. Wildfires in previous years had also increased flood risk, with more runoff from burned areas that could not absorb as much rain. Rivers and creeks overflowed their banks, dams overtopped, basements flooded, and roads were washed out. By September 12, 2013, conditions shifted from the extremely hot, dry conditions to flooding that stretched nearly 150 miles (225 km) from the central part of the Front Range in the south to the Wyoming border in the north (Scott 2013).

Rainfall continued through September 12, increasing the amount of rainfall in the heaviest hit areas to over 15 inches (380 mm). Rainfall is characterized by amount, intensity, and distribution in time. The rainfall amount is the depth of

A small mountain town in Boulder County was cut off due to massive flooding along the Colorado Front Range, 2013. (FEMA)

water in inch received during an event. Intensity is expressed in inch of water depth per hour (inch/hour). Heavy rainfall resulted in the total amount exceeding 8 inches (200 mm) in many locations and over 2 inches (50 mm) occurring up into the mountains, much of the water flowing down onto the plains through the canyons. In Boulder County, rainfall rates were sustained between 1 and 2 inches per hour (25 and 50 mm per hour), spreading northwest into the mountains toward Estes Park, up the Big Thompson Canyon (Gochis et al. 2015).

DEATH AND LOSS

Initially eight people were reported dead; the final number was revised to 10. Among the fatalities that occurred, seven were attributed to drowning and one to a mudslide. More than 18,000 people evacuated their homes, approximately 1,882 structures were destroyed, and thousands more were damaged (Colorado Division of Homeland Security and Emergency Management 2015). The flooding devastated 968 businesses, shutting down many for days, weeks and months, and damaged over 485 miles (728 km) of highways and roads, 150 miles (225 km) of railroad tracks, 54 bridges, and 27 dams. Governor John Hickenlooper declared an emergency disaster in 14 counties (Colorado Division of Homeland Security and Emergency Management 2015). The two hardest hit counties were Boulder and Larimer Counties. Many residents were unable to return to communities with widespread destruction, particularly those in low-income housing. As one example, hundreds of mobile home park residents in Evans, a city of 20,000 south of Greeley, were left homeless by the floods (Draper 2013/2016). In addition to urban infrastructure, flooding often has a substantial effect on the agricultural sector in terms of crop loss and damage to infrastructure. For example, irrigation ditches, common in eastern Colorado, are susceptible to damage from flooding, which occurred in the 2013 flood event. The estimated value of loss of agricultural production, not including infrastructure, was estimated between $3.4 and $5.5 million (Dalsted et al. 2019).

SECONDARY HAZARD

Any flood event can cause dispersion of sewage or chemical runoff from agriculture or urban applications. The 2013 Colorado Floods interfaced with two somewhat unique economic sectors. The northeastern corner of Colorado has 51,000 oil and gas wells. Of these, 1900 were shut down during the flooding and 15 releases of oil were identified. Testing by the Colorado Oil and Gas Conservation Commission (COGCC) indicated no impacts on surrounding rivers or streams. A 2015 statewide rule placed oil and gas extraction in floodplains under stricter guidance. While the wells seemingly did not cause significant impacts, many environmental groups voiced concern.

In Colorado, mining continues to evoke a sense of nostalgia along with environmental concerns that arise from a legacy of contamination. Mining flourished in northwestern Boulder County beginning in the 1860s and continued for many years before closing completely in the early 1990s. The region was typically mined for gold and silver, which contaminated the environment with heavy metals and other hazardous chemicals. The Captain Jack Mill at the headwaters of Lefthand Creek in Boulder County was placed on the Environmental Protection Agency's (EPA) Superfund National Priority List for cleanup in 2003. The EPA reported that remediation at the site performed well during the 2013 flood (EPA 2017). The 2013 flood affected the region of Lefthand Creek Watershed and ultimately could have had drastic consequences by disturbing and distributing contamination. Not all historical mines are well documented or necessarily well contained, creating a potential environmental hazard from flooding.

WARNING SYSTEM

More than a week prior to the flood event, the Climate Prediction Center under the National Oceanic and Atmospheric Administration (NOAA) indicated the weather pattern for Colorado would shift from dry to a potential of heavy rain. Then three days before the flooding, NOAA's Weather Prediction Center (WPC) forecasted the area was at increased risk for producing rainfall that could exceed flash flood levels. The conditions were suitable for flash flooding, but these centers could not predict localized details.

Weather Forecast Offices (WFOs) consist of 122 National Weather Service (NWS) field offices (NWS 2013). Only WFO Pueblo issued a flash flood watch, indicating conditions were possible for flooding and so people should be prepared, before the onset of flooding on the late afternoon and evening of September 11,

2013. The watch extended into the early morning hours of September 12. WFO Boulder did not have a formal team defined for hydrologic services as recommended from previous service assessments, which significantly limited the office's ability to respond to such a widespread and multifaceted flood event. As the flooding unfolded late on September 11, forecasters focused on issuing flash flood warnings, statements, and social media postings.

From September 11 to 15, 2013, WFOs Pueblo and Boulder collectively issued 78 flash flood warnings, meaning flooding was occurring and people should take protective action. The National Weather Service warned that any given storm cell was capable of producing multi-inch amounts in only a few hours. On September 12 in the morning, the NWS issued an alert, "(A) major flooding/flash-flooding event (is) underway at this time with biblical rainfall amounts reported in many areas," and flash flood warnings were issued for Adams, Denver, Larimer, Boulder, Jefferson, Arapahoe, Douglas, El Paso, Lincoln, Cheyenne, and Kit Carson counties. Although predicting flash flooding is challenging, the combination of warning messages and close collaboration with county EMs in the Pueblo and Boulder County Warning Areas (CWA) helped limit the number of fatalities (NWS 2013).

HISTORY OF FLOODING

How did this storm compare to other major rain/flood events in Colorado's history? In fact, Colorado has a long history of flash flooding. Considered the worst natural disaster in Colorado's recorded history, on July 31, 1976, between 12 and 14 inches (305 and 356 mm) of rain fell over a four-hour period in the mountains, producing a flash flood that killed143 people in the Big Thompson Canyon. While being the deadliest in Colorado history, this flood is representative of the flood risk in the canyons along the mountains. As an example, the town of Boulder is one of Colorado's most flood-risk communities. The city is situated at the mouth of a canyon, and the Boulder Creek flows through the middle of the town. Boulder experienced floods in 1894, 1896, 1906, 1909, 1916, 1921, 1938, and 1969. The most extensive of these occurred in 1894 and 1969 during the springtime corresponding to snowmelt runoff. The entire Front Range comprises canyon rivers that flow out of the mountains and onto the plains of eastern Colorado, representing a significant flood risk to those downstream. Many of the Front Range towns and cities are situated at the base of these canyons, and so flooding is a risk.

While the 2013 Colorado Flood was less deadly than the 1976 Big Thompson Flood, the 2013 Colorado Flood was more widespread than previous events. Another difference between the flood events is the 2013 Colorado Flood occurred in the fall, which is generally one of the drier months of the year. April and May are typically the wettest months followed by relatively lower precipitation in the summer months. However, precipitation is still common in this region in the summer months because of the North American Monsoon occurring between June and August. During this time, winds carry moisture from the Gulf of Mexico northward over the American Southwest. Monsoon storms can bring strong afternoon and evening thunderstorms to the state. September is typically drier with temperatures lower than summer highs, making the timing of the 2013 event somewhat unusual.

COMMUNITY RECOVERY

Many of the communities are still recovering from the 2013 floods, particularly those that were most affected. Many people were permanently displaced due to limited affordable housing availability, intervening job opportunities, or a desire to move from the affected areas. Vulnerable populations, like those renting in the former mobile home parks, have more limited opportunities during the recovery phase. Local-level decisions require infrastructure design, planning, and implementation along with the modification of former policies and/or new administrative approaches to serve community needs (Clavin et al 2017). Mitigation efforts (actions to reduce future impacts) implemented during recovery can reduce future flood impacts. These may include if, how, and where to rebuild; changes in land use and zoning practices; and public outreach programs.

Tonya Farrow-Chestnut and Deborah Thomas

Further Reading

Clavin, C. T., Z. E. Petropoulos, N. Gupta, and C. K. Tokita. 2017. *Case Studies of Community Resilience and Disaster Recovery from the 2013 Boulder County Floods.* Washington, DC: US Department of Commerce, National Institute of Standards and Technology.

Colorado Division of Homeland Security and Emergency Management, Department of Public Safety. 2015. After Action Report: State of Colorado 2013 Floods and Black Forest Fire. June 2015. http://hermes.cde.state.co.us/drupal/islandora/object/co%3A20487, accessed March 25, 2019.

Dalsted, N. L., J. Deering, R. Hill, and M. Sullins. 2019. Flood Damage Losses to Agricultural Crops in Colorado. http://www.wr.colostate.edu/ABM/Flood%20Damage%20Report.pdf, accessed March 1, 2019.

Draper, Electa. (2016) [2013]. Road Home Has Been Bumpy, Impassable for Flooded Mobile-home Residents. *The Denver Post.* November 26, 2013; updated April 19, 2016. https://www.denverpost.com/2013/11/26/road-home-has-been-bumpy-impassable-for-flooded-mobile-home-residents/, accessed May 16, 2019.

EPA (Environmental Protection Agency). 2017. Captain Jack Mill, Ward, CO: Cleanup Activities. https://cumulis.epa.gov/supercpad/SiteProfiles/index.cfm?fuseaction=second.Cleanup&id=0800892#bkground, accessed May 1, 2019.

Gochis, D., R. Schumacher, K. Friedrich, N. Doesken, M. Kelsch, J. Sun, and S. Matrosov. 2015. The Great Colorado Flood of September 2013. *Bulletin of the American Meteorological Society* 96(9): 1461–1487.

NWS (National Weather Service), U.S. Department of Commerce, National Oceanic and Atmospheric Administration. 2013. Service Assessment: The Record Front Range and Eastern Colorado Floods of September 11–17, 2013. https://www.weather.gov/media/publications/assessments/14colorado_floods.pdf, accessed April 2, 2019.

Scott, M. 2013. Historic Rainfall and Floods in Colorado, News and Features. *Climate-Watch Magazine.* https://www.climate.gov/news-features/event-tracker/historic-rainfall-and-floods-colorado, accessed December 1, 2018.

Cyclone Gorky, Bangladesh, 1991

The entire coast of Bangladesh is prone to violent tropical cyclones and their associated storm surges. The country is located in the humid tropics with the Himalayan Mountains lying to the north and the funnel-shaped coast touching the Bay of Bengal in the south. On average, every year about 12–13 tropical depressions develop in the Bay of Bengal, of which five attain cyclonic strength (i.e., wind speeds 74 mph [119 km/h] or over). Not all of the cyclones strike the Bangladesh coast with deadly force, but many do. The second deadliest cyclone, a category IV storm named Gorky, made landfall on the eastern coast of Bangladesh on the night of April 29, 1991 and affected all three coasts of the country (eastern, central, and southwestern). The cyclone was accompanied by storm surges up to 30 feet (9 m) high, and it battered the coastal area for at least 3–4 hours. Nine of the 19 coastal districts (the second largest administrative unit in Bangladesh) suffered severe losses from the event, but the most devastating losses were a direct outcome of the storm surge (Haque and Blair 1992).

The cyclone killed nearly 140,000 people and left as many as 10 million people homeless. Among the affected districts, loss of life and damage to property were the most severe in Chittagong and Cox's Bazar districts. It was estimated that 40–50 percent of all residents living on unprotected offshore islands died from the cyclone. This percentage ranges between 30 and 40 and between 20 and 30 for islands protected by embankment and mainland areas, respectively. Entire populations of several smaller islands were wiped out. A study reported that 63 percent of the deaths were children under age 10, who represented only 35 percent of the

An aerial view of coastal flooding in the aftermath of Cyclone Gorky, Bangladesh, 1991. (Department of Defense)

pre-cyclone population. Also, 42 percent more women died than men (Chowdhury et al. 1993). However, lack of proper sanitation and lack of safe drinking water supply caused a significant increase in the incidence of water-borne diseases. Nearly 7,000 people died from these and respiratory diseases during the post-disaster period. In addition, the cyclone caused more than 460,000 injuries (Haque and Blair 1992).

According to Government of Bangladesh estimates, a total of 522,000 houses were destroyed and another 431,000 were damaged. Three hundred schools and 655 health centers were either damaged or destroyed. Power, water, and communications lines to the affected areas were cut, and all modes of transportation were severely disrupted. More than 127 miles (190 km) of coastal embankments were destroyed and another 626 miles (940 km) sustained damage. In many affected areas, surface water was salinized and industries in and around the port of Chittagong suffered heavy damage. The agricultural and forestry sectors also sustained serious disruptions. In sum, the property damage caused by Cyclone Gorky was estimated at $2.4 billion (Haque and Blair 1992).

EARLY WARNING AND EVACUATION

The Bangladesh Meteorological Department (BMD), the public authority that prepares all weather forecasts and cyclone warnings, reported development of a tropical depression near the equator in the eastern Indian Ocean that turned into Cyclone Gorky on April 25, 1991, but it issued the cyclone warning only 15 hours before Gorky struck the northern part of the eastern coast of Bangladesh. Although the 15-hour warning was inadequate to evacuate all people from impending danger, after this cyclone, the BMD improved and upgraded its warning system. Now, the BMD issues a 48-hour warning that allows people to evacuate to safe public cyclone shelters and other safer buildings hours before a cyclone makes landfall. Cyclone shelters are generally multi-story buildings, often with open-structure foundations and reinforced concrete pillars. Available research suggests that nearly one-third of the total population of the areas that received cyclone warnings had evacuated prior to the landfall of Cyclone Gorky. This is consistent with the evacuation rate reported for Cyclone Sidr, which hit the southwestern coast in 2007. However, for Cyclone Gorky, just shy of only 3 percent of evacuees took refuge in designated cyclone shelters. The corresponding figure was 33 percent in the case of Cyclone Sidr (Paul et al. 2010).

One of the main reasons for the much lower rate of evacuees using public shelters in the case of Cyclone Gorky as opposed to Cyclone Sidr was the inadequate number of such shelters in 1991. The idea of building public cyclone shelters came after the deadliest cyclone that hit Bangladesh in 1970, also known as Bhola Cyclone. At that time, about 10 million people lived in the vulnerable, low-lying coastal belt and offshore islands of the country. Prior to the landfall of Cyclone Gorky, fewer than 500 cyclone shelters were distributed along the entire coast of Bangladesh; however, this number increased to 3,976 by the time of the Sidr landfall. Inadequate and relatively low utilization of public cyclone shelters was one of the major reasons that so many people died in 1991.

Another reason for the high number of deaths was that the cyclone hit in the dead of night, making it even more difficult for people to look for safe shelters. The high winds made the situation worse. This is because the landfall occurred around high tide, which was already 18 feet (5.5 m) above normal. Thus, walls of water, one after another, continued to pound the coastal areas for three to four hours. Only 3 percent of the houses were strong enough to withstand the onslaught of the tidal waves (Chowdhury et al. 1993). The storm surge submerged more than 107 miles (160 km) of the eastern coast from north of the port city of Chittagong to the tourist city of Cox's Bazar in the south, affecting also several densely populated offshore islands like Moheskhali, Kutubdia, and Sandwip. After making its way inland on April 30, 1991, the cyclone rapidly lost much of its power (Haque and Blair 1992).

RESPONSE

The Bangladesh Government's response to the 1991 cyclone was delayed. At that time, the ruling government expressed the view that extreme events were "acts of God" and no government has control over natural disasters. The government further claimed that nothing could reduce the impact of natural disasters. Perhaps the most outrageous excuse on the part of the sitting government was the suggestion that the 1991 cyclone was caused by failure of industrialized nations to limit emission of greenhouse gases. This was how the government shifted blame away from its inaction immediately before and after the landfall of Cyclone Gorky. Westerners, on the other hand, believed that the impact of the cyclone could have been reduced and many lives saved if the government had been adequately prepared for the event (Dove and Khan 1993).

Ultimately, many cyclone survivors remained marooned without shelter or in overcrowded temporary shelters for days after the storm while continuing rain and rough seas hampered relief operations. In view of the gravity of the situation, the government of Bangladesh ultimately issued an appeal for international assistance. Subsequently, the government and the international community launched a major response to the cyclone disaster. The government allocated $1.4 million for immediate relief and employed domestic military in the relief and rescue operations. In addition, 898 civil and military medical teams were dispatched to the affected areas. Each medical team consisted of two medical officers, three paramedics, and one technician. Relief from other countries was also forthcoming.

A United Nations task force estimated the cost of reconstruction and rehabilitation at $1.78 billion. The U.S. government alone gave almost $28 million to disaster response and recovery efforts, including 9,850 metric tons of wheat. In addition, under "Operation Sea Angel," one of the largest military disaster relief operations ever carried out on foreign soil, the U.S. Department of Defense sent 7,000 members of a naval task force to Bangladesh following Cyclone Gorky's landfall. This force was equipped with 15 ships that were returning to the United States after the Gulf War (August 1990–February 1991) and were subsequently diverted to Bangladesh. After the task force formally ended its operation, it left 500 military personnel, two

C-130 cargo planes, five Blackhawk helicopters, and four small landing craft to help complete relief operations in outlying districts.

The U.S. task forces played a decisive role in transporting relief items from Dhaka to cyclone-affected areas and delivered necessities to the hard-hit coastal areas and offshore islands by boats, helicopters, and amphibious crafts. The members of the task forces also quickly repaired bridges and roads and distributed preventive medical and water purification units among the cyclone victims. Their activities were integrated into the public relief operation. They maximized the civilian population's participation in the operation, coordinating all its activities with the various governmental and nongovernmental organizations (NGOs).

The people of Bangladesh and the government welcomed the task force because the country had no resources to reach many cyclone survivors, and ultimately, those relief efforts were credited with saving 200,000 lives. Other foreign countries such as the United Kingdom, China, India, Pakistan, and Japan also participated in disaster relief efforts. Because of the participation of both domestic and foreign military forces, the relief distribution was fair and effective.

CONCLUSION

Although the wind speed and height of storm surges were similar in both the deadliest and second deadliest cyclones, Gorky caused fewer casualties than the 1970 Bhola cyclone. Two reasons are attributed to this difference. For example, considerable improvement had been made since the Bhola Cyclone to improve the early-warning system and implement other cyclone preparedness and mitigation measures. During Cyclone Gorky, only about 60 percent of coastal residents heard a warning, with substantial variation among the affected areas. However, the 1991 warning systems were directed to the needs of ports and shipping, whereas the residents of the coastal districts also needed a separate warning system. At that time, the system contained up to 12 warning signals, which the coastal residents considered too many and confusing.

Following the Bhola cyclone, the Bangladesh government established a Cyclone Preparedness Program (CPP) in 1972 to execute cyclone warnings in the coastal areas. The CPP volunteers are stationed across coastal districts, and they are responsible for disseminating cyclone warnings using microphones, megaphones, and door-to-door contact. They also assist people in preparing to evacuate prior to cyclone landfall. At the time of the 1991 cyclone, 20,000 CPP volunteers were at work in the coastal districts, and their numbers consistently increased after Cyclone Gorky. Subsequent cyclones revealed that not all households received early warnings and not all people took preparatory actions, reinforcing the necessity of convincing people to take appropriate actions before a cyclone landfall.

As noted, following the Bhola Cyclone, the government of Bangladesh started to build public cyclone shelters in coastal districts, including offshore islands. At the time of Cyclone Gorky, an inadequate number of public shelters were dotted along the coast. Reports suggest that some of those who attempted to respond to the warnings in 1991 could not find public shelter. Notably, casualties were significantly lower in areas where people had made use of existing cyclone shelters (Chowdhury et al. 1993). After the 1991 cyclone, the number of public cyclone shelters significantly increased, yet it is still not sufficient to accommodate all the people at risk.

Although construction of coastal embankments started in Bangladesh in the 1960s to reduce flooding from high lunar tides and storm surges, the length of embankments was greater in 1991 than in 1970. This protective measure held back the initial thrust of the 1991 surge in many areas and thus reduced the fatalities. However, in several places, the embankment collapsed mainly because of poor maintenance. This negligence still exists.

Bimal Kanti Paul

Further Reading

Chowdhury, A.M.R., A. U. Bhuyia, A. Y. Choudhury, and R. Sen. 1993. The Bangladesh Cyclone of 1991: Why So Many People Died. *Disasters* 17(4): 292–304.

Dove, M. R., and M. H. Khan. 1993. Competing Constructions of Calamity: The April 1991 Bangladesh Cyclone. *Population and Environment* 16(5): 445–471.

Haque, C. E., and D. Blair. 1992. Vulnerability to Tropical Cyclones: Evidence from the April 1991 Cyclone in Coastal Bangladesh. *Disasters* 16(3): 217–229.

Paul, B. K., H. Rashid, M. Shahidul, and L. M. Hunt. 2010. Cyclone Evacuation in Bangladesh: Tropical Cyclone Gorky (1991) vs. Sidr (2007). *Environmental Hazards* 9: 89–101.

Cyclone Nargis, Myanmar, 2008

Myanmar, a self-isolated state, is considered by the United Nations (UN) as the "most at risk" country for natural disasters. The country is exposed to a wide range of hazards, including cyclones and associated storm surges, floods, earthquakes, landslides, and drought. Among these hazards, cyclones have historically caused the most destruction in the country. With winds up to 133 mph (200 km/h), Cyclone Nargis swept through the Ayeyarwady (Irrawaddy) delta region in Southern Myanmar on May 2, 2008. This cyclone was the worst natural disaster in the country's recoded history and triggered a sizable storm surge that flooded 25 miles (40 km) inland.

The tropical system that became Cyclone Nargis developed from a low-pressure system in the central area of the Bay of Bengal on April 27, 2008. Initially, it tracked slowly northwestward, and on April 29, it started to move steadily eastward. On May 2, 2008, Nargis rapidly intensified to reach peak winds of at least 105 mph (165 km/h), making it a weak Category IV cyclone. Then it moved ashore in the Ayeyarwady delta and passed near the former capital Yangon (Rangoon), gradually weakening until dissipating near the border of Myanmar and Thailand on May 3, 2008.

DEATH AND DESTRUCTION

On May 14, 2008, the Myanmar military government reported the death toll at 43,318, which was later increased to 84,500. However, independent observers claimed that the cyclone caused the deaths of more than 100,000 people and injured over 19,000. Many people died because the Myanmar government was not prepared for this disaster. It neither issued a cyclone warning nor evacuated a single person from the potential affected areas. In addition to the deaths and injuries, Cyclone Nargis displaced 800,000 people and severely damaged critical infrastructure, including electricity, communication and transportation networks, health facilities (75 percent destroyed), and schools (50–60 percent destroyed) in the affected areas. In all, around 2.5 million people were affected by the cyclone (Asia-Pacific Centre 2008; Leggett 2011).

Although the Myanmar government declared Yangon and the Ayeyarwady delta region disaster areas on May 6, 2008, the damage was most severe in the country's most populous delta region. The Irrawaddy delta is agriculturally most important for Myanmar. It is the main rice-growing area of the country, which makes Myanmar one of the leading exporters of the crop in the world. Unfortunately, Cyclone Nargis made landfall during a critical harvest period and thus caused significant damage to the crop. The UN's Food and Agricultural Organization (FAO) estimated that Nargis impacted 65 percent of the country's rice crop. All damages caused by the cyclone was estimated at $10 billion (UN 2008).

RESPONSE

Because of the massive scale of the humanitarian catastrophe, lack of resources and experience, and Myanmar's initial reluctance to allow international participants in relief efforts, its military government failed to adequately response to this deadly event. However, immediately after Cyclone Nargis, the government established an Emergency Committee, headed by the Prime Minister, and deployed armed forces and police units to the cyclone-affected areas to conduct rescue and cleanup operations and organize relief distribution. The government itself pledged $5 million for relief. On May 9, 2008, the UN made a flash appeal for $187.3 million for relief and rebuilding efforts in cyclone-affected areas. As the updated information was available, the flash appeal was revised to $481 million. Unfortunately, during the emergency phase of the relief operation, only 66 percent of that amount was raised (UN 2008).

The most serious problem with the Cyclone Nargis relief operation was the inflexible restriction of the Myanmar military government to give access to international agencies and civil society groups to provide emergency assistance to victims. Since the 1960s, the government has endorsed a self-reliance policy and thus avoided accepting aid from abroad, even if doing so caused more hardship to its people. This policy was a major factor behind the initial refusal of international assistance. Moreover, the policy was more strongly reinforced in 1988 when the international community enforced sanctions and trade embargos against Myanmar. These were aimed at forcing the military regime out of political power and bringing in a more democratic government. However, realizing the devastating consequences of Cyclone Nargis, the international community suspended the embargo temporarily in an attempt to provide the much needed emergency assistance to the survivors (Crisis Group 2008).

Before the storm, several international nongovernmental organizations (NGOs) such as Save the Children and Medecins sans Frontieres, or Doctors Without

Borders, had already been present in Myanmar, but because of restrictions imposed by the military government, these organizations were not able to pursue full relief operations. Other NGOs, UN agencies, and many foreign countries offered emergency assistance, but the Myanmar government was slow to issue visas and travel permits for workers and personnel of these organizations. Rather, the government insisted on distributing the relief assistance itself with as little help from external sources as possible (Asia-Pacific Centre 2008).

Responding to the grave situation, France, the United Kingdom, and the United States sent ships full of emergency supplies to Myanmar. Fearing that the humanitarian operation could be used as a pretext to overthrow the regime, the military government prevented these ships from entering the country's territory. Consequently, these ships anchored in international waters, awaiting permission to unload the much needed relief goods.

Several European countries considered the Myanmar response to Cyclone Nargis as a crime against humanity. This led the French government, with support from the United Kingdom and the United States governments, to propose that the UN Security Council on May 7, 2008, should invoke the "Responsibility to Protect" (R2P or RtoP) protocol to authorize the delivery of aid without the approval of the Myanmar government. The R2P is a global political commitment that was endorsed by all UN member states to prevent genocide, war crimes, ethnic cleansing, and crimes against humanity. However, Chinese and Russian governments rejected the proposal, arguing that the doctrine did not apply to natural disasters, and consequently the Security Council did not invoke.

Meanwhile, leaders of the Association of South-East Asian Nations (ASEAN) persuaded the Myanmar government to allow delivery of international emergency supplies. Myanmar is a member of this organization, and probably for this reason, it granted access for aid workers and other personnel of other member countries on May 12, 2008. Based on bilateral negotiation, the Myanmar government also allowed aid flights from Australia, China, India, Italy, and the United States. It granted visas for aid workers from Bangladesh, China, India, and Thailand as well as a small number of visas for aid workers of other countries. After intense diplomatic negotiations, on May 16, 2008, various UN agencies (e.g., the World Food Program, the World Health Organization, UN Children's Fund, and the Office for the Coordination of Humanitarian Affairs) and other international relief organizations were allowed to launch relief operations in the country.

Even after granting visas to international aid workers, the Myanmar government restricted internal movement of such workers, established military checkpoints in the worst affected areas, and confined workers to the immediate Yangon area.

RELIEF EFFORTS

The Myanmar military government's relief efforts suffered tremendously from lack of training and experience. For example, the relief operation confronted many barriers to effective aid delivery, including a weak telecommunications system, damaged road networks, lack of appropriate transportation and information, poor fuel supply, and lack of any banking systems (Leggett 2011). Oxfam and other aid organizations suspected that the emergency aid that did arrive in Myanmar was not being properly and adequately distributed. However, the strong local culture of sharing goods, especially in times of crisis, did help in overcoming relief distribution discrepancies.

Several international relief organizations also claimed that military officials diverted food away from the affected communities. It was also reported that military officers hoarded aid for themselves and for selling on the black markets. Officers kept high-energy and nutrition-rich biscuits supplied by foreign donors for themselves and distributed locally produced and low-quality biscuits in their place (Asia-Pacific Centre 2008). There is no doubt that Myanmar's governmental restrictions and intrusiveness, red tape, and corruption hampered relief activities. Still, international relief agencies provided emergency assistance to approximately 100,000 cyclone survivors.

The World Health Organization (WHO) distributed water purification kits and essential drugs in six of the seven most affected areas (Asia-Pacific Centre 2008). Meanwhile, the International Federation of Red Cross and Red Crescent Societies (IFRC) supported the Myanmar Red Cross Society (MRCS) in distributing relief supplies such as clean drinking water, clothing, plastic sheeting, bed nets to prevent malaria, and kitchen items. Additionally, the IFRC donated $189,000 to support the MRCS's relief effort. The IFRC also made an appeal for $5.9 million to support MRCS efforts to provide immediate relief in 30,000 households in terms of shelter, utensils, and basic household items (UN 2008).

Despite many difficulties, the international agencies tried their best to deliver emergency aid in an effective and accountable way. These agencies were able to provide new homes to 12,404 households and build 25 schools and 19 rural health centers. They also provided basic health care to more than 160,000 people and trained over 4000 people in community-based disaster risk management (IFRC 2011). Ultimately, the humanitarian works of the personnel of international relief organizations opened the prospect for future interactions among Myanmar government, society, and the international community.

Among the international communities, ASEAN enjoyed more access. In fact, the government along with representatives from the UN and ASEAN set up the Tripartite Core Group (TCG) as a means to jointly coordinate the relief effort. This group also prepared the three years Post-Nargis Recovery and Preparedness Plan (PONREPP). The plan required $700 million to complete the goals outlined in the plan; however, only $100 million were pledged by foreign countries. One major reason for this was the reluctance of donor governments to provide additional funding to the military government of Myanmar, and so the lack of adequate funding drastically reduced both relief and recovery efforts.

CONCLUSION

One year after the landfall of Cyclone Nargis, at least 500,000 people were still living without proper housing, and 350,000 required food donations from the WFP. This was the case despite the Myanmar military government having ordered an increasing number of cyclone survivors to return to their homes after it had declared the end of the relief phase on May 20, 2008. This approach was consistent with the belief that Myanmar and its citizens would do better if left on their own. The government wanted the cyclone survivors not to be too dependent on relief from international donors. For the same reason, the government was reluctant to issue visas for international relief workers even one year after the cyclone.

Despite the humanitarian crisis in the country, the Myanmar military government put its security and political agenda first, prioritizing the constitutional referendum over the delivery of assistance to the cyclone survivors. As scheduled, the referendum was held on May 10, 2008, in the nonaffected areas, and its sole purpose was to legitimize the military rule in the country. The referendum was postponed for two weeks in the cyclone-affected areas, but it still diverted much needed attention away from disaster relief operations. Moreover, in some affected areas, survivors who had taken shelter in schools and other public buildings were evicted to make room for voting booths. International media widely perceived this act as a means of the government pursuing its own agendas rather than addressing the needs of its citizens.

Given the mass fatalities caused by Cyclone Nargis, the Myanmar government did not pay any attention to the psychological problems of cyclone survivors, rescue workers, and the community as a whole. Many survivors, particularly those who lost one or more family members, were in urgent need of counseling and psychiatric care. Although up to 30 percent of households reported mental health problems as a result of the cyclone, they received no psychiatric care from the government (Leggett 2011). Fortunately, international relief agencies provided psychosocial support to 70,363 people.

Bimal Kanti Paul

Further Reading

Asia-Pacific Centre. 2008. *Cyclone Nargis and the Responsibility to Protect*. Brisbane, Australia: The University of Queensland.

Crisis Group. 2008. Burma/Myanmar after Nargis: Time to Normalise Aid Relations. https://www.crisisgroup.org/asia/south-east-asia/myanmar/burma-myanmar-after-nargis-time-normalise-aid-relations, accessed March 19, 2018.

IFRC (International Federation of Red Cross and Red Crescent Societies). 2011. Myanmar: Cyclone Nargis 2008 Facts and Figures. http://www.ifrc.org/en/news-and-media/news-stories/asia-pacific/myanmar/myanmar-cyclone-nargis-2008-facts-and-figures/, accessed March 19, 2018.

Leggett, S. 2011. Cyclone Nigris, Myanmar, 2008: A Critique of How the Government Managed the Crisis. https://studylib.net/doc/6966720/cyclone-nargis--myanmar--2008--a-critique-of-how, accessed January 25, 2020.

UN (United Nations). 2008. *Myanmar Tropical Cyclone Nargis: Flash Appeal*. New York: UN.

Cyclone Pam, Vanuatu, 2015

Vanuatu, along with many other small island developing states of the Pacific Ocean, is susceptible to extreme weather and geologic events such as tropical cyclones, floods, droughts, and earthquakes. The vulnerability of this maritime region is heightened by its proximity to the South Pacific Convergence Zone (SPCZ). On

Cyclone Pam in the South Pacific Ocean prior to making landfall in Vanuatu on March 13, 2015. (NASA)

average, about three tropical systems a year pass within 267 miles (400 km) of Port Vila, the capital city of Vanuatu. In 2015, Category 5 tropical cyclone, Pam, struck several countries of the South Pacific basin (e.g., Solomon Islands, New Caledonia, Tuvalu, Fiji, and New Zealand), but several major islands of Vanuatu were the hardest hit in this country that spreads north-south over 867 miles (1,300 km) of the south-west Pacific Ocean.

The cyclone entered the northern water mass of Vanuatu on March 12 and left its southern water mass on March 14, 2015. While slowly passing over a chain of about 80 islands of Vanuatu, the cyclone's estimated wind speed was 167 mph (250 km/h) and wind gusts peaked at around 213 mph (320 km/h). With respect to sustained winds, Cyclone Pam was the second most intense tropical cyclone of the South Pacific basin. On March 13, at around 11:00 p.m. local time, Cyclone Pam made landfall on the southeast corner of Efate Island, which is home to the capital city (GoRV 2015; Handmer and Iveson 2017).

FORMATION

Cyclone Pam formed on March 6, 2015, as a tropical disturbance east of the Solomon Islands and tracked slowly in a generally southward direction. Two days later on March 8, the disturbance reached tropical depression intensity, and on the next day, it was upgraded to tropical cyclone status when it was 667 miles (1,000 km) northeast of the Solomon Islands. Over subsequent days, Pam began to move a southward path toward Vanuatu, and on March 13, 2015, it reached Category 5 cyclone status on the Saffir–Simpson scales. After landfall, it began weakening in intensity and continued to move southeast along the western coasts of Erromango Island and Tanna Island. It dissipated on March 22, 2015.

DAMAGE

A combination of Cyclone Pam's high wind speeds, riverine runoff, and coastal flooding caused severe damage in Vanuatu. All sectors of the country's economy suffered significant damage, particularly tourism and agriculture; infrastructure and buildings were also heavily damaged. Tourism, which is the most important sector of Vanuatu, contributes about 60 percent of the GDP. Next, although farming and fishing employ 65 percent of the total labor force, agriculture accounts for slightly over 22 percent of the gross domestic product (GDP). In particular, food crops were badly damaged, and thus, the cyclone compromised the livelihoods of at least 80 percent of Vanuatu's rural, as nearly 96 percent of food crops were wiped out in the affected areas. The cyclone also destroyed vegetable gardens, coconut and banana plantations, and livestock (GoRV 2015).

Approximately 17,000 buildings, about 80 percent of the national housing stock, including schools and clinics, were either damaged or destroyed. Since most of the rural houses are made of palm, coconut and banana leaves, bamboo, and other local materials, the damage incurred in the housing sector was not high in monetary terms. However, the survivors faced difficulties in rebuilding houses because trees, which are the key material for construction, were extensively damaged by the cyclone.

Telecommunications ceased functioning in most cyclone-affected areas of Vanuatu as the cyclone destroyed all but one of the mobile towers, resulting in limited communication with the affected areas. It also severely damaged infrastructure, and water supplies were damaged or contaminated with saltwater, leaving nearly half of the 253,000 population in need of clean drinking water (OCHA 2015; Handmer and Iveson 2017). The storm damaged much of the ecosystem and medicinal plants of Vanuatu, particularly on Tanna Island, and in general caused severe and widespread damage on the larger islands of Tanna, Erromango, and Efate (GoRV 2015).

Approximately 15 people lost their lives either directly or indirectly as a result of Pam, which also injured many others. Eleven of these deaths were in Tafea and Shefa Provinces. An estimated 65,000 people were displaced from their homes, and overall, 188,000 people were affected in 22 of the 83 islands of Vanuatu. The total economic loss caused by tropical Cyclone Pam was estimated to be approximately $449.4 million, equivalent to 64.1 percent of the GDP, which indicates the scale of impact. The negative impact of the disaster on the overall economic conditions in the country would thus be felt for several years to come.

Tropical Cyclone Pam produced varied effects across the different economic and social sectors in Vanuatu. According to the Post-Disaster Needs Assessment (PDNA) reports, the housing sector sustained 32 percent of the total damage costs, followed by the tourism sector (accounting for 20 percent of all damage), the education sector (accounting for 13 percent of all damage), and the transport sector (accounting for 10 percent of total damage). The PDNA report further claimed that private sectors sustained 69 percent of the total damage, while the public sector sustained the remaining 31 percent of the damage. Damage differed not only by economic sectors but also regionally. The total damage was the greatest in Shefa Province, followed by Tafea Province, Penama Province, and Malampa Province, which combined experienced 99 percent of the total damage (GoRV 2015).

Before hitting Vanuatu, Cyclone Pam caused a damaging storm surge in Tuvalu, forcing a state of emergency declaration on March 13, 2015, after 45 percent of the nation's residents were displaced. Torrential rainfall occurred in the southeastern Solomon Islands, particularly in the Santa Cruz Islands such that the cyclone affected at least 3,000 households in the Solomon Islands. Cyclone Pam later brought heavy winds and rough surf to New Zealand's North Island during its weakening stages.

PREPAREDNESS

The death toll caused by Cyclone Pam was considered very low compared to its severity. One principal reason for this was that Vanuatu was well prepared. Because the cyclone moved relatively slowly as it headed toward Vanuatu, its government had enough time to provide early warning and evacuate people from its path. The cyclone made landfall at Port Vila of the islands of Efate at night, but it hit the southern islands of Tanna and Eromango during daylight hours. This helped residents of these two islands to monitor the path of the cyclone and shelter accordingly. Moreover, the initial warnings were issued two days before the cyclone reached Vanuatu, broadcast by radio and also sent through SMS to most of the at-risk population. The warnings included important information about the progression, location and direction of the cyclone, and the threat level. These two media also frequently updated the color code warning. At the time, there were about 66 mobile phone subscriptions per 100 people and 90 percent of Vanuatu was covered. Prior to the cyclone, aid organizations distributed disaster response materials and kits to Community Disaster Committee (CDC) members. The kits included solar power radios, water storage containers, and information on water and sewage hygiene (Handmer and Iveson 2017).

Apart from official early warnings, people also received warnings by observing nature, such as the abundance of a particular bird in the sky. This provided an indication of the approaching cyclone. Accordingly, people prepared for the event by storing food and water, cutting down banana leaves to prevent the trees from falling, and tying down roofs or evacuating to safer buildings. In fact, Vanuatu Red Cross identified safe houses for evacuation.

Almost all inland villages have at least one sturdy building to take shelter in during a cyclone. On March 10, 2015, the Red Cross requested churches to ensure they meet minimum standards for evacuation sites. Additionally, government and nongovernmental organizations (NGOs) evacuated people from the coastal areas, thinking that such areas could be affected by storm surges and/or flooding. In Port Vila, evacuation was targeted to residents of formal settlement as well as its relatively large numbers of residents of informal settlements. Although the government did not issue a code red alert, about 4,000 people were evacuated, one-fourth of whom were from Port Vila. As indicated, many local churches acted as evacuation centers. For instance, the Adventist Development and Relief Agency (ADRA) opened 10 evacuation centers in Port Vila (Handmer and Iveson 2017).

RESPONSE

More than four days after Cyclone Pam, much of the affected population of Vanuatu had yet to be provided with aid. As a result, most survivors in remote areas were forced to scavenge for food, and residents of Moso Island, located just north of Efatte, were forced to drink saltwater. By March 27, 2015, relief goods reached all of the affected islands. This delay was caused by a lack of airstrips and deep water ports. Additionally, the government and aid agencies struggled to provide immediate aid because communication networks were down. For official communication purposes, the government distributed satellite phones across the islands. Clearly, delivering relief aid was a challenge given the devastation.

Despite the slow response to emergency aid distribution, on March 15, the president of Vanuatu declared a state of emergency and called for international support. Many international and national NGOs and other humanitarian agencies from foreign countries responded and participated in Cyclone Pam relief efforts. ADRA responded by distributing shelter kits, water filtration kits, and food packages. In total, ADRA assisted more than 10,000 people in 2,586 households across three islands. On March 24, IsraAid distributed over 40 tons of rice, flour, and water to cyclone survivors in Tongoa in the Shepherds Islands. Furthermore, relief workers were sent from Australia, Britain, France, and New Zealand, and military assets were utilized to distribute government food packages to the affected people.

Two Australian Blackhawk helicopters were used to transport these packages. The Australian naval logistics/supply ship HMAS *Tobruk* operated water purification units to produce 26,417 gallons (100,000 liters) of safe drinking water a day. HMNZS *Canterbury* out of New Zealand was loaded with relief supplies on March 26 in Port Vila and departed for Epi Island. The French frigate *Vendemiaire* distributed relief supplies in Tanna by helicopter. On March 26, the Puma helicopter based on the *Vendemiaire* transported 10 tons of relief supplies to various locations. Also, the World Food Program (WFP) distributed high-energy biscuits (HEB) to approximately 38,000 people in nine islands (OCHA 2015). Moreover, Vanuatu received more than 70 containers of food items from different countries to help support the entire population as part of the relief supplies. However, 50 percent of these items were expired by the time they were about to be distributed among the cyclone survivors.

Before the disaster, many developed countries pledged funds to assist relief efforts in Vanuatu. For example, Australia pledged $3.8 million, the United Kingdom pledged $2.9 million, New Zealand pledged $1.8 million, and the European Union pledged $1.05 million. Other countries such as India and the French overseas territories of New Caledonia and French Polynesia also pledged immediate emergency aid. By the middle of April, donors had contributed more than $13.6 million toward the $29.9 million flash appeal launched by the government of Vanuatu (GoRV) and the UN on March 24 for humanitarian relief.

CONCLUSION

While distribution of emergency relief aid was delayed, recovery was relatively quick in several sectors as well as several affected islands. For example, sufficient repairs to the airport in Port Vila were completed on March 14, 2015, to allow the first flights from Australia carrying aid to arrive. Because Cyclone Pam provided evidence that traditional housing fared better and rural houses are made of local cheap materials, housing recovery did not take much time despite widespread incidence of poverty and shortage of trees. However, major NGOs reported that most of the rebuilt houses were not structurally stronger than the pre-cyclone housing. This meant that the new houses were not capable of withstanding cyclone force winds. Additionally, two years after the cyclone, a few schools and clinic were still under construction.

Agriculture in Tanna recovered within a year and supported cyclone survivors again in terms of food security. In other islands, it took more than one year because of the El Niño that brought drought and dry conditions. At the two years' anniversary, the Director of the National Disaster Management Office speculated that a full recovery from Cyclone Pam would take two more years. The office is still receiving donations toward the cyclone recovery, and at this stage, it prefers cash over material donations.

Bimal Kanti Paul

Further Reading

GoRV (Government of the Republic of Vanuatu). 2015. *Vanuatu Post-Disaster Needs Assessment: Tropical Cyclone Pam, March 2015*. Port Vila: GoRV.

Handmer, J., and H. Iveson. 2017. Cyclone Pam in Vanuatu: Learning from the Low Death Toll. *Australian Journal of Emergency Management* 32(2): 60–65.

OCHA (Office for the Coordination of Humanitarian). 2015. Vanuatu: Tropical Cyclone Pam Situation Report No. 12 (as of 26 March 2015). United Nations OCHA. https://reliefweb.int/sites/reliefweb.int/files/resources/OCHA_VUT_TCPam _Sitrep12_20150326.pdf, accessed December 9, 2018.

Cyclone Phailin, India, 2013

On October 12, 2013, Cyclone Phailin, the strongest storm to hit India in 14 years, made landfall on the coast of the Indian States of Odisha (formally Orissa) and Andhra Pradesh. The cyclone made landfall near Gopalpur in Ganjam district (the second largest administrative unit in India), Odisha. The sustained wind speed of the cyclone was between 133 miles/hour (200 km/hour) and 140 miles/hour (210 km/hour) with a maximum wind speed of about 147 miles/hour (220 km/hour) and was accompanied by a maximum 24-hour cumulative rainfall of about six inches (38 cm) in Cuttack district and eight inches (52 cm) in Bhubaneswar, the capital of Odisha. A maximum storm surge of approximately 11.5 feet (3.5 m) was reported in the Ganjam district of Odisha, and saline water inundation was reported up to 0.7 mile (1 km) inland from the coast. Besides Odisha and Andhra Pradesh, the cyclone also affected three inland states (Jharkhand, Bihar, and Chhattisgarh) to some extent (Singh and Jeffries 2013).

The storm originated in a remnant cyclonic circulation from the South China Sea. The circulation formed a low-pressure area off the Tenaserim coast, Myanmar, on October 6, 2013. It later moved over the North Andaman Sea as a well-developed low-pressure zone. It then intensified as a cyclonic depression on October 9, 2013, and later transformed into a cyclonic storm. On October 10, Phailin further intensified as a severe and then a very severe cyclonic storm when it made landfall along the Odisha coast as a Category 4 cyclone. Finally, as it moved inland, Phailin lost its strength (Mohanty et al. 2015).

LOSS AND DAMAGE

Cyclone Phailin affected around 13 million people and resulted in 44 deaths in the state of Odisha. Most of the deaths were caused by falling trees and/or collapsing houses. Deaths were also reported in other affected states. Compared to the magnitude of the storm, Cyclone

Cyclone Phailin in the Bay of Bengal prior to making landfall in Odisha, India, on October 12, 2013. (NASA)

Phailin accounted for far fewer deaths than would have occurred primarily because the Indian Meteorological Department (IMD) in Delhi provided early warnings of the cyclone, and the Indian and state governments evacuated to safer shelters, often by force, nearly 1 million people from the coastal areas before the storm's landfall. The warnings started five days before the storm's landfall as a red alert, which is the highest alert from the IMD, and the warnings specified where and what type of damage could be expected to houses and infrastructure. Another reason for far fewer deaths from this cyclone was that the landfall site was on a steep continental shelf, meaning fewer low-lying areas were vulnerable to storm surge (Samenow 2013).

After the deadly Orissa 1999 cyclone, which killed 10,000 people, the Indian government adopted a "zero-casualty approach." Following that cyclone, the IMD modernized its early-warning system and strictly monitored Cyclone Phailin's path, speed, and direction. However, while lives were spared, economic losses were very high. The cyclone and associated flood damaged 256,633 household units including 44,806 and 1,564 fishing and artisan households, respectively. A sector-wide Damage and Needs Assessment revealed that around 198,637 rural houses were partially damaged, and more than 57,000 thousand rural units were totally destroyed or severely damaged. Thus, the cost of reconstruction and provision of basic services accounted for about $477 million (Singh and Jeffries 2013).

In addition, more than 6,000 school buildings were damaged, along with thousands of other nonresidential structures like health centers/hospitals and government buildings. The estimated cost of reconstruction of the nonresidential and residential buildings came to about $105 million. Moreover, several historical monuments attesting to the rich history of Odisha were also damaged by Cyclone Phailin.

Meanwhile, the agricultural sector was hugely affected by the disaster. About 3.25 million acres (1.3 million ha) of agricultural and horticultural croplands suffered losses, including about 21 percent of Kharif croplands that were damaged in the three districts of Puri, Ganjam, and Khordah. Given the extent of damage to this sector, the estimated loss was about $287 million, including loss to crops and agricultural infrastructure. The cyclone and flood also caused widespread damage to the irrigation sector. About 725 miles (1,088 km) of canals, 8,152 minor irrigation projects, several minor lift irrigation projects, long stretches of embankments and drainage channels, and approximately 400 irrigation departmental buildings suffered damage owing to the storm. The total estimated cost of rehabilitation and/or strengthening of damaged irrigation infrastructure in the six districts of Ganjam, Puri, Khordha, Kendrapada, Jagatsinghpur, and Baleshwar was $77 million (Singh and Jeffries 2013).

The estimated livestock cost in the three most severely affected districts was about $2.23 million. More than 7 million livestock including large and small animals and poultry were affected by the cyclone, and 179,518 livestock were reportedly dead. Even though fishermen were called off from the sea and all kinds of movements near or into the sea were restricted, the fishery sector of Ganjam, Puri, and Khordah also suffered severe damage. The damage to and loss of fishing equipment and infrastructure, including boats, tanks, and nets, was estimated at $97 and $53 million, respectively (Singh and Jeffries 2013).

In Odisha, huge damage also occurred to surface communication systems, telecommunication, and power supply and water supply lines. Following the storm and heavy rainfall, flood conditions occurred in several rivers affecting the districts downstream. Reportedly 163 miles (245 km) of urban roads and long stretches of roadside drains were damaged, mostly by the post-cyclonic flooding and falling of trees due to severe wind speed. Additionally, 250 street lights and 21 miles (33 km) of water pipelines were also damaged. Thus, the cost of reconstruction of infrastructure included building more disaster-resilient structures to combat future disaster events such as raised embankments and concrete roads (given that concrete roads suffered the least damage). The estimated cost of reconstruction as of 2013 was $40.41 million for roads and $55,000 for reconstruction of roadside culverts (Singh and Jeffries 2013).

PREPAREDNESS

Unlike in the past, Odisha state was well prepared for Cyclone Phailin. As indicated, the IMD traced the formation of a cyclonic storm and its path on October 8, 2013, using statistical dynamic models to predict storms and storm surge. On the same day, the IMD reported the formation and intensification of the cyclonic storm to the Government of Odisha (GoO). This put the district authorities, the media, the National Disaster Response Force (NDRF), Police, and the Ministry of Home Affairs on high alert, and the GoO canceled the holidays of civil servants during the popular Hindu Dussehra festival to ensure proper functioning of the government and for close monitoring of the disaster (Padhy et al. 2015).

Later, on October 10, the government reported through an Orange Message the formation of a very severe cyclonic storm, subsequently called Phailin. The Odisha Disaster Management Authority (OSDMA), along with the National Disaster Management Authority (NDMA), started conducting mock drills in the multipurpose cyclone shelters. These drills acquainted people with disaster management skills and the equipment available in the cyclone/flood shelters. Because the number of cyclone shelters would not accommodate the population that was to be evacuated, the state government opened nearly 250 emergency shelters. These shelters were set up in sturdy buildings like schools, colleges, community halls, and public buildings. With the objective to minimize loss of lives, public officials of several districts of Odisha were directed to evacuate the coastal residents from the thatched temporary houses to the nearest flood/cyclone shelters by October 12, 2013, when the cyclone was predicted to make landfall. People were encouraged to move to shelters taking only a few clothes and important documents with them. As noted, the Indian government evacuated nearly a million people from coastal districts, of whom about 900,000 were from Odisha state (Padhy et al. 2015).

Measures were also taken to move livestock to safe places. In addition, 24-hour control rooms were set up in district offices and the district officials were equipped with satellite phones to ensure proper functioning of the evacuation and post-disaster relief operations. Bharat Shanchar Nigam Limited (BSNL), the telecom operating service of the Indian Government, was prepared with enough fuel to adequately operate the generators of the telephone exchange and the cell towers (Padhy et al. 2015).

As a proactive measure, NDMA placed 56 teams of the NDRF in vulnerable places. The state government also deployed disaster response teams with necessary equipment for rescue and relief operations. The state government ensured active and assertive coordination among various ministries and agencies involved in managing the event. A plan of action had been drawn up in detail such that all types of public and private vehicles were made available as needed, and adequate food supplies were stored near the potential cyclone-impact areas for immediate relief. Manpower from the less vulnerable districts was also directed to the highly vulnerable districts of Odisha. To enhance search operations, fire service personnel, other first responders, and civil volunteers were stationed at strategic points throughout the vulnerable districts.

For the most effective relief operations, the Government of Odisha requisitioned vehicles, including boats. Meanwhile, the Government of India deployed helicopters, fixed winged aircrafts, and warships from the Indian Air Force nearer to the potential impact area. The aircraft were requisitioned primarily for air-dropping operations if needed. The Indian Navy and the Coast Guard were also deployed to anchor fishing boats and ships in safe places. The Government of

Odisha prepared more than 1 million food packets and kept in reserve half a million tons of food grain along with drinking water. Medical facilities were also kept ready for relief operations, and fuel was stored for at least one week in the vulnerable districts (Padhy et al. 2015).

Arrangements for earth extractors were made to quickly clear roads right after the disaster had passed. Precautionary measures also included cancelation of trains and flights and cutting power supply in several coastal villages. Also, measures to lower water levels in the dams and the reservoirs, primarily to avert any flood situation, were taken in advance.

RELIEF OPERATION

In advance of Cyclone Phailin, National Disaster Response Force personnel along with Indian military were deployed in coastal areas with relief supplies and medical aid. Dry and cooked food and several essential materials were distributed to the people in the cyclone/flood shelters as post-disaster relief. Tankers and mobile vans ensured adequate potable water supply. Moreover, water pouches were distributed in the rural areas of the cyclone-affected places. Nearly half a million polythene sheets were also distributed to those whose houses were partially or totally damaged. To ensure post-disaster medical support, 185 medical teams and more than 300 medical relief centers were stationed in the villages against emergence of post-disaster epidemic. These teams were equipped with anti-snake venom, oral dehydration supplements, halogen tablets, and several other types of first aid. Several measures were also taken to prevent the spread of mosquito-borne diseases.

Ultimately, restoration of power supply in almost half of the affected villages was completed within one month of the disaster, and within six months, the entire power supply was restored. Adequate monitory compensation was provided to the families of the victims and the injured after verification, and loans were relaxed for the affected farming households (Padhy et al. 2015).

CONCLUSION

The teams of workers engaged in Phailin mitigation accomplished quite a lot. Joint efforts by the many organizations, including the Indian Army, the Indian Red Cross, and several philanthropic organizations, resulted in comparatively fewer deaths in Cyclone Phailin compared to Cyclone 05B in 1999, which claimed 10,000 lives. Consequently, the Chief Secretory of the OSDMA was commended at the International Conference on Humanitarian Logistics for the excellent mitigation efforts of the team. Furthermore, the Odisha chief minister, Naveen Patnaik, was commended for successful management of the disaster. Clearly, the joint management efforts witnessed during Cyclone Phailin should serve as a model for a future rapid-onset natural disaster.

Subarna Chatterjee

Further Reading

Mohanty, U. C., K. K. Osuri, V. Tallapragada, F. D. Marks, S. Pattanayak, M. Mohapatra, and D. Niyogi. 2015. A Great Escape from the Bay of Bengal "Super Sapphire-Phailin" Tropical Cyclone: A Case of Improved Weather Forecast and Societal Response for Disaster Management. *Earth Interactions* 19(17): 1–11.

Padhy, G., R. N. Padhy, S. Das, and A. Mishra. 2015. A Review on Management of Cyclone Phailin: Early Warning and Timely Action Saved Life. *Indian Journal of Forensic and Community Medicine* 2(1): 56–63.

Samenow, J. 2013. Major Disaster Averted: 5 Reasons Why Cyclone Phailin Not as Bad as Feared in India. *The Washington Post*. October 14, 2013. https://www.washingtonpost.com/news/capital-weather-gang/wp/2013/10/14/major-disaster-averted-5-reasons-why-cyclone-phailin-not-as-bad-as-feared-in-india/, accessed October 30, 2018.

Singh, D., and A. Jeffries. 2013. *Cyclone Phailin in Odisha, October 2013: Rapid Damage and Needs Assessment Report*. Washington, DC: World Bank.

Cyclone Sidr, Bangladesh, 2007

On the night of November 15, 2007, Bangladesh was devastated by tropical Cyclone Sidr. The cyclone struck the southwestern coastal areas and ripped through the heart of the country from the southwest to the northeast with 155 mph (230 kmph) winds. It originated from a tropical depression in the Bay of Bengal on November 11, 2007, and quickly reached peak sustained winds of 135 mph (215 kmph). This Category IV storm eventually hit offshore islands and made landfall on the southeastern part of the Sundarbans mangrove forest, and it was accompanied by tidal surges up to 20 feet (6 m) in some areas, breaching coastal and river embankments, flooding low-lying coastal areas, and causing extensive physical damage. Cyclone Sidr weakened into a tropical storm and dissipated on November 16, 2007. It claimed 3,406 lives in 28 of the 64 districts in Bangladesh, and more than 55,000 were injured by the event (Paul and Dutt 2010). A district is the second largest administrative unit in the country with an average population of 2,5 million.

Tropical Cyclone Sidr in the Bay of Bengal prior to making landfall on the Bangladesh coast on November 15, 2007. (NASA)

DAMAGE AND LOSS

More than 27 million people in 30 districts were affected by this cyclone; nearly 1.5 million homes were destroyed or damaged by Cyclone Sidr and subsequent storm surges. An estimated 1.87 million livestock and poultry perished, and crops on 2.5 million acres (1 million ha) suffered complete or partial damage. This translates into crop loss totaling nearly 1 million tons. The Joint Damage Loss and Needs Assessment (DLNA) Mission, led by the World Bank, estimated the economic damage and losses caused by Sidr at $1.7 billion, accounting for 3 percent of the total gross domestic product (GDP) of Bangladesh. The country's overall economic growth was affected by less than 0.5 percent in the 2007 financial year. When GDP and economic growth losses are considered, the impact of Cyclone Sidr was relatively moderate. Damage and losses were concentrated in the 12 coastal districts and in the housing sector. More than two-thirds of the disaster's effects were physical damages and one-third was economic losses. Both damage and losses were incurred in the private sector.

Despite being similar in severity, Sidr killed far fewer lives than Cyclone Gorky, a Category IV storm that struck Bangladesh in 1991 and killed an estimated 140,000 people. The relatively low number of fatalities as well as less-than-expected damage caused by Sidr was the result of a number of factors. First, following the cyclone of 1991, the Bangladesh government invested in Early Warning System (EWS) that provided timely weather forecasting and advance warning systems, resulting in the successful evacuation of people living on the southwestern coast. Additionally, after the cyclone of 1991, some 2,500 public cyclone shelters were constructed in the coastal regions of Bangladesh. During Cyclone Sidr, 3 million people were evacuated and 1.5 million were accommodated in public shelters (Paul 2009).

Second, Sidr struck the Sundarbans, the world's largest mangrove forest, before it reached populated areas. The forest's thick growth of mangrove trees served as a buffer, reducing the intensity of both the wind and the storm surges. The Sundarbans bore the brunt of Cyclone Sidr, thus saving residents near this area from more disastrous consequences. Fishermen who anchored their boats and remained in them along the small canals inside the Sundarbans probably saved their own lives. Most villages behind the trees were also saved. Third, Sidr made landfall during low tide, which resulted in relatively lower surge waves (GoB 2008). The length of a storm's surge is considered an important determinant of the number of deaths caused by a cyclone. Previous studies of cyclone-related mortality in Bangladesh clearly reveal that most cyclone deaths could be attributed to drowning or a combination of drowning and injuries sustained in the storm surge. The height of a storm's surge depends to a considerable extent on whether the surge occurs during a low or high tide.

Another reason for Cyclone Sidr's relatively low death toll was that the surge water remained on coastal land for fewer than three hours, and the affected areas experienced only one storm surge instead of a succession of surges. Finally, there is the coastal embankments factor (Paul 2009). Since the mid-1960, the Bangladesh government has constructed about 3,333 miles (5,333 km) of embankments in the coastal region to increase the agricultural production and protect the residents during cyclones and associated storm surges. In reality, these embankments saved many lives and property of coastal residents during the time of Cyclone Sidr.

CYCLONE WARNING AND EVACUATION

Five days before Sidr made landfall, the Bangladesh Government began to provide advance warning on radio and television, and it issued emergency evacuation orders and hoisted the highest danger signal almost 27 hours before landfall. Using megaphones, handheld bullhorns, and bicycle-mounted loudspeakers, 44,000 volunteers of the Cyclone Preparedness Program (CPP), local administrators, and emergency-management personnel warned people in coastal areas to leave their homes for public cyclone shelters and other safer places. A sample survey conducted in 2010 revealed that slightly over 78 percent of the respondents were aware of the cyclone warnings and evacuation orders prior to Sidr's landfall, primarily from CPP volunteers (Paul and Dutt 2010). Other sources of warning included friends, neighbors, relatives, local government officials, workers of nongovernmental organizations (NGOs), radio, and television.

The overall evacuation rate for Cyclone Sidr was reported as 33 percent, and approximately 20 percent of all respondents surveyed took refuge in designated cyclone shelters, and the remaining 13 percent in public buildings (e.g., government offices, mosques, and colleges), neighbors' houses that they perceived as being structurally stronger than their own, nearby embankments, roads, higher grounds, and large trees (Paul 2009). The Bangladesh government estimated that approximately 3.2 million (35 percent) of the 9.2 million coastal residents did evacuate after receiving the cyclone warning and evacuation orders. Those who did not evacuate provided a host of reasons, including no nearby public shelter, distance to public shelter, public shelters being too crowded, public shelters not in good condition, unsatisfactory attributes of public shelters (e.g., no running water, no toilet, or no separate room for women), not receiving warning, not believing the warning, evacuation orders being issued prematurely, warning message being incomplete, complicated warning signals, not having time, not realizing the danger, not having past experience, fearing burglary, feeling safe at home, and having a false sense of security.

Dhaka-based daily newspapers reported that some coastal residents were reluctant to evacuate to a cyclone shelter before putting their livestock in a safe place. Consequently, some of them were probably unable to reach a shelter before the cyclone's landfall. Additionally, in some areas, the agencies responsible for assuring safe and timely evacuations fell short of these important responsibilities. There were lapses in cyclone warning and evacuation procedures. The first evacuation order was issued 27 hours before the storm's landfall, and this was considered a premature action. One study reported that no fatalities occurred among people who opted to move to a public cyclone shelter (Paul 2009). All deaths occurred among people who did not comply with evacuation orders; were turned back from shelters (many of these facilities were already full and even overcrowded); or who took shelter in the structurally weak homes of neighbors, friends, and relatives.

RELIEF OPERATIONS

Immediately after Cyclone Sidr, the Armed Forces of Bangladesh initiated massive search and rescue and early relief operations. They also helped to bury the dead and remove dead livestock. Although the Bangladesh government did not make a formal international appeal for foreign assistance, the international community was quick to respond and pledged approximately $242 million in emergency aid as of December 17, 2007 (GoB 2008). The government, in collaboration with the United Nations and other relief organizations (e.g., CARE, Islamic Relief, Muslim Aid, Oxfam, World Food Program), undertook several humanitarian assessment missions and started relief operations focused on food, water and sanitation, shelter, and other emergency items. Medical teams and health workers were dispatched almost immediately after the cyclone to provide needed medical care. The nongovernmental organizations (NGOs) also opened makeshift hospitals. The government allocated $6.7 million for relief and housing construction purposes and sent medical teams to the affected areas. The Asian Development Bank (ADB) and World Bank contributed $150 and 100 million, respectively, to assist cyclone survivors (GoB 2008).

Although relief operations initially suffered from poor transportation and communications, as a result of severe damage to roads and telecommunication systems, subsequent distribution of emergency relief aid to cyclone survivors was conducted in a timely manner. With a few exceptions, the relief distribution was fair and unbiased. This was probably because the Bangladesh Army was primarily involved in disbursing relief goods and providing emergency services. The amount of emergency assistance, particularly food and clothing, provided to cyclone survivors was more or less adequate. Most affected people maintained that the government pursued an extensive relief program for cyclone survivors. It distributed 13,406 tents and 5,000 pieces of plastic sheets to build temporary shelters for those in need (Paul and Dutt 2010). However, Oxfam maintains that coordination gaps existed between local and national actors, resulting in poor-quality humanitarian response.

RECONSTRUCTION AND CONCLUSION

For quick and effective recovery, the Bangladesh Government formed the Early Recovery Cluster Coordination Group with 20 UN organizations, representatives from relevant governmental ministries, and international and national NGOs. The United Nations Development Program (UNDP) was appointed lead agency for the group. Using a "build back better" approach, the recovery program of Cyclone Sidr had two components: reconstruction and upgrading of damaged social and economic infrastructure and improvement of agricultural and nonfarm activities. The government decided to construct housing for 20,000 households; upgrade the transport network, damaged embankments, and health service infrastructure; rebuild damaged or destroyed marketplaces, schools, and public shelters; and reconstruct water supply and electric services. It also provided housing grants to cyclone survivors, but the amount was insufficient to build their homes, perhaps in part because many recipients spent their grant money on other emergency needs, such as food or winter clothing for children. However, several NGOs (e.g., Habitat for Humanity Bangladesh, Muslim Aid, Oxfam, and World Vision) and foreign governments (e.g.,

India, Kuwait, and Saudi Arabia) also participated in recovery efforts by providing transitional shelter and homes for Sidr survivors. For example, by the end of 2013, Muslim Aid constructed 800 transitional shelters and provided 2,000 households with construction materials to repair their homes in coastal districts.

Overall, the Bangladesh Government achieved success in short-term recovery efforts (e.g., providing food security, transitional shelter assistance, cash, and agricultural and fishing equipment; ensuring drinking water supply; and controlling infectious diseases). The NGO support was also remarkable in areas of providing nutrition for children and post-Sidr psychosocial support through recreational activities (e.g., songs, dance, and sports) in their schools. It was a large and welcome shift in disaster management in coastal Bangladesh when relief goods include items that addressed the specific needs of children.

However, both the government and NGOs did not achieve many of its long-term recovery goals because of lack of coordination and financial resources and competition among local NGOs to obtain funding. It was also reported that during relief distribution, NGOs favored their own microcredit borrowers. NGOs often demanded bribes when they distributed livelihood support such as livestock, nets, and boats. There was also no sharing of information among organizations that participated in relief operations. Primarily because of funding constraints and lack of roads to transport building materials in remote coastal districts, progress toward full recovery was slow. Not all damaged roads and embankments had been repaired, and many households were still living in temporary makeshift shelters even five years after the landfall Cyclone Sidr. Moreover, houses that were built or repaired in the affected coastal areas were not strong enough to withstand future cyclones.

Bimal Kanti Paul

Further Reading

GoB (Government of Bangladesh). 2008. *Cyclone Sidr in Bangladesh: Damages, Loss, and Needs Assessment for Disaster Recovery and Reconstruction.* Dhaka: GoB.

Paul, B. K. 2009. Why Relatively Fewer People Died? The Case of Bangladesh's Cyclone Sidr. *Natural Hazards* 50: 289–304.

Paul, B. K., and S. Dutt. 2010. Hazard Warnings and Responses to Evacuation Orders: The Case of Bangladesh's Cyclone Sidr. *Geographical Review* 100: 336–355.

The Dust Bowl, 1930s

The Dust Bowl was a series of human-environmental disasters in the 1930s in portions of the U.S. Great Plains states of Colorado, Kansas, New Mexico, Oklahoma, and Texas, where drought, heatwaves, rapid soil loss, dust storms, and mass soil re-deposition occurred. These negative environmental effects coupled with the ongoing Great Depression led to increased economic hardship, personal strife, and in some cases, mass farm closure and resettlement of agrarian people in the region. The causes of this long-term, spatially extensive series of disasters are largely attributed to both natural factors such as periodic drought and human factors such as intensive farming practices that were out of touch with the region's semiarid short grassland ecology. Thus, the Dust Bowl was a multifaceted human environmental disaster that spanned a broadly defined time and space (Porter 2012).

SCOPE AND CAUSES

Part of the difficulty in encapsulating the Dust Bowl is its temporal and spatial intermittence and broad scope. By that, the drought, erosion, dust storm events, and human responses that define it happened in different places at different times, roughly beginning at the close of the 1920s and ending in the late 1930s. Similarly, while other disasters such as floods, earthquakes, and volcanic eruptions and extreme storm events such as hurricanes and tornados occur at one moment, a few days, or a week in very localized geographic areas, the Dust Bowl lasted roughly a decade and had an approximate spatial footprint of 100 million acres (40.5 million ha). In essence, the Dust Bowl represents a time (1930s) in a region (southern Great Plains or U.S. Southwest) and the intermingling of economic and ecological disasters (McLeman et al. 2014).

Climatic conditions are some of the first to come to mind when considering the causes, especially natural ones, of the Dust Bowl. Drought, or below-average precipitation for a considerable period of time, creates the conditions where soil can be eroded by winds more easily. The broader southern Great Plains and Dust Bowl are bioregions that do experience regular cycles of drought. Paleoclimatology, or the science of understanding past climatic conditions, and the historical records provide evidence of these recurring periods of drought. The broad climatic drivers that control the drought periods there are the southerly migration of the jet stream, La Niña

This 1936 photo shows the destitution of farm families threatened by clouds of dust in the southern Great Plains during the American Dust Bowl. (Library of Congress)

conditions in the Pacific Ocean, and other, more fine-scaled regional oscillations. Large, persistent droughts hit the region especially hard in the 1500s and 1700s, and shorter droughts were common throughout the middle-to-late nineteenth century, when European-Eastern U.S. grain farming was brought to the area by white settlers. The 1930s, the decade of the Dust Bowl, was indeed a drought period.

The 1920s saw an increase of the mechanization and capitalization of agriculture across the Great Plains. Global grain shortages in the wake of the First World War (1914–1918) initially created the demand for higher yields, while mechanization allowed the expansion of agricultural areas at the expense of native shortgrass prairie ecology. At least 6 million acres (2.4 million ha) of prairie land were brought under cultivation prior to 1930, which should have never been touched by the plow. Additionally, the 1920s featured relatively wet climatic conditions that were ideal for grain farming. In the region that came to be known as the Dust Bowl, this period is referred to as the "Great Plow Up." The growth in the number of tractors, specialized implements, and harvester mechanization of this time allowed existing farmers to spatially expand their operations. Additionally, such technological introductions opened grain farming to newcomers and speculative operators called "suitcase farmers." High costs of machinery and other inputs combined with falling crop prices also forced farm operations to expand to recoup costs. The subsequent intensive plowing and the expansion of agriculture into previously unfarmed or marginal areas, which receives average rainfall of less than 20 inches (508 mm), vastly transformed the ecology of the region. Poorly cared for fields lay bare when not in production where previously native land cover had existed. The widespread replacement of native grass and forb communities featuring deeper root systems with the shallower annual wheat monoculture also worsened the wind erosion risk. All these were being set for the worst drought disaster to be experienced in this part of United States.

In short, economic boom combined with technological advances created the political economic preexisting condition for ecological ruin. According to Donald Worster, the preeminent environmental historian on the settlement of the Great Plains and the Dust Bowl (or "Dirty Thirties"), the ignorance of best farming practices in the drought-prone Plains was not at the root of human causes of the Dust Bowl. Rather, Worster places agency more in the entrepreneurial desire to capitalize on lucrative economic conditions in forcing the Dust Bowl. During the aforementioned "Great Plow Up," the southern Great Plains land base could thus be considered more a finite, extractable resource rather than a key component to a sustainable ecosystem or healthy agricultural economy. As such, one could consider the Dust Bowl as a disaster driven by the rapacious, extractivist logic of early-twentieth century rural capitalism. In this light, the economic collapse of the Great Depression that was the result of the unregulated and freewheeling Roaring Twenties' financial practices parallels the Dust Bowl ecological disaster that followed the expansive farming moment of the "Great Plow Up" (Worster 1979).

The ravaging of the land in the southern Great Plains continued until 1931, when a severe drought hit the region. Many places in the region experienced dry spells in 1933, 1934, and 1936. In 1934, many places in the Dust Bowl recorded their driest year on record, as well as one of the hottest. In terms of erosion and dust storm severity, 1935 was one of the worst years. In short, the climatic conditions for windblown soil erosion, or deflation, were there, but the native shortgrass prairie ecosystem had evolved to weather such climatic shifts. Like many disasters, natural causes coupled with anthropogenic or human-induced factors to create an enhanced disastrous effect (Cordova and Porter 2015).

Dust storms of epic size and frequency were remarkable events that came to characterize the disaster, referenced in the name "Dust Bowl." The loss of soil-anchoring native vegetation combined with drought allowed fast winds sweeping across the plains to mobilize vast amounts of topsoil in the air, creating exceptionally powerful dust storms throughout the 1930s. Most notable of these storms occurred on April 14, 1935, dubbed "Black Sunday," where 40–60 mile per hour (60–90 km/h) winds pulled so much soil into the air in the Oklahoma and Texas Panhandles that visibility was less than a foot (0.31 m). The dust darkened the skies of the region. Dusty and hazy conditions also expanded far outside the south Great Plains too.

During the mid-1930s, the time when drought and heatwaves were the most extreme, airborne dust from the Dust Bowl was seen as far away as in the U.S. East Coast. Soil loss from persistent winds and dust storms across the affected region was also intense. In many places across the Dust Bowl region, an estimated 75 percent loss of topsoil occurred, greatly affecting future agricultural productivity of the region, not to mention extant biological communities. The drought, dust storms, and soil loss over the course of the 1930s took significant tolls on the local human population in terms of economic loss, farm closure, and mass outmigration (NWS n.d.).

IMPACTS AND RESPONSE

Drought conditions throughout the 1930s caused massive crop loss for the largely agricultural economies of the southern Great Plains. The wheat and other grain monoculture that grew so successfully during the previous decade withered and died in the heat and dust of the 1930s. Farm animals also had no water, and herds dwindled while dust storms took their own toll. As the topsoil eroded away and persistent drought limited agricultural outlooks, land values also plummeted. The drought of the "Dirty Thirties" precipitated a cascading collapse of the region's rural economy as overextended farmers went bankrupt and credit lines dried up as rural banks went under. Combined with the effects on crops from drought, the loss of topsoil, and property damage from dust storms, thousands of many desperate, impoverished, and homeless people across the region left the affected areas. Some estimates put the homeless count as a result of the Dust Bowl as high as half a million. The aforementioned bankruptcy, farm closure, and property foreclosure; plummeting land values; lack of work; and precipitous economic collapse along with the sheer physical and emotional toll caused by years of economic and environmental struggle enhanced the uprooting effect and forced many people off the land and onto the highways.

The Dust Bowl saw hard-hit families from the region, especially Oklahoma, begin a mass exodus to nearby cities or the U.S. West Coast, especially California, in search of work in the truck farming sector. The term "Okie" was generally, and somewhat derogatorily, applied to these westward-bound Dust Bowl out-migrants. The entire out-migration during the 1930s and early 1940s across the broader Great Plains region is estimated between 3 and 4 million people (Worster 1979).

The U.S. federal government was largely helpless to stop the drought, which was a major natural causal factor in the Dust Bowl, but it did take proactive measures to curb other human causes and alleviate the suffering, poverty, and economic decline that resulted. Much of this federal response was part of a broader New Deal economic recovery strategy initiated by the Franklin D. Roosevelt administration in the wake of the Great Depression. To check and prevent further soil loss, the government established the U.S. Soil Conservation Service in 1935 within the Department of the Interior. Later, in 1936, soil conservation was broadened under the Soil Conservation and Domestic Allotment Act. This legislation provided $5 million to conserve fertile land, stop waste of soil, and develop flood control projects to reduce soil erosion.

On the other hand, the Soil Conservation Service, in conjunction with the Civilian Conservation Corps and other federal agencies, conducted publics works projects in the southern Plains, which ranged from massive tree plantings for wind breaks to farmer education and cash incentives for soil conservation farming techniques such as contour plowing, strip farming, crop rotation, and mulching practices. These techniques not only increase soil moisture and nutrients but also protect the soil from wind deflection. While the damage to much of the region's soil resources was permanent, the federal responses sought to preserve what was left. Other relief projects involved farm product purchases that helped out desperate farmers and ranchers while distributing foodstuffs to needy populations (Helms 1990).

CONCLUSION

The Dust Bowl remains to this day the worst human-induced ecological disaster in the history of the United States. This disaster radically changed human settlement, ecology, land use, and the economic landscape of the southern Great Plains. The Dust Bowl is instructive as it stands as a reminder for the need to collectively manage land resources from a systematic scientific perspective as well as focus on the ecological suitability of current economic activity. Although the federal government initiated reclamation and rehabilitation programs in response to the 1930s dust bowl, the country and the dust bowl region experienced many drought years, particularly in the mid-1950s and 1970s.

Andrew M. Hilburn

Further Reading

Cordova, C., and J. Porter. 2015. The 1930s Dust Bowl: Geoarchaeological Lessons from a 20th Century Environmental Crisis. *The Holocene* 25(10): 1707–1720.

Helms, D. 1990. Conserving the Plains: The Soil Conservation Service in the Great Plains. *Agricultural History* 64(2): 58–73.

McLeman, R., J. Dupre, L. Berrang J. Ford, K. Ford, K. Gajewski, and G. Marchildon. 2014. What We Learned from the Dust Bowl: Lessons in Science, Policy, and Adaptation. *Population and Environment* 35: 417–440.

NWS (National Weather Service). n.d. The Black Sunday Dust Storm of April 14, 1935. https://www.weather.gov/oun/events-19350414, accessed January 24, 2019.

Porter, J. 2012. Lessons from the Dust Bowl: Human-Environment Education on the Great Plains. *Journal of Geography* 111: 127–136.

Worster, D. 1979. *Dust Bowl: The Southern Plains in the 1930s.* New York: Oxford University Press.

East African Drought, 2011–2012

A severe drought occurred in the East African Region between July 2011 and August 2012. It is considered the worst drought in that region in several decades as it caused a severe food crisis in countries such as Somalia, Djibouti, Kenya, and Ethiopia, which affected around 9.5 million people (UN 2011). Severe food shortages led to increased rates of migration to neighboring countries, where people were bound to live in crowded, unsanitary conditions, and severe malnutrition, leading to a large number of deaths. Moreover, the destination countries such as Sudan, South Sudan, and some parts of Uganda were also affected by the food crisis but to a lesser extent than the origin countries of the migrants.

According to the Food and Agricultural Organization (FAO) of the United Nations, farmers in the south of Somalia were particularly affected by the food crisis. While the United Nations officially declared a famine on July 20, 2011, many believed that tens of thousands of people had already died in southern Somalia before the official declaration. Based on several studies, the declaration was late because Western governments were trying to weaken the Al-Shabaab militant group in that region by preventing aid from reaching the affected areas.

Additionally, internal fighting affected the delivery of emergency aid in some areas, but the relief operations were scaled up, and it resulted in significant reduction of malnutrition and mortality rates in Somalia by the middle of November 2011. According to the International Committee of Red Cross (ICRC), the food crisis in Somalia was no longer at the emergency level by January 2012. The UN officially declared the end of the famine on August 2012, and aid agencies shifted their focus to recovery efforts such as distributing seeds and digging irrigation canals. Several long-term strategies were adopted by the respective governments cooperating with development agencies to prevent the region from suffering future droughts and to make development activities more sustainable.

BACKGROUND OF THE DISASTER

The length of the main rainy season in the East African Region is from April to June, and most of the year's precipitation occurs in this period. In 2010, an unusually strong La Niña over the Pacific interrupted the seasonal rains for two consecutive seasons. Consequently, rainfall was 30 percent below the average of 1995–2010, and this led to crop failure and loss of livestock. In some places, the loss of livestock was around 40–60 percent (Perry 2011). This decreased the production of milk as well as aggravated a weak harvest. Since the whole economy of the region

This image shows severe drought in Somalia, Kenya, and southern Ethiopia during 2011–2012. The darker spots are an indication that plant growth slowed, with fewer photosynthesizing leaves than average. (NASA)

was based mainly on agriculture, the impact led to skyrocketed cost of cereals and livestock. Simultaneously, wages and purchasing power fell across the region. Rebel activity from Al-Shabaab group inside Somalia also compounded the crisis.

The head of the U.S. Agency for International Development (USAID) stated that the hotter and drier growing conditions in sub-Saharan Africa had reduced the resiliency of the communities and that climate change contributed to the severity of the crisis (UN 2011). But other professionals disagreed and said it was premature to blame climate change, as previous studies of climate models had predicted correctly the amount of rainfall in that region. Clearly, however, the unusually strong La Niña disrupted the pattern and contributed to the intensity of the drought. It is still debatable whether climate change is to blame as the relationship between La Niña and climate change is not well established.

EARLY WARNING FOR DROUGHT

The early-warning system for the crisis was criticized because of its failure to identify the gravity of the situation as suggested by the official declaration from the UN coming in July 2011 when tens of thousands of people were already dead.

Furthermore, the international community was late to react, which worsened the crisis, yet the Famine Early Warning System Network (FEWS NET) anticipated the crisis as early as August 2010, and the U.S. ambassador to Kenya declared the situation a disaster in early January 2011.

Several other development organizations criticized the Western governments for their late response to the disaster and their skeptical view about preventive action. It was one of the most severe food security emergencies of recent times, but the response was delayed due to the failure of the international community to heed the early-warning system. Early humanitarian response could have saved thousands of lives and minimized the severity of the disaster.

HUMANITARIAN IMPACTS

The center of the disaster was southern Somalia and its surroundings. In July 2011, the UN declared a famine in the Lower Shabelle and Bakool regions of southern Somalia. Later, in August, famine was declared in the Middle Shabelle, Mogadishu, and Afgooye areas. Famine spread to all eight regions of southern Somalia within four to six weeks due to a lack of humanitarian response. The food crisis primarily affected the farmers in southern Somalia, and staple prices were 68 percent higher over the five-year average of the entire region. Particularly, staple prices were 240 percent higher in southern Somalia, 117 percent higher in south-eastern Ethiopia, and 58 percent higher in northern Kenya. The UN estimated that 10 million people across the Horn of Africa region needed food aid in early July of 2011, and it conducted several airlifts of supplies in addition to supplying on-ground assistance. But the humanitarian response was inadequate because of the shortage of funding from international aid and lack of security in the crisis zone.

The crisis was expected to worsen, and many expected that large-scale assistance would be required until at least December 2011. Tens of thousands of people were internally displaced even as the Kenyan Red Cross indicated that the dire need of food supplies had increased the level of malnutrition to its highest ever. Schools were closed in the affected regions as there was no food for the children, yet about 385,000 children were already malnourished along with 90,000 pregnant and breastfeeding women.

The Taita-Taveta District of Kenya was reportedly affected by famine in August 2012, where an estimated 87,000 people were suffering due to drought and wildlife invasion. Apparently, large herds of elephants and monkeys overran farms in the lowland and highland areas of the district, ruining thousands of acres of crops, which contributed to the food crisis. The Karamoja and the Bulambuli districts of northern and eastern Uganda also reported food shortage, affecting an estimated 1.2 million people.

REFUGEE CRISIS

Alex Perry (2011) claims that in September 2011, at least 920,000 refugees from Somalia fled to neighboring countries, particularly Kenya and Ethiopia. The

United Nations High Commissioner for Refugees (UNHCR), based in Dadaab, Kenya, had opened three refugee camps in June 2011 for 440,000 people, though the maximum capacity for each camp was 90,000. The UNHCR also reported that 80 percent of the refugees were women and children who are more vulnerable to malnutrition. The daily mortality rate during the drought period was 7.4 per 10,000, which was more than seven times higher than during a typical "emergency" situation. As each camp was overpopulated, mostly with women and children, it was very difficult to manage resources. Ethiopia also hosted around 110,000 refugees from Somalia, which was reported in July 2011. According to the Lutheran World Federation, military activities in the conflict zones of southern Somalia and the eventual increased relief operations of humanitarian organizations decreased the rate of migration by December 2011.

HEALTH ISSUES

The inability of health service providers to render aid was a major issue in the refugee camps. As most of the refugees were women and children, severe impacts of malnutrition caused several diseases. For instance, measles broke out in the Dadaab camps in July 2011 such that Kenya faced a severe measles epidemic, as did Ethiopia. The World Health Organization (WHO) reported that due to crowded and unsanitary conditions, 8.8 million people were at risk of malaria and 5 million people of cholera in Ethiopia. Malnutrition rates among women and children also increased singnificantly in July to 30 percent in parts of Kenya and Ethiopia and over 50 percent in parts of southern Somalia before dropping in mid-September.

In early December 2011, the UN reported that the scaling up of relief operation had improved the severe acute malnutrition rates and decreased mortality rate in southern Somalia. The malnutrition rate fell to 20–34 percent from 30–58 percent, and the mortality rate was also decreased to 0.6–2.8 per 10,000 people per day from 1.1–6.1 per 10,000 people per day. The peak time of health crisis was from September to early November because of the higher migration rate and lack of resources to deal with a large number of refugees.

RESPONSE AND RECOVERY

Several factors delayed the response to this disaster. The international community was late in understanding the severity of the crisis and then faced security issues in distributing aid. Humanitarian agencies requested $2.48 billion to address the crisis, but they could not secure all of that despite several developed countries such as the United States, the United Kingdom, Canada, and Australia having contributed aid. The European Union (EU) also participated in emergency relief efforts. Meanwhile, Middle Eastern countries like Saudi Arabia, Iran, Lebanon, United Arab Emirates (UAE), Qatar, and Kuwait helped distribute supplies such as food and medicine.

International humanitarian organizations like Oxfam and Action Aid were involved in distributing aid and providing necessary supplies to refugee camps such as Dadaab, which had the largest refugee camp in the world at that time. Several countries and aid organizations supported the camps by providing food, clothes, medications, doctors, and security. Additionally, after the UN gave the official declaration of the end of famine in August 2012, a long-term recovery program has been undertaken by the concerned governments to rehabilitate the displaced population of this region. Thus, aid agencies shifted their focus to recovery and to helping the people through various activities such as digging irrigation canals and providing high-yielding seeds.

CONCLUSION

The drought and famine in the East African Region were a few of the major natural hazards in recent history. Thousands of people died, and millions were affected. The drought was triggered by natural climate variability, and the famine was the result of the drought. Poor economic infrastructure, weak government, and late response from the international community were the main reasons for the increased gravity of the disaster. The disaster, occurring in a politically unstable region with an ongoing conflict between government and militia groups, delayed the international response. This delay meant the disaster ultimately affected even more people.

The East African Drought resulted in one of the major food crises in the human history. Thousands of people were forced to migrate and suffered a major livelihood crisis. All the refugee camps were overpopulated with mostly women and children who were exposed to severe malnutrition and its lifetime impact. Drought recovery is a long-term process compared to recovery from other natural hazards. However, the impact of drought can be predicted, and early response can prevent the hazard from turning into a disaster. The drought of 2011–2012 in the East African Region is an example of the consequences of failed early response from the disaster management sector. If governments and international communities had been more proactive, thousands more lives could have been saved. People in that region are still recovering, and they are still vulnerable to this type of disaster due to continued weak infrastructure and lack of awareness.

Abu Sayeed Maroof

Further Reading

Lott, F. C., N. Christidis, and P. A. Stott. 2013. Can the 2011 East African Drought Be Attributed to Human-induced Climate Change? *Geophysical Research Letters* 40(6): 1177–1181.

Megersa, B., A. Markemann, A. Angassa, J. O. Ogutu, H. P. Piepho, and A. Valle Zaráte. 2014. Impacts of Climate Change and Variability on Cattle Production in Southern Ethiopia: Perceptions and Empirical Evidence. *Agricultural Systems* 130: 23–34.

Perry, A. 2011. Somalia: A Very Man-Made Disaster. *TIME.* August 8, 2011. http://world.time.com/2011/08/18/somalia-a-very-man-made-disaster/, accessed August 25, 2018.

UN (United Nations). 2011. *Final Report: Humanitarian Requirements for the Horn of Africa Drought.* https://reliefweb.int/sites/reliefweb.int/files/resources/Full_report_216.pdf, accessed August 25, 2018.

Edmonton Tornado, Canada, 1987

The Edmonton Tornado of 1987 was a powerful tornado that tracked through the eastern part of Edmonton, Alberta, Canada, causing major devastation to the city. The tornado was notable for its unusual intensity and duration, reaching wind speeds up to 260 miles (420 km) per hour and having remained on the ground for an hour. The tornado caused a significant number of fatalities, injuries, and destruction of property along its 27 miles (40 km) path, which stretched as wide as nearly one mile (1.5 km) in certain parts. The tornado was classified as an F4 on the Fujita F-scale. It is considered one of the worst natural disasters in Canada's history, and July 31, 1987, is commonly referred to as Black Friday by Edmontonians.

CHRONOLOGY AND DAMAGE

The meteorological conditions that caused the Edmonton tornado were typical of conditions that cause severe thunderstorms. A cold front in Western Alberta collided with a persistent warm air mass with high dew points, which was sitting over the central part of the province. This caused severe morning thunderstorms in the foothills region of the province (to the west of Edmonton), which moved eastward throughout the day and reached Edmonton by afternoon. Environment Canada warned of the high potential for unusually severe thunderstorms in the afternoon. The first touchdown of a tornado occurred east of the Leduc area, 21 miles (31.5 km) south of Edmonton. It was reported by a weather spotter at 2:59 p.m. The Edmonton Weather Office (i.e., part of Environment Canada) issued a tornado warning by 3:04 p.m. The tornado traveled northward along a 25-mile (37.5 km) path, passing through east Edmonton and neighboring Strathcona County. Along the way, it hit a major industrial area, Refinery Row, as an F4 tornado, killing 12 people before causing major destruction to 463 homes in the residential area of Clareview (APSS 1991). The tornado also struck the Evergreen Mobile Home Park as an F2/F3 tornado, killing15 people and destroying about 200 homes in the park.

The tornado left a trail of shattered buildings and trees, broken concrete and glass, twisted steel, and other debris, and it caused chemical spills, fires, and further hazards. Although the tornado caused specific devastation along the eastern boundary of the city, the associated wind, rain, and hail of the storm system collectively impacted most of the city. The size and swath of hailstones were very unusual, with well-documented evidence of walnut-to-tennis-ball-sized hail covering an area over 104 square miles (270 square km) (Charlton et al. 1998). This combined with significant flooding from the storm generated tens of thousands of insurance claims. There was substantial loss of life and properties, including 27 fatalities, hundreds of injuries, and 750 families left homeless, and more than $330 million in damage (APSS 1991). The storm system was also responsible for at least four other smaller tornados in the Edmonton region.

LACK OF PUBLIC AWARENESS

The Edmonton Tornado of 1987 serves as a notable disaster in Canadian history because at the time of the event the city was unprepared for a tornadic disaster of this magnitude. Although Canada ranks second in the world in terms of the number of tornados per year, the country witnesses substantially fewer tornadoes than the United States. For context, Canada averages 62 tornados per year (ECCC 2018b), while the United States averages 1,253 tornados per year (NCEI 2019). As such, tornados are comparatively rare in Alberta, with an average of 15 tornados per season for the entire province (ECCC 2018a). Given the large geographic extent of the province, most tornados that do form tend to strike in remote rural areas and consequently are often underreported. Further, when considering all known tornados in Alberta during the period of 1980–2009, the vast majority (95 percent) of them were relatively weak, either F0 or F1 strength (ECCC 2018a). The twin influences of a large spatial extent and a history of less impactful tornadoes contributed to a general lack of awareness and perceived risk concerning the threat of a major tornado hitting the city. In addition, Alberta lacked a robust radar network at the time, and consequently, Environment Canada relied heavily on trained weather spotters to confirm a touchdown. As of 2019, the Edmonton Tornado of 1987 is the only F4 ever to develop in the province (ECCC 2018a). The Alberta Public Safety Services refers to the Edmonton tornado as "unprecedented in Alberta's history because a long-path tornado happened to strike a major urban center" (APSS 1991, 14).

CHALLENGES FOR EMERGENCY RESPONSE

Although the police, fire department, and emergency medical services (EMS) in the city responded quickly and admirably to the disaster, the magnitude and suddenness of the event did create challenging circumstances. The City of Edmonton did have an Emergency Planning Committee (EPC), and it promptly established an Emergency Operations Centre (EOC) to help manage the crisis. While representatives from all departments and agencies (i.e., Disaster Services, Police, Fire, Water and Sanitation, Social Services) coordinated the activities, a lack of resources and technological limitations nonetheless hampered response efforts. Absent modern technology such as cell phone towers and social media communication, and the limited availability of portable radios and channels, hindered effective communication across the community of responders. For instance, responders were initially not even aware of the severe devastation in the Evergreen Mobile Home Park because of the damage to the phone exchanges in the park, leaving victims unable to call for help and forcing some of them to hike more than a mile to the Alberta Hospital (APSS 1991).

The hospitals in the city were overwhelmed with injuries of a magnitude they had never witnessed. Many people whom responders encountered were in a state of shock or disbelief (APSS 1991). Beyond threats to human welfare, the tornado's destruction at Refinery Row caused significant damage to industrial equipment. There was a high risk of explosion from industrial propane tanks, some of which were severely damaged and leaking (APSS 1991). During the event, it became clear that more detailed maps of various sections of the city were in demand and that evacuation methods could be improved. Nevertheless, the prompt response of emergency personnel averted numerous ancillary crises. For instance, when reports of potential natural gas leaks coincided with details of fallen power transmission towers and high tension lines, power was promptly shut off in the area (APSS 1991).

Despite the multitude of hardships, the people of Edmonton demonstrated resilience and compassion in their response to the disaster, and they are an example of how civilians and government can work together in an emergency. A civilian weather spotter first reported the tornado. Environment Canada responded quickly and issued a tornado warning within 5 minutes. Police, fire, EMS, and civilians worked together to search for survivors among the debris. Although over 400 people were brought to hospitals with injuries, less than 100 of those were delivered by ambulance—the others were brought in civilian vehicles commandeered by police or by civilians themselves. In their post-incident assessment of the assistance of on-site volunteers, the Alberta Public Safety Services (APSS) noted that the response to the tornado "confirmed that our most valuable resource is the civilian volunteer. Emergency organizations provide the overall structure and guidance while volunteers are able to provide additional support and expertise. This should be encouraged" (APSS 1991, 73).

CHANGES TO EMERGENCY RESPONSE PLANS

In the aftermath of Edmonton Tornado of 1987, a number of changes and improvements were implemented to facilitate emergency response for future incidents. The federal minister of environment initiated a review of the weather warning system with the aim to "analyze the existing system and recommend improvements that would result in more effective public warnings of dangerous weather conditions" (APSS 1991, 18). The review concluded that although the weather forecasts were excellent, the weather warning communication procedures were "too time consuming," and consequently public awareness and warning suffered (APSS 1991). The review prompted a number of changes to the weather warning system in the region.

One major response to the tornado was the implementation by Environment Canada of Alberta's first civilian Doppler radar station in Carvel (near Edmonton) in 1991, to improve forecasting and warnings (Charlton et al. 1998). Although the severe weather forecasting team in Edmonton had appropriately disseminated a tornado warning, it was determined that enhanced technological capability would improve regional storm detection and forecasting. The improved Doppler radar capability has also been complemented by a deeper investment in developing a more complete spotter network of qualified and experienced storm spotters. The potential for early detection of a tornado occurrence by Doppler radar data and timely ground verification from a storm spotter can result in increased warning lead time and be a lifesaving combination.

A second significant outcome was the implementation of the Alberta Emergency Public Warning System (EPWS) by the government of Alberta in 1992.

"The EPWS was the first rapid warning system of its kind to use media outlets to broadcast critical life-saving information directly to the public" (AEA n.d.a). The system has subsequently evolved into the Alberta Emergency Alert (AEA) since 2011 and now utilizes digital technology (including social media and a mobile app) to provide warnings and information to the public (AEA n.d.a).

The Black Friday tornado exposed "a number of weaknesses and deficiencies in the methods of communication of severe weather information to the public and in public awareness of Environment Canada's severe weather warning program" (APSS 1991, 22). The tornado made it clear that improvements to public education and outreach among Environment Canada, Emergency Preparedness Canada, provincial public safety agencies, media, and the public themselves would enhance the regional safety net in the event of another disaster. The devastating results of the tornado helped new partnerships to form between the Alberta government and local municipalities and served to strengthen commitments to Disaster Recovery Programs across the province. AEA now issues two types of alerts: Critical Alerts and Information Alerts. A Critical Alert is sent when there is an imminent life-threatening danger, and an Information Alert provides less critical information to help the public prepare for an emergency.

CONCLUSION

The Edmonton Tornado of 1987 left an imprint on the psyches of Albertans and Canadians that is still evident today. Black Friday stimulated an increasing consciousness about disaster preparation and risk management and highlighted the need for effective communication and public warning in the case of emergencies. The event motivated the Province of Alberta to become a leader in emergency management, with the Alberta Emergency Alert system preceding the national emergency notification system for Canada (Alert Ready) by several years. There is heightened recognition that urban growth in Alberta in general and in Edmonton in particular increases the population's susceptibility to severe weather. This recognition has generated more dedicated efforts to improve the ability to predict and forecast storm events and invigorated attempts to identify place-based vulnerabilities and to coordinate emergency planning across communities in the province.

Valerie Thomas and Robert Oliver

Further Reading

AEA (Alberta Emergency Alert). n.d.a. Program History. http://www.emergencyalert.alberta.ca/content/about/aboutprogram.html, accessed February 1, 2019.

AEA (Alberta Emergency Alert). n.d.b. Program History. https://www.emergencyalert.alberta.ca/content/about/mobileapp.html, accessed February 1, 2019.

APSS (Alberta Public Safety Services). 1991. *Tornado, A Report: Edmonton and Strathcona County, July 31st 1987*. Edmonton, AB, Canada: APSS.

Charlton, B. C., B. M. Kachman, and L. Wojtiw. 1998. The Edmonton Tornado and Hailstorm: A Decade of Research. *CMOS Bulletin* 26: 1–56.

ECCC (Environment and Climate Change Canada). 2018a. Alberta. All Verified Tornadoes by Fujita Scale (1980–2009), Including Tracks where Available. Canadian National Tornado Database: Verified Events (1980–2009)—Public. http://donnees.ec.gc.ca/data/weather/products/canadian-national-tornado-database-verified-events-1980-2009-public/AB_VerifiedTornadoes-TornadesVrifis_1980-2009.png, accessed February 18, 2019.

ECCC (Environment and Climate Change Canada). 2018b. Canadian Tornado Fact Sheet, Based on the 30-year (19880–2009) Data Set. Canadian National Tornado Database: Verified Events (1980–2009)—Public. http://donnees.ec.gc.ca/data/weather/products/canadian-national-tornado-database-verified-events-1980-2009-public/CanadianTornadoFactSheet.pdf, accessed February 18, 2019.

NCEI (National Centers for Environmental Information). 2019. U.S. Tornado Climatology. National Oceanic and Atmospheric Administration. https://www.ncdc.noaa.gov/climate-information/extreme-events/us-tornado-climatology, accessed February 18, 2019.

European Heat Wave, 2003

A heat wave is a prolonged period of excessive hot weather accompanied by high humidity and is generally measured relative to normal seasonal temperatures in a region. For example, the year 2003 was the hottest summer recorded in Europe since 1540, and many European countries experienced their highest temperatures on record. More than 35,000 people died in the 2003 heat wave, which was one of the 10 deadliest natural disasters on the continent in the previous 100 years. The situation was particularly severe in France, which alone experienced more than 14,000 deaths or 40 percent of the total human death toll. Along with France, several other European countries reported deaths: Italy (3,100), Portugal (2,100), the United Kingdom (2,000), the Netherlands (1,500), and Germany (300) (Brucker 2005; Pirard et al. 2005; Met Office 2015). In that year, Europe had extremely dry and warm weather from late May to the end of August. In some regions, during the day, the temperature remained above 100 degrees Fahrenheit in August, and even at night, temperatures exceeded the average. Overall summer temperature soared to 20–30 percent above the average during this extreme climatic condition, which particularly affected senior citizens as they were more prone to adversities. Ultimately, the losses from the heat wave are estimated at more than €13 billion (UNEP 2004).

The heat wave extended from northern Spain to the Czech Republic and from Germany to Italy. Anomalies in summer temperatures were also observed in Scandinavian countries. Meanwhile, the daytime high temperature broke the record in the United Kingdom on August 10, 2003, at above 101 degrees Fahrenheit (38.3 degrees Celsius); the following day, the record temperature in Switzerland was over 106 degrees Fahrenheit (38.3 degrees Celsius). Ultimately, June was the hottest month ever recorded in that country in 250 years of archives, while temperatures in France stayed above 100 degrees Fahrenheit (37.8 degrees Celsius) for two weeks. Moreover, combined with a lack of precipitation, the high temperatures created a water balance deficit of up to 15 inches (380 mm) in Southern Europe and of 8 inches (200 mm) over Western and parts of Eastern Europe (UNEP 2004).

ORIGIN AND BACKGROUND

May–August 2003 yielded anomalies of low-level stream function. Hence, through this period, Europe was covered by irregular anticyclone conditions, which prevented precipitation. From the Mid Atlantic to Eastern Europe, there was a displacement of the subtropical azure anticyclone, and a general strengthening flow, which normally covers the southern Scandinavian countries, was absent. For each subsequent summer month, anomalies of cyclonic or stream function struck almost the whole Atlantic sector. For the northern coast of South America, the stream function was low; for the Caribbean it was high, and the value was very high in continental Europe. During that period, there were also anomalies extending from Russia to the Pacific coast. Experts believe that these resulted in the worsening of weather conditions in Europe.

Additionally, heat accumulated on surfaces during the day, which were later gradually released at night; therefore, nights were warmer than average. However, due to gradual heat release in the form of radiation under clear skies at night, just before dawn, the temperature cooled down a bit. This created a strong ground flux, which contributed to exceptionally warm nights such that the sustained thermal stress led to increased mortality rates. In the summer of 2003, varied atmospheric circulations caused an imbalance of heat distribution under a clear sky, and subsequent absence of precipitation led to the warmest summer on record.

IMPACTS

A heat wave causes severe impact on humans, particularly on the elderly, infants, and children. People suffering from chronic illness, people on medication, people who spend a lot of time in the sun due to work or sport, and immobile people are also most at risk. Apart from deaths, major health impacts include heat stroke (a condition when the body fails to regulate the body temperature by sweating alone), dehydration (loss of necessary water from the body, which causes tiredness, breathing problems, and problems in heart rate), and sunburn (damage to the skin, which can be painful and can lead to cancer). Also, reportedly people have died due to drowning in pools while swimming because of extreme heat. Furthermore, heat waves cause huge amounts of air pollution from different sources such as wildfires (Met Office 2015).

The physical impact of the 2003 heat wave was immense, and it was observed ubiquitously across Europe. Lowered water surface in rivers and lakes was very common at the time. In Serbia, the River Danube fell to its record lowest level in 100 years. Weapons such as tanks and bombs submerged under water since World War II (1939–1945) were exposed. This caused danger to the people swimming in the rivers as well as to the local communities. Aquifers, reservoirs, and rivers, which were used for public water supply, dried up, and such situations were very common all over the Europe (Met Office 2015). Extreme heat caused snow and glacier melting in the European Alps, which led to rock and ice falls in the mountains. According to the World Glacier Monitoring Service (WGMS), a mass of Alpine glaciers decreased by up to 10 percent in 2003 during the heat wave. Simultaneously, temperatures increased above the elevation of 14,765 feet (4,500 m) for more than 10 days. This duration is considerable for shady rock walls in mountains and at high elevations. Ultimately, permafrost changes exposed land, which was locked for millions of years, while thaw penetration in the mountains was more than 6 feet (2 m) deep than in previous years, contributing to large landslides and rock falls.

Additionally, this heat wave decreased both the quality and quantity of harvests all over Europe; particularly in Southern and Central Europe, the situation was dire because while a large proportion of the harvest was under threat anyway, the heat wave increased production costs additionally. This is because it accelerated crop development by 10–20 days, causing early ripening and maturity. Added to that, the moisture level in the soil was very low, which affected winter crops as well. A very high amount of solar radiation and high temperatures began at the end of July until mid-August, which resulted in a very high rate of water consumption by the plants. This caused the water level to drop even more. Even in Switzerland, which is known as the "water tower" of Europe, water withdrawal for agricultural purposes from the river was banned from mid-July to mid-October. According to the Union of Swiss Farmers, production of potato and tobacco was very poor that year, and the loss amounted to $230 million (UNEP 2004).

The heat wave also seriously hampered wine production; however, the sectors most severely affected were fodder supply, arable and forestry sectors, and livestock. The fodder deficit ranged from 30 percent (Germany, Austria, and Spain) to 40 percent in Italy and 60 percent in France. The situation was so serious that Switzerland had to import fodder from Ukraine while cereal production decreased by more than 23 million tons. Livestock farmers suffered more due to lack of green fodder, and even in winter, that particular situation was not fixed. Due to the drought, some crops failed completely (UNEP 2004).

The European heat wave-induced wildfire raged across Western Europe and destroyed nearly 1.6 million acres (650,000 ha) of forestland. This wildfire was significantly widespread and deadly in comparison to any other past summer wildfires. More than 25,000 fires were reported all over Europe but mostly in Portugal, Spain, Italy, France, Finland, Ireland, Denmark, and Austria. In Portugal, forest fires burned 5.6 percent of the total forest area of the country, comprising 964,050 acres (390,146 ha), which is an area bigger than Luxemburg. Spain lost more than 315,114 acres (127,525 ha) of forestland. Wildfire also burned more than 109,028 acres (44,123 ha) of agricultural land. For Portugal, the incident was the worst of any in the previous three decades, and the financial impact was more than one billion Euros (UNEP 2004; Met Office 2015).

The 2003 heat wave created a huge problem in the energy sector. It caused great damage to France's nuclear plants, as there was severe shortage of water in the rivers for cooling down reactors. The country derives about 75 percent of its electricity from nuclear energy. Usually, nuclear power facilities in France return water to rivers after cooling down, which is environmental friendly and causes less environmental damage than other sources of energy production. However, in some places, water levels of rivers were so low that it became impossible to get sufficient water for running the power plant. As a consequence, some power plants were shut down just when demand for electricity peaked. The demand grew much higher as people were using air-conditioning and refrigerators more than usual. Thus, the supply of nuclear energy-derived electricity was greatly reduced. The situation was so bad that France cut its power export by more than half in order to conserve energy. Germany, which is less dependent on nuclear energy than France, had to close two nuclear stations (UNEP 2004).

The heat wave also caused great social and environmental impacts all over Europe. For instance, there was a drinking water shortage in some parts of the United Kingdom, and banning of hose pipes was ordered. Many parts of the United Kingdom saw increased numbers of tourists as people thought the temperatures were a bit cooler there than in the other parts of Europe, which made transport and traveling very difficult. Underground rail stations became unbearable, and some road surfaces melted completely and were unusable. Small rivers shrunk, and lack of water hampered travel routes. Moreover, the London Eye was closed for a day due to excessive temperatures as the cabin was too hot for people.

RESPONSES

Responses to the heat wave were somewhat insufficient and generated anger in part because nobody actually knew how to act in such a situation. France had no idea how to manage so many dead bodies and ended up collecting bodies in a room with refrigeration. This kind of action made people unhappy and angry. Lack of water and energy was widespread. France ended up applying for a fund from the European Union (EU) to deal with the situation. Scarcity of drinking water was a huge issue in the United Kingdom and Croatia. Television, newspaper, Internet, and radio were used to inform the masses to drink more water, wear cool clothing, and stay in shade during the middle of the day. Several countries, such as the United Kingdom, imposed speed limits on trains to avoid any kind of accident, while working hours for workers were rescheduled to avoid the worst of the heat (Met Office 2015).

To reduce the impacts of the heat wave, the National Health Service (NHS) in the United Kingdom advised people to stay out of direct sunlight and avoid going out at midday, when the sun was the hottest. It also advised people who wished to go outside in daytime to wear the following: 1) sun protection, 2) loose-fitting light clothing, and 3) a hat. Other advisories included staying well hydrated, avoiding alcohol, and having cold showers/baths. If any persons planned to travel, they were advised to carry adequate water.

CONCLUSION

The heat wave in 2003 was a great disaster for Europe. It took many lives and caused tremendous amounts of damage to property and the natural environment. In the meantime, it showed the world the consequences if people did not make some changes to their lifestyles, given the link that some experts found between the 2003 heat wave in Europe and the phenomenon of climate change. After the event, Europeans developed strategies to anticipate what to do if climate change were to bring more heat waves to Europe and how to reduce their number. These strategies centered on more research and fund gathering. In fact, after the 2003 event, two major heat waves struck Europe, but the damage was less. Clearly, at the cost of many lives, the heat wave of 2003 has made people more aware of environmental issues.

Md. Nadiruzzaman

Further Reading
Brucker, Gilles. 2005. Vulnerable Populations: Lessons Learnt from the Summer 2003 Heat Wave in Europe. *Eurosurveillance* 10(7–9): 147.
Met Office. 2015. The Heatwave of 2003. https://www.metoffice.gov.uk/weather/learn-about/weather/case-studies/heatwave, accessed September 21, 2018.
Pirard, Philippe, Stephanie Vandentorren, Mathilde Pascal, Karine Laaidi, Alain Le Tertre, Sylvie Cassadou, and Martine Ledrans. 2005. Summary of the Mortality Impact Assessment of the 2003 Heat Wave in France. *Eurosurveillance* 10(7–9): 153–156.
UNEP (United Nations Environment Programme). 2004. *Impacts of Summer 2003 Heat Wave in Europe: Environment Alert Bulletin 2*. Nairobi, Kenya: UNEP.

Eyjafjallajökull Eruption, Iceland, 2010

Eyjafjallajökull, an icecap-covered volcanic peak in the Eastern Volcanic Zone in southern Iceland, began to erupt early morning on April 14, 2010, ejecting a plume of volcanic ash at times several miles into the atmosphere. The plume rose above snow and cloud and created a shadow to the north. Horizontally, the plume extended roughly 40 miles (60 km) almost east shortly after noon, but by 2:25 p.m. local time, the plume reached more than 60 miles (100 km) from the eruption site. The vertical eruption caused widespread disruption of air travel in Northern and Western Europe. The first eruption, which occurred on March 20, 2010, came from an ice-free area on the north-east side of the volcano. This eruption produced lava but little explosive activity. The volcano remained relatively quiet until April 14, 2010, when eruption caused melting of large amounts of ice, leading to flooding in southern Iceland. The interaction of magma with water created a plume of volcanic ash and gas over 6.6 miles (10 km) high, which spread out and was carried by winds south-eastward towards Northern and Western Europe (BGS 2010).

On Sunday April 18, 2010, Eyjafjallajökull ejected about 750 tons of magma every second, and by April 20, most of the ice in the crater appeared to have melted; the plume was only reaching heights of up to about 3 miles (4 km), but the amount of material being ejected had increased significantly. As the amount of ice available to interact with the magma decreased, the volcano switched from producing ash to producing mainly fire fountains. By the end of April, explosive activity had virtually ended, and a weak plume remained, largely made up of steam, but flowing lava had advanced a few miles northward from the crater. In early May, explosive activity began to increase again, with more ash being ejected into the plume, which reached at times up to 6 miles (9 km) height. During early May, the activity became cyclical, and ash and gases started to rise from the

Mount Eyjafjallajökull in Iceland erupted in 2010. The Thorvaldseyri farm and fields are shown below the mountain. (Johann Helgason/Dreamstime.com)

volcano. However, thereafter, explosive activity gradually decreased, and by May 23, little or no ash was being ejected from the volcano (BGS 2010).

Iceland lies on the Mid-Atlantic Ridge (MAR), the boundary between the Eurasian and North American tectonic plates. As these two divergent tectonic plates move apart at a rate of 0.98 inch (2.5 cm) per year, magma from the mantle continually reaches the sea surface and creates new sea floor. This ongoing ridge of volcanism interacts with a mantle plume and has created the island of Iceland, which comprises many active volcanoes. This volcano-prone country experiences on average one volcanic eruption every three to four years. However, Eyjafjallajökull last erupted in 1821, and the previous eruptive cycles there had begun with basaltic eruptions on the flanks of the volcano, followed by the eruption of more silica-rich magmas from the volcano's summit. Preliminary ash analyses of the 2010 eruption followed a similar pattern (BGS 2010). Historically, Eyjafjallajökull eruptions have been followed by eruptions in its larger sister volcano, Katla, which indeed erupted on July 8–9, 2011.

SEISMIC AND VOLCANIC ACTIVITIES PRIOR TO ERUPTION

Growing seismic and volcanic activity in Eyjafjallajökull and in the surrounding region preceded the 2010 eruption. In February, the Icelandic Civil Protection Authority (ICPA) increased monitoring of the area, and on March 4, the ICPA declared an "uncertainty phase" for the area, the lowest of three emergency levels. On March 20, 2010, an eruption began in Fimmvörðuháls mountain pass between the Eyjafjallajökull and Mýrdalsjökull glaciers. At the time of eruption, the emergency level remained unchanged, reflecting the extreme difficulty of predicting volcanic eruptions. Moreover, the Fimmvörðuháls eruption took place near a popular hiking trail where many hikers were present, who were evacuated from the trail by Coast Guard helicopters. About 500 farmer households also were evacuated from the areas of Fljótshlíð, Eyjafjöll, and Landeyjar prior to eruption and flights to and from Reykjavík and Keflavík International Airport were postponed. The police temporarily closed the road to Þórsmörk. Finally, the Fimmvörðuháls eruption stopped on April 13, 2010.

On the evening of March 21, 2010, domestic and international air traffic was permitted again. Evacuees were allowed to return to their farms and homes on March 22, and authorities temporarily halted the evacuation plan. Additionally, roads and trails in the volcanic risk zones were reopened on March 29, 2010. When the second fissure appeared, the road was closed again because of the danger of flash floods, which could have developed if the fissure had opened near big ice caps or other snow reservoirs, but the road was again opened on April 1.

However, early morning of April 14, 2010, a far larger and more powerful eruption began beneath Eyjafjallajökull glacier. While the Fimmvörðuháls eruption was characterized by a steady flow of lava from the crater, the Eyjafjallajökull eruption violently threw magma and ash into the air. This eruption produced very little lava, but ejected huge quantities of glass-rich ash into the atmosphere. The ash plume ejected by the eruption covered farmland south of the glacier, causing significant disruption and hardship to farmers in the area. Not until June 23, 2010, did the Iceland Meteorological Office (IMO) and the University of Iceland, Institute of Earth Sciences (IES), stop issuing regular status reports of the eruption.

IMPACT

The eruption of Eyjafjallajökull dramatically affected air traffic in much of Northern and Western Europe. The volcano's explosive power was enough to inject ash directly into the jet stream, directly above the eruption site. The direction of the jet stream was unusually stable at the time of the explosive phase (April 18–May 4) of eruption, which happened beneath 660 feet (200 m) of glacial ice. The resulting meltwater flowed back into the erupting volcano, which created two specific phenomena: 1) the rapidly vaporizing water significantly increased the eruption's explosive power and 2) the erupting lava cooled quickly and created a dark cloud with glass-rich ash.

Naturally, airplanes need to avoid flying into ash plume, which is made up of large amounts of microscopic particles of volcanic rock and pulverized glass. These small particles can clog up jet engines and stop them from working. In such situations, planes are either rerouted or grounded to ensure the safety of passengers. Re-routing proves to be safer, but it is costly. Consequently, from April 15 to 21, 2010, over 300 airports in about two dozen countries closed their airspace to commercial jet traffic, causing massive impacts on air travel worldwide. That week, over 100,000 flights were canceled or delayed, causing the largest disruption to air traffic since the World War II (1939–1945) and costing airlines $1.7 billion in lost revenue. Iceland suffered economically because of a significant decrease in tourism, which had pulled the country out of a deep economic recession following the 2008 financial crisis.

At the time of the Eyjafjallajökull eruption, many suspected that the impact would lead to the collapse of the tourism industry in Iceland. Therefore, the government initiated an advertising campaign to encourage people to visit Iceland. Ironically, in fact, international media provided free advertising so that news of the eruption dominated the media across the globe for almost a month. The result was very positive; the number of foreign travelers visiting Iceland increased by 15 percent in 2011 compared to 2010, by 19 percent in 2012, and by 20 percent in 2013. By 2016, nearly 1.8 million foreign travelers visited the country (Iceland Magazine Staff 2017).

The Eyjafjallajökull eruption caused flooding in nearby areas because of the glacial meltwater that lay above the volcano. This flood and heavy ash fall damaged agricultural land. Furthermore, wet ash had become compact, making it very difficult to continue farming, harvesting, or grazing livestock, and it even poisoned animals on nearby farms. Additionally, roads were destroyed, and the ICPA advised people to stay indoors because of the danger from ash in the air. Apart from widespread disruption of air travel, European countries lost business trade because several sectors of economy that depended on air-freighted imports and exports were badly affected by the flight disruptions. For example, shortages of imported flowers, fruits, and electronic hardware were reported in the immediate days after the disruption.

Many people in Europe were not able to get to work because they were stranded. Perishable foods were wasted and drugs spoiled as they could not be transported. Several Asian and African countries also experienced similar economic losses. Sporting, entertainment, and many other events were canceled, delayed, or disrupted when individuals or teams were not able to make the trip to their destination. A number of world leaders and politicians had to postpone scheduled trips, or flights were diverted and delayed due to the closure of airports and airspace. The timing of the disruption during the Easter holidays was when levels of tourism typically are high.

Although no one died directly from the 2010 eruption of Eyjafjallajökull, the people who lived near the volcano site reported high levels of irritation symptoms even while their lung function was normal. Six months after the eruption, the residents of southern Iceland showed more respiratory symptoms than the people of northern Iceland, who endured no or much less ash fall. Similar respiratory problems were also reported in Europe, particularly in Scotland. Climatologists believed that the atmospheric effects of Eyjafjallajökull would not last long. There would be short-term impacts such as the glass-and-rock mixture being fatal to most plant species since it prevents photosynthesis; as a result, greenery in Iceland and parts of northern Europe fought to survive for several months until the ash simply blew away or fossilized to form new rock. It is worth noting that volcanic eruption does not emit much carbon to the atmosphere and thus has little impact on global climate change (Stone 2010).

RESPONSE

The ICPD evacuated people from volcanic risk zones. On April 18, 2010, the Icelandic Food and Veterinary Authority (IFVA) alerted all horse owners who kept their herds outside about the heavy ash fall, and where ash fall was projected to be significant, the IFVA directed all horses to be sheltered indoors. Based on in-depth interviews and questionnaire surveys conducted among governmental officials, rescue team members, and residents a few months after the eruption, some scholars maintain that Icelanders responded reasonably well to evacuation guidelines when the volcano Eyjafjallajökull erupted in 2010 (Bird and Gísladóttir 2018). Scholars also believe that the government was able to trigger a positive response to evacuation orders due to ongoing preplanning in response to seismic assessments and the trust and respect Icelanders had for the authorities.

Officials of the ICPD alerted individuals and households near Eyjafjallajökull based on evidence of a sudden increase in the amount of water in the area's streams and rivers. Combined with rising water temperature, the increase in streamflow signaled the onset of a flood caused by an eruption under a glacier. Thus, the officials contacted residents via an automated phone alert system as well as by going door to door to make sure that all residents had received the warning and accordingly taken safer shelter. However, only about 50 percent of people who received a warning actually evacuated while the remaining 50 percent stayed home to take care of older family members, others, or livestock (Bird and Gísladóttir 2018).

CONCLUSION

Eyjafjallajökull was not a big eruption, but still it caused severe disruptions to air traffic in Europe and the North Atlantic region. However, the positive impact of this eruption has been a notable increase in volcanic-cloud research and a subsequent burst of published scientific articles. Overall, this and subsequent eruptions in Iceland have prompted the aviation industry, regulators, and scientists to cooperate more closely to define hazardous airspace and improve the manner in which airspace is forecasted and communicated. There is another positive aspect: airplanes did not fly for a week, reducing the amount of carbon emission even after allowing for the amount that came from the eruption of Eyjafjallajökull.

Bimal Kanti Paul

Further Reading

BGS (British Geological Survey). 2010. Eyjafjallajökull Eruption, Iceland: April/May 2010. https://www.bgs.ac.uk/research/volcanoes/icelandic_ash.html, accessed December 20, 2018.

Bird, D. K., and G. Gísladóttir. 2018. *Responding to Volcanic Eruptions in Iceland: From the Small to the Catastrophic*. Palgrave Communications 4: 151. https://doi.org/10.1057/s41599-018-0205-6, accessed December 20, 2018.

Iceland Magazine Staff. 2017. Seven Years Ago Today: Eruption in Eyjafjallajokull, the Volcano with the Un-pronounceable Name. *Iceland Magazine*. April 14, 2017. https://icelandmag.is/article/seven-years-ago-today-eruption-eyjafjallajokull-volcano-un-pronouncable-name, accessed December 20, 2018.

Stone, Daniel. 2010. The Environmental Effects of Iceland's Volcano. *Newsweek*. April 20, 2010. https://www.newsweek.com/environmental-effects-icelands-volcano-70371, accessed December 21, 2018.

Grand Forks Flood, North Dakota, 1997

Flooding from spring snowmelt is an annual occurrence in the Red River Valley of the North, a long, narrow, prosperous, agricultural valley along the North Dakota-Minnesota border. The Red River of the North flows slowly northward along a sinuous path within a former glacial lake-bed, creating a natural flood conveyance problem during any spring with significant runoff due to snowmelt production.

Weather conditions during the winter of 1996–1997 produced a series of weather and flood disasters throughout the state. A total of nine blizzards, seven more than normal, led to the death of 123,000 cattle statewide. On April 4–6, 1997, Blizzard Hannah, the most extreme blizzard in fifty years, with a minimum central pressure of 974 mb (974 hPa) and maximum wind speeds of 80 mi/hr (130 km/hr), left half of North Dakota without electricity in sub-freezing temperatures for a period of days to weeks. Several smaller communities experienced spring flooding along the Red River, and the Fargo metropolitan area narrowly escaped severe flooding. The Grand Forks, ND–East Grand Forks, MN metropolitan area, was not as fortunate, however, and experienced what, at the time of the flood, was the costliest flood for a metropolitan area on a per capita basis in U.S. history. Estimates of the flood recurrence interval place the flood peak at about a 200–250-year event (Pielke 1999).

Grand Forks officials and residents knew the spring flood threat was extreme. Fall precipitation was up to 300 percent of the normal, reducing the ability of soils to absorb spring snowmelt. Wet soils froze to as much as three feet (1 m) depth, creating soil frost impermeable to the infiltration of spring snowmelt. Winter snowfall totaled 100 inches (250 cm), was heavy across the entire basin, averaging 200–300 percent of the normal, and was unusually dense, with 6–16 inches (15–40 cm) of liquid water equivalent. The early snowmelt period in March reduced the snow cover modestly across the region, but the 1.4–3.0 inches (3.5–7.5 cm) of liquid water equivalent from the blizzard of April 4–6 replaced all the melted snow cover. After Blizzard Hannah, air temperatures that had been below 32 degrees Fahrenheit (0 degrees Celsius) for nearly four months suddenly rose to daytime highs of 50 degrees Fahrenheit (10 degrees Celsius), with nighttime lows above 32 degrees Fahrenheit (0 degrees Celsius). Rapid and continuous snowmelt production occurred throughout the Red River Valley, overwhelming the capacity of the Red River of the North to convey river flows (Todhunter 1998).

EMERGENCY FLOOD PREPARATIONS

City emergency management officials had spent months preparing for the spring floods. The initial National Weather Service (NWS) numerical outlook of 49.0 feet (14.94 m) allowed city officials to protect the city to a numerical outlook of 52.0 feet (15.85 m). The 49.0 feet (14.94 m) outlook was the same river stage of the

The Sorlie Memorial Bridge between Grand Forks, North Dakota, and East Grand Forks, Minnesota, during the Red River of the North flood, April 21, 1997. (U.S. Geological Survey)

previous flood-of-record (FOR) in 1979, when emergency preparations by the city and a massive volunteer effort successfully protected the city from catastrophic flooding. Local government officials and city residents were confident that a similar emergency preparation and volunteer effort could successfully protect the city once again. This led to a narrow flood defense strategy that placed too much confidence in the 49.0 feet (14.94 m) forecast and too heavy a reliance on emergency flood-protection efforts, which included work by 12,000 volunteers to fill and deploy almost 1.4 million sandbags (Todhunter 2001).

The NWS maintained a numerical outlook of 49.0 feet (14.94 m) for 47 days. On April 14, 1997, the first numerical forecast was issued by the NWS based upon the more complex and physically based NWS River Forecast System. This numerical river forecast was raised six times within 100 hours to a forecast peak of 54.0 feet (16.46 m). The rapid change in the forecasted flood peak was in response to the extremely rapid snowmelt occurring across the basin. Raising flood protection from 49.0 feet (14.94 m) to 54.0 feet (16.46 m) in a week was simply an impossible task for city emergency managers and engineers (Pielke 1999).

THE FLOOD CATASTROPHE

On April 18, 1997, the first break in the emergency flood defense works occurred. This was followed in rapid succession by numerous other failures in flood defenses throughout the metropolitan area. Chaos ensued in the following

24 hours as the city rapidly transitioned from flood-protection to city-evacuation mode. Eventually, river discharge peaked at 137,000 cubic feet per second (3,880 cubic meters per second) on April 18, 62 percent greater than the FOR. River stage peaked at 54.35 feet (16.57 m) on April 22, 5.55 feet (1.69 m) higher than the FOR. Several fires, probably initiated by floodwaters, started in the downtown on April 19, which caused city officials to shut off electricity as individual sections of the city were evacuated. Flooding occurred from multiple sources: overland flow from the river, backward flow through the city's storm drain system, sewage backup in residential homes, and ground water flooding in basements from sump holes made nonworking due to the electrical shut-off. The Red River of the North was at the official flood stage from April 4 to May 19, 1997 (Todhunter 1998, 2001).

The flood affected approximately 60,000 people. About 55,000 residents—100 percent of East Grand Forks, MN residents, and 85 percent of Grand Forks, ND residents—evacuated their homes for periods of weeks to months. Grand Forks Air Force Base served as an emergency processing center for evacuees, who relatively quickly relocated to the homes of friends, relatives, and volunteers in the region. Grand Forks was without running water for 13 days and drinkable water for 23 days. In total, nearly 36,500 North Dakotans registered official damages with the Federal Emergency Management Agency (FEMA).

FLOOD IMPACTS AND FINANCIAL ASSISTANCE

Statewide, 2,200 sq. mi (5,700 sq. km) area was flooded, and the Red River was 7–10 miles (11–16 km) wide at its peak. An estimated 12,000 residential and commercial properties sustained damage; this included 694 dwellings and 493 other structures demolished because their damage exceeded 50 percent of property value. An additional 325 homes were bought out for flood mitigation purposes or construction of the floodwall/dike system. Also, 161 homes and 414 other structures were relocated following the flood. In addition, 11 buildings burned down from the fires of April 19. Infrastructure and property damage totaled $558 million, including $42 million to repair or replace sewer and water lines, street, and related infrastructure. At the University of North Dakota (UND), 70 buildings were affected by flooding, totaling $75 million in damages. In the aftermath of the flood, the population of Grand Forks fell by 3,000 people (6 percent), East Grand Forks lost 1,200 people (15 percent), and UND enrollment fell by 900 students (8 percent) (NDDES 2011).

Direct federal payments for emergency flood protection efforts, emergency operations, flood cleanup, and all other forms of flood assistance and recovery totaled $1.5 billion. City workers removed more than 60,000 tons of debris from the city. Statewide, 1,000 North Dakota National Guard and 1,886 Air and Army Guard personnel assisted in the emergency efforts. The National Flood Insurance Program paid $75 million in claims to city residents, although the number of purchased flood insurance policies was low prior to the flood. A total of 4,378 Small Business Administration loans totaling $180 million were administered to business and individuals to assist in repair efforts. FEMA provided $44.8 million to 21,846 applicants for Disaster Housing Assistance, and an additional $5 million in Disaster Unemployment Assistance benefits. Federal and state aid through the Public Assistance program to governmental and nonprofits for infrastructure repair reached $211 million. The U.S. Department of Housing and Urban Development allocated $201 million in grants to the state. FEMA Hazard Mitigation Grant Program funding of $13.7 million was awarded to the City of Grand Forks for various projects to mitigate against future flood damages (NDDES 2011).

Assistance from private, philanthropic, and nongovernmental organizations was also very significant. Most notable was the $15 million given by Joan Kroc, the McDonalds heiress, who was moved by images of the flood damages and community response. These funds were awarded in the amount of $2,000 to more than 7,500 applicants. The Ronald McDonald House Foundation contributed an additional $5 million toward this effort.

LONG-TERM FLOOD RECOVERY

Grand Forks is recognized by FEMA as one of the most successful long-term disaster recovery efforts in U.S. history. Several factors contributed to this success. The disaster occurred during a period of strong national economic growth, which allowed for significant federal disaster assistance. The state economy was also strengthening, which allowed the state to assist in the local cost share attached to the selected federal assistance. Although the disaster was very intense, it had a limited geographical scale. This enabled volunteer efforts to be effectively coordinated and targeted for disaster recovery. There was strong agreement among state and local representatives and the citizens of Grand Forks on recovery priorities, including the need for effective long-term flood mitigation, a more prosperous downtown, urban redevelopment, housing development, and economic growth. The limited size of the city and strong social connectedness of the region led to transparency in decision-making processes, helping the community to come together in the recovery process. Strong local leadership, combined with a well-placed and effective congressional delegation, led to very efficient integration of local, state, federal, and private agencies and services. The city residents possessed high levels of social, political, and economic capital and strong human resources, which made the community less vulnerable to the ravages of the flood catastrophe, more resilient to its aftereffects, and more focused on the long, slow path to recovery. Finally, the flood disaster occurred during a relative "news desert," which, when combined with some spectacular visual photographic images, helped the city to garner considerable assistance and support for its recovery efforts (Todhunter 2001).

The flood threat in Grand Forks has been effectively mitigated since the 1997 flood. Most of the subsequent city growth and expansion has been to the south and southwest, the highest elevation parts of the metropolitan area that were unaffected by the 1997 flood. More than 1,000 homes and structures were removed from the former 100-year floodplain. In 2007, the U.S. Army Corps of Engineers completed a $409 million levee and floodwall project that is eight miles (12.8 km) long and protects the city to a river stage of 60 feet (18.3 m). The State of North Dakota paid for the majority of the Grand Forks local cost share (NDDES 2011).

The $172 million in FEMA Community Development Block Grants were used to acquire land for flood mitigation and levee/floodwall construction (20.8 percent), invest in new housing construction (19.2 percent), rehabilitate existing housing (6.6 percent), repair and replace damaged infrastructure (23.3 percent), and promote economic development (23.6 percent). The downtown has experienced revitalization due to strong public and private investment and serves as a renewed center for financial, cultural, residential, and entertainment sectors. The Greater Grand Forks Greenway occupies what was once old residential housing and has become a focus of city life. The city population now exceeds pre-flood levels, as does the university student population; the city economy is also now stronger than at the time of the flood. The Alerus Conference Center, Ralph Engelstad Arena, Betty Engelstad Sioux Center, Cabela's, and King's Walk Golf Course, all built after the flood, offer amenities rarely found in a city of comparable size. Without question, the most significant long-term negative consequence of the flood has been its effect on housing affordability. The loss of nearly 1,000 homes due to flood damage, the geographical limitations on growth to any direction but the south, and the shortage of skilled workers in the building trades has produced a dramatic rise in median housing value. Housing affordability remains the most important long-term negative impact of the flood disaster.

Paul Todhunter

Further Reading

NDDES (North Dakota Department of Emergency Services). 2011. North Dakota Response to the 1997 Disasters. North Dakota Department of Emergency Services, Division of Homeland Security: Bismarck, ND. http://www.nd.gov/des/uploads%5Cresources%5C736%5Cn.d.-response-to-the-1997-disasters.pdf, accessed December 20, 2017.

Pielke, Roger A., Jr. 1999. Who Decides? Forecast and Responsibilities in the 1997 Red River Flood. *Applied Behavioral Science Review* 7(2): 83–101.

Todhunter, Paul E. 1998. Flood Hazard in the Red River Valley: A Case Study of the Grand Forks Flood of 1997. *North Dakota Quarterly* 65(4): 254–275.

Todhunter, Paul E. 2001. A Hydroclimatological Analysis of the Red River of the North Snowmelt Flood Catastrophe of 1997. *Journal of the American Water Resources Association* 37(5): 1263–1278.

Great Ice Storm of 1998, Canada

In January 1998, a major freezing rainstorm caused catastrophic losses across eastern Canada and the northeastern United States. Although ice storms in these regions are fairly commonplace, one study noted that "the quantity of ice accumulation and the persistence of the 1998 Ice Storm is often considered unprecedented" (Henson et al. 2007, 37). The 1998 ice storm was remarkable for its spatial extent, duration, and the severity of its short- and long-term impacts on human population, economy, environment, and the infrastructure and financial resources of the affected provinces and states. For this reason, it has popularly been called the Great Ice Storm of 1998. The ice storm devastated the electricity grid, leaving more than 15 percent of the Canadian population and more than 500,000 Americans without power (heat, light, and even water) during mid-winter. Twenty-eight Canadian and 17 American citizens lost their lives during the event (Lecomte et al. 1998). Most of the deaths occurred due to hypothermia and traffic accidents. Beyond human suffering, the ice storm caused significant monetary loss across numerous economic sectors that made the event the "costliest natural disaster to occur to date in Canada's history" (Lecomte et al. 1998, 37). While tragic, the event also prompted a more nuanced examination of the links between climate change and socio-spatial vulnerability and a reconsideration of technological resilience.

METEOROLOGICAL CONDITIONS

The ice storm of 1998 was the result of a temperature inversion. A series of low-pressure systems in the Southern United States brought warm, moist air from the Gulf of Mexico into southern Ontario and Québec. Meanwhile, a large stationary Arctic high pressure over northeastern Canada, coupled with a stationary Arctic front positioned just south of the affected area, caused "cold air to flow into the St. Lawrence and Ottawa River valley regions which the warm advected air could not dislodge" (Henson et al. 2007, 38). The synchronism of these air masses meant that as the precipitation from the warm moist air fell through the cold air it became super-cooled, and when it hit the cold surface, it froze instantly, coating everything in hard layers of ice.

Freezing precipitation began late on January 4, 1998, and continued to fall until January 10, 1998. Some areas received more than 3.2 inches (80 mm) of freezing precipitation during the event, while others experienced a mix of freezing precipitation and solid precipitation (i.e., ice pellets and snow). The accumulation of ice was extraordinary, more than doubling the amount recorded for historic severe storms in the region (Kerry et al. 1999). Although the storm was forecasted by Environment Canada and the National Weather Service (NWS) in the United States, the duration of the storm, the accumulation of ice, and the vast geographic area impacted were underestimated and served to expose vulnerabilities in the region (Murphy 2009). Apart from Ontario and Quebec, New Brunswick, Nova Scotia, and Newfoundland provinces were also affected by the ice storm. In the United States, several states such as Maine, New Hampshire, Vermont, and upstate New York incurred damage from this ice storm.

IMPACTS AND RESPONSE

The rapid formation of ice on all surfaces (i.e., tree branches, roads, wires, transmission towers) quickly became destructive. As the ice thickened, transmission towers collapsed, transformers failed, and electrical wires were severed. Falling trees and tree limbs caused further disturbance to the electrical, transportation, and communication networks. Impacts of the storm were magnified in part because it affected two of Canada's major urban centers, Ottawa and Montreal. Power grids were devastated in these metropolitan centers and in rural and urban communities throughout the region. In the province of Ontario alone, "2,800 kilometers of power wire had to be restrung, 10,750 hydro poles had snapped and had to be replaced, 84,000 insulators and 1,800 transformers had to be replaced, 300 steel transmission towers had been destroyed or damaged, and 40 percent of its electrical system had been destroyed or damaged" (Murphy 2009, 93). In Quebec, the damage was even greater, with Hydro Quebec, the province's main utility provider, reliant on a vast above-ground power distribution system (Kerry et al. 1999). Consequently, millions of people were left without power, some for more than a month. The situation quickly became dire as "an atmospheric disturbance of nature became a technological disaster" (Murphy 2009, 66).

The duration and spatial extent of the storm surprised officials (i.e., ministers of public security, meteorologists, emergency services personnel, and political representatives) and many were slow to respond. The miscalculation of risk, combined with a sizable portion of the population that was simply unprepared for the severe conditions or naive to potential dangers of the storm, resulted in social disorganization. The Province of Quebec, for instance, unlike Ontario and the affected states in the United States, did not declare a state of emergency during the storm because it did not have a history of making such declarations (Murphy 2009). Emergency response quickly scaled from the municipal level to senior orders of government, as many localities did not have suitable emergency plans in place to direct the delivery of resources and services. The storm did, however, reveal countless acts of courage, courtesy, and compassion by the citizenry to ease suffering where and whenever possible (Kerry et al. 1999).

The failings of the electrical grid had repercussions for other critical infrastructure networks (i.e., transportation, telecommunications, water supply, financials, and services). Road crews were unable to keep pace with the accumulation of ice and traffic congestion, and automobile accidents had cascading effects on hospital and other emergency service operations. Flights were delayed or canceled, train service interrupted, and numerous schools and businesses were closed. Three hundred sixty-two shelters were opened in Ontario and only 85 in Quebec. But all shelters struggled with issues of capacity and operability. Scores of individuals relocated to hotels and the houses of friends or family in search of power. In the absence of reliable power, communities were forced to use generators to power essential infrastructure. Thus, households were deprived not only of entertainment (i.e., television, and computers) and conveniences (i.e., microwaves, and stoves) but also of heat, light, and water (Murphy 2009).

The banking system also broke down, with people unable to access money because banks were closed and numerous automatic teller machines (ATMs) were inoperable. The ability to acquire basic goods became increasingly troublesome as the freezing rain persisted. The scope of the damage required a massive mobilization of the Canadian Armed Forces to provide humanitarian assistance and resulted in compassionate outreach from across the continent. While the buildup of ice was responsible for numerous injuries (i.e., traffic accidents, falling icicles, and slips and falls), deaths from carbon monoxide poisoning and fires caused by the improper use of fireplaces and other heat sources contributed to the tragic nature of the storm. The most affected people from this ice storm were the elderly who were dependent on medication and in need of special care during their stay in cramped shelters. Local hospitals treated more than the usual number of cases from falls on the slippery ice surface.

The consequences of lack of power quickly extended to the agricultural sector as dairy and other farming operations struggled to remain productive. Tens of thousands of farms, with varying degrees of preparedness, were affected. Reliant on electricity for powering milking machines, well pumps, manure-removal systems, and other forms of automated technology, many farmers were left in a precarious position. The unreliability of generators, combined with impassable roads, meant that petrol delivery was unpredictable and resulted in more than 20,000 cows dying in New York State (Murphy 2001). The death and injury to dairy cows and other animals (e.g., chickens, hogs, and piglets) and the inability to process, store, and distribute milk resulted in considerable economic loss in the agricultural sector (Kerry et al. 1999). In Ontario and Quebec, about 13.5 million liters of milk were discarded by approximately 5500 farmers, accounting a loss of income of the order of nearly $8 million. In New York State alone, over $4 million was lost due to this dumping.

The forestry sector was also severely impacted. The region is a key site of the maple syrup industry, whose taps were damaged as ice accumulated on trees. Sugar bush and orchard owners, tree farmers, and greenhouse operators all suffered loss and damage (Lecomte et al. 1998). Numerous other economic sectors, such as manufacturing, tourism (especially the alpine industry), communication, and retail, were also adversely impacted. The widespread damage to farms, trees, and personal property resulted in more than 840,000 insurance claims from policyholders in the United States and Canada. The recorded total economic loss exceeded $5 billion (Murphy 2009). However, the Great Ice Storm of 1998 produced in excess of 840,000 insurance claims in Canada and the United States, that is 20 percent more claims than claimed by victims of Hurricane Andrew (Lecomte et al. 1998).

RECONSTRUCTION AND RECOVERY

The Great Ice Storm of 1998 clearly illustrated that "interdependencies among essential infrastructures constituted one of the most fragile elements of modern society" (Murphy 2009, 158). The tight coupling of technology across the region and the linear design of the energy system meant that a much larger area was exposed to risk. While it is possible to construct transmission towers and lines to be more resilient, it is costly. Reconstruction of damaged infrastructure began immediately, with care being taken to protect the vulnerability of water treatment plants and water systems and to enhance redundancy in the electrical and communication systems. Nevertheless, cost-benefit considerations have resulted in the abandonment of further safeguards being implemented. Although further risk reduction is possible, the improbability of another storm of this magnitude occurring has resulted in a delicate negotiation by utility companies to try and strike a balance between system robustness and increasing consumer expenses.

Beyond triggering a massive expenditure of the public and private sectors to rebuild infrastructure, the storm prompted a number of organizational reforms. A series of cooperative agreements, as well as the enactment of new laws and regulations designed to increase regional disaster preparedness and resilience, emerged in the post-disaster period. A large number of emergency response agencies have developed or revised their disaster preparedness plans to address the various inadequacies of the management, communication, and other socially embedded issues that were exposed during the storm. Although various governments, agencies, and organizations have accepted responsibility for different failings, there has also been frequent reminders that such events require individuals to take responsibility for their health and safety.

CONCLUSION

Susceptibility to the ill-effects of the ice storm was due to a much larger-than-normal weather event and a centralized electrical grid, whose failure exposed and exacerbated social vulnerability (Murphy 2009). What began as a weather-induced disturbance quickly escalated into a socially manufactured crisis. When the intensity, duration, and scope of the storm exceeded expectations, human misery rose significantly. Although populations located in and near the Ottawa and St. Lawrence Valleys are familiar with harsh winter weather, the Great Ice Storm of 1998 revealed that a good deal of climatic comfort is predicated on the assumption that energy is prevalent and that blackouts are short-term interruptions.

Robert Oliver and Valerie Thomas

Further Reading

Henson, W., R. Stewart, and B. Kochtubajda. 2007. On the Precipitation and Related Features of the 1998 Ice Storm in the Montréal Area. *Atmospheric Research* 83(1): 36–54.

Kerry, M., G. Kelk, D. Etkin, I. Burton, I., and S. Kalhok. 1999. Glazed over: Canada Copes with the Ice Storm of 1998. *Environment: Science and Policy for Sustainable Development* 41(1): 6–11.

Lecomte, E, A. Pang, and J. Russell. 1998. *Ice Storm '98*. Boston/Toronto, ON, Canada: Institute for Catastrophic Loss Reduction and Institute for Business and Home Safety.

Murphy, R. 2001. Nature's Temporalities and the Manufacture of Vulnerability. *Time & Society* 10(2–3): 329–348.

Murphy, R. 2009. *Leadership in Disaster: Learning for a Future with Global Climate Change*. Montreal, QC, Canada: McGill-Queen's University Press.

Great Kanto Earthquake, Japan, 1923

With a magnitude of 7.9 on the Richter scale, the Great Kanto Earthquake struck the Tokyo-Yokohama metropolitan area just before noon on September 1, 1923. Because the earthquakes devastated these two cities, it is also called the

Tokyo-Yokohama earthquake, or simply the Tokyo earthquake. Both cities are located on the Kanto Plain of the Japanese main island of Honshu. The prefectures affected by this earthquake included Chiba, Kanagawa, Saitama, and Shizuoka. The earthquake's 14-second duration set a record of destruction and deaths to this day. For this reason, the anniversary of the Great Kanto Earthquake is now National Disaster Prevention Day in Japan. The earthquake destroyed more than half of the brick buildings and one-tenth of the reinforced concrete structures in the affected region. It claimed 142,800 lives, including about 40,000 who went missing and were presumed dead (Denawa 2014).

The epicenter of the Great Kanto Earthquake was in shallow waters, only 6 miles (9 km) below the floor of Sagami Bay, about 25 miles (37 km) south of Tokyo. A rupture of a part of the convergent boundary where the Philippine Sea Plate subducts beneath the small Okhotsk Plate along the line of the Sagami Trough caused the earthquake. The earthquake triggered a tsunami 39.5-feet (12 m) high at Atami on the Bay, and the tsunami waves swept away homes, railway tracks, bridges, and roads. Kamakura, ancient capital of Japan, was inundated by 19-feet (6 m) tall tsunami waves. This city is located almost 40 miles (60 km) from the epicenter, and the earthquake killed 300 people there. The north shore of Sagami Bay rose permanently by almost 6 feet (2 m), and parts of the Boso Peninsula, which forms the eastern edge of Tokyo Bay, moved laterally 15 feet (4.5 m) (Szczepanski 2017). Over the next three days, earthquake survivors experienced nearly 2,000 aftershocks. The comparable Japanese earthquakes in the twentieth century were at Kōbe on January 17, 1995, and the 2011 Tohoku earthquake.

IMPACT

Most of the deaths caused by the Great Kanto Earthquake were due to fire. At the time of the earthquake, many people in the affected region were preparing the noon meal on cooking stoves. Thus, the wood of the destroyed homes caught fire from the overturned stoves, and gas explosions also caused fires. These were the two sources of more than 130 major fires across Tokyo, which lasted two days and three nights. Although high winds spread the fire rapidly, fires remained clustered in eastern and central sections of the city, where they quickly got out of control, and the city lost its water supply. Many residents of Tokyo fled the city to escape widespread fire, while of those who stayed, many burned to death in streets, alleyways, parks, and in the few open spaces that existed in Tokyo. One of the open spaces was the Rikugun Honjo Hifukusho, once called the Army Clothing Depot where between 38,000 and 44,000 working-class Tokyo residents gathered in an area of 10 acres to escape fires. Flames surrounded them at about 4 p.m., and a 300-feet tall "fire tornado" roared through the area. All but only 300 of the people gathered there perished (Szczepanski 2017). People also sought refuge in Tokyo's Sumida River, dying in the hundreds from drowning when several bridges collapsed (Schencking 2013; Denawa 2014).

Fires also burned large swaths of Yokohama, Japan's largest port and its gateway to the West. Fire and the ground shaking together claimed 90 percent of the

The ruins of the Yokohama Specie Bank after the Great Kanto Earthquake in Japan, 1923. (Library of Congress)

homes in Yokohama (Szczepanski 2017). The earthquake not only destroyed two of Japan's largest cities and traumatized the entire nation but also provoked nationalist and racist zeal across the country. The survivors immediately started to search for reasons for the earthquake, tsunami, and firestorm. In the mid-afternoon on the day of the quake, false rumors quickly spread that the Korean immigrants in Japan had started the disastrous fires, looted shops and homes, poisoned wells, murdered women and children, and planned to overthrow the government. Individuals and members of certain neighborhood vigilante groups, police, and military units responded to these rumors with violence.

Armed with makeshift weapons, they looked for "lawless Koreans" to kill. "Approximately 6,000 unlucky Koreans, as well as more than 700 Chinese who were mistaken for Koreans, were hacked and beaten to death with swords and bamboo rods. The police and military in many places stood by for three days, allowing vigilantes to carry out these murders, in what is now called the Korean Massacre" (Szczepanski 2017). Racism, hatred, resentment, and criminal opportunism all contributed to this tragedy. In fact, prejudice and hostility toward Korean minorities started with Japan's colonization of Korea in 1910. On September 4, the police requested that every Prefectural Governor take firm measures to stop all punitive activities against Koreans and protect them in all possible ways. However, not a single member of the police or army force was prosecuted for the crimes they committed (Hammer 2011; Denawa 2014).

Forty-eight percent of all homes in Tokyo Prefecture were either destroyed or classified as uninhabitable as a result of the Great Kanto Earthquake, leaving 61 percent of Tokyo residents homeless (Szczepanski 2017). It also destroyed nearly 7,000 factories and 121 bank offices out of a total of 138 in the city. One hundred fifteen out of 172 hospitals in Tokyo and its suburbs were destroyed by the earthquake, along with 117 out of 196 primary schools.

RELIEF

Given the destruction, upheaval, and chaos that followed, the Japanese Government declared a state of martial law in all of Tokyo, Kanagawa, Chiba, and Saitama prefectures. The government also mobilized 52,000 troops from around the country for deployment to Tokyo and Yokohama to restore order, assist with relief and recovery efforts, and repair damaged infrastructure. It took 10 days for stability, peace, and public order to return, and this was possible due to deployment of nearly one in five members of Japan's entire standing army. After restoring order, the troops immediately started to provide food, clean drinking water, and other emergency items to earthquake survivors. Additionally, navy ships were commissioned to transport rice from military warehouses in Kobe, Osaka, Kure, and Sasebo. Unfortunately, providing medical assistance proved problematic as many hospitals and other medical facilities had been destroyed. Additionally, the heavy damage to Tokyo's infrastructure made the transportation of medical personnel and supplies from across the country exceedingly difficult, contributing to the delay in treating burn victims that cost many lives.

Troops were deployed to repair roads, bridges, and rail tracks. They rebuilt 57 miles (85 km) of rail track, constructed 27 temporary bridges, and removed over 3,000 damaged rail cars and trams left derelict in Tokyo. Navy personnel repaired damaged docks and rebuilt piers to accommodate domestic and foreign relief shipments. However, both Tokyo and Yokohama cities faced difficulties in providing temporary shelters for the earthquake survivors. Ultimately, city officials in Tokyo constructed temporary barrack housing to accommodate nearly 150,000 people by September, while more than half a million homeless individuals returned to where their houses once stood and constructed temporary makeshift houses. The situation was so grave that the Japanese government distributed food to needy sufferers in Tokyo until April 1924, and a supply of clean water to earthquake survivors continued up to the end of December of that year. Initially, Korean communities in Japan were not provided with emergency relief, but from September 11, 1923, the authority started to give Koreans rice, noodles, firewood, and medical supplies (Schencking 2013; Denawa 2014).

At the time of the earthquake, U.S. naval ships were anchored in Yokohama port. Navy officers helped to flash the news across the sea, which in turn galvanized an international relief effort led by the United States. Hearing the news about the Great Kanto Earthquake, President Calvin Coolidge immediately made an appeal for donations. He also ordered the U.S. naval fleet in Asia and troops stationed in the Philippines to carry out relief efforts in Japan. Within a week of the event, the United States sent dozens of warships with full relief supplies—rice, canned roast beef, reed mats, tents, gasoline, and medical supplies. The American

Red Cross also initiated a national relief drive, raising $12 million for Japanese earthquake survivors. J.P. Morgan, an American private financial institute, also provided loans to the cities of Tokyo and Yokohama.

According to the Japanese government, the total amount of cash donated to Japan following the 1923 earthquake amounted to nearly 22 million yen with the United States accounting for about 70 percent of the total aid. Of the total value of relief supplies shipped to Japan, the U.S. military donated approximately $9 million. Thus, the United States provided both cash and goods worth about $20 million, which was roughly 2.4 percent of America's Gross Domestic Product (GDP) at the time.

One principal reason for the United States to lead relief efforts in Japan was to improve deteriorating Japanese-American relationships. On April 19, 1906, San Francisco was struck by a devastating earthquake, and the city was largely destroyed in the fires that followed. At that time, relations between the two countries were deteriorating because of American opposition to Japanese immigration to the West Coast, including San Francisco. The Great Kanto Earthquake provided an opportunity for the United States to improve relations with Japan. Another reason was to "pay back" to the Japanese for their generosity after the San Francisco earthquake in 1906. Of the foreign donations, Japan provided the largest amount of aid, giving $46,000. Thus, the enormous amount of aid from the United States at the time of the Great Kanto Earthquake could be considered a repayment of sorts.

CONCLUSION

An examination of the government publications of stories of the survivors suggests that the Japanese government tried to cover up events following the Great Kanto Earthquake, which revealed morally unjust acts committed against Korean immigrants by both Japan's citizens and leaders. The truth is that the government failed to protect the lives of its minority citizens and subsequently permitted false rumors to spread in place of facts. As citizens of Japan, its minorities deserved the rights that the rest of the nation was guaranteed. Several lessons came from this tragic event. Firstly, taking action immediately after an extreme event or any time based on rumors is not a good idea. Secondly, the violent discrimination against the minorities revealed that the prejudice the Japanese had toward Koreans played a role in their readiness to believe false rumors.

Thirdly, the event led to the modernization of Tokyo in two sectors. Streets became wider after the 1923 earthquake. This is because of the realization that the narrow roads of the city posed difficulty for many people to evacuate to safer places. Also, many new buildings adopted additional, modern, earthquake-resistant codes. Admittedly, a number of buildings built to such codes did not withstand the 1923 event; therefore, because fire was the primary destructive force, buildings constructed after the earthquake introduced extra walls both inside and outside to improve strength and flexibility of the buildings. New buildings were also made of fewer flammable materials, and builders tried to avoid wood as a construction material. More sprinkler systems were introduced to reduce the risk of uncontained fire. Fireproof buildings, such as the Shirahige-Higashi apartment, were also constructed for similar effect. New open spaces were created to act as firebreaks, again as a containment measure. Similarly, Tokyo city officials initiated additional measures to protect municipal water storage and delivery systems to help in combating fire-related earthquake deaths.

One of the main reasons for the huge death toll in the Great Kanto Earthquake was because the residents of Tokyo were thoroughly taken by surprise by the event. Therefore, measures were taken after the 1923 earthquake to improve earthquake prediction, and instruments were installed to monitor and forecast earthquakes. To educate its residents, the city of Tokyo organizes earthquake drills every year on September 1. Schools throughout the city also regularly practice drills to educate the young on how to best survive earthquakes. Residents are advised to switch off stoves and ovens and not to keep things on high shelves, as falling objects are one of the major killers during an earthquake. Above all, the event initiated a massive effort in Japan to predict earthquakes and tsunamis.

Bimal Kanti Paul

Further Reading

Denawa, M. 2014. Behind the Accounts of the Great Kanto Earthquake of 1923: The Great Kanto Earthquake of 1923. Brown University Library Center for Digital Scholarship. https://library.brown.edu/cds/kanto/denewa.html, accessed July 11, 2018.

Hammer, J. 2011. The Great Japan Earthquake of 1923. *The New York Times*. March 13, 2011. https://opinionator.blogs.nytimes.com/2011/03/13/in-deadly-earthquake-echoes-of-1923/, accessed July 11, 2018.

Schencking, C. 2013. *The Great Kantō Earthquake and the Chimera of National Reconstruction in Japan.* New York: Columbia University Press.

Szczepanski, K. 2017. The Great Kanto Earthquake in Japan, 1923. https://www.thoughtco.com/the-great-kanto-earthquake-195143, accessed July 10, 2018.

Great Mississippi River Flood, United States, 1993

During the summer of 1993, the Mississippi River and its tributaries in the American Midwest underwent a catastrophic flooding event caused by an unusually long period of wet weather conditions. Primarily affecting the states of Minnesota, Wisconsin, Iowa, Illinois, North Dakota, South Dakota, Nebraska, Kansas, and Missouri, the flooding affected roughly 260,000 square miles (673,400 square km) of land and lasted as long as six months in some locations (Spence and Smith 2011). The 1993 flood caused as much as $20 billion in damage, making it the largest and economically costliest flood in the history of the United States up to that point (Phillips 1994).

ENVIRONMENTAL CONTEXT

The Mississippi River Flood of 1993 primarily affected the upper Mississippi River Basin. The largest tributary to the Mississippi is the Missouri river, which joins the Mississippi just north of the city of St. Louis, Missouri. As such, the Missouri river also experienced notably heavy flooding during this event. Most of the land within the floodplains of the Mississippi and Missouri rivers is dedicated to agriculture, although many towns and cities are also located within the floodplains. As small flood events occur with regularity, an extensive network of constructed levees and dams exists on the Mississippi and Missouri rivers to control the flow of water to protect agricultural land and population centers. However, the magnitude and duration of flooding in the summer of 1993 caused many of these flood control structures to fail.

Uncharacteristically wet weather conditions preceded the flooding. An anomalous dip in the jet stream crossing North America formed a low-pressure system over the Mississippi River Basin and set the stage for high levels of precipitation. This low-pressure trough drew humid, moist air masses from the Gulf of Mexico northward, which converged with cool, dry air masses moving south from Canada. As the jet stream remained stationary, this weather formation resulted in months of abnormally heavy precipitation, beginning around January and lasting through August. Many locations received two to three times the average amount of precipitation (Johnson et al. 2003). As snow from winter precipitation melted in the spring, rainfall persisted. Continued precipitation and snowmelt saturated the soils and filled tributary waterways throughout the upper Midwest, leading to high water levels and flooding events beginning in April and continuing through August.

The duration of flooding therefore lasted for a long time in many locations. High water in areas along the Missouri were reported to last for 100 days, and in some locations along the Mississippi, this length of time topped 200 days (Spence

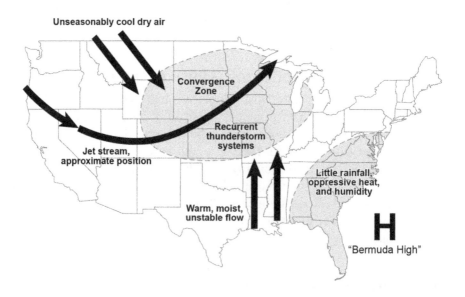

This map illustrates how cool, dry air from the northwest converged with warm, moist air from the south to produce the unusually heavy and steady rains that occurred during June and July 1993. (NOAA)

and Smith 2011). For Wisconsin, Iowa, and Illinois, the months of June and July were the wettest in their recorded history. Flooding reached its peak between the months of July and August, when flood records were set at dozens of monitoring stations along the Mississippi river. The flood event was the largest in 140 years of gauged measurements (Phillips 1994). It was significant not only for the record-setting levels of discharge and large extent of flooding but also for the duration of flooding that lasted for months in some locations.

DAMAGES

Over 150 major tributaries and rivers to the Mississippi river flooded. The volume of water overtopped levees in many locations, and the duration of flooding caused others to lose integrity as water seeped through. As a result, 1,000 levees—about two-thirds of the total number of levees on the Mississippi and Missouri Rivers—failed, and rushing water surged forth into agricultural fields and towns (Phillips 1994). In some cases, the water bursting through failed levees resulted in flash flood conditions that swept away structures and vehicles. The deaths of at least 52 people are attributed to the flooding, and most people drowned when they were trapped by floodwaters in their automobiles (Changnon 1996). Many locations affected by the flood, however, had warning systems in place, and local residents were able to safely evacuate. The deadliest flood from the Mississippi occurred in 1927, with 313 recorded deaths, and the improvement in forecasting and flood warning systems since then is thought to have saved hundreds of lives in 1993 (Changnon 2000).

The entire state of Iowa and counties in eight other states were declared a federal disaster area. The flood inundated parts of at least 75 towns, causing damage to roughly 150,000 residences and displacing large populations. Flooding impacted over 5,000 businesses, with estimates of physical damage and economic injury exceeding $334 million. Furthermore, public infrastructure, such as water treatment plants, was also heavily impacted. Water treatment facilities often draw surface water from rivers and are thus located in the floodplain nearby. In Des Moines, Iowa, floodwaters inundated the water treatment facility and caused water service shutdown for 19 days, leaving 200,000 people without potable tap water. Although the downtown area of Des Moines was not flooded, businesses closed due to the water shortage.

Wastewater facilities are also located on floodplains because they take advantage of the gravity flow at the lowest point in the town and discharge the treated effluent into rivers. Inundated wastewater treatment plants discharged raw sewage directly into waterways in many cities in Iowa (e.g., Cedar Rapids, Denison, Burlington, and Davenport) and other locations, causing concerns over potential outbreaks of water-borne disease. Standing floodwater also provided an ideal breeding ground for mosquitoes. Luckily, no significant increase in malaria was observed during or after the 1993 flood.

In addition to damaging property by direct inundation, the floodwaters also caused damage through processes of scouring and deposition. Rushing water is an effective erosional force. At areas where stream velocity increased, floodwaters removed soil, sand, and even gravel and carried it downstream as sediment. As water rushed through levee breaks, the water rapidly scoured soil from fields. In other locations, deposits of sand and gravel as deep as five feet were left as floodwaters receded and the sediment settled out (Phillips 1994). Despite these negative impacts, floodwater in 1993 carried nutrients that fertilized aquatic plants and washed fish into formerly isolated lakes. Available reports claim that catfish and other gamefish in lakes and other permanent water bodies significantly increased after the 1993 flood on the Mississippi and Missouri floodplains (Coch 1995).

IMPACTS TO AGRICULTURE

As agriculture is the most significant land use in the upper Mississippi River Basin, flooding severely affected farmlands. Heavy rains saturated soils and caused standing water in fields, which prevented planting or caused the root systems of crops to rot. Crops behind failed levees were almost completely lost as water broke through, inundated land, and rapidly washed away crops and eroded soils in the process of scouring. Roughly 10,250 square miles (27,000 square km) of cropland flooded during this event, costing approximately $8.9 billion in damages (Changnon 2000).

Further damage to agricultural land occurred as river floodwaters receded. The sediment-laden waters deposited large amounts of sand on farm fields and within ditches that were used for drainage. Thousands of acres were covered in sand deposits that were more than two feet thick, which subsequently required costly removal, with the U.S. Department of Agriculture (USDA) spending over $10 million to clear ditches of sand and debris (Changnon 2000). In addition to harming crops and livestock, the flooding also caused significant amounts of direct damage to farm infrastructure, destroying buildings, fences, machinery, and equipment.

IMPACTS TO TRANSPORTATION SYSTEMS

The 1993 Flood significantly disrupted the transportation sector and damaged infrastructure in the region. Transportation systems that were primarily affected included road systems, railways, airports, and river barges. Damages and revenue loss in the transportation sector totaled approximately $1.9 billion (Changnon 1996).

Road systems in the upper Mississippi River Basin typically follow waterways and frequently cross them. Inundation made many roads impassable for long stretches of time. Furthermore, floodwaters washed out sections of roads and highway systems, causing major damage. More minor damage occurred to other road infrastructure, including blowing out culverts. Floodwater swept away numerous bridges, and scouring floodwaters structurally undercut many others that subsequently required costly repairs or replacement. Road and bridge damage left numerous locations inaccessible and caused many businesses to temporarily shut down as customers and employees were unable to reach them.

Floodwaters also damaged freight railways' main lines that run parallel to both the Missouri and Mississippi rivers. While tracks are generally elevated on embankments that would withstand most flood events, over 800 miles (1,200 km) of track were flooded. In addition to tracks, the flooding physically damaged signals, communication lines, and locomotives. Many trains had to be rerouted along flooded sections, resulting in additional losses to railroads.

Airports, particularly in the states of Missouri and Iowa, were also affected by the floodwaters. Located in flood plains because of the flat terrain, 12 commercial airports were inundated by floodwaters, which temporarily closed, including the Spirit of St. Louis airport. Dozens of smaller airports were also damaged, many of which went out of business and never reopened (Changnon 1996).

The Mississippi and Missouri rivers themselves serve as a part of a major transportation network as barges move large quantities of materials. Barge traffic halted completely for about a month during the height of the flooding from July to August, and it was severely limited in the following months. The restricted navigation of the two rivers resulted in millions of dollars in lost revenue.

GOVERNMENT RESPONSE

Among the nine states affected by flooding, 532 counties received presidential disaster declarations that allowed them to receive federal aid to deal with the impacts of flooding (Changnon 1996). Over the next year, Congress approved $6.2 billion in flood aid (Changnon 2000). Other government expenditures included $1.6 billion in insured crop losses and payments of $301 million for flood insurance payments (Changnon 2000).

Following the 1993 food disaster, the federal government pursued efforts to strengthen against susceptibility to future flooding events. As a result, at least 12,000 buildings in high-risk areas for repeat flooding were elevated, acquired, and razed or relocated (Johnson et al. 2003). The federal government also bought 25,000 flood-prone properties in the Midwest to prevent repeat flood damage (Spence and Smith 2011). In some cases, lands in high risk of repeat flooding were set aside as public natural areas in the form of parks or wildlife habitat. In addition, the federal government installed more river sensors and funded further data collection and subsequent development on improved forecasting systems (Changnon 2000).

Although federal flood and crop insurance programs existed, this event demonstrated major gaps in the coverage of these programs. Areas that were affected by flooding typically only had 10 percent of insurable properties covered by the flood insurance program. For the crop insurance program, roughly only half of the insurable crop acres affected by the flood were insured against losses. Subsequently, policy changes encouraged the uptake of the programs, but with limited success. In fact, many properties in areas with risk of flood continue to rely on relief aid in case of future flood events (Changnon 2000).

CONCLUSION

The Great Flood of 1993 heavily impacted the lands and communities surrounding the Mississippi river and its largest tributary, the Missouri river. Although not the deadliest, it has thus far been the costliest river flooding event in the history of the United States. The flooding notably destroyed much of the levee system in place along the upper Mississippi and caused large losses to the agricultural sector and transportation networks. The disaster subsequently fueled debates over the role of infrastructure in mitigating flood damage and the effectiveness of disaster insurance programs.

Audrey Joslin

Further Reading

Changnon, S. A., ed. 1996. *The Great Flood of 1993: Causes, Impacts, and Responses.* Denver, CO: Westview Press.

Changnon, S. A. 2000. The Record 1993 Mississippi River Flood. In *Floods*, Vol. 1, edited by D. J. Parker, 288–301. New York: Routledge.

Coch, N. K. 1995. *Geohazards: Natural and Human.* Englewood Cliffs, NJ: Prentice Hall.

Johnson, G. P., R. R. Holmes, and L. A. Waite. 2003. *The Great Flood of 1993 on the Upper Mississippi River—10 Years Later.* United State Geological Survey. https://pubs.er.usgs.gov/publication/fs20043024, accessed September 23, 2018.

Phillips, Steven. 1994. *The Soil Conservation Service Responds to the 1993 Midwest Floods.* Washington, DC: United States Department of Agriculture.

Spence, Patric R., and Jacqueline S. Smith. 2011. Floods. In *Encyclopedia of Disaster Relief*, edited by K. Bradley Penuel and Matt Statler, 216–221. Newbury Park, CA: SAGE Publications.

Gujarat Earthquake, India, 2001

On the early morning of January 26, 2001, while India was celebrating its 52nd Republic Day, a powerful earthquake of magnitude 7.7–7.9 struck the northern Indian state of Gujarat at about 8.46 a.m. It was one of the strongest and deadliest earthquakes in India. The epicenter was 45 miles (65 km) east of Bhuj, in the district (the second largest administrative unit in India) of Kachchh (formerly Kutch). The earthquake originated at a depth of 10 miles (16 km), and the ground shook furiously for about 60 seconds, causing massive destruction of life and property. Bhuj was almost destroyed by the event, and for this reason, the event is also known as the Bhuj earthquake. In addition to neighboring Pakistan, the earthquake was felt as far away as Nepal and Bangladesh.

With a population of 1.26 million in 2001, Bhuj is the headquarters for the district of Kachchh and a center of important trade, commerce, and political activity. While Kachchh was the worst-affected district, earthquakes are not uncommon there. The district is part of the Mesozoic rift zone (a region where the crust has split apart), with the tectonic belt running in two almost parallel locations with several east-west trending tectonic faults. Two major fault lines are located within the district boundary, one in the barren lands of the north and the other in the populated areas of Kachchh. Seismologists analyzed the source region of the earthquake's origin and found that reverse faulting, which causes the rock above the fault plane to move up and over the rock below, was the dominant compressive stressor causing movement of the Indian plate. Resultant seismicity determines the region's classification as a high tremor risk zone where earthquakes have a magnitude of eight and above.

Kutch has experienced earthquakes ranging from 4 to 8 magnitude on the Richter scale. Of the 91 earthquakes experienced since the 1800s, 12 had magnitude between 5 and 8 and 79 were between 4 and 5. Two major intraplate earthquakes occurred in 1819 and 1956. The former is known as the Rann of Kutch earthquake, the magnitude of which ranged from 7.7 to 8.2 and created the Allah Bund (Dam of God). The latter is called the Anjar earthquake and had an estimated magnitude of 6.1.

CAUSE

The cause of the Gujarat earthquake was stress release from the Indian plate pushing north into the Eurasian plate, and it occurred far from the plate boundary. Thus, the earthquake was a continental intraplate earthquake, also called a mid-plate earthquake, but no primary fault rupture was identified as the source. Intraplate earthquakes occur on the interior of a plate rather than between two continental plates. They often occur due to sudden movement along the preexisting faults because of large stresses built up from regional plate tectonic forces. In this instance, the Gujarat earthquake happened along the east-west trending fault at a fairly shallow depth. While this earthquake caused a few surface cracks, it did not cause significant surface ruptures; however, there were instances of liquefaction, and riverbeds that had been dry for more than a century filled with volcanic mud flows as a result.

A surprising feature of the 2001 Gujarat earthquake was the lack of foreshocks before such a major earthquake. Although some seismologists argue that foreshocks were observed about 163 days before this earthquake, no one paid attention to the signals. The main shock was followed by about 638 aftershocks, which spread over 65 days. According to the Indian Meteorology Department (IMD), there were more than 500 aftershocks of magnitude 3.0 and above (Bolt 2004).

IMPACTS

The Gujarat earthquake caused massive loss of life and injury. About 19,727 people died and 166,000 suffered injuries in India, while another 18 died in southern Pakistan. The earthquake left 600,000 families homeless, destroyed 348,000 houses, and damaged an additional 844,000. It also destroyed civic infrastructures such as schools, hospitals, health clinics, public buildings, water supply systems, and communications and power. In these ways, the earthquake directly or indirectly affected 15.9 million people and reportedly killed more than 20,000 cattle. Loss of lives and property, including business interruption losses or loss of employment for residents of the damaged communities, was catastrophic. One month after the earthquake, an official estimate from the Indian Government placed direct economic losses at $1.3 billion. Other estimates indicate losses might have been as high as $5 billion (Bolt 2004).

Twenty-one out of 25 districts in Gujarat were affected by the 2001 earthquake, and it severely damaged Kachchh, Ahmedabad, Rajkot, and Jamnagar districts. Among the cities and towns most affected were Bhuj, Bhachau, Anjar, and Rapar;

in fact, Bhachau was destroyed. In addition, more than 200 villages were either severely damaged or leveled to the ground. Luckily, the epicenter of the 2001 earthquake occurred in a less densely populated area, thereby limiting the casualties. But the damage was widespread and severe because most of the buildings in the affected areas did not comply with seismic codes, which were not mandatory (Bolt 2004). For example, in Bhuj, 90 percent of buildings were affected, and Kachchh was cut off from the rest of the country for 24 hours. Additionally, over 70 percent of the buildings in the district were completely destroyed. Most of the earthquake-induced deaths also occurred there (Lecy 2007).

RESPONSE

The damage caused by the 2001 Gujarat earthquake was so extensive that search and rescue (SAR) operations were overwhelmed. In fact, initial SAR activities were done by family, friends, and neighbors. Trained SAR teams arrived within hours of the earthquake, including a team from a consortium of voluntary organizations called Sangh Parivar. For the first 48 hours, the Sangh Parivar undertook SAR operations despite the official machinery being paralyzed; eventually the SAR operations were undertaken jointly by Sangh Parivar and the army. Immediately after the earthquake, the Indian government deployed 22,500 troops to support the emergency effort by clearing rubble, unloading relief supplies, and performing medical services. Even though the army and the State Rapid Action Force promptly provided the manpower, they did not have either the expertise or the heavy equipment to undertake SAR operations. It took nearly three days before they could acquire heavy equipment and cranes. However, the Indian Army and the Sangh Parivar played a key role in recovering bodies from rubble and cremating of the dead. Meanwhile, many injured were treated on-site by Red Cross Societies of many countries, while seriously injured persons had to be flown to Mumbai and Pune for better medical treatment by the Indian Air Force.

The state and national governments and national and international communities responded quickly to this disaster with emergency relief and rehabilitation services. For example, the state government was quick to secure funds and able to disburse relief goods among earthquake survivors. Furthermore, some of the earthquake-affected towns and villages were adopted for relief and reconstruction by corporate firms, industrialists, and many governmental and nongovernmental organizations (NGOs) agencies across the globe (Mistry et al. 2001). Before fully devoting their resources to rehabilitation programs, the Government of Gujarat addressed the needs of specific households that suffered deaths or serious injuries. The family members of the deceased were compensated for their losses through benefits that ranged from $1,363 to $3,409, and individuals with serious injuries received up to $1,136. Over 97 percent of the applications for these funds subsequently were awarded benefits (Lecy 2007).

Although the economic sector seriously suffered from the 2001 earthquake, and many survivors of the affected areas lost their primary income source, the post-disaster stage provided income-generation opportunities for them. Many men were employed for demolishing structures, and many women removed rubble. Despite the state government's proactive policy, available reports suggest the public relief distribution suffered from discrimination against many on the basis of caste, religion, and political affiliation. The government was also criticized for its lack of disaster management policy and its failure to implement proper construction laws.

REHABILITATION

Within nine days of the earthquake, the state government established the Gujarat State Disaster Management Authority (GSDMA) with the initial responsibility to direct the government relief efforts and coordinate work and resources among relief agencies. However, the authority turned its attention from immediate relief interventions to the long-term rehabilitation of the affected communities. This shift was due to the fact that many earthquake survivors did not sleep indoors for more than three weeks for fear of falling buildings, which was exacerbated by many aftershocks. In addition, the state government realized that the most serious vulnerabilities stemmed from the massive loss of housing, and therefore the majority of the available funds were directed to housing reconstruction and rehabilitation. The government aimed to provide permanent housing for the 1.7 million homeless people and to repair the infrastructure necessary for normal economic activity (GSDMA 2007).

The Government of Gujarat received a modest amount of emergency aid from external sources for relief and recovery for the 2001 earthquake. For instance, it received $900 million from the World Bank and Asian Development Bank (ADP) and an additional $108 million from Overseas Development Authority (ODA), London, and the majority of these funds actually came as loans rather than grants (Mistry et al. 2001). Given this limited funding, the state government was extremely successful in housing reconstruction. Over 900,000 houses were repaired and nearly 200,000 were completely rebuilt. During the process, many homes received infrastructure upgrades. The percentage of houses with toilets increased from 32 percent to 53 percent, while plumbing increased from 30 percent to 34 percent. The government also rebuilt and repaired over 50,000 schoolrooms (Mistry et al. 2001), and public infrastructure was also upgraded. Another remarkable success was that most of these accomplishments were completed within two years of the earthquake.

In addition to support from a host of domestic and external sources, social capital played a crucial role in coping with the disaster impacts, particularly in the Kachchh district. Within two and a half weeks of the disaster, many affected communities had established their own temporary shelters with the aid of resources provided by members of social groups. In addition, many residents in Katchchh refused to put orphaned children up for adoption. These two examples reflect a high capacity of social capital following the disaster, which extended beyond the village level. The earthquake-affected districts were home to a sizeable business class with networks throughout the state. Also, a large diaspora in Mumbai and abroad was active during relief efforts through donations and other support (World Bank and Asian Development Bank 2001).

CONCLUSION

Despite some criticisms, the government of Gujarat was successful in recovering from the damage caused by the 2001 earthquake in a relatively short time. The government carried out an exhaustive selection of beneficiaries to ensure that the right people were receiving emergency relief aid and rehabilitation funds while organized citizen groups advocated for the interests of the needy. As a result, the reconstruction efforts were directed appropriately to the survivors. Public success in both response to and recovery from the event is attributable to a variety of factors, including strong leadership by the state government, a clear organizational structure provided by the GSDMA, effective relief disbursement, beneficiary participation in post-disaster activities, and demand of disaster survivors, which held government programs and NGOs accountable and transparent (Lecy 2007).

The Gujarat earthquake was a low-frequency, high-intensity event. To improve outcome of similar events, in 2003, the Science and Technology Department of the Government of Gujarat set up the Institute of Seismological Research. The mission of this organization is to safeguard life and property, create public awareness, carry out hazard assessment and mitigation, and investigate societal problems and industrial needs along with encouraging and promoting research and education in applied seismology. A combination of satellite imaging, modern digital technology, establishment of regional seismic networks, gaining assimilation and diffusion of information, database establishment, and gathering of bulletins from multiple sources is making monitoring more efficient. This is undoubtedly a right direction on the part of the state government to reduce the effect of future earthquakes; however, other earthquake mitigation measures, such as strict building code adherences, are also needed.

Avantika Ramekar

Further Reading

Bolt, B. A. 2004. *Earthquake*. 5th ed. New York: W.H. Freeman and Company.

GSDMA (Gujarat State Disaster Management Authority). 2007. Website of the Gujarat State Disaster Management Authority. http://www.gsdma.org/profile.htm, accessed March 16, 2007.

Lecy, J. D. 2007. Aid Effectiveness after the Gujarat Earthquake: A Case Study of Disaster Relief. *Journal of Development and Social Transformation* 4(1): 5–12.

Mistry, R., W. Dong, and H. Shah. 2001. *Interdisciplinary Observations on the January 2000, Bhuj, Gujarat Earthquake*. Singapore: World Seismic Safety Initiative (WSSI) and Earthquake and Megacities Initiative (EMI).

World Bank and Asian Development Bank. 2001. Gujarat Earthquake Recovery Program Assessment Report: A Joint Report by the World Bank and Asian Development Bank. https://www.gfdrr.org/sites/default/files/publication/pda-2001-india-gujarat.pdf, accessed June 21, 2018.

Haiti Earthquake, 2010

A catastrophic earthquake of 7.0 magnitude on the Richter scale hit the south-central part of Haiti at 4:53 p.m. on January 12, 2010. It had been 150 years since this part of Haiti last experienced an earthquake of any size. The epicenter of the earthquake was near the town of Léogâne, which is located approximately 16 miles (25 km) west of the capital of Haiti, Port-au-Prince. It occurred at a shallow depth of only 8.1 miles (13 km), and thus, the shaking was felt throughout Haiti and the neighboring Dominican Republic as well as in parts of nearby Cuba, Jamaica, and Puerto Rico (USGS 2015). The initial shock was soon followed by two aftershocks of magnitudes 5.9 and 5.5. Darkness fell within two hours of the earthquake, and most of the disaster area was without electricity. Residents of the affected areas camped out in streets and parks to avoid any structure that might collapse. At one point, a false rumor of a tsunami spread a new wave of fear through Port-au-Prince and caused widespread panic as people fled for higher ground.

BY THE NUMBERS

The 2010 Haiti earthquake is typically described as causing over 200,000 deaths, injuring 500,000 people, leaving up to 1.8 million people without permanent housing, affecting a total of 3.5 million people, destroying or critically damaging more than 200,000 structures, and creating $14 billion in damages. However,

The aftermath of a magnitude 7.0 earthquake that hit Port-au-Prince, Haiti, in 2010. (Frankbirds/Dreamstime.com)

depending on the source, the number of causalities varies by an order of magnitude, from a high of 316,000 by the Haitian government to 222,750 reported by the United Nations to a low of 46,000 determined by an epidemiologic survey. Following the quake, there have been at least three epidemiologic surveys that, either directly or indirectly, provide estimates of fatalities; these surveys—generally based on surveying individuals and households in Port-au-Prince—provide lower estimates that range from 46,000 to 158,000 deaths.

ESSENTIAL BACKGROUND

The 2010 Haiti earthquake occurred in the boundary region separating the North American plate (which includes Cuba and the Bahamas) and the Caribbean Plate, on which Haiti is located. These two tectonic plates are moving past each other at the speed of about 1 inch (2 cm) per year. This motion is accommodated by subduction zones in the ocean to the north (the North Hispaniola fault zone) and south of the island of Hispaniola (the Muertos trough) and two major strike-slip faults: the Septentrional-Oriente fault zone that trends east-west roughly along Haiti's north coast and the Enriquillo-Plantain Garden fault zone that runs approximately parallel through Haiti's southern peninsula. Between these two faults is the Gonâve microplate. Over time, the movement of the two major plates causes pressure to accumulate until a break occurs along a fault zones and produces an earthquake. The greater Port-au-Prince area had experienced earthquakes historically in 1860, 1770, and 1751, and Léogâne was destroyed by an earthquake in 1701, but there has been no earthquake disaster in living memory.

Seismologists originally assumed the earthquake to result from a strike-slip movement on the Enriquillo-Plaintain Garden fault (EPGF), but with additional evidence—including the absence of a surface rupture along the EPGF—they have located a previously unmapped subsurface thrust fault, the Léogâne fault, as the major source of the 2010 earthquake. The Léogâne fault dips to the north just above the EPGF; in the 2010 earthquake, the majority of energy was released by strike-slip motion and the remaining third by dip-slip motion. Had the rupture along the Léogâne fault transferred to the larger EPGF, the earthquake would have been of even greater magnitude. The fact that the more major fault, the EPGF, did not rupture suggests that larger earthquakes are yet possible.

VULNERABILITY TO EARTHQUAKES

Vulnerability can be seen in factors related to place, timing, and postdisaster response. Many of these proximate factors can, in turn, be traced to broader histories of Haiti's relationship with the wider region and world.

The earthquake's epicenter was not far from Haiti's major population center, the Port-au-Prince metropolis, with its 2.5 million people. Housing density increased loss of life, with the poorest neighborhoods being the most densely inhabited. Many Port-au-Prince residential areas are accessible only by foot, with narrow corridors between closely built houses; in neighborhoods like these, there are few safe places to shelter in an earthquake. In many cases, buildings were not designed and constructed to withstand earthquake hazards. Historically, buildings in Haiti were constructed of wood, wattle and daub, or thatch, but nearly all newer residential housing is made up of masonry or concrete block, often with roofs of the same. The masonry buildings are heavy and tend to have insufficient reinforcement to withstand a large earthquake.

In the 2010 quake, multistory buildings were more prone to collapse and more difficult to escape from. Site characteristics played an important role, too: while damage was widespread across the Port-au-Prince metropolis, some neighborhoods sustained much greater damage than others. It has long been known that earthquake waves can be amplified in valleys and low-lying areas filled with soft sedimentary rock, but the 2010 Haiti earthquake showed that topography—in this case, hills and steep, narrow ridges—can amplify seismic waves and increase damage.

In terms of time, the long period of seismic inactivity in Haiti meant many were unaware of the risk of earthquakes. The time of day when the earthquake struck was another factor: had the earthquake struck during the night or earlier in the day when more people would have been at school or work, the fatalities likely would have been even higher.

Problems related to postquake rescue and emergency efforts also contributed to increased loss of life and injury. The extent of the disaster; the lack of planning for such a disaster; the near total destruction of government buildings in the capital; the damage sustained by hospitals, roads, the electrical grid, and other infrastructures; and the fact that medical facilities and professional search-and-rescue services and equipment were insufficient even prior to the earthquake, all served to undermine the immediate response to the disaster. The vast majority of search-and-rescue operations was conducted by friends, family, and neighbors using their hands and commonplace tools. Timely, professional medical treatment was in drastically short supply; in too many cases, blood loss, infection, renal failure, and other treatable health problems led to death.

Starting in the 1970s, Haiti was loaned money from international sources and financial institutions on the conditions that Haiti would allow foreign investment—such as providing attractive conditions for foreign-owned garment manufacturing—and make it cheaper to import foodstuffs from abroad. These conditions, which Haiti had little choice but to accept, led to the fantastic expansion of Port-au-Prince's population—estimated at 75,000 new migrants per year over the decades leading up to the earthquake—which quickly outstripped the city's services and infrastructure.

Growth of Port-au-Prince happened simultaneously with pull factors—the lure of low-paying jobs in manufacturing factories based in the capital—and push factors—the base of the Haitian economy has always been agriculture, but cheap food imports eroded Haiti's farm economy and led to urban migration. For the United States, these conditions helped provide inexpensive clothing (Haitian workers are paid less than 50 cents/hour) and a market for the U.S. agriculture surplus (Haiti now imports the majority of its food, and the United States is the main supplier). But for Haiti, these conditions have built, over decades, increased vulnerability to earthquakes.

RESPONSES

The 2010 earthquake in Haiti attracted the world's attention. Journalists, relief workers, and others flocked to Haiti to offer assistance. The Haitian government, centralized in Port-au-Prince and having lost 14 out of 16 government ministries and the national palace in the quake, was overwhelmed. The number of bodies overwhelmed the morgues, cemeteries, and the ability of many families to pay for burials. Government dump trucks picked up bodies and dumped them in mass graves north of the city. Thousands temporarily left the capital for other parts of the country. Rural areas throughout Haiti experienced an influx in people and struggled to provide shelter, food, and care.

In the name of providing help, the U.S. military seized control of the Port-au-Prince international airport and seaport. Given previous U.S. invasions of Haiti in 1915 and 1994, many questioned why the U.S. humanitarian response was so militarized. The Interim Haitian Reconstruction Commission (IHRC), co-chaired by former U.S. President Bill Clinton and then Haitian Prime Minister Jean Max Bellerive, was formed to oversee the reconstruction. The IHRC was criticized for giving Haitians little say in the decisions made on reconstruction. After a year and a half, the IHRC had accomplished little and its mandate was not renewed.

In a more successful effort, the Haitian Ministry of Public Works, Transport, and Communications inspected over 400,000 structures and tagged each as red (unsafe, approximately 20 percent), yellow (restricted entry, 26 percent), or green (safe, 53 percent) and began the process of repairing yellow-coded buildings.

The most obvious evidence of the earthquake is now largely gone from the landscape. Gradually, the rubble was cleared. People returned to sleeping indoors. The central business district of downtown Port-au-Prince has largely relocated to areas, such as the suburb of Pétion-Ville, that sustained less damage, and the newly built government ministries are more dispersed throughout the city.

The international community pledged over $10 billion for earthquake relief and reconstruction efforts. Much of this money never reached the people who most needed it. Other funds went predominantly to nongovernmental organizations, donor military and civilian entities, and private contractors. Many of these entities lacked transparent ways of reporting how monies were spent. For instance, the American Red Cross, which raised almost half a billion dollars for the earthquake disaster, built only six permanent houses in five years (Elliot and Sullivan 2015). The government of Haiti received less than 1 percent of humanitarian funds (Ramachandran and Walz 2015). The international community, to the detriment of their success, rarely went beyond token input and participation of Haitians and Haitian institutions for designing and implementing projects.

DISPLACED PERSONS CAMPS

The most widespread need following the quake was shelter. Even people whose houses sustained no damage were fearful of sleeping indoors. Parks, empty lots, courtyards, and roads filled with temporary shelters built of tarpaulins, sheets, and cardboard, and later tents and other temporary structures. At their maximum, these displaced persons camps sheltered an estimated 1.5 million people. Some camps became small cities with services like restaurants, barber shops, and health clinics, and various governance structures. Often foreign nongovernmental organizations came to manage each camp and direct the distribution of aid. The camps had issues with security, and sexual violence became a problem, especially for women and girls. The camps rarely had the necessary latrines and water to serve their populations. And the camps tended to swelter in the sun and turn to pools of mud in the rain.

Through relocation, eviction, rental subsidy payments, or voluntarily, most of the camps have been vacated. An area coined Canaan, in an arid region north of Port-au-Prince, became a sprawling new city of 300,000 as many camp dwellers relocated here. Seven years after the earthquake, 37,000 persons were still living in temporary shelters in 21 sites in and around Port-au-Prince and Léogâne (Harlet 2017).

Anna Versluis

Further Reading

Elliot, J., and L. Sullivan. 2015. How the Red Cross Raised Half a Billion Dollars for Haiti and Built Six Homes. ProPublica and NPR. https://www.propublica.org/article/how-the-red-cross-raised-half-a-billion-dollars-for-haiti-and-built-6-homes, accessed May 8, 2019.

Harlet, J. 2017. Humanitarian Aid: 7 Years after the 2010 Earthquake, Who Remains Displaced? International Organization for Migration. https://haiti.iom.int/humanitarian-aid-7-years-after-2010-earthquake-who-remains-displaced#_ftn2, accessed May 9, 2019.

Ramachandran, V., and J. Walz. 2015. Haiti: Where Has All the Money Gone? *The Journal of Haitian Studies* 21(1): 26–65.

USGS. 2015. Earthquake Hazards Program Did You Feel It? Report for M 7.0 Haiti Region Earthquake of 2010-01-12 21:53:10 (UTC). https://earthquake.usgs.gov/data/dyfi/, accessed May 9, 2019.

Heat Wave and Wildfires, Russia and Eastern Europe, 2010

In mid-June 2010, an unusual meteorological phenomenon began to form in Eastern Europe. A very large volume of hot and dry airmass stalled over west-central Russia and northern Ukraine, and the system remained stationary for nearly eight weeks. During this time, many long-standing temperature records were broken (particularly in cities), wildfires occurred throughout rural districts of the Russia west of the Urals, and extreme heat and urban air pollutants and emissions from the wildfires were trapped by atmospheric inversions, leading to significant health threats.

High pressure settled in the Volga River valley near the start of summer and remained stationary for nearly two months. Intense high-pressure systems are not abnormal for the Northern Hemisphere; in fact, a similar long-lasting event occurred further west, over western and central Europe in 2003. This system was more intense however; it was the most intense of the last several centuries. Temperatures throughout western Russia, Ukraine, and Belarus during these weeks were 39.2–46.4 degrees Fahrenheit (4–8 degrees Celsius) above average. Numerous Russian cities experienced historic high temperatures: for instance, temperatures in Volgograd exceeded 105.8 degrees Fahrenheit (41degrees Celsius), in Moscow, they reached 102.2 degrees Fahrenheit (39 degrees Celsius) (eclipsing the previous high temperature set nine decades earlier [WHO 2010]), and in Saint Petersburg, the high was 98.6 degrees Fahrenheit (37 degrees Celsius). The temperature topped 86.0 degrees Fahrenheit (30 degrees Celsius) in Moscow for 33 days in a row.

Surface high-pressure systems are characterized by subsiding, warming, and desiccating air. The air warms adiabatically, caused by increasing atmospheric pressure closer to the surface, and its capacity to hold moisture expands. Evaporation draws moisture from soils and vegetation. Natural and human activities, like lightning and the burning of debris, can ignite the desiccated vegetation and even dry soils (like peat) to produce rapidly developing and long-lasting wildfires. The dry, hot winds produced by the fires intensify the conditions and promote the ignition and spread of fires.

Aside from the destruction caused by the burning of homes, property, and forestland, the wildfires generated significant amounts of airborne toxic chemicals and particulates that drifted into suburban and urban areas. Smoke, carbon monoxide, ozone, particulate matter, and even radionuclides (from the region of northern Ukraine, Belarus, and western Russia near the Chernobyl complex) were released from the burned-over districts and caused significant air-quality hazards.

VULNERABILITIES

The impacts of hazardous extreme events such as heatwaves and wildfires are magnified in some regions and for some populations, depending upon the characteristics that make the places or people more likely to suffer consequences from unhealthy conditions. These characteristics (of people, places, or systems) are known as vulnerabilities (the opposite characteristics are often referred to as adaptive capacities or as resilience), and they are unique to each hazard. Each of the following vulnerabilities of people and places were factors that affected the extent of life, health, and property losses during the catastrophe.

The people who are most vulnerable to extreme heat or heatwaves are the elderly; those who are dependent on the care of others (children and the disabled); and those with pulmonary, respiratory, or other physiologically stressful, weakening, or debilitating illnesses or diseases. Places or settings can exacerbate the vulnerabilities of people who live or work in them by having characteristics that lengthen or intensify the effects of heatwaves. These include cities (due to the added intensification of heat by the urban heat-island effect and lack of nighttime cooling), dwellings (urban or rural) with poor ventilation, dwellings lacking cooling or air-circulating devices (such as chillers, air-conditioning, or fans), areas that lack public access to indoor spaces that provide relief from the heat, and areas that lack environments (such as parks, public pools, or water features [rivers or lakes]) that are cooler and have more air movement, thus are less stifling and physiologically stressful. Private homes in Russia typically do not have air-conditioning due to high costs, low supply, and expensive permitting processes, while fans may be more widely used—stocks of both were exhausted in Moscow and the surrounding regions during the heat wave; more than 1.8 million air-conditioners were sold in 2010, and experts believe that more than four times as many could have been sold.

Spatial responses to find relief from heat often requires access to transportation (personal automobiles) or transportation infrastructure (buses or subways) and the financial capacity to travel, but the effort creates new physical stresses by adding excess financial burdens.

A somewhat unexpected hazard and its attendant human vulnerabilities were revealed by the public's efforts to find relief from the heat. Many sought comfort in or around pools, rivers, and lakes in the regions of extreme heat. A separate social problem—alcoholism and alcohol abuse—arose with attempts to cool down in these waterbodies. Adults who would consume alcohol to treat their heat stress often found themselves too drunk to escape from the water. Some of these adults also chaperoned their families in their recreational pursuits—but inebriation diminished vigilant visual attention to their children, and many children thus drowned as well. During the first three weeks of July alone, at least 1,244 people had drowned in Russia: most were inebriated and swimming without attendants. Heavy drinkers and their dependents were, therefore, most vulnerable to such tragedies. And waterbodies that lacked constant professional vigilance (i.e., lifeguards) were the most likely locales for the deaths.

Those susceptible to wildfire hazards (fire, heat, and smoke or air pollution) included the young, the elderly, those with mobility issues, those with pulmonary and respiratory problems or other illnesses, and smokers. Places that were more vulnerable to rural wildfires were those that were less accessible by emergency services (such as fire departments or firefighting brigades) and those with more combustible landscapes (parched forests or grasslands or in regions with drained swamp lands). Other vulnerable places may have limited egress (remote locations with few escape-path options to safety) and may have been settled or built up in more dangerous ways (i.e., structures were built within the urban-wildland interface zone that enveloped structures in high-risk conditions).

Peat bogs presented a real, human-made hazard for the residents of the Moscow region. Since the 1918 revolution, Soviet engineers had drained swamps to access and extract the submerged peat soils. The dried peat was burned in electrical power stations for much of the following 50 years. The use of peat continued until natural gas was discovered in Siberia during the 1950s. The bogs, no longer used for peat fuel, remained dry, and they spontaneously ignited during the period of hot and dry weather in 2010. By July, at least 43 peat bogs were burning in the vicinity of Moscow alone. Peat fires are difficult to both extinguish and contain as the fire depends on the depth of the peat deposit and its water content—some layers can be 16.5 feet (5 m) or deeper and can smolder for months. Such significant volumes of water were needed to fight a bog fire, so the Russian government constructed a 31-mile (50 km) long pipeline from the Oka River to a region east of Moscow.

Finally, by early August, the fires had begun to encroach upon the area where the Russian, Belorussian, and Ukrainian borders meet—near the site of the Chernobyl nuclear disaster. The fires began to burn the lands that were contaminated by radiation and toxic materials from the explosion in 1986. At least 28 fires covering 665 acres (269 ha) burned in the Bryansk region; at least 15 square miles (39 square km) had burned in areas that had vegetation and soils polluted with radiation. However, government radiation experts investigated and concluded that radiation levels beyond the contaminated region had not increased, and this seemingly eased the tensions.

IMPACTS

According to the Centre for Research on the Epidemiology of Disasters CRED), 55,736 people died from this heat wave, the wildfires, and the other related consequences of the calamity. The officially reported Russian government death toll was a mere 65. More than 26,000 wildfires were estimated to have burned more than 2 million acres (816,515 ha) across western and central Russia. More than 3,500 people lost their homes. Experts projected the damages probably exceeded $15 billion, and it may have exceeded $300 billion if lost timber was included in the calculation. An official government assessment has never reported an accounting of the damages. Russia's hottest summer destroyed more than a third of the Russian wheat crop, prompting the government to ban exports of wheat. And an estimated 32.9 million acres (13.3 million ha) of cropland in 41 provinces were impacted by the heat, drought, and fires.

GOVERNMENT RESPONSES

By the end of July 2010, the Russian president, Dmitry Medvedev, had declared a state of emergency in seven regions and ordered the Defense Ministry to deploy the military to battle the fires. The prime minister, Vladimir Putin, called for the resignation of local officials in 14 districts in central Russia because of their failures to act to control and end the emergencies and because they lost the trust of the people. At the time, it was reported that 25 had died, 1,257 homes (60 near Moscow) had burned, and more than 2,000 were without homes. Medvedev called for immediate construction of housing for those made homeless by the fires. The Russian Regional Development Minister Viktor Basargin vowed that those whose homes were destroyed by fires would receive new ones before the end of the year. And Putin pledged $33,000 for each family that suffered a death, $100,000 to reconstruct each home that was destroyed, and other compensation for the loss of property. U.S. President Barack Obama expressed condolences to Medvedev and offered firefighting equipment and expert assistance to combat the fires. While several other countries of the Commonwealth of Independent States and of the European Union were already assisting, the United States' offer was declined. The emergency phase of the disaster soon ended in the last weeks of August.

The Public Commission on Investigation of Causes and Consequences of the Wildfires in Russia in 2010 concluded that although the extreme nature of the meteorological conditions that settled over European Russia, Ukraine, and Belarus was the fundamental trigger of the disaster, it is clear that the Russian government's policies substantially worsened it. Due to poor organization; inadequate funding; and ineffective response to the wildfires, drought, and heat, the government's firefighting and emergency response forces were futile and could not prevent or contain the fires. Ad hoc brigades of ordinary Russian citizens, however, were more successful.

In the wake of the disaster, the Russian government initiated several policies to improve forest management in the short term. They committed to investing in fire suppression and prevention, reorganization of fire prevention and firefighting responsibilities in the central government, and improvements of transparency and accountability, particularly by making satellite imagery of fire-impacted territories publicly available and easily accessible via the Internet.

CONCLUSION

Extreme heat, drought, and forest fires are natural conditions that occasionally occur in virtually all mid- and high-latitude regions of the planet. The intensification of the extremes and frequencies of such events are now being influenced by the climate ramifications of global warming of the atmosphere. The occurrence of disaster from worsening and more frequent conditions is substantially determined by human actions and responses to the unusual conditions. Clearly, the lack of preparedness of the people in this region to cope with extreme heat is not surprising given the infrequent heat waves that were experienced in past decades. The more important matter is that the Russian government was unprepared for the outbreak of forest fires, fast-moving wildfires, and burning peat bogs, despite the relatively regular occurrence of these phenomena over the last century. No small part of the blame should be placed on forest managers whose campaigns of fire suppression contributed fuel availability to the hot, dry, and windy conditions.

The pledges made by the government in the wake of the fires amounted to more commitment to the same policies of the past. A disregard for the contribution of vulnerabilities (of people and places) to the disaster equation, combined with an apparent lack of reflection upon past national and regional policies, likely means that, even though more severe fire-years have occurred in the last decade, the prospects for future disasters remain high. No significant adjustments appear to have been made to mitigate the vulnerabilities and overarching policy remains unchanged.

John Paul Tiefenbacher

Further Reading

Maier, F., A. Obregón, P. Bissolli, C. Achberger, J. J. Kennedy, D. E. Parker, O. Bulygina, and N. Korshunova. 2011. Summer Heat Waves in Eastern Europe and Western Russia in State of the Climate in 2010. *Bulletin of the American Meteorological Society* 92: S210.

Sidortsov, R. 2011. Russia's Perfect Firestorm: Climate Adaptation Lessons from the Summer of 2010. May 13, 2011. https://ssrn.com/abstract=1851964, accessed June 8, 2019.

WHO (World Health Organization). 2010. Wildfires and Heat-wave in the Russian Federation—Public Health Advice. https://www.euro.who.int/__data/assets/pdf_file/0012/120090/190810_EN_Russia_wildfire_advisory.pdf?ua=1, accessed July 8, 2020.

Hurricane Andrew, United States and the Bahamas, 1992

Hurricane Andrew began forming in mid- to late-August 1992, off the western coast of equatorial Africa. Its progression from tropical wave to tropical storm, and ultimately to hurricane status, saw the system weaken initially and then regain strength as it continued moving toward the Bahaman Islands, leading to the first of three landfalls, one in the Bahamas and two more in the United States. After passing over the Bahamas, the storm regained strength and moved toward the U.S. Florida Keys and South Florida, where it caused extensive damage. The third portion of the storm's track pushed into the Gulf of Mexico, where it weakened, turned north, and made landfall in Louisiana, causing significant but less extensive damage. The system then turned northeast and moved into the continental United States, affecting an additional five states, but only inflicting light-to-moderate damage. The storm finally dissipated on August 28, 1992, over North Carolina.

METEOROLOGICAL SYNOPSIS

Before becoming a named storm, Andrew began as a tropical wave, forming off the coast of Equatorial West Africa on August 14, 1992. As it moved into the tropical North Atlantic, it gained strength and was upgraded to "Tropical Depression Three" on August 16. Continuing northwest, the depression continued to drop in pressure and increased in speed to 35 mph (56 kph). On August 17, Andrew moved closer to the Caribbean Basin, and it became a tropical storm (TS). At that time, its path was pointing toward the Lesser Antilles, east of Puerto Rico, the Dominican Republic, and Haiti, but a low-pressure system forced TS Andrew to veer north. Gaining strength from upper-level energy, the storm intensified on August 21, reaching 63 mph (104 kph) while preparing to shift its track from a northwest- to western-moving storm. A high-pressure system forced the storm to move west, an eye began to form, and on the morning of August 22, TS Andrew was upgraded to Hurricane Andrew.

Over the next 36 hours, Andrew's pressure decreased rapidly, contributing to the change in category on the Saffir-Simpson Hurricane Scale from 1 to 4—such a rapid drop in pressure is referred to as a "rapid deepening," a phenomenon in which a hurricane exhibits a 35 mph (56 kph, or 30 kt) increase in speed or a 40-millibar (mb) drop in pressure over a 24-hour period. Andrew made landfall as a Category 4 storm on Eleuthera island in the Bahamas on August 22, 1992. However, reanalysis suggests that the intensity may have been greater, causing Andrew to reach Category 5 upon making landfall in the Bahamas (Landsea et al. 2004). Andrew experienced moderate weakening while passing over the Grand Bahama Bank, but regained strength as it moved into the warmer waters near the Florida Coast. Vertical circulation, called convection, assisted in intensifying the hurricane on this segment of its path. As it moved into position for landfall on the United States, at Homestead Air Force Base, Florida, Andrew was a Category 5. Typically, hurricanes weaken over land, and Andrew followed this convention, dropping to a Category 4 and seeing an increase in pressure to 950 mb as it moved over the southern tip of the Florida peninsula.

Moving into the Gulf of Mexico as a Category 4, Andrew experienced only moderate intensification, reaching a low-pressure mark of 937 mb and wind speeds of 145 mph (233 kph). North of the storm, the high-pressure system that had been previously contributing to its intensity began to weaken, thus causing Andrew to decelerate as it approached the Louisiana coast. On August 26, Andrew made landfall in Louisiana as a Category 3 and began to weaken rapidly as it turned northeast. Within just 10 hours, Andrew was downgraded from a hurricane to tropical storm on August 27, and it moved through the states of Mississippi, Alabama, Georgia, Tennessee, and North Carolina, where its final remnants merged with a cold front. Andrew's last gasp of life occurred as a tropical depression on August 28, 1992, near the border between Tennessee and North Carolina (NOAA/NHC 2017a).

WARNINGS

Models predicting the path of Andrew performed well, displaying approximately 30 percent fewer errors than the average for the time. Lead times were 24 hours in advance for the Bahamas and 21–24 hours in advance for the United States. This provided ample time for coastal and inland residents to prepare and/or evacuate in the face of the anticipated high winds that are the hallmark destructive agents of a hurricane event.

Evacuations were executed on a large scale, with approximately 1.23 million people moving out of Florida locations, 1.3 million in out of Louisiana, and an additional 250,000 displaced from Texas. Hurricane watches were issued on August 22, 1992, for portions of the eastern and western coasts of Florida, with the western watches upgraded to warnings the same day, except for the northwestern watches, which were downgraded to tropical storm warnings. Given the short duration of the hurricane overland in Florida (approximately four hours), all watches and warnings were discontinued by August 24, 1992.

Preparations made in the Bahamas prior to landfall, given the extensive lead time, were adequate, and additionally, the hurricane struck the sparsely populated northern portion of the island, contributing to a very small threat to inhabitants. The Bahaman government issued a hurricane watch for the northwest region on August 22, 30 hours in advance of the storm, and upgraded the watch to a warning six hours later, giving the people in the predicted path up to 24 hours of lead time to prepare. Additionally, 15 hours of lead time was provided through the issuance of a second hurricane warning for the central Bahamas.

DAMAGE

In the Bahamas, Hurricane Andrew damaged an estimated 800 homes, affecting about 1,700 people. Damage to roads was also reported, and additional homes were destroyed in several smaller villages in the north. Storm surge and hurricane-spawned tornadoes contributed to the damage caused. Just three fatalities resulted in the Bahamas, largely attributed to the storm striking mostly unpopulated areas.

Florida shouldered nearly 97 percent of the total monetary damage caused on U.S. territory; of the $26.5 billion in damage, approximately $25.3 billion resulted from impacts sustained there. Wind speeds varied across the path of the storm, but several very substantial measurements were recorded. At the Turkey Point Nuclear Station, a top speed of 146 mph (235 kph) was experienced. Coral Gables saw gusts of 164 mph (264 kph), Key Largo reported a 13-minute sustained wind of 114 mph (183 kph), and Perrine recorded a 177-mph (285 kph) gust. These high winds tore roofs from buildings, as well as weakened interior walls, generating a large debris field that contributed to further damage. Most of the high winds, and consequently, the damage, were limited to a small portion of the state, namely in and around Key Largo. Homestead Air Force Base sustained heavy damage as well.

Damages in Louisiana made up the bulk of the remaining monetary impact. Heavy rain, high winds, flooding, and 14 tornadoes all caused damage, mostly along the coastline. Crop damage, as well as nearly loss of 1,000 homes, brought the total to just over $1.5 billion in the state. As Andrew continued inland, the remainder of the damage caused to other U.S. states resulted largely from tornadoes generated by thunderstorms associated with the weakening depression.

Destroyed buildings at Homestead Air Force Base in Miami-Dade County, Florida, after Hurricane Andrew, 1992. (PH2 Davis Tucker, USAF/Department of Defense)

Approximately $500 million in damage occurred outside the states of Florida and Louisiana.

Fatalities resulting from Andrew total 65, with 44 of those deaths occurring in Florida (15 direct, 29 indirect), 17 in Louisiana (8 direct, 9 indirect), and the remaining four occurring in the Bahamas (3 direct, 1 indirect). Efforts by the Centers for Disease Control (CDC) as well as a research team at Emory University state 462 injuries in the United States resulted from Andrew. The extent of injuries to people in the Bahamas was minimal.

RESPONSE AND RECOVERY

The American Red Cross, the United Kingdom, the United States, Canada, Japan, and the United Nations all responded to provide aid and assistance to the Bahamans in the aftermath of Andrew. As the damage was contained to a relatively small section of the island, cleanup efforts and rebuilding commenced almost immediately, and the extended impact of the disaster only affected tourism for a few seasons.

The impact on Florida and, to a lesser extent, Louisiana was much more extensive. The U.S. government issued a disaster declaration for the two states and approved a $11.1 billion aid package, of which $9 billion went to Florida. As it would be in several disasters lying in wait in the future, and had been most recently in the wake of Hurricane Hugo (1989), the Federal Emergency Management Agency (FEMA) was on the receiving end of widespread criticism for its lack of preparedness and response to the disaster area. The Florida Army National Guard and the U.S. and Canadian armies provided food, established tent cities, and repaired services to the affected areas. Crime rose almost immediately, as is a relatively common occurrence in large-scale disaster areas, but quickly slowed with the arrival of U.S. military troops.

Important lessons were learned concerning building codes in the wake of Andrew. The massive influx of insurance claims, brought about by very poor construction, ended the existence of several insurance companies and prompted the state of Florida to create new regulatory mechanisms to prevent a reoccurrence of such failures. The eventual result of those efforts was the creation of the Florida Building Code in 1998 (operational in 2002), a system that eventually supplanted the morass of local codes, creating a coherent and standard set of regulations applicable statewide. Studies of the damages caused by major hurricane events post-Andrew (e.g., Ivan, Katrina, Wilma) showed that buildings constructed under these new codes and subsequently affected by a hurricane sustained less damage than their pre-Andrew counterparts (Bradford and Sen 2004).

CONCLUSION

In terms of size, Andrew was not significant, but it ranks within the top 10 of a few major categories of hurricane records. Perhaps most notably is Andrew's inclusion on a very short list of just three hurricanes to make landfall in the United

States as a Category 5. There have been a total of 33 Atlantic hurricanes to achieve that level of strength, but only the Labor Day Hurricane (1935), Camille (1969), and Andrew (1992) have struck a U.S. coast while maintaining that strength. Andrew holds the number four position on the list of most intense (as measured by pressure) hurricanes that made landfall in the contiguous United States, with a low-pressure minimum of 922 mb; Katrina (2005) is third on that list (920 mb), with Camille (1969) coming in second at 900 mb, and the Labor Day Hurricane (1935) first at 892 mb.

Andrew was a very costly event, causing $26.5 billion in damage to areas struck in the United States (with an additional $250 million suffered by the Bahamas). At the time of the event, Andrew was the costliest hurricane in U.S. history, a record that was not surpassed for 13 years, when Katrina (2005) inflicted $108 billion in damage. As of 2018, Andrew is fifth on the list of costliest hurricanes all times (U.S. only), with Ike (number four, 2008), Sandy (number three, 2012), Katrina (number two, 2005), and Harvey (number one, 2017) ranking higher.

Andrew sits tied with 10 other Atlantic hurricanes at eighth place for highest one-minute sustained wind speed at 175 mph (282 kph). The duration of Andrew as a Category 5 is also notable, coming in at 23rd, having lasted for 16 hours at that strength (the longest Category 5 occurred in Cuba, 1932, and stretched over 78 hours, with the more recent Irma coming in at second, lasting for 75 hours at the top Saffir-Simpson level) (NOAA/NHC 2017b). The name "Andrew" was retired after the 1992 season, and this event will remain permanently fixed in the annals of hurricane history as one of the most important events on record.

Mitchel Stimers

Further Reading

Bradford, N. M., and R. Sen. 2004. Gable-End Wall Stability in Florida Hurricane Regions, 10-Year Review: Post-Hurricane Andrew (1992) to Florida Building Code (2002). *Journal of Architectural Engineering* 10(2): 45–52.

Landsea, C. W., J. L. Franklin, C. L. McCaide, J. L. Bevin II, J. M. Gross, B. R. Jarvinen, R. J. Pasch, E. N. Rappaport, J. P. Dunion, and P. P. Dodge. 2004. A Reanalysis of Hurricane Andrew's Intensity. *Bulletin of the American Meteorological Society* 85(11): 1699–1712.

NOAA (National Oceanic and Atmospheric Administration), and NHC (National Hurricane Center. 2017a. Historical Hurricane Tracks. https://www.coast.noaa.gov/hurricanes/, accessed January 3, 2018.

NOAA (National Oceanic and Atmospheric Administration), and NHC (National Hurricane Center). 2017b. Hurricane Research Division Database (HURDAT) version 2. http://www.nhc.noaa.gov/data/hurdat/hurdat2-1851-2016-041117.txt, accessed January 3, 2018.

Hurricane Charley, United States, 2004

Only the third hurricane of the 2004 Atlantic hurricane season, Hurricane Charley was already the second major hurricane. Florida is more affected by hurricanes than other U.S. states, with hundreds of recorded cyclones and billions of dollars in damage. However, Hurricane Charley became the first of four hurricane strikes on Florida that year, making it the first time in recorded history that the state was hit by four hurricanes in just one season (Smith and McCarty 2009). The storms that have affected Florida have ranged widely in power, but when it made landfall as a Category 4 hurricane after rapid intensification, Hurricane Charley was the most powerful storm to hit Florida since Hurricane Andrew's landfall in 1992. Hurricane Charley caused extensive damage in Florida, as well as later affecting states like South Carolina, North Carolina, and Virginia. It also affected other areas like Cuba and Jamaica.

PREDICTIONS, EVACUATIONS, AND THE HURRICANE'S PATH

On August 11, 2004, as Hurricane Charley passed near Jamaica, Florida Governor Jeb Bush issued a state of emergency for Florida as the National Hurricane Center (NHC) issued tropical storm and hurricane warnings for parts of the state. Massive evacuations began, with approximately 2 million people in western and southern Florida being encouraged to leave, including in the areas of Tampa and the Florida Keys. Large installations and organizations also began to prepare, including MacDill Air Force Base, the Kennedy Space Center (which encouraged the evacuation of normally on-site staff and secured equipment like the Space Shuttles), and amusement parks in the Orlando area (with relatively rare closures). States of emergency were also declared by Governors Sonny Perdue of Georgia and Mark Sanford of South Carolina.

As Hurricane Charley continued moving, it turned, increased its forward speed toward Florida, and rapidly intensified (Pasch et al. 2011 [2004]). This in turn led to questions about the storm's path and the warnings and predictions associated with it. Among the many pieces of information issued by the NHC and other organizations, the cone

Damaged boats lie in disarray after Hurricane Charley passed through Punta Gorda, Florida, on August 13, 2004. (FEMA)

of uncertainty is often shown by the media and followed by the general public as they follow the hurricane. The graphic shows the probable track of a hurricane over a five-day period and is generally accurate with roughly two-thirds of hurricane paths falling within their respective cones. The cone sometimes includes a central line focusing on a predicted forecast track of the center of the hurricane (often at specifically indicated times). However, the use of the cone can be problematic as the public may put too much confidence in it, as demonstrated by a study completed by the Federal Emergency Management Agency (FEMA), the U.S. Army Corps of Engineers, and the National Oceanic and Atmospheric Administration (NOAA) (Broad et al. 2007).

Importantly, in Hurricane Charley, the track of the hurricane shifted at a critical point. Graphics and information issued by the NHC between 5:00 a.m. on August 12 and 11:00 a.m. on August 13 put the center of the storm over Tampa Bay while the ones issued just hours later at 2:00 p.m. put the center over Charlotte County, 100 miles (150 km) further south (Broad et al. 2007). The hurricane ultimately made landfall about 70 miles (105 km) south of Tampa around 3:45 p.m., first on the barrier island Cayo Costa and then onto the mainland, before moving across Florida.

While there had been extensive evacuations in the Tampa Bay area, there had been far fewer evacuations in Charlotte County (south of Tampa and Sarasota), an issue potentially tied to the cone of uncertainty, people's belief in it, and the storm's path. As "FOX4 News meteorologist Dave Roberts stated, 'A lot of the folks here [in Charlotte County] weren't taking it [Charley] seriously. It was supposedly this big problem for Tampa Bay'" (Broad et al. 2007, 656). Others echoed similar concerns, but contextualized the cone and track within a broader media narrative, like WFOR News meteorologist Bryan Norcross, who "argued that much of the television coverage on Florida's west coast inappropriately urged the public to pay attention to the forecasted track line, leaving those within the cone, but not on the line, vulnerable to wobble in the hurricane's path" (Broad et al. 2007, 657).

This was supported by later research indicating that 27 percent of respondents to the survey did not evacuate, thinking they "could ride out the hurricane" (Smith and McCarty 2009, 134). Additionally, 26 percent thought the storm would make landfall elsewhere, 5 percent said they did not have time to evacuate, and 4 percent said they did not know the hurricane was coming (Smith and McCarty 2009). Further complicating this was Charlotte County's population. The area has a disproportionately high population over the age of 65, one of the highest percentages of such a population in a U.S. county at the time (Broad et al. 2007). Charlotte County has a higher proportion of retirees than most areas, and 4 percent of the people surveyed by Smith and McCarty (2009) indicated medical conditions as a concern related to their evacuation decision.

Looking back on the forecast data, the NHC had two main notes about Hurricane Charley: (1) at 96 hours and 120 hours out, the official track forecasts were significantly worse than the 10-year average but that (2) at 72 hours they were better than that 10-year average (Pasch et al. 2011 [2004]). They noted that while the official forecasts were off from the main predicted track, the hurricane actually made landfall in areas under hurricane watches and warnings and stayed within the cone (Pasch et al. 2011 [2004]). The director of the NHC, Max Mayfield, pointed out that "the Charlotte Harbor area was within the hurricane warning area for 24 hours. They were in the cone of uncertainty for four days" (Broad et al. 2007, 657). Additionally, the NHC was also issuing forecasts and advisories about its power, issuing a special advisory when it strengthened up to a Category 4.

As Hurricane Charley moved up the Florida peninsula, it triggered nine tornados throughout the state. By the time it left Florida, it was still at hurricane strength. Once back in the Atlantic, the storm briefly re-intensified. It then made another U.S. landfall near Cape Romain, South Carolina, at around 10:00 a.m. on August 14 as a Category 1 hurricane before moving back offshore again and making another U.S. landfall near North Myrtle Beach, South Carolina, around noon. As it traveled over North Carolina, it triggered five tornadoes and was downgraded to a tropical storm. While the storm continued to weaken, it began to move back into the Atlantic, passing near Virginia Beach, Virginia, as technically an extratropical system, causing two tornados in the state. Over the Atlantic, it continued to move toward the northeast, dissipating near southeastern Massachusetts by the morning of August 15, 2004.

THE EFFECTS

Hurricane Charley caused extensive damage throughout Florida and was the starting point of a lengthy and problematic hurricane season for the state. Hurricane Charley killed 10 people directly and an additional 25 indirectly in the United States, along with at least five additional direct deaths in Cuba and Jamaica combined (Pasch et al. 2011 [2004]). Most of these deaths were in Florida, including nine direct deaths and 24 indirect deaths. The direct deaths were caused by a series of issues, including damage to or destruction of buildings, flying debris, car accidents related to storm conditions, or drowning in related flood or storm water conditions (Pasch et al. 2011 [2004]).

In the wake of Hurricane Charley, President George W. Bush declared all Florida counties presidential disaster areas. Contextualized in the 2004 season, at the peak of needed response, over 16,000 temporary housing units were in use across the state and 10,000 were still in use by June 2005, including individuals and families housed in mobile homes or travel trailers (Tobin et al. 2006). While some were located on personal property, others were built relocation parks that also offered other forms of aid like a county mobile clinic and spaces for residents to meet with agencies like FEMA. Despite this, there were issues with the facilities such as a lack of nearby grocery stores and public phones (Tobin et al. 2006).

When Tobin et al. (2006) surveyed residents of a relocation park, they found that the most common concern about the impact of the hurricane among residents was their loss of housing, with 100 percent of respondents pointing to this as a concern. This was followed by loss of transportation (46 percent), loss of a job (39 percent), food (37 percent), family (26 percent), loss of everything (20 percent), stress (18 percent), and child care (16 percent). While FEMA and other agencies offered assistance for these issues to the residents, they remained concerns for people. When these same residents were asked how emergency response had met their needs, their average response, on a scale of 1 (not at all) to 5 (completely satisfied), was 3.36 (Tobin et al. 2006).

The total damage estimate for Hurricane Charley was revised in September 2011 up to $15.1 billion, including $6.8–$7.4 billion in insurance costs (Pasch et al. 2011 [2004]). At the time of the 2011 revision, Hurricane Charley was the sixth costliest hurricane in recorded American history behind Katrina (2005), Ike (2008), Andrew (1992), Wilma (2005), and Ivan (2004). However, Hurricane Charley has since dropped to the eleventh costliest American hurricane behind the storms already listed, as well as Harvey (2017), Maria (2017), Sandy (2012), Irma (2017), and Rita (2005), which include hurricanes from more recent seasons (especially the extremely costly 2017 season). This list does not, as of the time of this writing, include the likely-to-be costly 2018 season storms, including Hurricane Michael's impact on Florida. These lists of the costliest hurricanes, like the most powerful or largest hurricanes, change regularly over time, both as new hurricane seasons happen and as insurance and damage estimates or other data sets are reevaluated or revised.

CONCLUSION

With conflicting views over the cone of uncertainty and the hurricane track's accuracy and portrayal in the media, confusion seems to have arisen in people's understanding of where the hurricane might go and who should evacuate. This confusion is important to discuss because it can help us understand how people view, think about, and respond to hurricane information before the storms make landfall and thus help us improve communication in disasters. Later surveys have shown that 57–68 percent of people surveyed after Hurricane Charley said that "the cone of uncertainty graphic was a 'very important' factor in their evacuation decision" (Broad et al. 2007, 658). And with Hurricane Charley specifically, this confusion is part of the story of the storm, who it affected, and how it affected them. Hurricane Charley's devastation clearly demonstrates how disasters are the product of not only the hazard (the hurricane itself) but also the populations it affects, especially populations with potentially high vulnerability rates, like the elderly.

Jennifer Trivedi

Further Reading

Broad, K., A. Leiserowitz, J. Weinkle, and M. Steketee. 2007. Misinterpretations of the "Cone of Uncertainty" in Florida during the 2004 Hurricane Season. *Bulletin of the American Meteorological Society (BAMS)* 88(5): 651–668.

Pasch, R. J., D. P. Brown, and E. S. Blake. 2011 [2004]. *Tropical Cyclone Report: Hurricane Charley, 9–14 August 2004.* October 18, 2004 (Revised September 15, 2011). Miami, FL: National Hurricane Center.

Smith, S. K., and C. McCarty. 2009. Fleeing the Storm(s): An Examination of Evacuation Behavior during Florida's 2004 Hurricane Season. *Demography* 46(1): 127–145.

Tobin, G. A., H. M. Bell, L. M. Whiteford, and B. E. Montz. 2006. Vulnerability of Displaced Persons: Relocation Park Residents in the Wake of Hurricane Charley. *International Journal of Mass Emergencies and Disasters* 24(1): 77–109.

Hurricane Galveston, United States, 1900

Hurricane Galveston is also known as the Great Galveston Hurricane, the Great Storm of 1900, the Great Storm, the 1900 Storm, the Galveston Flood, or the Galveston Hurricane. Before striking the United States, the hurricane made landfall in the Dominican Republic, Hispaniola, and Cuba. But the Category 4 hurricane's devastation in the United States, especially Texas, marks it as one of the worst disasters in American history. Hardest hit was the low-lying Galveston Island, which sat just five feet above sea level.

GALVESTON BEFORE

The city of Galveston itself sits on Galveston Island in Texas. In 1900, prior to the hurricane, it occupied six square miles of the island. It included several public parks and markets, as well as schools, churches, and homes. Following the Civil War (1861–1865), the city experienced a population boom, growing over 400 percent in the 40 years before the hurricane made landfall to approximately 40,000 residents. Galveston relied on its location on the water, with two miles (3.2 km) of wharfs and more under construction when the hurricane made landfall, mostly designed to handle shipments of grain and cotton. Beyond this, the city claimed 306 factories as of 1900, as well as businesses such as banks, wholesale grocers, and wholesale liquor dealers (Coulter 1900).

THE GREAT STORM OF 1900

Wireless telegraph messages and sailors coming into port shared news of the storm's effects on Caribbean Islands in advance of its Texas landfall (Roth 2010). In addition, the report of United States Weather Bureau at Galveston after the storm noted that they had received the first message about the storm at 4:00 p.m. local time on September 4 when the hurricane was moving over Cuba, followed in the days after by updates that the bureau officials posted (Coulter 1900). After moving into the Gulf of Mexico on September 6, the storm strengthened into a hurricane. Beginning at 4:00 a.m. local time, rain and large waves hit Galveston Island. The hurricane flag was raised, warning local residents of the impending hurricane, resulting in the evacuation of 20,000 people at the time when the island's population was 30,000 (Roth 2010).

Hurricane Galveston reached Category 4 status on September 8. That day, as storm conditions increased on the island, people reportedly went to the beach to watch the increasingly severe waves (Roth 2010). Survivors of the hurricane later shared memories of themselves and others seeing rough water, a high tide, and water coming in toward the island. The Weather Bureau office workers deemed the entire island in danger from the hurricane by 10:10 a.m. Throughout the day, the Weather Bureau's office was inundated with people asking about the weather and the office's telephone line was busy until the lines went down (Coulter 1900).

While residents were aware of an incoming hurricane, they did not have the more nuanced predictions people have access to on the American Gulf Coast in the modern era. As one resident, William Mason Bristol, recounted: "We knew there was a storm coming, but we had no idea that it was as bad as it was" (Burnett 2017). Despite this, many evacuated, in large part due to the efforts of the workers at the Weather Bureau. As John Coulter included in his edited volume later that year, "The warnings of the United States Weather Bureau were the means of thousands of lives being saved through the hurricane" (Coulter 1900, 50).

Isaac Cline, of the Weather Bureau, and his brother rode through the area, warning people once more and encouraging them to evacuate, but for many, this was too late as bridges to the mainland were already being damaged and destroyed (Roth 2010). Late that night, before midnight, hurricane force winds began to hit the island, casting debris around the area, the storm surge increased, and 15 feet (4.6 m) of water pushed across the island. Overnight into September 9, this sent people fleeing toward the higher center of the island, while others attempted to take sanctuary in the Bolivar Lighthouse or the brick Rosenberg School. Others climbed higher up in buildings like homes and tied their families together to keep them from being separated. As people attempted to get away from the rising water, they were also at risk from floating debris and animals attempting to escape the floodwaters. The fact that the storm struck overnight and much of this behavior was happening in the dark further complicated the situation. Ultimately, 10 inches (254 mm) of rain fell on Galveston, setting a 24-hour record for September, and parts of the island were covered by anywhere from five (1.5 m) to 12 feet (3.7 m) of water (Coulter 1900; Roth 2010).

The storm began to dissipate on the island early in the morning on September 9. As it moved into the United States, it became an extratropical storm that moved through other parts of the United States before hitting Canada. Severe winds affected parts of Florida, Mississippi, and Louisiana. Areas across the Midwest, New York, and New England also experienced severe winds and rain. Six different Canadian provinces were affected, causing deaths, damage, and crop loss. However, in its most serious effects, it left behind massive destruction and large numbers of deaths on Galveston Island.

THE EFFECTS OF THE STORM

The hurricane killed an estimated 8,000 people. The death toll included a wide range of people who had remained on the island during the storm, such as 10 Catholic nuns and 90 children in their care at St. Mary's orphanage, trapped in the remains of the orphanage with the children tied to one another, seemingly in an attempt to prevent them from being separated by the floodwaters. The orphanage had been built close to the water and the beach under the belief that the fresh air would minimize the children's risk of yellow fever (Coulter 1900; Burnett 2017).

The death toll was so high that officials declared martial law and forced some residents, mostly black men, to gather and dispose of the corpses. Accounts from the time noted that able-bodied men were conscripted for this work and claimed that men who refused to do the work were threatened or shot (Coulter 1900).

At first, attempts were made to identify each corpse, but the number of dead and the need to deal with the bodies quickly due to public health concerns complicated this matter, reducing officials to adding descriptions to records and grave markers that amounted to listing sex, race, age, etc. The ground was so oversaturated that digging graves proved difficult. Initially, officials and responders attempted to dispose of the corpses at sea by taking them out on barges, weighing the bodies down, and throwing them into the water (Coulter 1900). When some bodies washed back onto the island, officials attempted to use funeral pyres to burn the remains instead (Roth 2010; Burnett 2017). Further complicating matters were reports of people refusing to handle the remains of black victims at all, to a greater extent than people not wanting to handle the remains of white victims (Coulter 1900).

Black men who survived were particularly more likely to be accused of looting or "despoiling the dead," based on accounts from that same year such as Coulter's edited volume of firsthand accounts *The Complete Story of the Galveston Horror*. At least some of those accused of such thefts were arrested and, without trial, shot immediately, despite any of their own claims of innocence or explanation (Coulter 1900). More broadly, the state police who had arrived on site also "arrested every suspicious character and the jail and cells at the police station were filled to overflowing" (Coulter 1900, 46).

Building and infrastructure destruction was widespread, including over 2,600 homes, all bridges to the mainland, telegraph and telephone lines, and 15 miles (23 km) of railroad tracks (Roth 2010). This widespread destruction resulted in approximately 10,000 people being left homeless (Burnett 2017). Contemporary accounts place the amount of destruction in the millions of dollars (Coulter 1900). The estimated $21 million at the time equates to over $631 million when adjusted for inflation today.

With the destruction of the bridges and railways, initial relief came in via boat from Texas City on Tuesday, September 11. This influx of relief not only provided much needed physical supplies but also potentially contributed to improving the mental health of local residents. However, later that same day martial law was declared by the mayor, purportedly in response to looting in the area. The Bolivar Point Lighthouse that served as a shelter for people on the island was also pushed into service after the hurricane passed, being used as a relief center and resuming its normal work as a lighthouse. Following this, tents were erected in Emancipation Park to help provide shelter, and roughly 1,000 inhabitants went to Houston, Texas, to stay out of the way while efforts began to clean and rebuild the city (Coulter 1900).

Reports were sent onto the mainland United States recounting the initial extent of the damage, asking for assistance, supported by local mayors, businessmen, and members of the Galveston Local Relief Committee. In response to the appeal, relief poured in, both in the form of material aid—like army rations to feed people

A house is tipped on its side following the hurricane and storm surges that hit Galveston, Texas, in 1900. Hurricane Galveston caused nearly 10,000 deaths, making it one of the worst natural disasters in U.S. history. (Library of Congress)

remaining in Galveston—and funds, all sent from individuals and communities across the United States, as well as from international locations such as England, Germany, and France. Starting in Chicago, a movement even began for students to send in their pennies to help.

THE SEAWALL AND GRADE RAISING

In the aftermath of the hurricane, efforts began on the island to protect structures and inhabitants from future storms and hurricanes. November 1901 saw a convention in Fort Worth, Texas, where Galveston's needs were discussed, leading eventually to a new municipal government in Galveston forming a board of engineers comprising Henry Martyn Robert, Alfred Noble, and Henry Clay Ripley to plan out ways to protect the island. The following January, they recommended the construction of a 3-mile-long, 17-foot-high concrete seawall, a recommendation that was approved and led to work beginning in the fall on the trench needed to support the wall. The large-scale seawall was constructed employing 40 men to build the wall in sections. It was completed in July 1904.

The city and its residents also worked on a "grade-raising," lifting 2,000 surviving buildings and everything around them, including utilities. Nearest the seawall, the city was raised over 16 feet (4.9 m), but in residential areas further inland from the wall, areas were only raised about eight feet (2.5 m). The grade raising was completed in August 1910, nearly a decade after the hurricane had made landfall. The raising, combined with the seawall, cost the town $3.5 million; but when another hurricane hit in August 1915, the improvements resulted in a much smaller death toll (Hansen 2007).

CONCLUSION

The Galveston Hurricane caused massive destruction on the island, and it remains one of the deadliest disasters in American history, even well over 100 years later. The extent of the storm left local leaders and residents coping with questions of how to help and shelter people in the area, as well as how to cope with the thousands of bodies. The effects of the storm led to efforts to send both aid and supplies into the affected community from around the world, but they also revealed preexisting problems in the United States, such as racism. As the city of Galveston began to move forward in recovery, they also turned to mitigation efforts to reduce the effects of future hurricanes and flooding, efforts that have remained part of their disaster planning efforts to the modern day.

Jennifer Trivedi

Further Reading

Burnett, J. 2017. The Tempest At Galveston: "We Knew There Was a Storm Coming, but We Had No Idea." National Public Radio. November 30, 2017. https://www.npr.org/2017/11/30/566950355/the-tempest-at-galveston-we-knew-there-was-a-storm-coming-but-we-had-no-idea, accessed January 23, 2019.

Coulter, J., ed. 1900. *The Complete Story of the Galveston Horror*. Sandy Springs, GA: United Publishers of America.

Hansen, B. 2007. Weathering the Storm: The Galveston Seawall and Grade Raising. *Civil Engineering* 77(4): 32–33.

Roth, David. 2010. *Texas Hurricane History*. Camp Springs, MD: National Weather Service.

Hurricane Harvey, Texas and Louisiana, 2017

The Gulf Coast of the United States is an area that frequently experiences hurricanes and the strong winds, heavy rain, and storm surge that come with them. From the months of May to November, the states bordering the Gulf of Mexico are at risk of feeling the impacts of a hurricane. Each year, an average of one to two hurricanes makes landfall in the continental United States. To be classified as a hurricane, the storm must have winds that exceed at least 74 miles per hour (111 km/h). On August 25, 2017, Hurricane Harvey made landfall in Rockport, Texas, as a Category 4 hurricane. Hurricane Harvey was the first major hurricane to make landfall in Texas since Hurricane Celia in 1970. At the time, Harvey ranked as the second-most expensive hurricane to impact the United States as it caused over $125 billion in damage.

FORMATION

On August 12, 2017, Hurricane Harvey originated from a tropical wave that emerged from Africa over the eastern Atlantic Ocean. By August 18, 2017, Harvey had reached tropical storm status as it passed over Barbados and St. Vincent before weakening again to a tropical wave on August 19, 2017. As a tropical wave, Harvey continued to move west through the Caribbean Sea and eventually over the Yucatan Peninsula on August 22, 2017. By August 23, 2017, the storm system regained enough strength to be classified again as Tropical Depression Harvey. Fueled by the warm waters of the Gulf of Mexico, Harvey began to rapidly intensify and turned north toward the Gulf Coast region of the United States. On August 24, 2017, Harvey reached maximum sustained winds of 80 miles per hour (120 km/h), thus, becoming a Category 1 hurricane. As the storm progressed through the warm waters of the Gulf of Mexico, Hurricane Harvey continued to increase in strength and sustained winds. Just prior to making landfall in Texas on August 25, 2017, Hurricane Harvey recorded maximum sustained winds of up to 130 miles per hour (195 km/h), classifying it as a Category 4 hurricane.

WARNING AND EVACUATION

The National Weather Service (NWS) issued a hurricane watch regarding Harvey during the afternoon of August 23, 2017, approximately 48 hours before hurricane-force winds hit the Texas coast. A hurricane warning followed on the morning of August 24, 2017, approximately 36 hours before Hurricane Harvey began to affect the coast.

Based on these warnings, 40,000 residents evacuated and took refuge from the hurricane in shelters throughout Texas and Louisiana. In response to Hurricane Harvey's predicted impact, the American Red Cross (ARC) opened the first of its kind, a "mega" shelter, in the convention center in downtown Dallas, Texas. During its peak usage, this "mega" shelter provided refuge for 5,000 evacuees.

Notably, Houston Mayor Sylvester Turner did not issue an evacuation order for the city. Instead, he encouraged the city's 6.5 million residents to prepare to ride out the storm in their residences. His rationale for this decision was partially based on the negative evacuation experiences of many Houston residents during Hurricane Rita in 2005. During that evacuation, traffic became gridlocked on major highways, leaving many to sit in traffic for over 20 hours, and over 120 evacuees died from the effects of accidents, a bus fire, and heat stroke. During Harvey, though many residents were able to either evacuate or ride out the storm safely in their homes, local, state, and federal emergency services officials, as well as private boat owners (including members of the "Cajun Navy"), performed 30,000 water rescues (FEMA 2018).

LANDFALL AND IMPACTS

After landfall, Hurricane Harvey continued to move northeast through Texas, impacting towns such as Holiday Beach, Port Lavaca, and Refugio. Residents of coastal Texas were threatened with 130 miles per hour (195 km/h) winds, heavy rains, and high storm surge. The combined effects of the storm surge produced by Hurricane Harvey and local tides resulted in maximum inundation levels of 6–10 feet (1.8–3.0 m).

Within the first 12 hours after landfall, Harvey rapidly weakened to be classified as a tropical storm. Winds decreased to around 40 miles per hour (60 km/h); however, the storm continued to produce great amounts of rain. With this weakening, Harvey became much slower moving, meaning that Houston and southeastern Texas experienced catastrophic, record-setting levels of rainfall. The highest reported total rainfall from Hurricane Harvey was 60.58 inches (1,539 mm) near Nederland, Texas. This amount made Harvey the most significant tropical storm rainfall event in U.S. history. Reports from the Houston metro area showed the total rainfall to range from 36 to 48 inches (914–1,219 mm) (Blake and Zelinsky 2018).

One possible explanation for the slow-moving nature of Harvey is climate change. Warmer global temperatures have caused the jet stream to move northward, meaning that the winds that may have at one time pushed Harvey to move faster were missing and allowed for the storm to slowly move across Texas and Louisiana (Coy and Flavelle 2017). Another unique aspect of Harvey was the size of the area in which heavy rainfall was felt. Heavy rainfall was experienced from the coastal area of Rockport to the inland city of Houston—a distance of nearly

A flooded highway in Houston, Texas, after Hurricane Harvey, August 31, 2017. (Karl Spencer/iStockphoto.com)

200 miles (300 km). In total, Harvey dumped over 27 trillion gallons of rain on Texas. This value makes Hurricane Harvey the wettest Atlantic hurricane on record (Huber et al. 2018).

The combined effects of the heavy rains and storm surge caused by Hurricane Harvey resulted in catastrophic flooding throughout the area impacted. The Houston metro area experiences frequent flooding and has historically spent billions in constructing dams and drains to protect the built infrastructure (Coy and Flavelle 2017). However, the rainfall from Harvey strained their flood-control reservoir systems. By August 29, two of the reservoirs had breached, resulting in increased water levels throughout the Houston metro area and causing sizable environmental, infrastructure, and property damage. A number of factors increased the city's vulnerability to flooding hazards, including excessive paving, reservoirs being too small to meet the needs of the fourth largest city in the country, and a lack of rigorous building codes (Coy and Flavelle 2017).

Harvey continued to move northeast, ultimately reaching the state of Louisiana as a tropical depression on August 30, 2017. In Louisiana, storm surge ranged from 2 to 4 feet (0.6–1.2 m) along the coast, and peak rainfall was estimated by radar to be 40 inches (1,016 mm) (Blake and Zelinsky 2018).

Casualty Estimates

It is challenging to accurately estimate casualties following natural disasters. The numbers included are the values reported by the National Oceanic and Atmospheric Administration (NOAA) in its Hurricane Harvey report. The impacts of Hurricane Harvey resulted in 68 direct deaths in the United States. Direct deaths refer to casualties that occurred as a result of the forces associated with the storm system, including wind-related events, lightning strikes, and drowning caused by storm surge, rough seas, and floodwaters. Hurricane Harvey also resulted in approximately 35 indirect deaths. Indirect deaths occur from events that happen during the storm, such as fires, vehicle accidents on wet roadways, effects from downed power lines, or heart attacks (Blake and Zelinsky 2018).

Damage Estimates

Hurricane Harvey is estimated to have caused $125 billion in damage in the United States. This estimate makes Harvey the second costliest hurricane behind only Hurricane Katrina in 2005. Damage was catastrophic and vast throughout southeastern Texas. Harvey affected approximately 13 million people. At least 300,000 structures and approximately 1 million cars experienced flood damage. Of the structures damaged, the Federal Emergency Management Authority (FEMA) estimated that up to 80 percent were uninsured residences or businesses (FEMA 2018). Additional structures that were damaged included schools throughout the state of Texas. Dozens of schools remained closed for more than a month as crews worked to repair damages from the rain, wind, and flooding. In the months following Harvey, the number of Texas residents filing for unemployment reached an all-time high, showing further the economic impacts of this storm (Huber et al. 2018).

The damage caused by Hurricane Harvey called into question the relaxed building and zoning codes throughout Texas. Though the state of Texas is one of the nation's most vulnerable states to flooding, their codes fail to reflect that reality. Subsequently, the number of residential and business properties built in flood-prone areas has increased without restriction. A consequence of this development scheme means that damage from storms such as Hurricane Harvey is often far worse than they need to be. In recent years, federal policy has pushed for the presence of greater local efforts targeting disaster resilience in order for states to be eligible to get maximum disaster relief funds (Coy and Flavelle 2017). In the case of Houston, creating and enforcing better building codes is one way the city can become more resilient following the aftermath of Harvey.

An estimated 336,000 customers lost power during Hurricane Harvey. Several oil and gas refineries located in the Gulf of Mexico along coastal Texas were forced to cease their operations for several days. While no damage was sustained at these facilities, this still caused gas prices in the United States to spike. Numerous highways and roads throughout the state experienced significant flooding and were closed both during and in the time immediately following the storm. In the metro Houston area, road flooding meant vital lifelines such as hospitals were inaccessible for days. Flooding also impacted 800 wastewater treatment facilities and 13 Superfund sites, which contaminated the floodwaters with sewage and toxic chemical waste. The contaminated floodwaters compounded with mold created significant challenges for property owners aiming to quickly begin the recovery process after the storm passed. It was also estimated that Hurricane Harvey resulted in over 200 million cubic yards of debris.

RECOVERY EFFORTS

On September 8, 2017, President Donald Trump signed a bill that approved $15.25 billion in aid for those impacted by Hurricane Harvey. Over 800,000 households applied for disaster assistance through at least one program offered by FEMA. Approved applicants received a variety aid, including temporary housing assistance, flood insurance claim payments, and low-interest small business disaster loans. Additionally, more than 300 voluntary organizations provided support to households and businesses impacted by Hurricane Harvey. These organizations performed a number of services including debris removal, feeding people, distributing supplies, physical and mental health services, and home repairs. A storm of the magnitude of Hurricane Harvey means that recovery efforts will be ongoing for members of the affected communities in Texas and Louisiana for years.

CONCLUSION

Hurricane Harvey caused significant impacts in portions of Texas and Louisiana for the next several days after the landfall. Once on land, Harvey caused record-setting rainfall in some locations. This rainfall made Harvey the most significant tropical storm rainfall event to ever occur in the United States.

In response to Hurricane Harvey, 40,000 residents sought refuge in shelters, and over 30,000 water rescues occurred in the days immediately following the storm. Over 100 direct and indirect deaths were attributed to Harvey. An estimated $125 billion in damage was caused, making Hurricane Harvey the second costliest hurricane for the United States. Recovery will continue in Texas and Louisiana for years to come as communities attempt to repair and rebuild the vast damage and losses felt from the storm. A storm of the magnitude of Hurricane Harvey raises the question of the ways Texas and Louisiana will recover in order to be more resilient in the face of the next storm.

Rachel A. Slotter

Further Reading

Blake, E. S., and D. A. Zelinsky. 2018. *National Hurricane Center Tropical Cyclone Report: Hurricane Harvey (AL092017)*. Washington, DC: National Oceanic and Atmospheric Administration.

Coy, Peter, and Christopher Flavelle. 2017. Harvey Wasn't Just Bad Weather: It Was Bad City Planning. *Bloomberg Businessweek*. August 31, 2017. https://www.bloomberg.com/news/features/2017-08-31/a-hard-rain-and-a-hard-lesson-for-houston, accessed August 31, 2017.

FEMA (Federal Emergency Management Agency). 2018. *2017 Hurricane Season: FEMA after Action Report*. Washington, DC: Department of Homeland Security.

Huber, C., H. Klinger, and K. J. O'Hara. 2018. *2017 Hurricane Harvey: Facts, FAQs, and How to Help*. World Vision. Last updated September 7, 2018. https://www.worldvision.org/disaster-relief-news-stories/hurricane-harvey-facts, accessed September 7, 2018.

Hurricane Ike, United States, 2008

While Hurricane Ike made landfall in the United States as a Category 2 storm, not even a "major" hurricane (a description generally reserved for hurricanes of Category 3 or higher), it caused widespread damage and destruction in the Caribbean, along the American Gulf Coast, especially in Texas and Louisiana, and much further inland, in states like Ohio. It also caused issues for a much larger number of Americans due to its impact on the oil and natural gas industries.

STORM FORMATION AND WARNINGS

Ike was upgraded to hurricane status on September 3, 2008, while near the northern Leeward Islands. By September 4, it reached its peak strength as a Category 4 hurricane. Following this point, the hurricane began to weaken before re-strengthening back to a Category 4 storm and then weakening again. Hurricane Ike made two landfalls on Cuba on September 8 and September 9, which disrupted the hurricane's structure.

After this, the wind field of the hurricane began to expand, meaning the storm covered a much larger area, ultimately reaching over 275 miles (413 km) away from the center of the storm. This larger size also led to the National Hurricane Center (NHC) issuing a tropical storm warning for parts of the Louisiana coast. This was unusual as the area had not first been issued a tropical storm watch. In general, a watch is usually issued before a warning, indicating that the conditions are right for an event like a tropical storm versus a warning's indication that a hazard is in the area or that the impact of a hazard is imminent. Around this time, officials in the United States began preparing resources and aid for potentially affected areas. Texas Governor Rick Perry declared 88 counties a disaster area on September 8, and on September 10, President George W. Bush issued an emergency declaration for Texas.

Hurricane Ike made landfall on Galveston Island, Texas, at 2:00 a.m. CDT on September 13 as a Category 2 hurricane. The storm then moved up Galveston Bay and across eastern Texas, where it weakened to a tropical storm by just 1:00 p.m. CDT that day. Although under a mandatory evacuation order, many people along the Texas coast and an estimated 25–40 percent of people on Galveston Island itself did not evacuate (Morss and Hayden 2010). Research has shown that people's views of hurricane risk can be impacted by a range of different factors, including things like their previous disaster experience, their trust in authorities, and their own perception of things like risk in general and vulnerability, issues that may complicate or discourage evacuation in disasters (Morss and Hayden 2010). This relative lack of evacuation was also complicated by the fact that some areas did not receive evacuation orders until 48 hours ahead of landfall. While this would be problematic in any hurricane evacuation, due to Hurricane Ike's size and strength, flooding began well in advance of landfall, meaning some people

Destruction along the Bolivar Peninsula in Galveston County, Texas, after Hurricane Ike, 2008. (National Weather Service/Galveston County OEM)

attempting to evacuate found their routes blocked by storm surge and floodwaters (Morss and Hayden 2010).

STORM SURGE AND FLOODING

Due to the sheer size of Hurricane Ike and the spread of its winds, higher water levels made landfall well in advance of the eye of the storm. This storm surge was also spread across the American Gulf Coast compared to the eye's landfall in Texas. The west coast of Florida and Key West, Florida, received approximately one to three feet (0.3–0.9 m) of water in the days before the storm's U.S. landfall. Parts of the Louisiana and Texas coasts, including Galveston Island, saw anywhere from 10 to 20 feet (3–6.1 m) of water, from up to 24 hours prior to and including landfall. Although some areas of the island were somewhat protected by a seawall built after the area's catastrophic hurricane in 1900, not all were protected, allowing water to come onto the island (Berg 2009; Morss and Hayden 2010).

Storm surge waters also went much further inland in Louisiana, up to 30 miles (45 km) into areas of lower Vermilion, St. Mary, and Terrebonne Parishes. The highest storm surge, however, affected areas of Texas between Galveston Bay and High Island and was estimated at between 15 and 20 feet (4.5 and 6.1 m) (Berg 2009). As with the unusual tropical storm warning for the Louisiana coast, Hurricane Ike's storm surge threat also led to an unusually early hurricane watch for parts of Louisiana and Texas in advance of the usual 36-hour window because of the threat of early storm surge. A hurricane warning for parts of Louisiana and Texas then followed about 24 hours in advance of the storm surge (Berg 2009).

Later research also found that while people in the affected areas expected Hurricane Ike to be a dangerous storm in general, there was some surprise about the extent of flooding triggered by the storm (Morss and Hayden 2010). When Morss and Hayden (2010) spoke to people affected by Hurricane Ike in parts of Texas, including Galveston, they noted how people discussed the destruction flooding had caused in their homes with some interviewees having lost everything in their homes and others having lost all the things that fell in the floodwaters (like furniture or sheetrock and wiring in their walls). This is not uncommon in hurricanes, especially those with large storm surges, but it serves to underscore the level of damage people can be left to deal with even if their house remains, technically, intact after landfall. It is also worth noting that two of their interviewees noted that they only had minor flood or water damage in their homes. These two people also noted that their homes had been raised about seven feet (2.1 m) off the ground after the historic 1900 hurricane (Morss and Hayden 2010, 179).

OIL AND NATURAL GAS PROBLEMS

As the storm moved forward in the Gulf of Mexico, the U.S. Department of Energy (2008) noted that in addition to evacuations on land, about 84 percent of the Gulf's manned crude oil and natural gas platforms had been evacuated, as well as personnel from 84 percent of oil rigs. While such evacuations were critical for personnel safety, and therefore absolutely necessary, when combined with the related need to shut in oil platforms, or cap production lower than the available output, they caused significant issues in oil and natural gas production and distribution. At least 14 oil refineries, two Texas Strategic Petroleum Reserve (SPR) sites, and one Louisiana SPR site closed due to the hurricane (U.S. Department of Energy 2008). According to the U.S. Department of Energy (2008), by September 13 over 99 percent of the Gulf's crude production and 98 percent of its natural gas production remained shut-in and 77 percent of major natural gas processing plants were closed, due to both the threat of Hurricane Ike and residual effects of Hurricane Gustav.

Once Hurricane Ike hit, it damaged 10 offshore oil rigs and several pipelines (Berg 2009). Closed and damaged pipelines had to be checked and fixed before being reopened. In the days following the storm, those that did reopen were sometimes operating at a reduced flow compared to their normal operations. The Louisiana Offshore Oil Port (LOOP) normally allows tankers to unload and temporarily store their crude oil, but it too faced limited operations in some facilities, further hampering oil production and distribution (U.S. Department of Energy 2008). This combination of damage, delays, and limited production and transportation led to a two-fold effect: rising gas prices and gas shortages (Berg 2009).

These problems affected areas across the United States, not just areas directly impacted by the hurricane, but further complicating the problem in Texas were power outages at gas stations. Fuel companies such as Chevron and ConocoPhillips began to respond to the situation, sending generators into the area for gas station use and prioritizing the need for gas along evacuation routes (U.S. Department of Energy 2008). The efforts to prioritize power restoration to these stations were important not only for consumers' daily fuel use but also for the transportation of evacuees and necessary responders and aid materials into the area.

WINDSTORMS

After interacting with another front, Ike became an extratropical storm and moved across parts of Arkansas, Missouri, and the Ohio Valley area before continuing into Canada in parts of Ontario and Québec in the following days. This caused damage, destruction, and various problems in a range of states, especially Ohio. The NHC included language in their warnings about the remains of Ike that noted the possibility of gale-force winds and extratropical cyclone formation as it moved over the United States and into Canada.

As the storm moved over Ohio, it resulted in wind gusts ranging from 63 to 78 miles (95–117 km) per hour, a so-called windstorm, which became a federally declared disaster. Losses amounted to $1.1 billion, the highest in Ohio since 1974 (when the Xenia tornado touched down as part of the 1974 Super Outbreak) (Schmidlin 2011). The storm caused seven deaths and over 600 injuries. Utility infrastructure was damaged by the high winds, resulting in wide-scale power outages nearly equal to the power outages that occurred in Texas and Louisiana when Hurricane Ike initially made landfall. Fallen trees blocked roadways and damaged buildings. Hospitals and public water supply systems had to resort to using backup generators. In addition to providing shelters for hurricane evacuees in Texas and

Louisiana, the Red Cross was ultimately responsible for operating 25 shelters in Ohio and 86 feeding stations (Schmidlin 2011). Governor Ted Strickland declared a state of emergency, and upon Strickland's request, President Bush declared 33 counties as affected by a major disaster, thereby allowing for the possibility of federal disaster aid funds (Schmidlin 2011).

CONCLUSIONS

Ultimately, Hurricane Ike killed at least 20 people in Texas, Louisiana, and Arkansas, as well as 28 in Tennessee, Ohio, Indiana, Illinois, Missouri, Kentucky, Michigan, and Pennsylvania as it moved over the United States. In addition to problems with the storm surge, areas experienced complications due to the winds associated with Hurricane Ike. The storm damaged several ports by the storm, including those in Galveston and Houston. Over 2 million people in Texas and Louisiana lost power, as did more than 2 million in Ohio (U.S. Department of Energy 2008). Damage estimates for Hurricane Ike, including for storm surge and inland flooding damage, ultimately reached $30 billion, making Ike the fourth costliest American hurricane at the time and currently (as of late 2018) the sixth costliest.

In addition to the scale of Hurricane Ike's destruction, both as a hurricane and as it dissipated, the storm raised complications nationwide in the United States. Areas as far away from landfall as Florida were forced to cope with early and heavy storm surges, areas as inland as Ohio had to deal with cleanup from high-force winds, and across the nation, but especially in Texas, had to deal with fuel shortages and increasing prices. Hurricane Ike is a clear example of the potential for widespread effects of a hurricane, even when landfall is located in a narrower area.

Jennifer Trivedi

Further Reading

Berg, R. 2009. Tropical Cyclone Report: Hurricane Ike, 1–14 September 2008. January 23, 2009. Miami, FL: National Hurricane Center.

Morss, R. E., and M. H. Hayden. 2010. Storm Surge and "Certain Death": Interviews with Texas Coastal Residents Following Hurricane Ike. *Weather, Climate, and Society* 2: 174–189.

Schmidlin, T. W. 2011. Public Health Consequences of the 2008 Hurricane Ike Windstorm in Ohio, USA. *Natural Hazards* 58(1): 235–249.

U.S. Department of Energy. 2008. Hurricane Ike Situation Report #2. September 14, 2008. Washington DC: Department of Energy Office of Electricity Delivery & Energy Reliability.

Hurricane Irma, Florida, 2017

Florida's geographic location means that the state and its residents are frequently exposed to hurricanes. From May to November, Florida residents are at risk of experiencing hurricanes and the strong winds, heavy rain, and storm surge that accompany them. Annually, one to two hurricanes make landfall in the continental United States. Storms must have winds that exceed 74 miles (111 km) per hour in order to become named hurricanes. Winds recorded at that speed mean those storms are classified as Category 1 hurricanes on the Saffir-Simpson Scale. Hurricane Irma made landfall on September 10, 2017, as a Category 4 storm in Cudjoe Key, Florida. Later that day, Irma again made landfall in Florida as a Category 3 storm in Marco Island, Florida. With the landfall of Irma in the continental United States, the 2017 hurricane season became the first time in a century during which two storms (the first was Hurricane Harvey) exceeding Category 4 on the Saffir-Simspon Scale made landfall within the same season.

FORMATION

On August 27, 2017, a tropical wave originated off the west coast of Africa and later upgraded to both a tropical depression and tropical storm three days later. The following day, August 31, 2017, Irma was officially categorized as a hurricane and was located about 400 miles (600 km) west of the Cabo Verde Islands. Notably, Irma upgraded from a tropical depression to a hurricane within 30 hours. Between September 1 and 4, 2017, Irma's wind strength fluctuated between Category 2 and 3 strength before ultimately reaching Category 5 strength on September 5, 2017. Hurricane Irma sustained Category 5 wind speeds for 60 hours, making the storm the second-longest sustained Category 5 hurricane on record behind the 1932 Cuba Hurricane of Santa Cruz del Sur. At the storm's maximum intensity, winds sustained at 185 miles (278 km) per hour. These winds extended up to 50 miles (75 km) from the center of the storm system with tropical storm force winds extending up to 185 miles (278 km) away.

LANDFALL

In total, Hurricane Irma made seven landfalls throughout the northern Caribbean and Florida. On September 6, 2017, Hurricane Irma made landfall three times as a Category 5 storm system on the Caribbean islands of Barbuda, St. Martin, and the British Virgin Islands with winds over 180 miles (270 km) per hour. On September 7, 2017, Hurricane Irma made another landfall in the Caribbean. This time Irma made landfall in the Bahamas as a Category 4 hurricane. After moving over the Bahamas, Irma intensified again as it moved over the Atlantic Ocean toward Cuba where the storm made its fifth landfall as a Category 5 hurricane on September 9, 2017, with winds up to 150 miles (225 km) per hour and waves at heights of close to 30 feet (9.1 m). Irma weakened significantly after its landfall in Cuba to a Category 2 storm system. However, the storm intensified again as it moved north toward Florida.

Irma made its sixth landfall, the first in the United States, as a Category 4, impacting the lower Florida Keys on September 10, 2017. The Keys received an average of 12 inches (305 mm) of rain and up to 8 feet (2.4 m) of storm surge as Irma moved over the islands. Ft. Pierce, Florida, reported the highest rainfall with

Flooding on Atlantic Avenue in Garden City, South Carolina, caused by Hurricane Irma, 2017. (John Erbland/U.S. Geological Survey)

22 inches (559 mm) falling in three days (NOAA 2018). Weakening in intensity, Irma continued to move north before making a seventh and final landfall in Marco Island, Florida, as a Category 3 hurricane. The combined effects of storm surge and tide yielded maximum inundation levels between 6 and 10 feet (1.8–3.0 m) in southwestern Florida.

Over the next two days, Irma continued to move north through Florida, impacting the areas of Naples and Ft. Myers as a Category 2 hurricane and the Tampa and Orlando metropolitan areas as a Category 1 hurricane. During this period, Irma's wind field continued to be significant. Tropical storm force winds extended over 350 miles (525 km) from the eye of the storm, causing impacts across the state. Significant flooding occurred throughout the state of Florida and in portions of Georgia and South Carolina. Irma also produced many tornadoes. In total, the National Oceanic and Atmospheric Administration (NOAA) confirmed 25 tornadoes with the strongest rated as an EF-2 on the Enhanced Fujita Scale. The EF-2 tornadoes occurred in the Florida cities of Mims, Crescent Beach, and Polk City.

By the evening of September 11, 2017, Irma weakened to a tropical storm and was located in northern Florida just west of Gainesville. Tropical Storm Irma also impacted the state of Georgia before the remnants of the storm system moved north into parts of Alabama and Missouri.

EVACUATION AND SHELTERING

In advance of Hurricane Irma making landfall in Florida, a record-breaking 6.8 million people were issued evacuation orders (FEMA 2018). Beginning on September 6, 2017, mandatory evacuation orders were issued throughout the state, starting with residents of the Florida Keys. In total, 75 percent of Florida Keys residents evacuated ahead of the storm making landfall.

Given the record number of people ordered to evacuate, sheltering was prominent in the days leading up to and immediately following the landfall of Hurricane Irma. In fact, the sheltering efforts in response to Hurricane Irma are one of the largest in the history of the United States. At its peak, over 191,000 people were sheltered in nearly 700 shelters in Florida (FEMA 2018). Within 30 days, the number of shelters opened had dwindled to 177, and by day 60, the last shelter had closed.

Considering the magnitude of this storm, closing all shelters within 60 days was impressive and differed from the sheltering experiences following the other major 2017 hurricanes that impacted the United States. Several reasons are attributed to this difference, including accommodation options and the levels of hurricane preparedness within the impacted areas. Specifically, Hurricane Irma caused low levels of damage to residences across the state, and local municipalities ensured noncommunal accommodations were available for impacted households to reside in. Additionally, the state of Florida and impacted municipalities across the state were prepared for housing needs of this magnitude given their prior experiences with major hurricanes, such as Hurricane Andrew in 1992.

EFFECTS

Casualty Estimates

It can be challenging to provide an accurate estimate of casualties following the occurrence of a natural disaster. According to NOAA's report, Hurricane Irma resulted in 10 direct deaths in the United States. The direct death is a term referring to casualties that occurred as a result of the forces associated with the storm system. In the case of Hurricane Irma, these deaths resulted from impacts of storm surge, rough seas, and wind-related events. Hurricane Irma was associated with 82 indirect deaths in the United States. Indirect deaths occur from events that happen during the storm, such as car accidents, fires, or electrocutions. Events that resulted in the indirect deaths associated with Hurricane Irma included vehicle accidents, carbon monoxide poisoning from generators, chainsaw accidents, electrocutions, and overheating.

Damage Estimates

Hurricane Irma is estimated to have caused $50 billion in damage in the United States, making it the fifth costliest hurricane at the time behind Katrina, Harvey, Maria, and Sandy. FEMA estimated that within the impacted areas 25 percent of buildings were destroyed and 65 percent were significantly damaged (FEMA 2018). Damage was most severe in Florida Keys. Most residential properties were badly damaged, completely destroyed, or uninhabitable. Additionally, over 1,300 boats were either badly damaged or destroyed in the Keys. Other significant damage occurred in the Marco Island area, where Hurricane Irma made landfall for the final time. Notably, 50 percent of the agricultural industry in Miami-Dade County was damaged, resulting in estimated losses at $245 million. This damage impacted the prices of orange juice and sugar and resulted in the losses of crops such as tomatoes, green beans, and cucumbers.

Hurricane Irma also caused significant power outages. In the days immediately after landfall, there were approximately 6 million people in Florida and 1.5 million people in Georgia without power. In total, 31 percent of Florida residents lost power and 16 percent of the state's hospitals were affected by power outages. Utility companies worked diligently to restore the power once the storm system had passed. Ten days after Irma had made landfall, the number of people in Florida without power was down to 75,000 (FEMA 2018). In many cases though, power outages lasted for a prolonged period of time. These prolonged power outages created difficulties for many residences, including nursing homes. In Hollywood Hills, Florida, the Rehabilitation Center at Hollywood Hills lost power due to the effects of Hurricane Irma. Their generator system also failed before power was restored, meaning that the residents of the nursing home were exposed to the extreme heat. In total, eight residents of the nursing home died due to indirect effects from the power outages caused by Hurricane Irma (Reisner and Fink 2017).

Effects on Policy

Concerns over the levels of disaster preparedness for health care facilities are not unique to Hurricane Irma. Following Hurricane Katrina in 2005, 215 people in hospitals and nursing homes died throughout the state of Louisiana. In response to these deaths, policy makers sought to rectify how ill-prepared health care facilities nationwide are for disasters. Efforts to create new health care rules took years of challenging debate. Ultimately, a new federal rule now requires nursing homes to have alternative means of maintaining temperatures to protect the safety and health of their residents. However, this policy is vague and does not mandate backup generators that are devoted specifically to air-conditioning systems. In the case of the Rehabilitation Center at Hollywood Hills, their facility did not have a generator devoted to the air-conditioning systems, and residents were put at severe risk following the widespread power outages (Reisner and Fink 2017). This Hollywood Hills example again raises concerns as to whether health care facilities have made adequate strides in the years after Hurricane Katrina to have become better prepared for disasters. Future policy will have to address these issues at greater length following the negative effects during Hurricane Irma.

CONCLUSION

Hurricane Irma made landfalls in Cudjoe Key, Florida, as a Category 4 storm and Marco Island, Florida, as a Category 3 storm on September 10, 2017. Once on land, Irma proceeded to cause significant impacts in Florida, Georgia, South Carolina, Alabama, and Missouri for the next several days. The strength of

Irma's winds caused significant damage and power outages throughout the impacted areas.

In advance of Hurricane Irma, over 6 million residents of Florida evacuated. Additionally, Irma proved to be one of the largest sheltering endeavors in the history of the Federal Emergency Management Agency. At its peak, over 191,000 residents were housed in nearly 700 shelters throughout the state of Florida; 92 direct and indirect deaths were attributed to Irma (NOAA 2018). An estimated $50 billion in damage was caused, a value that makes Hurricane Irma the fifth costliest hurricane in history for the United States. Damage was most severe in the Florida Keys and caused a significant portion of residents to permanently relocate.

Irma also resulted in prolonged power outages, which raised concerns once again over the disaster preparedness levels of health care facilities in the state of Florida. Though stricter legislation has passed since Hurricane Katrina, the eight heat-related deaths at the Rehabilitation Center at Hollywood Hills calls to question how effective this policy change is and how well it has been enforced. Following Hurricane Irma, Florida policy makers will have to consider how to better prepare these facilities in the future. This is of great concern considering the large elderly population in the state. In Hollywood Hills, restoring power to nursing homes was lower on the priority list for the local utility company. One solution may be to place all health care facilities, including nursing homes, at the highest priority level following a disaster. Doing so may ensure that future heat-related deaths are prevented.

Rachel A. Slotter

Further Reading

Amadeo, Kimberly. 2018. Hurricane Irma Facts, Damages, and Costs. *The Balance*. July 25, 2019. https://www.thebalance.com/hurricane-irma-facts-timeline-damage-costs-4150395, accessed January 11, 2019.

Cangialosi, J. P., A. S. Latto, and R. Berg. 2018. *National Hurricane Center Tropical Cyclone Report: Hurricane Irma (AL112017)*. Washington, DC: National Oceanic and Atmospheric Administration.

FEMA (Federal Emergency Management Agency). 2018. *2017 Hurricane Season FEMA After-Action Report*. Washington, DC: Department of Homeland Security.

NOAA (National Oceanic and Atmospheric Administration). 2018. National Hurricane Center Tropical Cyclone Report Hurricane Irma (AL112017) 30 August–12 September 2017. https://www.nhc.noaa.gov/data/tcr/AL112017_Irma.pdf, accessed January 11, 2019.

Reisner, N., and S. Fink. 2017. Nursing Home Deaths in Florida Heighten Scrutiny of Disaster Planning. *New York Times*. September 14, 2017. https://www.nytimes.com/2017/09/14/us/nursing-home-deaths-irma.html, accessed January 14, 2019.

Hurricane Katrina, United States, 2005

On August 29, 2005, Hurricane Katrina made landfall on the border between Louisiana and Mississippi. The coastal areas of Alabama, Mississippi, and Louisiana, some 90,000 square miles, were devastated by a Category 3 storm that included winds of 111–130 miles (167–195 km) per hour and a storm surge in the range of 9–12 feet (2.7–3.7 m) (Johnson 2006). Though the storm had reached Category 5 status in the Gulf of Mexico, it decreased in strength as it made landfall and proceeded inland. The decrease, however, did not limit the amount of death and destruction. It killed between 1,800 and 2,000 people and inflicted between $100 and 150 billion in physical damage (Burton et al. 2011).

PREPARING FOR THE STORM

On Friday, August 26, 2005, Hurricane Katrina made its way into the Gulf of Mexico. Hurricane models were predicting landfall on the Florida Panhandle. Katrina first affected southern Florida before making landfall in the Gulf Coast of the United States. In southern Florida, it impacted Miami-Dade and Broward counties with approximately 80 miles (120 km) per hour wind. The initial hit caused damage to infrastructure and left approximately 1.3 million Florida residents without power (Brown 2005). Over the course of the day, the storm shifted to the west, and predictions indicated that the new target was New Orleans, Louisiana (Johnson 2006).

With Katrina projected to make landfall around New Orleans, residents and government officials became concerned that this could be the anticipated "Big One." New Orleans had been fortunate in the past as storms had veered away from the city (Johnson 2006). Officials feared that a direct hit on the city could cause a break in the levees. The levees protect the city from Lake Pontchartrain on one side and the Mississippi River on the other side. Much of New Orleans lies below sea level. A complex system of canals and pumping stations protect the city from intense rainstorms.

In response to the updated forecasts, New Orleans began to implement evacuation plans. As the evacuation process began, those who had access to cars fled the city. Though interstate-highway lanes were converted to outbound only, the onslaught of people leaving clogged the highways days prior to the arrival of the storm (Glantz 2008). For many, the prospect of evacuation was not possible. Lack of access to a vehicle and limited public transportation left many residents stranded in areas of the city below sea level. There were also many residents in at-risk coastal areas who chose not to leave. For them, past evacuations that had resulted in false alarms led to the assumption that this was just another forecast with a low probability of hitting coastal Louisiana. It was estimated that there were some 150,000 people who either could not evacuate or refused to leave (Johnson 2006).

Much of the focus prior to landfall was on New Orleans and the Louisiana coast, but the governors in Alabama and Mississippi were also preparing their residents for the possibility of storm impacts. Haley Barbour, the governor of Mississippi at the time, warned of not only coastal impacts but also the possibility that the storm would cause destruction well inland. Initially, Bob Riley, the governor of Alabama, offered his assistance to those who were in the targeted areas (Brown 2005).

IMPACT AND RESPONSE

On August 29, 2005, Hurricane Katrina made landfall on the border between Mississippi and Louisiana. Once again, New Orleans avoided a direct hit. Though it appeared that the city had been spared, the ensuing days proved that Hurricane Katrina had created the impacts many would associate with the "Big One." The problems for New Orleans started early on August 29 as a storm surge pushed up the Mississippi River-Gulf Outlet. The storm surge provided enough force to overtop the earthen levees and breach the concrete walls along the Industrial Canal. The pumps still left in service were not able to deal with the disaster, and ultimately, 80 percent of the city of New Orleans was flooded (Johnson 2006; Glantz 2008).

Many residents were trapped in the city and had to seek refuge wherever they could. Some 25,000 refugees ended up living for several days in the Louisiana Superdome. Images and interviews revealed the social impacts of a disaster of this magnitude. People had none of the basic amenities—toilet paper, soap, diapers, clean clothes, bathing facilities, and bottled water (Glantz 2008). Images and videos released from the impacted areas of people wading through chest-high waters, seeking essentials and transporting family members by any means possible, revealed to the rest of America and the world the level of destruction and despair.

Within days of people taking refuge in the Superdome, problems started to arise in living conditions. Many had arrived at the facility without food or water. Eventually, hunger and dehydration set in. The smell of body odor and human waste became overpowering. Due to the large number of people in the facility,

Thousands of Hurricane Katrina evacuees from Louisiana rest inside the Astrodome in Houston, Texas, September 2, 2005. (FEMA)

toilets backed up. Because of the turmoil in the Superdome, buses began to arrive to rescue the evacuees. Many of the evacuees were transported to the Houston Astrodome. Helicopters were deployed to carry the seriously ill and injured (Brown 2005).

Katrina evacuees fled to many locations. Estimates based on applications for federal assistance received by the Federal Emergency Management Agency (FEMA) reveal that 40 percent fled to other parts of Louisiana, 30 percent to other parts of Mississippi, 12 percent to Texas, and 8 percent to Alabama (Glantz 2008). The largest number of evacuees from Louisiana ended up in Houston, Texas. The city opened its doors to the evacuees, but within a year wanted them to return home.

College students were hit hard by the storm. Universities from across the United States offered to assist the displaced students by offering free tuition or allowing the payment of fees once students received refunds from their original schools. Other storm victims became unemployed by the hurricane and lost their income. The human toll was also exacerbated by the evacuating process itself, whereby many families became separated, and due to the circumstances, it was difficult to reunite (Brown 2005).

Other images coming from New Orleans centered on those who had not evacuated. Many of these individuals were stuck in homes that were partially submerged. Several photos showed people sitting on rooftops, waiting to be rescued (Glantz 2008). Some had retreated to their attics to avoid the rising waters and had to chop holes in their roofs to escape. The United States Coast Guard became critical in the rescue effort. The Coast Guard rescued 12,533 people by air and 11,584 by boat (Johnson 2006).

National media attention focused on New Orleans, but other areas along the Gulf Coast were hugely impacted. The Mississippi coast was hit hard, with major destruction in Biloxi, Gulfport, and Bay St. Louis. Every building along the entire Mississippi coast was virtually destroyed. Beachfront homes were washed away, while others were flooded. One in five homes in Gulfport and Biloxi was destroyed by the storm. The destruction in Mississippi went well beyond the coast. Inland areas suffered extensive destruction, particularly to houses and mobile homes (Brown 2005).

Like New Orleans, rescue operations on the Mississippi coast were significant. Shortly after the storm passed, Mississippi National Guard troops entered the damaged areas and began search and rescue. Within a week, the United States Coast Guard rescued 1,700 residents of Mississippi. One of the major issues on the Mississippi coast in the days and weeks after the storm was associated with extensive power outages.

Although Alabama did not suffer a direct hit like Mississippi and Louisiana, there was still significant damage in the state. The most severe impact of the storm in Alabama was associated with the wind. One bridge had to be closed when a 13,000-ton oil platform was pushed onto it by the storm. The winds were so strong that the water in the Mobile River was forced upstream, leading to flooding in portions of Mobile. Dauphin Island, a barrier island connected by a bridge to the mainland, suffered extensive damage (Brown 2005).

Though Alabama sustained plenty of hurricane damage, Governor Riley offered housing assistance to victims in affected areas. After the storm subsided, National Guard troops rescued storm victims from flooded homes, and crews from Alabama and neighboring states worked to restore electricity. Within the week, contractors were hired to remove debris from cities in Mobile County. Community cooperation was particularly important in the debris removal process (Brown 2005).

Other prominent officials, including President George W. Bush, faced a great deal of criticism for their handling of the emergency in the immediate aftermath. On September 2, 2005, the president signed a federal relief bill that provided $10.5 billion of aid for victims of Hurricane Katrina, promising that more funding would be approved later. FEMA was another primary target of criticism. However, FEMA sent medical teams from Texas and Nevada into the affected areas within the week, dispatched hundreds of supply trucks, and provided shelters for evacuees (Brown 2005).

RECOVERY

In early September, many parts of New Orleans resembled a ghost town, and it was estimated that only one in five customers in the metro area had electricity. Water was scarce, and available water sources were contaminated, which required boiling before use. Public schools closed for the year, and many universities in the area canceled classes for the fall semester. Businesses shut down and public transportation came to a halt. Through the first half of September, police permission was required to enter most of the city (Johnson 2006).

By mid-September, some residents returned to their homes. Though the levee breaches had been fixed, residents were able to see the extent of flooding due to the water stain lines left on their homes. Many of the belongings found inside had been floating on water for days, and black mold was everywhere (Johnson 2006). Officials requested that residents returning to their homes and salvage through debris should wear masks and keep antibiotic ointment on hand to prevent infection.

New Orleans metro area was open to all residents in May 2006. Most areas had basic utilities and functioning traffic lights by this time. In many of the low-income areas where extensive flooding took place, governments were still contemplating when and how rebuilding should occur. For many of these residents, applying for a FEMA trailer and FEMA-issued blue tarps was the best they could do (Johnson 2006). Residents of the hardest-hit coastal communities in Mississippi were facing similar situations.

Many of the structures located along the Mississippi shoreline and inland a few blocks in Biloxi were destroyed by the storm surge. Recovery and reconstruction varied greatly within the community. Affluent homeowners and casinos initiated reconstruction quickly. In the lower income African-American and Asian-American communities, residents struggled to rebuild. Recovery during the first year was slow, but it increased significantly after that, reaching 90 percent during the observation period (Burton et al. 2011).

The displacement of people along the Mississippi coast was noticeable, but the city of New Orleans was hit particularly hard. Many residents of the city took refuge in new places, some going as far as Los Angeles. Polls taken in the year after the storm revealed that New Orleans could potentially lose many of those who had relocated. According to a USA Today/CNN/Gallup poll taken after the storm, 39 percent of those surveyed indicated that they either would not or probably would not return to New Orleans (Johnson 2006). Population numbers calculated shortly after the storm indicated that the city had lost approximately two-thirds of its pre-Katrina population of 484,674 (U.S. Department of Housing and Urban Development 2006).

One of the biggest issues for communities after catastrophic destruction is the ability to begin the reconstruction process, particularly the building of new homes. Almost 288,000 people were left homeless in Louisiana after the hurricane. Approximately 56 percent of the dwellings in Orleans Parish were severely damaged or destroyed. The prices for the remaining houses and rental properties skyrocketed. In addition, residents did not have a job to return to as many businesses and organizations were impacted by the storm. Over time, the job market rebounded, specifically for those associated with the construction industry (Johnson 2006).

CONCLUSION

In the aftermath of Hurricane Katrina, there was great concern as to how the communities impacted would be able to respond to the catastrophic damage. Many estimates suggested that full recovery would take over 11.5 years, though other estimates were in the range of 3–15 years (Johnson 2006). Twelve years after the storm, progress has been made. In heavy-hit areas such as the Lower Ninth Ward in New Orleans and lower income residential areas on the Mississippi coast, new structures are going up. There are still areas along the coast and a few blocks inland that have been slow to rebuild, and remnants of destruction remain in the form of foundations left behind.

The fact that the damage associated with Hurricane Katrina was so extensive should be a warning for low-lying coastal communities. The idea that climate change is impacting sea level rise is significant because future tropical events will have the potential to produce storm surges that move further inland. There is also the possibility that tropical storms may become more frequent and more intense (Johnson 2006). Based on this and the destruction witnessed after Hurricane Katrina, communities need to become more knowledgeable about coastal construction practices.

Vicki L. Tinnon

Further Reading

Brown, D. M. 2005. *Hurricane Katrina: The First Seven Days of America's Worst Natural Disaster.* Morrisville, NC: Lulu.com.

Burton, C., J. T. Mitchell, and S. L. Cutter. 2011. Evaluating Post-Katrina Recovery in Mississippi Using Repeat Photography. *Disasters* 35(3): 488–509.

Glantz, M. H. 2008. Hurricane Katrina as a "Teachable Moment." *Advances in Geosciences* 14: 287–294.

Johnson, M. L. 2006. Geographical Reflections on the "New" New Orleans in the Post-Hurricane Katrina Era. *Geographical Review* 96(1): 139–156.

U.S. Department of Housing and Urban Development. 2006. *Current Housing Unit Damage Estimates: Hurricanes Katrina, Rita, and Wilma.* Washington, DC: U.S. Department of Housing and Urban Development.

Hurricane Maria, Puerto Rico, 2017

Category 5 Hurricane Maria was the deadliest Atlantic Hurricane since Jeanne in 2004 and the costliest in Puerto Rican history. It formed from an African easterly wave that moved across the tropical Atlantic Ocean six days after it became a tropical storm on September 16, 2017, east of the Lesser Antilles. The hurricane reached Category 5 strength on September 18 just before making landfall on Dominica, becoming the first Category 5 hurricane on record to strike the island. Maria made landfall in Yabucoa, Puerto Rico, as a strong Category 4 hurricane on the morning of September 20, 2017. From Puerto Rico, it moved northeast of the Bahamas and close to the North Carolina coast. Moving slowly to the north, Maria gradually degraded and weakened to a tropical storm on September 28. Embedded in the westerlies, Maria accelerated toward the east and later east-northeast over the open Atlantic, becoming extratropical on September 30 and dissipating by October 3.

While deaths were also reported in Dominica, the Dominican Republic, Haiti, Guadeloupe, and parts of the United States, including the U.S. Virgin Islands, the majority of the deaths were reported in Puerto Rico, and their final total remains unclear. The U.S. government was widely criticized for its response to the disaster and research since the event has uncovered wider and more serious repercussions than was even first considered. Total losses from the hurricane were estimated at upward of $91.61 billion (2017 USD), mostly in Puerto Rico.

RESPONSE

Hurricane Maria made landfall on Puerto Rico on September 20, 2017, at 6:15 a.m. local time as a Category 4 hurricane, just two weeks after the area had been affected by Hurricane Irma. It was described by researchers at the Climate Impact Lab as a 1-in-3,000 event (Centro 2018). The storm brought severe storm surge, winds, heavy rainfall, and widespread flooding, including flash floods. Rain continued throughout the next day, and that, combined with the initial storm impact, resulted in the electrical grid being shut down, the closure of airports and ports, and the collapse of 95 percent of the cellular network (Centro 2018). Preliminary estimates noted major damage to or destruction of 250,000 homes, as well as significant damage to 300,000 homes (Centro 2018).

Roughly a week after the storm's landfall, nearly 100 percent of the population remained without power and 44 percent without water (Centro 2018). Just 15

A U.S. Geological Survey hydrologic technician flags a high-water mark in Comerio, Puerto Rico, from flooding on the Rio de la Plata caused by Hurricane Maria, 2017. (U.S. Geological Survey)

percent of hospitals remained open, leading to an additional lack of medical care that was not really offset by the limited deployment of the United States Naval Ship (USNS) *Comfort*, an American hospital ship, arriving nearly a week after the storm (Centro 2018). This lack of accessible medical care impacted the overall death toll of the storm. The lack of access to resources continued for months. By the six-month mark, 15 percent of Puerto Rican residents continued to lack electricity and 12 percent lacked potable water (Centro 2018).

With the long-term lack of power and other resources after Hurricane Maria, at least 35,000 people left Puerto Rico, traveling to areas like the mainland United States (Centro 2018; Kishore et al. 2018). Many relocated to areas like Florida and New York, a pattern of movement rooted in "existing population nodes," that is, people relocating to areas that already had significant Puerto Rican populations. Such movement is revealed through information sources like school enrollment data, which reveals that Florida was the most likely area for relocation and how over 11,000 Puerto Rican students enrolling in Florida school districts in the months after the storm also demonstrated that the relocation included whole families (Centro 2018).

HISTORICAL CONTEXT, ECONOMIC ISSUES, AND THE JONES ACT

Hurricane Maria caused approximately $90 billion in damages, largely in Puerto Rico and the U.S. Virgin Islands. It was the third costliest hurricane in U.S. history since 1900 (Kishore et al. 2018). Further complicating the economic impact of the storm was the historical colonial context of the island and legal problems like the Jones Act. Puerto Rico's history in the United States is distinct—it is an organized but unincorporated territory and has been since 1898 following the Spanish-American War. Puerto Rican residents have been American citizens since 1917, but they lack self-determination, full representation in Congress, and the ability to vote for the president from the island. This status means Puerto Rico is not fully a state but rather a territory while remaining in a colony-like status, lacking full representation and support from the U.S. federal government while coping with laws imposed on the island by that government. Laws like the Merchant Marine Act of 1920—or Jones Act—demonstrate some of the problems with this status. The Jones Act puts forward that Puerto Rican waters and ports are controlled by U.S. agencies and that non-U.S. ships and crews cannot engage directly in commercial trade with the island. In the aftermath of Hurricane Maria, this meant that international aid could not be brought directly to the island, but instead, it had to be rerouted through the U.S. mainland.

Moreover, Puerto Rico's pre-Hurricane Maria financial status was also caused by a unique colonial history. Laws that made Puerto Ricans U.S. citizens in 1917 also allowed them to raise money via tax exempt bonds. This led to the local government accruing massive debt, and with shifts in federal law and local government, eventually Congress stripped Puerto Rico of its ability to declare bankruptcy. This combination of federal control, lack of local power, and debt has left Puerto Rico owing $100 billion in bonds and unpaid pension debts—almost 70 percent of the island's gross domestic product (Rodríguez-Díaz 2018). It has also left Puerto Ricans with few options on how to respond to the growing financial crisis.

As of Centro's Puerto Rico Post Maria report, Puerto Rico has not yet received a $4.9 billion loan as part of the Congressional aid package. The Government Development Bank released an index of economic indicators that revealed a 20 percent decline in economic output in Puerto Rico after Hurricane Maria (Centro 2018). In the aftermath of Hurricane Maria, this meant the territory was dealing with (1) an ongoing general widespread economic crisis, (2) a further decline in the economy since the storm, and (3) a lack of aid funds. Such a combination is not only rooted in both Puerto Rico's colonial history and economic problems but also tied directly to Hurricane Maria and the major problems with recovery and aid. It also reveals the potential for future and ongoing economic and related problems. For example, the Climate Impact Lab estimated that "Maria could lower Puerto Rican incomes by 21 percent over the next 15 years—a cumulative $180 billion in lost economic output" (Centro 2018). In addition, Centro found that while from 2006 to 2016 Puerto Rico had lost 14 percent of the island's population to movement to the mainland United States, tied in part to the period's economic stagnation, the island could lose another 14 percent of their population from 2017 to 2019, creating additional problems for economic and other recovery (Centro 2018).

CONTROVERSIES

The initially reported death toll for Hurricane Maria was just 64 (Centro 2018; Kishore et al. 2018). However, this number was soon called into question, and research began to determine a more accurate total. The Centers for Disease

Control and Prevention (CDC) argued that tropical cyclone or hurricane deaths include not only those caused by event-related issues (like flying debris) but also those caused by unsafe conditions that lead to injury or illness or a lack of medical service (Kishore et al. 2018). The studies conducted on the Hurricane Maria death toll worked to include these deaths in their total count.

One project was commissioned by the Governor of Puerto Rico and had three key goals: (1) assessing the excess total mortality in Hurricane Maria, (2) an evaluation of the "implementation of Centers for Disease Control and Prevention (CDC) guidelines for mortality reporting in disasters," and (3) assessing the plans in place for and actions taken by the government of Puerto Rico, including from the perspectives of key informants (Milken Institute of Public Health 2018). The second and third components also included a goal of determining recommendations for potential improvements that could be made in that area. It was conducted by the Milken Institute of Public Health at George Washington University. Their findings included a key point that was widely discussed in the media after it became public, specifically that the likely death toll tied to Hurricane Maria, covering a period from September 2017 to February 2018, was 2,975 (Milken Institute of Public Health 2018, iii). This time frame, importantly, includes both the hurricane itself and its initial aftermath.

Researchers at Harvard University, working with other experts, also conducted a study attempting to better estimate the death toll from Hurricane Maria in Puerto Rico. Their estimate put the death toll even higher at 4,645 from September 20, 2017, to December 31, 2017 (again including both the disaster and its initial aftermath) (Kishore et al. 2018). Moreover, their estimate came with a caveat: "This number is likely to be an underestimate because of survivor bias" (Kishore et al. 2018, 162). In addition to this, the Harvard study notes that the mortality rate in Puerto Rico was not simply focused on the moment of impact of the hurricane, but it remained higher than expected throughout the end of the year, with a third of the deaths being "attributed to delayed or interrupted health care" (Kishore et al. 2018, 162). One study also found that Puerto Rican households went an average of 84 days without electricity and 68 days without water. In addition, 31percent of households surveyed reported an issue with access to medical services and care, including problems with access to medications, respiratory equipment (which also required power, something many people also lacked), closed medical facilities, absent doctor, or the inability to reach help via 911 (Kishore et al. 2018).

Other death toll estimates in the months after the storm included that of Center for Investigative Journalism at 985 deaths in September and October, of the *New York Times* at 1,052 deaths during those same months, and of CNN at 499 deaths between September 20, 2017, and October 19, 2017 (Centro 2018). While these death tolls differ from source to source, the point remains clear: Initial government-supported death toll numbers in Puerto Rico tied to Hurricane Maria vastly underestimated the real effects of the storm. Such underestimations can cause a series of problems for victims, ranging from recognition and closure to access to financial aid.

In addition to the controversy regarding the death toll from the storm, additional problems arose when a contract to rebuild power infrastructure on the island was awarded to Whitefish Energy Holdings, which, according to the press, the Federal Emergency Management Agency (FEMA), and members of Congress, had little experience and a questionable history relative to what was needed for the project. For many, this controversy pointed to larger problems with the lack of and delayed response to help Puerto Ricans. The U.S. Senate held a hearing on the matter and other recovery issues in Puerto Rico and the U.S. Virgin Islands in which Puerto Rican Governor Ricardo Rosselló participated. He ultimately canceled the $300-million contract (Centro 2018).

CONCLUSIONS

Hurricane Maria's impact, particularly on Puerto Rico, should not be underestimated. Striking an island already struggling economically, the scale of the storm's devastation and the massive problems and delays in recovery mark a historic point from which additional problems are likely to stem, including in terms of economic and population losses. Even basic components of disaster relief and response such as counting the dead were initially grossly mismanaged, requiring additional research and revisions in the months after. Delays in aid—both financial and physical, including medical assistance—also dramatically increased the death toll. Months of lacking power, telecommunications access, and water created a cascading series of complications for residents following the hurricane. Understanding the complexities of Hurricane Maria's effects on Puerto Rico ultimately requires a conversation about not only the hurricane and its effects but also the history of Puerto Rico and its place in the larger U.S. system.

Jennifer Trivedi

Further Reading

Centro. 2018. *Puerto Rico Post Maria*. New York: Center for Puerto Rican Studies, Hunter College, CUNY. https://centropr.hunter.cuny.edu/events-news/rebuild-puerto-rico/puerto-rico-post-maria-report, accessed December 16, 2018.

Kishore, N., D. Marqués, A. Mahmud, M. V. Kiang, I. Rodriguez, A. Fuller, P. Ebner, C. Sorensen, F. Racy, J. Lemery, L. Maas, J. Leaning, R. A. Irizarry, S. Balsari, C. O. Buckee. 2018. Mortality in Puerto Rico after Hurricane Maria. *The New England Journal of Medicine* 379: 162–170.

Milken Institute of Public Health. 2018. Ascertainment of the Estimated Excess Mortality from Hurricane María in Puerto Rico. Milken Institute School of Public Health at The George Washington University, in collaboration with the University of Puerto Rico Graduate School of Public Health. https://prstudy.publichealth.gwu.edu/sites/prstudy.publichealth.gwu.edu/files/reports/PR%20PROJECT%20REPORT%20FINAL%20Sept%2024%202018%20AGM%201325hrs.pdf, accessed December 17, 2018.

Rodríguez-Díaz, C. E. 2018. Maria in Puerto Rico: Natural Disaster in a Colonial Archipelago. *American Journal of Public Health* 108(1): 30–32.

Hurricane Matthew, United States, 2016

Hurricane Matthew affected multiple countries in 2016, including Haiti, Cuba, the Bahamas, and the United States. It was the first Category 5 hurricane in the Atlantic since Hurricane Felix in 2007. Matthew became a tropical storm on September

28 before escalating into a hurricane the next day. Following this, Hurricane Matthew underwent a period of explosive intensification, reaching Category 5 status on October 1. Its intensity fluctuated over the next few days as it made landfall in various areas. It left a path of destruction in its wake in several areas and ultimately also triggered widespread and expensive inland flooding in the United States. At least 585 people died in Hurricane Matthew, with over 500 of those deaths in Haiti. The damage, destruction, and death caused by Hurricane Matthew throughout the Caribbean and the United States was so extensive that the name "Matthew" was subsequently retired from the list of hurricane names for future use.

HAITI AND THE DOMINICAN REPUBLIC

In preparation for the hurricane, 340,000 people in Haiti and 8,500 in the Dominican Republic evacuated (Steward 2017). At approximately 7:00 a.m. local time (U.S. Eastern Time) on October 4, Hurricane Matthew made landfall near Les Anglais, Haiti. It was a Category 4 hurricane at landfall, the first major hurricane (Category 3 or higher) to make landfall in Haiti since Hurricane Cleo in 1964. Meanwhile the Dominican Republic experienced large areas of heavy rainfall. While moving over Haiti, Hurricane Matthew weakened slightly, but maintained its Category 4 status.

The storm's impact killed approximately only five people in the Dominican Republic, but 546 in Haiti. In addition, at least 439 people were injured there and 128 reported missing (Steward 2017). Further, extensive rainfall from Hurricane Matthew resulted in mudslides, flash floods, and river flooding in the Dominican Republic, contaminating multiple sources of drinking water, destroying homes, and causing damage or washing away roadways and bridges (Steward 2017).

Beachfront damage and destruction were more common in Haiti, although winds and rains also affected areas not directly on the waterfront. Some areas of Haiti lost up to 80 percent of their crops, as estimated by the United Nations Office for the Coordination of Humanitarian Affairs (Steward 2017). As in the Dominican Republic, heavy rain caused widespread mudslides, flash floods, and river flooding, destroying roadways, bridges, and homes; damaging and destroying crops, farmland, and livestock; and causing problems with the transportation of aid workers and relief supplies into the affected areas. In some areas, as much as 90 percent of the homes were heavily damaged or destroyed, resulting in at least 120,000 families losing their homes. In total, damage to Haiti was estimated at $1.9 billion.

Further complicating the situation, a cholera outbreak developed in the aftermath of Hurricane Matthew, in part due to the damage to Haiti's infrastructure systems, leading to nearly 10,000 cases of cholera (Steward 2017). Even prior to Hurricane Matthew, some Haitians had faced problems in accessing safe water, sanitation, and hygiene. Cholera epidemics are complex, involving the exposure of humans to the bacterium *Vibrio cholerae*, the outbreak of disease on a larger scale, and secondary transmission. Based on environmental conditions, Haiti is at a high risk of cholera outbreaks and public health policies in general, and after disasters, Haiti must take into account this risk, providing clean drinking water and access to sanitation (Khan et al. 2017). This must also be done preventatively in longer-term recovery and other periods, allowing for some protection against potential outbreaks in future disasters and in weather and environmental conditions that support outbreaks (Khan et al. 2017).

CUBA AND THE BAHAMAS

Prior to landfall, Cuba evacuated 380,000 people from high-risk areas. Around 8:00 p.m. local time on October 4, Hurricane Matthew made landfall near Juaco, Cuba. Although Hurricane Matthew had weakened to a Category 3 hurricane after moving over Cuba, it briefly strengthened again, making landfall near West End, Grand Bahama Island, around 8:00 p.m. on October 5. Following this strike, Hurricane Matthew turned north-northwest on October 7.

Cuba faced problems of damage to coastal areas related to storm surge flooding and wind. The city of Baracoa took the brunt of this destruction, having been hit by Hurricane Matthew's eyewall, leaving 90 percent of the area's homes damaged or destroyed. Most of the damage was within Guantánamo Province, estimated at about $2.58 billion (Steward 2017).

As in Cuba, certain areas of the Bahamas were more severely affected than others. Tropical storm or hurricane force winds caused widespread damage, limiting access to affected areas with trees and power lines downed across roadways and damaging homes, particularly roofs. The townships of Eight Mile Rock and Holmes Rock on western Grand Bahama Island were among the most severely affected, with residents left facing severe damage or destruction of 95 percent of homes there. Damage in the Bahamas as a whole was estimated at about $600 million (Steward 2017).

UNITED STATES

In advance of Hurricane Matthew, more than 3 million Americans evacuated from the coastal areas in Florida, Georgia, South Carolina, and North Carolina. Some cities, such as Savannah, Georgia, assisted in evacuation efforts by offering transportation to those who needed it from a predetermined location in the city to shelters further inland in areas such as Augusta, Georgia. Others, such as Tybee Island, Georgia, sent out text messages to local residents warning them of evacuation orders. After making landfall in the Bahamas, Hurricane Matthew began to travel up the side of Florida. It stayed just offshore, but the edge of the eyewall did hit NASA's Cape Canaveral launch facility with Category 2 force winds. Areas of Florida faced wind damage, as well as storm surge-related coastal flooding. Damage to beaches and dunes alone in some areas topped $50 million (Steward 2017).

As it moved north, Hurricane Matthew continued to weaken, but its wind field also expanded. Hurricane force winds hit both Georgia and South Carolina, especially impacting their barrier islands. Parts of Georgia flooded due to extensive rainfall, but record storm surges were also reported in coastal areas such as Tybee Island, and flooding from the Savannah River caused problems in the city of

Photos taken before Hurricane Matthew on September 6, 2014, and after on October 13, 2016, show that the storm cut a new inlet between the Atlantic Ocean and the Matanzas River near St. Augustine, Florida. (U.S. Geological Survey)

Savannah and surrounding areas. Trees toppled over by high winds resulted in damaged buildings and, in some cases, deaths as people inside those buildings were killed (Steward 2017).

Continuing north, Hurricane Matthew weakened into a Category 1 hurricane, making landfall around 11:00 a.m. near McClellanville, South Carolina, making it the first hurricane since 1954 (Hurricane Hazel) to make a U.S. landfall north of Florida in October. Storm surge flooding was also a problem throughout South Carolina's coast, and as it occurred in Georgia, wind gusts caused both direct structural damage and uprooting and knocking over of trees, damaging buildings. In some areas, the storm surge flooding reached as far as 1,000 feet (305 m) inland, and in many places, the surge was three to five feet above normal tide levels (Armstrong 2017; Steward 2017). Islands off the South Carolina coast and areas around Charleston were especially hard hit.

Hurricane Matthew then moved offshore and continued northward, following along the coast of North Carolina. While a large part of Hurricane Matthew remained offshore, its eyewall extended into both South Carolina and North Carolina, causing hurricane force winds, rain, and storm surge to coastal areas of the state. It was officially declared a post-tropical system on October 9. It then merged with another weather system that day while still east of North Carolina. Although no longer technically a hurricane, it maintained hurricane force winds, and upon turning northeast, it continued across the United States and into Canada over the following two days. Damage in the United States due to Hurricane Matthew, including inland flooding, was estimated at $10.3 billion (Armstrong 2017).

INLAND FLOODING

Some of Hurricane Matthew's heaviest rainfall came in North Carolina as the storm transitioned out of hurricane status and merged with other weather fronts. Heavy rainfall happened throughout parts of the state from October 7 through October 9, 2016. Near Evergreen, North Carolina, 18.95 inches (481 mm) of rain was reported on October 8 and 9. Other areas experienced up to 10–18 inches (254–457 mm) of rain throughout the affected states and primarily in areas of North Carolina. Further complicating this heavy rainfall was the fact that many of the same areas had just gotten 6–10 inches (152–254 mm) of rain from Tropical Storm Hermine, leaving waterways and rivers already full and the ground already saturated. While Hurricane Matthew had already caused major issues with flooding in other locations, there were questions about warnings and evacuation orders related to inland flooding in some areas of North Carolina.

Flooding was so extensive that, in addition to the closure of local roads, parts of Interstates 95 and 40 had to be closed. In total, more than 600 roads were partially or completely closed in North Carolina due to the flooding. After the flooding went down, more than 2,100 road repairs were necessary, including a range of fixes like correcting shoulder washouts, damage to drainage systems, and repairs to bridges. In addition to the damage done to roadways and bridges, estimates indicated that nearly 99,000 structures throughout North Carolina were damaged or destroyed by floodwater (Carroll et al. 2017).

Record levels of river flooding occurred in multiple locations. This combined with a relative lack of early inland evacuations, compared to coastal areas, resulted in the need for many emergency evacuations, conducted by emergency responders, the Coast Guard, and people trying to help their own neighbors, friends, and family members. But these efforts often took time; two days after Hurricane Matthew had passed, people in areas such as Lumberton, North Carolina, were still waiting for rescue. The majority of America deaths in Hurricane Matthew also resulted from inland flooding, mostly in North Carolina and mostly due to drowning, often in vehicles while attempting to travel or evacuate through floodwaters.

CONCLUSION

Hurricane Matthew caused widespread damage, destruction, and deaths throughout Haiti, the Dominican Republic, Cuba, the Bahamas, and the United States. In addition to hurricane force winds and storm surges, Hurricane Matthew's extensive rainfall and triggering of inland and river flooding marks it as distinct from many other hurricanes and an important lesson in the potential effects and necessary considerations in hurricane-related preparedness planning and evacuations. Residents of places such as inland areas of North Carolina did not expect to feel so much of the impact of the hurricane, although some had prepared for side effects like power outages. Widespread mudslides, flash flooding, and river flooding in multiple countries revealed far more extensive hurricane effects than those limited to coastal locations, and they are a key component in the story of Hurricane Matthew. Moreover, the aftermath of Hurricane Matthew with a cholera outbreak in Haiti points to the importance of focusing on public health and hygiene issues in the aftermath of disasters and in people's daily lives, including working to ensure ongoing access to clean water supplies.

Jennifer Trivedi

Further Reading

Armstrong, Tim. 2017. Hurricane Matthew in the Carolinas: October 8, 2016. National Weather Service, NWS Wilmington, NC Weather Forecast Office. https://www.weather.gov/ilm/Matthew, accessed September 21, 2018.

Carroll, Kathleen, Shawna Cokley, William Ellis, Brandon Locklear, Michael Moneypenny, Keith Sherburn, Brandon Vincent, Nick Petro, and Jonathan Blaes. 2017. Hurricane Matthew, October 2016. National Weather Service, Raleigh NC. http://www4.ncsu.edu/~nwsfo/storage/cases/20161008/, accessed September 21, 2018.

Khan, R., R. Anwar, S. Akanda, M. D. McDonald, A. Huq, A. Jutla, and R. Colwell. 2017. Assessment of Risk of Cholera in Haiti Following Hurricane Matthew. *American Journal of Tropical Medicine and Hygiene* 97(3): 896–903.

Steward, S. R. 2017. Hurricane Matthew (AL142016), 28 September–9 October 2016. April 7, 2017. Miami, FL: National Hurricane Center.

Hurricane Mitch, Central America, 1998

On October 29, 1998, Hurricane Mitch made landfall on the Caribbean Sea coast of Honduras, the second poorest country in the Western Hemisphere. The storm then began moving very slowly inland till November 3, causing high winds and torrential downpours at a rate of about 4 inches (100 mm) per hour in Honduras and Guatemala. In total, the coastal area of Honduras received more than 30 inches (750 mm) of rainfall, while the interior received 50 inches (1250 mm). In fact, Hurricane Mitch dumped a year's worth of rain on Central America in 48 hours. In time, the storm turned toward the northeast around Tegucigalpa, the capital city of Honduras. Before leaving the country on October 31, 1998, Mitch was downgraded to a tropical depression. Before reaching Florida as a tropical storm on November 5, Hurricane Mitch moved through several Central American countries such as Belize, Costa Rica, El Salvador, Guatemala, Nicaragua, and Panama. Panama and Nicaragua were also severely affected.

ORIGIN

Hurricane Mitch begun as a tropical depression in the Caribbean Sea near the Panama coast on October 22, and it attained hurricane force on October 24. On the same day, Mitch received its name. It was the 13th hurricane of the 1998 season (Ensor and Ensor 2009). By the afternoon of October 26, Mitch had intensified into a Category 5 hurricane—the highest rating on the Saffir-Simpson hurricane intensity scale—and set a record for staying continuously 33 hours at this category status (Ensor and Ensor 2009). After making landfall in Honduras and subsequently causing devastation in Nicaragua and Yucatan Peninsula, Hurricane Mitch traveled east-northeast, regaining its strength in the Bay of Campeche, Mexico. It made landfall near Naples, Florida, as a tropical storm on November 5 and finally dissipated over the Atlantic on November 9. Because of its disastrous impact, the World Meteorological Organization (WMO) removed Mitch from its list of Atlantic Ocean hurricane names in 1999.

IMPACT

Hurricane Mitch was the deadliest hurricane to hit the Western Hemisphere in more than 200 years, destroying crops and infrastructure with its attendant floods and mudslides. Moreover, intense rain-induced floods not only washed away all crops but also left them covered with sand in some areas, rendering the land unsuitable temporarily for any crop production. Many urban centers were also buried under mud contaminated with sewage, pesticides, chemical fertilizers, and even decomposing human and animal remains (Ensor and Ensor 2009). Both in Honduras and Nicaragua, agricultural infrastructure and production suffered severely. In Nicaragua, important export crops such as coffee, pineapple, and banana sustained widespread damage. In this country, more than 12,000 cattle were lost. Damage to basic infrastructure, agricultural production, and the industrial sector virtually destroyed more than two decades of progress in the Mitch-affected Central American countries.

Flood damage along the Choluteca River near Tegucigalpa, Honduras, caused by Hurricane Mitch, 1998. (Debbie Larson, NWS/NOAA)

Regarding infrastructure, the storm destroyed about 35,000 houses and damaged another 50,000, making up to 1.5 million people homeless, or about 20 percent of Honduras's population. Furthermore, it either destroyed or damaged 23 of the 30 hospitals. Additionally, over 75 rural and urban health centers were seriously damaged, and approximately 25 percent of the schools throughout the country were destroyed. Over 70 percent of the transportation infrastructure was damaged, mostly highways and bridges. Widespread areas lost power, and about 70 percent of the country lost water supply after the storm. Flooding in Tegucigalpa damaged many buildings, and the rains led to devastating landslides on nearby deforested hills and mountainsides. Hurricane Mitch affected all of Honduras's 18 territorial departments. Subsequently, on November 3, 1998, Honduras officials imposed a 15-day curfew from 9:00 p.m. to 5:00 a.m. and also temporarily restricted constitutional rights to maintain order.

Crime rates and domestic violence increased in Honduras immediately after the hurricane. About a million people left their homes, because their farms or their jobs could no longer sustain them. There were outbreaks of various diseases such as diarrhea, acute respiratory illness, dermatitis, malaria, and conjunctivitis. Sanitation facilities in both urban and rural areas were destroyed, and potable water sources as well as food supplies were contaminated. Nearly 2.9 million residents faced food and water shortages. Nine months after the hurricane struck, resettled people were still suffering from severe malnutrition (UN Interagency 1999).

Hurricane Mitch killed somewhere between 11,000 and 18,000 people. Most of the deaths occurred in Honduras and Nicaragua, and Honduras alone accounted for over 50 percent of total deaths. This huge death toll was exacerbated by human factors. Before the hurricane, two-thirds of Hondurans were living below the poverty line, and about half of them were living in extreme poverty and in marginal areas (Ensor and Ensor 2009). However, the single worst incident of the disaster took place in Posoltega, Nicaragua, where more than 2,000 people perished in a mudslide. The town of Casitas was also virtually wiped off the map by a mudslide. Meanwhile, in Honduras, the hurricane injured 12,272 people, and over 600,000 people were evacuated. Nearly a half million people took shelter in one of the 1,375 temporary shelters, most of them built in the municipalities of Tegucigalpa, Choluteca, and San Pedro Sula whose communities were the hardest hit, particularly those in the marginal areas of the urban centers. Overall, the hurricane caused more than $5 billion in damages (Ensor and Ensor 2009).

RESPONSE

Although hurricane warnings were issued in both Honduras and Nicaragua by radio, print, and television, the overall postdisaster response was slow and inadequate. The Honduran government did evacuate some 45,000 people on the Bay Islands and 55,000 from the mainland, while the Guatemalan government issued a hurricane warning, requested boats to stay in port, and evacuated some 10,000 people. The government also asked others to prepare for the approaching hurricane. In Belize, the government alerted people about the potential threat of Hurricane Mitch and asked citizens on offshore islands to evacuate to safer locations on the mainland. The government also evacuated people from Belize City. Finally, nearly 20,000 people were evacuated in the Mexican state of Quintana Roo.

However, in Honduras and Nicaragua, where a large number of dams failed, causing severe flooding, there was no warning for dam failures. One important barrier to quick response was lack of information due to poor emergency communication infrastructure. Whatever infrastructure existed prior to the hurricane was either destroyed or severely damaged. Moreover, there were acute shortages of personnel and equipment such as planes and helicopters to address the emergency. For these reasons, Honduras and foreign armed forces, including the Mexican military, were confronted with difficulties in rescuing people and distributing emergency relief goods to survivors of the storm.

Following the storm, the United Nations Office for the Coordination of Humanitarian Affairs (OCHA) made an initial appeal of more than $150 million to the international donor community to fund the emergency relief efforts for Hurricane Mitch survivors and to address the immediate rehabilitation requirements. In addition, the Food and Agricultural Office (FAO) of the UN made a consolidated appeal of $22.4 million to provide assistance to 50,000 farm families to buy seed, fertilizer, insecticide, and hand tools. President Flores of Honduras also requested international assistance on November 2, 1998, which ultimately totaled $2.8 billion over a period of several years. He also sent the armed forces and National Police to the affected areas to prevent looting and maintain order (Fuentes 2009).

International assistance started to arrive in the country immediately after both appeals with the United States and Spain taking an early lead in donating relief supplies. Although the Hondurans government formed a task force to receive and distribute foreign assistance, there was a concern about its ability to distribute aid fairly and efficiently to the survivors of the storm. In early 1998, Honduras had been rated the second most corrupt country in Latin America. Ultimately, the Catholic Church, and, to a lesser extent, some Evangelical churches, took on the responsibility. In November, the World Bank established a Central American Emergency Trust Fund to help Mitch-affected countries to channel disaster relief. Despite this, the gross domestic product began decreasing at the end of 1998 and contracted by 1.9 percent in 1999 (Ensor and Ensor 2009).

RECONSTRUCTION

After the emergency relief phase, medium- and long-term reconstruction projects began with assistance from foreign countries and international agencies. On November 5, 1998, the Honduras government began drafting a national plan for reconstruction in which selected civil society groups of the country participated. In December 1998, the Inter-American Development Bank (IADB) started coordinating longer-term recovery work. Unfortunately, favoritism and lack of transparency were widely reported as reconstruction projects were implemented. Apparently, reconstruction contracts in Honduras were being granted to friends and family members of high-level politicians. However, as the Honduran government estimated that it needed $4 billion to rebuild the country, ultimately it committed to a reconstruction plan that sought the following: (1) transparency in spending and managing the reconstruction projects, (2) increased participation of both local governments and civil society groups in public decision making, and (3) reduction in the country's ecological and social vulnerability (Fuentes 2009). The reconstruction phase officially ended on the third anniversary of Hurricane Mitch. While the official reconstruction period did not take that long, recovery from the hurricane was made more difficult by the severe droughts of 2001 and 2002 (Ensor 2009).

Donors and multinational organizations pledged approximately $2.8 billion to finance emergency relief, reconstruction, and transformation programs in Honduras and $6.2 billion for other affected Central American countries. In addition, the International Monetary Fund (IFM) and the World Bank agreed to relieve part of Honduras's debt. Despite promoting transparency and ensuring the participation of civil society in the reconstruction activities, nonetheless, reports of corruption were reported in a study funded by the World Bank. However, posthurricane reconstruction provided a window of opportunity for change in Honduras by the foreign donors and the members of civil society. They played a critical role in encouraging the country to undertake political reforms (Fuentes 2009).

The World Bank assisted the government of Honduras in developing a new national emergency strategy with large-scale flood diversion projects planned. The government also needed to develop early warning and mass evacuation protocols, as well as determine designated safe haven areas in every locality. Hurricane Mitch triggered a considerable measure of legislative reform, increased international investment, a strengthening of civil society, and other social and political changes. It also sparked an intense global debate on the morality of international debt and whether poor nations stricken by natural disaster should receive special help.

CONCLUSION

Natural disasters are never entirely caused by the forces of nature, and indeed, human factors were largely blamed for severely compounding the effects of Hurricane Mitch, particularly in Honduras and Nicaragua. Widespread mud slides occurred not only because of heavy rainfall but were also triggered by large-scale deforestation and the cultivation of marginal lands without any attempt to conserve the soil. Moreover, flooding was aggravated by a lack of adequate watershed management. In the Hurricane Mitch-affected countries of Central America, the poor bore the brunt of its devastating effects. Increasingly marginalized and affected by environmental degradation over many years of hardship because of restricted land rights, population pressure, rapid urbanization and international debt, prolonged civil strife, and adverse climatic conditions brought on by the El Niño phenomenon, poor small farmers were forced to cultivate marginal lands to survive.

Bimal Kanti Paul

Further Reading
Ensor, M. O., ed. 2009. *The Legacy of Hurricane Mitch: Lessons from Post-Disaster Reconstruction in Honduras.* Tucson: The University of Arizona Press.

Ensor, B. E., and M. O. Ensor. 2009. Hurricane Mitch: Root Cases and Response to the Disaster. In *The Legacy of Hurricane Mitch: Lessons from Post-Disaster Reconstruction in Honduras*, edited by M. O. Ensor, 22–46. Tucson: The University of Arizona Press.

Fuentes, V. E. 2009. Post-Disaster Reconstruction: An Opportunity. In *The Legacy of Hurricane Mitch: Lessons from Post-Disaster Reconstruction in Honduras*, edited by M. O. Ensor, 100–128. Tucson: The University of Arizona Press.

UN Interagency. 1999. *Transitional Appeal for Hurricane Mitch (Period December 1998 to May 1999). Resident Coordinator of Operational Activities of the United Nations System*. Tegucigalpa, Honduras: UN Interagency.

Hurricane Stan, Guatemala, 2005

Hurricane Stan caused widespread flooding and landslides throughout Central America, killing between 1,000 and 2,000 people both directly and indirectly. In Guatemala, Hurricane Stan and its aftermath killed at least 669 people, possibly up to 1,000, with an additional 1,400 minimum missing and presumed dead (IFRC 2005; Pasch and Roberts 2006). Calculating the death toll from Hurricane Stan has proven difficult, in part due to its effects in more remote areas and in part due to some confusion over what counts as a direct or indirect death from the storm.

Guatemala has a history with disasters, including hurricanes and flooding, as well as complications related to both. This history is further complicated by Guatemala's population—approximately 12.6 million people, 52 percent of whom live in rural areas (de Ville de Goyet 2008). These rurally located peoples may find themselves facing problems in receiving aid rapidly during disasters, due to both distance and a limited number of roadways for access, which themselves may be disrupted, damaged, or destroyed in events like flooding and landslides.

Tropical Storm Stan made landfall on the Yucatan Peninsula on October 2, 2005. As it moved across the peninsula, it weakened back down to a tropical depression, but once it hit the Gulf of Mexico and began to turn, it also began to strengthen again. On October 4, it reached hurricane status. Later that day, it made its second landfall near Veracruz, Mexico. Between October 1 and October 4, as it moved across and around the Yucatan Peninsula, Stan generated heavy rainfall, affecting not only Guatemala but also Belize, El Salvador, Honduras, and Mexico. October 3 saw the first reports of flooding damage in Guatemala. The hurricane itself remained relatively weak on the Saffir-Simpson Scale, never reaching above a Category 1 hurricane. However, this is not fully indicative of the destruction and damage it caused or the problems it revealed.

Before Hurricane Stan's landfall, on September 29, Guatemala's Institute for Seismology, Volcanology, Meteorology, and Hydrology, part of the nation's Office of National Coordination for Disaster Reduction (CONRED), had predicted that several of the country's departments might be affected by Stan (de Ville de Goyet 2008). However, more broadly speaking, there were serious issues with the accuracy of predictions and forecasts related to Hurricane Stan in this progression. While the system had initially been somewhat slow to develop, its escalation to a hurricane happened more quickly than expected, and its turn before its second landfall and the time at which it made this second landfall were difficult to predict exactly (Pasch and Roberts 2006).

The effects of Hurricane Stan also went beyond previous historic disasters in the area, like Hurricane Mitch in 1998. Hurricane Stan and the flooding and landslides it triggered also affected many of Guatemala's departments (administrative and political divisions within the country). Moreover, the flooding and landslides also caused more severe long-term effects, as crops in a range of areas were damaged or destroyed, causing long-term food access and economic problems. These effects included the loss of sugarcane in coastal areas and the loss of maize and beans in Altiplano areas (IFRC 2005). Guatemala's Ministry of Agriculture placed the damage to agriculture in the nation alone at $46 million, and the United Nations Economic Commission for Latin America and the Caribbean (ECLAC) estimated that 17,000 jobs or more were lost (de Ville de Goyet 2008). The hurricane particularly affected poor peasants and rural workers who depend on terrace farming, referring to a technique of using retaining walls and a series of tiered platforms in agriculture, predominately used in mountainous terrain for agriculture, including among Mayan areas in nations such as Guatemala (de Ville do Goyet 2008).

LANDSLIDES AND FLOODING

Hurricane Stan caused between five and ten days of rain in Guatemala, triggering flooding and landslides throughout the area that affected approximately one and a half million people. Over 6,000 people were evacuated in areas around the departments of San Marcos, Jalapa, Sololá, Santa Rosa, Huehuetenango, Jutiapa, Sacatepéquez, Escuintla, and Retalhuleu. Among these, San Marcos, Escuintla, and Huehuetenango were particularly hard hit, with at least 359, 258, and 100 communities affected in each department, respectively (IFRC 2005, 5).

The Humanitarian Aid Department of the European Commission noted that among the departments affected, Hurricane Stan hit the hardest in those that had low human development indexes and low-income levels, many areas of which overlapped with area of high indigenous (here often Mayan) populations (de Ville de Goyet 2008). The human development index is a composite index that factors in various issues such as life expectancy, education, and per capita income, and having a low human development index can point to problems people in an area may face in living long and healthy lives, accessing education, and having higher per capita incomes. This overlap between areas being more severely affected by the disaster and having predisaster low human development indexes and low-income levels points to broader issues in disasters, where vulnerable or less economically or socially powerful or stable people face severe disaster effects that compound the problems they face on a daily basis.

Rescue and response operations included both governmental and nongovernmental organizations (NGOs), such as volunteer firefighters and the Guatemalan

Red Cross, but such efforts were themselves also hampered by the ongoing effects of the disaster. Over 200 landslides throughout the country caused the destruction of communities and, importantly, of roadways that provided access to areas, including more remote locations with limited points of access. A lack of clear and useable roadways and bridges delayed rescue, aid workers, and resources from reaching the affected communities and limited the information flow about what those severely affected communities needed.

There was significant destruction in several Mayan communities and in areas with higher indigenous Mayan populations in general. This included communities such as Panabaj, Tzanchaj, and Santiago Atitlán, which were severely affected when ongoing rain, mudslides, and landslides spread throughout the mountains surrounding Lake Atitlan. These mudslides and landslides caused widespread damage and destruction, serious problems with rescue efforts, and in the most severe cases, the declaration of whole communities as mass graves. Outside aid was especially slow to reach such communities, meaning that it took several days at best for some areas to be reached by additional outside aid workers and resources. However, local residents and people with particular skill sets, such as doctors and nurses, responded quickly to help their neighbors as they could, a fairly common occurrence in disasters as people in a local community work as their own first responders.

MEDICAL ISSUES AND RELIEF EFFORTS

Further complicating the impact of the hurricane and landslides on small Mayan communities was the composition of the communities themselves. For example, in Santiago Atitlán, 75 percent of the households fall below the poverty line in Guatemala, and there are numerous long-standing health issues, including those related to a lack of clean drinking water (Peltan 2009). Hospitalito Atitlán opened to address some of these issues in April 2005, but it too was covered by landslides in Hurricane Stan. Physicians in the area rushed to help people affected by the landslide as quickly as possible, and they relied on their own and local efforts to treat survivors in the initial aftermath of the disaster. The hospital and its workers regrouped quickly, opening a new facility just over two weeks after the hurricane, and within a month after the hurricane, it was already functioning at the same levels it had been prior to the landslide (Peltan 2009).

Relief efforts in reachable communities ranged from the opening of emergency shelters, such as those at schools in areas like Tecún Umán, to the distribution of supplies such as blankets, hygiene kits, kitchen sets, water-related supplies, and mosquito nets to affected and evacuated people. CONRED focused on four key areas of response to the flooding and landslides: (1) water and sanitation, (2) shelter, housing, and social infrastructure, (3) food security and nutrition, and (4) health (de Ville de Goyet 2008). Within this fourth category, relief efforts also included medical outreach specifically to shelters and the distribution of fliers on

Emergency aid being unloaded from a UH-60 Black Hawk helicopter to survivors of Hurricane Stan in Guatemala City, Guatemala, October 15, 2005. (Petty Officer 1st Class Robert McRill/Department of Defense)

health-related behaviors. Such medical efforts included finding and treating conditions such as various forms of infections, skin conditions, diarrhea, and intestinal parasites; attempts to distribute medication such as cough treatments, antibiotics, eye drops, and vitamin supplements; and monitoring the growth and development of affected children.

Once outside aid sources were able to reach more affected rural communities, they did so in large numbers. A United Nations flash appeal raised $24.7 million in aid and a World Food Program (WFP) request for $14.1 million, resulting in a real contribution of $39.8 million (de Ville de Goyet 2008). The bulk of such aid came from five key sources: Sweden (18 percent), the United States (17.7 percent), the European Union (12 percent), the Netherlands (11.3 percent), and Norway (10 percent) (de Ville de Goyet 2008, 122). This total does not include the amounts raised by other nongovernmental organizations (NGOs). In addition, there were a number of governmental agencies involved in response, including but not limited to the Ministries of Agriculture, Health, the Environment and Natural Resources, Communications, and Education. CONRED was also able to secure equipment on loan, technical support, and other materials from the private sector with sources ranging from GBM (an IBM alliance company) to Cervecería Centro Americana (a beer and soft drink producer) (de Ville de Goyet 2008). The response to Hurricane Stan's flooding and landslides in Guatemala demonstrates clearly the complexities of disaster response efforts, incorporating a wide range of groups and resources.

CONCLUSION

Hurricane Stan's status as a Category 1 hurricane may be initially misleading, not fully conveying the level of destruction experienced by countries such as Guatemala or the extent of effects such as flooding and landslides on their people, particularly those in rural areas. While aid efforts were swift in the beginning at the local level, the extent of the damage to roadways and communities limited and delayed larger scale relief in the days after the disaster. Such issues were especially problematic in the aftermath of Hurricane Stan in Guatemala, but they also point to important lessons to be learned from in disasters as a whole—not only people and local communities could be their own first responders, a line often reinforced by emergency managers, but also a lesson in how less direct impacts of disasters can cause catastrophic damage, destruction, and death tolls. Thinking about hurricanes cannot be limited to just the hurricane itself and its initial landfall, but it must also consider subsequent effects such as rainfall, flooding, and landslides as the hurricane dissipate.

Jennifer Trivedi

Further Reading

de Ville de Goyet, Claude. 2008. The Use of a Logistics Support System in Guatemala and Haiti. In *Data Against Natural Disasters: Establishing Effective Systems for Relief, Recovery, and Reconstruction*, edited by S. Amin, and M. Goldstein, 83–142. Washington, DC: The International Bank for Reconstruction and Development/The World Bank.

IFRC (International Federation of Red Cross and Red Crescent Societies). 2005. *Operations Update on Central America, Mexico, and Haiti: Floods from Hurricane Stan*. Appeal Number 05EA021. Operations Update Number 03. Geneva, Switzerland: IFRC.

Pasch, R. J., and D. P. Roberts. 2006. Tropical Cyclone Report: Hurricane Stan, 1–5 October 2005. February 14, 2006. Miami, FL: National Hurricane Center.

Peltan, I. 2009. Disaster Relief and Recovery after a Landslide at a Small, Rural Hospital in Guatemala. *Prehospital and Disaster Medicine* 24(6): 542–548.

Indian Ocean Tsunami, 2004

The 2004 Indian Ocean tsunami (also known as the Great Sumatra-Andaman Earthquake and the Boxing Day Tsunami) ranks as one of the greatest natural disasters in modern history in terms of destruction, mortality, and number of people affected. In the early hours of December 26, 2004, a 9.1–9.3 magnitude earthquake struck at a 20-mile (30 km) focal depth about 107 miles (160 km) off the coast of Indonesia's Sumatra Island. The massive fault slip radiated tsunami waves at heights up to 99 feet (30 m) tall near the epicenter and 10–50 feet (3–15 m) tall in peripherally affected areas. The waves traveled at hundreds of miles per hour, impacting nations in Asia and Africa. In total, 15 countries were affected and deaths were recorded in 14 countries. Confirmed deaths total about 185,000, but estimates range from 230,000 to 300,000 when the missing people are included. The lack of predisaster warning systems, scale of impact, and amount and use of humanitarian aid ultimately led to a reevaluation of how to prepare for and respond to disasters.

WARNING AND PREPARATION

While tsunamis have impacted the Indian Ocean in the past (e.g., 1883 Krakatoa eruption, 1945 Balochistan Earthquake, and 1983 Diego Garcia Earthquake), a regional tsunami warning system was not in place prior to the 2004 tsunami. This lack of a warning system contributed to the severity of impacts. Thus, the tsunami waves took most populations by surprise even though there was a gap of several hours between the earthquake and when the waves arrived in the many affected areas. Due to gaps in warning and communication, the Indian Ocean Tsunami Warning System was established in 2005. The Indian Ocean Warning System, which operates as an intergovernmental group under UNESCO, used existing warning systems in Japan, Alaska, and Hawaii as models. Interestingly, the Hawaii Pacific Tsunami Warning System was heavily criticized for not issuing a warning to countries in the Indian Ocean, even though they had detected the 2004 earthquake and potential for a tsunami. Apart from the absence of a warning system, the other factors that contributed to vulnerability included a lack of structural mitigation measures (e.g., disaster-resistant housing and wave breakers) and the development of coastlines and natural buffers (e.g., mangroves, sand dunes, and sea wall) that could have mitigated the tsunami's effects.

IMPACTS AND RESPONSE

Fifteen countries were affected by the time the waves receded. The countries, which represent virtually the entire ring of the Indian Ocean, are (estimated mortalities in parentheses): Indonesia (167,540), Sri Lanka (35,322), India (18,050), Thailand (8,210), Somalia (290), Myanmar (500), Maldives (108), Malaysia (75), Tanzania (13), Seychelles (3), Bangladesh (2), South Africa (2), Yemen (2), Kenya (1), and Madagascar (no deaths) (TEC 2006). These numbers include at least 9,000 foreign tourists, mostly from Europe (e.g., Britain, France, Germany, Norway, and Sweden), who either died or were missing. Roughly 70 percent of all confirmed and estimated mortalities occurred in Indonesia alone, with the area of Banda Ache the most heavily affected. The other regions that experienced many fatalities include the east coast of Sri Lanka; Tamil Nadu and the Andaman and Nicobar Islands in India; and the west coast of Thailand. Beyond loss of life, an estimated 125,000 people were injured and 1.69 million were displaced (TEC 2006). Most of

Tsunami damage along the coast of Sumatra, Indonesia, near Banda Aceh. Considered one of the worst natural disasters in modern history, the tsunamis that followed a powerful undersea earthquake on December 26, 2004, killed more than 100,000 people in Indonesia alone. (U.S. Navy)

the displaced (about 95 percent) were in the heavily affected countries of Indonesia, Sri Lanka, and India.

Effects on the built, social, and natural environments were equally astounding. According to the Asian Coalition for Housing Rights, 720,000 houses were damaged or destroyed, and 175,000 fishing boats were lost. The destruction of boats, as well as fisheries and coastal aquaculture operations, represents a significant loss of capital and livelihood to coastal populations that rely on the fishing industry to support their families and local economies. More than 3,000 miles (4,500 km) of roads were damaged or destroyed. Furthermore, critical services such as electricity, water and wastewater treatment systems, and healthcare facilities were heavily impacted. For example, Indonesia lost 700 health professionals and 693 healthcare facilities, while Sri Lanka lost 35 hospitals and 57 healthcare facilities and India lost seven hospitals and 93 healthcare facilities (WHO 2005). Thus, aside from loss of life, survivors struggled with locating and obtaining the critical health services that they desperately needed to keep maladies such as diarrhea, dehydration, typhoid fever, malnutrition, hepatitis A, and leptospirosis at bay.

Additionally, access to dialysis and lifesaving medicines such as insulin and vaccines created grave situations and caused additional secondary deaths among the affected populations. Damage to the tourism industry was also great. The countries of Thailand, Malaysia, Sri Lanka, Maldives, and Seychelles all rely on tourism dollars to help prop up their national and local economies. Unfortunately, the number of tourists declined due to the massive loss of hotels, transportation infrastructure, utilities, and critical services required to support tourist activities. Collectively, these damages totaled to roughly $2.9 billion in Indonesia, $1.1 billion in Sri Lanka, and $500 million each in India, Thailand, and Maldives (TEC 2006).

Finally, the environment also suffered impacts from the tsunami. Beaches, sand dunes, mangroves, and other natural protective features were impacted, and waves led to the damage of terrestrial vegetation as well as saline and heavy metal deposits on land and in freshwater ecosystems. These damages are difficult to quantify because some environmental features recovered quickly (e.g., beaches) while others remained damaged for longer periods of time (e.g., soil quality), not to mention that some effects likely remained unnoticed and that the valuation of ecological features and ecosystem services is difficult to begin with.

The international response to the Indian Ocean Tsunami was overwhelming. Thirty-four foreign countries (including Australia, Germany, India, Japan, New Zealand, Singapore, the United Kingdom, and the United States) sent their armed forces to remove dead bodies, clear rubble, distribute emergency assistance, and provide immediate medical attention and psychological treatment. For proper distribution of disaster aid, the United Nations coordinated all foreign relief efforts in the tsunami-affected countries. The organization also quickly deployed its assessment teams in five severely affected countries of Asia.

Money from throughout the world poured in to address human and economic suffering. The funds, totaling approximately $13.5 billion (Telford and Cosgrave 2007), came from an array of supranational agencies (e.g., the United Nations and World Health Organization), national governments (e.g., United States, Germany, China, and Japan), and nonprofit organizations from every corner of the globe (e.g., Red Cross, World Vision, and Oxfam). While the much-needed funds certainly helped to alleviate suffering, several issues arose including timeliness of the funds, flexibility in usage of funds, tracking and monitoring of funds, and flow of financial information to affected populations. This has led some to contend that the massive amount of aid represented a "second tsunami" that when misused created a "disaster after the disaster" for some affected populations.

RECONSTRUCTION AND RECOVERY

Recovery after the tsunami focused on the concept of "build back better." The concept argues that while the disaster was devastating, the scenario should be leveraged as an opportunity to rebuild in a proactive manner that addresses past development failures while also leaving people less vulnerable and relatively better off than before the tsunami. The opportunity to "build back better" proved difficult as infrastructure, public utilities, and entire villages had to be rebuilt from the ground up. However, only half of the aid was donated for reconstruction and recovery (half earmarked for 2005 and half for 2006–2010), with the remaining half of aid dedicated to relief and emergency assistance. Thus, reconstructing numerous countries' coastlines proved challenging, as did supporting educational, vocational, micro-credit, and other livelihood programs to help get affected populations back on their feet. The result, which is not shocking given the large scale of the tsunami, was it led to both successes and failures across the affected countries.

In terms of successes, many affected populations are now less vulnerable to future tsunamis, cyclones, and coastal hazards. Coastal management zones were created in some countries (e.g., Sri Lanka and India), which resulted in the resettlement of affected communities away from the coast. Some of these managed areas were also populated with mangroves, pine trees, and other vegetation to absorb the impact of future coastal hazards. Furthermore, many affected families were provided concrete, disaster-resistant houses, and they gained formal ownership of land and/or property for the first time. Along with these settlements often came better access to piped water resources, toilets, and community infrastructure (e.g., on-site water treatment plants, community halls, libraries, and space to open shops and micro-enterprises). In addition to civil infrastructure, many affected populations were also provided with boats, micro-credit loans, and training for income-generation activities and were offered temporary employment in reconstruction activities.

In terms of failures, most reconstructed houses were prototypical in nature. These houses did not match local cultures and climates, and they were built for nuclear families and thus failed to accommodate joint families. Resettlement, which may have been wise from the standpoint of risk reduction, faced opposition in some locations (e.g., India) due to loss of coastal livelihoods (e.g., fishing) and access to the ocean. Further critiques include a lack of participatory approaches, lack of (timely) information flow to affected populations, and contractors running behind schedule due to the large scope of reconstruction and increases in demand for sand, cement powder, brick, and steel prices. Finally, some utilities were simply "patched up" to get them back online, but they were not overhauled to withstand future disasters.

Given the successes and failures, many lessons were learned and best management practices were identified. As argued by Olsen et al. (2005), we must avoid reconstruction mistakes of the past that tend to foster a "business-as-usual" pattern of disconnected projects and uncoordinated programs. Applying this to the 2004 tsunami, reconstruction actors should have aimed at respecting natural forces (do not ignore the dynamics of coastal systems); reducing human exposure to make communities safer; improving public services (reducing previous inequities and increasing access to potable water and sanitation); and promoting diversified and sustainable livelihoods that maximize community benefits while not degrading the environment. With these principles in place, national governments should have established clear goals (a set of specific measurable outcomes); decentralized decision making (a process in which local knowledge can mesh with technical expertise); celebrated successes (a process in which lessons learned are communicated); and promoted accountability (actors must recognize their limitations and failures). Collectively, these principles seek to produce enhanced outcomes not only for affected populations but also for reconstruction and recovery actors themselves. The goal is to create a positive feedback loop that helps to avert future failures in humanitarian aid provision.

CONCLUSION

The 2004 Indian Ocean Tsunami ranks as one of the greatest natural disasters in modern history. Affecting 15 countries and millions of individuals, the tsunami shook the globe in its geographical scale and need for postdisaster assistance. While there were certainly areas for improvement throughout the disaster cycle, the establishment of an Indian Ocean Warning System—which has successfully sent alerts after its inception—has brought much-needed disaster communication to coastal populations. Furthermore, thousands of families have been provided disaster-resistant houses, updated civil infrastructure, and livelihood support. The tourism industries have also been rehabilitated, which brings additional development monies to affected populations. Finally, many critiques of humanitarian aid have surfaced after the tsunami (see Korf 2006), and these critiques now inform many government, supranational, and nonprofit institutions.

Luke Juran

Further Reading
Korf, B. 2006. Antinomies of Generosity: Moral Geographies and Post-Tsunami Aid in Southeast Asia. *Geoforum* 38: 366–378.
Olsen, S. B., W. Matuszeski, T. V. Padma, and H.J.M. Wickremeratne. 2005. Rebuilding after the Tsunami: Getting It Right. *Ambio* 34(8): 611–614.
TEC (Tsunami Evaluation Coalition). 2006. *Joint Evaluation of the International Response to the Indian Ocean Tsunami: Synthesis Report.* London, UK: Tsunami Evaluation Coalition.
Telford, J., and J. Cosgrave. 2007. The International Humanitarian System and the 2004 Indian Ocean Earthquake. *Disasters* 31(1): 1–28.
WHO (World Health Organization). 2005. *Moving Beyond the Tsunami: The WHO Story.* New Delhi, India: WHO.

Iowa Flood, United States, 2008

Iowa experienced one of the worst floods in 2008, which started around June 8 and ended on July 1, and it affected the floodplains of most of the rivers in eastern and central Iowa. These rivers included the Upper Iowa River, the Iowa River, the Turkey River, the Maquoketa River, the Wapsipinicon River, the Cedar River (and its significant tributaries), and the Skunk River at its various forks. The Des Moines River of central Iowa had some minor flooding, breaching a levee in Des Moines. The Upper Mississippi River, which receives the outflow from all these rivers, remained at flood stage but did not cause significant damage. On June 13, 2008, floodwaters in the Cedar River crested at 31.12 feet (9.5 m) in Cedar Rapids, breaking the historic record of 20 feet (6.1 m),

The Des Moines River flooding over U.S. Route 30, west of Boone, Iowa, June 2008. (Adam Larsen/Dreamstime.com)

which was recorded in 1929, and it heavily impacted a wide geographic area of the Cedar River valley. On June 15, 2008, at 5:00 p.m. the Iowa River crested at about 31.5 feet (9.6 m) in Iowa City. The 2008 Flood was so damaging that it is often referred to as "Iowa's Katrina," and many cities along the affected rivers (e.g., Cedar Falls, Cedar Rapids, Iowa City, Waterloo, Waverley, and Oakville) were severely flooded.

CAUSES

Due to an unprecedented wet winter of 2007 and spring of 2008, Iowa experienced an extremely wet antecedent condition through the whole state (USGS 2010). In turn, the major cause behind the historic summer flood of 2008 is mostly associated with precipitation in the winter months of December 2007 to February 2008 that worked as a major catalyst for the flood event. Reports from across the region stated temperatures were below normal and that precipitation was above normal all across the state during this period. According to climate statistics from the beginning of February to mid-June in Mason City, Waterloo, and Des Moines, the below normal temperatures and above normal precipitation played a vital role in soil moisture content all across Iowa. Waterloo, Mason City, and Des Moines recorded seven individual daily precipitation records between April and mid-June in 2008, while the Waterloo area had a high monthly record in April 2008 of 11 inches (28 cm) of precipitation. Due to continuous heavy precipitation, a number of rivers experienced record water flow for several weeks, which caused levee failure at numerous locations. Due to anomalies in these weather patterns, both

Iowa and other Midwestern states' farmers had to delay planting their crops until May of 2008.

Due to this unusual delay, both rain and snow meltwater passed through creeks and watersheds without being absorbed. According to the weather service reports, snow remained in Waterloo and Des Moines on April 12, 2008, and in Mason City on April 25, 2008. The weather service also reported that in the first half of June, Iowa's average temperature was 62 degrees Fahrenheit (16.6 degrees Celsius) warmer than normal, and in the second half of the month, the average temperature was 61.7 degrees Fahrenheit (16.5 degrees Celsius) cooler than normal. Waterloo, Mason City, and Des Moines were prime examples of the average temperature remaining below normal for an extended period of time (NWS 2008).

IMPACTS

Both the Great Mississippi and Missouri Rivers Flood of 1993 and the 2008 Flood are considered the worst floods in the history of Iowa. However, in terms of geographic area, the 2008 Flood impacted a smaller area than did the 1993 Flood. The hardest-hit areas in 2008 were mainly confined to the Cedar and Iowa River basins of eastern Iowa. Due to the magnitude of the flood's impacts, 85 of Iowa's 99 counties were declared Federal Disaster Areas with total statewide damage estimated around $10 billion and over 40,00 people affected. An estimated 80 percent of 56,272 square miles (145,782 square km) of state land, 74 percent of the cities, and 87 percent of the counties in Iowa were declared disaster areas by the state governor. According to the National Weather Service (NWS), during the flood, an estimated 3 million acres (1.2 million ha) of corn and soybeans were underwater, and ultimately, about 1.3 million acres (0.53 million ha) of corn and 2 million acres (0.81 million ha) of soybeans were destroyed worth $2 billion (NWS 2008).

The Cedar River set records in terms of water discharge during the 2008 Flood in multiple areas where population density is higher than in other regions in the state, particularly in the Waterloo-Cedar Falls area and in Cedar Rapids, along the course of the Cedar River. Although statewide damage in 2008 was record high compared to that of past floods, two major cities (Cedar Rapids and Iowa City) experienced the hardest impacts. Indeed, the 2008 Flood overreached the 500-year floodplain boundary in many areas in Iowa.

As a result of the 2008 Flood, much of Iowa City's 500-year floodplain experienced mild to catastrophic effects of the rapidly flowing, polluted water. It was the city's largest flood on record, covering some 1,600 acres (648 ha). According to a University of Iowa report, up to 19 campus buildings were affected by rising waters. By using sandbags, volunteers constructed a massive levee around the university's main library, and extensive efforts were undertaken to move materials from the library to safer locations. Authorities at the University of Iowa abandoned nonessential operations, including classes, and evacuated flood-prone buildings on June 14, 2008.

Nearby, evacuation and sandbagging started in Iowa City on June 4, 2008. The city mayor, Regenia Bailey, issued a curfew restricting anyone except those authorized by law enforcement from being within 100 yards (91 m) of any area affected by the flood between 8:30 p.m. and 6:00 a.m. Meanwhile, Iowa City's Water Pollution Control (WPC) facility was underwater and out of service for several weeks. In Iowa City alone, an estimated 304 residences across the city were directly affected.

In Cedar Rapids, floodwater penetrated nearly 10 square miles (16 square km) or 14 percent of the city, impacting 7,198 parcels of land, or 1,300 city blocks. It damaged and/or destroyed 5,390 houses, displacing more than 18,000 residents, and damaged 310 city facilities, including 818 commercial properties and government buildings (FEMA 2009). Additionally, 45 registered day-care providers facilities were damaged, displacing 1,547 children, and three of four city collector wells and 46 vertical wells were disabled. Consequently, the city lost 1,360 jobs, and Cedar Rapids firefighters recused about 8,000 people from flood-affected areas using 423 boats. As for flood mitigation measures, volunteers started sand-bagging and constructing dirt levees around several buildings on June 8, 2008. Meanwhile, the Iowa National Guard arrived in Cedar Rapids to help the flood victims. Cedar Rapids alone experienced an estimated loss of $5.4 billion.

Besides the property and crop damages, the transportation network in the State of Iowa experienced severe damage due to the 2008 Flood, resulting in the closing of multiple roads and highways, including especially the local highways in eastern Iowa (NWS 2008, 2009). According to an Iowa state department official, an estimated cost for road damage reached $30 million; the damage was along an estimated 465-mile (748-km) state highway system and included 303 bridges and culverts. Nonmanufacturing, small, and intermediate businesses lost over $5 billion. In agriculture sectors, the estimated damage was over $3 billion for crops and grain, equipment, agricultural-related infrastructure, and storage and handling facilities.

RECOVERING AND REBUILDING

Due to the unprecedented impacts of the 2008 Flood event, several new initiatives were established to recover flood losses through state-wide programs such as the Rebuild Iowa Office (RIO) and Rebuild Iowa Advisory Commission (RIAC). Both programs were established by Governor Chet Culvert and coordinated state-wide short- and long-term flood recovery effort. The main purpose of RIO and RIAC formation was to address issues such as housing, agriculture, infrastructure, economic and workforce development, hazard mitigation, and floodplain management (RIO 2011). Specifically, after the 2008 Flood event, the City of Cedar Rapids considered several flood recovery action plan initiatives. For example, in 2009, the Cedar (and Iowa) River basin was designated as a Hydrology for the Environment, Life, and Policy (HELP) program by the United Nations Educational, Scientific, and Cultural Organization (UNESCO), and Cedar Rapids was selected as a pilot city by the National Academies and its Resilient America program.

The success story of Cedar Rapids became a national example for flood resilience and flood management programs. The city developed an interim flood response plan after the 2008 Flood event, which was finally approved in February 2009. This was done by Stanley Consultants, based in Muscatine, Iowa, who reviewed the flood protection measures and flood protection levels, and also the Corps of Engineers conducted product testing, which later became integrated into the Flood Response Manual in early 2015.

Cedar Rapids relied upon federal taxpayer emergency monies support from the Federal Emergency Management Agency (FEMA) and the U.S. Department of Housing and Urban Development (HUD) to acquire the damaged properties and to compensate the affected property owners. Another major program, called the buyout and acquisition process, was HUD's Community Development Block Grant (CDBG) Disaster Recovery program. These programs also supported other flood-affected communities in Iowa. Besides these programs, a voluntary buyout program was also initiated for providing economic assistance to property owners who voluntarily participated in removing many damaged floodplain structures vulnerable to high river flows and flooding in various areas including Cedar Rapids and Iowa City.

Like Cedar Rapids, another Iowa city was very successful in recovering from the 2008 Flood event. After the flood, the City of Cedar Falls, which is located about 60 miles (90 km) northwest of Cedar Rapids, passed new 500-year floodplain regulations. It is Iowa's first such floodplain ordinance. The city received necessary money from both federal and state governments to buy properties that were severely flooded by the 2008 Flood. The city bought over 200 homes severely damaged by the flood. In addition, the city provided nearly $2 million to rebuild 93 homes or to purchase houses outside the flooded area using state fund. In December of 2009, the City of Cedar Falls adopted a new floodplain ordinance to restrict any new or reconstructed projects in the Cedar River flood way and floodplain.

CONCLUSION

The 2008 Flood became the worst flooding event in Iowa's flood history since 1993, which occurred due to an unusual series of rain events from May 30 through June 14, 2008. During the first week of June 2008, historic flooding in southwestern Iowa was also partly associated with severe thunderstorm driven tornado events and damaging high winds and immediate prior cold winter and record snowfalls in the region. The flood of 2008 also reached 500-year flood levels.

The subsequent initiatives came in response in part to the clear lack of education, public information, and awareness about floodplains in Iowa. Due to the rapid development and significant population growth in all major cities, many developments had been built in recent years. In the case of Iowa, there was a lack of practical knowledge about floodplains among residents evidenced by many infrastructure and residential places built on floodplains; thus, public awareness and education is essential for reducing future flood-related impacts in the state of Iowa.

The flood in 2008 was devastating and generated extremely bad memories for many affected communities. Although the event opened many doors for better preparation on both federal and state levels, more cooperation between the two levels of government is needed to reduce future damage associated with such floods. Finally, the 2008 Flood offered numerous opportunities to improve the state of Iowa's floodplain management's postflood actions and outcomes, which may make Iowans more resilient compared to residents in other states in the United States.

M. Khaledur Rahman

Further Reading

FEMA (Federal Emergency Management Agency). 2009. Midwest Floods of 2008 in Iowa and Wisconsin. https://www.fema.gov/media-library-data/20130726-1722-25045-0903/fema_p_765.pdf, accessed February 8, 2019.

Miller, David L. Iowa Disaster 2008: Responding, Recovering, Rebuilding. Iowa Homeland Security and Emergency Management Division. http://www1.udel.edu/DRC/emforum/recordings/20100203.pdf, accessed February 8, 2019.

NWS (National Weather Service). 2008. Flooding in Iowa, National Weather Service. https://www.weather.gov/safety/flood-states-ia, accessed January 24, 2019.

NWS (National Weather Service). 2009. Central Iowa Floods of 2008: Late May through Mid June, 2008. https://www.weather.gov/media/dmx/SigEvents/2008_Central_Iowa_Floods.pdf, accessed January 25, 2019.

RIO (Rebuild Iowa Office). 2011. Quarterly Report & Economic Recovery Strategy. Des Moines: RIO. http://publications.iowa.gov/11071/1/2011-04_Quarterly_Report_FINAL.pdf, accessed January 25, 2019.

USGS (United States Geological Survey). 2010. Floods of May and June 2008 in Iowa. https://pubs.usgs.gov/of/2010/1096/pdf/OFR2010-1096.pdf, accessed January 24, 2019.

Izmit/Marmara Earthquake, Turkey, 1999

Earthquakes are the most frequent and devastating of natural disasters in Turkey in comparison with others. This is because 92 percent of the country is located in the active earthquake zone, and 98 percent of the people of this country live in an earthquake-prone area. In fact, earthquakes account for 66 percent of all damages caused by all natural disasters in Turkey, which means the average annual loss caused by earthquakes alone constitutes 0.8 percent of the national income in the country (TRCS 2006). The country is being squeezed sideways to the west as the Arabian plate pushes into the Eurasian plate at a rate of 0.71–0.98 inches (1.8–2.5 cm) per year. In the northern part of the country, the North Anatolian Fault (NAF) runs east-west along the transform boundary between the Eurasian Plate and the Anatolian Plate for a distance of over 1,000 miles (1,500 km).

The NAF slipped on August 17, 1999, 67 miles (100 km) east of Istanbul, causing an earthquake of 7.4 magnitude. This earthquake is known as the "Kocaeli" or "the Izmit," for both cities close to the epicenter. The earthquake was so powerful that it generated a local tsunami of about 2.5 feet (0.8 m) high within the enclosed Sea of Marmara. For this reason, the earthquake is also known as the Marmara earthquake; moreover, this was the strongest earthquake to strike northern Turkey since 1967. Its epicenter was in the Gulf of Izmit on the western segment of the NAF and about 7 miles (11 km) southwest of Izmit, or 53 miles (86 km) southeast of Istanbul. It occurred at a depth of 9.3 miles (15 km) and lasted for between 37 seconds and one minute.

The sudden breakage or rupture of the earth's crust along a western branch of the NAF caused the Izmit earthquake. The total length of the fault rupture was about 73 miles (110 km). The main earthquake struck the region along the southern part of the Marmara Sea at 3:02 a.m. local time, when people were sleeping at homes. Numerous aftershocks with magnitudes above 4.0 followed the earthquake. The first of the aftershocks (magnitude of 4.6) occurred 20 minutes later. Two moderate aftershocks occurred on August 19, 1999, about 50 miles (80 km) west of the original epicenter. Most of the aftershock activity was clustered near Akyazi and Izmit. Ultimately, large areas of a number of mid-sized towns and cities were destroyed. Subsequently, a second smaller but still major earthquake with a magnitude of 7.1 hit the same area on November 12, 1999.

LOSS AND DAMAGE

The 1999 Izmit earthquake affected the most densely populated and industrialized area of Turkey and caused the deaths of 17,480 people and injured another 43,953. High casualty figures were reported especially in the towns of Golcuk, Derince, Darica, and Sakarya (TRCS 2006). Nearly 120,000 poorly engineered houses were damaged beyond repair, 30,000 houses were heavily damaged, 2,000 other buildings collapsed, and 4,000 other buildings were heavily damaged. Nearly 214,000 residential units and 30,500 business units were lightly to heavily damaged by the earthquake, and 43 schools were destroyed while 377 other schools and 11 hospitals were seriously damaged. The earthquake left 300,000 people homeless, requiring approximately 121 tent cities to accommodate displaced people (Bolt 2004). Most of the casualties occurred in poorly built, masonry buildings. So many casualties and building collapses were caused by liquefaction (a process in which water-saturated sands jostled by an earthquake rearrange themselves into a closer packing arrangement), which occurred in parts of Izmit. Moreover, the Avcılar district in Istanbul experienced heavy damage because this part of the city was built on relatively weak ground mainly composed of poorly consolidated Cenozoic sedimentary rocks.

Other important factors in the collapse of and damage to property were the long duration and strong velocity of the earthquake, which triggered, for example, a disastrous fire at the Tüpraş petroleum refinery. Attempts to extinguish the fire were unsuccessful because the event rendered the water pipelines inactive. Therefore, aircraft were called in to douse the flames with foam, and people were evacuated from within three miles of the refinery. The fire spread over the next several days, warranting the evacuation within three miles. The fire was brought under control five days later. Although the tsunami waves killed 155 people, they did not cause extensive damage. Because of the economic significance of the earthquake-affected region, the country experienced labor losses and production and market losses due to the relatively large number of deaths and injuries. Outmigration also played a role in such losses, and ultimately, property damage reached $7 billion (Hyndman and Hyndman 2006).

RESPONSE

Immediately following the earthquake, local residents engaged in search and rescue (SAR) operations. Also, within 24 hours of the earthquake, Turkey's national civilian voluntary search and rescue organization began SAR activities. The international community provided rapid and generous support with rescue teams from 24 foreign countries joining the operation within 48 hours. These countries also donated emergency assistance for the earthquake survivors as well as immediate grants directly to the Turkish authorities. Then, on the third day after the earthquake, the Turkish Red Crescent Society (TRCS) and the Turkish army participated in rescue and relief operations. For the first two days, the SAR operations were slow because of the lack of heavy machinery, access to which in turn was caused by damaged road ways. From the third day onward, the government deployed heavy machinery to start SAR activities as well as to clean rubble. The central government deployed heavy machinery according to damage estimates by the governors of the provinces affected.

However, due to the absence of a systematic plan, distribution of emergency items such as drinking water and food was slow, and as a result, large quantities of spoiled food piled up uneaten, creating mounds of garbage. The Turkish government did appeal for international emergency aid, firefighting troops, recovery, and reconstruction assistance. Also, a report published by the International Federation of Red Cross and Red Crescent Societies claims that in total 50,000 people were rescued from beneath rubble, but local residents performed 98 percent of these rescues (IFRC 2002).

Prior to the earthquake, relations between Greece and Turkey were volatile. However, the Greek government, citizens, and NGOs offered immediate support. In fact, Greece was the first foreign country to pledge aid for the earthquake survivors. The Turkish government could not disregard the Greek initiatives, and as a result, the adversarial relationship between the two neighboring countries thawed. In the summer of 1999, Greece and Turkey initiated diplomatic conversations, leading to improved political relations between the two countries.

Immediately after the Izmit earthquake, the Greek Ministry of Foreign Affairs contacted its counterparts in Turkey by sending envoys. Two days after the earthquake, the Greek Ministry of Public Order sent a rescue team of 24 people along with two trained rescue dogs. The ministry also sent fire extinguishing planes to help put out the fire at the Tüpraş petroleum refinery, and the Greek Ministry of Interior Affairs sent a fully equipped medical team of 11 people as well as tents, ambulances, medicine, drinking water, clothes, food, and blankets. The Greek Ministry of Health set up three units for blood donations on August 18, 1999, and the same day sent aid to Turkey. On August 19, 1999, the Ministry of Foreign Affairs set up three receiving stations in Athens, Thessaloniki, and Komotini, Greece, to collect the spontaneous donations from its citizens. The Greek Red Cross, the Athens' Medicine Association, and other NGOs, also sent emergency aid, including the Greek departments of the Médecins Sans Frontières (Doctors Without Borders).

Along with rescue and distribution of emergency assistance for the earthquake survivors, the government of Turkey initiated a damage assessments process employing 1,200 technical personnel. They assessed the damage of 334.000 properties in three categories (moderately damaged, badly damaged, and destroyed) in 20 days. Owners of the moderately damaged properties received a reparation loan of 2 billion Turkish Lira (TL). Most of the owners of badly damaged and/or completely damaged properties were provided a house constructed by the government (TRCS 2006).

RECONSTRUCTION

Given the extensive physical and social damage, a major reconstruction effort and recovery plan was needed, in addition to a mechanism to reduce the costs of future earthquakes in Turkey. Along with the Turkish government, the World Bank initiated its assessment to outline the likely impact of the earthquake on the economy and the cost of reconstruction and recovery. The bank estimated the total fiscal burden arising from the earthquake to be in the range of $1.8–2.2 billion. The largest direct cost to the budget, estimated to be in the range of $740 million–$1.2 billion, came from reconstruction and repair of damage to the housing stock. Costs for infrastructure replacement and rehabilitation were estimated to add a further $450 million (World Bank 2006).

As part of the comprehensive response to the August 17 Izmit earthquake, the Turkish Government initiated three reconstruction projects, namely, the Marmara Earthquake Emergency Reconstruction Project (MEER), the Emergency Earthquake Recovery Loan (EERL), and the Turkey Earthquake Rehabilitation and Reconstruction Assistance Project (TERRAP). The first project provided funding for reconstruction and interventions that would contain damage in the case of a similar future event. The government allocated $505 million, and the project had three main components. The first component outlined a comprehensive emergency management structure that coordinates and integrates risk reduction strategies, preparedness, response, and recovery. It also created an insurance mechanism by establishing and expanding national catastrophic risk management and risk transfer capabilities for resident owners of damaged or destroyed buildings; reducing government fiscal exposure; ensuring the financial solvency of the pool; and reducing government dependency on the World Bank and other donors. Finally, this component planned to establish a land information system while updating and improving obsolete registers and maps. The second component of MEER was to fund the development of a trauma program for psychological support for adults, and the third component comprised methods of construction of permanent housing in earthquake-affected areas.

The EERL was allocated $252.53 million and the MEER $737.11 million (of which the World Bank provided $505 million as a loan). The EERL was used for short-term purposes, while the MEER was used for short-term and medium-term projects. The TERRA project was implemented to finance the reconstruction of earthquake-affected areas, including urban infrastructure, using the loan provided by the European Investment Bank. Consequently, the Turkish government constructed 14,068 permanent settlements in several cities affected by the 1999 earthquake. In addition, it built 2,000 settlements in rural areas (TRCS 2006). Because these three projects required a large amount of money, assistance was also provided by other financial institutions and aid organizations besides the World Bank. Notable among them was the European Investment Bank.

CONCLUSION

The immediate public response to the August 1999 earthquake was slow because of a combination of several factors: the huge extent of the stricken area, inadequate planning, and lack of personnel experience, resources, and effective leadership. The gravity of the event pointed out the need for upgrading the existing emergency response system, the lack of effective enforcement of Turkey's building codes, and the inadequate coverage of earthquake insurance in the housing sector. The 1999 earthquake was a lesson for the Turkish government to reduce the impact of future earthquakes. One noteworthy outcome of this earthquake was that it facilitated a better political relationship between Turkey and Greece.

Bimal Kanti Paul

Further Reading
Bolt, B. A. 2004. *Earthquake*. New York: W.H. Freeman and Company.
Hyndman, D., and D. Hyndman. 2006. *Natural Hazards and Disasters*. Belmont, CA: Thomson.
IFRC (International Federation of Red Cross and Red Crescent Societies). 2002. *World Disasters Report: Focus on Reducing Risk*. Geneva, Switzerland: IFRC.
TRCS (Turkish Red Crescent Society). 2006. *International Disaster Response Law: 1999—Marmara Earthquake Case Study*. Ankara, Turkey: TRCS.
World Bank. 2006. *Implementation Completion and Result Report (IBRD-45170) on a Loan in the Amount of US$ 294 Million to the Republic of Turkey for Marmara Earthquake Emergency Reconstruction Project*. Ankara, Turkey: World Bank.

Johnstown Flood, Pennsylvania, 1889

Floods result from both natural and anthropogenic (resulting from human activity) sources. The latter source includes failure of artificial structures (e.g., dams, dykes, levees, and embankments) constructed primarily to control streamflow and/or minimize the effects of floods. Dam failure is a common source of human-made floods in many developed as well as developing countries of the world. It occurred in the past and is also occurring now. Excessive rainfall, poor and/or lack of maintenance, and poor location such as a geologically unstable setting are the main causes of the collapse of a dam. As a consequence, the dam releases a large volume of water that abruptly inundates areas downstream, damages property, washes away farmland, and kills and injures many people. One such a disaster happened in Johnstown, Pennsylvania, at 3:10 p.m. on May 31, 1889.

DEATHS AND DESTRUCTION

The 1889 Johnstown flood directly killed 2,209 people and caused massive destruction. However, the death toll did not include those who died of injuries or exposure in the first day after the disaster. Without those numbers, the flood recorded the largest loss of life in the United States from any natural disaster at the time and was the worst flood to hit the country in the nineteenth century. Among the deaths, 99 households lost all their members, and 396 children aged 10 years or younger were killed in the flood. One-third of the dead were never identified. The flood widowed 124 women, orphaned 98 children, and left 198 men without wives. In the affected area, the flood killed about one person out of every 10 and one out of nine in Johnstown itself. In all, it destroyed 1,600 homes (McCullough 1968). Ultimately, the flood damage was estimated at $17 million, which is equivalent to about $463 million in 2017 (FRBM 2018).

CAUSES OF THE FLOOD

The collapse of a large earth-fill dam primarily caused the 1889 flood. This event resulted from three factors: human neglect, poor judgment, and a two-month period of unusually heavy rainfall (Ebert 1993). The earthen dam above South Fork and Johnstown was constructed to impound water for the canal from Johnstown to Pittsburgh, known as the Western Division of the Pennsylvania state's "Main Line" canal. Prior to the dam, shortages of water nearly every summer interrupted the traffic of this canal. Another purpose of the construction was to protect against the area's susceptibility to floods. Construction of the dam began in the mid-1820s and was completed in 1853.

The early success of the Erie Canal in New York spurred the Main Line canal, which was a part of a cross-state canal system. The Commonwealth of Pennsylvania constructed all the canals in the state to facilitate water transport. The earthen dam constructed above Johnstown was 72 feet (22 m) high and 931 feet (284 m) long. The depth of the reservoir, Lake Conemaugh, ranged from 40 to 100 feet (12–31 m) and had a maximum capacity of 17.7 million cubic yards (13.5 cubic m) (Ebert 1993). The claim in 1889 was that it was the largest earth dam in the United States, and it created the largest human-made lake (McCullough 1968).

The dam was constructed approximately 14 miles (21 km) upstream of Johnstown near the confluence of the Little Conemaugh and Stony Creek Rivers. The town is 60 miles (90 km) east of Pittsburgh in a valley near the Allegheny Mountains. It is located on a floodplain that has been subjected to frequent flooding. In 1889, when the devastating flood occurred, Johnstown was home to 30,000 people, many of whom were employed nearby in the steel industry.

As railroads superseded the canal as a means of transporting goods, the Commonwealth abandoned the canal and later sold both the dam and the reservoir to the Pennsylvania Railroad, which then sold the two properties in 1857. In the late 1870s, these properties were sold to the South Fork Fishing and Hunting Club, a sporting club formed by wealthy patrons from Pittsburgh. The club converted the reservoir into a private resort lake for their wealthy associates, many of whom were connected through business and social links to Carnegie Steel. The club also built a road across the length of the dam and placed a fish screen in the spillway, which allowed water to spill into the stream below the dam. Its members also built cottages and a clubhouse to create the South Fork Fishing and Hunting Club, an exclusive and private mountain retreat.

A tree trunk protrudes from a toppled house during the Johnstown flood in western Pennsylvania, May 1889. (Library of Congress)

The dam and the reservoir suffered many years of neglect, and by 1889, they were in dire need of repair. Since its completion, the reservoir had manifested several leakages around the main drain pipes. Additionally, after purchasing the dam, the club built a grating over the spillway outlet to prevent the fish from escaping the lake, whereupon the outlet became clogged because a large amount of floating material was trapped in it (Ebert 1993).

EVENTS ASSOCIATED WITH THE FLOOD

Heavy rainfall occurred continuously for several days in the vicinity of the dam prior to its collapse. A day before the collapse, the most extensive rainfall of the century occurred. Around six to eight inches (15–20 cm) of rain fell in 24 hours, accelerating the melting of snowfall in the nearby Allegheny Mountains (McCullough 1968). Then a spillway at the dam became clogged with debris that could not be dislodged, putting the dam under tremendous pressure from the swollen waters of Lake Conemaugh. The lake's water level passed over the top of the dam. Realizing that the dam's collapse was imminent, an engineer of the South Fork Fishing and Hunting Club considered cutting through the dam's end, where the pressure would be less, but he decided against it as that would have ensured the failure of the dam.

The engineer then rode a horse to warn the residents of nearby towns. Another source claims that he went to nearby town to telegraph warnings to Johnstown people, explaining the critical nature of the eroding dam. A telegraph warning message of flood came to Johnstown around 2:45 p.m. However, flooding was recurrent in the valley, and hardly any one took the warning seriously because residents had received such alarms, many false, many times in the past (McCullough 1968).

The central portion of the dam broke, causing water to rush from Lake Conemaugh down the valley toward Johnstown at 40 miles (60 km) per hour. The lake at the club's house was about 450 feet (140 m) in elevation above the town. The flood hit the town of South Fork first, which had a population of 1,500 people. The town was on high ground, and most of the people ran to the nearby hills to escape the flood. Four people were killed, and nearly 30 houses in South Fork were either destroyed or washed away. The flood then headed to the Conemaugh Viaduct and the small town of Mineral Point, one mile (1.6 km) below the Conemaugh Viaduct. About 30 families lived on the village's single street, and the flood killed 16 in that town. Before reaching Johnstown, the flood hit the Cambria Iron and Steel Works located at the town of Woodvale, near Johnstown, sweeping up railroad cars and barbed wire in its moil. Three hundred fourteen of the 1,100 residents died in the flood (McCullough 1968).

On the way to Johnstown, rushing floodwaters carried debris, boulders, trees, houses, barns, farm animals, and pets. People in the path of waters were often crushed as their homes and other structures were swept away. Thirty-three train cars were pulled into the raging waters, creating more hazards. Telephone lines were downed, and rail lines were washed away. Waves of rushing water, estimated at 30–40 feet (9–12 m) high and nearly half a mile wide, reached Johnstown at 4:07 p.m. The flood swept away the northern half of the city, destroying some 1,500 buildings and drowning 2,000 people. Four square miles (6 sq. km) of downtown Johnstown were completely destroyed.

The floodwater rose in the downtown area as high as 10 feet (3.0 m). A few survivors were washed up several miles down the valley, while few others took to the top floors of the few tall buildings in town. At the old Stone Bridge in downtown Johnstown, debris piled 40 feet (12 m) high and caught fire, which lasted for three days. As a result, some 80 flood survivors burned to death. After floodwaters receded, the pile of debris at the bridge covered 30 acres (12 ha) and reached 70 feet (21 m) in height. It took workers three months to remove the debris.

RELIEF AND RECONSTRUCTION

Flood survivors in Johnstown had no food, money, or dry clothing, and many of them did not eat anything for 24 hours or more. Some people suffered from early stages of pneumonia. The carcasses of drowned cows, horses, pigs, dogs, cats, and other animals posed an imminent threat of epidemic. However, people from hillsides and farmers from neighboring nonaffected areas came to Johnstown with food, water, and clothing. The city opened a temporary hospital to treat injuries, and the survivors along with other people collected and identified those who were killed by the flood (McCullough 1968).

The first relief train from Pittsburgh reached Johnstown on June 2, 1889. People across the United States sent tons of relief supplies and donated $3.6 million. Eighteen foreign countries, including Australia, France, Germany, Great Britain, Russia, and Turkey, also donated emergency assistance in the value of $141,301, making a total of $3.7 million (McCullough 1968). The unprecedented response from the American people was the result of the record news coverage. Also, most members of the South Fork Fishing and Hunting Club contributed financially for both disaster relief and recovery efforts in Johnstown. After the flood, industrialist and philanthropist Andrew Carnegie built a new library in the town. At its peak, 7,000 relief workers from across the United States participated in emergency aid distribution.

One of the American Red Cross's first major peacetime relief operations took place in the immediate aftermath of the flood. Clara Barton, founder and president of the American Red Cross, arrived with 50 doctors and nurses five days after the disaster to lead the relief efforts and stayed for five months. The Red Cross constructed shelters for homeless residents and provided furniture and supplies for flood survivors. It also distributed new and used supplies valued at $211,000, and the organization helped some 25,000 people. The Philadelphia Red Cross also participated in the disaster relief efforts by establishing a headquarters in Johnstown separate from that of the American Red Cross. The former specialized in medical relief, working out of the Cambria Iron Company's hospital.

Despite the great tragedy, flood disaster reconstruction of the devastated community began almost immediately and lasted for five years. Meanwhile, recovery of the dead continued for months. Dead bodies were found as far away as Cincinnati (600 miles or 900 km) and as late as 1911. In just 14 days, the Pennsylvania Railroad had rebuilt 20 miles (30 km) of track and bridges, allowing it to reopen its line to the east. Six weeks after the flood, the town was well on its way to recovery; the river channels had been cleared and huge piles of wreckage had been removed. Although Cambria Iron and Steel's facilities were heavily damaged by the flood, the company returned to full production within a year and a half.

CONCLUSION

As in the past, Johnstown also experienced deadly floods in 1894, 1907, 1924, 1936, and 1977. However, some people blamed and criticized the South Fork Fishing and Hunting Club particularly for the 1889 flood; ultimately, the club successfully defended the court cases and paid nothing to anyone. Like several church ministers, the court claimed that the dam break was an "Act of God." At the confluence of the Stony Creek and Little Conemaugh Rivers in Johnstown, an eternal flame burns in memory of the 1889 flood victims. The library built by Carnegie is now used as the Johnstown Flood Museum. It features a display about Barton and the American Red Cross, including examples of some of the relief items she distributed, documents, photographs, and more. In fact, the 1889 flood played an important historical public relations role for the American Red Cross. Also, portions of the Stone Bridge in downtown Johnstown have been made part of the

Johnstown Flood National Memorial, established in 1969 and managed by the National Park Service. The 1889 deadly flood has been the subject of or setting for numerous histories, novels, and other works as well.

Bimal Kanti Paul

Further Reading

Ebert, C.H.V. 1993. *Disasters: Violence of Nature Threats by Man*. Dubuque, IA: Kendall/Hunt Publishing Company.

FRBM (Federal Reserve Bank of Minneapolis). 2018. Consumer Price Index (Estimate) 1800. January 2, 2018. https://www.minneapolisfed.org/community/financial-and-economic-education/cpi-calculator-information/consumer-price-index-1800, accessed January 2, 2018.

McCullough, D. G. 1968. *The Johnstown Flood*. New York: Simon and Schuster.

Joplin Tornado, Missouri, 2011

A multiple-vortex EF-5 (on the Enhanced Fujita Scale) tornado struck Joplin, Missouri, on the afternoon of May 22, 2011. At nearly a mile, the tornado generated a six-mile path through the community, destroying approximately 7,000 homes and damaging hundreds more. Almost every business along a main east-west road through the city suffered either heavy damage or was completely destroyed. Gas leaks in some locations caused injuries and overnight fires across the city. The total damage area covered approximately 1,800 acres, or nearly one-quarter of the total area of Joplin, and caused an estimated three billion dollars in insured losses (Paul and Stimers 2012).

METEOROLOGICAL SYNOPSIS

The Joplin tornado was part of the larger May 21–26 outbreak that recorded a total of 261 confirmed tornadoes. On May 22, the risk of severe weather was deemed moderate by the National Weather Service's Storm Prediction Center (SPC), but a unique mix of atmospheric conditions converged across southern Missouri on the afternoon of the event. Usually, when two storm cells collide, they produce one larger but weaker supercell, as the circulation patterns of both are disrupted. In the case of the supercell that eventually produced the Joplin tornado, the newly combined larger supercell gained strength, and the circulation pattern did not break down. The cell over southern Missouri was weaker than the approaching cell from northern Oklahoma, and as the two merged, it developed into a pattern typical of a classic tornado-producing supercell.

At 5:17 p.m., the SPC issued a tornado warning, and at 5:24 p.m., the supercell merged with a third additional smaller cell, gained strength, and quickly developed the tornado that touched down just outside the western city limits as an EF-1 ten minutes later. The tornado rapidly increased in intensity, forming a multiple-vortex classic wedge structure, and began its devastating trek through the community (FEMA 2013).

DAMAGE

The tornado was approximately a half-mile wide as it approached Schiffendecker Avenue, causing EF-2 to EF-3 damage, and growing to a width of three-quarters of a mile between 26th and East 20th Street. The funnel stayed south of downtown Joplin, but in the southern section of the city, it destroyed four schools, including Joplin High School, and damaged six others. Major damage was also inflicted upon several industrial buildings in the area. The total damage swath encompassed nearly 7.5 square miles (19.4 square km) of the total 32 square miles (83 square km) that make up the community. According to the 2010 census, 13,547 people—27 percent of the population—lived in the 500 census blocks directly affected by the tornado.

The Joplin tornado completely destroyed 4,000 homes and caused damage to 3,000 others; at least 9,200 people were displaced as a result. Further, it destroyed 553 businesses, affecting nearly 4,000 jobs. The tornado heavily damaged one of the community's two hospitals, St. John's Regional Medical Center, which hindered emergency medical capacity in the immediate wake of the disaster. The hospital was eventually deemed a total loss and demolished. Buildings located on Freeman Hospital West campus, the other hospital in Joplin, also sustained damage, but they remained operational. Other significant buildings destroyed included the Full Gospel Church, Harmony Heights Baptist Church, and the Greenbriar Nursing Home (several fatalities were associated with the nursing home).

Near South Duquesne Road extending near the Interstate 44 and 249/71 junction, the funnel began shifting right and tracking to the south-east across I-44, and it began to lose strength; yet it remained strong enough to flip vehicles near the U.S. Route 71. As the tornado continued to break down, it moved into rural areas of south-eastern Jasper county and north-eastern Newton County—damage in

This car was found among the debris after an EF-5 tornado hit Joplin, Missouri, in 2011. (Amynbrady/Dreamstime.com)

those areas was moderate to minor, which is consistent with EF-1/EF-2 damage indicators. The tornado finally lifted east of the community of Diamond at 6:12 p.m., having laid down a total track length of at least 22 miles (36 km).

DEATH AND LOSS

The 2011 Joplin tornado was the deadliest single tornado event in the United States since modern recordkeeping began in 1950. Previous to Joplin, the Flint, Michigan, tornado in 1953 on June 8 claimed 116 lives. Taking into account the sheer size of the funnel and the magnitude of the event, and the population density of Joplin, the large death toll was not an unusual scenario. Over the past 30 years, the United States typically sees around 55 deaths per year, on average, resulting from tornadic activity; thus, the record number of fatalities caused by the Joplin tornado was far higher than the average number of annual tornado deaths in the United States in the past three decades.

Adequate warning cannot be effective if the population at risk does not have access to tornado shelters. Houses that lack basements and/or the inability to reach a nearby shelter are important factors when considering response to tornado warnings. The local geology of Jasper county is such that the water table is close to the surface, and the substrate is comprised largely of rocky and difficult-to-penetrate ground. Located near the border of four states (Missouri, Kansas, Oklahoma, and Arkansas), the area is home to former lead and zinc mines, most of which were last operational in the mid-1950s. Many tracts of land are located over old mine shafts, which has resulted in problems with subsidence (depressions in the ground due to a lack of subsurface support). These physical conditions have created a situation in which basements are not practical for most building schemes, and thus, nearly 80 percent of homes in Joplin did not have a basement included as part of the construction of the home.

In addition to lack of subsurface construction, the aging housing stock of the community was built according to the common building codes and standards of the time, which by comparison are far less demanding and extensive than what contractors encounter when building homes in the twenty-first century. According to Keith Stammer, Director of the Jasper County Emergency Operations Center (JCEOC), many older homes in the community are not secured to their foundation by the very common modern method of attaching sill plating to the foundation, and some homes were constructed with no foundation at all (Paul and Stimers 2012). The majority of respondents to a survey indicated that they were unsure whether or not their home was properly anchored to its foundation or whether hurricane roof straps had been installed. (Hurricane roof straps secure the rafter to the joist via a metal plate, as opposed to simply nailing those two components together, and they have been found to be extremely effective at keeping the roof connected to the house during extreme wind events, including tornadoes. Damage to structures is greatly lessened when the roof remains intact, as the weakened walls are not exposed to high winds, under which they can easily fail.) After the tornado, the Joplin City Council encouraged residents to use hurricane straps to strengthen any new housing construction.

It is typical for a larger percentage of deaths resulting from a tornado striking a community to occur in residential locations, which, however, was not the case with the Joplin event; the largest percentage of deaths occurred in business facilities (e.g., hospitals, nursing homes, restaurants, churches, gas stations, and retail stores). When comparing Joplin to U.S. tornado statistics by location of death, 35 percent more fatalities occurred in businesses during the Joplin tornado than usual, as the tornado moved through major commercial districts—no other tornado is known to have inflicted such damage on the business sector of a community.

Compounding the path of the tornado through areas with many business locations is the fact that the event occurred on a Sunday, a day when many cities see an increase in shopping and general outdoor activity. Although many of these facilities did have designated safe areas, those areas did not offer a high level of protection against an FE-5 tornado. It is worth noting that the high school graduation had taken place earlier in the day; Joplin High School held its graduation ceremony, but the ceremony did not take place at the high school, which was destroyed. The ceremony instead took place at a college on the northern border of the city. Had the tornado struck during the event, just a few hours earlier, and had the event been held in its typical location, the death toll would have undoubtedly topped 300 (Paul and Stimers 2012).

TORNADO WARNING-TIME PRECEDING THE EVENT

Lack of warning can be a major contributor to deaths resulting from rapid-onset natural disasters, such as flash floods and tornadoes, but data shows that approximately 90 percent of those surveyed were aware of warnings before the tornado hit Joplin. The public received warnings about the Joplin tornado with a 17-minute lead time, and the entire path of the tornado through Joplin was contained within the warning polygons. Authorities activated the siren in Joplin twice, with the initial three-minute alert sounding at 5:11 p.m. and the second three-minute alert sounding at 5:38 p.m., when the tornado set down in southwest Joplin.

Further, 86 percent of respondents to the same survey indicated that the NWS did an adequate job of warning the public. For these reasons, insufficient warning of the tornado threat was not a strong contributor to the high death toll. Among the 11 percent who were not aware of the warnings, the most common reason specified was that they were inside their homes but not watching television or listening to a radio. They also reported that they did not hear any warning sirens because they were indoors. (Emergency warning sirens are not designed to be heard inside buildings, particularly when doors and windows are closed.)

Roughly one-third of residents who claimed to have received a warning did not comply or took no action in regards to that warning. A key reason for this was the indifferent attitude of some respondents toward tornado warnings, as they had proven to be inaccurate based on past uses of those systems. Statistically, tornado occurrence is above the average for the state of Missouri and is 161 percent above the average for the United States as a whole—the implication being that tornadoes are not uncommon events in this area. A survey revealed that 75 percent of respondents in the city of Joplin had experienced a tornado in the past, as residents of either Joplin or another close-by community. Most tornadoes in this region, however, have been of low magnitude (EF-0–EF-2).

Concerning response to the warnings provided, residents who had no previous experience with a tornado exhibited a higher rate of compliance (voluntary action taken) compared to those who had experienced at least one tornado in the past. Further data from a survey showed that most of the residents who took action had been living in the city for a shorter period of time than those who opted to not heed the warnings. This suggests that long-time residents were less likely to take action for a variety of reasons, including a perceived understanding of severe weather and a lack of serious attention paid to warnings as a result.

In addition to perceptions of tornado-based risk, some Joplin residents felt that the warnings did not convey a sufficiently high level of risk. For many residents, the reality of the tornado threat only came when the warning sirens were activated. Authorities activated sirens twice in a short time span, and some residents thought the second siren's blast was an indicator that the tornado had passed, while others viewed the second siren as an indicator of the seriousness of the warning, and that prompted them to get to a nearby shelter or to the best shelter available. Several residents registered complaints concerning how often the warning sirens are activated in Joplin. Too frequent activation of a warning system is known to cause residents to become somewhat immune to the intended effect of a warning (Paul et al. 2014).

Finally, some Joplin residents demonstrated what is termed *optimism bias*, a phrase that describes the state in which diminished perceptions of threats exist, and consequently, lesser actions are taken in the face of potential danger. The effect is essentially the product of not wanting to believe that a tragedy will befall the individual or the area in which they live, and thus, the person remains optimistic in their judgment that they are correct and, thus, safe.

COMMUNITY RECOVERY

Even though the tornado caused an incredible amount of destruction, the community has since recovered and rebuilt much of what was lost. Almost all residential structures destroyed or damaged by the tornado were rebuilt or repaired. Businesses and civil infrastructure have been rebuilt as well, including a new hospital and a new high school.

Mitchel Stimers

Further Reading

FEMA (Federal Emergency Management Agency). 2013. Mitigation Assessment Team Report: Meteorological Background and Tornado Events of 2011. https://www.fema.gov/media-library-data/20130726-1827-25045-5594/tornado_mat_chapter2_508.pdf, accessed September 29, 2017.

Paul, B. K., and M. J. Stimers. 2012. Exploring Probable Reasons for Record Fatalities: The Case of the 2011 Joplin, Missouri, Tornado. *Natural Hazards* 64(2): 1511–1526.

Paul, B. K., M. J. Stimers, and M. Caldas. 2014. Predictors for Survivor Responses to the 2011 Joplin, MO, USA, Tornado Warning. *Disasters* 19(1): 108–124.

Kashmir Earthquake, Pakistan, 2005

Kashmir, the northernmost geographical region of the Indian subcontinent, is disputed territory controlled in part by Pakistan and India. It is located in the subduction zone of active faults where the Indian plate collides with the Eurasian plate, which makes the region prone to intense seismic activity. On the morning of October 8, 2005, this region experienced a massive earthquake with a magnitude of 7.6 on the Richter scale. The epicenter of the earthquake was 12 miles (18 km) northeast of Muzaffarabad, the capital of Pakistan-administered Kashmir (PAK), and 62 miles (100 km) north-northwest of Islamabad, the capital of Pakistan. The earthquake originated at a depth of 18 miles (26 km) (United Nations System 2005).

Because of the epicenter's location in PAK, and because the earthquake caused the most extensive damage in PAK and the North West Frontier Province (NWFP) (now called Khyber Pakhtunkhwa Province) in Pakistan, it is also known as the Pakistan earthquake. However, it also affected two other neighboring countries of Pakistan: Jammu and Kashmir (J&K) administered by India and Afghanistan. Because of its geographic coverage, the earthquake is often called the South Asia earthquake. It was also felt in Tajikistan and in Xinjiang province in China.

DAMAGE AND DEATH

The 2005 earthquake affected over 11,583 square miles (30,000 square km) of treacherous Himalayan terrain. Of all the affected regions, the hardest hit in terms of casualties and property damage was the PAK area, where Muzaffarabad suffered the most. The devastation was tremendous as the earthquake completely destroyed many hospitals, schools, and rescue services, and it paralyzed communication systems in severely affected areas. Over 12,000 schools and colleges were either destroyed or damaged, and more than 32,000 buildings collapsed in Kashmir alone (Ozerdem 2006). The earthquake destroyed a total of 203,579 units of housing and damaged another 196,574 units. This translates to 84 percent of housing damaged or destroyed in PAK and 36 percent in NWFP. Also, it produced 200 million tons of debris. Thus, total property damage in Pakistan was estimated at $5 billion (United Nations System 2005).

According to the official estimates, 73,000 people died in Pakistan, and more than 128,000 were injured while approximately 3.5 million were displaced (UNICEF n.d.). The death toll varied not only from one affected country to another but also within a given country. As noted, the overwhelming majority of the deaths in Pakistan occurred in PAK, and the Muzaffarabad district (the second largest administrative unit in Pakistan) accounted for more than 70 percent of those casualties. The remaining 30 percent occurred in eight other districts of northern Pakistan. Indian government-controlled Jammu and Kashmir experienced at least 1,369 deaths, with about 6,266 injured. Fortunately, only four people died and 14 were injured in Afghanistan. Additionally, an estimated 250,000 farm animals were killed. The severity of the damage and the high number of fatalities were exacerbated by poor construction in the affected areas (United Nations System 2005).

An almost equal number of men and women died in the 2005 Pakistan earthquake. This could be explained by the fact that both male and female adults were either sleeping after taking a meal before sunrise or resting at their homes as the earthquake occurred in the month of Ramadan when Muslims keep daytime fasting. However, the mortality estimates differs by age and gender. According to a UNICEF estimate, 17,000 children died, most of them due to widespread collapse of school buildings since this earthquake occurred on a regular weekday (Saturday) shortly after 8:50 a.m. local time (UNICEF n.d.), while children were in school. Nearly two-thirds of fatalities that occurred among pre- and grade-school children were boys. This was not surprising because school enrollment was higher among boys than in girls. Ultimately, the earthquake orphaned 42,000 children in Pakistan, disabled 23,000 children, and widowed an estimated 17,300 women (UNICEF n.d.).

RESPONSE AND RELIEF

After this event, numerous powerful aftershocks occurred in the affected region of which more than 1,200 had a minimum magnitude of 4.0 or greater; some of these aftershocks were close to 6.0 on the Richter scale (United Nations System 2005). Naturally, the aftershocks hampered the response and relief efforts. In addition, aftershock-induced landslides and falling rocks damaged highways and mountain roads and made large parts of the affected region inaccessible for several days. Besides severe weather conditions, another reason for delayed relief was that the Pakistani military government was not prepared for a disaster of this magnitude. The last earthquake of similar magnitude occurred in 1935, which is known as the Quetta earthquake. However, after initial delay, domestic response to the disaster was extensive, particularly in Pakistan.

The Government of Pakistan set up the President's Relief Fund for earthquake relief operations and established a Federal Relief Commission (FRC) to coordinate the massive rescue and relief operation several days after the earthquake. The commission also took the responsibility to conduct relief operations in collaboration with central and provincial governments and nongovernmental organizations (NGOs). The government issued a flash appeal for international aid three days after the disaster, which was later revised to request some $550 million for humanitarian assistance to earthquake survivors for the following six months (United Nations System 2005). The government requested for blankets, tents, medicine, warm clothes, food, and many more emergency supplies.

The government deployed 50,000 troops into the earthquake-affected northern part of the country, including PAK, to perform search and rescue (SAR) operations and relief distribution. Given the rugged mountain terrain and road blockade by landslides, helicopters were the only effective means to successfully complete

SAR operations and relief distribution. Unfortunately, the Pakistan army had only 15 helicopters, and conveyance of additional helicopters offered to Pakistan was delayed. Therefore, relief items arrived in some remote locations no sooner than one week after the earthquake.

In a display of goodwill, the governments of Pakistan and India opened five crossing points along the line of control (LoC), the military demarcation line between the Indian- and Pakistani-administered parts of Kashmir, to facilitate rescue efforts and the flow of relief goods. However, the Pakistan government refused an offer of Indian army helicopters (Ozerdem 2006). Nevertheless, Pakistan International Airlines (PIA) transported relief goods from all over the world to Pakistan, and Turkey made a similar offer to transport goods. The Pakistan government also provided cash to compensate for loss of life, injuries, and damage to property. Given that the earthquake occurred in early winter, the government built "warm rooms" in the affected areas for the survivors. To reduce the cost, some residential dwellings were converted to such rooms. Consequently, in Indian-administered Jammu and Kashmir, the country's government initiated rescue and disaster relief operations.

Pakistanis from all provinces and walks of life generously donated relief supplies in both cash and kind for the earthquake victims. Given the magnitude of the disaster, the government alone was unable to provide relief to all the people affected by this earthquake. Luckily, the earthquake occurred during the holiest month of the Islamic calendar, when followers of Islam are especially generous to the poor and donate significantly to charities. Overseas Muslim communities, particularly Pakistanis and Kashmiris living in Britain, were also extremely generous in providing emergency aid to the earthquake victims. In fact, British volunteers of Pakistani and Kashmiri origin came to the earthquake-affected areas to distribute emergency relief goods (Ozerdem 2006). In another example, a UK-based charity, Kashmir Relief and Development Foundation (KRDF), set up relief camps in all the major cities of PAK. Their volunteers worked around the clock to provide emergency aid to the earthquake victims.

Like the domestic response, the global response was also delayed initially. In addition to domestic military forces, several foreign forces, including troops of the North Atlantic Treaty Organization (NATO), took part in disaster relief operations. Various international aid and UN agencies participated in the operation and played significant roles in rescue and relief efforts. For example, NATO troops airlifted nearly 3,500 tons of emergency supplies for earthquake survivors in Pakistan and deployed engineers, medical units, and specialist equipment, including helicopters, to assist in relief operations. The supplies included thousands of tents, stoves/heaters, blankets, mattresses, and sleeping bags necessary for protection from the severe cold to nearly 1 million survivors who were sleeping under the open sky in early winter.

NATO also provided medical supplies. Using helicopters, the NATO troops transported the seriously injured from remote areas to the hospitals of Rawalpindi and Islamabad every day. Meanwhile, the NATO field hospital treated nearly 5,000 patients and conducted 160 major surgeries. In addition, its mobile medical units treated some 3,424 patients in the remote mountain villages. These services helped prevent the outbreak of diseases in the earthquake-affected areas. In such ways, NATO and other foreign troops worked closely with the Pakistani army. However, lack of coordination was often a concern for successful completion of the relief mission. This was not unexpected since more than 100 international and national aid agencies participated in search and rescue operation and disaster relief distribution. Foreign aid agencies came from many countries such as China, Japan, Russia, Turkey, and the United Kingdom. These agencies often could not function in timely and efficiently ways because of inaccessibility to the disaster-affected region and severe weather (Ozerdem 2006).

The extensive use of both domestic and foreign military forces in the rescue and relief work was criticized by some politicians as well as national and local Islamist groups, notably the Jamaat-e-Islami (JeI) and the Jamaat-ud-Dawa (JuD). These groups also participated in emergency relief distribution to the survivors. In particular, they established field camps, mobile hospitals, and tent villages in the affected areas. In all, more than 1 million people were reached in the first year after the earthquake (IFRC 2012).

RECOVERY AND REBUILDING

Typically, the relief stage of a major disaster lasts for the first three to six months after an extreme event. Consequently, the official relief phase for the 2005 Pakistan earthquake ended in April 2006 (UNICEF n.d.). Before the end of this phase, the government of Pakistan initiated recovery and rehabilitation activities in the region, first creating the Earthquake Rehabilitation and Reconstruction Authority (ERRA) to support medium- to long-term rebuilding efforts. Several foreign countries such as Turkey, Japan, Saudi Arabia, and United Arab Emirates (UAE) helped in these efforts. For example, Japan prepared a land-use plan for Muzaffarbad city.

The Pakistani government with the support of the UN and international community formulated an early recovery plan with a set of 10 principles: focus on the most vulnerable and socially disadvantaged groups; restore institutional capacities to manage the recovery process; reestablish people's livelihoods; secure human development gains and progress in poverty reduction; reduce disaster risk; engage the civil society and private sector; maintain independence and self-sufficiency; have transparency and accountability; maintain subsidiarity and decentralization; and support aid coordination. The early recovery also emphasized providing cash for work (e.g., rubble clearance), transitional shelter, microfinance schemes for the vulnerable, and capacity building for local governance (UN Systems 2005). The recovery phase of the operation was completed in December 2009, with almost all activities in the earthquake-affected areas finalized (IFRC 2012).

CONCLUSION

The Pakistan Army controlled all postearthquake field activities. Because the military follows top-down orders, some critics complained that military response was slow. However, the Pakistani army surprised everyone with its accomplishments, and without its hard work, the death toll and suffering of those who survived would have been much higher. The 2005 Kashmir earthquake set a record in the history of success in global rescue, relief, and recovery efforts. The cluster approach was applied for the first time following the earthquake. Flash appeal is generally divided into categories or sector, which are called clusters. The government of Pakistan adopted ten sectoral cluster groups to channel and coordinate international humanitarian communities, who would, in turn, channel and coordinate their own efforts. The clusters provided a forum for bringing together a broad range of UN and non-UN humanitarian agencies.

M. Khaledur Rahman and Bimal Kanti Paul

Further Reading

IFRC (International Federation of Red Cross and Red Cross Societies). 2012. *Final Report: Pakistan Earthquake*. Geneva, Switzerland: IFRC.

Ozerdem, A. 2006. The Mountain Tsunami: Afterthoughts on the Kashmir Earthquake. *Third World Quarterly* 27: 397–419.

UNICEF. n.d. UNICEF Pakistan Earthquake Response: Overview. www.unicef.org/pakistan, accessed March 5, 2018.

UN Systems. 2005. *Pakistan 2005 Earthquake: Early recovery Framework with Preliminary Costs of Proposed Interventions*. Islamabad, Pakistan: UN Systems.

Kerala Floods, India, 2018

The state of Kerala on India's west coast is highly vulnerable to several natural disasters. Of these disasters, flood is the most common with 14.5 percent of the state's land area prone to it and almost 50 percent of some districts (administrative units in India, which are equivalent to counties in the United States) vulnerable to floods. Kerala occupies 1.18 percent of land area of India, but it accommodates 2.76 percent of total population. Located in the southwest corner, the state is divided into three geographical regions, east to west: High-, mid-, and lowland. Each of these geographical divisions has distinct economic functions and agricultural produce. Historically, the state has been called the "Spice Coast of India." Typically, Kerala experiences two rainy seasons, the Southwest Monsoon from the end of May to September and the Northeast Monsoon in the middle of October. Even during the monsoon season, the rainfall has interludes with sunny skies, allowing routine activities to continue. Occasionally it rains for hours, but sunshine is not too far away. But in 2018, Kerala experienced a one-in-a-thousand-year flood event.

CAUSE

Kerala experienced unprecedented rainfall between June 1 and August 18, 2018, which resulted in the worst flood in a century for the state, accumulating 92.4 inches (2,346.6 mm) of rainfall, which was 42 percent above normal. In particular, Idukki district received 92 percent above normal rainfall while Palakkad district received 72 percent above normal rainfall. Kerala is drained by 44 rivers; 41 of them are westward flowing while 3 are eastward flowing. All the rivers are entirely fed by the monsoon rains. Not surprisingly, the state has 43 total dams with 35 major reservoirs. During the 2018 rainfall event, the capacity of each was close to full reservoir level (FRL). With no buffer storage to accommodate heavy inflows from August 10, 2018, onward and with exceptionally heavy rainfall continuing, 170 percent above normal in catchment areas, the authorities were compelled to release water from the dams downstream into the rivers. Consequently, the overflowing of all riverbanks led to widespread flooding all over the state.

IMPACT

All 44 rivers that run across Kerala overflowed, and floodwater washed away low-lying areas and riverbanks. Moreover, 10 of the 14 districts reported nearly 342 landslides. The devastating floods and landslides caused extensive damage to houses, roads, and infrastructure; washed away crops and livestock; and affected the lives of millions of people. Ultimately, 12 of the 14 districts were impacted by the floods, and seawater rushed into coastal areas threatening homes (*Hindustan Times* 2018; *Times of India* 2018). Moreover, about 111,197 acres (45,000 ha) of farmland were submerged under water. Meanwhile, the raging waters claimed more than 350 lives in 15 days, destroying 27,000 houses, displacing over 100,000 people, and damaging 60,000 miles (96,561 km) of roadways and 134 bridges. This disaster is classified as a Level 3 disaster, the highest in the country. The only

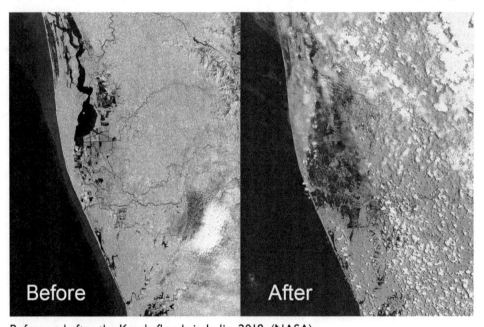

Before and after the Kerala floods in India, 2018. (NASA)

disaster that Kerala probably experienced on a similar scale was in 1924, but the lessons learned were lost in the intervening years

Kerala suffered huge economic losses because of the floods. Conservative estimates suggest approximately 2.6 percent of Kerala's gross state domestic product (GSDP) got washed away by floods. The havoc threatened the agriculture and tourism industries. Kerala authorities estimated that the state's roads, public buildings, and other infrastructure suffered a loss of $2.7 billion. According to the World Bank report, in the aftermath of the calamity, upon the request of the Ministry of Finance of the Government of India, a Joint Rapid Damage and Needs assessment (JRDNA) was conducted in September 2018 by the Kerala Government, the World Bank, and the Asian Development Bank. The estimated recovery needs amounted to approximately $3.56 billion for priority sectors alone. The assessment identified transportation, rural and urban infrastructure, tourism, and housing to be the worst affected sectors (PDNA 2018).

The impact of disasters went beyond the physical damage to homes and infrastructure in that it stressed both affected and unaffected populations. In addition to assistance for physical damages, communities and individuals also needed psychosocial support.

RESPONSE

Any emergency requires a quick response, and typically the local administration is the first responder with assistance from the state government. Since 2005, India has had in place an institutional and coordination mechanism for effective disaster management at local, state, and federal levels. In 2016, India adopted the National Disaster Management Plan (NDMP) as an action framework designed to be consistent with United Nation's Sendai Framework for Disaster Risk Reduction 2015–2030. According to the NDMP, disaster management is considered the responsibility of the state governments, meaning they are primarily responsible for undertaking rescue, relief, and rehabilitation measures in the event of a natural disaster. The damage assessment and all phases of recovery and reconstruction, short- to long-term, are the responsibility of the state governments with the federal government supplementing their efforts with logistical and financial support during severe disasters as requested by the state government. Following the 2018 flood, the state government's Disaster Management and Mitigation Unit (DMMU) formed a Rapid Spot Assessment team that visited relief camps and villages and collaborated with local partners (Christian Agency for Rural Development (CARD) and Marthoma Church of India (MCI) social wing) to identify a base from which to work. Relief packages for households included hygiene kits, kitchen utensils, disinfectants, bleaching powder, and sanitation brushes (Reliefweb 2018).

Postflood, short-term, and long-term rehabilitation plans and programs were the biggest challenges the state government faced. The state police, fire force, and armed forces engaged in rescue operations to house a total of 1,028,000 people in 3274 relief camps (numbers vary depending on the source) in the flood-hit districts (*Times of India* 2018). Two immediate scares were, first, the spread of water-borne diseases, communicable diseases, and dead animal carcasses and, second, unstable structures that could injure individuals. As people started returning to their flood-hit homes, many faced the tough task of repairing their homes, including restoring electricity, plumbing, removing the dirt brought by the water, and even dealing with reptiles such as poisonous snakes. Several states sent tons of relief material to Kerala. Moreover, individuals sent monetary help and material to the Chief Minister's Disaster Relief Fund (CMDRF). For instance, the state of Maharashtra sent 30 tons of aid including packaged food, milk powder, blankets, bed sheets, and clothes. The armed forces including army, navy, and air-force joined in the search and rescue and relief operations. The Indian Air Force arranged for medical assistance in relief camps and set up a Rapid Action Medical Team in one of the worst impacted districts. Indian Railways transported over 20,00,000 liters of water in special water wagons along with arranging delivery of around 200,000 bottles of drinking water.

Kerala's economy depends extensively on fishing, agriculture, tourism, and on remittances from guest workers. This state has a huge diaspora outside the country, and thus, the flood attracted the attention of both diaspora and government humanitarian agencies in foreign countries. In fact, international aid came pouring in as soon as the world got news of Kerala's flood. For instance, Kerala's contribution to the economy of the Middle East is undeniable with 80 percent of workers in United Arab Emirates (UAE) being Keralites. To this point, the federal government allocated $100 million, and the Kerala state government and other states contributed $30 million. Additionally, the United Arab Emirates offered $100 million for flood relief aid while Qatar offered $500,000 and Maldives $50,000 (Rai n.d.).

But the Indian Government declined help from these nations, although the Kerala Government was keen to accept donations from such sources, especially the UAE. As a matter of policy, the Government of India does not issue any appeals for foreign aid following a disaster and has not done so for 18 years since the then Prime Minister Manmohan Singh declared it. In 2005, this stand was adopted as policy, and it sparked debates during the recent Kerala floods. In fact, the chief minister of Kerala said the state government had planned to approach the federal government about allowing donations. This federal-state conflict is quickly turning into a question of apathy and ignorance. On the one hand, the policy reflects India's desire to change its international image from that of a poor country to one capable of dealing with its own problems, while on the other hand, many question whether the government is being indifferent to and ignorant of Kerala's needs.

REHABILITATION

The rains finally stopped on August 20, 2018, and the water started receding, allowing displaced people to return home (where possible). Finally, search and rescue operations were called off on August 22 after over 20,000 stranded people were evacuated. As people started moving back into their homes, the state moved from the relief to the recovery phase of the disaster. At this time, the Government of Kerala commissioned the Post Disaster Need Assessment in collaboration with international development partners such as the European Union, the World Bank, and the UN to effectively assess the damage and loss and to estimate recovery needs. After the flood, the immediate challenge was to clean the homes, prevent water-borne diseases, and rehabilitate impacted populations. One of the major issues for the people on returning home was witnessing the absolute devastation caused by over 1.6 feet (50 cm) of mud within their houses. Additionally, families were at risk of secondary disasters such as mudslides and collapsing structures due to prolonged inundation. Accordingly, health authorities warned people and prepared defense forces against the spread of diseases in the state.

Implementing reconstruction and rehabilitation measures is a long and expensive process. The World Bank extended its support of up to $500 million to the Government of Kerala for its recovery efforts and for building a resilient future. Under the proposed plan, the World Bank would help the Government of Kerala financially and by building resilient institutions, systems, and infrastructure, in accordance with the government's aim of "Build Back Better." The proposed recovery and resilience program by the JRDNA had the following phases of development. The first phase helped the state to raise funds from domestic and international sources; phase two included policy and institution strengthening for water resources, transportation, social protection, environmental protection, and disaster risk reduction (PDNA 2018). In phase three, the government itself would cautiously reconstruct the houses, restricting, for instance, construction of houses in areas that had experienced landslides to avoid future damage due to unstable slopes.

CONCLUSION

Floods are the most common natural hazard in Kerala, and a disaster of the 2018 magnitude can take up to a decade for full recovery. Kerala suffered huge economic losses on account of these floods, whose magnitude has underscored the need for research and knowledge-generation activities. In the future, the availability of state-of-the-art geographical information system (GIS) technologies can help create maps and disseminate information to local government agencies and authorities, enabling effective early-warning systems and risk information communication to reduce loss of life and property.

Avantika Ramekar

Further Reading

Hindustan Times. 2018. The Worst Is Over: Kerala Shifts Focus to Rehabilitation, Rebuilding Flood-hit Areas. August 24, 2018. https://www.hindustantimes.com/india-news/the-worst-is-over-kerala-shifts-focus-to-rehabilitation-rebuilding-flood-hit-areas/story-wFAyULgRQL09S5iZhCFAfP.html, accessed February 12, 2019.

PDNA (Post-Disaster Needs Assessment). 2018. Kerala Post Disaster Needs Assessment Floods and Landslides-August 2018. https://www.undp.org/content/dam/undp/library/Climate and Disaster Resilience/PDNA/PDNA_Kerala_India.pdf, accessed March 23, 2019.

Rai, S. n.d. Kerala Flood: Repercussions of the Institutional Dysfunctionalism. https://www.academia.edu/38660937/Kerala_Flood_Repercussions_of_the_Institutional_Dysfunctionalism, accessed March 8, 2019.

Reliefweb. 2018. Overall Update on Kerala Flood Relief 2018. September 3, 2018. https://reliefweb.int/report/india/overall-update-kerala-flood-relief-2018-3rd-september-2018, accessed September 9, 2018.

Times of India. 2018. Kerala Floods: Over 1 Million in Relief Camps, Focus on Rehabilitation. August 21, 2018. https://timesofindia.indiatimes.com/india/kerala-floods-over-1-million-in-relief-camps-focus-on-rehabilitation/articleshow/65478086.cms, accessed January 23, 2019.

Kobe Earthquake, Japan, 1995

The Kobe earthquake, also referred to as the Great Hanshin or Hanshin-Awaji earthquake, occurred on January 17, 1995, at 5:46 a.m. local time, in the Osaka-Kobe (Hanshin) metropolitan area of western Japan, which is located in the southern part of Hyogo Prefecture. With 11 million inhabitants, this densely populated metropolitan area is Japan's second largest urban area. This urban earthquake registered a magnitude of 7.2 on the Richter scale, and it was among the strongest, deadliest, and costliest to ever strike that country. It lasted for 14–20 seconds, and its epicenter was located near the northern edge of Awaji Island in the Inland Sea, only 12.5 miles (20 km) south of the coast of the port city of Kobe, which is one of the largest container ports in the world. The hypocenter was shallowly situated in a depth just about 10 miles (16 km) below the earth's surface. The tectonic plate boundary between the Philippine Sea plate and the Eurasia plate was about 133 miles (200 km) away from the epicenter (Bolt 2004).

CAUSE

Most of the major earthquakes in Japan are caused by subduction of either the Philippine Sea plate or the Pacific plate, with release of energy within the subducting plate or the accumulation and sudden release of stress in the overlying plate. The Kobe earthquake of 1995 was a third type, known as an "inland shallow earthquake." Four foreshocks preceded the earthquake, and 50 aftershocks of magnitude 4.0 or higher followed within the week. Kobe is located further away from a plate margin than other cities in Japan, and thus, this devastating earthquake was not directly connected to the nearby subduction zone of the Philippine Sea plate. Instead, the earthquake was generated along a local active fault system formed by the intersection of the Nojima fault with the Suma fault (Koketsu et al. 1998).

The Nojima and Suma faults cut across Awaji Island and run from the coast through Kobe city to Itami city, which is located just north of Osaka city. Because the fault shifted, this great earthquake occurred with sharply enhanced shaking as the Nojima fault rapture headed straight toward and then underneath Kobe. After the earthquake, the actual fault appeared on the surface of the ground in Awaji city. Two parts of the ground slipped about 3.3–6.6 feet (1–2 m) from each other, and one part rose 1.6–3.5 feet (0.5–1.2 m). The length of the ruptured Nojima Fault extended horizontally for up to 33 miles (50 km) (Bolt 2004). This fault was later designated a natural monument, and a museum has been built on part of the fault.

DAMAGE AND DESTRUCTION

The 1995 Kobe earthquake surprised many Japanese seismologists because they had not identified the Nojima Fault as an imminent threat and thus didn't predict the earthquake. Japan is known globally for its comprehensive earthquake prediction program with thousands of seismometers and monitoring stations across the country. However, an earthquake had not occurred in Kobe in 400 years, and so the city contained less prediction equipment than other more prone areas of Japan. While mild earthquakes were common in the Kansai region where Kobe is located, nonetheless, no one expected an earthquake of this magnitude.

The 1995 Kobe earthquake is one of the largest earthquakes to strike a dense urban area. In Japanese history, the Kobe earthquake is the second deadliest earthquake after the 1923 Great Kanto earthquake, which impacted Yokohama-Tokyo and killed almost 100,000 people and injured 150,000 others (Kobayashi and Kazuki 2005). The Kobe earthquake caused the deaths of 6,400 people, wounded 40,000 people, made more than 300,000 people homeless, and damaged in excess of 240,000 homes. Millions more homes in the affected region lost electricity and water service. Consequently, many homeless people spent days on the roads of Kobe out of fear of aftershocks. They tried to keep warm against the 36-degrees Fahrenheit (2.2-degrees Celsius) cold by making fires on the sidewalk with pieces of their damaged or destroyed homes and furniture.

Great Hanshin earthquake ruins in Kobe, Japan, 1995. (Gyro/iStockphoto.com)

A one-mile (1.6 km) long portion of the Hanshin Expressway, linking Osaka and Kobe, which was built to withstand an 8.1 earthquake, was completely destroyed, and other parts of the expressway were heavily damaged during the earthquake. Moreover, trains derailed so that only 30 percent of the rail lines on the Osaka-Kobe rail network were operational. Additionally, emergency assistance vehicles needed to use the damaged roads, but many of the roads were destroyed during the earthquake. Raised motorways collapsed, and other roads were damaged, limiting rescue, search, and relief operations. Many small roads were inaccessible because of fallen debris from buildings or cracks and bumps caused by the earthquake. Most collapsed freeways were built in the 1960s and 1970s, and after their construction, no retrofit was done to strengthen these freeways (Bolt 2004).

Telephones and other communication services were put out of action, making communication slow and difficult. Electricity and water supplies were badly damaged over large areas. Many factories, commercial buildings, and bridges were damaged. Twenty of the 150 wharfs built parallel to the shoreline were destroyed. Because the earthquake occurred early morning, regular train service had not begun, saving many lives. However, 7,100 dwellings burned down because the earthquake occurred when people were cooking breakfast, causing over 300 fires, which took over two days to extinguish. Broken gas pipes also caused fires, and road damage and lack of adequate water hampered firefighting capabilities (Bolt 2004).

With 4,571 fatalities and more than 14,000 injured, the earthquake hit Kobe the hardest, and thus the event was named after this city. The earthquake damaged more than 120,000 of about 500,000 structures in the city, and more than half of those fully collapsed. The central part of Kobe was the worst affected area, including the main docks and port area. This area is built on soft and easily moving rocks, especially the port itself, which is built on reclaimed land. Here, the ground actually liquefied due to intense shaking of water-saturated sediments and became like thick soup, allowing buildings to collapse, sink, or topple sideways, and disabled the port, which in turn crippled the Japanese economy for months. Only about 20 percent of buildings in the downtown area of Kobe were useable (Bolt 2004). So many buildings collapsed because the city was directly in the path of fault orientation and rupture direction, which produced intense shaking and consequently severe damage.

Many buildings in Kobe were constructed prior to the 1960s and were made primarily of wood, which caught fire, causing widespread structural collapse. On the other hand, buildings, schools, and factories built after the 1960s in Kobe, as well as in the rest of Japan, are earthquake proof (necessary by law) with counterweights on the roofs and cross steel frames. This is particularly true for new buildings and roads constructed 20 years prior to the earthquake. However, many older buildings either collapsed or caught fire. The traditional houses in the eastern part of Kobe, typically characterized by their wooden columns and heavy roofs, were reduced to dust, and nine in every 10 deaths recorded in this part of the city were caused by failure of these heavy roofs during the earthquake.

The Kobe earthquake seriously impacted the Japanese economy with the damage amounting to an estimated $102.5 billion. A major decline in the Japanese

Stock Markets followed (Bressan 2012). For instance, on the day following the earthquake, the Nikkei 225 index plunged by 1,025 points. It took nine months for the index to return to its pre-earthquake level. Many residents of the affected areas also lost their jobs for varying lengths of time, and private consumption fell by about 5 percent after the earthquake and did not return to normal levels for four months.

RESPONSE

It is well known that Japan is an earthquake-prone country, which is why its modern buildings and other infrastructure are earthquake proof. However, the Kobe earthquake proved that the country's superior earthquake-resistance construction was inadequate. Numerous supposedly earthquake-resistant buildings did not withstand the earthquake, particularly older wooden frame houses. Rail lines, elevated highways, and other transportation networks were either completely or nearly completely paralyzed, and the earthquake also exposed the inadequacy of national earthquake preparedness. The Japanese government and the Kobe city authorities were heavily criticized for their slow and ineffectual response, as was the government's initial refusal to accept international offers of emergency aid. Fire departments, police departments, and the self-defense force operated independently, which implied lack of coordination among emergency responders.

Despite overall slowness of public response, homeless people were provided for reasonably quickly. Similarly, patients were successfully evacuated and transported from the affected hospitals in the disaster area to backup hospitals. Unfortunately, the managing capacity of the regular emergency medical care system was overburdened because major roads and freeways, life-lines (including water, gas, and electricity), and many hospitals were severely damaged. Thus, many affected hospitals remained completely isolated for days. Moreover, there was no coordinated effort to monitor patient flow from one hospital to another. Because of a lack of any central disaster management agency in Japan, such as the Federal Emergency Management Authority (FEMA) in the United States, emergency responders such as fire departments, police departments, and the Self-Defense Force operated independently without any coordination.

During the first three months after the earthquake, around 1.2 million volunteers participated in the relief work. No other previous earthquake had received so many volunteers to provide emergency services to disaster survivors in Japan. For example, several big retail firms stepped in and used their existing supplies to provide basic amenities to those affected by the event. Similarly, mobile networks such as NTT and Motorola provided free telephone service to the earthquake survivors.

RECOVERY

Water, gas, electricity, and telephone services were fully working by July 1995, and the railways were back in service by August 1995. However, clean freshwater was in short supply until April 1995. A year after the earthquake, 80 percent of the port in Kobe was working, while less than 15 months afterward, manufacturing activity in greater Kobe was at 98 percent of its estimated prequake level (*The Economist* 2011). By January 1999, 134,000 housing units had been constructed, but some people were still living in temporary houses. The Japanese government did not immediately pass any law to make buildings and transport structures even more earthquake proof, but more seismic instruments were installed in the affected areas to monitor earthquake movements.

Within a decade, Kobe fully recovered from the disaster, but the Japanese government had to spend a huge amount of money to rebuild the city, approximately $6.8 billion on rebuilding the port and $3 billion to put the freeway back in place. Roads, bridges, and buildings were reinforced against another earthquake, and the national government revised its disaster response policies. The Hyogo prefectural government also devised an emergency transportation network and set up evacuation centers and shelters in Kobe.

CONCLUSION

The magnitude of the 1995 Kobe earthquake can be considered moderate, but it was a highly destructive event because of the earthquake's shallowness, and it occurred along active faults. The huge loss of life in the Kobe earthquake can be attributed to two causes: firstly, it occurred in one of the most populated areas of the world, and secondly, it took everyone by surprise. The Kobe earthquake was a major wake-up call for the Japanese government as well as Japanese disaster prevention authorities. After this event, additional efforts were initiated to improve the accuracy of earthquake forecasting, and building codes were also improved to facilitate structures that could withstand such natural disasters. Finally, the Japanese government improved the already strong guidelines for responding quickly in the case of future extreme events.

Bimal Kanti Paul

Further Reading

Bolt, B. A. 2004. *Earthquakes*. New York: W.H. Freeman and Company.

Bressan, D. 2012. January 17, 1995: The Kobe Earthquake and Early Antiseismic Architecture. *Scientific Americans*. January 12, 2012. https://blogs.scientificamerican.com/history-of-geology/january-17-1995-the-kobe-earthquake-and-early-antiseismic-architecture/, accessed May 10, 2018.

The Economist. 2011. Economics Focus: The Cost of Calamity, March 19–25. https://www.economist.com/weeklyedition/2011-03-19, accessed May 10, 2018.

Kobayashi, R., and K. Kazuki. 2005. Source Process of the 1923 Kanto Earthquake Inferred from Historical Geodetic, Teleseismic, and Strong Motion Data. *Earth, Planets and Space* 57(4): 261.

Koketsu, K., S. Yoshida, and H. Higahihara. 1998. A Fault Model of the 1995 Kobe Earthquake Derived from the GPS Data on the Akashi Kaikyo Bridge and Other Datasets. *Earth Planets Space* 50: 803–811.

Loma Prieta Earthquake, California, 1989

On Tuesday October 17, 1989, around 5:00 p.m., an earthquake of magnitude 7.1 on the Richter scale struck central California, including the San Francisco Bay area. The quake was made famous for occurring immediately prior to Game 3 of that year's Major League Baseball World Series between the San Francisco Giants and the Oakland Athletics, which was taking place at Candlestick Park in San Francisco. The epicenter of this earthquake was on the San Andreas Fault, which is roughly 10 miles (15 km) northeast of Santa Cruz and 56 miles (84 km) south of San Francisco. Using the tradition of naming earthquakes after the most prominent geographic feature near the epicenter, the earthquake was named Loma Prieta—for the highest peak in the Santa Cruz Mountains at an elevation of 3,790 feet (1,160 m). However, the earthquake is also known as the World Series or San Francisco-Oakland earthquake.

The Loma Prieta earthquake was felt as far away as western Nevada and San Diego. It occurred in a segment of the San Andreas Fault zone, which had been designated by geologists as a seismic gap, meaning a section of a fault that had produced earthquakes in the past but was then quiet. Such a gap is useful in forecasting future earthquakes. For example, several days before the Loma Prieta earthquake, scientists recorded changes in the magnetic field and in radio signals in the Santa Cruz Mountains. In fact, the U.S. Geological Survey (USGS) predicted an earthquake of magnitude equal to or greater than 6.5 in the area because of the lack of a seismic event along the San Andreas Fault since 1906—the year when an earthquake of 8.3 magnitude struck the San Francisco area.

SELECTED PHYSICAL/GEOLOGIC CHARACTERISTICS

Though the earthquake lasted only for 15–20 seconds, the Loma Prieta earthquake was one of the worst to have hit the San Francisco Bay area. A slip along the San Andreas Fault, which forms the boundary between the Pacific and the North American Plate, caused the quake. Relative to the land on the east, the land to the west of the fault had been moving in a northwest direction at an average rate of two inches (5.1 cm) per year. During the 1989 earthquake, the Pacific Plate moved 6.2 feet (20 m) to the northwest and 4.3 feet (14 m) upward over the North American Plate. Unlike the typical California earthquake focal depth of 4–6 miles (6–9 km), the Loma Prieta earthquake was recorded at a depth of 11 miles (16 km).

Following the main shock, an aftershock of 5.2 magnitude occurred about 2.5 minutes later. A total of 20 aftershocks of magnitude 4.0 (or greater) and more than 300 aftershocks of magnitude 2.5 (or greater) were recorded. The main earthquake and its aftershocks triggered more than 1,000 landslides and rockfalls. These occurred in the epicentral zone in the Santa Cruz Mountains. One landslide disrupted traffic on the State Route 17 for over a month. The aftershock zone extended 25 miles (37 km) from south of Watsonville near Highway 101 to north of Los Gatos near Highway 17.

The aftershocks map shows that part of the San Andreas Fault ruptured in the main earthquake, which began at a depth of 13 miles (19 km), stopped at a depth of 2 miles (3 km), and stretched about 26 miles (42 km). It ruptured the southernmost part of the break that had caused the 1906 San Francisco earthquake, and rupturing lasted only 11 seconds. This duration was unusually short for a 7.1-magnitude earthquake. If rupturing had lasted for 20–30 seconds, the expected duration for a 7.1-magnitude earthquake, numerous other structures would have been destroyed (Abbott 2008). Furthermore, the rupture direction was bilateral; that is, the fault rupture proceeded from the focus in both directions along a fault strike, away from the source to the northwest and southeast (Brumbaugh 1999; Bolt 2004).

The 1989 Loma Prieta earthquake is a good example of the complex nature of slip on a fault. Most of the slip occurred between a depth of 6 and 11 miles (9 and 16 km) and was concentrated in two main areas to the northwest and southeast of the focus (Brumbaugh 1999). Slip is the relative motion of one face of a fault relative to the other. Thus, the Loma Prieta earthquake in 1989 uplifted the west side of the Santa Cruz Mountains about 3.3 feet (1 m) (Bolt 2004). Similar to the subsequent 1994 Northridge earthquake, it did not produce fault scarps (a cliff or steep slope formed by displacement of the ground surface). However, liquefaction was one of the worst effects of the Loma Prieta earthquake. Notable liquefaction occurred in the Marina district of San Francisco Bay, along the Pacific coastline with sandy soil and near the epicenter (Bolt 2004). Liquefaction is caused by constant shaking of ground. Given this shaking in conjunction with the weight of the buildings, water is squeezed out of the soil and thus causes the soil to temporarily develop a liquid consistency, which causes buildings to topple and collapse.

DEATH AND DAMAGE

Although the Loma Prieta earthquake released energy over 30 times less than that of the 1906 San Francisco earthquake, the damage caused by the former earthquake was one of the worst in the history of the country. It killed 63 people, injured nearly 3,800 others, and caused an estimated $6 billion in property damage. It left 8,000 people homeless (Bolt 2004; Abbot 2008). Of the total deaths, 41 died because of collapse of a one-mile segment of the double-decked Interstate 880 freeway (also called the Cypress Viaduct) in Oakland, where travelers were trapped in cars and trucks between levels of the roadway. Inadequate reinforcing of the freeway caused these deaths. However, the death toll is considered to be relatively low for two reasons. First, the source area of the earthquake was sparsely populated because of its rugged topography. Residents there experienced severe shaking but not many fatalities. Also, timing kept the death toll low as the earthquake occurred at about 5:00 p.m. A lot of people drive back to their homes using freeways around that time. However, because of a World Series game, the freeways were relatively empty. Most people left work early to watch the game at home on television. If there had been no game at 5:00 p.m. on October 17, collapse of freeways in rush hour would have killed many more.

Collapsed and burned buildings in the Marina District of San Francisco, California, following the 1989 Loma Prieta earthquake. (U.S. Geological Survey)

More than 1,300 houses were completely destroyed, and 20,000 houses were damaged. Similarly, 400 businesses were destroyed, and an additional 3,500 were damaged (Bolt 2004). The most significant damage to buildings occurred along the waterfront in San Francisco, called the Marina district. The structures in this district were built on bayfill materials consisting of unconsolidated sand and mud. These loose materials amplified the ground shaking owing to their poor elastic response, and thus, they suffered a great deal of ground shaking and consequently caused severe damage to buildings. Here also fires broke out early in the evening of October 17.

However, the city of Watsonville, located about 13 miles (20 km) southeast of Santa Cruz, and older buildings in downtown Santa Cruz were badly damaged. About 90 percent of structural damage in Watsonville was because of the failure of unreinforced masonry buildings and wooden structures that were not properly bolted to foundations. There was one death in this city. The Pacific Garden Mall of Santa Cruz was virtually destroyed. Four people died here as did five in the Marina district of San Francisco. The city of Oakland also experienced considerable damage. Many old masonry structures throughout the city suffered structurally. In all, 18 bridges were closed for several days after the earthquake (Bolt 2004).

The San Francisco Bay Area was distant from the epicenter, but the loose sediments magnified the effects of ground shaking and thus caused considerable damage at large distances from the source (Brumbaugh 1999). However, the rocky parts of the hills around the Bay experienced much less damage than did the Marina district. There, the horizontal shaking of the ground lasted for a second or two, while the district and other soft soil areas of the San Francisco Bay were subject to three times the forces for five times the duration. Thus, pockets of San Francisco and Oakland experienced Modified Mercalli (MM) intensities of IX while other earthquake-affected areas experienced MM intensities of VII and VIII (Bolt 2004). However, widespread disruption of transportation, utilities, and communication caused economic hardship of the people in all of the earthquake-affected areas.

RESPONSE AND RECOVERY

In the first hour, power outages, a breakdown in communication systems, and release of hazardous materials slowed public response and delayed moving critical emergency staff from home to business. Moreover, much of the telephone system was not working. Despite the problems, the Oakland Fire and Police Departments and several area ambulance providers responded with supplies and resources to the Cypress freeway collapse. Other states and nearby affected city emergency agencies and personnel helped evacuate many victims from the freeway and transported them to local hospitals, primarily by helicopters. People of the earthquake-affected areas also evacuated on their own and took shelter with relatives and friends. They did so not because of physical damage caused by the earthquake, but from fear of aftershocks and further damage and concern about the safety of their children. In addition to trained search and rescue (SAR) teams, local volunteers participated in search and rescue operations while trained paramedics and other health workers began administering medical care. By the morning of October 18, 1989, primary SAR operations were completed on the Cypress freeway (USDI 1994).

Victims of the Loma Prieta earthquake received emergency and recovery assistance not only from state and federal government programs but also from numerous religious groups and volunteer organizations. Low-income households and residents of Santa Cruz Mountains did not have earthquake insurance and were aided only by nonprofit volunteer organizations. For this reason, public agencies, such as the Federal Emergency Management Authority (FEMA), were criticized. According to a 2006 report published by the Public Policy Institute of California (PPIC), insurance companies paid approximately $570 million to earthquake survivors. The average claim was valued between $9,000 and $18,000 while FEMA and the Small Business Authority (SBA) provided loans and grants totaling over $560 million. In addition, the California Disaster Assistance Program (CALDAP) provided $44 million in loans to owners of rental property and $43 million to homeowners for repair and rebuilding. In addition, the American Red Cross donated $13 million for housing recovery after the Loma Prieta earthquake.

The residents of central California raised many questions after the 1989 Loma Prieta earthquake, one of which was: How safe are California's freeways and bridges? The government of California conducted several inquiries and concluded that with the exception of several main freeways, others were seismically safe. Subsequently, the government issued an executive order in which seismic safety was given priority consideration in the allocation of resources for transportation and for construction of all state and public buildings (Bolt 2004). After the earthquake, the state's Office of Emergency Services has established a heavy search and rescue capability sector and installed a new satellite communications system, which links key state agencies with county emergency operation centers (Tobin 1994).

After the Loma Prieta earthquake, the California legislature considered over 300 seismic bills. In addition to a prepaid residential recovery fund, the legislature created a new seismic hazard mapping program and an earthquake deficiency disclosure requirement for residences and certain commercial buildings. The legislature also required the California Department of Transportation (CALTRANS) to retrofit existing vulnerable bridges (Tobin 1994). Similar measures were also taken by the cities of San Francisco, Los Angeles, Berkeley, and Oakland to retrofit certain buildings and improve emergency response capabilities. Statewide, many cities and counties accelerated their efforts to comply with the state law, requiring the identification of unreinforced masonry buildings, the notification of owners, and the adoption of mitigation programs. On their own initiatives, private sector businesses throughout the state have acted to retrofit their buildings, diversify their production, and prepare for their emergency response and business recovery.

CONCLUSION

After the Loma Prieta earthquake, scientists predicted a 50 percent chance for one or more earthquakes (of magnitude 7.0) in the San Francisco Bay Area in the next 30 years. Will the lessons learned and experienced gained from the 1989 earthquake be enough for tackling such a future event? In fact, involvement of disaster managers and relevant public officials in past earthquakes in California did prove to be successful in limiting the aftermath effects of the Loma Prieta event. This earthquake was invaluable for them to prepare a more effective response, to enhance preparedness, and to improve mitigation efforts for future events of similar or higher magnitude.

Bimal Kanti Paul

Further Reading
Abbott, P. L. 2008. *Natural Disasters*. Boston: McGraw Hill.
Bolt, B. A. 2004. *Earthquake*. New York: W.H. Freeman and Company.
Brumbaugh, D. S. 1999. *Earthquakes: Science and Society*. Upper Saddle River, NJ: Prentice Hall.
Tobin, T. 1994. Legacy of the Loma Prieta Earthquake: Challenges to Other Communities. In *Practical Lessons from the Loma Prieta Earthquake*, edited by the National Research Council (NRC), 19–27. Washington, DC: NRC.
USDI (United States Department of Interior). 1994. *The Loma Prieta, California, Earthquake of October 17, 1989—Fire, Police, Transportation, and Hazardous Materials*. Washington, DC: USDI.

Mexico City Earthquakes, Mexico, 1985

Geography and allied disciplines provide an ideal lens through which to understand the impact of disasters. Its attention to the connections of space, place, environment, and society provides a broad view of the causality and outcomes before, during, and following disasters. When considering the catastrophic September 1985 earthquake that hit Mexico City, such an event initially appears as a geologic concern caused by the inevitable forces of plate tectonics. Yet with the addition of knowing that the earthquake epicenter was over 200 miles (300 km) away, a more comprehensive, geographical approach is warranted to understand how land-use changes and unique environmental factors resulted in the deaths of an estimated 10,000–15,000 people, 14,000–20,000 injuries, and between three and four billion dollars in damages. Additionally, the repercussions of the 1985 earthquake, when contextualized with the political and economic circumstances of Mexico in 1985, precipitated great social and political change across the country.

PHYSICAL AND URBAN LANDSCAPES OF MEXICO CITY

The Valley of Mexico, the geophysical region that largely encompasses the Mexico City metropolitan area, is seismically active. Additionally, owing to long-term population dynamics and historical land-use changes, the geologic earthquake hazard is enhanced. On a large scale, this highland valley is located in the southern portion of the Trans-Mexican Volcanic Belt (TMVB), a physiographic region that runs in an east-west trending, 667-mile (1,000 km) long belt from approximately 18.5°N to 21.5°N latitude across Mexico from its Pacific to Gulf Coast. The TMVB is the result of the subduction (downward diagonal movement) of the medium-sized Cocos and tiny Rivera plates under the southern edge of the massive North American Plate. Rising magma and crustal uplift on the leading edge of this subduction has made the TMVB a volcanic region for the past 20 million years, and it remains active even today (Anderson et al. 1986).

Prior to the colonial and modern eras, a system of lakes covered the Valley of Mexico. For the past 500 years, efforts by colonial governments and the modern state of Mexico to control flooding and to reclaim land have drained the majority of lakes, leaving a highly plastic and seismic-wave enhancing sedimentary foundation for much of the city. As of the 2010 census population estimates, the entire Mexico City metropolitan area, extending roughly 5,333 square miles (8,000 square km), held a massive 20.4 million people, about one-sixth of the entire population of Mexico. Like today, this area has long been a focus of human settlement since the first arrivals of human populations in the region, about 12,000 years ago, through the first instances of complex agricultural societies leading to the Aztec Triple Alliance (1420–1521) that held political and economic dominance over the region upon contact with Europeans.

At time of the arrival of the Spaniards, conservative population estimates for the Valley of Mexico sit at around 1 million people. Following the conquest by the Spanish, led by Hernán Cortés, the ancient Aztec capital of Tenochtitlán was converted into the Spanish vice regal capital of Mexico City, serving as the head of the viceroyalty of New Spain for 300 years, from 1521 to 1821. Following a catastrophic decline of the native population of the Valley of Mexico, largely due to disease, warfare, and colonial abuses, estimates indicate only 10 percent of the roughly 1 million people there resided a century after Cortes's conquest. The population only bounced back to its original million by the first decade of the twentieth century, leading to the roughly 14 million that lived there in late 1985.

Owing to the long occupation of intensive human settlement in the Valley of Mexico, populations have lived with the hazard and risk of earthquakes there. The largest earthquake in the modern record to strike Mexico City before the 1985 events was the one in July of 1957, which registered a magnitude of 7.9, resulting in widespread damage and over 300 deaths, and another 7.6-magnitude event in March of 1979, which resulted in fewer than 40 deaths but did considerable damage. While these two prior historical events did progressively impact building codes and emergency preparedness in the Mexico City region, such preparations failed to account for the strength and persistence of the 1985 events.

GEOLOGIC EVENTS OF SEPTEMBER 19–20, 1985

In general terms, the 1985 Mexico City Earthquakes is often referred to as one single event. Yet the human-environmental disaster was a combination of two earthquakes, an 8.1-magnitude event on September 19 and another 7.5-magnitude event just 36 hours later on September 20, with tremors between them and afterward. The first earthquake began just before rush hour at 7:14 a.m. on Thursday, September 19, 1985. The epicenter of the earthquake occurred approximately 233 miles (350 km) to the southwest of Mexico City, on Mexico's Pacific Coast in Michoacan State. The area where the earthquake occurred was a well-known active continental plate boundary where the Cocos Plate, wedged between the Pacific Plate to its west and the North American Plate to its east, subducts or is pushed underneath the North American plate at about a 15-degree angle. Specifically, the area where the earthquake originated is known as the Michoacan seismic gap. High-resolution seismic data pointed to the Michoacan Gap as a potential source of major earthquakes before the September 1985 events. The September 19 earthquake was categorized as a subduction thrust or slip-dip earthquake, or one where the upper plate is violently thrust upward over the subducting one. While the area on the Pacific Coast closest to the epicenter experienced intense movement, the horizontal acceleration energy decreased significantly as it passed through the hard rock over the 233 miles (350 km) to Mexico City (Anderson et al. 1986).

However, as the shockwaves passed through the soft lake sediments of the Valley of Mexico, their acceleration and amplitude increased. Recent modeling of the vibration of the sediments under much of Mexico City indicates such surface conditions can amplify the shaking up to 100 times (Cruz-Atienza et al. 2016). Additionally, the unique properties of the lake sediments can cause shaking to endure nearly 300 percent longer than the surrounding hard rock areas. The most powerful and persistent shaking, some of which lasted over three minutes, affected the portions built on lake deposits (Chandler 1986).

DEATHS AND DESTRUCTION

A definitive tally of the cost in terms of human lives of the 1985 Mexico City Earthquake does not exist due to methodology, the chaos of counting the dead in a disaster zone, and political viewpoints of the accounting agencies. In all reality, due to the immense destruction across one of the most densely populated places on the planet, any count is nothing more than a vague estimate. Bodies can be miscounted, informally recovered, or lost completely to the rubble, not to mention the attribution of secondary causality to deaths that happened less directly from the destruction, such as precipitated health effects due to chronic illness and the lack of health care during a disaster. Thus, a range is often given to quantify the toll in lives for the earthquakes. Many estimates place the number of lives lost to these events between 10,000 and 15,000. Estimates of the injured and hospitalized are also disputed. A ballpark figure would be between 14,000 and 20,000, but any real figure is truly unknown or unknowable (Davis 2004).

The physical damage was most pronounced in areas where the shaking intensity was the highest and the duration longest, correlating to places situated on the old lakebed. In spite of the fact that Mexico City had one of the most progressive earthquake building codes at the time, in over 250 buildings some portions were damaged or they entirely collapsed while tens of thousands others suffered significant structural damage. Over 100,000 housing units were damaged, sending an estimated 2 million residents to shelters, relatives' homes, or the streets in the days, weeks, and months following the quakes. The quake especially hit the central historical portion of the city hard, destroying or damaging national monuments and government offices, as well as key commercial, health care, and education infrastructure. About 20 percent of the population lived in the hard-hit downtown area that also served as the premier economic zone of the city where

The General Hospital of Mexico collapsed after the Mexico City Earthquake in 1995. (Department of Commerce)

even greater proportions of *chilangos* (Mexico City residents) worked, shopped, and conducted business. Collapsed bridges, debris-filled streets, and damaged roadways hindered road and foot transportation all over the city. The earthquake also rendered the metro subway useless with destroyed tunnels, collapsed stations, and offline rails (Davis 2004).

RECOVERY EFFORTS, ADAPTATION, AND SOCIAL EFFECTS

Rescue and recovery began immediately after the shaking stopped. As in the immediate time following any earthquake, municipal authorities and everyday citizens alike began to organize rescue parties to pull survivors and cadavers from the wreckage. However, due to Mexico's top-down hierarchical and single-party federal system at the time, government institutions responded slowly to the disaster. The disaster also occurred amid a major debt and currency crisis, and concerns over debt repayment are believed to have hindered recovery. Also, making matters worse, a lack of liquid funds to speed the recovery led to a delayed response. A full 36 hours had passed after the first earthquake before President Miguel de la Madrid addressed the nation, and it was broadly perceived that the federal response for recovery was disorganized and slow. Also, many citizens complained that repairing key infrastructure for restarting Mexico's economy took precedent over taking care of the basic needs of its citizens. The federal government either did not seek or dismissed foreign aid outright. As a result, citizens created many neighborhood-based grassroots recovery organizations and began recovery and redevelopment plans of their own, many of which later garnered government support (Davis 2004; Adler 2015).

In the Mexican political landscape, the 1985 earthquake and the perceivably lackluster government response was a watershed moment. The PRI or Partido Revolucionario Institucional (Institutional Revolutionary Party) had enjoyed single-party control of the government since the mid-1920s and through decades of economic growth and prosperity following the revolution (1910–1921). Over this time, the PRI had developed a hierarchical, unilateral patron-client state centered in the capital. However, the political rifts from mounting debt, recent economic crisis, and what was largely a nondemocratic, authoritarian state were split wide open by the earthquake and laid bare for the Mexican public and international community to see (Adler 2015). While the 1985 earthquake was not the only agent of change, growing dissatisfaction with PRI governance led to the rise of substantive political challenges, eventually leading to the 2000 presidential election where a non-PRI candidate won that office for the first time in 75 years.

CONCLUSION

Through a geographic lens, one can see a much broader set of circumstances that led to the cataclysmic disaster that struck Mexico City on September 19–20, 1985. From the city's situation within the seismically Trans-Mexican Volcanic

Belt, it has an exceptionally high risk of earthquake disaster. Human effects such as long-term land-use changes and expanding urbanization led to a large, key portion of the city to be located on top of a drained lakebed. The subsoil of this anthropogenic (human-made) landscape is thus composed of highly plastic sediments that amplify the strength and duration of shockwaves, greatly enhancing the potential for damage, death, and injury for even seemingly far-off earthquake epicenters. In this sense, Mexico City is an urban metaphor for the mutuality of social and environmental aspects of hazards, risks, and disasters. Going beyond the event itself, the 1985 earthquake in Mexico City is a marquee example of how disasters are often critical political moments that both summarize the grievances and problems between citizens and their government while pointing the way to potential avenues of change.

Andrew M. Hilburn

Further Reading

Adler, D. 2015. The Mexico City Earthquake, 30 Years On: Have the Lessons Been Forgotten? *The Guardian*. September 18, 2015. www.theguardian.com/cities/2015/sep/18/mexico-city-earthquake-30-years-lessons, accessed September 11, 2018.

Anderson, J., P. Bodin, J. N. Brune, J. Prince, S. K. Singh, R. Quaas, and M. Onate. 1986. Strong Ground Motion from the Michoacan, Mexico, Earthquake. *Science* 233(4768): 1043–1049.

Chandler, A. 1986. Building Damage in Mexico City Earthquake. *Nature* 320(10): 497–501.

Cruz-Atienza, V., J. Tago, J. D. Sanabria-Gómez, E. Chaljub, V. Etienne, J. Virieux, and L. Quintanar. 2016. Long Duration of Ground Motion in the Paradigmatic Valley of Mexico. *Nature, Scientific Reports* 6(38807): 1–9.

Davis, D. 2004. Reverberations: Mexico City's 1985 Earthquake and the Transformation of the Capital. In *The Resilient City: How Modern Cities Recover from Disaster*, edited by T. Campanella and L. Vale. New York: Oxford University Press.

Millennium Drought, Australia, 2001–2012

During the 2000s, southern and southeast Australia were gripped by the longest and worst drought in 100 years, which is widely known as the Millennium Drought. In fact, the drought began in 1999 and continued until late 2009. The emergence of La Niña weather conditions in 2010 rapidly ended the drought with record-breaking rainfall and flooding, although officially it ended in 2012. The Millennium Drought was not a continuous one. For example, it appeared to break in early 2007, but returned again in August. However, the year 2006 was the driest on record, and large parts of Australia experienced drought. At its peak, 78.6 percent of the area of the eastern state of New South Wales experienced drought, while 100 percent of the farmland in Victoria experienced drought.

Australia is no stranger to drought because it is the driest landmass in the world. Mean annual rainfall is only 18.5 inches (470 mm), about much less than the world average of 28 inches (711 mm). However, coastal areas receive relatively higher rainfall (40 inches or 1,016 mm) while the interior receives 10 inches (254 mm) or less. Because of arid and semiarid climate of the interior, 25 percent of Australians live within 2 miles (3 km) of the coast, and 75 percent live within 27 miles (40 km) of a beach. About 40 percent of Australians live in the two great metropolitan areas of Sydney and Melbourne, and 70 percent live in the 10 largest cities.

Although the Bureau of Meteorology in Australia uses rainfall deficiencies to identify regions that are under drought conditions, drought is recurrent for other reasons too, such as shifting of the rainfall belt to the central Pacific, low humidity, high wind speeds, high temperatures, and large numbers of sunshine days. Some of these factors are influenced by climate drivers such as El Niño and La Niña. These drivers are a part of a natural cycle known as the El Niño–Southern Oscillation (ENSO) and are associated with a sustained period (many months) of warming (El Niño) or cooling (La Niña) in the central and eastern tropical Pacific.

The 2000s drought affected all of Australia, but its devastating effects were primarily confined to the populous southeastern part of the country, home to the four largest cities of Sydney, Melbourne, Brisbane, and Adelaide. All told, the Millennium Drought affected an area of 23,160 square miles (60,000 square km). The water supply of a large number of major cities (e.g., Adelaide, Brisbane, Canberra, Melbourne, Perth, and Sydney) was severely interrupted. In fact, the drought overwhelmingly affected the urban people.

Although several El Niño weather patterns affected the Millennium Drought, the Australian Bureau of Meteorology concluded that climate change exacerbated the extent and severity of the drought. However, some researchers claim that it is difficult to make a link between human-induced climate change and drought for this particular event, which lasted for several years. They argue that the link is not easy because Australia has a much more variable climate than many other parts of the world.

DROUGHT DEFINITION IN AUSTRALIA

Although there are many definitions of drought, the event is simply defined as an extended period of rainfall deficit during the growing season. However, the actual period of rainfall deficit differs by country. For example, in Southern Canada, this period refers to no rain for 30 continuous days, but this duration is meaningless in Australia because as a dry continent, it experiences no rainfall for at least one 30-day period per year. In fact, in most parts of Australia, even in coastal regions, the dry season lasts several months without any rain. Therefore, in this country, drought is generally defined as a calendar year in which rainfall is in the lowest 10 percent on record (Bryant 1991). Apart from rainfall deficiencies, other factors (e.g. low humidity, high wind speed, and great amounts of sunshine) are often used to describe drought. All of these factors increase water loss from soils and plants.

IMPACTS

The impacts of the Millennium Drought were wide-ranging and felt by everyone in Australia. It devastated rural and urban communities, industries, and the environment, which all depend on water from the Darling River and the River Murray to prosper. In addition to urban demand, these two rivers supply water for agriculture. Naturally, the supply was severely restricted for irrigators, putting pressure on agricultural and horticultural industries and on regional communities. Because of severe shortage of water, 33 wetlands were temporarily disconnected to help save water, risking long-term damage to the ecosystem. For example, the drought triggered major bushfire events in 2003 and 2009 (van Dijk et al. 2013).

With irrigation curtailments and no rainfall, the drought severely affected the agriculturally rich Murray-Darling river basin, Australia's largest river system, which produces one-third of Australia's food supply. The country's agriculture depends on irrigation as grain crops dominate its agriculture. However, crops were planted in half of the land area usually devoted to growing such crops, and Australia's cotton production decreased during the drought period. This threatened the livelihood of 10,000 people who were directly employed by the cotton industry. Meanwhile, farmers faced difficulty in finding enough fodder to feed cattle and sheep, and dairy producers were hit particularly hard by the drought.

Environmental impacts of the Millennium Drought included lowering of regular stream flows to 80–90 percent, and the water flow of the Lower River Murray in South Australia was recorded at the lowest level in over 90 years of records. The Darling River had no flow for nearly one year. During the period of low or no flow, water in the Lower Lakes (Lake Alexandrina and Lake Albert) fell below the mean sea level, reversing the usual positive hydraulic gradient from the lake to the sea. As a result, the Murray Mouth closed, and saline seawater started to enter inland, severely affecting soil fertility. Thus evapoconcentration, lack of flushing, and increased resuspension resulted in extremely poor water quality. Further, falling water levels in the Lower River Murray and Lower Lakes resulted in acidification of soil and ground water, threatening water supplies for people and livestock.

This significant reduction of flow also restricted supply for critical water needs of urban dwellers. Drought-affected cities had to implement severe water restrictions and other measures, including, in some cities, a complete ban on outdoor water use for years. For example, the Millennium Drought forced residents of greater Melbourne, a city of 4.3 million people, to find innovative ways of increasing water supply and decreasing water demand. People of drought-affected areas experienced many health problems related to dirty or poor quality water or no water. Because of the drought, over 70 percent of the floodplain forest died, and several species were newly threatened with extinction. Also, Aboriginal communities suffered due to the exposure of ancient burial grounds. Additionally, drought contributed to increase in electricity costs.

RESPONSE

The Australian government response to the Millennium Drought was manifold. The most important water policy response of this drought was the declaration of the National Plan for Water Security by then Prime Minister John Howard in 2007. This led to the passage of the Water Act (2007) by the Commonwealth Parliament, the formation of the Murray-Darling Basin Authority in 2008, and the publication of the Murray-Darling Basin Plan in 2012.

In addition to these actions, federal, state, and local governments all prepared in advance of the event to diversify water supplies and to scale up efficiency and conservation programs. They enforced rules about the use of domestic, agricultural, and industrial water, and those who did not follow the restrictions were fined. At the domestic level, these restrictions were associated with the use of washing machines, toilets, cooling towers, and showers. Accordingly, these measures reduced consumption of water per capita in the affected cities and rural areas. For example, in southeast Queensland, these measures contributed to reducing each person's water use to 87 gallons (140 liters) per day. At the same time, several innovative public measures were also initiated, which included recycling waste water from commercial buildings, factories, and sports fields and recovering water from the sewage system for recycling (The Conversation n.d.).

Individual cities, such as Melbourne, also introduced innovative measures to reduce water demand during the drought years. The city implemented residential and industrial water conservation programs by using recycled water. In place of surface water, the city also experimented with recycled water to support agriculture in the Werribee Irrigation District. To some extent, the Millennium Drought provided Melbourne with the opportunity to develop a more integrated approach to water management (Low et al. 2015).

Both federal and state governments asked homeowners to install tanks to save and collect water, and federal and state governments provided health services and public works for those affected by the drought. The federal government constructed a controversial desalination of seawater plant at Kurnell in southern Sydney at a cost of $1.4 billion (A$1.9 billion). The construction was completed in 2010, and yet the plant shut down in 2012 because the drought ended, and the region's dams were full. However, the plant was controversial because it was expensive and energy-intensive to operate.

Additionally, to prevent acidification of soil, lake, and groundwater, the government constructed an embankment and arranged to pump water along it using aerial limestone dosing. To ensure salt and other pollutants could flush out of the River Murray, the government had to initiate around-the-clock dredging. Apart from these actions, the government made comprehensive investments in water conservation and efficiency involving households, businesses, and local governments. These investments helped cities cope with the Millennium Drought and also reduced vulnerability to future droughts (The Conversation n.d.). The drought also threatened the city water supply systems. For example, in Adelaide, pipelines were built to deliver drinking water to the Lower Lakes communities and sustain valuable horticultural industries.

Research has shown that suicide rates might increase among farmers who had lost their crops and those who had experienced financial problems. To counter this, the government invested in improving mental health services during the drought period of 2001–2007 (Guiney 2017). Meanwhile, the federal government have provided aid assistance to the affected cities and people worth A$4.5 billion since 2001. Most of this aid was for concessional loans, a part of which was interest free for five years. The affected farmers received these loans that helped them to switch from higher interest commercial loans and saved them at least $20,000 a year on average. The federal government also provided the Farm Household

Allowance (FHA) and grant to organizations such as the Foundation for Rural and Regional Renewal (FRRR). State governments also provided drought grants in the form of transport subsidies and waivers on a range of state government fees including council rates and car registrations. In addition, the New South Wales government waived Local Land Services (LLS) levies and allocated about one billion dollars to cover drought preparedness, emergency relief, and mental health services. Major Australian banks also participated in relief efforts by donating to the Red Cross and other humanitarian organizations.

CONCLUSION

As a creeping disaster, Australia's Millennium Drought lasted for almost a decade, and as a consequence, it had devastating effects on agriculture, the environment, and water supplies. This drought affected more urban than rural people, but farmers and agriculture were the hardest hit by the event. As with previous droughts, Australia's cities and rural areas survived the Millennium Drought, but they had to implement restrictive and innovative measures. These measures can provide important lessons for drought-prone areas of the world. Although Australian governments at two levels were successful in responding to the disaster, several initiatives and decisions did not work so well. As noted, the desalination project was not a successful response to the drought. However, the important factors in the overall success were government and utility programs that rallied community support for lowering household water demand.

Bimal Kanti Paul

Further Reading

Bryant, E. A. 1991. *Natural Hazards*. Cambridge, MA: Cambridge University Press.

Guiney, R. 2017. Farming Suicides during the Victorian Drought: 2001–2007. *Australian Journal of Rural Health* 20: 11–5.

Low, K. G., S. B. Grant, A. J. Hamilton, K. Gan, J.-D. Saphores, M. Arora, and D. L. Feldman. 2015. Fighting Drought with Innovation: Melbourne's Response to the Millennium Drought in Southeast Australia. *WIREs Water*. https://doi.org/10.1002/wat2.1087.

The Conversation. n.d. What California Can Learn from Australia's 15-Year Millennium Drought. https://theconversation.com/what-california-can-learn-from-australias-15-year-millennium-drought-55300, accessed November 17, 2018.

van Dijk, A.I.J.M., H. E. Beck, R. S. Crosbie, R.A.M. de Jeu, Y. Y. Liu, G. F. Podger, B. Timbal, and N. R. Viney. 2013. The Millennium Drought in Southeast Australia (2001–2009): Natural and Human Causes and Implications for Water Resources, Ecosystems, Economy, and Society. *Water Resources Research* 49: 1040–1057.

Mozambique Flood, 2000

Mozambique is one of the most flood-prone countries in Africa, its geography being the principal cause of frequent flooding. It has the fourth largest flood plain in Africa and experiences some flooding almost every year, with a major flood occurring every 5–10 years. Mozambique has 10 major rivers, including the Zambezi and the Limpopo Rivers, and the former is the largest. Nine out of the 10 rivers originate in the adjacent seven countries that cross Mozambique from west to east and drain into the Indian Ocean along its 1,667 miles (2,500 km) coastline. The catchment areas of these rivers drain water from vast swaths of southern Africa, stretching into Botswana.

Since the beginning of this century, Mozambique has experienced three devastating floods: the first in 2000, the second in 2007, and the third most recently in 2013 (UNDP 2014). Of these three major floods, the flood in February 2000 was the worst in 50 years. It affected the southern half of the country from the Limpopo River basin to Maputo, the capital of Mozambique, causing the deaths of 699 people, displacing tens of thousands, and severely damaging housing, agriculture, public buildings, schools, hospitals, roads networks, railways, and communications (World Bank 2000).

CAUSES OF THE FLOOD

The 2000 flood in Mozambique was triggered by exceptionally heavy and persistent rainfall within the country as well as in the surrounding countries of South Africa, Botswana, and Swaziland. The rainfall started on February 9, 2000, and continued for five weeks. During this period, Mozambique recorded over 43 inches (1,100 mm) of rainfall, which accounted for more than 75 percent of the country's annual rainfall. Prior to this rainfall, Mozambique experienced a long dry spell, which made the surface impermeable and facilitated the chances of a flood by increasing the rate of surface runoff into its major rivers. Notably, drought is also a constant threat in Mozambique. However, not only did these rivers receive a massive amount of rainfall, but after the rainfall had stopped, they also continued to receive floodwaters from the neighboring countries as those rivers flow through Mozambique into the sea.

As noted, the capital of Mozambique was flooded, displacing tens of thousands of people. Further north, the flood also made many people homeless in Gaza province. The flood was exacerbated by two tropical cyclones, which brought additional rain and strong winds to the flood-affected areas. As Mozambique is a coastal country, just as with floods, the country is also vulnerable to tropical cyclones. Sixty percent of the population lives along the coastline. Cyclones Eline and Gloria hit the coast and inland on February 22, 2000, and March 5, 2000, respectively, and thus the 2000 flood began in February and ended in March 2000. Every major valley south of Beira was affected, as rivers burst their banks. For the first time in recorded history, simultaneous flooding occurred in all of the major rivers that flow into the Indian Ocean through Mozambique (UNDP 2014).

FLOOD DAMAGE AND IMPACTS

The 2000 Mozambique flood affected about 540 sq. miles (1,400 sq. km) of arable land in southern and central parts of the country, which was 11 percent of

Flood-damaged areas around the Save River in Mozambique, March 16, 2000. (TSGT Cary Humphries, USAF/Department of Defense)

the total cultivated area. It also affected 90 percent of the irrigated land in the five flood-affected provinces of Maputo, Gaza, Inhambane, Sofala, and Manica. The largest impact on arable occurred in Gaza (accounting for 43 percent of the flooded cultivated land), followed by Maputo (31 percent) and Sofafa (18 percent). This translated into a loss of up to 21 percent of expected agricultural production in the affected provinces (World Bank 2000).

According to the government of Mozambique, one-third of the country's staple crop, maize, was destroyed. Also, almost 4.5 million people or some 25 percent of the total population were seriously affected, with half needing food aid. Meanwhile, approximately 700 people died and 20,000 head of cattle were lost while almost 1 million people lost their homes. Additionally, about 6,000 fishermen lost 50 percent of their boats and gear. Consequently, the national Gross Domestic Product (GDP) fell from 7.5 percent in 1999 to 1.6 percent in 2000 (World Bank 2000), and the estimated overall damage was $620 million in 2014 USD (UNDP 2014).

With very little fertile ground being available for cultivation in Mozambique, the lack of crops because of the flood meant a lack of food and subsequent lack of income. The deaths of cattle also reduced the income of flood-affected rural residents. Furthermore, the flood damaged many businesses, creating a high unemployment rate in the postflood period. Moreover, disease and malnutrition were common in the months following it due to lack of clean water, food, and medical supplies. After the flood, lack of food caused widespread malnutrition in the affected areas. Additionally, when the flood receded, pools of stagnant water were left behind, which are ideal breeding grounds for mosquitoes. Floodwater also contaminated many sources of water supplies in rural areas. As a consequence, many flood survivors suffered from water-borne diseases, including malaria, cholera, and diarrhea.

RESPONSE AND RECOVERY

One day before the landfall of Cyclone Eline, the Mozambique government made an initial appeal to the international community for $76.5 million for rebuilding the infrastructure and providing emergency materials to flood survivors who were at risk of disease. This amounted to around $15 per head to cover health care, food, and accommodation. Later, the government increased the appeal amount to $160 million. According to Wikipedia, by 17 March, a number of countries (e.g., Australia, Canada, Denmark, Ireland, Italy, the Netherlands, Portugal, Spain, the United States, and Sweden) had pledged $119 million to Mozambique, and by mid-May 2001, 93 percent of the appeal had been met. By 4 March, 39.6 tons of various relief goods reached the country, which nearly overwhelmed the small airport at Maputo. Some of these countries (e.g., the UK and Italy) canceled their $650 million debt in March.

The United States distributed the resettlement cash grant, amounting to $9.7 million, among 106,000 flood survivors from December 2000 to April 2001 for flood recovery. Cash grants were provided directly to families whose homes and farms were destroyed in the flood. Each affected family received about $92 (USAID 2002). The grant was delivered to the woman of each flood-affected household, with the intention that she would utilize the money in the best interests of the whole household. Grant-recipient families therefore were able to choose for themselves which goods or services were their highest priority or whether saving was their priority. Postgrant evaluation reveals that the resettlement grant was successful in achieving its objective of implementing an alternative to traditional relief programs. The evaluation further shows that a cash transfer program is an efficient and cost-effective way to help relieve the affected households quickly.

However, in addition to Western countries, several Asian and African countries, such as Botswana, Ghana, Mauritania, and Saudi Arabia, also responded to the appeal made by the Mozambique government. Several international relief agencies, such as the World Food Program (WFP), UNICEF, International Federation of Red Cross and Red Crescent Society (IFRC), Doctors without Borders, and the Bill and Melinda Gates Foundation also responded to the appeal. Over 49 countries and 30 international NGOs provided humanitarian assistance either in the form of cash or kind (e.g., food, medicine, and tent). Thus, massive national and international relief operations avoided greater loss of life and suffering. Ultimately, the UN Disaster Assistance Coordination team (UNDAC) was responsible for overall coordination of the flood operation; as part of this team, the WFP is the logistics coordinator for the delivery of food and nonfood items to flood survivors.

The major roads of Mozambique were under water during the 2000 flood, and the primary north-south highway was flooded in three locations. Because of breakdown in the transportation system, public and private response was relatively late—it started at the end of February. Evacuation and rescue operations were also delayed because Mozambique had a limited capacity for conducting such operations. Therefore, boats and helicopters were used to evacuate flood victims from the affected areas. South Africa provided 12 planes and helicopters to operate search and rescue missions as well as airdrop food. They were assisted by two helicopters from Malawi, six from the United Kingdom, and 10 from Germany. By March 7, the fleet of 29 helicopters had rescued 14,204 people (McGreal 2000).

To accommodate people who were displaced by the flood, the Mozambique Government established 121 temporary flood shelters across the country. This arrangement often separated family members and friends in different shelters, causing psychological problems. Also, the conditions at these shelters were poor—no sanitation, little food, and lack of medical supplies. Compounding all this, poor infrastructure made it difficult to transport emergency supplies.

Flood response in Mozambique was not only delayed but also inadequate for several historical reasons. Prior to the 2000 flood, disaster management in Mozambique was not a proactive policy, mainly due to political instability caused by 17 years of civil war, which ended in 1992. Then development programs received priority, and these programs were not integrated with disaster prevention and preparedness programs. Additionally, ongoing tensions remained between the Renamo areas of the north and center and the Frelimo areas of the south, which also impeded flood relief and recovery efforts in Mozambique. North-south tensions were further exacerbated by the difficult road communications in such an elongated country, with the major roads running east-west. Despite all these barriers, relief and recovery efforts ran somewhat satisfactorily because donor countries and international relief agencies have been active in Mozambique since its independence in 1975. Additionally, the country seemed to receive adequate funding from external sources, and it managed this funding largely through the national budget system, which avoided complex and multiple funding arrangements. This system facilitated rapid disbursement of external funding and strengthened national accountability and transparency mechanisms (UNDP 2014).

CONCLUSION

Mozambique adopted its National Disaster Policy one year prior to the 2000 flood event, which marked the beginning of proactive measures for disaster management in the country.

The policy included introduction of early disaster warning systems with community involvement, allocation of funds for contingencies, and support of livelihood recovery through labor-intensive strategies. It promotes mainstreaming of disaster risk reduction through national and sectoral development planning and emphasizes the importance of inter-sectoral coordination in disaster prevention and response (UNDP 2014). The same year (2000) the country formed the National Disasters Management Institute (Portuguese: Instituto Nacional de Gestão de Calamidades, INGC) with operational responsibility for the coordination of disaster management activities, including multi-sectoral prevention, risk reduction and mitigation, awareness campaigns, disaster response, and postdisaster relief and rehabilitation and reconstruction. The institute operates under the Ministry of State Administration (MAE).

Before the 2000 flood event, flood forecasting and early-warning systems were in the early stage of development, and probably for this reason, the government failed to warn and evacuate people before the event affected the country. As indicated, poor early warnings and flood control systems for Mozambique are a regional issue that involves close collaboration with neighboring countries. However, by the 2007 major flood, which killed only 117 people as opposed to the 699 who died in the 2000 flood, all these were in place. This sharp decrease in fatality rates and the impact of flooding indicates that floods are being managed more effectively. This reflects lessons learned from the prior decade, including flood forecasting, early-warning systems, and trans-boundary cooperation.

Most importantly, national investment and international cooperation in strengthening institutional capacity and readiness have noticeably reduced the devastating impact of disasters on human lives in the country. Also, the management of water flows from two major dams, the Cabora Bassa and the Kariba, has had a major impact on reduction of flood risks in Mozambique. The 2000 situation, when human and financial resources were limited both in preparing for the flood and in recovering from it, no longer exists. There is no doubt that the government of Mozambique was more successful in responding to and recovering from the subsequent major floods in 2007 and 2013 than the 2000 flood. This trend should continue in future events.

Bimal Kanti Paul

Further Reading

McGreal, Chris. 2000. Mozambique: Amid the Filth, the First Sign of Disease. *Mail and Guardian*. March 7, 2000. https://reliefweb.int/report/mozambique/mozambique-amid-filth-first-sign-disease, accessed June 3, 2018.

UNDP (United Nations Development Program). 2014. *Mozambique - Recovery from Recurrent Floods 2000–2013: Recovery Framework Case Study*. New York: United Nations.

USAID (U.S. Agency for International Development). 2002. *Mozambique Flood Resettlement Grant Activity*. Washington, DC: USAID.

World Bank. 2000. *A Preliminary Assessment of Damage from the Flood and Cyclone Emergency of February–March 2000*. Washington, DC: World Bank.

Nepal Earthquakes, 2015

Within 16 days, two devastating earthquakes struck Nepal in 2015—one on April 25 and the other one on May 12. The first earthquake had a magnitude of 7.8 and the second of 7.3. These events together affected 31 of the country's 75 districts, which comprise the second largest administrative unit in Nepal. Considering the extent of damage, the Nepalese government declared 14 of the 31 affected districts as "crisis-hit" (GoN 2015). Of the five regions of Nepal, most of the earthquake-affected districts are located in the Western and Central regions. However, the epicenter of the April 25, 2015, event was Barpak VDC (Village Development Committee) in the

Gorkha District, about 50 miles (80 km) northwest of the capital city of Kathmandu. The country had not experienced an earthquake of this magnitude for over 80 years (GoN 2015). This powerful earthquake damaged several World Heritage sites and caused numerous landslides and rock/boulder falls in the affected districts. As a consequence, many roads in these districts were blocked, which delayed both rescue operations and recovery activities.

The epicenter of the May 12, 2015, earthquake was near Kodari, about 113 miles (170 km) southeast of the epicenter of the April earthquake or about 50 miles (80 km) northeast of Kathmandu. After the first earthquake, the country experienced many aftershocks, 421 of them with a magnitude of greater than or equal to 4 (GoN 2015). The two earthquakes combined killed more than 8,857 people and injured 22,304 others. The earthquakes affected about one-third of Nepal's population and destroyed or damage over 885,000 houses (GoN 2015). Sindhupalchok district experienced the highest number of deaths both in absolute and relative terms. The earthquake impacted an area similar to the size of the state of New Jersey and generated a total economic loss of $10 billion (GoN 2015).

SEARCH AND RESCUE OPERATIONS AND EARTHQUAKE RELIEF EFFORTS

Although search and rescue (SAR) operations started immediately after the April 25 earthquake, particularly in the Kathmandu valley area, both SAR and

The Japanese army takes part in the rescue effort at Sankhu village after the Nepal earthquake, 2015. (Prabhat Kumar Verma/Pacific Press)

relief operations were delayed for other affected areas. Nonetheless, nine out of 10 Nepalese troops and almost all police officers participated in SAR operations. In addition, 134 international SAR teams from 34 countries participated in rescue operations. Moreover, 300 U.S. military personnel assisted in SAR operations in Nepal. Overall, 4,236 helicopter flights were used for SAR, and teams recued 7,558 persons by air and 4,689 persons by land. Additionally, 4,000 public and private health workers participated in SAR and relief operations (GoN 2015). Foreign physicians and health workers also provided necessary treatment for those injured in the earthquake.

At the start of relief operations, the Nepalese government confronted several serious problems. However, the relief efforts gained strength within 10 days of the first major earthquake, providing emergency aid to earthquake survivors despite the country having suffered from ineffective governance since the mid-1990s. Between 1996 and 2006, the country experienced a Maoist insurgency aimed to end the monarchy system and establish a popular democracy. Then on June 1, 2001, the crown prince killed himself along with all his family, including his parents. The king's younger brother became the new king of Nepal. However, the general public of the country started agitating to end the monarchy system. The new king resigned in 2008, and the leader of the former Maoist rebels came into power.

Political instability in Nepal continued even after the end of the monarchy. Since 2008, several governments have come into power, but none has lasted. In 2015, agitation by the Madhesi, the majority population of the southern plain, negatively impacted the economy of Nepal, and the protesters blocked oil imports from India. Nepal relies completely on Indian oil. The conflict between Madhesi and Kathmandu manifested itself after the promulgation of Nepal's new constitution. When the marginalized Madhesi community found out that the earlier promise of the Nepalese government was not reflected in the new constitution, they protested and blockaded a key route that was used for transporting Indian oil.

In addition to political problems, the initial relief operation was delayed because of Nepal's topography. The earthquakes affected the country's mountainous and hilly regions, where severely damaged access roads were nearly impassable due to collapsed bridges, earthquake-triggered landslides and avalanches, and rubble from road-side buildings destroyed by the events. Damage to infrastructure further hampered search and rescue operations, particularly in remote locations. The country's bureaucracy also contributed to the delay in relief operations. To transport relief goods to remote affected areas, the Nepalese government had an inadequate number of trucks and drivers, so it had to recruit both from the private sectors. For transporting the goods, the government also needed earthmovers and cranes to clear roadblocks caused by landslides and rock falls. Unfortunately, it did not possess any such equipment prior to the April earthquake (Paul et al. 2017).

Another reason for delay was the air congestion caused by the Nepalese government opening Kathmandu airport as the staging ground for international relief goods, which meant many planes were unable to land on time. One month after the earthquake, the government imposed several rules and regulations for bringing relief items to the country, including custom inspections and import taxes. These checks were necessary because government authorities did not know what was coming into the country. However, while these actions certainly delayed disaster response, they probably discouraged the sending of unnecessary items from abroad (Paul et al. 2017).

According to a sample survey conducted in Nepal, earthquakes survivors in Nepal received many emergency items from both domestic and foreign public and private sources. These materials included food, drinking water, clothing, cash, housing materials, kitchen ware, tent, dignity kits (e.g., towels, sanitary napkins, soaps, toothbrushes and toothpaste, and combs) for girls and women, medicine, and other necessary items. As immediate relief, the Nepalese government allocated 15,000 Nepalese rupees (NPR) (approximately $150.00) and then 10,000 NPR (approximately $100.00) as winter relief to each earthquake-affected household (Paul et al. 2017).

Disaster relief distribution was source and item specific. For example, the Nepalese government provided cash; nongovernmental organizations (NGOs) distributed about 92 percent of all tents; the World Food Program (WFP) of the UN delivered food items, such as rice, beans, and high protein biscuit; and the United Nations Population Fund (UNPF) distributed dignity kits (GoN 2015). In the immediate aftermath of the earthquake, for food and basic medical services, the survivors of remote and isolated affected areas depended entirely on the local people who had experienced little or no earthquake damage. Once roads were cleared of debris and rubble, the emergency assistance arrived relatively quickly. Unlike government agencies, NGOs were selective in providing disaster relief to the earthquake-affected areas such that relief distribution was restricted to the areas where NGOs had been working for a long time. Notably, among all developing countries, Nepal has one of the highest number of NGOs per capita (Paul et al. 2017).

The national and international press depicted the relief efforts to be going on poorly, but some researchers (e.g., Paul et al. 2017) note that the operation ran smoothly and efficiently. According to these researchers, one primary reason for a satisfactory assessment of relief operations was that the international community provided overwhelming support to the earthquake-affected people of Nepal. The Nepalese government also contributed to make the relief effort a success. It claims that "overall, 22,500 civil servants, 65,059 staff of the Nepal Army, 41,776 staff of Nepal Police and 24,775 staff of the Armed Police Force, as well as 4,000 government and private health workers were mobilized to aid rescue and relief efforts" (GoN 2015, xii).

However, distribution of relief was not consistently fair in part because Nepalese societies have been deeply segregated in terms of caste and ethnicity. In some instances, unequal distribution of relief goods and systematic exclusion of people belonging at the bottom of the caste hierarchy (known as Dalit or "untouchable") and other marginalized groups (e.g., poor, elderly, female, and disabled) was reported by Amnesty International. Also, in some districts emergency aid distribution was politically motivated (Paul et al. 2017). The organization further reported that the Nepalese government was distributing to each earthquake-affected household 22 lbs. (10 kg) of rice without considering the level of damage or economic conditions. This government action offended small and landless farmers and members of the Dalit caste because their need for rice was much greater than that of other relatively affluent households. A report published by the Dalit Civil Society claims that more than three-fourths of the Dalits in earthquake-affected areas were displeased with the way authorities distributed relief goods among their communities.

CONCLUSION

Numerous domestic and international actors participated in emergency relief operation after Nepal was hit by two major earthquakes in 2015. Despite a lack of coordination among these actors, and logistical, political, and other problems, the operation ultimately gained momentum. The Nepalese government introduced a "one-door policy" for smoothly running the relief operation. According to this policy, all earthquake donations had to be deposited to the Disaster Relief Fund established by the prime minister. These donations were then sent to nearby army headquarters, and finally VDC/Municipality authorities distributed donations among the earthquake survivors. Note that the government agencies disbursed only two items: equal amount of cash and rice to all earthquake-affected households. However, this policy was criticized by some donors, many of whom doubted the even distribution of relief items because Nepal was not known for its effective public service.

The Nepalese government's one door policy and the distribution of restricted numbers of relief items does seem to have worked well. The one-door policy did seem to block the flow of unnecessary and even useless goods from abroad. Additionally, the import taxes enforced by the government and the policy of checking disaster relief items arriving from foreign countries also worked well. However, the people who lived far away from the VDC/Municipality headquarters or offices of local political parties believed that the distribution was neither even nor always justified (Paul et al. 2017). Additionally, many of those people had no adequate access to information about emergency aid, which acted as a barrier to receive such aid.

Although the government of Nepal outlined the medium- and long-term recovery needs and priorities of all four selected sectors (social, productive, infrastructure, and cross-cutting) on June 25, 2015, recovery efforts, nonetheless, seemed to be very slow in the country, particularly in the housing subsector of social sectors (GoN 2015). With nearly 1 million houses being either fully or partially damaged, the rebuilding would continue for several years. After outlining a housing reconstruction policy that included adaptation of earthquake-proof standards for all rebuild and repair houses, the Nepalese government has only recently begun to distribute a housing grant to earthquake-affected households. This is an important step, which can substantially reduce fatalities and injuries as well as housing collapses due to future earthquakes in the country, provided the government becomes successful in enforcing building standards in the earthquake-affected areas.

Bimal Kanti Paul

Further Reading

GoN (Government of Nepal). 2015. *Nepal Earthquake 2015: Post Disaster Needs Assessment, Vol. A: Key Findings*. Kathmandu: GoN.

Paul, B. K., B. Acharya, and K. Ghimire. 2017. Effectiveness of Earthquake Relief Efforts in Nepal: Opinions of the Survivors. *Natural Hazards* 85: 1169–1188.

Pakistan Flood, 2010

Flood is a perennial problem in Pakistan, particularly in the Indus plain. In the summer of 2010 (July and August), this plain experienced a catastrophic flood considered one of the largest floods in the history of Pakistan. The 2010 flood in Pakistan killed about 2,000 people, while over 1.7 million households were either destroyed or damaged, and more than 18 million people were seriously affected by this event. In terms of the lives affected, the impact of the 2010 flood in Pakistan was greater than that of the 2004 Indian Ocean tsunami, the 2005 Kashmir earthquake, and the 2010 Haiti earthquake combined (UNOCHA 2010). With 20 percent of Pakistan's total area covered by water, the 2010 flood caused extensive damage to property, crops, and infrastructure. A total of 78 of the 156 districts were affected by the 2010 flood. A district is the third-order administrative unit in Pakistan, below provinces and divisions, but it forms the first-tier of local government. The most affected province was Sindh, where approximately 18 percent of the total population was affected (Lahiri-Dutt 2010).

CAUSES

The catastrophic 2010 flood started in mid-July following heavy monsoon rains that led to flash floods and landslides in the northwest and eastern parts of Pakistan. Provinces affected by the flood included Khyber Pakhtunkhwa, Sindh, Punjab, Azad Kashmir, Gilgit Baltistan, and Baluchistan. The intensity of the rains was very high with about four months' worth of rainfall falling in just a couple of days. Some areas in Northern Pakistan received over three times their usual annual rainfall in just 36 hours. The high-intensity rainfall continued for the rest of July and persisted into August. This led to flooding in the Indus River Basin, which traverses Pakistan from north to south (Lahiri-Dutt 2010).

The excessive runoff from the rain led the Indus River and its tributaries to rise above their banks and flood the surrounding areas. As of mid-August, the floodwater had moved along the Indus River from the northern region (e.g., Punjab and Khyber Pakhtunkhwa provinces) downstream to the south, causing flooding in new areas of densely populated areas in southern provinces such as Sindh (Kronstadt et al. 2010).

Several factors contributed to the 2010 Pakistan flood event. Inadequate water resource planning and development of infrastructure played an important role as the Indus River was unable to drain the excess waters into the Arabian Sea. In recent decades, there have been a lot of diversions of the river flow for irrigation. Levees have been built along the banks of the river by (mostly influential) farmers

Aerial view of floods in Pakistan, August 25, 2010. (Cpl. Jenie Fisher/Department of Defense)

not only to protect their farms from flooding but also for irrigation. This affects the natural system of the river when excess sediments are deposited on the riverbed causing it to rise above nearby plains. This, in turn, makes the river channel less efficient and thus enhances flood risk (Lahiri-Dutt 2010). Also, there has been concern that climate change during the last half century has contributed to the extent and severity of the flood event by triggering heavier rains than usual. However, separating the climate change effects from anthropogenic factors contributing to flooding is difficult.

DAMAGE AND IMPACTS

The 2010 Pakistan flood event started in July 2010 in northern Pakistan and persisted into August after reaching the south, and then it extended to new areas on the eastern and western banks of the Indus River in Thatta district in September. As the flood progressed southward, the people affected in the north started recovering from the impacts. But the floodwaters were stagnant in the south, causing delay of flood recovery until November 2010. As the flood lingered, the flood damage and its impacts substantially increased. According to the United Nations Office for the Coordination of Humanitarian Affairs (UNOCHA 2010) report, the flood affected more than 20 million people in Pakistan. About 1.8 million houses were damaged, forcing some people to move into internally displaced persons (IDP) camps and some out of their home districts for several weeks. In all, nearly 2,700 people were injured because of the flood.

The 2010 flood also caused significant and widespread damage to various types of infrastructure and several economic sectors of the country. The United Nation's

Food and Agriculture Organization (FAO) estimated that about 8.1 million acres of standing crops, including rice, corn, cotton, fruit orchards, and vegetables, were damaged or lost completely due to flooding. Also, 1.2 million livestock and 6 million poultry died. Many dikes and embankments were destroyed. This impact increased the potential for (long-term) food shortages and increased the prices of staple goods (Kronstadt et al. 2010).

Moreover, the 2010 flooding had a huge impact on the health sector in part because the flood significantly impeded access to safe drinking water. Stagnant water from the floods, widespread damage to sanitation infrastructure, and overcrowding caused by the displaced population led to increase in outbreaks and spread of water-borne and vector-borne diseases such as gastroenteritis, diarrhea, cholera, malaria, and skin diseases. The major concern was the reduction in access to health services and medications as large numbers of health facilities in flood-affected areas were damaged or destroyed. Also, about 11,000 schools were damaged or destroyed. Roads, railways, and bridges were also damaged, and electricity was severely disrupted (Kronstadt et al. 2010).

The Asian Development Bank (ADB) and the World Bank estimated that the 2010 flood caused a total of $10 billion in damage. Long- and short-term economic impacts included inflation as prices of items and services increased, increased government fiscal expenditure, and lower GDP growth compared to that of the previous year.

RESPONSE AND RECOVERY

The government of Pakistan through its Initial Floods Emergency Response Plan (PIFERP) requested $460 million on August 11, 2010, to assist flood-affected people. Following the initial request, about 35 percent of the requested funds were either already spent on recovery or committed by humanitarian organizations. The plan was later revised in September, and an additional $1.6 billion was sought to enable adequate support for the Pakistani government in addressing residual relief and early recovery needs of flood-affected families for the following year (UNOCHA 2010).

The PIFERP first focused on rescue and immediate relief for the flood victims; then it shifted focus to help people displaced by the floods to return to their damaged homes and helped in repairing and rebuilding those homes. Based on strategic objectives, the response plan covered various humanitarian-clustered projects that included water and sanitation, health, shelter, agriculture, food, community restoration, protection, education, nutrition, logistics, coordination, and camp coordination/management. Funds were allocated to each cluster, and activities for each were clearly outlined. Evaluation of the progress of each cluster was built into the plan. All of this meant fund requirements increased substantially during the response plan revision, and noteworthy funding gaps in clusters such as agriculture, community restoration, and education were revealed. Also, several organizations were listed as coordinators for each cluster. For example, the United Nations Food and Agriculture Organization (FAO) was designated for the agriculture cluster, the United Nations Development Program (UNDP) for community rebuilding, and International Rescue Committee (IRC) for water and sanitation needs (Kronstadt et al. 2010).

Despite the coordination by several international organizations, ultimately, Pakistan's National Disaster Management Agency (NDMA) was responsible for the overall coordination of disaster response efforts by working closely with federal ministries, government departments, the military and the international community to organize, collect and distribute relief goods, while the Economic Affairs Division (EAD) coordinated donors (Kronstadt et al. 2010). With a lot of donors, the initial funding for the response plan was immediate, resulting in a total of $307 million (about 67 percent of the initial request) by the end of August. Donors ranged from international organizations such as UNOCHA, DHL International, Asian Development Bank, AT&T Foundation, World Bank, Exxon Mobil, and UPS Foundation to countries such as the United States, the United Kingdom, China, Bangladesh, Jordan, Australia, Denmark, Canada, and Cyprus and private donors such as Bill and Melinda Gates.

The United States was one of the largest donors for relief and recovery efforts related to the 2010 Pakistan flooding with a total of about $250 million; moreover, including in-kind civilian and military support, such as the prefabricated steel bridges, halal meals, blankets and air transport, amounted to another $89.1 million (Kronstadt et al. 2010; UNOCHA 2010). The United Kingdom donated about $200 million in response to the UN flood appeal. Additionally, it provided new bridges to replace some of those destroyed by the floods. China claims it was the first to donate and had provided about $47 million including tents, electricity generators, medicines, and water purification tablets.

The European Union donated a total of about $450.9 million in both cash and kind aid, including about $210.4 million from the European Commission Humanitarian Aid Department and the remaining coming from more than 20 countries. For example, Germany donated about $34 million, Denmark about $24.4 million, Sweden about $25.5 million, and France about $3 million. Japan also donated a total of $55 million in cash and relief goods, Australia provided $75 million, and Canada donated about $30 million with an additional $1.4 million from a Canadian humanitarian coalition. Morocco also provided about $ 2 million (UNOCHA 2010).

All these donations covered almost all immediate relief requirements, including providing temporary shelters and safe drinking water. However, relief work was also hindered to some extent by the difficult logistical terrain, destruction of infrastructure, and the threat of terrorist attacks against aid agencies. These made emergency work and volunteering hard to carry out, and it was mostly done in an uncertain and challenging working environment. The Pakistani people blamed the government for a slow and unsystematic response to the flood, which led to several riots with attacks on and raiding of aid convoys (Kronstadt et al. 2010).

CONCLUSION

Pakistan experienced one of the most devastating flood events in 2010, garnering attention and help from the entire world. While Pakistan had experienced severe and dangerous flooding in the past, this event affected more people than had similar events in the past 40 years. Extreme weather (precipitation), which is the dominant climate condition for Pakistan, coupled with the Asian monsoons and poorly developed infrastructure, increased flood risk for the country. And the frequency of flood events raises questions about whether improving flood resiliency of the country is possible and if so how. Since it is difficult to control extreme weather, effective early-warning systems and creating or revising intrinsic plans for response are critical.

After the 2010 flood event, Pakistan developed and updated a well-equipped flood early-warning system at the upper Indus basin to help prevent or reduce financial, human, and property losses caused. In collaboration with the Pakistan Council of Research in Water Resources (PCRWR) and the Pakistan Meteorological Department (PMD), UNESCO, using financial assistance from the government of Japan, worked on the improvement and strengthening of the flood early-warning system and management capacity of Pakistan. With these put in place, the number of fatalities has decreased, and so has the impact of flooding after the year 2010. The 2011, 2012, and 2013 floods killed about 361, 100, and 80 people, respectively, as opposed to the over 2,000 fatalities in the 2010 flood. This type of planning might be an indication of more effective preparation for and management of floods through the strengthening of the country's institutional capacity and with international cooperation.

Hilda Onuoha

Further Reading

ADB (Asian Development Bank). 2010. ABD-World Bank Assess Pakistan Flood Damage at $9.7 Billion. Press Release. October 14, 2010. https://www.worldbank.org/en/news/press-release/2010/10/14/adb-wb-assess-pakistan-flood-damage-at-97-billion, accessed May 20, 2018.

Kronstadt, K. A., P. A. Sheikh, and B. Vaughn. 2010. *Flooding in Pakistan: Overview and Issues for Congress—CRS Report R41424*. Washington, DC: Congressional Research Service.

Lahiri-Dutt, K. 2010. Special Essay: Pakistan Floods. http://www.globalwaterforum.org/2010/09/06/special-essay-pakistan-floods/, accessed May 20, 2018.

UNOCHA (United Nations Office for the Coordination of Humanitarian Affairs). 2010. Pakistan—Revised Floods Relief and Early Recovery Response Plan. https://reliefweb.int/report/pakistan/pakistan-floods-relief-and-early-recovery-response-plan-revision-november-2010, accessed May 20, 2018.

Sichuan Earthquake, China, 2008

China is located between the world's two most active earthquake belts—the Circum-Pacific Belt or commonly referred to as the Pacific "Ring of Fire" in the east and the Alpide Belt, which stretches from the Mediterranean to the Himalayas and Indonesia. China therefore is a hotbed of seismic activity, and there have been many devastating earthquakes. The Sichuan earthquake in 2008, also known as the Wenchuan earthquake, is considered the most destructive earthquake in China's recent history. It is also the deadliest temblor since the 1976 Tangshan earthquake that killed more than 242,000 people.

The massive earthquake struck a mountainous area in southwest China's Sichuan province at 2:28 p.m. local time on May 12, 2008. Its epicenter was near Yingxiu, a town of about 9,000 people in Wenchuan county of Sichuan province, about 50 miles (80 km) northwest of the provincial capital—Chengdu. The magnitude of the quake was initially measured at 7.9 on the Richter scale and later revised to 8.0 by the Chinese government. It killed 69,227 people and injured 374,643 more, with 17,923 people missing as of September 18, 2008. At least 5 million people were left homeless. Estimated total economic losses were about $125 billion, making this earthquake one of the costliest natural disasters in Chinese history.

This earthquake hit less than three months before China was to host its very first Summer Olympic Games in Beijing, a much anticipated "coming-of-age" event for the world's most populous and rapidly-industrializing country. The Chinese government, aware of the heightened worldwide interest in China, reacted quickly to the quake. Its unprecedented rescue effort was widely lauded by the international community. That a large number of children perished because their school buildings collapsed during the quake caused many to question the reasons behind the shoddy construction quality in this earthquake-prone region and whether corruption among government officials contributed to it, but the government concluded that the sheer destructive power of the quake, not the subpar construction, was to blame for the collapse of schools.

OVERVIEW OF THE QUAKE

The Sichuan earthquake occurred on the well-known and active Longmenshan, or Dragon's Gate Mountain, fault system on the northwestern margin of the Sichuan Basin. The strike of the Indo-Australian Plate onto the Eurasian Plate at a rate of 2 inches (5 cm) per year results in the uplift of the Tibetan Plateau and its east-northward movement, generating high levels of stress in the Longmenshan fault system. Seismic activities are common in this area. A 7.5-magnitude earthquake on August 25, 1933, in Diexi, about 56 miles (90 km) northeast of the 2008 one, killed about 9,300 people. Another earthquake struck the area in 1976.

With a focal depth of 12 miles (19 km), the 2018 earthquake ruptured the Beichuan fault in the Longmenshan system more than 190 miles (306 km) from southwest to northeast with surface displacements of up to 30 feet (9 m) and at least another fault. The rupture lasted for nearly two minutes, but most of the energy was released in the first 80 seconds. The relatively firm terrain in southwest and central China meant that the seismic waves would travel a long distance without losing their power. The tremors were felt in much of China, as far away as Beijing (930 miles, or 1,500 km to the north) and Shanghai (1,060 miles, or 1,700 km to the east). Shaking also occurred for several minutes after the main quake in Hong Kong, Taiwan, and in a number of South and Southeast Asian countries. Japanese seismologists concluded that the seismic waves of the Sichuan earthquake traveled around the earth six times.

While no foreshocks were reported before the earthquake, a large number of aftershocks hit the area and did not subside until months after the main quake. Within 72 hours of the main quake, 104 major aftershocks with magnitudes ranging from 4.0 to 6.1 were recorded. By January 14, 2009, there were a total of 42,719 aftershocks, 8 of which registered above 6 in magnitude, with the strongest earthquake measuring 6.4. The strong aftershocks caused further damage and casualties.

DAMAGES

The shallow earthquake brought devastation to the severe shock zone on the fault, with a maximum intensity of XI on the Chinese seismic intensity scale, which is similar to the Modified Mercalli intensity scale. An elongated zone extending from Yingxiu—the town closest to the epicenter—near the southwest end, to the town of Beichuan in the northeast, which was struck repeatedly by strong aftershocks, suffered the most severe damages (intensity scale of XI). Almost 80 percent of the buildings in Beichuan County were destroyed. The town was later abandoned and preserved as part of the Beichuan Earthquake Museum to remind people of the terrible disaster. The area that sustained damages of Level VI or above covered 170,056 square miles (440,442 sq. km) in an oval-shaped region that is 582 miles (936 km) long and 370 miles (596 km) wide. The powerful temblor knocked out electricity and communication networks. All highways into the earthquake-stricken areas were damaged or blocked by landslides. Other infrastructure was also devastated. For example, 391 dams, most of them small, were damaged. The earthquake also killed 12.5 million livestock and caused considerable other damages to agriculture.

The massive property damage is attributed to the fact that China did not have an adequate seismic design code until after the 1976 Tangshan earthquake. This

Collapsed buildings in Sichuan, China, after the earthquake in 2008. (Yekaixp/Dreamstime.com)

means buildings constructed before 1976 were not likely to have been designed to withstand earthquakes. In rural areas, in particular, even recently constructed homes did not necessarily follow earthquake codes.

The most heartbreaking part of this devastating natural disaster is probably the deaths of a large number of elementary and high school students. When the earthquake struck at 2:28 p.m., the schools were in session. At least 7,000 school buildings throughout the province collapsed, killing 5,335 students. At the Beichuan Middle School alone, 1,300 students and teachers died.

It appeared from the crumbled school buildings that many were not reinforced with steel beams. Many grieving parents demanded investigations into school collapses and potential responsibilities for shoddy construction qualities. They suspected that corruption among government officials who oversaw the construction projects and unscrupulous builders who cut corners was to blame. But the issue was largely ignored by the state-controlled media.

GEOHAZARDS

The Sichuan earthquake triggered what has characterized as the largest number of geohazards ever recorded. There were more than 60,000 landslides in an estimated area of 316 square miles (811 sq. km), which created 501 new lakes because landslide debris blocked rivers completely, and in 327 other cases, landslides partially obstructed or diverted river channels without forming lakes. The earthquake lakes posed grave dangers to villages and cities downstream because the blockages could crumble under the weight of the rapidly rising water behind the dams and hence cause serious floods and debris flows. To ward off catastrophic breaches, engineers and soldiers dug canals to drain the lake gradually. As a precautionary measure, hundreds of thousands of people were evacuated to safe grounds.

Landslides also blocked roads and mountain passes, which delayed the arrival of rescue workers. Many buildings that survived the shaking were demolished by massive boulders rolling down the mountains. In some places, landslides actually did more damage than the earthquake. The geohazards might have been responsible for at least a third of the human deaths in this earthquake.

RESPONSE

The Chinese government mounted a rapid and full-scale disaster response. The China Earthquake Administration (CEA), the government agency charged with earthquake monitoring and prediction, issued a preliminary forecast of at least 7,000 deaths only 30 minutes after the earthquake. The death toll was considerably underestimated. Premier Wen Jiabao flew to the earthquake area only 90 minutes after the earthquake to oversee the search and rescue work. On the same day, the Ministry of Health sent 10 emergency medical teams to the hardest-hit Wenchuan County, and the Chinese military dispatched 50,000 troops and armed police there to help with disaster relief. However, aftershocks, landslides, falling rocks, persistent heavy rain, and destroyed roads made it difficult for the soldiers and other relief workers to reach the earthquake-damaged villages and towns. Some of them had to get there on foot.

On May 12, the fourth day after the earthquake, China's National Disaster Relief Commission enacted the Level I Emergency Contingency Plan, designed for the most serious class of natural disasters. Twenty helicopters were deployed to deliver food, water, and other relief supplies as well as to evacuate the injured and serve reconnaissance purposes. By May 13, more than 15,600 troops from the Chengdu Military Region had joined the rescue force in the earthquake-stricken areas. The next day, 15 Special Operations Troops from the Chinese military parachuted into the inaccessible northeast part of Wenchuan County to help stranded locals. By May 15, 90 more helicopters joined the rescue work. This was the largest peacetime airlifting operation ever conducted by the People's Liberation Army. One rescue helicopter carrying earthquake survivors crashed in fog on May 31, killing all passengers and the crew. The Chinese military arguably played a major role in the rescue effort.

The Chinese government initially did not accept international help, but on May 13, it accepted a donation of some 100 tons of relief supplies from the Tzu Chi Foundation in Taiwan, and on May 14, China formally requested the support of the international community to respond to the needs of disaster relief. Rescue teams from Japan, Russia, Singapore, South Korea, and Taiwan arrived in the earthquake area shortly after that. The United States shared satellite images of the disaster area in addition to donating tents and generators.

Many Chinese companies, foreign companies operating in China, celebrities, and the general public all over the country donated money or goods for earthquake relief. The Red Cross Society of China stated that the disaster areas needed tents, medical supplies, drinking water, and food, but it suggested that cash instead of relief supplies be donated, as it was difficult to reach the disaster areas due to damaged roads, bridges, and blockages from landslides. On May 16, China stated it had received $457 million in donated money and goods.

Immediately after the earthquake, mobile and terrestrial telecommunications were cut to the affected areas due to power disruption and severe traffic congestion. They were restored shortly after that. The Internet was extensively used to share information related to rescue and recovery. The official Xinhua news agency set up an online rescue center to identify the blind spots in disaster relief. Several websites were set up by volunteers to share contact information for victims and evacuees. The State Council of China declared a three-day period of national mourning for the earthquake victims from May 19 to May 21, the first time a national mourning period was declared for reasons other than the passing of key government or Communist Party leaders. The international media and the International Federation of the Red Cross generally praised the Chinese government for reacting swiftly and openly to the earthquake. Some contrasted that with its less well-coordinated response to the 1976 Tangshan earthquake.

CONCLUSION

The Sichuan earthquake devastated the mountainous areas around Wenchuan, but they have made an impressive comeback, thanks to the assistance from the

Chinese central government and provincial governments throughout the country. In June 2008, the Chinese government issued regulations for the recovery and rebuilding of the earthquake-ravaged areas. The Chinese State Council devised a paired support system that called on 19 eastern and central provinces and municipalities with large economies to each support a severely affected country in Sichuan province (i.e., "one province helping one county"). The arrangement would work for three years, and each assisting province or municipality was expected to devote no less than 1 percent of its total annual budget for this purpose. On November 6, 2008, the Chinese government announced it would spend about $147 billion over the next three years to rebuild the earthquake-damaged areas.

The plan apparently worked. In 2012, the deputy governor of Sichuan province Wei Hong announced that the restoration and reconstruction were completed, with $137.5 billion spent. Local governments had helped more than 12 million people repair their homes and relocated 200,000 farmers whose farmland was lost in the earthquake. Presumably the new buildings followed higher construction standards. In addition to helping in rebuilding those places, the Chinese government has also been promoting more sustainable livelihoods in the fragile mountain ecosystem, including ending farming on steep slopes to reduce landslide risks. It is good to have something positive coming out of a major natural disaster.

Max Lu

Further Reading

Chen, Y., and D. C. Booth, eds. 2011. *The Wenchuan Earthquake of 2008: Anatomy of a Disaster.* New York: Springer.

The EERI-GEER Team. 2008. *The Wenchuan, Sichuan Province, China, Earthquake of May 12, 2008.* EERI Special Earthquake Report. October 2008. https://www.eeri.org/site/images/eeri_newsletter/2008_pdf/Wenchuan_China_Recon_Rpt.pdf, accessed October 28, 2019).

Wang, Z. 2008. A Preliminary Report on the Great Wenchuan Earthquake. *Earthquake Engineering and Engineering Vibration* 7(2): 225–234.

Xing, H., and X. Xu. 2011. *M8.0 Wenchuan Earthquake.* New York: Springer.

Sulawesi Earthquake and Tsunami, Indonesia, 2018

On Friday, September 28, 2018, a 7.5-magnitude earthquake struck Palu on the Indonesian island of Sulawesi just before dusk. The epicenter of this earthquake was only 48 miles (77 km) north of Palu, a short distance from the island's western coast at a depth of only six miles (10 km) below the surface. The earthquake triggered a devastating tsunami. Located 1,100 miles (1,650 km) northeast of Jakarta, Palu, with a population of about 340,000, is the capital of the province of Central Sulawesi. The city is situated on a long and narrow bay at the mouth of the Palu River. The Sulawesi earthquake was more powerful than a series of earthquakes that killed 500 people on the Indonesian island of Lombok in a span of one month between July and August 2018. The most powerful of the Lombok earthquakes struck the island on Sunday, August 3, 2018, measuring 7.0 on the Richter scale. A tsunami warning was also issued in the immediate aftermath of the Lombok earthquake.

However, the Sulawesi earthquake was felt over a wider area; noticeable shaking was reported as far as Malaysia. The strongest shaking was felt in the city of Donggala, where a maximum intensity of IX (violent) was recorded on the Mercalli intensity scale. This city is closer to the epicenter than Palu, where the maximum intensity was VII (very strong). The Indonesian Agency for the Assessment and Application for Technology (BPPT) estimated that the energy released by the earthquake was 200 times than that released in the nuclear bombing of Hiroshima.

CAUSE

Several small foreshocks had been reported throughout September 28, 2018, in the affected area. However, in the early evening, the Palu-Koru fault suddenly slipped and caused a major earthquake followed by a deadly tsunami. This very active fault, where the two plates are moving over each other, is straddled by the city of Palu. The area is at high risk of both earthquakes and tsunamis; for instance, in the past 100 years, 15 events of over 6.5 magnitude have occurred in the area. The largest was a 7.9-magnitude event in January 1996, about 66 miles (100 km) north of the 2018 earthquake. Geologists are not sure whether the tsunami on September 28, 2018, was caused by the movement on the fault rupture from the 2018 earthquake or from underwater landslides within Palu bay caused by the shaking from the earthquake (Cunneen 2018).

A collapsed bridge lies submerged in Palu, Indonesia, after a magnitude 7.5 earthquake triggered a tsunami that slammed into the island of Sulawesi, October 11, 2018. (Heyfajrul/Dreamstime.com)

TSUNAMI WARNING

The Sulawesi earthquake triggered a tsunami, reaching about 20 feet (6 m) high and traveling at a speed of 250 mph (375 kmph). However, geologists had predicted the tsunami height would not exceed the maximum height of 6.6 feet (2 m). About six minutes after the earthquake, an early warning of the potential for a tsunami was issued by the meteorology and geophysics agency (BMKG). However, BMKG officers could not contact the officers in the Palu area because of power outages. The earthquake destroyed the local mobile network in Palu, and no information could get in or out of the affected areas. The tsunami struck the shore of Palu within 20 minutes of the earthquake. To date, the September earthquake has been followed by a total of 150 aftershocks, ranging in magnitudes from 2.9 to 6.3. Fourteen aftershocks of magnitude 5.0 occurred within the first 24 hours of the main earthquake.

BMKG authorities lifted the tsunami warning too soon, only 34 minutes after it was issued, drawing harsh online criticism. Contributing to the premature decisions, the tsunami occurred in an area with no tide gauges to provide useful information about the height of the waves. The closest gauge to Palu is around 125 miles (200 km) away, and the decision to lift the tsunami warning was based on data obtained from this distant gauge. The sensor of the gauge was attached to one of the 22 buoys connected to the seafloor sensor, which transmits advance tsunami warnings (AP 2018b). Some scientists believe that even if a tsunami sensor or a tide gauge had been placed at the mouth of the bay, the tsunami generated by a submarine landslide would not enable detection of the tsunami waves before it struck the shore in Palu.

It is worthwhile mentioning that Indonesia is one of the world's most earthquake- and tsunami-prone countries. The country is located in the area of intense seismic and volcanic activity, known as the "Pacific Ring of Fire." After the 2004 Indian Ocean Tsunami (IOT) that killed 167,540 people, which accounted for more than half of the deaths caused by this extreme event (Paul 2019), a concerted international effort was launched to improve tsunami warning capabilities in the country. Through this effort, a network of 134 tidal gauge stations augmented by land-based seismographs, sirens in about 55 locations, and a system to disseminate warmings by text message was introduced. However, this network has been disabled by vandalism or theft or just became inactive due to lack of funds for maintenance (AP 2018b).

IMPACTS

The earthquake's impact was magnified because of the thick layers of sediment on which the Palu city lies. Thus, the Sulawesi earthquake caused widespread liquefaction, which means the soil behaves like a liquid. This, in turn, causes foundations of buildings and other structures to sink into the ground. As a result, the earthquake destroyed thousands of homes and other structures. Then the tsunami caused extensive damage to the main highway, which was cut off by a landslide, and a large bridge was washed away by the tsunami wave, making it hard for rescue and aid workers to reach the earthquake victims. Landslides also severely disrupted communication networks. Additionally, in Palu, more than half of 560 inmates in a jail were able to flee, and in Donggala, 100 inmates managed to escape because of fire in a jail immediately after the earthquake. Apart from Palu and Donggala, the city of Sigi and the surrounding rural areas were also devastated by the earthquake.

The earthquake and tsunami effectively cut off much of Palu and Donggala from the outside world for several days, and strong aftershocks were reported on the island the day after the earthquake. As noted, electricity and telecommunications were cut, causing severe disruption to air traffic in Palu airport and its seaport, on which the region relied for fuel supplies. As of October 12, 2018, more than 2,000 people have been confirmed dead, and 680 people were still missing while a further 2,500 were injured. Ultimately, the earthquake affected 1.5 million people, displaced 100,000, and put about 200,000 people in urgent need of assistance, about a quarter of them children. It damaged or destroyed more than 65,000 houses, leaving some 330,000 people without adequate shelter (UNOCHA 2018). In terms of number of deaths, this earthquake was the deadliest to strike the country since the 2006 Yogyakarta earthquake.

RESPONSE

After the earthquake and tsunami, local people immediately began efforts to rescue people trapped in the rubble of collapsed buildings and provide urgent assistance to survivors. Later, they were joined by search and rescue teams from the Indonesian Red Cross (PMI), the National Research and Rescue Agency (BASARNAS), the Indonesian National Armed Forces (TNI), and local government agencies. Around 700 military and police personnel were deployed to provide emergency assistance to earthquake victims. However, as with the Lombok earthquake of August 3, 2018, the government response was slow and late. Meanwhile, a large number of local and national aid agencies made appeals to raise funds to provide emergency assistance to earthquake victims. These agencies delivered thousands of shelter kits, solar lanterns, and water purifiers to the disaster zone in addition to trucks and power generators to help get them to where they were needed.

After two days of earthquake and tsunami, on October 1, the government of Indonesia through the national disaster management agency (BNPB) and Ministry of Foreign Affairs agreed to accept international relief aid for the survivors of the events. This surprised some because after the 2018 Lombok earthquake, the government refused to accept such offers from the international community. The BNPB considered that that earthquake did not rise to the level of a national emergency and thus the government did not receive international assistance (AP 2018b). It is customary, particularly for developing countries, to appeal for external help after each major disaster. However, the agency claimed that Indonesia had sufficient resources and substantial experience in handling natural disasters and that the country was proud that it had not declared a national disaster since the IOT in December 2004 (AP 2018a).

However, since October 1, relevant United Nations (UN) agencies and a number of international humanitarian nongovernmental organizations (NGOs) have been working closely with the government (UNOCHA 2018). In fact, one such international NGO, the UK's Disasters Emergency Committee (DEC), made an appeal immediately after the events to help the survivors of the Indonesia earthquake and tsunami. DEC and its local partners worked closely with the Indonesian authorities to provide emergency assistance to those who urgently needed it. Indonesian NGOs also played an important role in providing initial support at a time when broader humanitarian assistance had not yet reached many of the affected areas.

One week after the devastating earthquake and tsunami, the Indonesian Humanitarian Country Team (HCT) in consultation with the Indonesian government launched the Central Sulawesi Earthquake Response Plan, seeking $50.5 million for an initial period of three months (October 2018–December 2018). The government provided $15 million to support the activities included in the plan, which had seven major activities: logistics, water and sanitation, camp management, health, shelter, protection, and food security and livelihoods. The World Food Program (WFP) was committed to support logistics and food security and livelihood activities.

BNPB and the regional disaster management agency (BPBD) have been coordinating the public response to the earthquake and tsunami under the overall leadership of the Coordinating Minister for Political and Security Affairs. The ASEAN Coordinating Centre for Humanitarian Assistance on disaster management (AHA Centre) and other relevant regional agencies and personnel served as the conduit for offers of international assistance for the Sulawesi earthquake and tsunami response (UNOCHA 2018).

As of October 4, 2018, the Indonesian government had received emergency assistance from 29 countries. Initially, the United States provided $3.7 million in humanitarian assistance, including materials for emergency shelter that will help more than 100,000 survivors. In addition, teams of experts from the U.S. Agency for International Development (USAID) have been in Palu to assess needs and logistics. Other countries that provided assistance include Australia, European Union (EU), Malaysia, New Zealand, the Philippines, South Korea, Singapore, Taiwan, and the United Kingdom. By October 21, 2018, most of the health facilities in Palu, Donggala, and Sigi were operational.

CONCLUSION

The Sulawesi earthquake response posed a serious question about the effectiveness of Indonesia's tsunami warning system, which was developed after the 2004 IOT. The failure of tidal sensors prior to the tsunami highlights the weakness of the existing tsunami warning system as well as low public awareness about how to respond to tsunamis, and yet tsunamis are not uncommon in Palu. Still many people in the city are not aware of the risk of a tsunami following the earthquake. Whether tsunami warnings are issued or not, after earthquakes, residents of coastal area should take shelter on higher ground and stay there for at least a couple of hours.

Available reports suggest that instead, hundreds of people gathered on Palu's beach to celebrate an annual festival, the Palu Nomoni Festival. They saw the waves were coming, but did not take shelter on higher ground. This clearly suggests that they were not aware of what to do following an approaching tsunami.

Bimal Kanti Paul

Further Reading

AP (Associated Press). 2018a. Indonesia: Lombok Quakes Don't Meet International Disaster Status. August 20, 2018. https://apnews.com/4a6092bf4b464b95a80499d157ca0143/Indonesia:-Lombok-quakes-don't-meet-national-disaster-status, accessed August 24, 2018.

AP (Associated Press). 2018b. Indonesian Tsunami Warning System "Stuck in Testing Phase": Experts. *The Sydney Morning Herald.* October 1, 2018. https://www.smh.com.au/world/asia/indonesian-tsunami-warning-system-stuck-in-testing-phase-experts-20181001-p5070a.html, accessed August 21, 2018.

Cunneen, Jane. 2018. Would a Better Tsunami Warning System Have Saved Lives in Sulawesi? *The Conversation.* October 2, 2018. https://theconversation.com/would-a-better-tsunami-warning-system-have-saved-lives-in-sulawesi-104223, accessed December 1, 2018.

Paul, B.K. 2019. *Disaster Relief Aid + Changes & Challenges.* Gewerbestrasse, Switzerland: Palgrave Macmillan.

UNOCHA (United Nations Office for the Coordination of Humanitarian Affairs). 2018. Central Sulawesi Earthquake Response Plan (Oct 2018–Dec 2018). October 5, 2018. https://reliefweb.int/report/indonesia/central-sulawesi-earthquake-response-plan-oct-2018-dec-2018, accessed August 21, 2018.

Summer Floods, United Kingdom, 2007

Torrential downpours in May through July 2007 caused widespread surface and river or fluvial flooding across the United Kingdom (UK). During these three months, many cities in England and Scotland were under floodwater as the country experienced the wettest summer of the previous 250 years (Environment Agency 2007). The official records suggest that more than double the usual rain fell in May, June, and July in England and Wales. The location and strength of the jet stream, as well as unusually high surface temperatures in the Atlantic Ocean, caused more intense depressions near the United Kingdom, resulting in unusually heavy rainfall. For much of the 2007 summer, the jet stream was further south than usual, resulting in more rain-bearing depressions crossing southern and central parts of the United Kingdom.

The summer floods of 2007 differed in scale and type from most severe floods that occurred in recent decades. The main source of the flooding was surface water rather than river water. The Environment Agency (2007) estimated that overflowing rivers accounted for one-third of the flooding; problems with surface drainage systems, particularly in cities such as London, accounted for the other two-thirds. Surface water flooded nearly 1,400 properties in London. In the South-East and in the Yorkshire and Humberside regions, surface water flooded about 70 percent of damaged properties. Moreover, "just over half the properties flooded in the East and West Midlands and South-West regions were from surface water flooding" (Environment Agency 2007, 14).

The surface water caused flooding because drains, culverts, sewers, and ditches overflowed with rainwater. Local authorities, water companies, and other infrastructure providers, the Highways Agency, and landowners were responsible for maintaining these facilities. One distinguishing feature of this nonfluvial flooding in 2007 was that it inundated a very high proportion of properties and commercial premises. On the other hand, the river floods also were of a greater magnitude than the design limit of some flood alleviation schemes and many urban drainage systems. However, some localities were subjected to more than one separate flood episode (Marsh and Hannaford 2007).

The rain in May and early June saturated the soil so that little water could infiltrate from the rain that fell in June and July across much of England and Wales. By late July, soils were at their wettest for that time of year for at least 50 years. In urban areas, paved surfaces behaved like saturated soil. In such areas, if the rain is very heavy and its duration is long, the drains will not be able to cope, causing flash flooding. In several flooded areas such as South Yorkshire, Sheffield, and Rotherham, the 2007 summer flood was a 200-year event, meaning a flood of similar magnitude is expected to occur on average once every 200 years, or that a 0.5 percent chance of a flood of this magnitude exists for any one year (Environment Agency 2007).

IMPACTS OF THE FLOODING

Widespread and severe flooding in the June-August timeframe has been very uncommon in the United Kingdom during the last 100 years. Thus, summer flooding does not offer a substantial threat to lives and livelihoods. However, small and short-lived floods on a local scale are not uncommon in the summer. Meanwhile, with near-saturated soils and prolonged heavy rainfall in the summer of 2007, extensive flooding was inevitable. As a result, many rivers and their tributaries in southern England and Wales exceeded peak river flows and thus previous records by wide margins. This, in turn, caused substantial damage of property and suffering of people in the affected areas. The worst affected areas were in northern England (particularly urban areas in South Yorkshire and Humberside), the south Midlands, and parts of the upper Thames basin (Marsh and Hannaford 2007).

Over 55,000 homes and 6,000 businesses were flooded, and all 19,000 homes flooded by rivers were in the floodplain. A survey conducted by the Environment Agency (2007) indicated that about 28 percent of these were built in the previous 25 years. Tragically, 14 people died, and thousands of people stayed in temporary shelters and/or were left without power. In Gloucestershire, flooding of the Mythe Water Treatment plant left 350,000 people relying on bottled water for more than two weeks. Flooding affected over 100 sewage treatment plants in the Midlands, forcing many shut downs. Floods also caused structural damage to transport facilities as many motorways were closed as were many railway stations and lines. About 10,000 people were left trapped on the M5 motorway while others were left stranded at railway stations. Moreover, flooding affected more than 300 schools in

Yorkshire and Humberside and caused serious damage to many schools in Hull. The floods caused insured losses at $6 billion, and other losses were estimated at around $2 billion (Environment Agency 2007).

Many people in flooded and low-lying localities had to be evacuated. Near Rotherham, the threat of the Ulley Dam failure following the late June rainfall prompted the evacuation of around 1,000 people and closure of the nearby highway for 40 hours. Subsequently, emergency services rescued around 7,000 people. Meanwhile, more than 200 people were evacuated from floodwaters in Sheffield, and 20 were airlifted from the roof of a building in the Lower Don Valley (Boyd 2017). Civil and military authorities described the 2007 summer flood rescue effort as the biggest in Britain since the World War II (1939–1945). Flooding seriously affected commercial activities in many towns and cities. For example, in Sheffield, the Meadowhall shopping center closed for a week, and the region sustained significant local impacts on tourism.

The flood also caused widespread damage to maturing crops (e.g., wheat, barley, and fodder) and vegetables (e.g., broccoli, carrots, peas, and potatoes), while in Gloucestershire and Lincolnshire, farmers moved their livestock to higher ground (Marsh and Hannaford 2007). Luckily, there was no outbreak of flood-related illness. Of course, the Health Protection Agency advised people to avoid coming into direct contact with floodwater.

Regarding mitigation infrastructure, the floodwaters affected about 667 miles (1,000 km) of flood defenses (e.g., walls, embankments, flood storage areas, and pumping stations). About half of them were overtopped because of the huge amount of water flowing through the rivers. The sheer scale of the floods far exceeded the design standards for flood defenses, overwhelming those defenses and accounting for nearly 20 percent of the properties damaged by flooded rivers. Additionally, the force of the floodwater breached a small section of defenses. Flood defenses failed at just nine sites, but fortunately, these structural failures did not make the flooding worse at any one of the sites. At five sites, embankments failed only after they were overtopped. Those flood defenses were designed to withstand floods with a 1-in-100 probability of occurrence. The Environment Agency (2007) estimated that the flood defenses successfully protected over 100,000 homes and businesses during the summer of 2007. In the United Kingdom, most flood defenses are maintained by the Environment Agency. The rest are maintained by local authorities, internal drainage boards, and businesses, and individuals.

FLOOD WARNINGS AND GOVERNMENT RESPONSE

Effective warnings for river floods were sent directly to over 34,000 homes to help residents prepare for and cope with the floods. Unfortunately, there was no specific effective warning service for surface water floods, which affected over 35,000 homes and businesses. However, the warning service on river flooding was effective, timely, and accurate, with the exception of 4,100 properties where the authority failed to provide warning because of the technical limitations of flood forecasting systems (Environment Agency 2007). A total of over 500 "flood" and "severe flood" warnings were issued in different localities during the summer of 2007. About 80 percent of the warnings were issued two hours in advance of flooding. In other cases, more than two hours were needed, while in a very few cases, problems with river measuring gauges delayed flood warnings.

Flood warnings were widely issued through the radio network as well as by other means such as vehicle-mounted loudspeakers and sirens. In flood-affected areas, the Environment Agency provided a free service called Floodline Warnings Direct. This service informed people in flood-prone areas about the risk of flooding via a telephone call, text, or fax message. People utilized this free service, and they sent a flood warning to their family members, relatives, and friends in good time. In addition, the official website of the Environment Agency provided flood information. At the height of the floods on July 23, more than 4 million people sought over 10.7 million pieces of information from the website. On the same day, the website received ten times the normal number of requests for information (Environment Agency 2007).

In addition to official attempts to issue flood warnings through various means, many radio and television stations on their own initiatives provided good public service throughout the event. These electronic media broadcasted frequent informative updates, which undoubtedly helped inform the public about the risk associated with the event and the measures they could take to reduce the impact.

The public response to the floods in 2007 was immediate and sufficient. For instance, the British government allocated more funding for additional measures to protect people against future flooding. Since 2007, the government has constructed 1,176 new flood defenses to better protect more than 480,000 properties (Boyd 2017). The government also provided aid to flood-affected councils, supplied bottled water for flood victims, and completed several reports on flooding and response to the flooding. In April 2010, the government passed the Flood and Water Management Act 2010, which provided more power and responsibility to the Environment Agency and local authorities to plan flood defenses coordinated across catchment areas and the wider country.

However, some of the government actions were criticized by the public. For example, the government distributed flood-related responsibilities across four departments rather than via one ministry. Thus, there were too many organizations each with too many competing priorities, which restricted proper planning and management of flooding. As indicated, river and coastline flooding is the responsibility of the Environment Agency; drainage is in the hands of local authorities; water on main roads is the remit of the Highways Agency and Local Authorities; and sewer flooding is the domain of underwater companies. However, the British government was also criticized for not employing the army to assist flood victims. In fact, the Met Office (the United Kingdom's national weather service) warned the British government in the spring that summer 2007 flooding would be likely because the El Niño phenomenon had weakened. But the government failed to take any action to prepare for the anticipated flood.

CONCLUSION

In the contexts of extent and severity, the 2007 summer flooding in the United Kingdom was remarkable. Given the exceptionally heavy rainfall, which has no close modern parallel in the summer, the widespread flooding was expected and anticipated. Although summer rainfall surpassed past record, climatologists believed that it was not a result of global warming. A report by the Centre for Ecology and Hydrology concluded the rain was a freak event, not part of any historical trend. However, many settlements, both on well-populated flood plains and widely distributed urban centers, experienced a wide range of flood types, from the local effects of surface runoff to extensive inundations from river overflows. The 2007 floods underlined the United Kingdom's continuing vulnerability to climatic extremes, but the floods also confirmed the exceptional rarity of the hydrological conditions experienced that summer.

A number of lessons were learnt from the 2007 summer floods in the United Kingdom. First, the country is not immune to flooding. Next, the existing flood defense structures were not adequate, and those structures needed better design to withstand future floods. Further, as noted, there was no flood forecasting and warning system for the surface water flooding. The British government should have developed warnings for this type of flooding. Because surface flooding accounted for more than two-thirds of total damage, public and private authorities should have taken proper measures to reduce risk of surface flooding. Ultimately, the floods in 2007 demonstrated the need for flood risk management to be properly coordinated and to incorporate all sources of flooding.

Bimal Kanti Paul

Further Reading

Boyd, E. H. 2017. Ten Years from the Summer Floods of 2007. July 31, 2017. https://environmentagency.blog.gov.uk/2017/07/31/ten-years-on-from-the-summer-floods-of-2007/, accessed May 24, 2018.

Environment Agency. 2007. *Review of 2007 Summer Floods*. Bristol: Government of UK.

Marsh, T., and J. Hannaford. 2007. *The Summer 2007 Floods in England & Wales: A Hydrological Appraisal*. Oxfordshire, UK: Centre for Ecology & Hydrology.

Superstorm Sandy, United States, 2012

Hurricane Sandy, also known as "Superstorm Sandy," was a Category 1 hurricane that hit the eastern coast of the United States in October 2012. The storm began as a tropical wave (a trough of low pressure) in the warm waters of the Caribbean Sea on October 19, 2012, but it quickly developed into a tropical depression, then a tropical storm, and eventually a hurricane with maximum sustained winds of 74 mph (111 kmph) on October 24, 2012. The hurricane made landfall in three countries before hitting the United States: in Jamaica on October 24, in Cuba on October 25 as a Category 3 storm, and in the Bahamas as a Category 1 storm.

Hurricane Sandy made landfall in the United States as a Category 1 storm on October 29, 2012, near Brigantine, New Jersey, at approximately 8:00 p.m. At the time of landfall, the hurricane had maximum sustained winds of 80 mph (120 kmph). An above-normal high tide amplified the storm surge, causing much greater devastation than that expected of a typical Category 1 storm.

The storm is estimated to have caused over 147 deaths worldwide, including 71 in the United States (NOAA 2017). Within the United States, the hurricane damaged over 650,000 homes and left 8.5 million people without power. The storm was estimated to have caused over $70 billion in economic loss, which makes it the fourth costliest disaster to ever hit the United States (NOAA 2017). The hurricane was so significantly impactful that the World Meteorological Organization has permanently retired "Sandy" from the official list of Atlantic hurricane names.

DISASTER PREPAREDNESS AND RESPONSE

In anticipation of landfall, cities and states along the East Coast undertook several disaster preparedness and response measures. President Barack Obama declared a state of emergency in New York, Maryland, Washington, DC, Pennsylvania, North Carolina, New Jersey, Connecticut, Massachusetts, Rhode Island, and West Virginia and a Limited Emergency in Maine. Given the magnitude of the storm, and in a break from the usual procedure, the governors of New York, New Jersey, and Connecticut made the declaration requests verbally, and the presidential approval process was expedited to meet their needs.

The presidential emergency declarations activated actions and funding for preparedness and response measures, such as evacuation and preemptive service closures at the federal, state, and local levels. In New Jersey, Governor Chris Christie ordered evacuation of residents and casinos in the barrier islands from Cape May to Sandy Hook. In New York City, Mayor Michael Bloomberg ordered similar evacuations from low-lying areas as well as the closure of all public schools. New York Governor Andrew Cuomo also mobilized the army and Air National Guards in anticipation of the disaster.

Service and utility providers also took similar preventative measures. In the two days before landfall, Amtrak canceled all services originating and terminating in East Coast stations, while the New York Metropolitan Transit Authority (MTA) and the Port Authority of New York and New Jersey (PANYNJ) suspended all subway, commuter rail, and bus services. The Southeastern Pennsylvania Transportation Authority and the Metro in Washington, DC, suspended all services, and over 20,000 flights were canceled. New York's LaGuardia Airport and John F. Kennedy International Airport as well as Newark's Liberty International Airport halted operations, while the United Nations Headquarters in New York and U.S. federal offices in Washington, DC, closed in anticipation of the disaster.

HURRICANE IMPACT IN NEW YORK AND NEW JERSEY

Hurricane Sandy made landfall late evening on October 29, 2012, near Brigantine, NJ, just north of Atlantic City, causing extensive flood and wind damage and losses to coastal communities along the eastern seaboard. In New York City, a combination of heavy rains and a nearly 14-foot (2.3 m) storm surge caused the Hudson River, New York Harbor, and the East River to flood Lower Manhattan and several MTA subway tunnels. Over 11 million commuters in New York were left without service during this time. The New York MTA is estimated to have experienced approximately $5 billion dollars in losses from the hurricane, including $4.75 billion in infrastructure damage and $246 million in lost revenue and increased operating costs. More than 2 million customers in New York City were left without power during the storm, of which 8,200 remained without power until January 2013. This caused additional hardships for disaster-affected households during the cold winter months.

Power failure during the storm also led to the evacuation of multiple hospitals in Manhattan and Coney Island. The 18-story high New York University (NYU) Langone Medical Center had to evacuate over 300 patients, including 45 critical-state patients and 20 babies, who had to be carried down multiple flights of stairs. In the Breezy Point neighborhood of Queens, an electrical fire driven by high winds destroyed over 110 homes and damaged an additional 20. The incident was one of the worst residential fires recorded by New York's Fire Department since its inception in 1865. The storm damaged over 100,000 homes in Long Island alone. Power outages and flooding also forced the New York Stock Exchange to suspend trading operations for two days, its longest weather-related closure since 1888 (CNN 2018).

In coastal New Jersey, the storm destroyed Atlantic City's world-famous Boardwalk, boardwalks of Seaside Heights and Belmar, and damaged over 72,000 structures and 346,000 housing units in the region. Over 10,000 structures were slated for demolition due to disaster damage, and over 1,000 sites would need remediation after hazardous material discharge. About 75 percent of New Jersey's small businesses were directly or indirectly affected by the storm, causing over $8.3 billion in business losses. Over 2.4 million customers lost power during the disaster, and almost 20,000

Aerial photograph of damage caused by Hurricane Sandy at Mantoloking, New Jersey, on November 2, 2012. (Greg Thompson/USFWS)

residents of Hoboken, New Jersey, were isolated in their homes after water from the Hudson River overtopped its seawall. The storm also damaged over 51 schools and caused six schools to shut down for the whole school year (Smith 2013).

In addition to these impacts, the disaster-affected region also faced severe gas shortages owing to power loss and storm damage. Only 25 percent of New York and 40 percent of New Jersey gas stations were operational in the weeks following landfall, leading to the institution of gas rationing in both states. In New York, this plan required that drivers whose license plates ended in an odd number, a letter, or other character could only purchase gas or diesel fuel on odd-numbered days, while those with plates that ended in even numbers, including the number zero, could purchase gas only on even-numbered days. The rationing in New Jersey extended for about two weeks and almost a month in New York (CNN 2018).

DISASTER RESPONSE, RELIEF, AND RECOVERY IN NEW YORK AND NEW JERSEY

Formal disaster relief and response activities began as soon as there was landfall in all the affected states. This included rescue of endangered residents, shelter provision, medical aid, and financial assistance. A week after the disaster, the White House reported that 164,000 Connecticut, New York, and New Jersey residents had applied for federal assistance from the Federal Emergency Management Agency (FEMA). More locally, several prominent hospitals in New York City had to be evacuated after the disaster, including the NYU Langone Medical Center, the Bellevue Hospital Center, Coney Island Hospital, and Palisades Medical Center. A total of 6,700 National Guards actively participated in these rescue and relief activities (CNN 2018).

Aid organizations, both big and small, played a critical role in providing disaster relief, aid, and shelter. It is estimated that over 9,000 people across 13 states spent the first night of the storm in 171 American Red Cross-operated shelters. According to the American Red Cross, the organization disbursed over $312 million in donations to disaster-affected families and individuals after Hurricane Sandy (CNN 2018). At the other end of the spectrum, Occupy Sandy was a grassroots relief operation that functioned on the philosophy of mutual aid. Occupy Sandy volunteers collected clothes, blankets, and food, which were distributed through multiple decentralized "hubs" in disaster-affected neighborhoods. In addition to these activities, Occupy Sandy volunteers also coordinated a registry on Amazon's website to solicit material donations for hurricane relief and aid. Occupy Sandy organizers estimate having received almost 36,000 individual items valued at $718,000 during this donation drive (Heindl 2012).

On January 28, 2013, the U.S. Congress passed $50.5 billion "Hurricane Sandy relief bill," which provided funding for recovery efforts in all states affected by the disaster. New York City received the largest share, $13 billion, of which $9 billion was allocated through FEMA grants and $4.21 billion allocated through the Department of Housing and Urban Development (HUD). Private claim insurance in New York City alone was estimated to be over $18 billion (Finn et al. 2016). Most of the federal and insurance monies were used toward repair and reconstruction of homes, businesses, and public infrastructure damaged or destroyed in the disaster.

Several recovery planning activities were initiated at the federal, state, and local levels in the aftermath of the disaster. In 2013, HUD funded and oversaw implementation of "Rebuild by Design" (RBD), which solicited resilience-building projects for the Sandy-affected areas. Over 148 teams applied from across the world, of which 10 were selected to proceed to the design stage. Of these, six were eventually selected by HUD for implementation, including a series of artificial barrier islands along the New York and New Jersey harbors, a comprehensive resiliency plan for the Hunts Point section of New York City, and an integrated storm-water management program for parts of the New Jersey coast (Finn et al. 2016).

The state of New York initiated other recovery planning activities such as the New York Rising Community Reconstruction (NYRCR) program, which assisted communities across the state in the planning and reconstruction activities. The program was particularly beneficial for rural and small urban communities lying outside of the New York City metro area and was funded through the Community Development Block Grant-Disaster Relief (CDBG-DR) grant allocated by HUD to the state. In addition, the state offered tax relief measures to allow individuals to claim personal property loss and aid donations in their income tax filings. In New York City, the NY Economic Development Authority implemented the city-wide Special Initiative for Rebuilding and Resilience (SIRR), which focused on long-term strategies for disaster reconstruction as well as climate mitigation and adaptation. The final SIRR report released in June 2013 outlined various actions on infrastructure, the built environment, critical systems, and public awareness, such as coastal protection measures, building retrofits, and construction of resilient utility systems (Finn et al. 2016).

In New Jersey, the state's Department of Community Affairs (DCA) created the "reNew Jersey Stronger" initiative and implemented the New Jersey Post Sandy Planning Assistance Grant Program (PSPAG). Operated in phases, the program provided financial assistance to disaster-affected communities to create broad recovery strategies as well as more specific plans (Finn et al. 2016). The New Jersey Economic Development Authority also initiated the Storm Recovery Loans program, which provided low-interest loans to small businesses impacted by Sandy.

CONCLUSION

Although Super Storm Hurricane Sandy affected 24 states, it caused severe damage in New York and New Jersey. Since disaster recovery is a long-term prospect, local communities along the East Coast will continue to recover from the effects of Hurricane Sandy for many years to come.

Divya Chandrasekhar

Further Reading

CNN Libraries. 2018. Hurricane Sandy Fast Facts. CNN. October 29, 2018. https://www.cnn.com/2013/07/13/world/americas/hurricane-sandy-fast-facts/index.html, accessed November 1, 2018.

Finn, D., D. Chandrasekhar, and Y. Xiao. 2016. Planning for Resilience in the New York Metro Region after Superstorm Sandy. In *Spatial Planning and Resilience Following Disasters: International and Comparative Perspectives*, edited by S. Greiving, M. Ubaura, and J. Tesiliar, 196–314. Bristol, UK: Policy Press.

Gerhardt, Mary Beth. 2012. Hurricane Sandy—October 28–31, 2012. https://www.wpc.ncep.noaa.gov/winter_storm_summaries/event_reviews/2012/Sandy_Oct2012.pdf, accessed January 9, 2019.

Heindl, Matt. 2012. Occupy Sandy's Happy Hackers. Razorfish Buzzcut (Blog). January 25, 2012. https://razorfishbuzzcut.tumblr.com/post/41564190632/occupy-sandys-happy-hackers, accessed November, 2018.

NOAA (National Oceanic and Atmospheric Administration). 2017. *Costliest U.S. Tropical Cyclones Tables Updated*. Miami, FL: National Hurricane Center. http://www.nhc.noaa.gov/news/UpdatedCostliest.pdf, accessed January 9, 2019.

Smith, Christopher H. 2013. Floor Statement on Sandy Supplemental. https://chrissmith.house.gov/uploadedfiles/floor_remarks_on_sandy_jan_2_2013.pdf, accessed November 6, 2018.

Tangshan Earthquake, China, 1976

China lies between two major tectonic plates. The Indo-Australian plate thrusts northward along the Himalaya mountains in Southwest China while the Pacific plate pushes westward against the Eurasian plate from the east. The forces from the tectonic movements not only result in frequent seismic activities along the collision zones but can also reactivate faults deep within the Eurasian plate. China is therefore earthquake-prone.

On July 28, 1976, a 7.8-magnitude earthquake hit the northern Chinese city of Tangshan, located about 96 miles (155 km) east of the Chinese capital—Beijing, in the northeastern part of Hebei province. The powerful temblor nearly razed this coal-mining and industrial city of one million residents and killed a quarter million people, making it the second deadliest earthquake of the twentieth century, after the Haiyuan earthquake in 1920, in North China's Ningxia province. The sheer destruction caused by this earthquake shocked the world, but soon afterward, the Chinese government mobilized the resources to rebuild the city. Tangshan was rebuilt completely on its ruins in a decade. Today Tangshan is a bustling city of 8 million people. For its impressive achievements in rebuilding and reinventing itself, Tangshan earned the moniker "The Brave City of China."

THE EVENT

The main shock of the Tangshan earthquake struck at 3:42 a.m. local time on July 28, 1976, and lasted for approximately 14–16 seconds. Its epicenter was in the southern part of Tangshan, roughly 9 miles (15 km) above the earthquake's focus. The impact was felt within a 680-mile (1,100 km) radius of Tangshan, in 14 provinces and autonomous regions of the country. Beijing and another nearby city—Tianjin, which is 63 miles (101 km) to the southwest of Tangshan—endured considerable damages.

The earthquake took place along the shallow Tangshan fault—a near vertical right-lateral strike-slip fault unknown at the time, which runs approximately east-northeast right under the center of the city. It is one of the three faults in the Changdong fault zone. Rock strata in a strike-slip fault are usually displaced in a horizontal direction along the fault line. The main shock in the Tangshan earthquake resulted in the block on the southeast side sliding about 10 feet (3 m) to the southwest, which caused a 75-mile (120-km) subsurface rupture and many surface faults.

A series of aftershocks followed the main shock, with 12 of them at a magnitude of 6 or greater. The first major aftershock, with a magnitude of 6.2, occurred at 7:17 a.m. on the same day, just three and a half hours after the initial tremor. An even bigger aftershock (magnitude of 7.1) struck at 6:45 p.m. that afternoon near the city of Luanxian, located about 43 miles (70 km) to the northeast of Tangshan and just south of the northeastern end of the Tangshan fault. Most aftershocks took place between the two ends of the Tangshan fault, stretching about 87 miles (150 km). The strong aftershocks caused many remaining structures to collapse and more lives to be lost. They also hampered initial rescuing effort.

THE DAMAGE

The Tangshan earthquake is one of the deadliest earthquakes in recorded history. Its death toll is officially reported at 242,769 people, but the number is for the city of Tangshan only. Some estimates put the casualties as high as 700,000. In 7,218 households, every member perished. In addition, about 164,851 people in Tangshan were severely injured. More than 60,000 families were left homeless and 4,204 children became orphans.

People cross the crumpled Chengli Bridge in Tangshan, China, on July 28, 1976, following a massive earthquake. (U.S. Geological Survey)

Property damage was extensive. Tangshan endured category XI (extreme) damages in the modified Mercalli scale, the second strongest on the Chinese earthquake intensity scale. About 93 percent of residential buildings, unreinforced houses, and multistory buildings and 78 percent of industrial buildings collapsed. Estimated total economic losses exceeded $10 billion yuan. A large swath of the surrounding area (approximately 3650 square miles or 9,500 square km) suffered various degrees of damages. Beijing suffered level VI intensity shaking, damaging about 10 percent of the buildings in the city; at least 50 people lost their lives. Damage in Tianjin varied from categories VI to IX, and because of its much larger size, Tianjin's total property damage reportedly exceeded Tangshan's.

The quake knocked out water pumping stations, water pipes, sewage pipes, and electricity. The communication network in and around Tangshan was destroyed. Bridges collapsed, railroad tracks were severely damaged, and all but one of the roads in Tangshan were wrecked and impassable.

Several factors contributed to the great loss of human lives and severe property damages due to the Tangshan earthquake. First, the earthquake was very strong but shallow, which means much of the enormous energy released by the earthquake was converted to surface shaking. Tangshan is located in the northeastern part of the great North China Plain—an alluvial plain formed by sediments eroded from the nearly Yanshan mountains over thousands of years. The thickness of the sediment layer overlying the limestone and sandstone below ranges from several feet to around 2,000 feet (600 m). This type of ground is particularly vulnerable to earthquake shaking. To make things worse, the earthquake resulted in liquefaction and sand blows that accentuated its destructive power. Liquefaction is the transformation of the sandy soil in the region into a fluid-like mass due to violent shaking, which destabilized the ground and made the earthquake more destructive to buildings and infrastructure. Sand blows resulted from the increased weight brought on by slumping layers of soil upon waterlogged sand below, causing the outpouring of wet sand. This was particularly a problem in the area south of Tangshan, where sand blows silted wells and irrigation ditches and caused much damage to crops.

Second, seismic risks in the area around Tangshan were grossly underestimated. Chinese seismologists at the time did not anticipate a major earthquake there any time soon. Most buildings in the city were constructed without consideration for earthquakes and few were anchored to the bedrock. In fact, no earthquake building standards existed at all before the 1976 earthquake. Not surprisingly, most unreinforced concrete and brick buildings pancaked in the earthquake. In addition, the timing of the earthquake likely made the human loss greater. At that early morning hour, most people were asleep and were therefore indoors when the disaster befell them.

THE RESCUE EFFORT

The earthquake virtually flattened the whole city, trapping many people in the debris. Those who managed to crawl out to the open were stunned by the disappearance of their homes and city, but they quickly began to dig through the rubble using their bare hands and makeshift tools to answer muffled calls for help and look for survivors. For some time, survivors lacked clean drinking water, food, and medical supplies. Many people had to drink contaminated water from pools and other sources. Some injured people died from lack of timely medical care. Residents constructed temporary shelters for themselves from whatever materials they could find.

The Chinese government was initially unable to pinpoint the epicenter of the earthquake or find out about the scale of destruction because the communication facilities in Tangshan were destroyed. A messenger dispatched by the local government drove an ambulance to Beijing to deliver the news. However, once the scope of the disaster was clear, the central government reacted quickly by using military helicopters and planes to drop food and medical supplies. Some 100,000 soldiers, 30,000 medical personnel, and 30,000 construction workers went to Tangshan. The operation was initially chaotic.

Many soldiers had to march on foot hundreds of miles from where they were stationed to their destination. In their rush to reach Tangshan, relief workers accidentally clogged the only remaining road, causing themselves and their supplies to be stuck in traffic for hours. The rescue effort was also hampered by lack of heavy equipment and proper training as well as by the dangers of aftershocks. As time went by, the operation became much better coordinated. Local governments established Headquarters for Resisting the Earthquake and Relief Work to direct the rescue and relief efforts. Medical personnel set up makeshift clinics with very rudimentary equipment. It was estimated that 80 percent of the people trapped under rubble were saved. The Chinese government emphasized self-sufficiency and rejected all offers of foreign relief aid. Because the earthquake occurred during the heat of midsummer, the dead had to be buried quickly, mostly in shallow graves near their destroyed homes. This soon became a public health problem after rain exposed the decaying bodies. Many corpses had to be located and reburied outside of the city.

THE CONTROVERSIES AROUND PREDICTION

The Chinese government claimed the Tangshan earthquake was unexpected, and seismologists generally also agree that it was not predicted, but information made available later showed that it might not have been the case. In fact, it was predicted by seismologists at two earthquake monitoring stations, but their warnings were either not taken seriously or set aside because key government officials at the time were worried about the political ramifications of a false alarm in the capital region of the country. To be fair, earthquake prediction is still very much a scientific challenge even today, but one could only imagine the number of lives that could have been saved had the authority heeded the warning of an impending disaster at the time.

China has prided itself on a long tradition of trying to predict earthquakes. The Chinese invented the first functioning seismometer in 132 AD and has documented extensively anomalous natural phenomena that may be considered precursors to earthquakes. The Tangshan earthquake occurred near the tail end of China's Cultural Revolution (1966–1976), during which scientific research was generally neglected, but earthquake monitoring and prediction received much attention out of necessity. China established a nationwide earthquake prediction and reporting network, and Chinese seismologists claimed to have successfully predicted the 1975 Haicheng earthquake about 250 miles (400km) northeast of Tangshan, in Liaoning Province.

In July 1976, a seismologist in Tangshan and another in Beijing independently made short-term predictions of an imminent earthquake in the general vicinity of Tangshan, between July 22 and August 5, but they were eventually rejected by decision-makers at the State Seismological Bureau. The reasons include a belief at the time that major earthquakes were more likely to occur in western China than in the east where Tangshan is and the bureaucratic concerns over large-scale social and economic disruptions an earthquake warning may cause in the politically sensitive capital area.

While there were no foreshocks before the Tangshan earthquake, it was documented that nature gave a series of other advance warnings in the days before the earthquake, some of which were taken into consideration in the short-term predictions made by the two seismologists mentioned earlier. Well water in a village outside of Tangshan rose and fell several times the day before the earthquake. Gas reportedly spouted out of a well in another village on July 12 and increased, leading up to the earthquake. Many other wells in the area surrounding Tangshan showed cracking. There were also many reports of anomalous animal behaviors ahead of the earthquakes. Chickens at a farm would not feed and ran around frantically chirping. Mice and weasels came out of their nests in broad daylight and appeared panicked. The night before the earthquake, on July 27, many residents reported seeing flashes of colorful lights and roaring fireballs in the sky. However, since no official earthquake warnings were issued, no precautionary measures were taken, and with the exception of one county, people were completely unprepared when the disaster befell them.

CONCLUSION

The 1976 earthquake flattened Tangshan, but efforts to rebuild the city started shortly after the rubble was removed. More than 3,000 experts from across the country contributed to the planning for a new Tangshan. Large-scale construction started in the second half of 1979 and ended officially on July 28, 1986, when Tangshan commemorated the tenth anniversary of the earthquake. The rebuilding focused on factories, mines, and infrastructure first and residential housing later, but by October 1988, all residents in temporary earthquake shelters had moved into new housing. The construction was done following the first national earthquake-resistant building standards issued by the Chinese government following the Tangshan earthquake.

The 108-feet (33 m) tall earthquake memorial and the Tangshan Earthquake Relic Site Memorial Park remind visitors of the devastation and enormous human losses the city suffered, but they also attest to human triumph in fighting against powerful natural disasters.

Max Lu

Further Reading

Palmer, J. 2012. *Heaven Cracks, Earth Shakes: The Tangshan Earthquake and the Death of Mao's China*. New York: Basic Books.

Paltemaa, L. 2016. *Managing Famine, Flood, and Earthquake: Tianjin, 1958–1985*. London, UK: Routledge.

Qian, G. 1989. *The Great China Earthquake*. Beijing, China: Foreign Language Press.

Yong, C., T. Kam-Ling, F. Chen, Q. Zou, and Z. Chen, eds. 1988. *The Great Tangshan Earthquake of 1976: An Anatomy of Disaster*. New York: Pergamon Press.

Thomas Fire, California, 2017–2018

Wildfires are traditionally understood as unplanned and unintentional fires, including unauthorized human-caused fires, originating in undeveloped wildland areas. As they are unplanned and unintentional, wildfires are differentiated from prescribed fires, also referred to as prescribed burns or controlled burns. When conducted appropriately, prescribed fires are one of the most indispensable fire management tools.

Wildfires are steadily increasing in frequency, duration, and magnitude, prolonging the traditional fire season in many regions of the United States and becoming more destructive in terms of loss of life and economic impacts. Researchers attribute the intensification of wildfires to a combination of contributing factors, such as climate change, increased development along the wildland-urban interface (WUI), and invasive species. The devastating impact of wildfires has become exceedingly apparent in California during the last three decades. Sixteen of California's 20 largest wildfires, determined by the total acres burned, have occurred since 1999. As of January 2018, the largest California wildfire on record was the Thomas Fire.

The Thomas Fire ignited on December 4, 2017, near Thomas Aquinas College in Ventura County California, about 50 miles (75 km) northwest of Los Angeles. Regardless of speculation, the official cause of the fire remains unknown. The Thomas Fire rapidly spread through Ventura and Santa Barbara Counties, largely due to strong Santa Ana winds and persisting drought conditions that dried vegetation, amassing in a fuel bed to further feed the wildfire. During the incident, 103,253 residents evacuated and 934 sheltered in place. The fire caused the deaths of one firefighter and one civilian. By the time the U.S. Forest Services declared it to be 100 percent contained on January 12, 2018, the Thomas Fire had resulted in the destruction of 1,063 structures and damage to an additional 280 structures. Destroyed and damaged structures included both residential and commercial property. The Thomas Fire burned a total of 281,893 acres (114,080 ha) in Southern California, briefly becoming the largest wildfire in California history until being surpassed only a few months later in July 2018, when the Mendocino Complex fire burned 459,123 acres (185,804 ha) in Northern California (CWERT 2018).

The area of Ventura and Santa Barbara counties in Southern California consumed by the Thomas Fire has an active history of recurrent wildfire incidents. Since 1983, approximately 66 percent of the Thomas Fire area has experienced one or more wildfires. Since 2008, that same area has experienced 10 major fires that have directly threatened heavily populated areas, destroyed infrastructure, and resulted in debris flows (CWERT 2018).

Flames burn the mountains above Carpinteria, California, during the Thomas Fire, December 2017. (Erin Donalson/Dreamstime.com)

CONTRIBUTING FACTORS

Wildfires generally start naturally, typically sparked by lightning strikes. However, human-caused changes to woodland areas and climate change are major contributors to intensified wildfire activity (FCAT 2018). Some examples of human-caused sources include debris burning, campfires, arson, discarded cigarettes, sparks from operating equipment, and above-ground power lines. Once ignited, the spread and behavior of the fire is influenced by the type and prevalence of fuel available, weather conditions, and topography. Factors that contributed to the rapid and vast spread of the Thomas Fire included: climate change, persisting drought conditions, Santa Ana winds, the topography of the region, invasive species, and rampant beetle infestations.

Early in 2017, Southern California experienced very wet conditions, causing vegetation to flourish. Leading up to the Thomas Fire, drought conditions persisted for months, causing the abundant vegetation to become dry and brittle, an ideal fuel source. The dominant types of vegetation within the Thomas Fire perimeter were chaparral-southern coastal scrub, brush, and tall grass (CWERT 2018). The variance of vegetation height allowed flames to spread from the soil surface into the canopy of the Los Padres National Forest via "ladder fuel." In addition to drought conditions, woodland areas in western parts of North America have been dying off due to bark beetle infestations. Bark beetles cause trees to die, and they fall to the forest floor, contributing to an already substantial fuel bed (FCAT 2018).

As flames climb from the soil surface into the canopy, wind distributes embers, spreading wildfires at a faster rate over a greater distance. The Santa Ana winds occur from September through May in southwestern California. These hot and dry winds are regularly responsible for exacerbating wildfire conditions and knocking down power lines, which can also spark fires. In 2017, Santa Ana winds worsened drought conditions, further drying vegetation and producing more fuel for the Thomas Fire.

Santa Ana winds move westward toward the coastline of southern California, encountering the Sierra Nevada, San Gabriel, and the San Bernardino mountain ranges. Within the Thomas Fire perimeter, elevation ranges from sea level to 6,000 feet (USDA-Forest Service 2018). Downslopes in topography increase wind speed, further drying vegetation as winds distribute embers and gust downhill. Additionally, embers are pushed uphill and further dispersed with the rising heat.

COMPOUNDING DISASTERS

Wildfires strip landscapes of vegetation, increasing the potential for flood hazards. The critical concerns with the loss of vegetation in the Thomas Fire area were an increased probability for hillslope erosion, alluvial fan flooding, and debris flow hazards (CWERT 2018). According to the U.S. Forest Service Burned Area Emergency Response (BAER), Burned Area Report, most of the Thomas Fire burn area had an estimated vegetative recovery period of three to five years (USDA-Forest Service 2018).

On January 9, 2018, Santa Barbara County experienced a severe rain event. Accumulation was reported at a rate of 6.48 inches (165 mm) (an hour for the initial five-minute downpour and 3.44 inches or 165 mm) an hour for the subsequent 15-minute downpour (CWERT 2018). With the Thomas Fire almost contained, rapid accumulation triggered a debris flow event resulting in loss of human life, displacement of private and commercial residents, and property destruction. Debris flows can occur with little warning and overtake objects in their path with a devastatingly destructive force. Therefore, it is critical that residents living in and downstream from wildfires maintain situational awareness and be ready to evacuate if necessary (CWERT 2018).

Mandatory and voluntary evacuation notices went out at the outbreak of the fire and again in January when the storm moved in, and debris flow was a present danger. Due to the wording used (such as "warning" versus "order") and the timing of the notices, many residents found instructions confusing or did not know where to get proper information. In the weeks and months to come after, the affected counties and state government recognized that their notification system and timeline needed more adjustments to ensure information is clear and presented in a timely and consistent manner.

RECOVERY

After any natural disaster, recovery occurs in the form of rebuilding and re-establishing normal daily life. However, when certain areas suffer loss year after year due to wildfires, it may be time to question whether rebuilding homes is justifiable. With the rebuilding of houses or entire neighborhoods in WUI, which have burned in the past, come two primary concerns. Firstly, will homes built to the newest standards (such as the use of fire-safe materials or significant presence of defensible-space) be safe if another disaster occurs? Secondly, the prudence of maintaining residential development in wildlands, especially without policies, legislation, and practices to better care for the forests. Before addressing longer-term concerns, debris and hazardous material need to be removed safely.

On December 5, 2017, Ventura County Environmental Health Department requested a disaster declaration from the state. In securing that declaration, the county was able to begin a hazardous material removal program immediately. Nine days into the fire, a local assistance center opened to assist local homeowners with services, and several other centers opened in neighboring cities. Throughout the affected area, community meetings were held to educate locals on debris-removal programs, assistance programs, and mudslides and/or debris flows that could follow the fire. On December 8, 2017, California received a Presidential Disaster Declaration, which in turn triggered public and individual assistance, as well as Small Business Administration (SBA) disaster loans. That declaration was then amended to account for the damage done by flooding, mudslides, and debris flow.

Following full containment, the state began the Fire Debris Removal Program, which comprises two phases. Phase one is the removal of household hazardous waste. Complete debris removal is required before residents can acquire permits for rebuilding or beginning demolition. Phase two consists of debris and ash removal. At this point, residents can either qualify for the California Office of Emergency Services Fire Damage Debris Clearance Program (provided they have Debris Removal Right-of-Entry Permit and insurance coverage) or hire private contractors. In the six months following the fire, CalRecyle (a state program that conducts debris removal) assisted over 600 homeowners.

Environmental agencies also began recovery efforts as quickly as the fire started. Due to the location of the fire, both the U.S. Forest Service and Natural Resources Conservation Service were active in containment and recovery. The U.S. Forest Service Burned Area Emergency Response (BAER) team immediately began identifying threats to human life and safety, property, and critical natural or cultural resources. The primary goal of assessment is to prescribe emergency stabilization efforts such as installing erosion measures, and eventually, the forest service can incorporate long-term rehabilitation efforts such as re-growing the forest. The Watershed Emergency Response Team also deployed and began rapid evaluation of postfire hazards to life and property. Their recommendations are passed along to respective agencies with the intention of mitigating impacts of postfire secondary disasters (CWERT 2018).

LESSONS LEARNED

In response to the January 9, 2018, debris flows, Santa Barbara County opened the Montecito Center for preparedness, recovery, and rebuilding as a central source of information and services, as well as contributing input and housing a true community recovery process. The center and the Recovery Strategic Plan are two critical tools in shaping recovery and resiliency efforts.

Establishing resilience amid climate change, further encroachment of development on wildlands, and incomprehensive preparedness practices is difficult. After both the Thomas Fire and subsequent debris flow, the state has begun to take stock of what they could face in years to come, which requires actively addressing climate change and enacting preventative measures. In the summer following the event, the state released the California Forest Carbon Plan—a collection of goals to be implemented by 2030 as part of the larger 2017 Climate Change Scoping Plan (FCAT 2018).

If residents continue to live and rebuild in the same areas, city planning and warning/evacuation procedures must be carefully examined and adjusted to prevent loss of life and property. The California Environmental Quality Act (CEQA), Forest Practice Act and Rules, and National Environmental Policy Act (NEPA) contain some of the most important land management laws, advocating for WUI defensible-space, strategically placed fuel breaks, and ecological restoration (FCAT 2018). In addition to examining fire code building standards, agencies recognize that their warning system and evacuation language need to be adjusted to clearly communicate the present condition of an ongoing event and possible danger later. When residents are confused about the expected behavior, they may fail to protect their property and safety in the case of a fire or subsequent disasters.

The most significant obstacle that both California government and residents face in recovery and mitigation is a lack of time. There was not enough time to reflect and act accordingly because many residents had nowhere to go and needed housing. Then in July (less than six months following the Thomas Fire), the Mendocino Complex took the title for the largest wildfire and left more than a hundred homes destroyed and another hundred damaged. While measures such as the Carbon Plan will take time, other measures need to be considered quickly and implemented almost immediately to counter the next fire.

Christine Beste and Samantha Clements

Further Reading

CWERT (California Watershed Emergency Response Team). 2018. Thomas Fire Watershed Emergency Response Team Final Report. February 26, 2018. http://cdfdata.fire.ca.gov/admin8327985/cdf/images/incidentfile1922_3383.pdf, accessed January 3, 2019.

FCAT (Forest Climate Action Team). 2018. California Forest Carbon Plan: Managing Our Forest Landscapes in a Changing Climate. http://resources.ca.gov/wp-content/uploads/2018/05/California-Forest-Carbon-Plan-Final-Draft-for-Public-Release-May-2018.pdf, accessed January 2, 2019.

USDA (U.S. Department of Agriculture)—Forest Service. 2018. Draft: Thomas Burned Area Report. January 6, 2018. https://www.fs.usda.gov/Internet/FSE_DOCUMENTS/fseprd570172.pdf, accessed January 4, 2019.

Tohoku Earthquake and Fukushima Tsunami, Japan, 2011

Japan is located on the eastern edge of the Eurasian Plate under which lies the Pacific Plate, an oceanic plate, to the east of Japan. Because of its location at the juncture of two plates, Japan is the nation with the most recorded tsunamis in the world. On Friday, March 11, 2011, at 2:46 p.m. local time, a powerful earthquake measuring 9.0 on the Richter scale occurred along a subduction zone in the Pacific Ocean 43 miles (70 km) off the northeast coast of the Oshika Peninsula of Tohoku, Japan, 18 miles (29 km) below the sea surface. The nearest city to the epicenter was Sendai, 86 miles (130 km) away. Tokyo, the capital of Japan, was 248 miles (373 km) from the epicenter. The shaking of the earthquake lasted for about six minutes. It was the most powerful earthquake ever recorded in Japan and the fourth most powerful earthquake in the world since modern record-keeping began in 1900 (Tania 2011).

The earthquake resulted from a collision of the Pacific plate and the Eurasian plate, and it was so powerful that the earthquake knocked the whole planet off its axis by a foot (0.3 m) and increased earth's rotational speed, which shortened the length of a day by about a microsecond. The earthquake moved Honshu, the country's main island, 8 feet (2.4 m) east, and the effects of the earthquake were felt around the world. In addition to being called the 2011 Great Japan Earthquake, it is also known as Tohoku Earthquake and the 3/11 earthquake (following the Americans' labeling of the September 11, 2001, terrorist attacks as 9/11) (Karan 2016). The main earthquake was preceded by a number of large foreshocks, with over 1,000 aftershocks reported of which 80 were over magnitude 6.0.

The Great Japan Earthquake triggered deadly tsunami waves that reached heights of up to 133 feet (40.5 m) in Miyako in Tōhoku's Iwate Prefecture, which exceeded the record set by the 1896 Meiji earthquake and Sanriku tsunami. The tsunami waves traveled up to 6 miles (10 km) inland in the Sendai area. The waves

An aerial view of damage caused by the magnitude 9.0 Tohoku earthquake and Fukushima tsunami in Japan, 2011. (U.S. Navy/Petty Officer 3rd Class Alexander Tidd)

also traveled across the Pacific at a colossal speed of 533 miles (800 km) per hour, reaching from Alaska to Chile. Alaska Emergency Management reported a 5.1-foot (1.55-m) wave at Shemya. Also, up to eight feet (2.4 m) tsunami surges were recorded in California and Oregon, while in Chile, the wave height was 6.6 feet (2 m) (Oskin 2017). Japan, on average, experiences one tsunami every 6.73 years, the 2011 tsunami being the second largest such event in Japanese history; the largest occurred on June 15, 1896.

The crashing waves cracked the mighty tsunami-defense seawalls along the Japanese coast and caused meltdown of three reactors in the Fukushima Daiichi Nuclear Power Plant complex after the plant's cooling system failed from the loss of electrical power. For this reason, this tsunami is called the Fukushima Tsunami. The Tokyo Electric Power Company (TEPCO), the owner of the nuclear power plant, wanted to solve the cooling problem by pumping in saltwater from the sea. The Japanese Prime Minister initially opposed such a solution, but three days later approved it.

Meanwhile, the tsunami waves destroyed thousands of homes and industries and swept away hundreds of cars, boats, and household goods. Waves also wiped out towns and cities along an entire stretch of the northeastern Honshu coast. The events on March 11, 2011, created an estimated 24–25 million tons of rubble and debris. The tsunami waves also carried an estimated 5 million tons of debris out to sea, some of which washed up on North American coastlines years later. Because the earthquake subsequently triggered tsunami and nuclear accidents, these three events together are also called Japan's Triple Disaster of 2011.

DAMAGE

The Great East Japan Earthquake caused enormous damage, mostly by the earthquake-induced tsunami, which hit Japan's coastline less than an hour after the earthquake. According to an official estimate, 15,945 people died and 6,156 were injured because of the events immediately following the 2011 earthquake. While slightly over 23 percent of the Japanese population was 65 years and older in 2011, this age group accounted for 56 percent of all deaths (HelpAge International 2013). The earthquake also displaced 340,000 people, and as of June 10, 2016, more than 2,500 people were still reported missing. Moreover, more than 120,000 buildings were destroyed, and over 1 million buildings were damaged (Oskin 2017). Eleven hospitals were destroyed, and another 300 were damaged. The earthquake and resulting tsunami caused collapse of a dam and widespread fires due to broken gas lines, the largest occurring at the Cosmo Oil Refinery in Ichihara city in Chiba Province.

An estimated 230,000 cars and trucks were either damaged or destroyed, and the earthquake and tsunami together caused severe and extensive structural damage to roads and railways in northeastern Japan. Physical infrastructure was essentially nonfunctional for over two and a half months. Because the earthquake and tsunami affected the coastal region, the fishery industry also suffered significant loss. Additionally, around 4.4 million households lost electricity, and 1.5 million households remained without water for several weeks. Many coastal towns along the affected coast were either completely or partially destroyed.

The 3/11 tsunami caused considerable damage to agricultural crops, particularly in the Sendai Plain, where it flooded an estimated 217 square miles (561 square km). It destroyed farming machinery, warehouses, and agricultural canals, which were contaminated with saline water. It also destroyed the countryside adjacent to the built-up area. In the Sendai Plain, the Fukushima Daiichi nuclear power plant meltdowns created desperate emergency efforts that lasted months. This caused mass fear over radioactive contamination as well as worry about the safety of nuclear energy in Japan as well as around the world. The people of western Japan were reluctant to buy crops grown in the affected areas because of the fear of radioactive contamination. The same was true for the fish caught off the eastern coast of Japan. Subsequently, sushi restaurants suffered, and breaks in refrigeration caused by power cuts reduced the availability of frozen products. Ultimately, radioactive materials were even detected off the shore of the North American coast.

Finally, the total direct financial cost of the triple disaster was estimated at $219 billion with the cost of cleanup alone estimated at $50 billion (Karan 2016). This disaster also disrupted global markets in manufactured goods, especially automobiles, steel, and chemicals. It also caused fluctuations in global currency markets for several weeks following the event.

TSUNAMI WARNING

Japan has the best earthquake and tsunami warning system in the world. While the system is run by the National Research Institute for Earth Science and Disaster Prevention, the Japan Meteorological Agency issues both types of warning. Also, private early-warning systems exist at offices and factories. On March 11, 2011, within less than a second of the historic earthquake, a seismometer located on land closest to the epicenter detected signals to determine an alert was necessary. Quickly, the alert was automatically issued to factories, schools, TV networks, radio stations, and mobile phones. As it happens, Japan's scientists had forecast a smaller earthquake and tsunami would strike the northern region of Honshu, but in the decade before the 2011 earthquake, a number of Japanese geologists had predicted a large earthquake and tsunami would hit northern Honshu. However, their warnings were not considered seriously by officials responsible for the country's earthquake hazard assessments.

In effect, the tsunami warning system did not work properly because of more than one human error. First, tsunami warnings took longer than they should have because more calculations were involved than for an earthquake. Thus, a regional tsunami warning was issued nine minutes after the earthquake struck. In the areas severely affected, residents probably received only about 15 minutes of warning. In reality, the interval between tsunami warning and the arrival of the tsunami waves to the coast took about 40 minutes. Second, many coastal residents who took shelter at higher elevations mistakenly assumed either that no tsunami had occurred or that it had already arrived at the coast. Therefore, many of these residents returned to their homes to collect their belongings (Karan 2016).

The prediction of the height of the tsunami was also wrong because it was based on erroneous data. The Meteorological Agency of Japan wrongly predicted that the tsunami wave height would be between 9.8 and 10.6 feet (3–6 m). As a result, many residents of the affected areas took shelter on the rooftops of three-story buildings for safety as such buildings could survive a 10.6 feet (6 m) tsunami. As indicated, in many affected places, the tsunami waves were higher than that (Karan 2016). For this reason, Japan's Meteorological Agency was criticized.

Regarding evacuation, in some areas, such as Miyagi and Fukushima, only 58 percent of people sought shelter on higher ground (Oskin 2017). As indicated, older people in Japan were the most affected by the events on March 11. A study found that 69 percent of older people stayed at home when the earthquake struck, and their evacuation time was slowed because of power failures, lack of reliable information, physical debilities, and being alone (HelpAge International 2013). Meanwhile, the Japanese government ordered residents within a 12-mile (20 km) radius of the Fukushima Daiichi Nuclear Power Plant and a 6.2-mile (10 km) radius of the Fukushima Daini Nuclear Power Plant to evacuate mandatorily because the plants' systems were unable to cool the nuclear reactors. A 20-mile (13 km) voluntary evacuation was also issued, resulting in the displacement of 300,000 people. People also evacuated from other parts of the northeastern coast; for example, 20,000 people were ordered to evacuate after tsunami warnings.

RESPONSE AND RECOVERY

Initially, the Japanese government downplayed the Fukushima nuclear accident to prevent public panic. With that intention in mind, the government declared the triple disaster a Level 4 disaster. A week later, it was upgraded to a Level 7, only the second ever since Chernobyl in 1986. The national and international response to the triple disaster was overwhelming. One hundred sixteen countries and 28 international organizations participated in providing disaster assistance. Thus, Japan had received $5.1 billion in donations by March 2012, with the United States having donated the most followed by Taiwan. The disaster relief operation ran more or less satisfactorily. However, as with most other postdisaster relief operations, the survivors of the triple disaster received many nonessential emergency items as well as too many of the same items. For example, too many blankets and too much clothing were sent from abroad as disaster relief items.

The Japanese government pledged to spend $206 billion for reconstruction and radiation cleanup (Karan 2016), and indeed, it took three years to clean the debris, but the affected residents are still recovering. As of February 2017, about 150,000 people were still living away from their homes, and 50,000 of them were living in temporary housing (Oskin 2017). Recovery took taking a long time. The government itself believed that full recovery would take 10 years, one of the principal reasons for this being government red tape. Moreover, misuse and diversion of recovery funds were reported, along with reports that some people had taken advantage of tsunami victims by misusing government recovery funds (Karan 2016). However, some sectors, such as the fishing industry, have already recovered.

CONCLUSION

Many lives were saved because of the combination of stricter building codes and early-warning systems. After 2011, Japan worked on improving its mitigation and preparedness measures. For examples, the country installed a new and updated tsunami warning system, and the government asked tsunami experts from around the world to assess the history of past tsunamis, to better predict the country's future earthquake risk (Oskin 2017). The government also created the Investigation Committee on the Fukushima Nuclear Power Plant accident, which found TEPCO had failed to prevent the accident because of its reluctance to spend money in protecting the people. The committee also found that the company had a history of poor safety standards and cover-ups in the past. Nuclear power still provides 30 percent of the country's electricity, but movements against this source of energy have emerged across Japan since March 2011.

Finally, Japan observes Disaster Prevention Day every September as a reminder of the importance of earthquake preparation. On that day, all public and private facilities across the country practice earthquake and fire drills.

Bimal Kanti Paul

Further Reading

HelpAge International. 2013. *Displacement and Older People: The Case of the Great East Japan Earthquake and Tsunami of 2011.* Chiang Mai, Thailand: HelpAge International.

Karan, P. P. 2016. After the Triple Disaster: Landscape of Devastation, Despair, Hope, and Resilience. In *Japan After 3/11: Global Perspectives on the Earthquake, Tsunami, and Fukushima Meltdown*, edited by P. P. Karan and U. Suganuma, 1–42. Lexington: The University Press of Kentucky.

Oskin, B. 2017. Japan Earthquake & Tsunami of 2011: Facts and Information. *Live Science.* September 13, 2017. https://www.livescience.com/39110-japan-2011-earthquake-tsunami-facts.html, accessed March 11, 2018.

Tania, B. 2011. Tsunami, Earthquake, Nuclear Crisis—Now Japan Faces Power Cuts. *The Guardian.* March 13, 2011. https://www.theguardian.com/world/2011/mar/13/japan-tsunami-earthquake-power-cuts, accessed March 12, 2018.

Tri-State Tornado, United States, 1925

On March 18, 1925, the Tri-State tornado moved through Missouri, Illinois, and Indiana, and for three and a half hours, it laid down a track that eventually covered 219 miles (352 km) and took the lives of 695 people (Johns et al. 2013). As of 2017, the Tri-State event remains the single deadliest tornado in U.S. history. In addition to the enormous death toll, the tornado also injured 2,027 people (Maddox et al. 2013) and inflicted an estimated $2.1 billion in damage (inflation adjusted to 2017). Although the Fujita Scale was not introduced until 1971, Thomas Grazulis retroactively assigned the event an F5 rating, indicating wind speeds ranging from 261 to 318 mph (419–512 kph). Wind speeds for this event were estimated at 300 mph/483 kph, with a maximum forward speed of 77 mph (124 kph). The tornado was part of the March 18, 1925, outbreak that recorded 12 tornadoes and killed 747 people in total. The devastation that this event caused can be attributed to several factors, including sheer size, duration on the ground, path width, lack of adequate warning systems at the time, and conditions of the tornado itself, which in essence disguised the actual funnel over much of its path. The 1925 Tri-State tornado serves as a point in meteorological history in which it was recognized that public awareness of tornadoes needed to be increased.

METEOROLOGICAL SYNOPSIS

The storm system that set up the tornado (and the outbreak in general) is not considered to be meteorologically unusual, in that it exhibited characteristics typically found in large-scale weather patterns that develop as a result of cold fronts overtaking warm fronts in the Central and Southern United States. The tornado began at 1:01 p.m. CST, born of very low cloud heights and at what is termed the "triple point"—the intersection of the cold front, the warm front, and the dry line or a dew point line (Johns et al. 2013). The tornado stayed very close to this junction for most of its life, veering just slightly more east while the triple point moved slightly more north. While the meteorological set up was not unusual, perhaps an unusual phenomenon of the system was the close proximity of the track to the surface low—it is more typical for a violent tornado to form in the warmer southeastern section of the storm. Early-morning thunderstorms (approximately 7:00 a.m. CST) had developed over southeastern Kansas and northeastern Oklahoma, carrying with them hail, high winds, and possibly small unrecorded tornadoes. The low-pressure system associated with these storms was positioned over northeastern Oklahoma at the time and flanked on the east by a warm front. The dry line formed in front of the surface low, and that low-pressure system was evident in southern Missouri by 12:00 p.m. CST, setting the stage for the system that would produce the Tri-State tornado (Johns et al. 2013).

DAMAGE

Shannon County, Missouri, three miles north-northwest of Ellington, is recorded as the initial point of touchdown. Like many large tornadoes, the Tri-State began as a small funnel, but grew rapidly. Preceded by fair weather, the event took many residents by surprise. Moving along an east-northeast path with an average speed of about 60 mph (96 kmh), the funnel struck the Missouri communities of Ellington, Annapolis, Leadanna, Biehle, and Corwall (and crossed near several others), destroying homes, schools, and killing 11 people. Observers of the event across its Missouri path noted that hail and/or rain generally led the tornado, and clear skies and sunny weather followed, which is typical of supercells. At a point near Biehle, the single-funnel tornado spun into two separate funnels, but reformed into a single funnel before crossing the Mississippi River.

By 2:30 p.m. CST, the tornado moved across the state line and into Illinois, where the first community that it struck was Gorham. It essentially leveled the entire community and took the lives of 34 people. Golf ball-sized hail was recorded in Gorham, and the forward speed of the storm increased to an average of 73 mph (117 kph). The mile-wide funnel continued along its east-northeast path, violently striking Murphysboro, Illinois, causing major damage to a substantial portion of the community. At the city level, Murphysboro holds the record for the largest death toll suffered by a single city, with 234 people killed—which represents 33.6 percent of the total for the entire length of the tornado's path. Several other cities suffered major damage and heavy casualties as the tornado continued through Illinois, including De Soto, West Frankfort, and Parrish, which saw 69, 152, and 22 fatalities, respectively (35 percent of the total). The school in De Soto was filled with students and staff seeking shelter, but regardless of the strong brick construction, 33 children died as the building tumbled onto them.

Before exiting Illinois, the tornado killed 65 more people, bringing the death toll for that state to approximately 613, which represents an incredible 88.2 percent of the total number of fatalities for the entire life of the storm. Throughout its path across Illinois and Indiana, the supercell may have shifted to what is termed a *high precipitation* (HP) supercell; these types of formations are not as prolific in the production of F3–F5 tornadoes as often as are typical (referred to as "classic") supercells, yet the Tri-State tornado remained at F5 strength.

The final state to be affected was Indiana; two communities stand out as the hardest hit: Griffin and Princeton. Striking Griffin, and causing heavy damage, the tornado killed 25 and caused more damage as it moved through farmland on its way to Princeton, where an additional 45 lives were lost. Shortly after striking Griffin, it is reported that the single funnel split into a tri-funnel formation. Southwest of Petersburg, Indiana is recorded as the point of dissipation for the event, which occurred at 4:38 p.m. CST. By the end of the event, 15,000 homes were destroyed in 19 communities and rural areas, leaving thousands of residents without shelter.

WAS IT MORE THAN ONE TORNADO?

Initially, and for many years after, it was assumed that the tornado was one singular event, and even now it is referred to as *The* Tri-State *tornado*, rather than some other moniker that might indicate a multiple-tornado supercell. Meteorologists now recognize that large supercells often produce multiple and distinct tornadoes, that is, one ends and a new funnel forms, within the life span of the cell, referred to as a tornado family. Tornadoes that cover extensive and connected paths are termed "very long-track tornadoes," or VLTs. When the damage path of the Tri-State VLT was observed post event, the gaps in the damage path were thought to be the result of "skipping," which was assumed to have occurred when the tornado "skipped" over an area and touched down at another point along the path, with the structural integrity of the funnel remaining intact. Comparisons with more recent tornado-producing supercells suggest that the Tri-State tornado may have actually been more than one distinct funnel. Although a segment of

174 miles (280 km) from Madison County, Missouri, to Pike County, Indiana, was found to have no evidence of data gaps in the damage record, it is possible, but unconfirmed, that a small portion at the beginning of the path and the last 45 miles (67.5 km) of the damage path were caused by multiple tornadoes (Johns et al. 2013). Regardless, the Tri-State tornado still holds the official record for the longest damage path and the longest duration for a single tornado.

EMERGENCY RESPONSE

With a death toll of 695, it is not hard to imagine the stress such an event would put on existing response, rescue, and health-care systems, in addition to overall social security and law enforcement (postevent looting and theft were major problems). It is well known within the emergency management community that citizens respond with heightened degrees of altruism following a disaster event in their local area, and the Tri-State event demanded such a response. In addition to locals, doctors, nurses, and necessary medical supplies were dispatched to the hardest-hit locations, and 13 relief centers were established along the damage path, with the headquarters located at Murphysboro, Illinois. Within the next year, Red Cross personnel handled over 6,800 cases, providing services including medical care, food, transportation, and building supplies. The surrounding areas that were not affected by the disaster contributed to the care of the injured and ongoing recovery of the communities struck.

WHY SO MANY WERE KILLED?

The sheer size of the event and with the fact that it scored direct hits on multiple communities are the obvious reasons for the high number of casualties. But there are many eyewitness reports of the event that provide indications that people may have been unaware of the fast-approaching tornado. The cloud base of the supercell (the height from the bottom of the clouds to the ground, what meteorologists term *lifted condensation level* or LCL) was very low, which may have obscured the funnel from distances that normally would provide adequate warning time. Further, given the supercell's transition to a high-precipitation event, the funnel was most likely "rain wrapped" for much of its life through Illinois and Indiana—the tornado would have been partially or completely obscured by rain, making visual identification difficult if not impossible.

A severe weather phenomenon known as "downburst" also accompanied the main funnel along much of the damage path. Downbursts occur when cool air rush into a void left by rising warm air, collides in a central location, and descends rapidly, causing additional and often major damage in areas adjacent to the main funnel. Additionally, many tornado formations will produce vortices along the edges of the main vortex—smaller funnels along the main funnel; with such a large main funnel, many additional vortices were likely produced, enhancing the damage potential of the tornado.

An overall deficiency in knowledge and understanding of tornadoes was also a major contributor to the high death toll. John Park Finley, a lieutenant in the U.S.

Army signal corps, made attempts to predict tornadoes in 1884 and 1886 as a part of test studies, and while he was able to predict thunderstorms with some degree of accuracy, based on observations of typical conditions when storms moved in, he was not able to predict tornadoes very well. The word *tornado* was even banned from forecasts for fear that it might cause public panic—a ban that remained in place until the early 1950s.

The U.S. Weather Bureau had been placed under the control of the Department of Agriculture in 1891, and very little research was conducted on severe storms and tornadoes after that, not starting up again in earnest until the late 1940s. The first successful forecast of a tornado came in 1948, when meteorologists Major Earnest Fawbush and Captain Robert Miller, stationed at Tinker Air Force Base in Oklahoma City, predicted a tornado after similar weather patterns had produced one five days earlier.

Today there is an extensive and advanced network of radar, trained storm spotters, warning systems, and heightened public awareness of severe storm activity, none of which existed in 1925; television had not yet been invented, and even radio was only just becoming a prominent feature in large cities. By the last decade of the twentieth century, the average number of people killed annually by tornadoes in the United States stood at around 50; much of that relatively low number can be attributed to research and awareness, lacking before the 1950s, and certainly a major contributor to the incredible loss of life that resulted from the 1925 Tri-State tornado.

Mitchel Stimers

Further Reading
Johns, R. H., D. W. Burgess, C. A. Doswell III, M. S. Gilmore, J. A. Hart, and S. F. Piltz. 2013. The 1925 Tri-State Tornado Damage Path and Associated Storm System. *Electronic Journal of Severe Storms Meteorology* 8(2): 1–33.

Maddox, R. A., M. S. Gilmore, C. A. Doswell III, R. H. Johns, C. A. Crisp, D. W. Burgess, J. A. Hart, and S. F. Piltz. 2013. Meteorological Analyses of the Tri-State Tornado Event of March 1925. *Electronic Journal of Severe Storms Meteorology* 8(1): 1–27.

Tropical Storm and Floods, Yemen, 2008

Yemen is a disaster-prone country where extreme events such as floods, cyclones, earthquakes, tsunamis, landslides, droughts, and volcanic eruptions frequently occur. Every other year, the country experiences a tropical storm that is associated with heavy rainfall, which in turn causes (flash) floods. On October 23, 2008, a tropical storm, designated Deep Depression ARB 02 or Tropical Storm 03B, struck two less densely populated eastern provinces (Hadrsmout and Al-Mahrah) of Yemen, producing heavy rainfall in the typically arid region. In 30 hours, the coastal areas received about 4 inches (90 mm) of rainfall, almost 18 times greater than the typical rainfall of 0.2–2.4 inches (5–6 mm). In some inland areas such as Wadi Hadramout valley, the storm dropped possibly as much as 8 inches (200 mm) of rain over a 7,700 square mile (20,000 square km) area. This translated to about 528 billion gallons (2 billion cubic m) of water, more than twice the capacity of most valleys and waterways (*wadis*), which the waterways were unable to contain. This led to severe flooding that swelled water levels to 59 feet (18 m). Some claim that the magnitude of the flooding was the greatest in 600 years, attributing the scope to climate change (Wiebelt et al. 2011).

Tropical storm 03B was the second worst natural disaster on record in Yemen following the deadly floods in 1996. The storm was the sixth tropical cyclone of the 2008 North Indian Ocean cyclone season and the second tropical cyclone in the Arabian Sea that year. Most of the tropical cyclones that hit the Arabian Peninsula originate in the Arabian Sea, the portion of the Indian Ocean north of the equator and west of India. As indicated, tropical cyclone-caused damage in the Arabian Peninsula is primarily due to flooding. The 2008 cyclone formed on October 19 off the west coast of India about 450 miles (725 km) southeast of the Yemeni island of Socotra. Moving generally westward toward the Gulf of Aden, the depression failed to intensify much due to its meeting dry air. It reached maximum sustained winds of only 35 mph (55 km/h), and the storm dissipated immediately after the landfall (IMD 2009).

IMPACTS

Overall, the storm and subsequent flooding killed 183 people in Yemen (GoY 2009), and the damage was extensive, particularly in Hadramout province, where most homes were built of mud bricks. More specifically, Wadi Hadramout was the worst hit region, sustaining 67.5 percent of the total damage and loss, with 16 of its 19 districts reporting damages. Several Wadi Hadramount villages, listed as United Nations Educational, Scientific and Cultural Organization (UNESCO) World Heritage sites, sustained heavy damage. The 2008 storm and flood together destroyed 2,826 houses and damaged another 3,679, leaving about 25,000 people homeless. These two events together affected about 700,000 people. Total damage was estimated at $1.6 billion, which was equivalent to roughly 6 percent of Yemen's 2008 gross domestic product (GDP).

The floods damaged infrastructure to the value of $113 million in 2008 dollars, most of which related to damaged roads. Major roads, communications, power, and water supply networks all sustained major destruction and damage. As a result, access to the affected areas was very difficult in the early response period, and it took close to one week to reach some of them. At the Sayun Airport, floods damaged runways and other facilities. Widespread irrigation systems were also damaged, including 359 dams, 65 reservoirs, 1,241 wells, and 1,229 water pumps. The floods also damaged 170 schools, as well as many health facilities, and along the coast many fishing boats and equipment (Wiebelt et al. 2011). The storm sent a plume of moisture throughout the Arabian Peninsula, contributing to dust storms as far north as Iraq. Because of flooding, hundreds of residents were trapped in their homes, while businesses and schools were shut down. The homeless people took temporary shelter in mosques and schools or stayed with friends and relatives.

Among all the economic sectors, with nearly 64 percent of total damages, agriculture was the most affected. However, impact on agriculture was unevenly distributed with some districts sustaining much more damage and loss than others. For example, the flood affected 75 percent of the farmers in Hadhramaut. It also damaged 22,902 acres (9,268 ha) of cultivated lands as well as 51,455 acres (20,823 ha) of uncultivated lands due to soil erosion. In addition, the flood affected crops, vegetables, and forage crops; killed about 58,500 livestock; and wrecked 309,103 honey beehives. The flood washed away soil from the fields and uprooted 550,000 palm trees, 161,449 fruit trees, and 16,587 citrus tree cells. Despite the heavy agricultural damage, the floods did not disrupt the national food supply (GoY 2009).

This tropical storm caused the Yemen floods, October 23, 2008. (NASA)

RESPONSE

The government of Yemen sent search and rescue teams into the flood-affected two eastern provinces of the country to help stranded residents, but because of strong winds, these efforts were not fully successful. For instance, in Seiyun, six soldiers died because of lightning strikes while attempting to rescue trapped residents, and ultimately the affected governorates of Hadramout and Al-Mahara were declared disaster areas on October 27, 2008. The government of Yemen also requested assistance from the international community after declaring the preceding two provinces as disaster areas. On the same day, the government of Yemen, represented by the Ministry of Planning and International Cooperation (MOPIC), requested the international community's support to assess the damages, losses, and postdisaster needs and to join reconstruction and recovery efforts (GoY 2009).

The Yemeni government prepared for the donors a list of urgent items required by the flood victims: food, tents, blankets and utensils, water pumps, water purification equipment, mobile health units, and electricity generators. The last item was listed because 70 percent of the population was without electricity. One serious problem was that the hardest-hit areas had poor infrastructure, which caused difficulties in reaching and distributing relief goods quickly. To overcome the transport issue, the government sent the military to assist storm victims and sent aircraft with tents, food, and medicine to the worst hit areas. Red Cross and Red Crescent Societies (IFRC) of Yemen provided meals and water to about 21,000 people. The agency also provided school kits for 4,500 students whose facilities were damaged.

Various agencies of the United Nations provided emergency and recovery supports for storms and flood victims. For example, the World Health Organization (WHO) provided medical kits to the worst hit areas. It also sent an Inter-Agency Diarrhoeal Disease Kit (IDDF) and medical supplies to the Al-Mahara province. This kit benefited nearly 10,000 people for three months. Further, WHO supported the Ministry of Public Health and Population (MOPHP) to enhance the disaster surveillance system, control vector-borne diseases, and bolster a campaign to prevent communicable disease transmission. Moreover, the United Nations Development Program (UNDP) provided construction materials to displaced storm and flood victims while the World Food Program (WFP) assisted 43,000 people with food and other emergency supplies.

A number of international relief organizations responded to the disaster immediately, such as Islamic Relief, Oxfam, Turkish Red Cross, USAID, and Adventist Development and Relief Agency International (ADRA). The Organization of the Islamic Conference collected funds for the relief of flood victims. WHO, along with other UN agencies and local authorities, coordinated relief aid distribution. The Red Cross and Red Crescent Societies of Yemen assisted over 70,000 storm victims through health programs and also helped residents cope with stress and hygiene issues. Probably for these reasons, there was no outbreak of any disease in the affected areas (GoY 2009).

To raise money for disaster recovery, the Yemeni government cut one day of salary for all workers, equating to $4.25 million, and the government provided another $100 million from its annual budget; meanwhile, local charities and residents collectively raised $8.5 million. Many Middle Eastern countries, including Saudi Arabia and the United Arab Emirates, sent money and supplies to help rebuild damaged infrastructure and houses. Saudi Arabia made a pledge of $100 million in assistance, and the United Arab Emirates (UAE) pledged $30 million for the flood victims. The Arab Fund sent $135 million, including $35 million for road reconstruction. Many non-Middle Eastern countries (e.g. Japan, Germany, and Italy) also pledged relief supplies and funds to Yemen. Japan sent items such as tents, blankets, plastic sheets, sleeping mats, and water purifiers and also provided funding to rebuild shelters for 700 displaced Al-Akhdam people. Additionally, Singapore sent relief supplies worth $20,000.

As indicated, WFP and other United Nations Agencies provided food and logistical support. Relief efforts continued to make progress until 2011 when Yemen underwent a political uprising. However, disaster relief efforts suffered initially from many problems. First, in some areas, relief item distribution was duplicated due to lack of coordination, while in some areas flood victims did not receive any help. Yemen's Deputy Prime Minister for Internal Affairs coordinated the public relief efforts in conjunction with the governors of the most affected areas; however, the scale of the disaster proved too great for some ministries to handle. Conversely, the Ministry of Public Works relatively quickly reopened roads and repaired the damaged Sayun Airport. Within two months after the flood, most roads, power systems, hospitals, and communication services were restored.

RECOVERY

The estimated recovery and reconstruction needs from the storm and floods in Yemen were grouped into three categories: (1) urgent actions under an immediate intervention program; (2) reconstruction and recovery needs over the short term and medium term, spanning an estimated four years for reconstruction and recovery to predisaster level; and (3) disaster risk management needs, to reduce the vulnerability and risk of occurrence of a similar disaster in the future. The total reconstruction and recovery needs for the two affected provinces assessed by the Damage, Loss and Needs Assessment (DLNA) team were $929 million. The team disaggregated that amount: $872 million for Hadramount and the remaining $57 million for the Al-Mahara province. The largest needs were found in three sectors that are listed in descending order: (1) the agriculture sector ($496 million), (2) the housing sector ($115 million), and (3) the transport sector ($112 million).

However, the World Bank estimated the cost of recovery at $1.046 billion, mostly for rebuilding houses, re-growing crops, and restoring social services. The World Bank also provided $41 million for the Yemen Flood Protection and Emergency Reconstruction, which rebuilt vital infrastructure. At an international donor conference, various individuals and countries pledged $301 million to help with reconstruction. The UAE Red Crescent assisted in the reconstruction work, sending $27.3 million to rebuild 1,000 houses, 750 of which were completed by December 2009.

Despite all the help and funding, Yemen has never fully recovered from the destruction caused by the 2008 storm and floods. For many households, the effects of the storm and floods lasted several years due to insufficient assistance. By 2010, about 40 percent of the overall recovery cost was met by international donations, although funding was halted after the political uprising began in 2011. However, the affected areas' economy largely returned to preflood levels by 2010.

CONCLUSION

One useful outcome of the 2008 Yemen floods was that the country prepared its Floods Response Plan (FRP). A flood protection master plan was also prepared by Wadi Hadramout. The FRP recommended constructing major flood protection infrastructure such as storm-water control for Mukalla and other urban areas, updating land-use plans and building regulations in urban areas, and developing a program for upgrading the standards of construction for roads. It also recommended an early-warning system for storms and floods. Meanwhile, the master plan urged the government to construct water diversion structures, embankments, and sedimentation traps. Although the 2008 tropical storm and floods led to one of the worst adverse natural events ever to affect Yemen, these events helped the country to better prepare for such extreme events.

Bimal Kanti Paul

Further Reading

GoY (Government of Yemen). 2009. *Damage, Losses and Need Assessment: October 2008 Tropical Storm and Floods, Hadramount and Al-Mahara, Republic of Yemen.* Sana: GoY.

IMD (Indian Meteorological Department). 2009. Report on Cyclonic Disturbances over North Indian Ocean During 2008. New Delhi. http://www.rsmcnewdelhi.imd.gov .in/images/pdf/publications/annual-rsmc-report/rsmc-2008.pdf, accessed December 10, 2018.

Wiebelt, M., C. Breisinger, O. Ecker, P. Al-Riffai, R. Robertson, and R. Thiele. 2011. *Climate Change and Floods in Yemen*. Washington, DC: International Food Policy Research Institute (IFPRI).

Typhoon Haiyan (Yolanda), Philippines, 2013

Typhoon Haiyan, known in the Philippines as Typhoon Yolanda, made landfall in the central coast of the Philippines as a Category 5 tropical cyclone in the early morning of Friday, November 8, 2013. With winds of 195 mph (315 km/h) speed, this "super typhoon" was a powerful and destructive natural disaster that struck not only the Philippines but also other Southeast Asia countries such as Vietnam, China, and Taiwan. It was associated with storm surges that were 16 feet (5 m) high in Tacloban, a coastal port city of San Pablo Bay in Leyte province, directly in the path of the typhoon. The enclosed nature of the bay magnified the storm surges. Much of the city and towns to its south, Palo and Tanauan, were inundated because their elevation is less than the height of the storm surge. In total, 73 square miles (191 square km) of coastal area was flooded. It originated from a broad low pressure in the western Pacific, to the east of Micronesia on November 2, 2013. The following day it developed into a tropical depression, and on November 4, 2013, it turned into a tropical storm. It rapidly strengthened into a typhoon and moved west toward the central Philippines.

IMPACT

Typhoon Haiyan is the deadliest typhoon on record in the Philippines. It killed 6,300 people in the country, causing 28,688 injuries and 1,062 to go missing. After the typhoon, President Benigno Aquino III called for an investigation into why so many people died from the event. However, the devastation clearly was widespread. The typhoon damaged 1.1 million houses of which 551,000 were destroyed; it affected 16 million people from over 3.4 million households, and it caused $10 billion in economic losses, equivalent to 4 percent of the Philippines' yearly gross domestic product (Yamada 2017).

The humanitarian impact of Typhoon Haiyan was enormous; huge numbers of people were left homeless. In the worst-affected areas in coastal communities in Leyte province and the southern tip of Eastern Samar, the typhoon knocked out power, telecommunications, and water supplies. It swept directly through six of 80 provinces in the central Philippines and affected 38 other provinces. Ultimately, more than 10 percent of the nation's population of 105 million people suffered from Typhoon Haiyan with nearly 5.6 million people requiring food assistance. Airports, vital links to the rest of the archipelagic country, were damaged in typhoon-affected areas. With a population of 222,000, the city of Tacloban, capital

A young boy stands among the rubble caused by Typhoon Haiyan in Tacloban City, Leyte, Philippines, 2013. (Hrlumanog/Dreamstime.com)

of Leyte province, was the hardest hit by the event and the place of the most concentrated destruction and death (CRS 2014). There, because of delayed response and delivery of emergency supplies such as food and shelter by the Philippine government, many storm survivors abandoned their homes for other parts of the country.

Moreover, Typhoon Haiyan affected some of the poorest parts of the Philippines, where people are either farmers or fishers. Several crops such as sugarcane, coconut, and pineapple grow in the affected areas; however, damage to these crops had little impact on the livelihoods of the affected people. Additionally, because the region affected was one of the country's least developed, impact on the Philippines' manufacturing production was negligible. Nonetheless, although the affected areas were not the country's largest sources of rice, the extensive damage to rice crops in the region was suspected to have reversed the Philippines' trend toward rice self-sufficiency (CRS 2014).

WARNING AND EVACUATION

With the storm's landfall anticipated along the central part of the east coast of the Philippines, government officials in Tacloban and Manila began to prepare for the typhoon. Three military C-130 aircraft with supplies, along with 32 helicopters, were made ready to tackle the emergency response immediately after landfall. Twenty Philippines navy ships were also put in a state of heightened readiness.

The city officials in Tacloban devised plans to shelter many of the city's more than 220,000 residents, designating 23 evacuation centers, of which 22 were school buildings. Hundreds of officials went door to door to encourage residents of Tacloban to take shelter in an evacuation center. Electronic media also urged residents in the storm's path to take safe shelters. However, no mandatory evacuation order was issued, although the president of the Philippines in a television address urged people to evacuate low-lying areas prior to landfall of the typhoon.

Unfortunately, the evacuation rate was very low; only about 15,300 people of Tacloban's 220,000 residents took shelter in the evacuation centers. This translated to about 7 percent of the city population. This rate was even lower among vulnerable residents, and many of them stayed at home. At least three reasons were responsible for the low rate of evacuation. First, residents were reluctant to leave their homes, wanting to guard their possessions. Many others were doubtful about the danger of the impending disaster. Finally, a tsunami warning had been issued, which subsequently had initiated mass evacuation about 10 months earlier but turned out to be a false alarm. Thus, in this November 2013 instance, the government officials did not initiate mass evacuation because of fears of triggering violence.

Available reports suggest that the government officials did provide a storm surge warning along with an advanced typhoon warning. However, another reason for not heeding the typhoon warning was because the term *surge* was unfamiliar even to those who had lived for years with fierce storms. Thus, the term's implications were not understood by the populace (Yamada 2017). Rather, they were familiar with the term *tsunami*, which is triggered by earthquakes occurring in the sea, but not with the term *tropical storms*. Although designated evacuation centers had sheltered many in past typhoons, a considerable number of residents of the affected areas as well as some officials thought that public shelters opened by the government and city authorities would not protect the evacuees. The metal sheet roofs of most of the designated shelters were not tightly tied to the walls, and many shelters were also not on elevated land and thus were liable to inundation from storm surges. In fact, of the 300 people who sought safety in one of the shelters in Tacloban, at least 23 of them died by drowning when storm surge water inundated the shelter. Ultimately, the majority of the deaths in the affected areas were caused by either drowning or by flying debris (CRS 2014; McPherson et al. 2015). However, people bought necessary supplies before the typhoon's landfall.

DOMESTIC AND INTERNATIONAL RESPONSE

The domestic response was initially delayed for almost two weeks because of many factors. First of all, the Philippine government failed to anticipate the extent of the damage and miscalculated its own capacity to respond quickly and appropriately. Additionally, downed trees and other debris blocked roads in the affected areas. Other factors included a general lack of transportation, extremely limited communications systems, seriously disrupted government services, shortage of available government workers, and heavy rains. These factors also contributed to widespread looting in the typhoon-affected areas, including the city of Tacloban, causing desperate residents to take the law into their own hands. They broke into grocery stores, drugstores, and shopping malls, looting milk powder, rice, noodles, peanuts, soap, sardines, candy, and other items ranging from books to furniture. Local authorities in Tacloban reported that at least eight people were crushed to death when crowds looted rice from a government warehouse in the nearby town of Alangalang.

Search and rescue (SAR) operations were also delayed for the given reasons. Additionally, the employees of the affected cities were unable to turn up to work and were unreachable. For example, only about 10 percent of Tacloban's 293 police force members appeared for duty. There was also no effort to clean rubble after the typhoon, and bodies lay decaying in the streets for days. Ultimately, hundreds of corpses were put into mass graves. The situation was so serious that President Aquino ordered deployment of soldiers from their base near Catbalogan City, about 67 miles (100 km) north of Tacloban, but because of the road conditions, the soldiers had to walk to Tacloban and arrived in the city about 20 hours later. A few hours later, the first Philippine Air Force C-130 transport plane landed at Tacloban airport. On November 11, 2013, the Aquino administration declared a state of calamity and allocated more resources for relief efforts (McPherson et al. 2015).

In addition to various United Nations agencies, numerous international organizations, nongovernmental organizations (NGOs), and foreign donors actively participated in Typhoon Haiyan relief efforts. As of January 31, 2014, foreign donors had contributed about $663 million to the efforts. The United States contributed approximately $87 million in disaster aid, and its private citizens and private corporations contributed an additional $59 million (CRS 2014). Among all foreign countries, the United States played the most significant role in typhoon relief efforts in the Philippines. It deployed the aircraft carrier USS *George Washington* with its 5,000 sailors and 80 aircraft, 12 navy ships, and 66 military aircraft, along with nearly 1,000 military personnel. U.S. military assistance included clearing roads, transporting aid workers and emergency goods, delivering 4 million pounds of equipment and relief supplies, evacuating over 20,000 people from impacted areas, and logging 2,400 hours of flight time (Yamada 2017).

Also, the U.S. government granted Temporary Protected Status to eligible Philippine nationals, which allowed them to stay and work in the United States and thereby support their families in storm-affected areas. The United States Agency for International Development (USAID) and Catholic Relief Services (CRS) worked in partnership with local officials to help restore normalcy and also provided shelter to 3,275 families in the city of Tacloban. For those who lost their homes completely, the program provided new shelters and built up new communities for relocation. The program also trained the residents of these new homes to expedite their recovery, engage in their new community, and reclaim their right to a life of dignity.

Other nations also contributed to the relief efforts by providing financial assistance as well as sending military resources. For example, the Canadian government deployed members of the Disaster Assistance Response Team (DART) to conduct search and rescue operations in affected areas, provided humanitarian aid, and helped repair damaged infrastructure. The government also sent military equipment, and the Canadian Red Cross opened a field hospital. The British and Japanese governments sent naval destroyers. China sent its naval hospital ship. Korean and Taiwanese landing ships also joined the relief mission. Philippine expatriates from all over the world, as well as Philippine guest workers in the Middle East, made generous donations to the Typhoon Haiyan relief funds. They also financially helped their family members and relatives living in the Philippines to recover from the impact of the event quickly.

After Typhoon Haiyan, the Philippine government provided temporary housing for people living in tent cities, particularly for those who formerly lived close to the sea. In January 2014, the government announced that it would build 60,000 permanent housing units over a two-year period (CRS 2014). In the same month, it developed a four-year typhoon rebuilding plan for the affected area known as Reconstruction Assistance in Yolanda (RAY). The government allocated $8.2 billion for this plan. The UN Humanitarian Country Team (HCT), in partnership with UN agencies and NGOs and international organizations, prepared a Strategic Response Plan (SRP) to support the Philippine government's initiatives in providing emergency assistance and its reconstruction goals (CRS 2014). Prior to development of the rebuilding plan, the Philippines government launched a web portal, the Foreign Aid Transparency Hub (FAITH), that enabled the public to track distribution of international disaster relief funds.

CONCLUSION

The Philippines government initially failed to respond adequately and in a timely way in the aftermath of Typhoon Haiyan. In response to the request by the government, the United States deployed military assets to provide critical relief supplies to the typhoon survivors. However, as the national government had to share power with the governments of provinces, it slowed the relief and recovery programs. Media reports criticized the national government for lack of adequate preparation and coordination among the various public agencies (Palatino 2015). Media and other critics also pointed out misuse of emergency funds and donations provided by foreign countries. Two years after the event, many typhoon survivors continued to live in either temporary shelters or tent cities with inadequate services and livelihood opportunities. By now, however, most of them have returned to their permanent homes, and rehabilitation is near complete.

Bimal Kanti Paul

Further Reading

CRS (Congressional Research Service). 2014. *Typhoon Haiyan (Yolanda): U.S. and International Response to Philippines Disaster.* Washington, DC: CRS.

McPherson, M., M. Counahanb, and J. L. Hallb. 2015. Responding to Typhoon Haiyan in the Philippines. *Western Pacific Surveillance and Response Journal* 6(S1): 1–4.

Palatino, M.. 2015. Typhoon Haiyan Two Years Later: The Philippines Is Still Recovering. *The Diplomat.* November 11, 2015. https://thediplomat.com/2015/11/typhoon-haiyan-two-years-later-the-recovery-continues/, accessed June 23, 2018.

Yamada, S. 2017. Hearts and Minds: Typhoon Yolanda/Haiyan and the Use of Humanitarian Assistance/Disaster Relief to Further Strategic Ends. *Social Medicine* 11(2): 76–82.

Valdivia Earthquake, Chile, 1960

On May 22, 1960, at 3:11 p.m., the largest recorded earthquake in the twentieth century struck Chile, one of the most seismically active nations in the world. The country lies over the Nazca Plate and the South American Plate. The 1960 seismic event resulted from the release of mechanical stress between the subducting Nazca Plate and the South American Plate. The 9.5-magnitude earthquake was the result of building tension in this zone, and it originated approximately 100 miles (160 km) off the coast of southern Chile, parallel to the city of Valdivia. For this reason, the event is called the Valdivia or Great Chilean earthquake. It lasted approximately 10 minutes and generated the largest tsunami in the Pacific region for at least 500 years. The tsunami was destructive not only along the coast of Chile but also across the Pacific in Hawaii, Japan, and the Philippines. Tsunami-induced waves up to 82 feet (25 m) high severely battered the Chilean coast in three large waves. The first wave reached the coast about 15 minutes after the earthquake. The third wave was the largest, but it traveled at only half the speed of the second wave. Further inland, giant landslides buried many Chilean villages and towns. The epicenter of this megathrust earthquake was near Lumaco, approximately 350 miles (570 km) south of Santiago, the nation's capital. It occurred at a depth of 21 miles (33 km). Many aftershocks followed through November 1 with five of them being of magnitude 7.0 or greater.

Between May 21, 1960, and the major shock, a series of three foreshocks hit Chile, which provided early warnings of the major earthquake. The first foreshocks measured 8.3 magnitude on the Richter scale and severely damaged the coastal town of Concepción, causing over 125 deaths and ruining one-third of the town's buildings (BSL 2015). The second and third foreshocks occurred on May 22 and measured 7.1 and 7.5 on the Richter scale, respectively. The last two foreshocks did not cause any fatality. These three successive foreshocks are together called the 1960 Concepcion earthquakes. The main event, the Valdivia earthquake, generated a 620-mile (1,000 km) long, 167-mile (200 km) wide rupture zone along the stretch of the Nazea Plate (Pallardy n.d.). Moreover, another earthquake also occurred in Chile 70 miles (105 km) away from Concepcion on February 27, 2010. This earthquake measured 8.8 on the Richter scale.

More than 2,600 miles (4,184 km) long, the coastline of Chile is dominated by the collision between the Nazca Plate and the South American Plate. The former plate is pushing eastward at a rate of approximately 2 inches (5 cm) per year. Because the Nazca Plate carries mostly oceanic crust, it is denser than the mostly continental South America Plate. The Valdivia earthquake even caused a slight setback to the unstoppable rotation of the earth, making the days a few milliseconds shorter (BSL 2015).

DEATHS AND DAMAGE

The number of deaths associated with both the earthquake and tsunami in Chile was never fully resolved. However, estimated fatalities range between 490 and 5,700, which represent less than 0.1 percent of the country's population given that the country had 7.3 million people in 1960. Although the world's largest earthquake on record, it did not generate the greatest death toll. This is because a large foreshock that struck only 30 minutes before the main event sent many people from their homes to the street. It saved them from the main shock, and their abandoned houses ultimately were destroyed during the main shock (Brumbaugh 1999). However, over 1,000 deaths in the 1960 Chile earthquake were caused by tsunami, which affected most of the coastal cities of the country, including Lebu, Puerto Aisen, Puerto Montt, and Valdivia. Nearly half of the buildings were rendered uninhabitable in the latter two cities, which experienced an earthquake intensity ranging from X to XI on Modified Mercalli Intensity (MMI) scale. The scale value ranges from I (usually not felt) to XII (devastating). About 3,000 people were injured in these coastal cities.

The earthquake left 2 million people homeless. While the Chilean government reported that 58,622 homes were demolished, in many cases, the homes that had been designed to resist earthquakes generally performed well during the 1960 earthquake. However, building damage was extensive when affected by soil subsidence, or small fault movements, with concrete buildings being the most affected structures. Many of these structures collapsed because they were not built using earthquake-resistant technology. Conversely, traditional wooden houses fared better than concrete buildings. Additionally, houses constructed on high ground suffered relatively less damage than those on the lowlands, which absorbed great amounts of energy. However, a huge landslide blocked the San Pedro River, which drains Lake Rinihue above Valdivia. This block caused flooding along the course of the river where about 100,000 people lived. Some of these people were evacuated while others left the affected areas. It took days to cut through this barrier and open a channel (Brumbaugh 1999). During this time, several other landslides also blocked the outflow of water, interrupted railway and highway transportation, and destroyed bridges and telecommunication systems in southern Chile. Meanwhile, the tsunami that struck the southern coast of Chile flooded agricultural farms and killed many people and livestock, the latter deaths crippling the rural economy of the affected countryside.

The earthquake and tsunami caused damage amounting to $550 million (1960 dollars) in Chile. In Hawaii, where tsunami waves arrived 15 hours later, it caused 61 deaths, 43 injuries, $23.5 million in damage, and millions of additional dollars in damage at Hilo Bay on the main island of Hawaii. Twenty-two hours later, tsunami waves reached the main Japanese island of Honshu and caused 139 deaths and destroyed or washed away almost 3,000 homes. Also, at least 21 people died in the Philippines due to the tsunami. Additional damage of $1 million, two deaths, and four injuries resulted on the U.S. west coast from 3.3 to 6.6 feet (1–2 m) waves (NOAA 2015).

Two days later on May 24, 1960, the Cordón Caulle volcano in Chile's Lake District erupted from a 1,000-feet (305 m) long fissure after nearly 40 years of inactivity. This event was thought by some seismologists to be linked to the earthquake, which also caused widespread ground deformation in Chile, with up to 6.6 feet (2 m) of subsidence in the coastal mountains, while the foothills of the Andes were raised over 1.5 feet (0.5 m) or half a meter in places. Some offshore islands in Chile were raised as much as 20 feet (6 m). The vertical changes in elevation covered an estimated 17,000 square miles (44,030 square km). This pattern of ground movement is similar to that caused by the Alaska subduction earthquake in 1964 (Bolt 2004). Ultimately, the global extent of the tsunami caused by the Valdivia earthquake led to the creation of the International Tsunami Warning System of the Pacific (ITSU) in 1965 (NOAA 2015).

RESPONSE

After the 1960 earthquake, the Chilean government introduced stringent building codes, which have been revised many times over the last several decades. As a seismic-prone country, Chile's government introduced a law that holds building owners accountable for any loss in the first 10 years of a building's existence from inadequate compliance with the building code. It has also significantly improved its emergency preparedness methods. Since the 1960 earthquake, the government has been educating its residents to familiarize themselves with disaster preparedness plans, practicing countless earthquake drills every year, and running through evacuation routes time and time again. Even young children are taught how to react if an earthquake hits the country. They learn from their school teachers what to do in case of an earthquake. After the 1960 earthquake, the government introduced a comprehensive tsunami warning system with a battery-operated backup in coastal areas, and it has improved the system over time, particularly after the 2010 earthquake. It also developed and upgraded a national strategy to improve the ability of buildings to withstand a 9.0-magnitude earthquake. Importantly, it developed a seismic design code for new buildings, and homes and offices are now constructed to sway with seismic waves rather than try to resist them. The building can crack, tilt, and even be declared unfit for future use, but it must not collapse.

Immediately after the earthquake, the Chilean government formed an emergency committee to solve the problems caused by the earthquake. At the same time, the government also asked the United States for emergency assistance, including medical aid. The U.S. government responded quickly and sent two field hospitals and 61 nurses on May 25, 1960. These hospitals were located in Puerto Montt and Valdivia, where the local hospitals had been destroyed. Meanwhile Chilean relief workers, army nurses, and emergency staff established a tent city and field kitchen for the survivors. They established a clinic and treated infections and minor injuries of the survivors. They also provided tetanus shots for relocated persons. In addition, the United States Geological Survey (USGS) prepared the first geologic map of Valdivia following the earthquake.

The Chilean government provided grants to earthquake-affected farmers for rehabilitation. For example, the dairy farmers and dairy industry in the affected zone received government subsidies after the earthquake. The dairy industry also received state support through a long-term policy. In addition, an international technological cooperation program was established in the dairy sector by the German and Danish governments. The Chilean government received $136.4 million from abroad in the form of donations, and it spent $292.6 million of its own money on emergency response and recovery.

RECOVERY

The 1960 Valdivia earthquake was a blow to Chile's economy. Many small farm towns, particularly along the coast, were never able to return to their pre-earthquake state and were forced into economic decline. Many city blocks of the affected urban centers with destroyed buildings in the city center remained empty until the 1990s and 2000s; some of them still are used as parking lots. The earthquakes also damaged an area that had suffered a long period of economic decline, which began with shifts in trade routes due to the expansion of railroads in southern Chile and the opening of the Panama Canal in 1914. These changes slowed the recovery efforts in the affected region. However, after the earthquake, the government created the "Ministry of Economy, Development and Reconstruction," which was in charge of recovery. Within two years, the province of Valdivia had recovered, but it took several decades to come back to what it was.

CONCLUSION

Located in a subduction zone between Nazca Plate and the South American, Chile is one of the highest risk countries in the world for earthquakes. The country experiences a major earthquake frequently and as such is well prepared for such events. Similar to the 1960 Chile earthquake, earthquake also occurred in 1575 in Valdivia, which caused both landslides and tsunami. But the 1960 Valdivia or Great Chilean earthquake is most remembered for two reasons. It was the largest on record, and it forced the Chilean government to make the country resilient in the face of earthquake events. After the 1960 event, the government started long-term investment in earthquake mitigation and preparedness measures, including better urban planning and public education on how to react to such crises.

The event also generated other benefits. For example, land subsidence in Corral Bay caused by the 1960 earthquake improved navigability as shoal banks, produced earlier by sediments from Madre de Dios and other nearby gold mines, sank and were compacted. The earthquake also initiated useful research on long-term economic effects, health outcomes and other consequences, and recovery of regional geomorphologic system. Finally, a score of geologic studies were conducted after the 1960 earthquake in Chile.

Bimal Kanti Paul

Further Reading
Bolt, B. A. 2004. *Earthquakes*. New York: W.H. Freeman and Company.
Brumbaugh, D. S. 1999. *Earthquakes: Science and Society*. Upper Saddle River, NJ: Prentice Hall.

BSL (Berkeley Seismology Lab). 2015. Today in Earthquake History: Chile 1960. May 22, 2015. https://seismo.berkeley.edu/blog/2015/05/22/today-in-earthquake-history-chile-1960.html, accessed August 12, 2019.

NOAA (National Oceanic and Atmospheric Administration). 2015. May 22, 1960 Southern Chile Earthquake and Tsunami. https://www.ngdc.noaa.gov/hazard/data/publications/1960_0522.pdf, accessed August 12, 2019.

Pallardy, R. n.d. Chile Earthquake of 1960. https://www.britannica.com/event/Chile-earthquake-of-1960, accessed August 12, 2019.

Vietnam Flood, 1999

Vietnam is one of the countries that are most affected by flooding. For instance, during the 1990s to the early 2000s, the country experienced numerous floods, particularly in its central and southern regions. Located in Southeast Asia, bordered by China in the north, Laos and Cambodia in the west, and the South China Sea in the east, the country has 2,860 small and large rivers with a total flow of about 867 billion cubic meters per year. Moreover, Vietnam is located at the end of the Red River in the north and the Mekong River in the south, the latter being one of the 10 largest rivers in the world. All of these rivers carry an estimated 300 million tons of eroded mud per year, which is one of the reasons Vietnam frequently experiences floods (ODM 2005).

The central region of Vietnam is narrow, mountainous, and close to the coastline. Many rivers pass through the region, which had a total population of 7.5 million in 1999. These rivers originate in the western mountain range and flow into the South China Sea. Moreover, Vietnam is tropical, and so the country's northern and north central regions are influenced by a monsoon type of climate; thus, these regions receive heavy rainfall during the summer. Because of its unique topography and tropical climate, the country is prone to annual flooding.

At the beginning of the twenty-first century, Vietnam experienced several severe floods, and the 1999 flood in the northcentral region was the worst flood the country had experienced in a century. It inundated seven central provinces, killed more than 700 people, injured more than 500, and made at least 55,000 people homeless. In all, about 1.7 million people were affected by the flood, and it caused $780 million damage to the region (UNOCHA 1999). Furthermore, the impact of the flooding was compounded by a number of human-made factors. The region is the country's poorest, and it was devastated during the Vietnam War (1955–1975) by the U.S. military's bombing, shelling, and use of chemicals (James 1999).

CAUSES

In October and November 1999, the central region of Vietnam experienced more than typical rainfall due to a series of storms. The first storm to hit was Tropical Storm Eve on October 19, and the main event occurred from November 1 to November 6. Provinces from Ha Tinh to Quang Ngai experienced heavy rains. On October 20, 1999, these provinces received from 4 to 18.5 inches (100–470 mm) of rainfall, exceeding their monthly averages. Floodwater levels on rivers in Quang Binh and Quang Tri provinces, and downstream of the Thu Bon River (Quang Nam Province) as well as Thua Thien Hue Province, reached their highest levels. Also, the floodwater level on the Kien Giang River (Quang Binh) reached 6.7 feet (2.05 m) above alarm level II; the Quang Binh River reached 8.8 feet (2.69 m), above level I; and the Thu Bon River reached 5.9 feet (1.79 m), above level I (ADPC 2003).

From November 1 through 6, 1999, the combined effect of low-pressure air over the South China Sea and the tail of a cold front caused heavy rain over the central region, severely flooding provinces from Quang Binh to Binh Dinh. During this week, these provinces received 23.6–39.4 inches (600–1,000 mm) of rainfall, occurring on a large scale and concentrated in a short time period. In Hue City, the average rainfall reached over 51 inches (1,300 mm), the highest level in the city since 1886 (ADPC 2003).

According to the Central Committee for Flood and Storm Control (CCFSC), on December 8, 1999, additional flooding caused by heavy rainfall from December 1 through 6 once again attained historical levels in some rivers in the central provinces from Thua Thien Hue to Khanh Hoa. These provinces' rainfall average reached 78.7 inches (2,000 mm) and caused destructive inundation (UNOCHA 1999).

DAMAGE AND IMPACTS

The 1999 flood in the central region in Vietnam damaged or destroyed thousands of schools and health clinics and hundreds of miles of roads. The rising waters also destroyed thousands of acres of newly planted rice as well as affected several vast areas of secondary crops such as sugarcane and coffee. Thousands of head of cattle, pigs, and other valuable livestock were lost, along with 3,000 shrimp and fish ponds and over 1,200 fishing boats (James 1999). After the double floods in November and December, some paddy fields were covered with more than three feet (1 m) of sand and mud, while the soil was completely washed away in some areas. Essential irrigation networks and dykes were destroyed also. Officials reported that crops replanted after November's flood were washed away and many temporary relief shelters were destroyed.

The complexity of the terrain in the central region contributed to the flood impact. For example, for the lowland population, the largest blow to the household economy was the loss of and damage to rice that was stored in houses, depriving residents of both food security and main income for the year. People lost between one and five tons of rice per household. Moreover, the dyke and canal system were damaged, which increased production insecurity. The second main source of income, animal husbandry, was badly affected also. Almost all households lost pigs and poultry in the flood and in the epidemics immediately after the event.

In the highland of the western part of the central region, the inundation damage remained severe and caused large losses of garden crops, such as pepper, fruit trees, and cassava. The loss of income from pepper constituted the largest loss for many households, and cattle died both during and after the flood; however, the loss of the cassava crop had the most serious short-term impact. Most of the cassava was grown on flat land close to the river, instead of on the hill slope as in previous generations. Therefore, most of the cassava areas were inundated, and the root rotted and become inedible (ADPC 2003). Banana was another important crop damaged by the flood because the continuous rain caused more banana pests than normal. In general, the food security situation was thus more threatened by the 1999 flood than by floods in previous years. Notably, this mountainous part of Vietnam experienced flash flooding rather than river flooding.

Naturally, all affected people were very vulnerable to the failure of the rice harvest. There was hardly any surplus from rice production in 2000 because of the high production costs for drainage and replanting. Additionally, heavy rains in January and February caused inundation of the newly planted fields, and people had to replant two times and in some areas even three times. Surprisingly, the rice price in 2000 was lower than normal, but then the crop in May 2001 was down to 30 percent in terms of quantity and quality. Also, almost all households immediately bought new piglets, but they died. People invested again, and the pigs died again. It took about a year before the animal health situation stabilized. In part, this was because land available for production decreased due to the floods and to the government's restrictions on the use of hill land.

RESPONSE AND RECOVERY

The ministers of Defense, Health, Transport and Communication and Agriculture and Rural Development implemented life-sustaining and production recovery measures under the guidance of the government. Through helicopters, the Vietnamese military dropped emergency supplies of food to thousands of flood survivors. All of these entities participated in emergency disaster relief efforts, providing many necessary items such as instant noodles, lifeboats, raincoats, sets of clothing, blankets, canoes, water purification powder, medicine and food items, and shelter materials such as tents, plastic sheeting, and canvas. With support from the International Federation of Red Cross and Red Crescent Societies (IFRC), the Vietnamese Red Cross (VNRC) was not only active in evacuating people from flood-risk areas but also participated in relief distribution and assessment at the onset of the disaster. The VNRC, through its extensive network, high level of experience, and local knowledge, was able to meet the emergency needs of the flood survivors. It also received the full support and cooperation of the Vietnamese government and the armed forces of the country.

The IFRC launched a revised appeal on November 10, 1999, for $4.6 million for 258,000 households; the first appeal had been for $3.1 million. Besides its own emergency fund of $210,000 for the flood, UNICEF committed an additional $60,000 to flood relief in December. Meanwhile, the World Food Program (WFP) procured 850 tons of rice, which was distributed with assistance from VNRC/ IFRC. The Food and Agricultural Office (FAO) also prepared emergency relief packages including rice seeds and fertilizer along with $400,000 for flood victims in Quang Tri (UN OCHA 1999).

National and foreign governments such as the United States, France, Thailand, Australia, Hong Kong, China, Malaysia, and the EU countries and multilateral and nongovernmental organizations (NGOs) provided both cash and in kind assistance to the flood survivors for immediate support as well as during the rehabilitation and recovery period. For instance, rice was widely distributed in several batches. The flood-affected households were classified into three groups depending on how badly they had been affected. Accordingly, each household received between 330 and 551 lbs. (150–250 kg) of rice (ADPC 2003). Vegetables were planted immediately with seeds provided by the public authorities. Many households, especially in the lowland communes, borrowed rice from neighbors or relatives to be repaid after harvest.

The government and donors provided materials support for reconstruction, and the community provided labor. In many cases, migrant laborers from nonaffected provinces participated in reconstruction. Irrespective of the employment status, bank loans were available for all flood survivors to repair and make houses flood-proof. In addition, the poorer flood survivors received housing grants from the government, the VNRC, and NGOs. Finally, the Red Cross distributed medicine and disinfectants to purify the water in public supplies, as well as to all types of health facilities, including hospitals and private clinics.

As noted, educational institutions were seriously affected by the 1999 flood. However, the committees at different levels of administration quickly invested in repairing school buildings and in replacing furniture and school materials. Teachers worked extra hours without pay to help students catch up, and they all shared books and helped each other to dry or replace wet books. The Agriculture Ministry distributed seed and planting materials for sweet potatoes, vegetables, and other short-term crops. At the local level, the People's Committee made plans to enlarge the capacity of the agriculture cooperative to do this task. The Vietnam Bank for Agriculture and Rural development (VBARD) allocated $1.3 million of credit to the Hai Lang District, which was distributed as a one-year credit with 0.3 percent interest for rice farmers for short-term production recovery. The decision to emphasize the short-term needs was partly an issue of administrative convenience (ADPC 2003).

CONCLUSION

Vietnam is a long, narrow country that is frequently subjected to floods and tropical storms. Severe storms with strong winds are often accompanied by heavy rains causing river water levels to rise and subsequently the areas to flood. Moreover, where a storm or tropical depression combines with a cold front, it can result in long and torrential rains, causing serious flooding over river basins of the central region. In response, the strategy of the Vietnamese Government on disaster management for central Vietnam is to promote flood and storm prevention measures with the policy of pro-active prevention, mitigation, and adaptation. Management and mitigation measures include construction of upstream reservoirs and dykes in the plain areas. Combining such infrastructure with irrigation systems for stabilizing and enhancing agricultural production, the government could adopt a more proactive role to prepare for and reduce death, loss, and vulnerability.

Vu Vo

Further Reading

ADPC (Asian Disaster Preparedness Center). 2003. Local Institutions Response to 1999 Flood Event in Central Vietnam. http://www.fao.org/docrep/007/ae080e/ae080e04.htm, accessed October 20, 2018.

Hays, J. 2014. Devastating Floods in Vietnam in 1999, 2000, and 2001. Facts and Details. http://factsanddetails.com/southeast-asia/Vietnam/sub5_9h/entry-3491.html, accessed November 2, 2018.

James, G. 1999. Floods Devastate Central Vietnam. World Socialist Web Site, December 31, 1999. https://www.wsws.org/en/articles/1999/12/viet-d31.html, accessed October 24, 2018.

ODM (Open Development Mekong). 2005. National Report on Disaster Reduction in Vietnam. https://www.unisdr.org/2005/mdgs-drr/national-reports/Vietnam-report.pdf, accessed November 2, 2018.

UNOCHA (United Nations Office for the Coordination of Humanitarian Affairs). 1999. *Vietnam: Floods*. OCHA Situation Report No. 2. New York: UNOCHA.

Yangtze River Flood, China, 1931

The Yangtze River (Chang Jiang) in central and eastern China periodically experiences destructive floods. It is the longest river in Asia and third in the world in terms of length and discharge. It is one of the world's major waterways, carrying about two-thirds of all goods shipped on China's inland waterways. The 4,200 mile (6,300 km) long river originates in the very high and mountainous plateau of Tibet and travels generally eastward along a meandering course until it empties into the East China Sea near Shanghai on the coast. Flooding is very common in the lower course of the Yangtze River, downstream of the Three Gorges Dam, where the river flows through low-lying terrain dotted by lakes and marshes. Among the most destructive flood events in Yangtze River valley were those of 1870, 1931, 1954, 1998, and 2010. The 1931 flood was perhaps the worst natural disaster of the twentieth century as it killed 3.7 million people directly or indirectly. However, estimated death tolls vary greatly from source to source. The 1931 flood started on August 18, 1931, and lasted more than three months. The river runs through one of the most populated areas on earth, which dramatically increased deaths and injuries. Most people on the floodplain lived at a subsistence level, depending on the river for water for their personal and farming needs.

CAUSES

Usual weather patterns and decades of poor river management primarily led to the 1931 flood. Many earthen levees and dikes were constructed along the Yangtze

River banks before the 1931 flood to restrict the flow of water within the bank; however, these resulted in heavy deposition of sediments on the river bottom, which in turn reduced the river's water carrying capacity and thereby caused the flood. In addition, many of the lakes that had once acted as part of a flood control system either were cut off from the river by levees or were converted into cropland. Deforestation further reduced the capacity of the river to handle intense rains, which created more runoff and hence floods.

The Yangtze River has a drainage basin area of 700,000 square miles (1.8 million square km) that lies in the subtropical monsoon region. As a result, when the lower Yangtze basin receives heavy rains, the consequences are catastrophic. In April 1931, the lower river-basin area received far-above-average rainfall. When torrential rains came again in July, the stage was set for disaster. As the waters of the river continued to rise in the first half of August and even more rain fell, the rice fields that dominated the surrounding landscape were flooded, destroying the crop.

The weather and environmental conditions prior to 1931 were also responsible for the historic Yangtze River flood. From 1928 to 1930, China was afflicted by a long severe drought. The lack of rainfall made the river flow lower, and in some areas, the river floor dried up. However, the winter of 1930 was particularly harsh, dropping a huge amount of snow, followed by extremely heavy East Asian monsoon rains in the spring in the Yangtze lower basin. These rains melted the remaining snow and added even more water to the river. In 1931, seven cyclones hit the central Pacific coast of China during the month of July alone, far more than typical two cyclones that hit the coast annually in summer. In addition, political instability, economic turmoil, and incessant civil war were also responsible for the devastating impacts of the 1931 flood.

IMPACTS

The 1931 Yangtze River flood inundated more than 30,000 square miles (77,700 square km), including the cities of Nanjing and Wuhan. The Yellow River and Huai River also flooded in 1931. Overall, floods caused by three rivers inundated approximately 120,000 square miles (180,000 square km), an area equivalent to the size of the states of New York, New Jersey, and Connecticut combined. The Yangtze River flood directly killed 300,000 people and made 40 million homeless. Thus, it affected about a tenth of the Chinese population. The flood damaged crops, washed out grain storage facilities, and destroyed homes and roads. In the Yangtze River Valley, around 15 percent of the rice and wheat was destroyed. The flood also caused a rapid increase in the prices of vital commodities, and subsequently a famine in the flood-affected areas.

Many people died due to flooding for several reasons: drowning, being crushed by debris, starving, or falling victim to waterborne and other diseases such as cholera, typhoid, and dysentery. At least 150,000 people died because of drowning and diseases during and after the flood; millions were also injured. Although rural communities bore the brunt of the disaster, the 1931 flood was so extensive that people in cities also suffered from flooding and starvation because of crop damage.

In late July, the dykes that encircled Wuhan collapsed, enabling the water that had been held back to enter the city at a rapid rate. The flood inundated the 32-square-mile (83 square km) area of the city, submerging it under many feet of water for close to three months. Thus, the streets of the city transformed into canals. Thousands of people living in houses constructed from timber and earth drowned or were buried alive. Those who survived left their homes and began to search for safer refuge. It was difficult not only for the people of Wuhan but also for all flood-affected people to find refuge because hardly any dry place was left. In part, this is because most of the buildings in Wuhan were only one-story high. Thus, about 30,000 people were forced to live on a railway embankment in central Hankou. With little food and a complete breakdown in sanitation, thousands soon succumbed to diseases (Courtney 2018a).

Meanwhile, a huge fire broke out in the city's Texaco oil depository, and the fire brigade could not come close to the site because it was too hot to approach. The fire burned for three days, spewing flaming oil onto the surface of the water and noxious smoke into the atmosphere. Additionally, more than 50 people were killed by electric shocks. Overall, no place was more profoundly affected than Wuhan, a bustling metropolis in the middle Yangtze valley, which is known as "China's Thoroughfare" or "the Chicago of China" (Courtney 2018b).

In Wuhan, the local traffic police struggled to maintain order in the streets, which were overrun by boats, locally called sampans. At first, police stood on boxes and tried to keep boats in lanes, but as the floodwaters rose, police were forced to climb into the branches of trees. With even greater rise in floodwater, they could do little but sit back and watch as large cargo junks sailed into the city center. In the meantime, floodwater mixed with overflowing sewers, causing boat traffic to run into polluted water, which in turn was responsible for both mortality and morbidity from water-borne diseases. The telegraph office, telephone exchange, and airport in the city were closed; large stretches of the railway line washed away, stopping trains from approaching Wuhan.

The city of Nanjing was also severely affected by the 1931 Yangtze River flood. At the time of the disaster, the city was the capital of Republican China. One of the most tragic single events during the flood occurred on August 25, 1931, in the city of Gaoyou. Without any warning, floodwater rushed into the city through the Grand Canal when the dikes near Gaoyou Lake were washed away, whereupon some 2,000 people drowned in their sleep (Pietz 2002).

RESPONSE

During the 1931 Yangtze River flood, China was in the midst of a communist insurgency (1927–1936), so the central administration was very weak and ineffective at providing emergency assistance to flood survivors. Realizing this, the business community in Wuhan raised substantial funding for flood relief, which was used to construct a wooden walkway to allow pedestrians to walk over the flooded streets. Funds were also used to open gruel kitchens to feed the flood survivors. In addition, the community purchased boats to distribute boiled water and dispatched crews to dispose of corpses. Buddhist monks participated in flood relief efforts by turning their temples into refugee camps and by providing food and other necessary items. Traditional medical practitioners massaged the stomachs of starving people and distributed herbal medicine to prevent or reduce exposure to disease (Courtney 2018a).

Although the government in Nanjing falsely claimed that it saved many lives from starvation, it were the philanthropic agencies that fed more than half a million people for months before shipments of wheat arrived in Wuhan city. However, flood survivors of rural areas solely depended on outside assistance. The flood destroyed their crops, but it also provided opportunities for alternative foods such as lotus, water chestnuts, and wild rice that grow in floodwater. The flood also brought plenty of fish, which was consumed by both rural and urban dwellers, while frogs, ducks, and turtles that swam into inundated homes became important sources of food for survivors. The floodwater also brought mosquitoes and snails that thrived in vastly expanded territories and caused epidemics of malaria and schistosomiasis.

Reeds were another flood-resistant species that proved invaluable for disadvantaged flood survivors. After knitting together and stretching reeds over bamboo frames, survivors built temporary shelters that were not only cheap but also portable, allowing survivors to escape quickly when water levels rose. While some dwellers of Wuhan treated displaced people with great generosity, others saw them as a threat to political and economic stability. In some cases, some elite people and foreign residents requested the city authorities to have the flood survivors removed from their temporary shelters to preserve the hygienic well-being of themselves.

On the other hand, the local military were convinced that communists were using the flood disaster as a way to infiltrate Wuhan. Therefore, they declared martial law and began patrolling the flooded city streets in sampans with mounted machine guns. Anyone suspected of looting or other subversive activities was executed on the spot. But even such draconian treatment could not quell the paranoia, and eventually soldiers expelled flood survivors from the city center, relocating them to camps built on the outskirts of the city, where thousands died from diseases. While natural disasters often bring communities together, the 1931 flood in Wuhan divided the community into two as if people were living through two separate floods: one that caused mild inconvenience and another that destroyed lives (Courtney 2018b).

The 1931 flood was one of the first major tests for China's Nationalist government. As the magnitude of the disaster became clear during the middle of the flood, the government established the National Flood Relief Commission and organized emergency relief distribution. The League of Nations provided assistance to this commission. Many prominent figures in the United States such as aviators Charles Lindbergh and his wife Anne Lindbergh provided donations for the survivors. In fact, humanitarian organizations across the globe helped with the relief effort. Overseas, Chinese communities in Southeast Asia generously supported the relief efforts. However, the effort became much more difficult following the Japanese invasion of Manchuria in the autumn of 1931. Eventually, the government managed to secure wheat and flour from the United States (Courtney 2018a).

CONCLUSION

The 1931 Yangtze flood was the most lethal in China's history. However, much of the disaster could have been averted if flood-control measures had been followed closely. The Yangtze carries large amounts of sediment, which accumulates in certain areas of the river and must be cleared regularly. However, with much of the area's resources devoted to civil war at the time, the river was neglected. After the 1931 flood, more effective levees and other flood protection structures were built. Still, the floods of 1954 and 1998 were nevertheless highly destructive and killed some 30,000 and 3,650 people, respectively. Furthermore, one of the major objectives of the Three Gorges Dam project was to alleviate flooding on the Lower Yangtze. The dam did prove effective during the 2010 flood by holding back much of the resultant floodwater and thus minimizing the impact of flooding downstream. The flood killed only 392 people, far fewer than the 1954 and 1998 floods. However, the dam had to open its floodgates to reduce the high water volume in the reservoir, and thereby it caused extensive property damage.

Bimal Kanti Paul

Further Reading

Courtney, C. 2018a. *The Nature of Disaster in China: The 1931 Yangzi River Flood.* New York: Cambridge University Press.

Courtney, C. 2018b. Picturing Disaster: The 1931 Wuhan Flood. China Dialogue. September 11, 2018. https://www.chinadialogue.net/culture/10811-Picturing-disaster-The-1931-Wuhan-flood/en, accessed January 18, 2019.

Pietz, D. 2002. *Engineering the State: The Huai River and Reconstruction in Nationalist China 1927–1937.* New York: Routledge.

Bibliography

Aitsi-Selmi, A., S. Egawa, H. Sasaki, C. Wannous, and V. Murray. 2015. The Sendai Framework for Disaster Risk Reduction: Renewing the Global Commitment to People's Resilience, Health, and Well-Being. *International Journal of Disaster Risk Science* 6: 164–176.

Albala-Bertrand, J. M. 1993. *The Political Economy of Large Natural Disasters: With Special Reference to Developing Countries.* New York: Oxford University Press.

Alexander, D. 1993. *Natural Disasters.* New York: Chapman & Hall, Inc.

Alexander, D. 2000. *Confronting Catastrophe: New Perspectives on Natural Disasters.* New York: Oxford University Press.

Armenian, H. K., A. Melkonian, E. K. Noji, and A. P. Hovanesian. 1997. Deaths and Injuries Due to the Earthquake in Armenia: A Cohort Approach. *International Journal of Epidemiology* 26(4): 806–813.

Bakir, P. G. 2004. Proposal for a National Mitigation Strategy Against Earthquakes in Turkey. *Natural Hazards* 33: 405–425.

Balluz, L., L. Schieve, T. Holmes, S. Kiezak, and J. Malilay. 2000. Predictors for People's Response to a Tornado Warning: Arkansas, 1 March 1997. *Disasters* 24(1): 71–77.

Barker, J. C. 2000. Hurricanes and Socio-Economic Development on Niue Island. *Asia Pacific Viewpoint* 41(2): 191–205.

Barnes, L. R., E. C. Gruntfest, M. H. Hayden, D. M. Schultz, and C. Benight. 2007. False Alarms and Close Calls: A Conceptual Model of Warning Accuracy. *Weather and Forecasting* 22: 1140–1147.

Barnett, B. L. 1998. US Government Natural Disaster Assistance: Historical Analysis and a Proposal for the Future. *Disasters* 23(2): 139–155.

Betzold, C. 2015. Adapting to Climate Change in Small Island Developing States. *Climatic Change* 133: 481–489.

Billon, P. L., and A. Waizenegger. 2007. Peace in the Wake of Disaster? Secessionist Conflicts and the 2004 Indian Ocean Tsunami. *Transactions of the Institute of British Geographers* 32(3): 411–427.

Birkmann, J., P. Buckle, J. Jaeger, M. Pelling, N. Setiadi, M. Garschagen, N. Fernando, and J. Kropp. 2010. Extreme Events and Disasters: A Window of Opportunity for Change? Analysis of Organizational, Institutional and

Political Changes, Formal and Informal Responses after Mega-Disasters. *Natural Hazards* 55: 637–655.

Blaikie, P. M., and R. Lund, eds. 2010. *The Tsunami of 2004 in Sri Lanka: Impacts and Policy in the Shadow of Civil War*. London: Routledge.

Borden, K. A., and S. L. Cutter. 2008. Spatial Patterns of Natural Hazards Mortality in the United States. *International Journal of Health Geographics* 7: 64. https://doi.org/10.1186/1476-072X-7-64.

Boruff, B. J., J. A. Easoz, S. D. Jones, H. R. Landry, J. D. Mitchem, and S. L. Cutter. 2003. Tornado Hazards in the United States. *Climate Research* 24: 103–117.

Boulter, S., J. Palutikof, D. J. Karoly, and D. Guitart, eds. 2013. *Natural Disasters and Adaptation to Climate Change*. New York: Cambridge University Press.

Bradshaw, S. 2013. *Gender, Development and Disasters*. Cheltenham, UK: Edward Elgar.

Bressan, D. 2012. March 27, 1964: The Great Alaskan Earthquake. *Scientific American*. March 27, 2012. https://blogs.scientificamerican.com/history-of-geology/march-27-1964-the-great-alaskan-earthquake/, accessed February 21, 2020.

Burrus, R. T., Jr., C. F. Dumas, C. H. Farrell, and W. W. Hall, Jr. 2002. Impact of Low-Intensity Hurricanes on Regional Economic Activity. *Natural Hazards Review* 3(3): 118–125.

Burton, C. G. 2010. Social Vulnerability and Hurricane Impact Modeling. *Natural Hazards Review* 11(2): 58–68.

Burton, I., R. W. Kates, and G. F. White. 1978. *The Environment as Hazard*. New York: Oxford University Press.

Chakraborty, J., T. W. Collins, C. Marilyn, M. C. Montgomery, and S. E. Grineski. 2014. Social and Spatial Inequities in Exposure to Flood Risk in Miami, Florida. *Natural Hazards Review* 15(3): 04014006.

Changnon, S. A., ed. 1996. *The Great Flood of 1993: Causes, Impacts, and Response*. Boulder, CO: Westview Press.

Chen, J., S. Chen, and P. F. Landry. 2013. Migration, Environmental Hazards, and Health Outcome in China. *Social Science and Medicine* 80: 85–95.

Collins, A. E. 2009. *Disaster and Development*. London: Routledge.

Cross, J. A. 2002. Megacities and Small Towns: Different Perspectives on Hazard Vulnerability. *Environmental Hazards* 3: 63–80.

Cunny, F. 1983. *Disasters and Development*. New York: Oxford University Press.

Cutter, S. L., ed. 2001. *American Hazardscapes: The Regionalization of Hazards and Disasters*. Washington, DC: Joseph Henry Press.

Cutter, S. L., L. Barnes, M. Berry, C. Burton, E. Evans, E. Tate, and J. Webb. 2008. A Place-based Model for Understanding Community Resilience to Natural Disasters. *Global Environmental Change* 18: 598–606.

Cutter, S. L., B. J. Boruff, and W. L. Shirley. 2003. Social Vulnerability to Environmental Hazards. *Social Science Quarterly* 84(2): 242–261.

Cutter, S. L., and C. Finch. 2008. Temporal and Spatial Changes in Social Vulnerability to Natural Hazards. *Proceedings of the National Academy of Sciences* 105(7): 2301–2306.

Davis, I., and D. Alexander. 2016. *Recovery from Disaster.* London: Routledge.

de Bruycker, M., D. Greco, M. F. Lechat, I. Annino, N. de Ruggiero, and M. Triassi. 1985. The 1980 Earthquake in Southern Italy—Morbidity and Mortality. *International Journal of Epidemiology* 197: 113–117.

Donner, W. 2007. The Political Ecology of Disaster: An Analysis of Factors Influencing U.S. Tornado Fatalities and Injuries, 1998–2000. *Demography* 44(3): 669–685.

Dow, K., and S. L. Cutter. 2000. Public Orders and Personal Opinions: Household Strategies for Hurricane Risk Assessment. *Environmental Hazards* 2: 143–155.

Downey, D. C., ed. 2016. *Cities and Disasters.* Boca Raton, FL: CRC Press.

Eckel, E. B. 1970. *The Alaska Earthquake March 27, 1964: Lessons and Conclusions.* Washington, DC: United States Government Printing Office.

Ellis, K. N., L. R. Mason, and K. L. Gassert. 2019. Public Understanding of Local Tornado Characteristics and Perceived Protection from Land-Surface Features in Tennessee, USA. *PLOS ONE* 14(7): e0219897. https://doi.org/10.1371/journal.pone.0219897.

FEMA (Federal Emergency Management Authority). 2005a. *Hurricane Charley in Florida: Observations, Recommendations, and Technical Guidance.* Washington, DC: FEMA.

FEMA (Federal Emergency Management Authority). 2005b. *Hurricane Ivan in Alabama and Florida: Observations, Recommendations, and Technical Guidance.* Washington, DC: FEMA.

Fernandez, G., and I. Ahmed. 2019. "Build Back Better" Approach to Disaster Recovery: Research Trends since 2006. *Progress in Disaster Science* 1: 100003. https://doi.org/10.1016/j.pdisas.2019.100003.

Fothergill, A., and L. A. Peek. 2004. Poverty and Disasters in the United States: A Review of Recent Sociological Findings. *Natural Hazards* 32: 89–110.

Freedman, L. *Bad Friday: The Great and Terrible 1964 Alaska Earthquake.* Kenmore, WA: Epicenter Press.

Fussell, E. 2015. The Long Term Recovery of New Orleans' Population after Hurricane Katrina. *American Behavioral Science* 59(10): 1231–1245.

Gaydos, J. C., and G. A. Luz. 1994. Military Participation in Emergency Humanitarian Assistance. *Disasters* 18(1): 48–57.

Ghorai, D., and H. S. Sen. 2015. Role of Climate Change in Increasing Occurrences Oceanic Hazards as a Potential Threat to Coastal Ecology. *Natural Hazards* 75: 1223–1245.

Gunawardena, T., N. Tuan, P. Mendis, L. Aye, and R. Crawford. 2014. Time-Efficient Post-Disaster Housing Reconstruction with Prefabricated Modular Structures. *Open House International* 39(3): 59–69.

Gupta, K. 2009. Cross-cultural Analysis of Responses to Mass Fatalities Following the 2009 Cyclone Aila in Bangladesh and India. Quick Response Report #216. Boulder: The Natural Hazards Center, University of Colorado.

Hall, J. F., ed. 1994. *Northridge Earthquake January 17, 1994, Preliminary Reconnaissance Report.* Oakland, CA: Earthquake Engineering Research Institute.

Hamblyn, R. 2014. *Tsunami: Nature and Culture*. London: Reaktion Books Ltd.

Haney, T. J. 2018. Move Out or Dig In? Risk Awareness and Mobility Plans in Disaster-Affected Communities. *Journal of Contingencies and Crisis Management* 1–13. https://doi.org/10.1111/1468-5973.12253.

Hewitt, K., ed. 1997. *Regions of Risk: A Geographical Introduction to Disasters*. London: Longman.

Hoekstra, S., K. Klockow, R. Riley, J. Brotzge, H. Brooks, and S. Erickson. 2011. A Preliminary Look at the Social Perspective of Warn-on-Forecast: Preferred Tornado Warning Lead Time and the General Public's Perceptions of Weather Risks. *Weather, Climate, and Society* 3: 128–140.

Hulme, M. 2009. *Why We Disagree about Climate Change*. Cambridge: Cambridge University Press.

Iannone, G., ed. 2014. *The Great Maya Droughts in Cultural Context: Case Studies in Resilience and Vulnerability*. Boulder: University Press of Colorado.

Ito, H., W. Wisetjindawar, and M. Yokomatsu. 2014. Improving the Organizational Efficiency of Humanitarian Logistics in the Aftermath of a Large-Scale Disaster. *Journal of Integrated Disaster Risk Management* 4(2): 142–155.

Jonkman, S. N., and I. Kelman. 2005. An Analysis of the Causes and Circumstances of Flood Deaths. *Disasters* 29(1): 75–97.

Joyce, K. E., S. E. Belliss, S. V. Samsonov, S. J. McNeill, and P. J. Glassey. 2009. A Review of the Status of Satellite Remote Sensing and Image Processing Techniques for Mapping Natural Hazards and Disasters. *Progress in Physical Geography* 33(2): 183–207.

Kameyama, Y., A. P. Sari, M. H. Soejachmoen, and N. Kanie, eds. 2008. *Climate Change in Asia: Perspectives on the Future Climate Regime*. Tokyo: United Nations University Press.

Kapur, A. 2010. *Vulnerable India: A Geographical Study of Disasters*. Los Angeles, CA: SAGE.

Kent, R. C. 1987. *Anatomy of Disaster Relief: The International Network in Action*. London: Pinter Publishers.

Kirsch, T. D., E. Leidman, W. Weiss, and S. Doocy. 2012. The Impact of the Earthquake and Humanitarian Assistance on Household Economies and Livelihoods of Earthquake-Affected Populations in Haiti. *American Journal of Disaster Medicine* 7(2): 85–94.

Kozu, S., and H. Homma. 2014. Lessons Learned from The Great East Japan Earthquake: The Need for Disaster Preparedness in the Area of Disaster Mental Health for Children. *Journal of Emergency Management* 12(6): 431–439.

Kurita, T., M. Arakida, and S.R.N. Colombage. 2007. Regional Characteristics of Tsunami Risk Perception among the Tsunami Affected Countries in the Indian Ocean. *Journal of Natural Disaster Science* 29(1): 29–38.

Laframboise, N., and B. Loko. 2012. *Natural Disasters: Mitigating Impact, Managing Risk*. Washington, DC: International Monetary Fund.

Lamond, J., C. Booth, F. Hammand, and D. Proverbs, eds. 2012. *Flood Hazards: Impacts and Responses for the Built Environment*. Boca Raton, FL: CRC Press.

Le Billion, P., and A. Waizenegger. 2007. Peace in the Wake of Disaster? Secessionist Conflicts and the 2004 Indian Ocean Tsunami. *Transactions of the Institute of British Geographers* 32: 411–427.

Lindell, M. K., J. E. Kang, and C. S. Prater. 2011. The Logistics of Household Hurricane Evacuation. *Natural Hazards* 58: 1093–1109.

Lindell, M. K., C. S. Prater, H. C. Wu, S.-K. Huang, D. M. Johnston, J. S. Becker, and H. Shiroshita. 2016. Immediate Behavioural Responses to Earthquakes in Christchurch, New Zealand, and Hitachi, Japan. *Disasters* 40(1): 85–111.

Luther, L. 2008. *Disaster Debris Removal after Hurricane Katrina: Status and Associated Issues.* Washington, DC: Congressional Research Service.

Mahue-Giangreco M., W. Mack, H. Seligson, and L. B. Bourque. 2001. Risk Factors Associated with Moderate and Serious Injuries Attributable to the 1994 Northridge Earthquake, Los Angeles, California. *Annals of Epidemiology* 11: 347–357.

Marano, K. D., D. J. Wald, and T. I. Allen. 2010. Global Earthquake Casualties Due to Secondary Effects: A Quantitative Analysis for Improving Rapid Loss Analyses. *Natural Hazards* 52: 319–328.

Margesson, R. 2005. *Indian Ocean Earthquake and Tsunami: Humanitarian Assistance and Relief Operations.* Washington, DC: Congressional Research Service.

Middleton, N., and P. O'Keefe. 1998. *Disaster and Development: The Politics of Humanitarian Aid.* London: Pluto Press.

Mileti, D. S. 2000. *Disasters by Design: A Reassessment of Natural Hazards in the United States.* Washington, DC: Joseph Henry Press.

Mirza, M.M.Q., R. A. Warrick, N. J. Erickson, and G. J. Kenny. 2001. Are Floods Getting Worse in the Ganges, Brahmaputra and Meghna Basin? *Environmental Hazards* 3: 37–48.

Mitchell, J. K., ed. *Crucibles of Hazard: Mega-Cities and Disasters in Transition.* Tokyo: United Nations University.

Miyake, H., S. N. Sapkota, B. N. Upreti, L. Bollinger, T. Kobayashi, and H. Takenaka. 2017. The 2015 Gorkha, Nepal, Earthquake and Himalayan Studies: First Results. *Earth Planets Space* 69(12). https://doi.org/10.1186/s40623-016-0597-8.

Morgan, O., M. Tidball-Binz, and D. V. Alphen, eds. 2006. *Management of Dead Bodies after Disasters: A Field Manual for First Responders.* Washington, DC: PAHO.

Mustafa, D. 2003. Reinforcing Vulnerability? Disaster Relief, Recovery, and Response to the 2001 Flood in Rawalpindi, Pakistan. *Environmental Hazards* 5(1): 94–105.

National Research Council. 2015. *Affordability of National Flood Insurance Program Premiums: Report 1.* Washington, DC: The National Academies Press.

Niederkrotenthaler, T., E. M. Parker, F. Ovalle, R. E. Noe, J. Bell, L. Xu, M. A. Morrison, C. E. Mertzlufft, and D. E. Sugerman. 2011. Injuries and Post-Traumatic Stress Following Historic Tornados: Alabama. *PLOS ONE* 8(12): e83038. https://doi.org/10.1371/journal.pone.0083038.

Noji, E. K., G. D. Kelen, H. K. Armenian, A. Oganessian, N. P. Jones, and T. Sivertson. 1990. The 1988 Earthquake in Soviet Armenia: A Case Study. *Annals of Emergency Medicine* 19(8): 891–897.

Oliver-Smith, A., and S. M. Hoffman, eds. 1999. *The Angry Earth: Disaster in Anthropological Perspective*. London: Routledge.

Paul, B. K. 1998. Coping with the 1996 Tornado in Tangail, Bangladesh: An Analysis of Field Data. *The Professional Geographer* 50(3): 287–301.

Paul, B. K., and D. Che. 2011. Opportunities and Challenges in Rebuilding Tornado-Impacted Greensburg, Kansas as "Stronger, Better, and Greener." *Geojournal* 76: 93–108.

Pelling, M., ed. 2003. *Natural Disasters and Development in a Globalizing World*. London: Routledge.

Pelling, M. 2011. *Adaptation to Climate Change: From Resilience to Transformation*. London: Routledge.

Phillips, B. 2014. *Qualitative Disaster Research*. New York: Oxford University Press.

Phillips, B., D. M. Neal, and G. Webb. 2011. *Introduction to Emergency Management*. Boca Raton, FL: CRC Press.

Platt, R., ed. 2000. *Disasters and Democracy: The Politics of Extreme Natural Events*. Washington, DC: Island Press.

Pradhan, E. K., K. P. West, Jr., J. Katz, S. C. LeClerq, S. K. Khatry, and S. R. Shrestha. 2007. Risk of Flood-Related Mortality in Nepal. *Disasters* 31(1): 57–70.

Prater, C. S., and M. K. Lindell. 2000. Politics of Hazard Mitigation. *Natural Hazards Review* 1(2): 73–82.

Quarantelli, E. L. 1998. *What Is a Disaster?* London: Routledge.

Rahman, M. S., R. Yang, and L. Di. 2018. Clustering Indian Ocean Tropical Cyclone Tracks by the Standard Deviational Ellipse. *Climate* 6: 39; https://doi.org/10.3390/cli6020039.

Schmidt, A., J. Wolbers, J. Ferguson, and K. Boersma. 2018. Are You Ready2Help? Conceptualizing the Management of Online and Onsite Volunteer Convergence. *Journal of Contingencies and Crisis Management* 26: 338–349.

Sharma, K., K. C. Apil, M. Subedi, and B. Pokharel. 2018. Challenges for Reconstruction after M_w7.8 Gorkha Earthquake: A Study on a Devastated Area of Nepal. Geomatics, *Natural Hazards and Risk* 9(1): 760–790.

Siddique, A. K., A. H. Baqui, A. Eusof, and K. Zaman. 1991. 1988 Floods in Bangladesh: Pattern of Illness and Causes of Death. *Journal of Diarrhoeal Disease Research* 9(4): 310–314.

Simmons, K. M., and D. Sutter. 2011. *Economic and Societal Impacts of Tornadoes*. Boston, MA: American Meteorological Society.

Simmons, K. M., and D. Sutter. 2012. *Deadly Season: Analysis of the 2011 Tornado Outbreaks*. Boston, MA: American Meteorological Society.

Smith, K. 2001. *Environmental Hazards: Assessing Risk and Reducing Disaster*. London: Routledge.

Sobel, A. 2014. *Storm Surge: Hurricane Sandy, Our Changing Climate and Extreme Weather of the Past and Future*. New York: Harper Wave.
Spiegel, P. B. 2005. Differences in World Responses to Natural Disasters and Complex Emergencies. *Journal of American Medical Association* 293: 1915–1918.
Stern, G. 2007. *Can God Intervene? How Religion Explains Natural Disasters*. Westport, CT: Praeger.
Stough, L. M., and D. Kang. 2015. The Sendai Framework for Disaster Risk Reduction and Persons with Disabilities. *International Journal of Risk Science* 6: 140–149.
Telford, J., and J. Cosgrave. 2007. The International Humanitarian System and the 2004 Indian Ocean Earthquake and Tsunamis. *Disasters* 31(1): 1–28.
Thomas, D., B. Phillips, W. Lovekamp, and A. Fothergill. 2013. *Social Vulnerability to Disasters*. Boca Raton, FL: CRC Press.
Tobin, G. A., and B. E. Montz. 1994. *The Great Midwestern Floods of 1993*. Fort Worth, TX: Saunders College Publishing.
UNDP (United Nations Development Program). 2004. *Reducing Disaster Risk: A Challenge for Development*. New York: UNDP.
Waker, P. 1992. Foreign Military Resources for Disaster Relief: An NGO Perspective. *Disasters* 16(2): 152–159.
White, G. F., ed. 1974. *Natural Hazards: Local, National, Global*. New York: Oxford University Press.
Wiharta, S., H. Ahman, J.-Y. Haine, J. Lofgren, and T. Randal. 2008. *The Effectiveness of Foreign Military Assets in Natural Disaster Response*. Solna, Sweden: Stockholm International Peace Institute.
Wisner, B., P. Blaikie, T. Cannon, and I. Davis. 2004. *At Risk: Natural Hazards, People's Vulnerability and Disasters*. London: Routledge.
Wu, H., and D. A. Wilhite. 2004. An Operational Agricultural Drought Risk Assessment Model for Nebraska, USA. *Natural Hazards* 33: 1–21.

About the Editor and Contributors

EDITOR

BIMAL KANTI PAUL, professor of geography and geospatial sciences, Kansas State University

CONTRIBUTORS

SOUMIA BARDHAN, assistant professor in the Department of Communication, University of Colorado, Denver, Colorado

PALASH BASAK, PhD student in the School of Environmental and Life Sciences, University of Newcastle, Ourimbah, Australia

CHRISTINE BESTE, planner (terrorism preparedness) at the Delaware Emergency Management Agency and has a master's degree in disaster science and management from the University of Delaware, Newark, Delaware

DIVYA CHANDRASEKHAR, assistant professor in the Department of City and Metropolitan Planning at the University of Utah, Salt Lake City, Utah

SUBARNA CHATTERJEE, ABD in the Department of Geography and Geospatial Sciences at Kansas State University, Manhattan, Kansas

BEN CLARK, graduate student in the Department of Civil and Environmental Engineering at Virginia Tech, Blacksburg, Virginia

MATTHEW CLARKE, Alfred Deakin Professor and head of the Department of the School of Humanities and Social Sciences at Deakin University, Melbourne, Australia

SAMANTHA CLEMENTS, DAFN (disability, access, and functional needs) emergency manager in Brooklyn, New York

JILL S. M. COLEMAN, associate dean of the College of Sciences and Humanities and Interim Department Chair of Anthropology at Ball State University, Munchie, Indiana

KELSEY COMPTON, student in the Pamplin School of Business at Virginia Tech, Blacksburg, Virginia

SCOTT CURTIS, distinguished professor in the Department of Geography, Planning, and Environment at East Carolina University, Greenville, North Carolina

TONYA FARROW-CHESTNUT, graduate student in the Department of Geography and Earth Sciences at the University of North Carolina, Charlotte, North Carolina

JOHN A. HARRINGTON JR., emeritus professor in the Department of Geography and Geospatial Sciences at Kansas State University, Manhattan, Kansas

LISA M. B. HARRINGTON, professor in the Department of Geography and Geospatial Sciences at Kansas State University, Manhattan, Kansas

SYED A. HASNATH, freelance researcher and scholar, Boston

ANDREW M. HILBURN, assistant professor in the Department of Social Sciences at Texas A&M International University, Laredo, Texas

ASIF ISHTIAQUE, postdoctoral research fellow in the School of Environment and Sustainability at University of Michigan, Ann Arbor, Michigan

TRISHA JACKSON, faculty at Cloud County Community College, Gary County Campus, Junction City, Kansas

AUDREY JOSLIN, assistant professor in the Department of Geography and Geospatial Sciences at Kansas State University, Manhattan, Kansas

LUKE JURAN, associate professor in the Department of Geography at Virginia Tech, Blacksburg, Virginia

WILLIAM P. KLADKY, freelance writer based in Lutherville, Maryland

MAX LU, professor in the Department of Geography and Geospatial Sciences at Kansas State University, Manhattan, Kansas

IAN MACNAUGHTON, undergraduate student in meteorology at Virginia Tech, Blacksburg, Virginia

ABU SAYEED MAROOF, PhD student in the Department of Geography and Geospatial Sciences at Kansas State University, Manhattan, Kansas

RICHARD MARSTON, university distinguished professor emeritus in the Department of Geography and Geospatial Sciences at Kansas State University, Manhattan, Kansas

ROSE G. MICKE, undergraduate student in the Department of Geography and Geospatial Sciences at Kansas State University, Manhattan, Kansas

SALIM MOMTAZ, associate professor in the School of Environmental and Life Sciences, University of Newcastle, Ourimbah, Australia

LIPON MONDAL, doctoral candidate in Sociology at Virginia Tech, Blacksburg, Virginia

About the Editor and Contributors

MD. NADIRUZZAMAN, research fellow in the Institute of Geography, University Hamburg, Hamburg, Germany

KATHERINE S. NELSON, assistant professor in the Department of Geography and Geospatial Sciences at Kansas State University, Manhattan, Kansas

ROBERT OLIVER, associate professor of Geography at Virginia Tech, Blacksburg, Virginia

HILDA ONUOHA, ABD in the Department of Geography and Geospatial Sciences at Kansas State University, Manhattan, Kansas

ANJANA PAUL, freelance writer based in Overland Park, Kansas

M. KHALEDUR RAHMAN, assistant professor in the Geology & Geography Department at Georgia Southern University, Statesboro, Georgia

VAHID RAHMANI, assistant professor in the Department of Biological and Agricultural Engineering at Kansas State University, Manhattan, Kansas

AVANTIKA RAMEKAR, ABD in the Department of Geography and Geospatial Sciences at Kansas State University, Manhattan, Kansas

JALAL SAMIA, postdoc in Laboratory of Geo-Information Science and Remote Sensing at Wageningen University, Wageningen, the Netherlands

ROBERT M. SCHWARTZ, professor in the Department of Disaster Science and Emergency Services at the University of Akron, Akron, Ohio

KATHLEEN SHERMAN-MORRIS, professor in the Department of Geosciences at Mississippi State University, Starkville, Mississippi

RACHEL A. SLOTTER, PhD candidate in the Department of Disaster Science and Management at University of Delaware, Newark, Delaware

ANDREW STEVENS, student in the Department of Geography at Virginia Tech, Blacksburg, Virginia

MITCHEL STIMERS, president and CEO, Compita Consulting, Junction City, Kansas

AMENEH TAVAKOL, recently received her PhD degree from the Department of Biological and Agricultural Engineering at Kansas State University, Manhattan, Kansas

ARNAUD TEMME, associate professor in the Department of Geography and Geospatial Sciences at Kansas State University, Manhattan, Kansas

DEBORAH THOMAS, professor and chair in the Department of Geography and Earth Sciences at the University of North Carolina, Charlotte, North Carolina

VALERIE THOMAS, associate professor in the Department of Forest Resources and Environmental Conservation at Virginia Tech, Blacksburg, Virginia

JOHN PAUL TIEFENBACHER, professor in the Department of Geography at Texas State, San Marcos, Texas

VICKI L. TINNON, assistant professor in the Department of Chemistry and Geosciences at Jacksonville State University, Jacksonville, Alabama

PAUL TODHUNTER, professor in the Department of Geography at University of North Dakota, Grand Forks, North Dakota

JENNIFER TRIVEDI, assistant professor in the Department of Anthropology and a Core Faculty of Disaster Research Center at University of Delaware, Newark, Delaware

ANNA VERSLUIS, associate professor in the Department of Geography, Environmental Studies, and LALACS at the Gustavus Adolphus College, St Perter, Minnesota

VU VO, completed an undergraduate degree in the Department of Geography and Geospatial Sciences at Kansas State University, Manhattan, Kansas

Index

Note: **Bold** indicates volume numbers; page numbers followed by *t* indicate tables and *f* indicate figures.

ActionAid International, **1**:263–267
  areas of work, **1**:264–266
  on climate change and climate justice, **1**:264–266
  on climate change-induced migration, **1**:264–265
  federal model of governance, **1**:266–267
  on food security and farming, **1**:265
  headquarters, **1**:263
  on mining supply chains, **1**:266
  mission and purpose, **1**:263
  on natural resources and land rights, **1**:266
  on sustainable agriculture, **1**:265–266
American Red Cross, **1**:267–272
  Armed Forces Emergency Services, **1**:268
  Barton, Clara, founder, **1**:267, 268–269
  Biomedical Services, **1**:268
  blood donation program, **1**:268, 270, 271
  Board of Governors, **1**:268
  chapters, **1**:268
  debate and controversy, **1**:271
  Disaster Services, **1**:268
  First World War and, **1**:269
  foundation and history, **1**:267–271
  Haiti earthquake and, **1**:270, 271
  Health and Safety Services, **1**:267–268
  Hurricane Katrina and, **1**:270, 271
  International Services, **1**:268
  natural disasters and, **1**:270
  organizational structure, **1**:268
  Second World War and, **1**:270
  September 11, 2001, terrorist attacks and, **1**:270
Avalanches, **1**:1–11
  classification of snow avalanches, **1**:5
  climate change and, **1**:8–10
  components of snow avalanches, **1**:3–5
  debris avalanches, **1**:6, 238
  definition of rock avalanches, **1**:1
  definition of snow avalanches, **1**:1
  earthquakes and, **1**:2, 6
  frost wedging (freeze-thaw cycle) and, **1**:9
  glacial outburst floods and, **1**:10
  Huascarán avalanches, Peru (1960 and 1970), **1**:6
  humans and, **1**:7–8
  major events, **1**:6–7
  moisture content, **1**:3
  overloading, **1**:2–3
  physical characteristics, **1**:1–2
  Plurs avalanche, Switzerland (1618), **1**:7
  primary factors in rockslides, **1**:2–3
  rockslides and climate change, **1**:9
  Rogers Pass avalanche, British Columbia (1910), **1**:7
  snow avalanches and climate change, **1**:9–10
  snowpack and, **1**:4
  terrain and, **1**:4
  types of snow avalanche, **1**:5
  weather and, **1**:4–5
  Wellington avalanche, Washington (1910), **1**:7
  White Friday, Dolomite Mountains, Italy (1916), **1**:6–7

Bam earthquake, Iran (2003), **2**:1–6
  aftershocks, **2**:2
  cause of, **2**:1
  damage and loss, **2**:1, 2–3
  damage to health facilities, **2**:2–3
  damage to infrastructure, **2**:2–3
  damage to sanitation system, **2**:3
  food security and, **2**:3
  impact on children, **2**:3
  Iran Red Crescent Society and, **2**:3
  magnitude, **2**:1
  nongovernment organizations (NGOs) and, **2**:3
  reconstruction, **2**:4–5
  relief, **2**:3–4
  United Nations Children's Fund and, **2**:3–4
  United Nations High Commissioner for Refugees and, **2**:4
  World Food Program and, **2**:4
  World Health Organization and, **2**:4
Bangladesh flood (1998), **2**:6–10
  aman rice and, **2**:8
  causes of, **2**:6–7
  economic losses, **2**:7–8
  loss of life and extent of damage, **2**:7–8, 9
  nongovernment organizations (NGOs) and, **2**:8
  post-flood epidemics, **2**:7
  recovery, **2**:9
  relief works, **2**:8
  Vulnerable Group Feeding (VGF) program, **2**:8
Bengal famine (1943–1944), **2**:10–14
  background and history, **2**:10
  causes and consequences, **2**:11–12
  deaths from, **2**:10
  as a "holocaust," **2**:13
  private and public relief efforts, **2**:13–14
Bhola Cyclone, Bangladesh (1970), **2**:14–18
  deaths from, **2**:14, 15
  equivalent of Category 3 hurricane, **2**:14
  impact, response, and aid, **2**:15–17
  impact on fishing industry, **2**:15
  land-use patterns and, **2**:15
  preparation and mitigation, **2**:14–15
  recovery and reconstruction, **2**:17
Big Thompson Canyon flash flood, Colorado (1976), **2**:18–23
  American Red Cross and, **2**:22
  Army Corps of Engineers and, **2**:22
  causes of, **2**:19–20
  damage from, **2**:20
  deaths from, **2**:18
  economic losses, **2**:20
  response, **2**:21–22
Black Saturday bushfires, Australia (2009), **2**:23–27
  cause of, **2**:23–24
  damage and impact, **2**:24–25
  damage on human communities, **2**:24–25
  damage to property and infrastructure, **2**:24
  deaths and injuries from, **2**:23
  response and recovery, **2**:25–26
Blizzards, **1**:11–22
  costs of winter storms in U.S., **1**:12
  damage to infrastructures, **1**:17–18
  definition and historical elements, **1**:12–14
  definitions of "heavy snow," **1**:13–14
  disaster declarations, **1**:18–19
  Eastern winter storm (2014), **1**:12
  economic losses, **1**:17–18
  emergency management, **1**:19–21
  etymology and use of the term "blizzard," **1**:12
  fatalities, **1**:19
  geographic distribution, **1**:15–17
  Groundhog Day Blizzard (2011), **1**:12
  impacts, **1**:17–19
  injuries, **1**:19
  Iran blizzard (1972), **1**:15
  naming winter storm events, **1**:11–12
  New England Blizzard (1978), **1**:12
  physical formation, **1**:14–15
  preparedness, **1**:19–21
  response and mitigation, **1**:21
  Storm Data (U.S. weather events database), **1**:19
  Superstorm/Storm of the Century (1993), **1**:12
  United States' blizzard variability by decade, **1**:17
  U.S. climatology and, **1**:16–17
  winter storm components, **1**:14
  winter weather terms (U.S. National Weather Service), **1**:13
  See also Buffalo Blizzard, New York (1977)

Index

BP Deepwater Horizon Oil Spill, United States (2010), **2**:27–32
  cleanup methods, **2**:29
  deaths from, **2**:28
  impact and response, **2**:29–30
  impact on fishing industry, **2**:29–30
  impact on wildlife, **2**:29
  litigation and settlement, **2**:30–31
  preparation and mitigation, **2**:28–29
  recovery and consequences, **2**:30–31
  RESTORE Act and, **2**:30
  worker injuries from, **2**:30
Brisbane and Queensland flood, Australia (2011), **2**:32–36
  causes of, **2**:32–33
  damage and impact, **2**:33–34
  deaths from, **2**:32, 33
  inefficient dam management and, **2**:33
  insurance claims, **2**:35
  litigation, **2**:35
  response and recovery, **2**:34–35
  size of disaster zone, **2**:33–34
Buffalo Blizzard, New York (1977), **2**:36–40
  Army Corps of Engineers and, **2**:39
  cause of, **2**:37
  deaths from, **2**:38
  economic losses, **2**:38
  food shortages, **2**:39
  impacts, **2**:37–38
  response, **2**:39–40
Bush, George W., **2**:153, 165, 168, 177

California drought (2012–2016), **2**:40–45
  damage to agricultural sector, **2**:41–42
  damages, **2**:41–43
  damages to ecosystems, **2**:42–43
  impact on forest trees, **2**:42
  impact on hydropower production, **2**:42
  preparations for the future, **2**:44
Carter, Jimmy, **2**:39
Catholic Relief Services (CRS), **1**:272–276
  agriculture program, **1**:275
  education program, **1**:275
  environmental and agricultural programs, **1**:273
  guiding principles, **1**:274
  health projects, **1**:275
  Integral Human Development (IHD) and, **1**:273
  microfinance programs, **1**:275
  mission and purpose, **1**:273–274
  natural disasters and, **1**:276
  origin and background, **1**:272–273
  programs and activities, **1**:275–276
  water and sanitation programs, **1**:275
Chicago Heat Wave, Illinois (1995), **2**:45–50
  causes of, **2**:46–47
  daily temperatures, **2**:46
  deaths from, **2**:45, 47–48
  impact on human health, **2**:47–48
  impact on infrastructure, **2**:48
  lessons learned, **2**:49
  response, **2**:48–49
Chi-Chi earthquake, Taiwan (1999), **2**:50–54
  aftershocks, **2**:50
  damage, **2**:50–52
  damage to property and infrastructure, **2**:50–51
  deaths from, **2**:51, 52
  landslides, **2**:52
  magnitude, **2**:50
  recovery, **2**:53–54
  response, **2**:52–53
Christchurch earthquake, New Zealand (2010–2011), **2**:54–59
  aftershocks, **2**:56, 57, 58
  cause of, **2**:55–56
  deaths from, **2**:56
  impact, **2**:56–57
  magnitude, **2**:54
  Maori population and, **2**:58, 59
  recovery, **2**:58–59
  response, **2**:57–58
Climate change, **1**:8–10, 22–33
  ActionAid International and, **1**:264–266
  adaptation options, **1**:32
  anthropogenic global warming (AGW), **1**:23
  avalanches and, **1**:8–10
  Black Saturday bushfires, Australia (2009) and, **2**:23, 26–27
  causes of, **1**:23–26
  climate hazards, **1**:27–29
  climate sensitivity and feedbacks, **1**:26
  coastal erosion and, **1**:34–35, 36–37, 39, 42
  desertification and, **1**:44–45, 47–48, 51–52

Climate change (*cont.*)
  drought and, **1:**28, 59, 63, 205; **2:**97, 245
  erosion and, **1:**78, 100, 101, 102, 103–105, 106
  flooding and, **1:**28; **2:**259, 300
  food insecurity and, **1:**51
  food safety and natural ecosystems and, **1:**30–31
  freshwater resources and, **1:**31
  GHG emissions and, **1:**25–26
  human health and economy and, **1:**31
  hurricanes and, **1:**130, 137–138, 178; **2:**161
  impacts, **1:**30–31
  International Organization for Migration and, **1:**298–299
  IPCC definition of, **1:**22
  Islamic Relief Worldwide and, **1:**303
  long-term future of, **1:**30
  Lutheran World Federation and, **1:**306, 307–308
  mitigation options, **1:**31–32
  near-term future of, **1:**29–30
  precipitation and, **1:**27–28
  Refugees International and, **1:**325–326
  salinization and, **1:**172, 173, 175, 176
  sea level rise and, **1:**36–37, 42
  socio-spatial vulnerability and, **2:**119
  storms and, **1:**28
  temperature extremes and, **1:**29, 199, 200, 201, 206, 208–209; **2:**108–109
  transportation and, **1:**206
  warming and, **1:**27
  wildfire and, **1:**29
  wildfires and, **1:**257, 258, 260; **2:**286, 287, 290
  See also Intergovernmental Panel on Climate Change (IPCC)
Clinton, Bill, **2:**49, 140
Coastal erosion, **1:**33–43
  adaptation responses, **1:**40–41
  Bruun model or rule, **1:**38
  causes and types of, **1:**35–36
  climate change and, **1:**34–35, 36–37, 36–37, 39, 42
  defense responses, **1:**40
  definition, **1:**33–34
  impact on future of coastal communities, **1:**39–40
  impact on infrastructure, **1:**39
  impact on people and communities, **1:**39
  impact on sediment, **1:**38–39
  impacts, **1:**38–40
  managed retreat/realignment responses, **1:**41
  measurements and tracking, **1:**37–38
  rapid-onset coastal erosion, **1:**36–37
  responses, **1:**40–42
  sea level rise and, **1:**36–37
  slow-onset coastal erosion, **1:**36–37
Colombia floods (2010–2011), **2:**59–64
  damages, **2:**60–61
  deaths from, **2:**60, 61
  environmental context, **2:**60
  government response, **2:**62–63
  health effects, **2:**62
  impacts to agriculture, **2:**61–62
  impacts to livestock, **2:**62
  impacts to public infrastructure, **2:**61
Colorado flood, United States (2013), **2:**64–68
  community recovery, **2:**67–68
  death and loss, **2:**65–66
  event summary, **2:**64–65
  history of flooding, **2:**67–68
  mining industry and, **2:**66
  secondary hazard, **2:**66
  warning system, **2:**66–67
Concern Worldwide, **1:**276–281
  activities, **1:**278–281
  aid to famines and refugees, **1:**280
  climate-smart emergency programs, **1:**279
  conflict-driven emergencies and, **1:**279
  displaced persons and, **1:**278–279, 280
  employees, **1:**280
  famines and, **1:**278–279
  fundraising, **1:**278
  mission and purpose, **1:**276–277
  natural disasters and, **1:**278–279, 280
  O'Loughlin-Kennedy, John and Kay, founders, **1:**277
  origin and history, **1:**277–278
Cooperative for Assistance and Relief Everywhere, **1:**281–286
  development, **1:**282–283
  founding, **1:**282
  fundamental principles, **1:**284–285
  Marshall Plan and, **1:**283
  mission and goals, **1:**281–282
  organization and source of funding, **1:**285
  Peace Corps and, **1:**283
  World War II and, **1:**282

Cyclone Eline, **2:**249, 251
Cyclone Gloria, **2:**249
Cyclone Gorky, Bangladesh (1991), **2:**69–73
  damage to infrastructure, **2:**70
  deaths from, **2:**69–70, 71
  early warning and evacuation, **2:**70–71
  nongovernment organizations (NGOs) and, **2:**72
  "Operation Sea Angel" and, **2:**71
  response, **2:**71–72
  United Nations task force and, **2:**71–72
Cyclone Nargis, Myanmar (2008), **2:**73–77
  Category 4 cyclone, **2:**73
  death and destruction, **2:**73–74
  Doctors Without Borders and, **2:**74–75
  international response, **2:**75
  Myanmar military and, **2:**74
  nongovernment organizations (NGOs) and, **2:**74–75
  relief efforts, **2:**75–76
  response, **2:**74–75
  Save the Children and, **2:**74
  World Health Organization and, **2:**75, 76
Cyclone Pam, Vanuatu (2015), **2:**77–82
  Category 5 cyclone, **2:**78
  damage to economic and social sectors, **2:**79–80
  damage to property and infrastructure, **2:**78–79
  deaths from, **2:**79, 80
  nongovernment organizations (NGOs) and, **2:**80–81, 82
  preparedness, **2:**80
  response, **2:**80–81
  warnings, **2:**80
Cyclone Phailin, India (2013), **2:**82–87
  Category 4 cyclone, **2:**83
  damage to agriculture and livestock, **2:**84
  damage to property and infrastructure, **2:**83–84
  deaths from, **2:**83, 86
  National Disaster Management Authority (NDMA) and, **2:**85
  "zero-casualty approach" and, **2:**83
Cyclone Sidr, Bangladesh (2007), **2:**87–91
  Category 4 storm, **2:**87
  damage and loss, **2:**87–88
  damage to Sundarbans (mangrove forest), **2:**88
  deaths from, **2:**88, 89
  nongovernment organizations (NGOs) and, **2:**89, 90–91
  reconstruction, **2:**90–91
  relief operations, **2:**89–90
  warning and evacuation, **2:**89

Desertification, **1:**43–53
  agriculture and, **1:**50
  climate change and, **1:**44–45, 47–48, 51–52
  conflict and, **1:**52
  definition of, **1:**43
  dust storms and, **1:**51–52
  ecoregions and, **1:**44
  food security and, **1:**50–51
  Gobi Desert, China and Mongolia, **1:**49
  human systems and, **1:**50–51
  Index of Human Insecurity (IHI) and, **1:**52
  Koeppen-Geiger climate classification and, **1:**44
  major problem areas, **1:**47–50
  major processes in, **1:**44–47
  net primary production (NPP) and, **1:**47
  physical characteristics of, **1:**44
  Sahel Region, Africa, **1:**48–49
  salinization, **1:**46
  secondary natural disasters and, **1:**51–52
  soil compaction, **1:**46–47
  soil degradation, **1:**44–45
  vegetation degradation (deflation), **1:**45–46
Doctors Without Borders, **1:**286–290
  in Africa, **1:**289
  disaster relief, **1:**288–289
  displaced persons and, **1:**289
  establishment and organizational structure, **1:**287–288
  Kouchner, Bernard, and, **1:**287
  missions, **1:**288
  Nicaraguan earthquake (1972) and, **1:**288–289
  principles, **1:**288
  in Southeast Asia, **1:**289

Droughts, **1:**53–64
  agricultural drought, **1:**56
  agricultural responses, **1:**61–62
  causes of, **1:**58–59
  climate change and, **1:**28, 59, 63, 205; **2:**97, 245
  cooling facility responses, **1:**62
  as "creeping disasters," **1:**54, 62
  definitions and types of, **1:**54–57
  economic impacts, **1:**59–60
  energy impacts, **1:**60
  health and mortality impacts, **1:**61
  hydrological drought, **1:**56–57
  impacts, **1:**59–61
  indices of, **1:**58
  meteorological drought, **1:**55–56
  National Drought Mitigation Center (NDMC), **1:**54–55
  Palmer Drought Severity Index (PDSI), **1:**58
  positive drought impacts, **1:**63
  responses, **1:**61–62
  social aid responses, **1:**62
  social impacts, **1:**61
  socioeconomic drought, **1:**57
  water restriction responses, **1:**62
  wildlife impacts, **1:**60
  See also California drought (2012–2016); East African drought (2011–2012); Millennium drought, Australia (2001–2012)
Dust Bowl, the (1930s), **2:**91–96
  agricultural mechanization and practices and, **2:**93
  climate conditions and, **2:**92–93
  impacts and response, **2:**94–95
  New Deal and, **2:**95
  scope and causes, **2:**92–94
  Soil Conservation Service and, **2:**95

Earthquakes, **1:**64–75
  aftershocks, **1:**67–68
  avalanches and, **1:**2, 6
  collapse earthquakes, **1:**71
  definition of, **1:**64–65
  earthquake magnitude and intensity comparison, **1:**69*t*
  explosion earthquakes, **1:**71
  geographic distribution of, **1:**69–70
  measures of magnitude, **1:**68–69
  myths, **1:**65–66
  National Earthquake Information Center (NEIC), **1:**68
  prediction, **1:**72–74
  preparedness, **1:**74
  Richter scale, **1:**68, 70, 71
  selected parameters of earthquakes, **1:**65*t*
  seismic waves and depth, **1:**66–68
  tectonic earthquakes, **1:**70
  types of, **1:**70–72
  volcanic earthquakes, **1:**70–71
  See also Bam earthquake, Iran (2003); Chi-Chi earthquake, Taiwan (1999); Christchurch earthquake, New Zealand (2010–2011); Great Kanto earthquake, Japan (1923); Gujarat earthquake, India (2001); Haiti earthquake (2010); Izmit/Marmara earthquake, Turkey (1999); Kashmir earthquake, Pakistan (2005); Kobe earthquake, Japan (1995); Loma Prieta earthquake, California (1989); Mexico City earthquakes, Mexico (1985); Nepal earthquakes (2015); Sichuan earthquake, China (2008); Sulawesi earthquake and tsunami, Indonesia (2018); Tangshan earthquake, China (1976); Tohoku earthquake and Fukushima tsunami, Japan (2011); Valdivia earthquake, Chile (1960)
East African drought (2011–2012), **2:**96–100
  ActionAid International and, **2:**99
  background of, **2:**96–97
  deaths from, **2:**96
  early warning, **2:**97–98
  humanitarian impacts, **2:**98
  impact on health, **2:**99
  Oxfam and, **2:**99
  refugee crisis, **2:**98–99
  response and recovery, **2:**99–100
Ecoregions, **1:**44, 76
Edmonton tornado, Canada (1987), **2:**101–105
  challenges for emergency response, **2:**102–103
  changes to emergency response plans, **2:**103–104
  chronology and damage, **2:**101

damage to property and infrastructure, **2**:101
deaths from, **2**:101
lack of public awareness, **2**:102
Erosion, **1**:75–87
  badlands and, **1**:79–80
  characteristics of river patterns and their relative stability, **1**:81*t*
  climate change and, **1**:78, 100, 101, 102, 103–105, 106
  cropland agriculture and, **1**:76–77
  forest removal and, **1**:77
  glacial erosion, **1**:82–83
  grazing and, **1**:77
  groundwater erosion, **1**:83–85
  gullies and, **1**:79
  land surface erosion, **1**:77
  natural rates and rates accelerated by human activities, **1**:75–78
  off-road vehicles and, **1**:77
  rills and, **1**:79
  rivers erosion, **1**:80–82
  sheetflow erosion, **1**:78–79
  subsurface tunnels (pipes) and, **1**:79
  urbanization and, **1**:78
  wind erosion, **1**:85–86
European heat wave (2003), **2**:105–109
  deaths from, **2**:105
  impact on agriculture, **2**:107
  impact on energy sector, **2**:107–108
  impact on property and infrastructure, **2**:106–107
  impacts, **2**:106–108
  origin and background, **2**:106
  social and environmental impacts, **2**:108
  wildfires and, **2**:107
Expansive soils, **1**:87–97
  alfisols, **1**:91, 92
  aridisols, **1**:91, 92
  Atterberg limits, **1**:93–94
  clay minerals, **1**:88, 89–90
  COLE (coefficient of linear extensibility), **1**:94
  engineering solutions, **1**:94–95
  environmental solutions, **1**:95
  geographic distribution of, **1**:88, 91–92
  as high shrink-swell soils, **1**:88
  measurement of, **1**:88, 93–94
  mollisols, **1**:91–92
  natural hazard and, **1**:88, 92–93

octahedral sheet and, **1**:90
Plasticity Index (PI), **1**:94
soil colloids and, **1**:89
soil orders and, **1**:88, 90–91
solution of building more structurally sound foundations, **1**:94–95
solution of control moisture changes, **1**:94
solution of mixing expansive and nonexpansive materials, **1**:95
vermiculite clay group, **1**:90
vertisols, **1**:91–92
Extinction, **1**:97–107
  Alvarez, Luis, on, **1**:99
  asteroid impact theory, **1**:99–100
  Chicxulub Impact event and, **1**:99, 103
  complexity of causes of mass extinctions, **1**:105–106
  Cuvier, Georges, on, **1**:98–99
  definition of, **1**:97
  early and contemporary understanding of mass extinctions, **1**:98–100
  fifth mass extinction (K-T event), **1**:99, 102–104
  first mass extinction, **1**:100
  fourth mass extinction, **1**:102
  possible sixth mass extinction, **1**:98, 104–105
  present-day extinction rate, **1**:104
  second mass extinction, **1**:101
  Smit, Jan, on, **1**:99
  third mass extinction, **1**:101–102
Eyjafjallajökull eruption, Iceland (2010), **2**:109–114
  flooding and, **2**:112
  health impacts, **2**:112
  impact on air traffic, **2**:111–112
  impact on tourism industry, **2**:112
  impact on transportation, **2**:112
  seismic and volcanic activities prior to eruption, **2**:110–111

Federal Emergency Management Agency (FEMA)
  American Red Cross and, **1**:270
  disaster declarations and, **1**:18
  earthquakes and, **2**:233, 238
  floods and, **1**:108, 117; **2**:116–118, 206
  hurricanes and, **2**:149, 152–153, 163, 173, 176, 177, 182, 279
  ice storms and, **1**:147

Floods, **1:**107–119
  agricultural and livestock impacts, **1:**115
  benefits of, **1:**115–116
  channelization and, **1:**117
  clean water impacts, **1:**115
  climate change and, **1:**28; **2:**259, 300
  coastal floods, **1:**110–111
  dams, **1:**117
  deaths from, **1:**114
  definition of, **1:**107–108
  definition of flood frequency, **1:**111
  demography and, **1:**114
  in developed and developing countries, **1:**114
  dikes and, **1:**116–117
  flash floods, **1:**109–110
  flood insurance and aid, **1:**118
  forecasting and warning systems, **1:**115
  frequency and impacts of floods by continents, **1:**112*t*
  geographic distribution of, **1:**111–113
  health impacts, **1:**113–114
  impacts, **1:**113–115
  national characteristics and, **1:**114
  number of deadly floods that killed at least 51 people by continents, **1:**113*f*
  physical dimensions of, **1:**111
  prevention and mitigation strategies, **1:**116–118
  river floods, **1:**108–109
  sandbagging and, **1:**117
  stilt houses and, **1:**117
  types and causes of, **1:**108–111
  See also Bangladesh flood (1998); Big Thompson Canyon flash flood, Colorado (1976); Brisbane and Queensland flood, Australia (2011); Colombia floods (2010–2011); Colorado flood, United States (2013); Grand Forks flood, North Dakota (1997); Great Mississippi River flood, United States (1993); Iowa flood, United States (2008); Johnstown flood, Pennsylvania (1889); Kerala floods, India (2018); Pakistan flood (2010); Summer floods, United Kingdom (2007); Tropical storm and floods, Yemen (2008); Vietnam flood (1999); Yangtze River flood, China (1931)

Grand Forks flood, North Dakota (1997), **2:**114–118
  damage to property and infrastructure, **2:**116
  emergency flood preparations, **2:**114–115
  events, **2:**115–116
  impact on agriculture and livestock, **2:**114
  impact on population statistics, **2:**116
  impacts and financial assistance, **2:**116–117
  long-term recovery, **2:**117–118
Great Ice Storm of 1998, Canada, **2:**118–122
  deaths from, **2:**119, 120–121
  impact on banking system, **2:**120
  impact on electrical grid and energy sector, **2:**119–120, 121
  impact on forest sector, **2:**121
  impacts and response, **2:**119–121
  meteorological conditions, **2:**119
  miscalculation of risk and, **2:**120
  reconstruction and recovery, **2:**121–122
Great Kanto earthquake, Japan (1923), **2:**122–127
  aftershocks, **2:**123
  American Red Cross and, **2:**125–126
  cash donations, **2:**126
  damage to property and infrastructure, **2:**124–125
  deaths from, **2:**123
  fires and, **2:**123–124
  impact, **2:**123–125
  impact on nationalism and racism, **2:**124
  magnitude, **2:**122
  relief, **2:**125–126
Great Mississippi River flood, United States (1993), **2:**127–132
  damage by scouring and deposition, **2:**129–130
  damages, **2:**129–130
  deaths from, **2:**129
  economic losses, **2:**127
  environmental context, **2:**127–129
  government response, **2:**131
  impacts to agriculture, **2:**130
  impacts to transportation systems, **2:**130–131
  wastewater facilities and, **2:**129
Gujarat earthquake, India (2001), **2:**132–136

Index

aftershocks, **2:**133, 135
  cause of, **2:**133
  deaths from, **2:**134
  impacts, **2:**133–134
  magnitude, **2:**132
  nongovernment organizations (NGOs) and, **2:**134, 136
  rehabilitation, **2:**135–136
  response, **2:**134–135

Hail, **1:**119–129
  car damage from, **1:**124
  crop damage from, **1:**124
  damage and destruction from, **1:**123–125
  death and injury from, **1:**124–125
  definition of, **1:**119
  economic losses from, **1:**125
  forecasting and warning systems, **1:**126
  global distribution of, **1:**121–122
  hail damage ($ million) in the United States, **1:**125*t*
  hail insurance, **1:**125–126, 127
  hail loss claims 2013–2015 by top five hail loss policy types, **1:**127*t*
  hailstone criteria, **1:**120
  preparedness and response, **1:**126–128
  property damage from, **1:**123–124
  residential property damage from, **1:**123–124
  size and layering, **1:**119–121
  top five states for major hail events, **1:**123*t*
  in United States, **1:**122–123
Haiti earthquake (2010), **2:**137–141
  aftershocks, **2:**137
  deaths from, **2:**137–138
  displaced persons camp, **2:**140–141
  essential background, **2:**138
  international relief, **2:**140
  responses, **2:**140
  U.S. military and, **2:**140
  vulnerability to earthquakes, **2:**138–139
Heat wave and wildfires, Russia and Eastern Europe (2010), **2:**141–146
  deaths from, **2:**143, 144
  government responses, **2:**144–145
  impact to agriculture, **2:**144
  impact to property, **2:**144
  impacts, **2:**144
  vulnerabilities, **2:**142–144

Hurricane Andrew, United States and the Bahamas (1992), **2:**146–150
  damage, **2:**148–149
  deaths from, **2:**149
  meteorological synopsis, **2:**146–147
  response and recovery, **2:**149
  Saffir-Simpson Hurricane Wind Scale category, **2:**146–147, 150
  tornadoes and, **2:**148
  warnings, **2:**147
Hurricane Charley, United States (2004), **2:**150–155
  deaths from, **2:**153
  economic losses, **2:**154
  effects, **2:**153–154
  predictions, evacuations, and hurricane's path, **2:**151–153
  Saffir-Simpson Hurricane Wind Scale category, **2:**151, 153
  tornadoes and, **2:**153
Hurricane Galveston, United States (1900), **2:**155–159
  background, **2:**155
  deaths from, **2:**156–157
  effects of, **2:**156–158
  impact on property and infrastructure, **2:**157
  looting and martial law, **2:**157
  Saffir-Simpson Hurricane Wind Scale category, **2:**155
  seawall and grade raising, **2:**158–159
Hurricane Harvey, Texas and Louisiana (2017), **2:**159–164
  casualty estimates, **2:**162
  damage estimates, **2:**162–163
  deaths from, **2:**162, 164
  formation, **2:**160
  landfall and impacts, **2:**161–163
  recovery efforts, **2:**163
  Saffir-Simpson Hurricane Wind Scale category, **2:**159, 160
  warning and evacuation, **2:**160
Hurricane Ike, United States (2008), **2:**164–168
  deaths from, **2:**167
  oil and natural gas problems, **2:**166–167
  Saffir-Simpson Hurricane Wind Scale category, **2:**164
  storm formation and warnings, **2:**164–166
  storm surge and flooding, **2:**166
  windstorms, **2:**167–168

Hurricane Irma, Florida (2017), **2**:168–173
    casualty estimates, **2**:171
    damage estimates, **2**:171–172
    deaths from, **2**:171, 172, 173
    effects, **2**:171–172
    effects on policy, **2**:172
    evacuation and sheltering, **2**:172–173
    formation, **2**:169
    landfall, **2**:169–170
    Saffir-Simpson Hurricane Wind Scale category, **2**:169, 172
    tornadoes and, **2**:170
Hurricane Katrina, United States (2005), **2**:173–179
    aftermath, **2**:178
    American Red Cross and, **1**:270, 271
    deaths from, **2**:174
    impact and response, **2**:175–177
    landfall, **2**:173–175
    preparing for the storm, **2**:174
    recovery, **2**:177–178
    Saffir-Simpson Hurricane Wind Scale category, **2**:173–174
    storm surges and, **1**:187; **2**:174, 175, 177, 178
    Superdome and, **2**:175–176
Hurricane Maria, Puerto Rico (2017), **2**:179–183
    controversies, **2**:181–183
    deaths from, **2**:179, 180, 181–182, 183
    historical context and economic issues, **2**:180–181
    impact on power and water supplies, **2**:179–180
    Jones Act and, **2**:181
    relocation and, **2**:180
    response, **2**:179–180
    Saffir-Simpson Hurricane Wind Scale category, **2**:179
Hurricane Matthew, United States (2016), **2**:183–188
    Cuba and Bahamas, **2**:185
    deaths from, **2**:184, 186, 187
    Haiti and Dominican Republic, **2**:184–185
    inland flooding, **2**:187
    Saffir-Simpson Hurricane Wind Scale category, **2**:183–184, 185–186
    United States, **2**:185–187
Hurricane Mitch, Central America (1998), **2**:188–193
    deaths from, **2**:190
    impact, **2**:189–190
    impact on agriculture, **2**:189
    impact on crime rates and domestic violence, **2**:190
    impact on infrastructure, **2**:190
    international assistance, **2**:191
    origin, **2**:188–189
    response, **2**:190–191
    Saffir-Simpson Hurricane Wind Scale category, **2**:189
Hurricane Patricia, **1**:132, 134, 135
Hurricane Rita, **1**:135, 136, 170; **2**:154, 160
Hurricane Stan, Guatemala (2005), **2**:193–197
    deaths from, **2**:193, 197
    landfall, **2**:193–194
    landslides and flooding, **2**:194–195
    medical issues and relief efforts, **2**:195–197
    nongovernment organizations (NGOs) and, **2**:194–195, 196
    Office of National Coordination for Disaster Reduction (CONRED) and, **2**:193, 195, 196
    Saffir-Simpson Hurricane Wind Scale category, **2**:193, 197
Hurricane Wilma, **1**:134, 135–136, 270; **2**:149, 154
Hurricanes, **1**:129–138
    category five events, **1**:136
    climate change and, **1**:130, 137–138, 178; **2**:161
    Coriolis Effect and global wind patterns, **1**:131
    costs and death records, **1**:136
    deadliest storms in South Asia, **1**:137
    eye and eyewall, **1**:130–131
    formation of, **1**:129–130
    Hadley Cell and, **1**:131
    hurricane season, **1**:133
    landfall and effects, **1**:131–133
    major Atlantic hurricanes and records, **1**:135–136
    naming, **1**:135
    rainfall and, **1**:132–133
    rotation, **1**:131
    Saffir-Simpson Scale, **1**:134
    Tri-Cell Model of Global Air Circulation and, **1**:131
    tropical cyclones and cyclones, **1**:129, 137
    typhoons, **1**:129, 137

warnings, **1:**133–134
*See also individual cyclones,
hurricanes, and typhoons*

Ice storms, **1:**139–151
  Arkansas and Oklahoma (December 2000), **1:**148–149
  atmospheric conditions and, **1:**139–141
  Canada and northeastern United States (January 1998), **1:**146–147
  catastrophic freezing rain events, **1:**144–145
  forecasting, **1:**141–142
  forest industry and, **1:**145
  ice accumulations on telegraph wires, **1:**145
  Kentucky and Gulf Coast (January 2009), **1:**147
  northern Idaho (January 1961), **1:**149
  notable ice storms, **1:**145–149
  power lines and, **1:**143
  regional frequency of, **1:**142–143
  safety tips during ice storms, **1:**149–150
  societal impacts, **1:**143–145
  southeastern ice storm (February 1994), **1:**146
  United States (December 2002), **1:**147–148
  vehicle accidents and, **1:**143–144
  See also Great Ice Storm of 1998, Canada
Index of Human Insecurity (IHI), **1:**52
Indian Ocean tsunami (2004), **2:**197–202
  deaths from, **2:**198, 199
  Great Sumatra-Andaman Earthquake and, **2:**197, 198
  impact on access to health care and medicine, **2:**199–200
  impact on built, social, and natural environments, **2:**199
  impact on environment, **2:**200
  impacts and response, **2:**198–200
  international response, **2:**200
  reconstruction and recovery, **2:**200–202
  warning and preparation, **2:**198
Intergovernmental Panel on Climate Change (IPCC)
  on adaption options, **1:**207
  on air temperature, **1:**23, 200, 209, 307
  definition of climate change, **1:**22
  on desertification, **1:**45, 47–48
  on displaced persons, **1:**325

Fifth Assessment Report (AR5), **1:**23
  on flooding, **1:**28
  Representative Concentration Pathways (RCPs), **1:**29–30
  on sea level rise, **1:**298, 325
International Federation of Red Cross and Red Crescent Societies (IFRC), **1:**290–299
  Disaster Relief Emergency Fund (DREF), **1:**293
  disaster reports and activities, **1:**294
  Dunant, Henry, founder, **1:**291
  founding, **1:**291–292
  fundamental principles, **1:**292–293
  humanity principle, **1:**292
  impartiality principle, **1:**292
  independence principle, **1:**292
  mission and purpose, **1:**290–291
  neutrality principle, **1:**292
  organization and sources of funding, **1:**293
  Red Crescent symbol, **1:**291
  unity principle, **1:**293
  universality principle, **1:**293
  voluntary service principle, **1:**293
  World War I and, **1:**291
International Organization for Migration (IOM), **1:**295–299
  climate change programs, **1:**298–299
  Cluster Approach and, **1:**296–297
  disaster rebuilding efforts, **1:**297–298
  Disaster Risk Reduction (DRR) program, **1:**298
  disaster-related activities and participation, **1:**296–298
  Hurricane Mitch and, **1:**296
  One Room Shelter Project, **1:**297
  structure, **1:**295
Iowa flood, United States (2008), **2:**202–207
  causes of, **2:**203–204
  impact on property and agriculture, **2:**204–205
  impact on transportation, **2:**205
  impacts, **2:**204–205
Islamic Relief Worldwide, **1:**299–304
  advocacy initiatives, **1:**303
  controversies, **1:**303–304
  developmental efforts, **1:**302–303
  global strategy, **1:**301
  governance, **1:**300–301
  humanitarian efforts, **1:**302
  vision, missions, and values, **1:**302

Izmit/Marmara earthquake, Turkey (1999), **2**:207–211
  aftershocks, **2**:208
  deaths from, **2**:208
  loss and damage, **2**:208–209
  magnitude, **2**:208
  nongovernment organizations (NGOs) and, **2**:209–210
  reconstruction, **2**:210–211
  response, **2**:209–210
  tsunami and, **2**:207, 208

Johnstown flood, Pennsylvania (1889), **2**:211–216
  causes of, **2**:212–213
  deaths and destruction, **2**:212
  events associated with, **2**:213–214
  relief and reconstruction, **2**:214–215
Joplin tornado, Missouri (2011), **2**:216–221
  community recovery, **2**:220
  damage to property and infrastructure, **2**:217–218
  death and loss, **2**:218–219
  meteorological synopsis, **2**:216
  tornado warning, **2**:219–220
  total damage area, **2**:216, 217

Kashmir earthquake, Pakistan (2005), **2**:221–225
  aftershocks, **2**:222
  damage and death, **2**:221–222
  magnitude, **2**:221
  nongovernment organizations (NGOs) and, **2**:222
  recovery and rebuilding, **2**:224
  response and relief, **2**:222–224
Kennedy, John F., **1**:283
Kerala floods, India (2018), **2**:225–230
  cause of, **2**:225–226
  impact, **2**:226–227
  rehabilitation, **2**:228–229
  response, **2**:227–228
Kobe earthquake, Japan (1995), **2**:230–234
  aftershocks, **2**:230, 231
  cause of, **2**:230–231
  damage and destruction, **2**:231–233
  damage to property and infrastructure, **2**:231–232
  deaths from, **2**:231, 232
  economic impact, **2**:232–233
  impact on communication sector, **2**:232
  magnitude, **2**:230, 234

Koeppen-Geiger climate classification, **1**:44

Landslides, **1**:151–162
  Bridge of the Gods, Oregon (1450), **1**:156–157
  causes of, **1**:153–154
  classification of landslide types by material and movement type, **1**:151
  earthquake and landslide near Villach, Austria (1348), **1**:156
  falls and topples, **1**:152
  flows, **1**:153
  geographic distribution of, **1**:155–156
  Haiyuan earthquake and landslides near Ningxia, China (1920), **1**:157
  hazard assessments, **1**:161
  impact on buildings and towns, **1**:158
  impact on crops, **1**:158–159
  impact on landscapes, **1**:159–160
  impact on roads and other infrastructure, **1**:158
  impacts, **1**:157–160
  risk assessments, **1**:161
  rockfall into Lake Geneva (563 CE), **1**:156
  rockfall of Mont Granier, France (1248), **1**:156
  secondary effects of upstream and downstream flooding, **1**:159
  size of, **1**:154–155
  slides, **1**:152
  spreads, **1**:152–153
  susceptibility assessments, **1**:160–161
  types of, **1**:152
Lightning, **1**:162–172
  causes of, **1**:163
  in Central Africa, **1**:168
  cloud-to-ground lightning, **1**:165
  contact or conduct injuries from, **1**:167
  deaths from, **1**:166
  features of, **1**:164
  fires from, **1**:167
  geographic distribution of, **1**:168–169
  ground-to-cloud lightning (triggered lightning), **1**:165
  heat damage from, **1**:167
  heat lightning, **1**:165
  in the Himalayan Mountains, **1**:168
  impact of lightning strike, **1**:166–167
  in Indonesia, **1**:168
  injuries from, **1**:166–167
  inter-cloud lightning, **1**:164

intra-cloud lightning (sheet lightning), 1:164
in Pamas of southeastern Argentina, 1:168
safety tips, 1:169–170
side flash injuries from, 1:166
in South America, 1:168
spider lightning, 1:165
types of, 1:164–165
in the United States, 1:168–169
wildfires and, 1:250, 252–253, 256
Loma Prieta earthquake, California (1989), 2:235–239
aftershocks, 2:235–236, 238
death and damage, 2:236–238
magnitude, 2:235
response and recovery, 2:238–239
selected physical and geologic characteristics, 2:235–236
Lutheran World Federation, 1:304–309
climate change, 1:307–308
disaster relief, 1:306–307
five core commitments, 1:305–306
founding and history, 1:305
partnerships, 1:306–307
sources of funding, 1:308
vision, purpose, and values, 1:304–305

Mass movement events, 1:1; 2:57. *See also* Avalanches; Landslides
Mennonite Central Committee (MCC), 1:309–314
activities, 1:312–313
cyclone response, 1:312
Democratic Republic of Congo emergency response, 1:313
headquarters, 1:310
organizational structure, 1:310–311
refugee response, 1:312–313
vision and mission, 1:310
Mexico City earthquakes, Mexico (1985), 2:239–244
deaths and destruction, 2:241–243
geologic events, 2:241
magnitude, 2:241
physical and urban landscapes of Mexico city, 2:240–241
recovery efforts, adaptation, and social effects, 2:243
Millennium drought, Australia (2001–2012), 2:244–248
drought definition in Australia, 2:245

environmental impacts, 2:246
impacts, 2:246–247
response, 2:247–249
Mozambique flood (2000), 2:248–253
causes of, 2:249
damage and impacts, 2:249–251
impact on agriculture and livestock, 2:249–250
impact on food security and health, 2:250–251
nongovernment organizations (NGOs) and, 2:251
response and recovery, 2:251–252

Nepal earthquakes (2015), 2:253–258
aftershocks, 2:254
deaths from, 2:254
magnitude, 2:253
nongovernment organizations (NGOs) and, 2:256
relief operations, 2:255–257
search and rescue operations, 2:254–255

Obama, Barack, 2:29, 30–31, 144, 277
Oxfam International, 1:314–319
focus areas, 1:315
founding and history, 1:314
fundraising, 1:316–317
member organizations, 1:315
mission and objectives, 1:314
organizational structures, 1:315–316
relief efforts and criticism, 1:317–318

Pakistan flood (2010), 2:258–262
aid from European Union, 2:261
aid from United States, 2:261
causes of, 2:258–259
damage and impacts, 2:259–260
deaths from, 2:258
Initial Floods Emergency Response Plan (PIFERP) and, 2:260
response and recovery, 2:260–261
Pan American Health Organization (PAHO), 1:319–323
activities, 1:321–323
Haiti earthquake (2010) and, 1:322
history, 1:319–320
Hurricane Sandy (2012) and, 1:322
measles eradication program, 1:321
publications program, 1:321
services to refugees and other disadvantaged persons, 1:321–322

Refugees International, **1**:323–327
  activities, **1**:324–327
  Climate Displacement Program, **1**:326
  internally displaced persons and, **1**:324
  international field missions, **1**:326
  mission and purpose, **1**:323
  Morton, Sue, founder, **1**:323
  programs for women and girls, **1**:326
  refugees and, **1**:323–324
  Rohingya Crisis and, **1**:326–327
  stateless people and, **1**:324
  target population, **1**:323–324
Rohingya Crisis, **1**:280, 281, 283, 284, 289, 303–304, 326–327, 335
Roosevelt, Franklin D., **2**:95

Salinization, **1**:172–181
  Aquarius/SAC-D (NASA satellite instrument), **1**:179, 180
  Atlantic Meridional Overturning Circulation, **1**:178–179
  climate change and, **1**:172, 173, 175, 176
  definition of, **1**:172
  density-temperature-salinity balance, **1**:178
  evaporation and precipitation in the water cycle, **1**:177–178
  ocean salinization, **1**:177–180
  river lake, aquifer, and geothermal salinization, **1**:176–177
  soil salinization, **1**:173–176
  sources of soluble salts in river basins, **1**:176
  SPURS-1 (first Salinity Processes in the Upper Ocean Regional Study), **1**:179–180
  SPURS-2 (second Salinity Processes in the Upper Ocean Regional Study), **1**:180
  variations in ocean surface salinity, **1**:178
Save the Children, **1**:328–332
  contemporary activities, **1**:329–330
  in Ethiopia, **1**:330
  formation, **1**:328–329
  in Haiti, **1**:330
  humanitarian responses, **1**:330–331
  Jebb, Eglantyne, and, **1**:328, 331
  in Mediterranean, **1**:331
Sichuan earthquake, China (2008), **2**:262–267
  aftershocks, **2**:263–265
  damages, **2**:264–265
  deaths from, **2**:265
  geohazards, **2**:265
  magnitude, **2**:263
  overview of, **2**:263–264
  response, **2**:265–266
Storm surges, **1**:181–189
  Ash Wednesday storm (1962) and, **1**:186
  in Bangladesh, **1**:186–187
  definition of storm tides, **1**:181
  extratropical storms and, **1**:182–183, 184
  fetches and, **1**:182
  forecasting and measurement, **1**:183–185
  freshwater flooding and, **1**:187
  Great Hurricane of 1900 and, **1**:186
  high water marks, **1**:184
  Hurricane Katrina (2005) and, **1**:187
  impacts and mitigation, **1**:185–186
  notable examples, **1**:186–187
  Saffir-Simpson (SS) scale and, **1**:182
  Sea, Land, and Overland Surges from Hurricanes (SLOSH) model, **1**:183
  tide gauge stations, **1**:184
  tropical cyclones/hurricanes and, **1**:182–183, 185, 186, 187
  wetlands and mangrove forests as defenses against, **1**:185
Subsidence, **1**:189–198
  Arecibo Observatory, Puerto Rico, and, **1**:191
  definition of, **1**:189
  dissolution and, **1**:190
  Global Positioning System (GPS) technology and, **1**:196
  groundwater extraction and, **1**:192–193
  human-induced subsidence, **1**:192–193
  interferometric synthetic aperture radar (InSAR) and, **1**:196
  in Jakarta, Indonesia, **1**:195
  karst and pseudokarst topography, **1**:190–191
  in Mexico City, Mexico, **1**:195
  mining of underground natural resources and, **1**:193
  Panel on Land Subsidence (NRC), **1**:196
  plate tectonics and, **1**:189–190
  sinkholes and, **1**:190–191
  soil and, **1**:193–195

subsurface void and, **1:**190
in Tokyo, Japan, **1:**195
in the United Kingdom, **1:**195–196
in the United States, **1:**189, 190–195
Sulawesi earthquake and tsunami, Indonesia (2018), **2:**267–272
   aftershocks, **2:**269, 270
   cause of, **2:**268
   deaths from, **2:**269, 270
   impacts, **2:**269–270
   magnitude, **2:**268–269
   nongovernment organizations (NGOs) and, **2:**271
   response, **2:**270–271
   tsunami warning, **2:**269
Summer floods, United Kingdom (2007), **2:**272–276
   deaths from, **2:**273
   impacts, **2:**273–274
   warnings and government response, **2:**274–275
Superstorm Sandy, United States (2012), **2:**276–281
   impact in New York and New Jersey, **2:**277–279
   landfall, **2:**276, 277, 279
   response, relief, and recovery in New York and New Jersey, **2:**279–280
   Saffir-Simpson Hurricane Wind Scale category, **2:**276

Tangshan earthquake, China (1976), **2:**281–286
   aftershocks, **2:**282, 284
   controversies, **2:**284–285
   damage, **2:**282–283
   event, **2:**281–282
   magnitude, **2:**281
   rescue effort, **2:**283–284
Temperature extremes, **1:**199–210
   acclimatization and, **1:**202
   adaption and, **1:**206–207
   advisories, watches, and warnings, **1:**207
   air temperature and trends, **1:**200
   apparent temperature, **1:**201
   arctic amplification and, **1:**208
   California (2012–2015), **1:**204
   climate change and, **1:**29, 199, 200, 201, 206, 208–209; **2:**108–109
   cold waves, **1:**201–202
   cooling centers, **1:**207
   drought hazard from, **1:**205
   Europe (2003), **1:**203
   Europe (2012), **1:**204
   examples of extreme temperature hazard, **1:**203–204
   extreme marine temperatures, **1:**202
   Florida (2010), **1:**204
   heatwaves, **1:**201
   human health/death hazard from, **1:**205
   IPCC models and future scenarios, **1:**208–209
   North Africa and Middle East (2016), **1:**204
   observed changes in number of record highs versus record lows, **1:**208
   Russia (2010), **1:**203
   St. Louis, Missouri (1980), **1:**203
   terminology, **1:**200–202
   transportation hazard from, **1:**206
   U.S. Midwest (1995), **1:**203
   wildfire hazard from, **1:**206
   worldwide annual average surface temperature anomalies, **1:**200
   See also Chicago Heat Wave, Illinois (1995); European heat wave (2003); Heat wave and wildfires, Russia and Eastern Europe (2010)
Thomas Fire, California (2017–2018), **2:**286–290
   compounding disasters, **2:**288
   contributing factors, **2:**287–288
   deaths from, **2:**286
   lessons learned, **2:**289–290
   recovery, **2:**288–289
Tohoku earthquake and Fukushima tsunami, Japan (2011), **2:**290–295
   aftershocks, **2:**291
   damage, **2:**292–293
   deaths from, **2:**292
   magnitude, **2:**291
   response and recovery, **2:**294–295
   tsunami warning, **2:**293–294
Tornadoes, **1:**210–220
   cone tornadoes, **1:**215
   directional sheer, **1:**213
   distribution of, **1:**215–217
   Doppler radar and, **1:**212
   drill bit tornadoes, **1:**215
   in early history, **1:**210
   Enhanced Fujita Scale (EF-Scale), **1:**217–218
   fatalities and damage from, **1:**218

Tornadoes (*cont.*)
  formation and physical characteristics, **1:**212–215
  Fujita, Dr. Tetsuya (Ted), on, **1:**212
  gustnado, **1:**214
  multiple-vortex tornadoes, **1:**215
  outbreaks, **1:**216, 217, 219
  rainfall and, **1:**213
  records, **1:**219–220
  research in the United States, **1:**211–212
  rope tornadoes, **1:**215
  speed sheer, **1:**213
  stovepipe tornadoes, **1:**215
  types of, **1:**215
  wall cloud, **1:**214
  waterspouts, **1:**214–215
  wedge tornadoes, **1:**215
  wind shear, **1:**213
  See also Edmonton tornado, Canada (1987); Joplin tornado, Missouri (2011); Tri-State tornado, United States (1925)
Tri-State tornado, United States (1925), **2:**295–299
  damage, **2:**296–297
  deaths from, **2:**295, 297, 298–299
  emergency response, **2:**298
  meteorological synopsis, **2:**296
  question of multiple tornadoes, **2:**297–298
Tropical storm and floods, Yemen (2008), **2:**299–304
  deaths from, **2:**300
  impacts, **2:**300–301
  recovery, **2:**303
  response, **2:**301–302
Truman, Harry, **1:**282
Trump, Donald, **1:**317; **2:**31, 163
Tsunamis, **1:**220–230
  coastal tide gauge data and, **1:**223
  deaths and injuries from, **1:**225–226
  deep-ocean data and, **1:**223
  as distinct from tidal waves, **1:**220
  earthquakes as triggers for, **1:**221, 222, 223, 228
  evacuation, **1:**224
  far-source-generated tsunamis, **1:**222
  formation and physical characteristics, **1:**221–222
  improved responses to, **1:**228
  International Tsunami Information Center (ITIC), **1:**224
  landslides as triggers for, **1:**221
  mental health effects from, **1:**227
  mid-source-generated tsunamis, **1:**222
  monitoring agencies, indigenous knowledge, and warnings, **1:**223–224
  near-source-generated tsunamis, **1:**222, 225, 229
  potential impacts, **1:**225–227
  response and recovery efforts, **1:**227
  seaquakes as triggers for, **1:**222
  seismic data and, **1:**223
  soil and vegetation contamination from, **1:**226
  tracking and predicting, **1:**222–223
  volcanic eruptions as triggers for, **1:**221
  water depth and, **1:**222
  See also Indian Ocean tsunami (2004); Sulawesi earthquake and tsunami, Indonesia (2018); Tohoku earthquake and Fukushima tsunami, Japan (2011)
Typhoon Haiyan (Yolanda), Philippines (2013), **2:**304–309
  Category 5 tropical cyclone, **2:**304
  deaths from, **2:**304, 306, 307
  domestic and international response, **2:**306–308
  impact, **2:**304–305
  nongovernment organizations (NGOs) and, **2:**307, 308
  warning and evacuation, **2:**305–306

United Nations
  Bam earthquake, Iran (2003), and, **2:**3–4
  Colombia floods (2010–2011) and, **2:**62
  Cooperative for Assistance and Relief Everywhere and, **1:**285
  Cyclone Gorky, Bangladesh (1991), and, **2:**71
  Cyclone Sidr, Bangladesh (2007), and, **2:**90
  East African drought (2011–2012) and, **2:**96
  Hurricane Stan, Guatemala (2005), and, **2:**196
  Indian Ocean tsunami (2004) and, **2:**200
  Lutheran World Federation and, **1:**308

Mennonite Central Committee and, **1:**310
Millennium Development Goals (MDGs), **1:**303, 334, 340, 343
Oxfam International and, **1:**317
Pan American Health Organization and, **1:**319–322
Sulawesi earthquake and tsunami, Indonesia (2018), and, **2:**271
Sustainable Development Goals (SDGs), **1:**266, 282, 303, 334
tropical storms and floods, Yemen (2008), and, **2:**302
See also International Organization for Migration (IOM); Pan American Health Organization (PAHO); World Food Program (WFP); World Health Organization (WHO)
United Nations Children's Fund (UNICEF), **1:**332–337
Child Survival Revolution, **1:**334
Development era, **1:**333–334
founding and history, **1:**332–333
functional areas, **1:**335
fundamental principles and mission, **1:**334–335
headquarters, **1:**335
organization and sources of funding, **1:**335–336
revenue, **1:**336
United Nations Development Program (UNDP), **1:**285; **2:**302
United Nations Economic Commission for Latin America and the Caribbean (ECLAC), **2:**194, 300
United Nations Educational, Scientific and Cultural Organization (UNESCO), **1:**224; **2:**5, 198, 205, 262, 300
United Nations Environmental Programme (UNEP), **1:**48
United Nations Food and Agriculture Organization (FAO), **2:**260
United Nations High Commissioner for Refugees (UNHCR), **1:**305, 324, 325; **2:**98–99
United Nations Office for the Coordination of Humanitarian Affairs, **2:**184, 191, 259
United Nations Population Fund (UNPF), **2:**256

Valdivia earthquake, Chile (1960), **2:**309–313
aftershocks, **2:**309
deaths and damage, **2:**310–311
magnitude, **2:**309
recovery, **2:**312
response, **2:**311–312
Vietnam flood (1999), **2:**313–317
causes of, **2:**313–314
damage and impacts, **2:**314–315
deaths from, **2:**315
nongovernment organizations (NGOs) and, **2:**316
response and recovery, **2:**315–316
Volcanic activity, **1:**230–240
Aa lava, **1:**233
active volcanoes, **1:**231
Baikal Rift Valley, **1:**236
Blocky lava, **1:**234
cinder cone volcanoes, **1:**232
composite volcanoes, **1:**232–233
deaths from, **1:**230
dormant volcanoes, **1:**231–232
East-African Rift Valley, **1:**235–236
explosive (central) volcanoes, **1:**233
extinct volcanoes, **1:**232
formation of magma, **1:**230–231
geographic distribution of, **1:**234–236
hazards from directed blasts, **1:**236
hazards from molten rock released into atmosphere, **1:**237–238
hazards from pyroclastic surges, **1:**238–239
hazards from volcanic structural collapse, **1:**238
landslide hazards from, **1:**238
lava dome volcanoes, **1:**233
lava flow hazards from, **1:**238
Pacific Ring of Fire, **1:**234–235
Pahoehoe lava, **1:**233–234
Pillow lava, **1:**234
positive effects of, **1:**239
quiet (fissure) volcanoes, **1:**233
Rio Grande Rift, **1:**236
shield volcanoes, **1:**232
tephra hazards from, **1:**236–237
types of lava, **1:**233–234
types of volcano, **1:**231–233
volcanic ash hazards from, **1:**237
volcanic earthquakes, **1:**236

Volcanic activity (*cont.*)
   volcanic materials and eruptions, **1**:230–231
   West Antarctic Rift, **1**:236
   See also Eyjafjallajökull eruption, Iceland (2010)

Waterspouts, **1**:240–250
   Cottage City waterspout, Massachusetts (1896), **1**:248
   Dark Spot Stage of, **1**:246
   Decay stage of, **1**:247
   definition of, **1**:241
   early history of, **1**:241, 242
   fair-weather (nontornadic) waterspouts, **1**:240–241, 243–244, 245
   formation of, **1**:242–243
   gustnados, **1**:245
   history of, **1**:241–242
   International Centre for Waterspout Research (ICWR), **1**:249
   landspouts, **1**:245
   life cycle of, **1**:246–247
   as marine hazards, **1**:247–248
   Mature Vortex Stage of, **1**:247
   research and forecasting, **1**:248–249
   snownados, **1**:245–246
   Spiral Pattern Stage of, **1**:247
   Spray Ring (Incipient Spray Vortex) stage of, **1**:247
   tornadic waterspouts, **1**:244
   twin tornadic waterspouts, **1**:245
   types of, **1**:243–245
   under-recording of, **1**:242
Weather Channel (U.S.), **1**:11–12
Wildfires, **1**:250–261
   adaptive capacity and, **1**:258–259
   Butte Fire, California (2015), **2**:26
   Camp Fire, California (2018), **1**:251–252
   definition of, **1**:250
   Elk Complex fire, Idaho (2013), **1**:256, 257
   fire in the environment, **1**:254
   fire weather, **1**:255–256
   Forest Fire Danger Index (FFDI), **2**:26
   Great Smoky Mountains fire (2016), **1**:252
   human exposure to, **1**:258
   human response to, **1**:259–260
   human-related wildfire ignition, **1**:251–252, 253
   lightning and, **1**:250, 252–253, 256
   natural wildfire ignition and support, **1**:252–253
   pyrocumulus or flammagenitus clouds and, **1**:254
   Rim Fire, California (2013), **2**:26
   Rush Fire, California (2012), **2**:26
   sensitivity to, **1**:258
   smokejumpers, **1**:255
   suppression and firefighting, **1**:255, 259–260
   temperature extremes and, **1**:199–200, 203, 206
   Valley Fire, California (2015), **2**:26
   wildland-urban interface zones and, **1**:251–252, 258
   winter wet, summer dry climates and, **1**:258–259
   See also Heat wave and wildfires, Russia and Eastern Europe (2010); Thomas Fire, California (2017–2018)
World Food Program (WFP), **1**:337–341
   criticism, **1**:340–341
   founding and history, **1**:337
   funding, **1**:339–340
   mission statement, **1**:337–338
   organization and activities, **1**:338–339
   World Food Program (WFP) USA, **1**:339
World Health Organization (WHO), **1**:341–346
   activities, **1**:342
   criticism and controversy, **1**:344–345
   founding principle, **1**:341
   funding, **1**:341–342
   health emergencies program, **1**:344
   history and accomplishments, **1**:343–344
   International Health Regulations, **1**:343–344
   organizational response to emergencies, **1**:342–343
   public health emergency of international concern (PHEIC), **1**:342–343, 344, 345
   Special Programme for Research and Training in Tropical Diseases, **1**:343
   universal health care and, **1**:344

World Vision International (WVI),
   1:346–349
  budget and funding, **1:**346
  extreme hunger and, **1:**347–348
  founding and history, **1:**346
  humanitarian work, **1:**347–349
  Pierce, Bob, founder, **1:**346
  Zika virus and, **1:**348–349

Yangtze River flood, China (1931),
   2:317–321
  causes of, **2:**317–318
  controversy, **2:**319–321
  deaths from, **2:**317
  fire and, **2:**319
  impacts, **2:**318–319
  response, **2:**319–321